ANIMAL DISPERSION
in relation to Social Behaviour

ANIMAL DISPERSION

in relation to Social Behaviour

V. C. WYNNE-EDWARDS

Regius Professor of Natural History
University of Aberdeen

OLIVER AND BOYD

EDINBURGH AND LONDON

OLIVER AND BOYD LTD

Tweeddale Court
Edinburgh

39A Welbeck Street
London W. 1

First published 1962

Printed in Great Britain by
Oliver and Boyd Ltd., Edinburgh

Preface

The theory presented in this book links together the subjects of population and behaviour. It applies to animals in general, which gives it an exceedingly wide scope. During the last seven years it has provided me with a novel and, it has often seemed, commanding viewpoint from which to survey the everyday events of animal behaviour; and some of the most familiar activities of animals, the purpose of which has never properly been understood, have readily been seen to have important and obvious functions. It has turned out to be an agreeable and characteristic feature of the theory not to keep butting against widely held, pre-existing generalisations, but to lead instead into relatively undisturbed ground. Needless to say, the reader is confronted at the beginning with two or three fundamental principles that, on account of their unfamiliarity alone, he may be expected to eye with a certain amount of scepticism, until they can by degrees be critically appraised in the light of each succeeding chapter of evidence.

The theory has been developed more or less freely, although this has led into some fields of knowledge where I have had to explore as a novice as I went along. Given an apparently rational, single explanation of the origin of social behaviour, for instance, one of the interesting corollaries has been to interpret the zoological background of conventional behaviour in man. This is not without its philosophical implications, although actually these have only very lightly been touched upon; seldom, as a matter of fact, could one find any clearer indication than emerges here of the closeness of man's kinship with his fellow animals. Our human experience as social beings turns out to be of frequent use in gaining an understanding of sociality in other species and, with reasonable discretion, it can be drawn upon without any serious fear of anthropomorphism. The evident loss by man, almost within the historic period, of the means of limiting population growth, which he formerly possessed like other animals, stands out with disturbing clarity.

A large book has been required in order to contain enough factual evidence to support the theory in each of its widespread ramifications. The literature that could and ideally should have been consulted is too vast to be within the compass of one person in any reasonable time. Though fairly full documentation for what has been included is generally attempted, it is too much to hope that nowhere will key references have been overlooked; and to those authors whose illuminating works I have inadvertently passed over I offer my apologies. During the final revision it has been practicable to incorporate only a few of the most recent publications, which have poured out in an ever-growing stream since the various chapters were first written, in some cases several years earlier. There must be places also where enthusiasm has led me rashly into conclusions that experts with their better-digested knowledge will regard as unwarranted. Such errors and omissions, whether numerous or few, will I hope be found to relate in the main to incidental points, and do nothing to undermine the solidarity of the central theory itself.

A general idea of the subject and how it is developed can be obtained without necessarily reading the whole book from cover to cover. The first chapter is intended to provide both the foundations of the theory and a preliminary review of it. Thereafter all the way through the book there are summaries at the end of the chapters. These are usually fairly full, though their condensation is naturally bound to make them tougher meat than the book itself. They may show the reader where to skip, where to pick out for detailed reading the subjects likely to interest him most, and where to turn to the main text when it seems essential to find corroborating evidence; they may also supply a brief reminder of the chief points that have emerged at each stage, after the chapter itself has been read. It is appropriate to note here that a previous, very condensed five-page sketch of the theory of animal dispersion was published in 1959 (*Ibis*, **101**: 436-41).

I am indebted for information and discussion to many more friends and correspondents than can be mentioned individually by name. I have abundant cause to be grateful to my colleagues in Aberdeen for their knowledge, criticism and encouragement, especially Doctors George Dunnet, Robert M. Neill, Philip Orkin, David Jenkins, Adam Watson and Guy Morison; the same is true of Dr William M. Clay and my other colleagues in the Department of Biology at the University of Louisville, Kentucky, where I spent four rewarding months between the writing of Chapters 19 and 20.

I must expecially thank Mr Ben Feaver, of Haverfordwest, and Miss Mabel Slack, of Louisville, for allowing me to use their coloured photographs for plates IV and IX. The butterfly specimens in plate VIII were kindly lent by the Royal Scottish Museum, and were selected and photographed under the skilled supervision of Dr D. R. Gifford, of Edinburgh University. For plate II I am grateful to Dr Robert S. Simmons, of Baltimore, Maryland; and for plates III (upper photo) and V, to Dr Robert Carrick, of Canberra, Australia. Acknowledgments are also due to the following for kind permission to reproduce illustrations already published elsewhere, namely Akademische Verlagsgesellschaft, Leipzig (fig. 6); Editor of *Ardea* (fig. 47); Editor, *Biological Bulletin Wood's Hole* (fig. 5); Cambridge University Press (figs. 28, 48, 49); Dieterich'sche Verlagsbuchhandlung, Wiesbaden (fig. 7); Entomological Society of America (fig. 40); Messrs Walter de Gruyter, Berlin (fig. 18); Editor of the *Ibis* (pl. XI); Messrs Macmillan & Co. (figs. 30, 31, 45 and 46); Marine Biological Association of the United Kingdom (figs. 33 and 36); Royal Irish Academy (fig. 32); the Royal Society (fig. 12); Messrs Oliver & Boyd (figs. 24 and 37-8); Sears Foundation for Marine Research, Yale University (fig. 29); Smithsonian Institution,

Washington (figs. 13 and 14); Superintendent of Documents, U.S. Government Printing Office (figs. 2 and 41-4); University of California Press (fig. 34, pls. VI and VII); University of Chicago Press (fig. 50); Messrs H. F. & G. Witherby (fig. 22); and the Zoological Society (pl. X).

Finally it is a pleasure to acknowledge the immense help I have received from my wife.

Contents

Illustrations

Chapter 1

An outline of the principles of animal dispersion

> ' Most of the singing and elation of spirits of that time [the amorous season] seem to me to be the effect of rivalry and emulation; and it is to this spirit of jealousy that I chiefly attribute the equal dispersion of birds in the spring over the face of the country.' Gilbert White, *The Natural History of Selborne*, Letter XI to the Hon. Daines Barrington, 8 February 1772.

1.1. *Introduction*

Animal dispersion may be defined as comprising the placement of individuals and groups of individuals within the habitats they occupy, and the processes by which this is brought about. It is something we are inclined to take largely for granted. People who observe these things know well that animals can be expected to occur more plentifully in habitats that contain an abundance of the kinds of resources they need, and that they will generally be scarcer or absent altogether where the environment is less well provided. Experience teaches us also the sort of habitat preferred by each species; and the inference is clear enough that animals must often have complex adaptations—either in the form of innate instincts or acquired responses—to enable them to select their habitats appropriately. Inference may carry our thoughts a step further, suggesting to us that habitat selection and dispersion must be essentially internal or domestic affairs that each species arranges for itself as best it can. But seldom do we get very far beyond the point of accepting dispersion as something that automatically happens, or even dimly begin to realise the gigantic, universal organisation it entails. Lack (1954*a*, p. 264) has rightly said that it ' presents a remarkable, and until now largely unappreciated problem '.

Lack devoted the final chapter of his book, *The natural regulation of animal numbers*, to dispersion. Like Gilbert White (quoted at the head of this chapter) he realised that it must be due, in part at least, to the responses and behaviour of the animals themselves. ' Dispersion [of birds] in winter ', he says, ' presents no particular problem and seems to be adequately explained by supposing that, within their favoured habitats, the individuals or flocks avoid areas where food is short and tend to settle where it is abundant' (loc. cit., p. 264). And further on (p. 269), 'the problem of dispersion was obviously in Howard's mind when he propounded the territory theory, though he rightly linked territory with other matters, particularly pair formation, and, also rightly in my opinion, linked dispersion with the food requirements of the young'. We shall find as we proceed that we have to

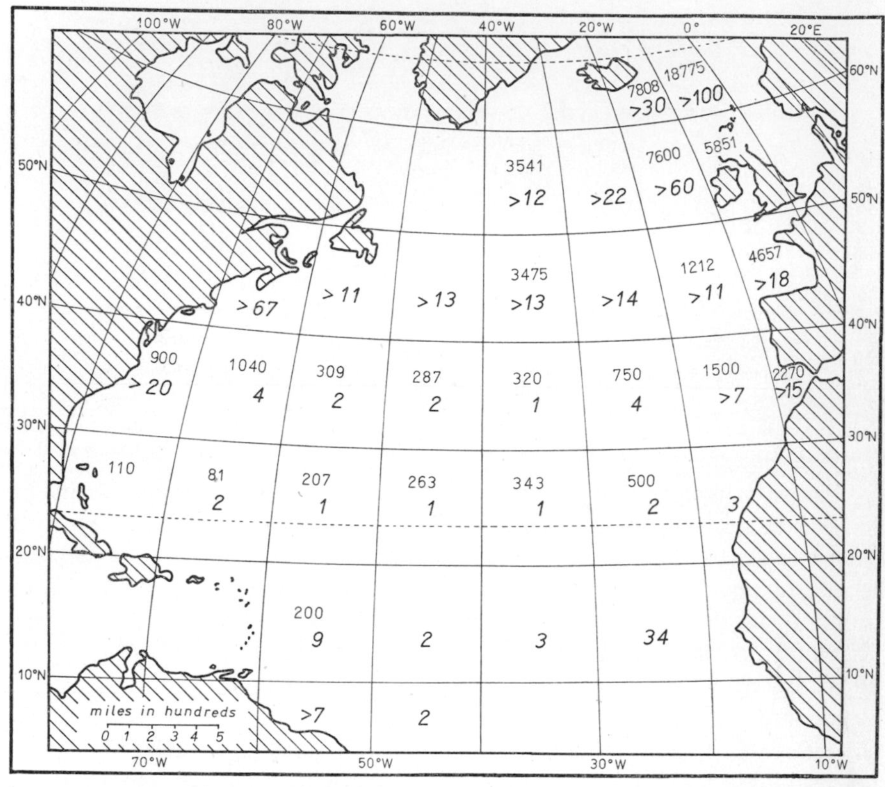

FIG. 1. Jespersen's correlation between the number of pelagic birds and the abundance of plankton in the North Atlantic. In each 10° sector the upper figure gives the average volume of macroplankton in cc in a standard haul, and the lower figure (in larger type) the average number of birds recorded per day. The correlation coefficient between them is +0·85. (Redrawn on an equal-area projection from the data in Jespersen's 'Dana' report, 1930.)

travel a long way from this initial viewpoint; but it is a very appropriate place to start, because it was from Lack's stimulating book and in particular the final chapter that the title of this one and the inspiration for writing it came.

A good illustration of what is involved in dispersion, all the better because it is quantitative, was provided by Jespersen (1924, 1930) as a result of the Danish oceanographic expeditions made in search of the birthplace of freshwater eels (*Anguilla*) between 1913 and 1928. Standard hauls of macroplankton were taken at scores of stations all over the North Atlantic, and these, measured by bulk, could be used as a broad index of the richness of the surface waters, in terms of food for pelagic birds; the hauls were grouped and averaged to give a single figure for each 10-degree sector of the ocean. A log was kept of the number of birds seen each day; and the counts for all the days spent in each 10-degree sector were similarly combined and averaged, to give an index of the population-density of birds. In spite of the collective character of the data and the differences in season at which the different areas were sampled, the correlation that emerges between bird density and abundance of plankton is very strong (*see* fig. 1): in numerical terms it can be expressed by a coefficient of $+0{\cdot}85$. The probability that such a situation could have arisen by chance alone is negligible ($P \leqslant 0{\cdot}001$).

This is a dispersion that the birds must have brought about by their own efforts; and, as far as can be judged from the averages of grouped samples such as we are given here, it results in an efficient *pro rata* allocation of birds to the available food-resources. The population-density in fact appears to be graded so that in every area it bears about the same constant average relation to the amount of plankton present; and this certainly appears to be a situation that we should no longer dismiss as ' presenting no particular problem '.

Another correlation of the same kind, in which the relationship between eater and food is simpler and more direct, and on a smaller geographical scale, was demonstrated by Hardy and Gunther (1935, p. 273), between the population-density of whalebone whales round South Georgia (determined from the kills by commercial whalers) and that of their principal food-species, the krill *Euphausia superba*.

Close adjustments between animals and their essential resources are not confined to birds and mammals, but can be found even in the Protozoa. There is usually of course at least a residual element of chance in dispersion, and of the influence of extraneous factors beyond the animals' control; but freely-moving individuals endowed with the necessary perceptions always have the means of combating chance and seeking to improve their situation. If we look, for instance, at the dispersion and settlement on the floor of the sea of free-swimming larvae, after they have drifted in the plankton apparently like seeds borne on the wind, we find it is far from being a purely random scatter; in fact their metamorphosis can generally be postponed for some considerable time while they search for a suitable substratum (*cf.* Wilson, 1952, p. 118). Thus oyster larvae tend to be gregarious on settlement, and prefer to attach themselves when they have found surfaces of suitable texture on which other oyster larvae have already settled, at least up to a certain density (Cole and Knight-Jones, 1949); and practically the same has been found to apply to larvae of the barnacle *Elminius modestus* (Knight-Jones and Stevenson, 1951). The complexity of the responses involved in dispersion, even in comparatively young and lowly animals, is very apparent.

1.2. *Optimum density in relation to resources*

Ideally the habitat should be made to carry everywhere an optimum density, related to its productivity or capacity, without making any parts so crowded as to subject the inhabitants to privation, or leaving other parts needlessly empty. It seems quite possible, on the basis of general experience, that something approaching these conditions may in fact often be realised. We ought therefore to consider what factors would be likely to determine the optimum density of a population; and this brings us to one of the fundamental principles underlying the theory of animal dispersion that we are setting out to examine.

Before proceeding, however, it should be stated that in the great majority of species of animals, just as in the sea birds, the critical resource, as far as population-density is concerned, is food. There are of course many other requirements to be satisfied in making a habitat habitable at all—it must be accessible, and within certain limits of tolerance as far as physical conditions are concerned; it must be able to provide whatever shelter is required, and be free from incompatible organisms of other kinds. But, granted that every essential need is met at least minimally, and life for a particular species is consequently supportable at all, food so generally becomes the factor which ultimately limits population increase that we can afford to regard any other ultimate limiting factors—mere ' standing room ', for instance, in the case of some of the sessile barnacles (*Balanus* etc.) in the intertidal zone, or nest-holes for breeding populations of pied flycatchers (*Muscicapa hypoleuca*) or purple martins (*Progne subis*)—as being special exceptions to a general rule. It is with the rule and not with the exceptions that we need at this stage to be concerned.

Our best approach to the subject of optimum density is to study man's own experience in exploiting natural recurrent resources for consumption as food and for other purposes—resources consisting of ' wild ' species whose numbers are sustained and renewed by allowing nature to take its course. Man the fisherman, in particular, still acts essentially as a predator exploiting natural prey. He goes away to sea with his nets and gear and comes home with the spoils; and hitherto he has been effectively prevented (except on a negligible scale) from undertaking any kind of cultivation or husbandry to increase the natural rate of replenishment of the stock. His experience is therefore closely relevant to the study of the relationship between other species and their food-resources.

Events of the last hundred years and more have proved beyond doubt that fishery resources are not inexhaustible. There is no more striking illustration of this than the one presented by the history of whaling. As early as the 16th Century the Basques had begun to make serious inroads into the stock of the North Atlantic right whale (*Balaena glacialis*). By the 17th Century the centre of the fishery had shifted north to the seas and fiords of Spitzbergen, and from there it moved still later to Baffin Bay. In these Arctic waters the principal prey was the Greenland whale (*Balaena mysticetus*).

The northern fishery reached a high peak of prosperity in the early years of the 19th Century: thousands of people were then directly or indirectly deriving their living from it, and in Britain alone the ports of Hull, Whitby, Leith, Dundee and Peterhead thrived as never before. Capital and enterprise poured into the industry, and there can have been little or no thought of moderating the catch and preventing overfishing. Although the invention of steamships and the harpoon gun about 1860 made it possible to catch the faster-moving blue and fin whales (*Balaenoptera*) and to stave off the ultimate disaster for another fifty years, before 1914 the whole North Atlantic fishery had crumbled in ruins. The stocks of the two right whales have never recovered, and the population of Greenland-whaling men and of those who ministered to them has become effectively extinct.

A similar though less disastrous history has followed the pursuit of all the chief commercial fishes, such as the cod (*Gadus callarias*), haddock (*G. aeglifinus*), plaice (*Pleuronectes platessa*), halibut (*Hippoglossus*), and at last apparently even the herring (*Clupea harengus*). What first opened our eyes to the reality of overfishing was the spectacular recovery made by some of these heavily exploited species in the North Sea, as a result of the respite they got in the 1914-18 war when vessels and crews were otherwise engaged and the Dogger Bank was covered by minefields. The ability commonly shown by partially depleted fisheries to regenerate in this way is, incidentally, a heartening feature in a generally depressing prospect, and ecologically it is of the highest importance.

Since 1918 the subject of overfishing has attracted increasing study by fishery scientists, and an understanding has been gained of the principles underlying the conservation and optimal exploitation of fishery resources (*cf.* Beverton and Holt, 1957). (An account of a very illuminating experiment on fishery exploitation by Silliman and Gutsell (1958) will help to bring these principles home in a later chapter; *see* p. 499). In its simplest terms the empirical position is that the effort devoted to exploiting what can be thought of at the outset as a virgin fishery can be steadily increased, with corresponding increases in the total catch, up to a certain critical maximum. If the fishing intensity is stepped up beyond this level, the stock begins to suffer depletion and the annual catch in time starts getting smaller; a position is thus reached where increasing effort produces a smaller total return.

This critical level of fishing intensity is the one that gives the greatest possible sustained landings of fish, both absolutely and in terms of unit effort expended, and also the greatest cumulative total catch. We are concerned here as simply as possible with the production of food, and can afford to neglect everything to do with variations in the details and efficiency of exploitation, for instance through altering the mesh-sizes of nets, and also ignore the question of market values and financial returns. The maximum is represented graphically in figure 2 as the highest point on the curve relating sustained yield (the same as 'equilibrium yield') to fishing intensity. This is more likely to be a rounded summit than a sharp culmination, so that in practice it is usually reached without anyone realising it; and it can actually be revealed only by careful long-term analysis or experiment.

By the time commercial fisheries unknowingly begin to exceed the optimum level they are generally paying handsomely. Increased effort still gives temporarily increased returns. The yield per man-hour tends of course to drop, at first very slightly, but this is likely to be masked for some time by natural irregularities in productivity from year to year. In any case there is a strong tendency to shut one's eyes to the unwelcome signs of overfishing, so that depletion may go a long way before any pinch is actually felt or the cause of it admitted by those directly involved.

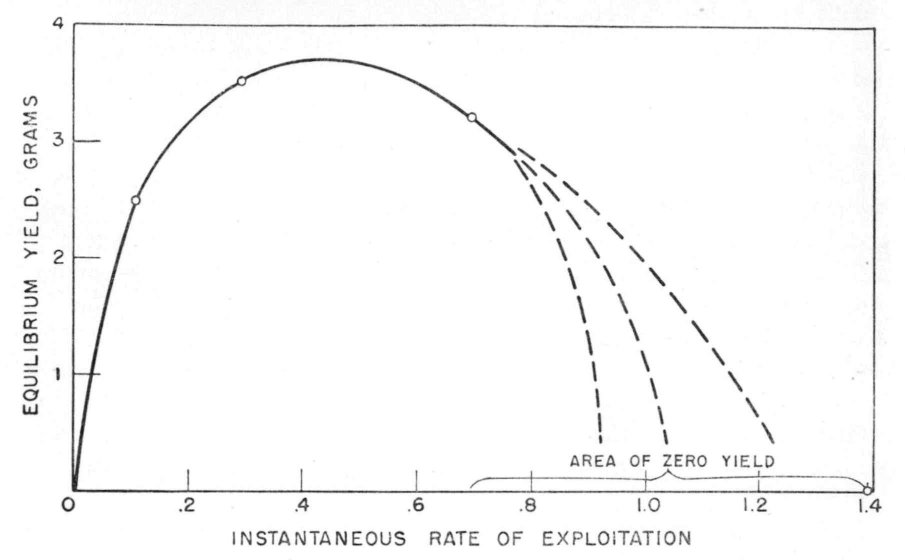

FIG. 2. Graph of the relation between the rate of exploitation (horizontal axis) and the equilibrium yield of fish (vertical axis) in Silliman and Gutsell's experiment. Under these particular conditions, the highest sustained yield would have been obtained by a 35-40 per cent exploitation rate, and any exploitation above about 55-60 per cent would have led to the extinction of the stock. (From Silliman and Gutsell, 1958.)

Having understood the problem it becomes possible to see a remedy: this is to determine the approximate optimum yield and limit the effort to match it. The slogan ' fish less and catch more ' puts the immediate aim in a nutshell. Politically it is extremely difficult to achieve, because it requires all participants to reach an agreement on some workable method of limitation, and thereafter scrupulously to abide by it. What is fundamental to conservation is that the system of free enterprise, with every man or company for himself, must be exchanged for a common code of rules.

This has nevertheless been agreed to in a promising number of cases in different parts of the world, among them that of the southern whale fishery. There the International Whaling Commission has for some years successfully controlled both the number of whaling expeditions taking part and the size of the combined catch for the year in terms of so many ' blue-whale units '; the moment the agreed total has been reached, it declares the fishery closed for the season. Admittedly the Commissioners' task is extremely difficult, and the results imperfect. For one thing, each expedition now tends to be expensively equipped to kill whales at the maximum possible speed in order to get the biggest obtainable share of the quota before the season is closed. The different expeditions are thus still in intense competition, and it would be much better if each country or outfit could be assigned an individual quota, allowing it freedom to take its harvest economically and in its own time, and entirely banishing the competitive element from the whaling grounds. A century ago, indeed, in the worst days of the North Pacific fur-seal fishery, things got so bad that vessels had to mount cannon to defend themselves and international incidents and bloodshed occurred; this was finally prevented and the present extremely successful scientific management made possible, by the North Pacific Sealing Convention of 1911—perhaps the first international treaty of the kind.

There are in fact five lessons to be learnt from this phenomenon of overfishing, the importance of which it would be difficult to exaggerate. The first is that *overfishing reduces both the yield per unit effort and the total yield;* in some circumstances, if sufficiently severe, it can damage the stock beyond recovery and even lead to its extermination. This actually happened in the 19th Century through the commercial exploitation in North America of the passenger pigeon (*Ectopistes migratorius*)—' the most impressive species of bird that man has ever known ' (Schorger, 1955, p. viii; *see also* below, p. 451). Secondly, the size of the optimum catch is not self-evident, and can be determined and adjusted only in the light of experience. Thirdly, the participants must all come to a common agreement or convention to limit their catch, preferably *while the stock is still at or near its maximum abundance and productivity*, and must forego any immediate personal advantage in favour of the long-term benefit of the community as a whole. Fourthly, argument about who shall participate and how much he shall take ought always to be transferred to some higher court, so that direct destructive and wasteful competition is entirely eliminated from the catching of the actual fish. Fifthly, no profitable fishery is immune from over-exploitation; the consequences are certain to follow wherever the optimum rate of cropping is persistently exceeded. Exactly the same is true of any other kind of profitable, renewable, living resource, whether of game, fur-bearers, or—if nutrients are not returned to the soil, or plants are killed or regeneration impaired—even of natural vegetation.

The importance of these inferences lies, of course, in the fact that there is no difference in principle between man exploiting fish or whales and any other predator exploiting any other prey. All predators face the same aspects of exactly the same problem. It is impossible to escape the conclusion, therefore, that *something must, in fact, constantly restrain them, while in the midst of plenty, from over-exploiting their prey.* Somehow or other ' free enterprise ' or unchecked competition for food must be successfully averted, otherwise ' overfishing ' would be impossible to escape: this could only result in lasting detriment to the predators and the risk, if they persisted in it, that the prey might be exterminated altogether.

That some such restraint often exists in nature is most easily comprehended, perhaps, in a situation where a population of animals has to depend for a period on a standing crop, which must be made to last out until the season comes round for its renewal or until alternative foods become available. Many small northern birds, for instance, are provisioned ahead for four or five months in the winter by the standing crop of seeds of a few species of trees or herbs, or by a finite stock of overwintering insects. It would be fatal to allow birds to crowd in freely in autumn up to the maximum number that could for the moment be supported by the superabundance of food, without regard to later consequences. The optimum population-density, on the contrary, would generally be one that could be carried right through till the spring on the same stock of food, supposing the birds were winter-resident species such as tits (*Parus*), robins (*Erithacus*) or woodpeckers (Pici). To achieve this optimum it would clearly be necessary to put a limit on the population-density *from the beginning*, while the resource was still untapped.

The need for restraint in the midst of plenty, as it turns out, must apply to all animals whose numbers are ultimately limited by food whether they are predators in the ordinary sense of the word or not. It is commonly the only way in which plenty can be conserved and maintained. It applies in general as forcibly to herbivores as to carnivores. Ruminants, for instance, are notoriously capable of impairing fertility by long-continued overgrazing; as is well known, this has resulted, in the brief span of history, in reducing large tracts of primeval forest and natural grassland in the Near and Middle East and other ancient centres of pastoral civilisation almost to the bare stones. The habitat generally contains only a finite total quantity of those nutrient elements, such as nitrogen, phosphorus, potassium and calcium, on the repeated circulation of which the continuity of life depends. One of them usually sets a limit to primary productivity. It seems probable, in closed communities at any rate, that selection will favour an optimal balance in the rate of circulation and in the proportion of the total that is held at any given moment by (i) the animal biomass, (ii) the plant biomass, and (iii) the remainder of the system. The hoarding of precious nutrients within the animal biomass, which is likely to be accentuated by over-population, may, especially if it is followed by emigration on a large scale, seriously prejudice the continued productivity of the system as a whole.

Where we can still find nature undisturbed by human interference, whether under some stable climax of vegetation or proceeding through a natural succession, there is generally no indication whatever that the habitat is run down or destructively overtaxed. On the contrary the whole trend of ecological evolution seems to be in the very opposite direction, leading towards the highest state of productivity that can possibly be built up within the limitations set by the inorganic environment. Judging by appearances, chronic over-exploitation and mass poverty intrude themselves on a mutually-balanced and thriving natural world only as a kind of adventitious disease, almost certain to be swiftly suppressed by natural selection. It is easy to appreciate that if each species maintains an optimum population-density on its own account, not only will it be providing the most favourable conditions for its own survival, but it will automatically offer the best possible living to species higher up the chain that depend on it in turn for food.

Such *prima facie* argument leads to the conclusion that it must be highly advantageous to survival, and thus strongly favoured by selection, for animal species (1) to control their own population-densities, and (2) to keep them as near as possible to the optimum level for each habitat they occupy.

Regarding the first of these conditions, the general hypothesis of self-limitation of animal numbers has been growing rapidly in favour among animal ecologists in recent years (*e.g.*, Kalela, 1954; A. J. Nicholson, 1954; Errington, 1956; Wynne-Edwards, 1955; Andrewartha, 1959); so far as it goes the evidence given already in the introductory section of course confirms it. To build up and preserve a favourable balance between population-density and available resources, it would be necessary for the animals to evolve a control system in many respects analogous to the physiological systems that regulate the internal environment of the body and adjust it to meet changing needs. Such systems are said to be homeostatic or self-balancing, and it will be convenient for us to use the same word. Physiological homeostasis has in general been slowly perfected in the course of evolution, and it is thus the highest animals that tend to be most independent of environmental influences, as far as the inward machinery of the body is concerned. Population homeostasis, it may be inferred, would involve adaptations no less complex, and it might therefore be expected that these would similarly tend to reach the greatest efficiency and perfection in the highest groups.

We are going to discover in the concluding chapters of the book that such homeostatic adaptations exist in astonishing profusion and diversity, above all in the two great phyla of arthropods and vertebrates. There we shall find machinery for regulating the reproductive output or recruitment rate of the population in a dozen different ways—by varying the quota of breeders, the number of eggs, the resorption of embryos, survival of the newborn and so on; for accelerating or retarding growth-rate and maturity; for limiting the density of colonisation or settlement of the habitat; for ejecting surplus members of the population, and even for encompassing their deaths in some cases in order to retrieve the correct balance between population-density and resources. Not all types of adaptation have been developed in every group, though examples of parallel evolution are abundant and extraordinarily interesting. Indeed this newest manifestation of homeostasis in the processes of life seems unlikely to remain long in doubt.

1.3. *The existence of natural conventions*

At this point, however, we must leave the subject of homeostasis temporarily in order to consider the second conclusion of our *prima facie* argument, namely that it would be advantageous to be able to keep the population-density at or near the optimum level. We are still concerned with those species of animals—knowing them to be in the great majority—whose numbers are ultimately limited by the resource of food.

It should first be recognised that, unless they are on the verge of extinction, all animals and plants have a great latent power of increase. Physiological provision is made, even in the slowest breeders, for the production of offspring in excess of what is needed merely to sustain the population. In order to prevent a geometrical progression of increase, therefore, some kind of brake must be applied; and if this is to allow the increase to proceed freely when the population-density is low and economic conditions permit it, but to prevent density going any higher once a particular limiting threshold has been reached, then the application of the brake requires to be ' density-dependent '. That is to say, that when the density is low, multiplication will be relatively unhindered, but as it mounts towards a ceiling the checks on increase will become progressively more severe, until population losses through death and emigration catch up with the gains from reproduction and immigration, and the increase is brought to a halt.

When an experimental population is set up in some kind of a confined universe with a finite but regularly renewed supply of food, whether the population consists of *Drosophila*, flour-beetles, *Daphnia*, *Lebistes* (guppies) or mice—to name some of the commonest experimental animals—this is exactly what happens. As we shall see later (*see* Chapter 21, p. 495), the experiments can be replicated time and again, and if the universe is made the same, the population will reach the same predictable ceiling of numbers each time within a narrow margin of error. It will be a matter of special remark when the experiments come to be described that the ceiling is never imposed by starvation, except perhaps indirectly in mice where the mothers often run out of milk at high population-densities, with profound effects on the production of recruits. On the contrary, the ceiling is normally imposed, and the level indefinitely maintained, while the members of the population are in good health—sometimes actually fat—and leading normal lives. Guppies, for instance, still breed and viviparously produce young, but after the ceiling is reached and there are no vacancies in the population, the young are gobbled up by their elders within a few minutes of their birth (*see* p. 543).

It is not necessary here to go into the subject of external density-dependent checks on population—caused by agents such as predators, parasites and pathogens that are likely to take a mounting percentage of lives as population-density rises and economic conditions become more severe. This will be discussed sufficiently fully in the appropriate place later on (p. 546). It will suffice for the present to say that these external checks, while they may sometimes be extremely effective in preventing population-density from rising, are on the whole hopelessly undependable and fickle in their incidence, and not nearly as perfectly density-dependent as has often been imagined. They would in most cases be incapable of serving to impose the ceilings found in nature; what is more, experiment generally shows that they are unnecessary, and that many if not all the higher animals can limit their population-densities by intrinsic means. Most important of all, we shall find that self-limiting homeostatic methods of density-regulation are in practically universal operation not only in experiments, but under ' wild ' conditions also.

Towards the fringe of its range the existence and population-density of any particular species of animal is often overwhelmingly dictated by the physical conditions of the environment—heat, cold, drought, shelter and the like—and by such biotic factors as the presence of better-adapted competitors or the absence of requisite vegetational cover; and it is frequently one of these factors that precludes any further geographical extension of its range. If we ignore the fringe, however, and confine our attention to the more typical part of the range, dispersion in the great majority of animals reflects the productivity of the habitat in terms of food, in just the same way as we saw in the special case of Jespersen's correlation between pelagic birds and plankton. That this is so is nowhere in serious dispute; and if we take care to exclude the minority of species where the food correlation is weak or absent, we can repeat once more that food is generally the ultimate factor determining population-density, and the one that predominates far above others.

We have already the strongest reasons for concluding, however, that population-density must at all costs be prevented from rising to the level where food shortage begins to take a toll of the numbers—an effect that would not be felt until long after the optimum density had been exceeded. It would be bound to result in chronic over-exploitation and a spiral of diminishing returns. Food may be the *ultimate* factor, but it cannot be invoked as the *proximate* agent in chopping the numbers, without disastrous consequences. By analogy with human experience we should therefore look to see whether there is not some natural counterpart of the limitation-agreements that provide man with his only known remedy against overfishing—some kind of density-dependent convention, it would have to be, based on the quantity of food available but ' artificially ' preventing the intensity of exploitation from rising above the optimum level. Such a convention, if it existed, would have not only to be closely linked with the food situation, and highly (or better still perfectly) density-dependent in its operation, but, thirdly, also capable of eliminating the direct contest in hunting which has proved so destructive and extravagant in human experience.

It does not take more than a moment to see that such a convention could operate extremely effectively through the well-known territorial system adopted by many kinds of birds in preparation for the breeding season. Instead of limiting the number of expeditions and fixing the total annual catch like the International Whaling Commission, conventional behaviour in this case limits the number of territories occupied in the food-gathering area. Birds will not, of course, submit to being overcrowded beyond a certain density, nor to the reduction of their territories below a basic minimum size, so that territorial behaviour is perfectly capable of imposing a ceiling on population-density (as C. B. Moffat realised in 1903). In its action it is completely density-dependent, allowing the habitat to fill amid mounting rivalry up to a conventional maximum density, after which any further intrusion is fiercely repelled. Where territories retain this simple primitive character as feeding areas, and where enough information is available, the evidence indicates that minimum territory size is inversely related to the productivity of the habitat: in other words, it is closely linked with the presumptive food-supply. Finally, our last condition is completely met, in that the contest among the participants is all for the possession of territory, and once they have established their claim to the ground they can do the actual food-getting in perfect peace and freedom, entirely without interference from rivals.

The substitution of a parcel of ground as the object of competition in place of the actual food it contains, so that each individual or family unit has a separate holding of the resource to exploit, is the simplest and most direct kind of limiting convention it is possible to have. It is the commonest form of tenure in human agriculture. It provides an effective proximate buffer to limit the population-density at a safe level (which is obviously somewhere near the optimum though we can only guess in animals how nearly perfection is attained); and it results in spreading the population evenly over the habitat, without clumping them in groups as we find in many alternative types of dispersion.

Much space is devoted in later chapters to studying the almost endless diversity of density-limiting conventions, and only the briefest indication of their range can be given here. What must first be appreciated is their artificial or slightly unreal quality. The food-territory just considered is concrete enough—it is the very place where food is found and gathered. But some birds have territories in which they nest but do not feed; and some have only token territories which are the nest-sites they possess and defend in a breeding colony. These, as we shall see later, perform exactly the same function, because the number of acceptable nest-sites in the colony is ' artificially ' limited by the birds' traditional behaviour. In a rookery, for instance, the fullest membership of the community, conferring the right to breed, belongs only to those pairs that can secure a nest-site, construct and defend a nest. Supernumerary ' unauthorised ' nests are constantly pillaged, and there are almost always in consequence a good many non-breeding adults present. These latter, however, may be accepted members of the rookery population, all of whom have the free right to all the resources of the communal territory of the rookery (*see* p. 158). Communal feeding-territories have been closely studied also in a few other birds, such as the

Australian 'magpie' *Gymnorhina tibicen* (Dr. Robert Carrick, orally), and in *Crotophaga ani* and *Guira guira* in Cuba and Argentina respectively (Davis, 1940a and b); but it will appear later that the phenomenon is evidently a general one, applying, not in the breeding season alone, to many birds that flock and to gregarious mammals at any time. In solitary mammals such as foxes the home-ranges of individuals freely overlap; but each *bona fide* resident must have the established use of a traditional earth or den in order to be tolerated and allowed the freedom of the local resources; and so it goes on (*cf.* fig. 8, p. 100).

In most cases the personal status of the individual with respect to his rivals assumes a great importance; and in fact conventional contests frequently come to be completely divorced from tangible rewards of property, and are concerned solely with personal rank and dominance. All who carry sufficient status can then take whatever resources they require without further question or dispute. We shall find that, although widespread and common, these abstract goals of conventional competition are especially characteristic of gregarious species.

The subject is evidently a complex one, that will require much amplification. Enough introduction has perhaps been given, however, to reveal that conventional competition really exists, and to suggest the forms the prizes or goals can take and the way they can be made to serve as dummies or substitutes for the ultimate goal that should never be disputed in the open—the bread of life itself.

1.4. *Social organisation*

In any homeostatic system there are necessarily two component processes. One is the means of bringing about whatever changes are required to restore the balance when it is disturbed, or to find a new balance when this becomes necessary through changes in conditions external to the system. It has already been briefly indicated in the previous section that populations (among the higher animals at any rate) do have the necessary powers to adjust their own population-densities. The other essential component of homeostasis is an input of information, acting as an indicator of the state of balance or imbalance of the system, that can evoke the appropriate corrective response. A stimulus is required that will check and reverse the trend of the system when the balance sways in one direction, in order to bring it back into equilibrium. A device of this kind is familiar to electronic engineers in the design of stable electrical systems, and is described by them as negative feed-back.

In the balance that we are considering here it is postulated that population-density is constantly adjusted to match the optimum level of exploitation of available food-resources; and as food-supply 'futures' change with the changing seasons the population-density must be adjusted to match—in so far as this is possible in existing circumstances. What is needed in the way of feed-back, therefore, is something that will measure or reflect the demand for food, assessing the number of mouths to be fed in terms of present and prospective supplies; to use another analogy, it has to undertake the instrumentation, and to respond to population-density and economic conditions in the same general way as a thermometer is used in a thermostatic system.

Free contest between rivals for any commodity will readily provide such an indicator. The keener the demand the higher the price in mettle and effort required to obtain the reward: the tension created is thus ideally density-dependent. One of our guiding first principles, however, is that undisguised contest for food inevitably leads in the end to over-exploitation, so that a conventional goal for competition has to be evolved in its stead; and it is precisely in this—surprising though it may appear at first sight—that social organisation and the primitive seeds of all social behaviour have their origin. This is a discovery (if it can be so described) of the greatest importance to the theory.

Any open contest must of course bring the rivals into some kind of association with one another; and we are going to find that, if the rewards sought are conventional rewards, then the association of contestants automatically constitutes a society. Putting the situation the other way round, a society can be defined for our purposes as *an organisation capable of providing conventional competition*: this, at least, appears to be its original, most primitive function, which indeed survives more or less thinly veiled even in the civilised societies of man. The social organisation is originally set up, therefore, to provide the feed-back for the homeostatic machine.

It might easily be assumed that male birds competing for territories (to return to the same illustration) were the direct opposite of a society, being all at enmity with each other; but this would be a completely false conclusion. As Kalela (1954) first pointed out, they are in fact strongly coordinated together, and often conspicuously coherent with non-random contiguous territories—essentially a static flock with a higher ' individual distance ' than we commonly associate with the use of that word. They sing in emulation of each other with the primary object of being heard and recognised by their rivals, particularly at the purely conventional singing-hours at dawn and dusk (this convention is dealt with in Chapter 15); and they are in personal even though militant contact with all their neighbours. This two-faced property of brotherhood tempered with rivalry is absolutely typical of social behaviour; both are essential to providing the setting in which conventional competition can develop. The remaining essential characteristic of a society, it may be repeated, is that it is concerned with convention: indeed conventional competition and society are scarcely distinguishable facets of a single phenomenon.

All the higher animals—especially the vertebrates and arthropods—have evolved organised societies that come within our definition. The lower down we go in the evolutionary scale the less conspicuous and elaborate does the homeostatic machinery become; but we shall find later on that wherever secondary sexual characters have appeared there is almost sure to be a recognisable conventional society (*cf.* Chapter 12). The existence of such societies is usually most clearly revealed, however, by the possession of methods of mutual communication and recognition, through conventional signs and signals given and received, which form an indispensible link in social integration. In fact the best starting point for the more detailed development of the theory has seemed to be a study of the nature and occurrence of methods of social communication in animals; and to this therefore the next several chapters will be devoted.

The reader is already aware that conventions and conventional behaviour, with which we are so pre-eminently concerned, are in the nature of artefacts, more or less widely divorced from the real sanctions by which their observance is ultimately enforced. The more remote they are from absolute reality, the more do they become symbols arbitrarily endowed with a meaning (just as are the conventional signs that make up a printed page); and the more they are likely to differ from ordinary ' real ' adaptations by being somewhat bizarre or even extravagant in nature. It is particularly desirable to keep this in mind at this stage. The mode of their evolution, which may already have begun to arouse questions in the reader's mind, will be dealt with a little later, in the next section of the chapter.

The actual regulation of population-density is largely a matter of exercising control over recruitment and loss in the population. In some kinds of animals there is a fairly continual dissipation of numbers through mortality, whereas in others discrete generations may arise to succeed each other year by year. In either case there is usually an annual breeding season, the basic function of which is in the first case to make good the losses of the preceding year, and in the second to create a new generation to succeed its progenitor. We have already briefly referred to experiments which demonstrate with great clarity that recruitment (at least in the species experimented with) is density-dependent; and it is obvious therefore that the feed-back part of the machinery will have to be especially active just before breeding commences, in order to elicit the required response from the breeding stock and produce the quota of recruits that current economic conditions dictate.

The same will be true whenever there is about to be some kind of self-inflicted loss to the population—for instance by the voluntary or forced emigration of part or all of its members, as in migratory birds before they leave their summer or winter quarters. Conventional competition is likely to be conspicuous also if the dispersionary equilibrium is upset from outside, for example by an unforeseen failure of the food-supply. As far as bird-migration is concerned, we are going to find evidence to suggest that all the way along the route it is subject to automatic traffic control, so that orderly millions of individuals can complete the successive hops of their long flights without ever encountering dangerous traffic-jams and locally exhausted fuel-resources (*see* Chapter 14, p. 292).

It is well known that the territorial convention in birds tends to be most active and vigorous at a particular one of these special junctures in homeostatic regulation, namely just before mating and egg-laying take place; in practice it often begins to build up some time before, and continues until the process of reproduction is concluded, but the climax is most often reached just when theory would predict it, immediately before breeding begins. At this time conventional competition can assume a variety of forms. Not only is there direct contest over the definition of territorial boundaries, but the males usually spend a large amount of time in aggressive display, by spirited singing or conspicuous flight or both. These displays are, of course, abstract conventions, endowed with a meaning that broadly conveys the threat of physical reprisals upon any rival that dares to dispute the occupant's claim to possession.

Symbolic displays are largely wasted if there is no audience to receive and interpret them; and because of the ordinary necessities of life such as feeding and sleeping they cannot often be indulged in all day long. The conventional, social nature of bird song is in no way more clearly revealed than by the almost universal tendency for all the individuals concerned to do it together, at times when, for a brief convenient period, it becomes the dominant communal activity. This usually occurs at the two periods of day when synchronisation is easiest, namely when the rate of change of daylight is most rapid, at dawn and dusk. Then for some minutes at least there is generally a chorus in which all the rival members of the local society take part. Nothing could be more perfectly adapted to indicate the population-density than such a synchronised vocal display.

Specially-timed communal displays (Chapter 15) occur in every group of the higher animals—in the dancing of gnats and midges, the milling of whirligig-beetles, the manoeuvres of birds and bats at roosting-time, the choruses of birds, bats, frogs, fish, insects and shrimps,—even, if we are not mistaken, in the vertical migration and surface-assembly of innumerable species in the plankton (Chapter 16). They are very commonly synchronised at dawn or dusk or both; but in fact they need not in certain circumstances (as when the participants have little else to do, or when feeding is a communal activity anyway) be confined to special hours at all. They form, as a class, a tremendously important and hitherto completely unexplainable component of social behaviour. It is essential to have a term to designate them, and we shall call them *epideictic* displays—signifying literally ' meant for display ', but connoting in its original Greek form the presenting of a sample. Though rather a rare word in its previous technical senses it already has a place in many English dictionaries.

Epideictic displays are especially evolved to provide the necessary feedback when the balance of population is about to be restored, or may need to be shifted, either as a seasonal routine or as an emergency measure. They generally involve conventions that have evolved away from the direct primitive contest, which is liable to end in bitter bloodshed and even killing of participants, and have come to assume a highly symbolic quality, not even directly implying threat in many cases, but producing a state of excitation and tension closely reflecting the size and impressiveness of the display. In many cases not only is a special time of day set aside, but also a traditional place, to which all the participants resort for the purpose; and this is undoubtedly the underlying cause of almost all types of communal roosting and hibernation, and many other gregarious manifestations among normally solitary animals (*see* Chapters 11, 14 and 19).

The study of epideictic phenomena leads inevitably to a reappraisal of the epigamic displays that characterise the marital relations of the sexes and typically culminate in fertilisation. We shall find that much that has been regarded hitherto as epigamic is unquestionably epideictic. In the prenuptial period, when dispersion is often organised on the basis of the mated pair, it is normal for the male to assume the whole dispersionary task, and to participate exclusively in epideictic displays as the representative of a mated pair (and sometimes at a later stage of a whole family party). The male is then the epideictic sex, and many of his conventional displays and adornments have their primary significance for other males with whom he is in epideictic competition, and only quite secondarily, if at all, do they affect or concern the female. Considerable attention will be given, in Chapters 12 and 21 especially, to unravelling the epigamic and epideictic aspects of sexual displays and the functions and evolution of sexual adornment, dimorphism, polygamy and the like.

Situations of this kind quite commonly arise, in which conventional competition concerns only one section of the population—the larvae, for instance, or the breeding adults as a group, or one sex alone; and in that case the particular category concerned is generally distinctively recognisable as a separate social caste. A special development of this has apparently led to the phase systems of locusts and some other species, in which the migratory phase is uniquely involved in the intensive epideictic displays that culminate in swarming and exodus flights. A discussion of this subject forms part of Chapter 20; and it is in the same chapter that consideration is given to the well-known fluctuations in numbers and density that occur in many populations, seeming as they do to negate any sweeping assumption of a universal evolutionary tendency towards homeostasis.

It is equally possible to discover an opposite situation, in which, instead of having a single species divided into a number of distinctive social castes, two or more species for the time being merge their identities and collaborate in a common dispersionary and social system. A good example of this is found in certain groups of weaver-birds (Ploceidae) in Africa, in which at breeding time all the males assume a highly specific livery and each species is more or less independently dispersed; but during the rest of the year neither sex nor species is readily recognisable, and mixed flocks are formed that share a common life. Sharing food-resources with other species is very common in nature; and it is obviously of no avail for one species to evolve a conventional code for the conservation of its resources if this is freely violated by competing species. To an important extent in some groups, in consequence, interspecific conventions have been built up, and a mutual ' understanding ' reached between competitors for the same resource, just as normally occurs at the intraspecific level. These are refinements of extraordinary interest from the evolutionary point of view, perhaps confined to the highest groups but appearing here and there in birds, mammals and other vertebrates, and in insects such as the social Hymenoptera. They present a radical antithesis to Gause's hypothesis that animals of similar ecology cannot survive together in the same habitat.

For the most part, of course, the book like the theory it presents is concerned with economic affairs *within* the species, at the population level, but Chapters 17 and 18 are devoted to the kind of interspecific relationship just mentioned and some others like mimicry, where the mingling of models and mimics may make the assessment of the population-density of either species alone exceedingly difficult; and in this situation it will be remembered that there are quite a number of mimetic Lepidoptera in which the males—no doubt as usual the epideictic sex—are non-mimetic and retain their visually distinctive specific characters.

1.5. *Social evolution and group-selection*

It is part of our Darwinian heritage to accept the view that natural selection operates largely or entirely at two levels, discriminating on the one hand in favour of *individuals* that are better adapted and consequently leave more surviving progeny than their fellows; and on the other hand between one *species* and another where their interests overlap and conflict, and where one proves more efficient in making a living than the other. Selection at the individual level is often designated as intraspecific, and that at the higher level interspecific. The latter covers a broad range of relationships; it is frequently concerned not so much with ecological overlaps between closely allied species in the same genus as with the mutually conflicting needs of two independent predators seeking the same prey, or two unrelated contestants for the same micro-habitat.

Neither of these two categories of selection would be at all effective in eliciting the kind of social adaptations that concern us here. We have met already with a number of situations—and shall later meet many more—in which the interests of the individual are actually submerged or subordinated to those of the community as a whole. The social hierarchy, or ' peck-order ' and its equivalent, is a common and important product of conventional competition, and its function is to differentiate automatically, whenever such a situation arises, between the haves and the have-nots (*see* Chapter 8). For those high enough in the scale the rewards—space, food, mates—are forthcoming; but when food, for instance, is already being exploited up to the optimum level, the surplus individuals must abide by the conventional code and not remain to contest the issue if necessary to the death. It is in the interests of survival of the stock and the species that this should be so, but it ruthlessly suppresses the temporary interests of the rejected individual, who may be condemned to starve while food still abounds.

The same applies to conventions in general. They are usually in fact extremely important adaptations, potentially of great survival value to the species: but it takes a group of individuals to maintain a convention, and if the individuals are isolated from one another it falls to the ground. Human conventions as we know are characteristic of a place, a people, a creed, a profession; and in general the characteristics and functions of a society belong to it collectively, and cannot be completely represented or discharged by any solitary member.

There are a great many important characters of this kind, not only in animals but in plants also, that are in the nature of collective attributes, all possessing the common quality of contributing to the welfare and survival of the group as such, and when necessary subordinating the interests of the individual. One of these is the reproductive rate. If intraspecific selection was all in favour of the individual, there would be an overwhelming premium on higher and ever higher individual fecundity, provided it resulted in a greater posterity than one's fellows. Manifestly this does not happen in practice; in fact the reproductive rate in many species, and recruitment of adults in others, is varied according to the current needs of the population. Restrictions can be imposed in a number of ways, for instance by each female laying fewer eggs, or eating more of them, or by allowing fewer females to breed—all adaptations that cut directly across the interests of prolificity in the individual. In the extreme case, in various social insects, it has been possible to evolve castes of sterile individuals, something that is inconceivable in a world where the most successfully fecund were bound to be individually favoured by selection and the infertile condemned to extinction; it could only have evolved where selection had promoted the interests of the social group, as an evolutionary unit in its own right.

It has become increasingly clear in recent years, not only that animal (and plant) species tend to be grouped into more or less isolated populations, due very largely to the physical discontinuities of the habitat, but that this is a very important feature from an evolutionary standpoint in the pattern of their distribution (*cf.* Sewall Wright, 1938; Dobzhansky, 1941, p. 166 *et seq.*; Carter, 1951, p. 142). A great weight of evidence will be forthcoming in the course of the book to confirm this conclusion. At least in terrestrial habitats the food-resources on which animals depend are of a strictly localised, immobile character, depending ultimately on a stationary or very slowly changing pattern of vegetation. The local stock of any given animal species, exploiting its resources, consequently tends to adopt many conventions of a strictly localised or topographical character—for example the traditional sites of breeding places. Other conventions rely equally strongly on a procession of mutual relationships among the individual local inhabitants. Above all, the local stock conserves its resources and thereby safeguards the future survival of its descendents; and no such conservational adaptation could have evolved if the descendents did not normally fall heirs to the same ground. Thrifty exploitation today for the benefit of some randomly chosen and possibly prodigal generation of strangers tomorrow would make slow headway under natural selection.

Thus it is clearly of the greatest importance in the long-term exploitation of resources that local populations should be self-perpetuating. If confirmation were needed of this conclusion, it could be found in the almost incredible faculties of precise navigation developed in all long-distance two-way migrants whether they are birds, bats, fish or insects, to enable them to enjoy the advantages of two worlds, and still retain their life-long membership in the same select local stock. Ideally, localisation does not entail complete reproductive isolation however; we shall have to consider later the pioneering element also—in most species relatively small—that looks after colonisation and disseminates genes. These are matters to be discussed in Chapter 20.

Strict localisation endows each local population with potential immortality. The population can therefore undergo adaptation as such, in the same way as the subspecies, the species, or any other group. The extent of its differentiation from other populations will be limited by the amount of pioneering and interchange of membership that occurs with neighbouring populations; but we shall see that local conventions and traditions tend to be long-perpetuated, and that social organisations, with their collective conventions, are sometimes extremely persistent.

Some such local groups may in practice maintain their identity for centuries, and even, as we shall see later, for thousands of years. Others are not so fortunate and suffer a constant turnover of colonisation and extinction, especially in the less permanent types of habitat. None of them last for ever in the same place, on account of the secular changes in climate and geology. Survival is the supreme prize in evolution; and there is consequently great scope for selection between local groups or nuclei, in the same way as there is between allied races or species. Some prove to be better adapted socially and individually than others, and tend to outlive them, and sooner or later to spread and multiply by colonising the ground vacated by less successful neighbouring communities.

Evolution at this level can be ascribed, therefore, to what is here termed group-selection—still an intraspecific process, and, for everything concerning population dynamics, much more important than selection at the individual level. The latter is concerned with the physiology and attainments of the individual as such, the former with the viability and survival of the stock or the race as a whole. Where the two conflict, as they do when the short-term advantage of the individual undermines the future safety of the race, group-selection is bound to win, because the race will suffer and decline, and be supplanted by another in which antisocial advancement of the individual is more rigidly inhibited. In our own lives, of course, we recognise the conflict as a moral issue, and the counterpart of this must exist in all social animals (*see* p. 131).

Group-selection is not by any means a new concept, though it has never been accorded the general recognition that its importance deserves. Particular attention has been paid to it in the present work, because it is fundamental to the dispersion hypothesis; the principal allusions to it in the chapters that follow have all been assembled in the index for ease of general reference. As being the only possible method of evolving sterile castes in social insects it has been recognised by Sturtevant (1938, p. 75) and O. W. Richards (1953, pp. 145-6); and the general action of natural selection on integrated social units as such has been explicitly pointed out by Allee (1940). References to the same general ideas are quite widely scattered in the literature (*cf.* v. Haartman, 1955, with respect to the evolution of clutch-size in birds); and, as described in the following section, they formed an essential part of Carr-Saunders' principle of the Optimum Number.

1.6. *Carr-Saunders' principle of the Optimum Number*

It is probably plain from the context that, after reaching the end of the book, the author turned back to this chapter and completely revised it. In the original draft there was no sixth section, because it was only relatively late in the day that I discovered that, albeit in a rather special and restricted context, the theory of dispersion through conventional behaviour had actually been published before.

This was in fact a very rewarding discovery. In its previous form it had been conceived as applying to man—and indeed primitive man—alone; and its author, Sir Alexander Carr-Saunders (1922), clearly perceiving that it depended on social evolution through group-selection, had assumed its origin as an evolutionary process to date from roughly the lower Palaeolithic, at which time, in his view, ' we should look for the origin of social organization ' (loc. cit., p. 239). Had he realised that social organisation goes back probably to the lower Cambrian, and is adumbrated in the Protozoa and in the plant kingdom, he could as easily have extended the principle as we have done here.

In this pioneer work, *The Population Problem*, Carr-Saunders was at great pains to show first that unrestricted nomadism did not exist in primitive races, but that all of them were without exception essentially territorial; so-called nomads simply travelled about within their tribal areas instead of having permanent settlements (*i.e.* they all tended to be locally self-perpetuating): next that they all limited their fertility by a variety of practices, the most important of which are abstention from intercourse, abortion and infanticide. Later (pp. 200, 213) he comes to the point that every population has an optimum number, or an optimum density, that enables the greatest income per head to be earned; and above this density returns diminish.

He sees that it is fatal to depend on starvation to eliminate the surplus, because it inevitably leads to social instability and to making useless all the special skills in food-getting on which the healthy population depends (p. 214). In this the wishes of the parents may have to be overridden in the interests of the community; and strict conformity with social practices is enforced by social pressure (p. 216).

Finally, this cannot be brought about without group-selection. 'Those groups practising the most advantageous customs will have an advantage in the constant struggle between adjacent groups' (p. 223); but group-evolution, and the optimum regulation of numbers in this way 'are clearly only applicable to races among whom social organization has become established' (p. 239).

There the matter rests. The several foundations of the present theory are all comprehended. Nothing could have given me greater reassurance than the knowledge that so distinguished a student had earlier pioneered the road. A fuller discussion will be found in Chapter 21 (p. 493).

Chapter 2

The integration of social groups by visible signals

2.1. *Signals used for social integration and other purposes*

A social organisation that is evolved to provide the basis for conventional competition presupposes that its members will be able first of all to recognise one another, and then to comprehend the bond of interest that unites them—a bond that arises from seeking the same conventional rewards, and turns under stress into rivalry. Recognition and comprehension imply communication, and this is a two-way process of transmitting and receiving, of signal and perception. The signals may evoke one or sometimes more than one of the senses, commonly either visual, auditory, chemical or tactile. They tend to be conventional signs, for the most part arbitrary and without directly intelligible reference to the response they elicit in the percipient. To take an example not connected with dispersionary behaviour, nestling birds may beg for food by stretching their necks and opening their mouths as wide as possible, or in some species by gripping the parent's bill, their signals in either case being closely related to the response released; but they may equally well solicit food by making vocal sounds that are purely symbolic, with a meaning understood only by an established convention, and not inherent in the sounds themselves. The same applies to the warning colour-patterns of black with yellow or red that occur in insects, amphibia and snakes, for instance; the meaning of these, if common opinion is not mistaken, relates conventionally to the distasteful or venomous qualities of the bearers.

The use of signals for social purposes naturally demands the possession of sense-organs capable of perceiving them. Animal groups containing many highly-coloured species are in some cases—among the insects, bony fishes and birds, for example—known to possess colour-vision; and similar correlations often hold in respect of other senses, as in the amphibia where sound production and the development of the tympanic membrane go together, or in the mammals where the multiplicity of scent glands coincides with a highly developed olfactory sense. But there are, of course, many other animals with brilliant colour patterns which they themselves apparently cannot perceive, such as sea-anemones, pennatulids, corals, tubicolous and other polychaete worms, and echinoderms; so that it is always necessary when considering the part that any of these potential signals could play in social communication to take account of their reception as well as their emission.

It is noteworthy that the types of signal that can be instantly perceived at a distance are on the whole commoner, and probably therefore more effective under most conditions, than those that rely on chemo-reception and the tactile sense.

Considerably greater attention has been given to studying the development and uses of colours, patterns and visual signals generally than to those involving any of the other senses; but in all cases the functions to be discharged in communication are essentially the same, and it is not particularly difficult to devise an analytical scheme applicable to signals in general along the same sort of lines as the familiar one drawn up by Poulton (1890) for the uses of animal colours.

From our point of view the primary subdivision in the functions of communication should clearly fall between signals directed at other members of the same species and those (like warning colours) directed at other kinds of animals; these divisions are respectively intra- and inter-specific. It is naturally with the former that we are generally concerned in the study of social integration. Intraspecific signals are obviously varied in purpose; those that are essentially social in function are capable of expressing a wide range of intentions, needs, moods and emotions that could if necessary be systematically analysed. They include expressions of threat, warning, fear, pain, hunger, and—at least in the highest animals—such elemental feelings as defiance, well-being, superiority, elation, excitement, friendliness, submission, dejection, and solicitude. Poulton's classification was hardly concerned with these at all, and out of his various categories of coloration that can reasonably be regarded as signals, only the 'episematic' (recognition marks) and 'epigamic' (courting colours) belong to our intraspecific division. It suits our purpose to continue to recognise epigamic signals, which can with advantage be restricted to communications between members of opposite sexes concerned more or less directly with mating and marital life. For the rest, however, it would lead us unnecessarily deep into ethology to attempt any detailed analysis of the purposes of social and familial communications at this stage. Some of the special purposes will emerge as we go along, but for the present all intraspecific signals that are not epigamic can be uncritically lumped together as social, since there appear to be none which this term cannot reasonably be stretched to cover.

Inter-specific signals, on the other hand, have no primary message for members of the signaller's own species, and are not directly concerned therefore with the subject of dispersion; but it is desirable to consider them briefly, in order to be clear later on regarding the different purposes for which signals are emitted. They seem to fall into four main categories, not rigidly separated, including (*a*) warning or 'keep-off' signals, intended to save any assailant or disturber-of-the-peace from the formidable consequences of molesting an animal protected by unpleasant taste, sting or odour; (*b*) intimidating signals, intended to frighten or shock the assailant into leaving the owner alone, or giving the latter a moment's pause in which to effect its escape: to this Poulton's ' pseudaposematic ' category corresponds in part; and (*c*) decoying signals, intended to draw the approach or rivet the attention of a member of another species, either to enable it to be seized for food (as in the angler-fishes belonging to the order Pediculati), or to divert it from attacking the eggs, the defenceless offspring, or the vital parts of the signaller itself. A fourth category (*d*) has developed from the fact that it is greatly to the advantage of most animals to suppress entirely the emission of signals most of the time, in order that their presence and identity may be overlooked by other animals; and to this end some animals have not merely adopted negative or cryptic colours and attitudes that enable them to disappear into their surroundings, but they give out actively dissembling or masking signals, which cause them, for example, to be mistaken by predators for unpleasant species when they are not, as in the case of Batesian mimics, or by their prey for innocuous species when they are in fact dangerous, as in the case of some ant-like spiders (for fuller details, *see* Cott, 1940; Tinbergen, 1953, Chap. 6).

Animals that are permanently protected by their nauseating taste, poison and defensive weapons may emit continuous warning signals, through their conspicuous permanent colour patterns, which are effective to visual predators whenever there is light enough for them to be seen. But the majority of signals made either for intra- or inter-specific communication are produced *ad hoc* in the appropriate context. This almost always applies to categories (*b*) and (*c*) in the preceding paragraph. Shock signals in particular owe their effect to the element of surprise; thus the cryptically-coloured eyed hawk-moth (*Smerinthus ocellatus*), resting on the bark of a tree-trunk, when touched by a sharp object such as a bird's bill, suddenly spreads its fore-wings to reveal a pair of dark staring ' eyes ' on the hind-wings, which are waved slowly back and forth (Tinbergen, loc. cit., p. 94). The pistol-shrimp (*Alphaeus californiensis*) is supposed for similar reasons to emit its unexpected loud detonation (MacGinitie, 1949, p. 276), and the tube-dwelling polychaete worm *Chaetopterus* to turn on its floodlights (Nicol, 1952, p. 429). Decoy signals may similarly require special postures and movements, such as the ' broken-wing act ' or injury-feigning which is widely developed among birds.

There is not always any certain criterion for distinguishing inter- from intra-specific signals: warning and threat, for example, are common to both, and may use the same conventions in each case, although the ulterior object is totally different. Usually, however, it is the circumstances that make it clear to what kind of audience the signals are being directed.

It is generally recognised that, instead of the visual colours and patterns so often evolved for communication, any of the other media for signalling can be effectively substituted. Thus some species such as rattlesnakes (*Crotalus*) apparently make use of warning sounds, and some of frightening sounds, like the pistol-shrimps already mentioned or the common dor-beetle

(*Geotrupes*) which crepitates when seized; others use intimidating or repellant odours, as do the devil's coach-horse (*Staphylinus*) and the ground-beetle (*Carabus*), or the caterpillars of swallowtail butterflies (*Papilio*) which have a foul-smelling forked horn or osmeterium that can be conspicuously everted during the threat display.

TABLE I

Summary analysis of animal signals

Category of signal	*Directed at*	*Poulton's nearest equivalent*
1. Intra-specific		
a. social	other members of own species or social group	episematic
b. sexual	other sex	epigamic
2. Inter-specific		
a. warning	predators and other assailants	aposematic
b. intimidating	,, ,, ,, ,,	(none)
c. decoying	either prey or predators	pseudepisematic
d. dissembling or masking	,, ,, ,, ,,	pseudaposematic

The various media of communication—visual, acoustic, electrical, chemical and tactile—are considered in turn in the following chapters: but it should be noted that the more highly organised animals, including the insects and vertebrates, seldom rely on communications of a single sensory type. They usually combine the use of alternative methods, for example visual and acoustic, acoustic and olfactory, or even three or four of them, although one often tends to predominate.

2.2. *Visual signals and perception*

Signals intended for visual perception depend either on reflected light, in that case being effective only in sufficiently well-lighted environments, and consequently commonest in diurnal animals; or on bioluminescence, effective only in the dark. The animal must in either case be living in a medium transparent to light, and not underground, in dense vegetation, or in turbid water, where sound or scent is likely to be a more effective means of making contact. Visual methods, however, tend on the whole to be favoured wherever possible above all other means of communication; for not only can animals display infinitely varied patterns, but these can be rendered especially conspicuous, and their meaning altered or emphasised, by movements associated with their exhibition.

Social integration, as already mentioned, requires the recognition of other members of the group and species; and among animals constantly in motion each recognition-contact may depend more on the characteristic general visual appearance of the moving object than on any particular mark or pattern. The movement itself may supply the chief component of the signal. In the Introduction (p. 1), for example, the dispersion of pelagic birds was shown to be closely correlated with the abundance of plankton in the surface waters of the ocean. In such a population of constantly moving members, density adjustment must be a continuous process, and it would appear likely to demand the frequent attention of each particular species of bird concerned, in order that they could govern as efficiently as they do their dispersion in relation to the food available. The most abundant and characteristic oceanic birds, the order Tubinares or petrels, appear to be largely nocturnal in their feeding habits, taking their food when it becomes accessible to them at the surface during the hours of darkness, as a result of its diurnal vertical migration. They are nevertheless for the most part solitary rather than gregarious, and silent, at sea; and their olfactory powers, while perhaps greater than in most birds, are almost certainly insufficient for maintaining contact among thinly-scattered populations such as theirs. Social integration can therefore fairly safely be assumed to be visual; and this accords well with the remarkable (and not otherwise explained) amount of time spent by albatrosses, shearwaters and petrels, in apparently aimless cruising in the daylight hours. Numerous voyagers have noted their tireless flight and ceaseless rise and fall, as they roam the waves hour after hour; they alight rather seldom for any purpose (unless in the wake of a ship), and generally appear not to be seriously engaged in feeding. The function of this diurnal flight could well be ' dispersive ', providing the individual with a frequency sample of visual recognition contacts, which could be integrated with the current abundance of food, and result in compensatory movements and the adjustment of density. Without some such adaptive mechanism, and on a basis merely of uncoordinated chance alone, it would clearly be impossible to achieve the high correlation between density and food-supply that has been shown to exist.

This type of population density determination, the method of sampling by travel and encounter, is perhaps rather general in free-moving animals living in a continuous medium, especially those in, or on the surface of, large bodies of water. Individuals cannot return to, or defend as their own, a fixed position in an unlimited uniform fluid environment, because they have no access to permanent solid points of reference. Either they can each travel independently, achieving the appropriate average density through their variable degree of tolerance or intolerance of the presence of others; or else all those in a given volume or area of water can unite into a school or flock, and, if desired, maintain an ordered spatial pattern with respect to one another as they travel in a body, in due course encountering and reacting to other similar bodies. This amounts to the same thing, only the cruising unit is not an individual but a gregarious group. The two systems are not mutually exclusive and can easily be intermixed. Social integration must be postulated as occurring in all cases through mutual recognition and response; but of course it need not be maintained visually. Examples will be given later of tactile, acoustic and lower-frequency vibration signals employed for the same purpose, and it is likely that chemical signals (including the so-called 'extrametabolites') are of importance also in these environments.

Travel and encounter can be used by very mobile animals in any medium when they are free to range over an undivided territory or habitat. Butterflies, for instance, may avail themselves of it, through their bright colours and patterns, which are often concealed except during their conspicuous and leisurely flight. The frequency of contacts will depend on three factors—the population-density, the speed of movement and the maximum range of recognition; so that the slower an animal moves the less frequent contacts it will make; and there is a tendency among slower-moving species, such as terrestrial mammals, or silk-spinning insects and spiders, or snails, for example, to leave a persistent chemical or tactile signal wherever they go, which must greatly increase the frequency of 'contacts'; but further consideration of this must be deferred to the appropriate place (p. 98). It is sufficient to point out that inverse correlations would be expected, in an unlimited fluid medium, between the population-density, the cruising speed, and the maximum range of recognition; that is to say that slow-moving animals must remain close together, or else have far-carrying or persistent signals, in order to maintain integration, whereas those which travel at higher speed can be spaced at greater intervals without losing the necessary contact.

Visual recognition within the species appears to be made possible in the great majority of cases by the display of conspicuous patterns and colours. Where there is nothing to be lost by continuous advertisement, as in large birds like swans, storks, albatrosses or gannets, which can protect themselves adequately from predators when adult, and need no concealment in hunting, we find both sexes with their plumage as conspicuous as possible. Probably on account of the high degree of efficacy of their visual advertisement, a number of species in these families have dispensed almost entirely with alternative methods of recognition-signalling; and, unlike the majority of birds which as a group are the most vocal of animals, they tend to be silent, except in sexual or social-dominance displays at close quarters (*e.g. Cygnus olor*, *Ciconia ciconia*, *Sula bassana* and many other Steganopodes, almost all Tubinares, etc.). The same limitation applies equally to some of the great raptorial birds, which render themselves highly conspicuous by soaring hour after hour; a good many are entirely silent in flight, and some may be virtually mute at all times (including the California condor *Gymnogyps californianus*, certain eagles, *e.g. Aquila chrysaetos*, *Haliaetus albicilla*, and especially vultures, *e.g. Aegypius monachus*, *Gyps* spp., *Pseudogyps* spp., *Torgos tracheliotus*, *Neophron percnopterus*, etc.).

The visual recognition signals typical of most birds are of much value to ornithologists as a means of species identification, and even where there is considerable sexual dimorphism they may be equally developed in both sexes and even in the immature stages. Examples of this are the specula of the surface-feeding ducks, produced by lustrous, coloured secondaries in the wings; or the wing-bars and coloured rumps of various finches, such as the pine grosbeak (*Pinicola enucleator*), white-winged crossbill (*Loxia leucoptera*), chaffinch (*Fringilla coelebs*) and brambling (*F. montifringilla*). It is very common for visual recognition signals to appear only when the bird moves or flies or otherwise deliberately exposes them. As long as it remains still its identity is concealed visually, alike from predators and members of its own kind. When it moves it becomes a relatively conspicuous object anyway, and this conspicuousness need not always be much increased by the recognition-label. Identity-marks include the white and coloured rumps on many passerines including the finches just mentioned; special wing-patterns, for instance in the lapwing (*Vanellus vanellus*) and many other Limicolae; white, coloured or patterned tail-feathers, as in the pipits (*Anthus* spp.), wagtails (*Motacilla* spp.), redstarts (*Phoenicurus*) and related forms. The number of examples which could be given is almost unlimited, because visual recognition signals are found in almost all birds.

Similar visual signals occur in most other groups of animals with highly developed eyes, and it is unnecessary to give more than representative examples. The combination of cryptic or concealing coloration in the motionless or resting animal, and recognition marks that appear as soon as it moves, just noted in birds, is exceedingly common. In orthopteran insects it is to be seen among those widely distributed grasshoppers that have their hind-wings resplendent with black and yellow, red or blue; very similar patterns are found in phasmids and various Hemiptera. The Carolina locust (*Dissosteira carolina*) may suffice as an example. When stationary the insect generally passes unnoticed. When it flies, it flashes the oversize hind-wings, coloured black bordered with brilliant yellow, and at the same time often emits a crackling sound. (This is not nearly so loud, however, as that of its relative and compatriot the cracker-locust *Cercotettix verruculatus*, which, with much less conspicuous yellowish-green hind-wings, instead relies principally on the sound signals consisting of a series of loud clicks like a burst from a police-rattle.) 'It is generally believed that so-called "flash colours" serve to confuse or misdirect an enemy in the pursuit of prey. . . .' (Cott, 1940, p. 376). This is a theory that has been attributed originally to C. Hart Merriam (*cf.* Chapman, 1928, p. 96); and Cott himself has no better alternative to suggest. But he says lower down: 'It must be admitted that in the present state of our knowledge the precise biological meaning of flash colours is not clearly understood.'

There are fairly convincing reasons for thinking that misdirection of pursuers is at most a secondary effect. In *D. carolina* none of the nymphal stages would be so protected, since the coloured wings and the ability to stridulate appear only in the adults. Flights are taken spontaneously, and the insects may frequently be seen to hover in the air, apparently displaying themselves. 'They rise at first about three or four feet making a light purring or beating sound and then, rising higher, change the motion of the wings when a curious, sharp, seesawing sound is produced' (Townsend, 1891). 'The flight consists of two positions, one with a very fast wing-flutter while hovering in the air. Then a change of position usually occurs and a further, longer period of hovering during which the flutter of the wings is at a distinctly slower rate' (Pierce, 1948, p. 258-259). Furthermore, red-headed woodpeckers (*Melanerpes erythrocephalus*) have been observed in southern Ontario perching on poles, from which they were making short flights to snatch flying insects; 'the most conspicuous insects taken were adults of the Carolina locust. . . . The locusts would . . . hover in the air with rapidly beating wings or dance up and down in flight above a particular spot. . . . During this hovering flight the locusts were easily snapped up by the foraging woodpecker.' On another occasion two or three woodpeckers were seen similarly engaged, and 'in some cases the bird would follow the locust down to the ground and then carry it to a pole before devouring it' (Judd, 1956). The flash colour theory is clearly contradicted in this case, since the woodpeckers succeeded in following the locusts even when they 'disappeared' by folding their wings and dropping to the ground.

It seems probable that the locusts are actually performing song-flights, and their dispersion is somewhat akin to territory in birds, as in the case of many other Orthoptera. Male rivalry and fighting have actually been observed in close association with the flights in this species, *Dissosteira carolina* (Snodgrass, 1925, p. 413) (*see also* p. 48). Professor Pierce of Harvard, the physicist who undertook important pioneer work in analysing the sounds produced by insects, has the following comment (1948, p. 262): 'These aerial evolutions in [*D. carolina*] take place prominently at the height of the mating season, and the flight . . . is a hovering over a nearly fixed spot . . . . It is described by various entomologists as a "sexual maneuver by the male in the presence of one or more females on the ground." I have not been able to determine that there was a female present to admire the male's antics. In fact, it seemed to me equally plausible to assume that the flight of the male takes place for the benefit of other males, rather than as an effort to charm an admiring female.' This detached observation by a scientist unhampered by preconceived biological ideas should not pass unnoticed.

In the Lepidoptera visual recognition marks revealed only by flight or deliberate movement are very common also, and often there could be no serious contention that they are developed as 'flash colours' to confuse enemies. More or less cryptic patterns on the undersides of the wings can be found in most of those families of butterflies which fold the uppersides together when at rest, and thus hide their bright signal-patterns; they are extremely common in the Satyridae (graylings, browns, arguses, etc.), Nymphalidae (fritillaries, tortoiseshells, etc.), Lycaenidae (hairstreaks, coppers and blues) and Pieridae (whites and sulphurs). We may note here in passing a North American satyrid, the pearly eye (*Lethe portlandica*) that has been found, like the Carolina locust, to exhibit a type of territorial behaviour; the ' males are quite likely to adopt a particular tree trunk as a " territory ", return to it day after day and chase other butterflies away from it. Combats between males are thus frequent ' (Klots, 1951, p. 66).

In moths, the forewings usually conceal the underwings in the resting insect; bright patterns of white, red or yellow offset with black, resembling those of the Orthoptera, are found on the upper surface of the underwings in a relatively small number of genera, including especially the well-known handsome species of *Catocala* (Plusiidae), and various others among the Caradrinidae, Brephidae, Arctiidae and Sphingidae. Rather few of these, other than the tiger-moths (Arctiidae), are abroad voluntarily in daylight, but some of them at least appear to have a well-synchronised period of flight at dusk, whilst the light intensity is still adequate for sight-recognition. Specific distinctions of colour and pattern are, as usual, the rule in these genera, though *Catocala* contains several groups of species based on rather minute differences.

Turning again to the vertebrates, it is well established that various teleost fishes can distinguish colours, and that many have developed visual recognition marks. Leaving aside luminescent signals for the present, we find visual patterns best developed in species in shallow water where illumination reaches a high intensity. Most of the trout, such as *Salmo trutta* and *Salvelinus fontinalis*, have brightly-coloured ocelli on the sides of the body. In the North American freshwater family Centrarchidae, the numerous species of sunfishes have developed a special ear-like flap on the operculum, covered by pigmented integument and differing sharply in size and colour-combinations in the various species. Even in some apparently ' difficult ' families, such as the freshwater Cyprinidae or the freshwater and marine Gobiidae, it usually does not take very long to learn to recognise the species at a glance by external characters alone. As we shall see again with the photophores of luminescent fishes, recognition patterns and silvery reflecting surfaces are almost always developed on the flanks and underside, where they are invisible from above, and where, on deliberately turning the body, they will flash out clearly against a contrasted background (*cf.* Ward, 1919, pp. 16-22).

Quite similar conditions obtain in the amphibia, the diagnostic colours of which, as well as their nuptial colours, are often developed on the flanks (*e.g.* the lateral row of scarlet, black-ringed dots on the common North American newt *Triturus viridescens*) or on the belly (*e.g.* many species of *Triturus*, or the fire-bellied toad *Bombinator igneus*). (Many amphibia have brilliant patterns over the upper surface as well, *e.g.* in *Salamandra* and *Ambystoma*, or the neotropical tree-frogs *Dendrobates* spp.; but these types are probably all poisonous and their colours are serving primarily to give warning to potential predators (Cott, 1940, p. 264).) In aquatic reptiles particularly the freshwater turtles, brightly coloured lateral and ventral patterns are again common.

The remarkable arboreal lizards belonging to the genus *Draco* have patagial fans of great size supported by ribs, and these fans can be extended at will to display their brilliant pattern of black and red, orange or yellow. Cott associates these with the ' flash colours ' of insects, at the same time calling attention to the fact that they are used in display, that is, in signalling to others of the same species. ' On a branch . . . they [the lizards] greatly resemble butterflies, especially since they have the habit of opening and folding their pretty wings ' (Gadow, 1901, p. 517). Evans (1938) describes the male of the lizard *Anolis sagrei* in Cuba as occupying a look-out post on a tree-stump or fence-post within its territory and displaying visually to intruders by erecting the dewlap and crest. Climbing and arboreal reptiles may well be better situated for making visual displays than the more strictly terrestrial forms of lizards and snakes, where the combination of general cryptic coloration and signals visible at any worthwhile distance may be very difficult to achieve. Various alternative methods of social integration are therefore indicated in the latter, including the use of audible and olfactory signals, and group-aggregations associated with reproduction or hibernation or both.

In many mammals visual recognition is subordinate to other, especially olfactory, methods of signalling, in accordance with the nocturnal habits of the majority of species. A number of mammals use the tail as a semaphore, though not necessarily for social contact; the dog, cat and squirrel serve well to illustrate its versatility in the expression of emotion. However, in two well-known groups of ruminants, namely the deer and antelopes, and in the rabbits and hares, the tail is usually part of a recognition system which is most frequently brought into action when the animal is moving, and thus falls into the category just illustrated in birds and insects.

In the deer ' the tail and disk are totally different in each species ' (Seton, 1909, p. 73); the white-tailed deer (*Odocoileus virginianus*) gets its name from the large foot-long tail, white on the ventral side, which is raised upright like a flag, and sometimes switched from side to side, as the deer retreats. The antelopes belong to a very distinct family, the Bovidae, but their adaptations converge with those of the Cervidae to such an extent that the two are freely confused in the popular mind. In the prairie-dwelling prong-horned antelope (*Antilocapra americana*) the white buttock patches, which Seton calls the ' discograph ', seem 'at first like the rest of his spots—a mere patch of white coat; but it is found to be specialised for an important service. It is composed of hair graded from short in the centre, to long at the front edges. Under the skin of the part is a circular muscle by means of which the hair can, in a moment, be raised and spread radially into two great blooming twin chrysanthemums, more or less flattened at the centre. When this is done in the bright sunlight they shine like tin pans, giving flashes of light that can be seen farther than the animal itself, affording a conspicuous identification mark that must be of great service to the species ' (Seton, 1909, p. 224). Musk-glands are situated in the centre of the spots. Erectile hair is similarly found in the buttock-patches of a number of deer.

The springbuck of South Africa (*Antidorcas euchore* (Forst.) = *A.*

PLATE I

Springbuck in ' pronking ' attitudes as they jump to avoid a bullet that has struck the ground. (From J. G. Millais, 1895, p. 16.)

marsupialis (Zimm.)) provides another good example of this sort of display. Like most antelopes, it has a bold specific colour pattern on the hind quarters and face; but in both sexes there also extends forward along the back, from the root of the tail to the saddle, a long double fold or narrow pouch of skin (whence Zimmermann's name *marsupialis*), lined with snow-white hairs 4 to 6 inches in length. ' The name springbok arose from the habit of these bucks when startled or at play, of leaping high into the air with body curved, legs held stiff and close together, and head down; at the same time the line of long white hair is displayed like a fan by the action of certain skin muscles ' (Fitzsimons, 1920, III, p. 94).

Leaping into the air, apparently for the sake of display, is especially associated with these same families of mammals: springbuck clear seven to twelve feet vertically, and impala (*Aepyceros melampus*) the same; the Indian antelope (*Antilope cervicapra*) also ' has the habit of occasionally springing into the air, all the members of a herd bounding, one after the other. This is done . . . before they are much frightened, when the herd is first moving off ' (Blanford, 1889-91, p. 523). The black-tailed jack-rabbit (*Lepus californicus*), ' when running, every so often . . . leaps much higher than usual '; Anthony (1928, p. 493) believes this is ' probably to enable it to watch the back track and the whereabouts of the pursuer ', but it could very easily be a display of the springbuck type. Of the white-tailed jack-rabbit (*L. townsendii*) Coues (1876) wrote: ' The first sign one has usually of a hare . . . is a great bound into the air, with lengthened body and erect ears. The instant it touches the ground it is up again, with a peculiar springy jerk, more like the rebounding of an elastic ball than the result of muscular exertion. It does not come fairly down, but seems to hold its legs stiffly extended, to touch only its toes, and rebound by the force of the impact. The action is strikingly suggestive of the "bucking" of a mule. . . .' The springs gradually grow weaker until it finally squats in its tracks. This behaviour very closely resembles that of the springbuck, and likewise appears to be demonstrative rather than inquisitive.

Rabbits and hares in fact have adaptations for visual signalling rather similar to those of antelopes; they all possess some variety of contrasting white or black ' scut ', which is revealed in retreat, and many of them have expressive ears as well. Of the white-tailed jack-rabbit in Manitoba Seton observes that ' in noting the hare's methods of intercommunication we must not forget its uniform, so distinctive with its blazonment of markings black and white—markings that plainly advertise to all the world the wearer's identity ' (1909, p. 661). In the white-sided jack-rabbits (*Lepus alleni*, *L. callotis* and related forms) of Mexico and adjacent Arizona, there is a specialisation of the cutaneous muscle every bit as remarkable as those already described in the pronghorn and springbuck. As E. W. Nelson wrote: ' By means of muscles the skin of either side can be drawn over the back at will. In this manner the buffy or brown dorsal area is shifted more or less completely to one side and the white on the opposite side is drawn nearly or quite to the median line. This habit has been observed when the rabbits were standing, or moving along at moderate speed, usually after they had been driven from their forms. This enlargement of the white area is always on the side turned toward the chance intruder, and accordingly alternates from side to side as the animals zigzag away. In the bright sunlight the snowy white side flashes brilliantly, attracting attention from afar, and affording a fine example of directive coloration. . . . I am inclined to think this flashing of the white is most frequent during the rutting time' (Nelson, 1909, p. 115).

A number of illustrations have now been given, taken from the insects and vertebrates, of colour patterns which are suitable for visual recognition signalling, and in fact appear to be used for intraspecific communication. In some of the cases, possibly, the same patterns may be turned secondarily to the diversion of predators. It is a feature of many of them that, in order to combine an overall cryptic coloration with signals visible at a distance, the signal patterns are concealed and have to be exposed by special flights, muscular movements or postures; moreover, as our own experience with signals shows, flashing patterns command attention more effectively than those that are continuously exposed. This is a general principle, applying with possibly still greater force to light and sound signals.

Before leaving this section it seems desirable to draw attention to the fact that, unless our own powers of visual recognition are in some essential way defective, there are groups of the higher animals which are notably wanting in specific visual recognition marks. The various species of *Clupea*, including the pilchard, herring and sprat, which in their immature stages frequently consort together in the same sheltered habitats, are very difficult indeed to differentiate by eye at that age. Among birds there are many notorious cases, of which three will suffice, namely the palaearctic warblers of the genus *Phylloscopus*, the nearctic tyrannid flycatchers (*Empidonax* spp.), and, above all, the females and eclipse-plumaged males of numerous African species of weaver-birds, bishop-birds and widow-birds (*Ploceus*, *Euplectes* and *Coliuspasser*). In butterflies, to take another different group, many boreal and arctic species of pearl-bordered fritillaries (*Brenthis*), yellows (*Colias*) or blacks (*Oeneis*) require close attention to detail and even the use of a lens to differentiate them; moreover their uppersides, which normally carry the recognition marks in butterflies and are obviously showy in *Brenthis* and *Colias*, are remarkably uniform in each genus and more or less devoid of useful taxonomic characters. In most of the species of the widespread genus of common social wasps, *Vespula*, the queens and drones can be identified specifically only by small details visible at very close quarters, and the workers may be considerably more difficult to tell apart; it is common to find two or three almost indistinguishable species occurring abundantly in the same locality, such as *V. vulgaris*, *V. germanica* and *V. rufa* in Britain (to which the somewhat more distinct *V. norvegica* and *V. sylvestris* may be added over large parts of the country). The resemblances appear to be a good deal closer than can convincingly be attributed to ' Müllerian mimicry '. Numerous other examples will doubtless be apparent to systematists familiar with particular groups.

It is necessarily true that all these species have some means of identifying their own kind, either audible or olfactory perhaps, at least in the breeding season; but they frequently stand out as exceptional members of the orders to which they belong, and it must consequently be considered whether visual recognition has in them been suppressed at the species level, and instead refers to the genus as a whole. There are indications at least in some of the groups mentioned that for part of their lives specific distinctions are little observed: this is true of the mixed groups of young clupeids and of bishop-birds; and, as with the social wasps just mentioned, three or four species of fritillaries may sometimes be found flying in the same habitat. It is to be noted that in the case of most of the birds the males are perfectly distinct in the breeding season, either by their songs in *Phylloscopus* and *Empidonax*, or by their showy nuptial plumages in *Euplectes*; but in winter their specific characters become more or less completely suppressed, and they are then scarcely easier to identify than the females and immatures. In some species of *Ploceus* even the male nuptial plumages are distinguishable only with difficulty, and White (1951) has recorded that at Lake Mweru he ' found *P. intermedius*, *P. collaris* and *P. melanocephalus* literally breeding side by side over a front of some two hundred yards on the lake shore, an unusual phenomenon to occur with three species of a single genus. In the case of the two latter, nests were completely mixed, with no segregation of the species '. They are of course sociable breeders. In the specimens collected (number not stated) the stomach contents were different in the three species.

The possibility that specific barriers may sometimes be broken down, and that groups of species may be dispersed on a common basis, as a mixed population unit, especially in the non-breeding season, may appear at first sight very unlikely, if not impossible on theoretical grounds. Further discussion of it must, however, be deferred for the present, but it will be considered again in Chapter 17, which is devoted to problems connected with the dispersion of siblings and mimics.

2.3. *Bioluminescence and visual signalling*

Pelagic animals, as mentioned already, require to maintain their social co-ordination in the absence of any fixed landmarks which could give a stable and persistent topographical pattern. Many more of them, of course, inhabit the well-lighted surface layers than the deeper levels, and ordinary visual recognition may often be sufficient, especially for smaller organisms living at higher population densities. In the bathypelagic habitat, on the other hand, the total absence of light, and rather low productivity or ' carrying capacity ', make the whole problem of intraspecific contact extremely formidable; and some deep-sea organisms belonging to the higher groups such as Crustacea, cephalopods and fishes, are blind, having evidently come to rely entirely on auditory, vibrational, tactile or olfactory signals.

Many pelagic and bottom-living species are provided with light sources, which appear to be capable of fulfilling most of the normal functions of the visual colours and patterns of diurnal animals. Bioluminescence is apparently used, by many of the more primitive invertebrates especially, as a signal to other species, in order to give warning of stings, and poisonous or unpleasant taste. All the Coelenterata (Cnidaria) are protected by stinging threads, and phosphorescence has appeared sporadically in all three classes, Hydrozoa, Scyphozoa and Anthozoa, into which the phylum is divided. Some of the luminescent forms are free-living pelagic medusae and siphonophores, and others, including several hydroids and pennatulids, are sedentary members of the bottom-fauna. Many of them (and of their non-phosphorescent relatives) are highly coloured in addition. Eyes may be absent altogether, or exist as more or less simple ocelli. Among the Annelida the polychaete worm *Polycirrus* is not only coloured bright red by reflected light, but produces a violet phosphorescence, and is known to be distasteful to certain fish (Benham, 1896, pp. 294-5). All of these protected organisms light up on stimulation, often with a somewhat persistent glow, and examples of spontaneous flashing do not appear to be found. The same is true of the well-known phosphorescent dinoflagellates, including *Noctiluca*, and a number of Ctenophora, though it is not known whether and in what respect these are harmful to other animals.

Light appears to be used also to frighten predators, and this may apply to some of the groups just mentioned, including the last. There are certain luminous millipedes (Cl. Myriapoda) which ' become brighter on irritation ' (Haneda, 1955, p. 360). The well-known polychaete *Chaetopterus* spends its whole life in its tube, but produces on stimulation a weird and vivid pattern of light from many parts of its body; its own eyes are minute and simple. Yet another function, not previously noted in connection with reflected-light coloration, is fulfilled in some of the polynoid polychaetes, which, when provoked, autotomise their segmental scales by means of a specially developed muscular mechanism, and these scales, being luminous, continue to glow after being cast off; they serve most likely as a ' sacrificial lure ' (Nicol, 1953, p. 82), rather comparable with the lizard's tail. (According to Nicol *Chaetopterus* under stress will autotomise the anterior end of the body, perhaps for the same purpose: it can subsequently be regenerated.) The deep-sea squid *Heteroteuthis dispar* and its close relative *Stoloteuthis* (*cf.* Haneda, 1955, p. 354), and several prawns including *Heterocarpus*, can eject luminous puffs or clouds, the effect of which in the dark is no doubt the same as that of the ink-cloud of *Sepia* in daylight, and covers their get-away (Harvey, 1952, pp. 281 and 346).

Still other uses of animal light with which we are not directly concerned here include its supposed employment as a lure to attract food-organisms. This is presumed to occur in the deep-sea angler-fishes belonging to the

Ceratioidea, at least the females of which carry a photophore in their illicia or ' fishing-baits '. Lastly, there is a suggestion that it may sometimes be used, for instance in the myctophid fish *Diaphus*, like a miner's head-lamp to illuminate the visual field of the owner (Marshall, 1954, p. 290).

In various higher groups, however, with well developed vision, including in the sea the Crustacea, cephalopods and fish, and on land the insects, there are examples of luminescent adaptations which appear to be exactly homologous with the recognition and other intraspecific signal-marks that depend on colours and patterns visible by reflected light, already reviewed in the previous section.

One of the principal authorities on luminous fishes, August Brauer (1908), points out that the number and position of the light-organs are specific characters. ' He advocates the idea that in the ocean the light-organs replace the specific colour-markings of terrestrial animals ' (Murray and Hjort, 1912, p. 679). In some even the young have light organs (a condition paralleled in the terrestrial fireflies), and there may be sexual differences, for example in the position and size of the bright tail photophores of *Myctophum* and *Diaphus*, or in the size of the cheek photophore in *Idiacanthus*: but we have already noted exactly parallel cases in colour patterns of birds, and there is no lack of similar examples in the coloration of fishes.

Beebe (1944) also recognises the function of specific recognition, and so does Marshall (1954, p. 284), at least as far as the typical lantern fishes (Myctophidae), with their rather feeble lights, are concerned. The same suggestion has been made by Welsh and Chace (1937) with reference to the luminous prawns, such as euphausiids and sergestids. In the cephalopods, moreover, 'each species has a slightly different arrangement, and in the various families the distribution may be quite different ' (Harvey, 1952, p. 287). As with the recognition colour-marks of fishes and other aquatic vertebrates, photophores are most commonly placed on the flanks and undersides in all these diverse groups, and special movements of the body may therefore be similarly involved in their display. Finally, it may be noted that in the numerous kinds of fireflies, the flash is characteristic for each species in intensity, duration, rate, interval and colour (Harvey, 1940, p. 74).

Of all the methods of signalling in the dark the use of flashing lights seems the most effective. During his bathysphere descents Beebe (1935, pp. 207 and 222) observed a continual succession of lights and flashes, varied in colour, some pale yellow and others pale bluish, some irregular and some blinking, and of an infinite variety as regards size and juxtaposition. At one stage he estimated that there were never less than ten visible at a time.

In warm coastal waters in many parts of the world there are various syllid worms, including the celebrated fire-worms of the West Indies (*Odontosyllis enopla*), which have mass displays at regular times of spawning. They swim at the surface for a period of twenty or thirty minutes, beginning an hour or less after sunset, at dates that are closely correlated with the lunar period. The males emit sharp flashes, while the larger females produce a strong continuous glow as they circle during ovulation, attracting the males to the discharge of their eggs. This is a fairly typical communal breeding display, many other examples of which will be given in Chapter 11. Though it may appear on the face of it solely concerned with the mutual attraction of the sexes, we may later be prepared to recognise that these and certain similar phenomena also provide an exhibition of population-density at the onset of reproduction.

Many luminescent creatures, though they may react to physical disturbance as well, can be observed to flash spontaneously—among them very notably of course the terrestrial glow-worms and fireflies. In the prawn *Sergestes* light is produced in intermittent flashes, in rapid sequence in the different photophores, the whole discharge lasting perhaps one or two seconds. In the numerous planktonic euphausians, which constitute the ' krill ' on which especially the southern whalebone whales largely depend, all the species, and usually some of their larval stages, are phosphorescent; it was observed that the northern *Meganyctiphanes norvegica* ' gave out short flashes of light from time to time as they swam restlessly back and forth in incessant activity ' (Harvey, 1952, p. 343). The Japanese cephalopod *Watasenia scintillans* is known as the firefly squid for the same reason; and, according to Walls (1942, p. 396), ' the pelagic fish *Anomalops* swim in schools, flashing their lights like so many fireflies. Despite the proximity of the organ to the eye, it is probably only a social signal '.

Some of the animals that flash their lights can be stimulated to do so in response to a flashing stimulus. One of the most spectacular of these is *Pyrosoma*, a colonial planktonic tunicate living in tropical seas; the colonies of one of the species sometimes attain the enormous length of four metres, and it may be remembered that the officers of H.M.S. *Challenger* traced their names on one of these cylindrical monsters with their fingers, in order to see their signatures brightly illuminated for several seconds. Polimanti and Burghause (1914) showed that *Pyrosoma* could be stimulated to respond to the turning on of an electric lamp, and also that the light of one colony would excite a response in another colony in the same vessel (cited from Harvey 1952, p. 487). The small ostracod crustacean *Cypridina noctiluca,* formerly collected by the Japanese to yield a luminous powder, has been seen to respond whenever a flashlight was shone on the water (Harvey, loc. cit.. p. 302); and the lantern-fish *Myctophum affine* was observed by Beebe to flash in response to light from his luminous watch-dial (Marshall, 1954, p. 285).

It appears in these latter cases that a light signal from one individual could in proper circumstances evoke a responding signal from other members of the species near enough to perceive it. Luminescent signals are clearly capable of procuring intraspecific recognition, and thereby of serving as the means of social integration. They may sometimes be employed to assemble the sexes at the season of reproduction; but a purely sexual or epigamic function cannot always be assigned to them. In *Pyrosoma*, for instance, the numerous individuals which form the floating colony are, like other ascidians, hermaphrodite, and the growth of the colony itself takes place asexually by budding. Just as with the visual recognition characters of birds, photogenic organs are sometimes found in the immature stages, both in insects, Crustacea and fishes; and, when we come to consider in a later chapter the remarkable communal displays of fireflies (p. 198), we shall find that in some species it involves only the males, ' and seems to have nothing to do with mating ' (Harvey, 1940, p. 74).

As with the reflected-light patterns and colours already discussed, there is again in the use of bioluminescence a strong emphasis on specific distinctions, alike in insects, Crustacea, cephalopods and fishes; and an exactly analogous development is, of course, to be found in the use of animal sound, which forms the subject of the next two chapters. It appears that any one of the different types of signal source is capable of providing an alternative means of mutual recognition among rival members in the society to which they belong. Some signals may be common to all individuals, some confined to males at the inception of the reproductive cycle, some to immature larvae, or to any other restricted category of potential competitors. Recognition may in some cases be sufficient in itself to set up the competitive stress that provides the index of population-density and elicits the appropriate response from the animal's homeostatic machinery, without any further display of open threat or actual contest.

2.4. *Summary of Chapter* 2

1. A social organisation presupposes communication between its members; and the most fundamental exchange of information is mutual recognition. An animal's identity is generally made known by signs or signals, very often of a conventional character, which have to be observed and comprehended by the percipient. Any of the sensory media can be used for this kind of communication, but signals that are instantly perceptible at a distance are in most circumstances more effective than those that rely on chemo-reception or touch.

In most animals, naturally, communication fulfills an immense range of other functions also. In making an elementary analysis of these functions, the first division falls between (1) communications that are intraspecific, i.e. addressed to other members of the same species, and (2) those that are interspecific. The latter have relatively little bearing on dispersion; they are divided into four fairly distinct categories, namely (a) warning, (b) intimidating, (c) decoying and (d) dissembling (*see* Table I, p. 26). Of the intraspecific functions of communication no simple analysis can be offered, but they are provisionally subdivided into (a) social and (b) sexual or epigamic.

2. Signals intended for visual reception rely on reflected daylight in the case of diurnal species, and on bioluminescence in those that are active in dark but transparent media such as air and water. In social integration,

visual methods are the ones most widely used. Not only can the signal patterns be infinitely varied in form and colour, but they can be actively displayed and transmitted by special *ad hoc* movements (or by flashing).

The nature of the movements, or the general appearance of the moving body, may in certain cases be the chief component by which animals identify each other; but often there are special recognition marks.

Animals constantly on the move have an opportunity to assess their population-density by direct observation through what may be called the method of ' travel and encounter '. This could apply rather widely to species living in limitless uniform environments such as large bodies of water or their surfaces, where for want of fixed landmarks it is impossible to identify and defend a home base or to maintain a stationary dispersion. It could be used by cruising pelagic birds which, though they are largely nocturnal feeders, nevertheless constantly make themselves conspicuous by their tireless diurnal ranging flight. It could be used also, for example, by cruising non-territorial butterflies.

Some typical visual recognition marks are mentioned, first in birds and then in insects, especially the Orthoptera. ' Flash colours ' are believed with reason to belong to this category of signal, and not to have any primary connection with evading predators; they are frequent in insects and also in vertebrates including fishes. In mammals the rump and tail ' semaphores ' of deer, antelopes, hares and rabbits are again typical social signal devices, often requiring for their full display elaborate special movements, reaching a climax in the antelopes and jack-rabbits.

Attention is drawn again to some anomalous groups of diurnal animals that are difficult to identify specifically for want of typical recognition marks, for instance the several species of yellow-jacket wasps (*Vespula*) or some of the African weaver-finches (Ploceidae). Here it seems possible that several species may share a common dispersionary system, at least locally or temporarily; but further consideration is deferred to Chapter 17.

3. Not by any means all bioluminescence subserves social recognition, but it is a medium well suited to this purpose and is so adapted, notably in marine and bathypelagic animals, including many Crustacea, cephalopods and fishes; description of its wonderful development in the terrestrial fireflies is deferred to Chapter 11. Of all the methods of signalling in the dark none seems more effective than the use of flashing lights. Though often especially associated with adult life and reproductive swarming, social bioluminescent displays do not necessarily have any connection with mating, and will in certain cases appear later on to be largely or entirely epideictic.

Chapter 3

Social integration by the use of sound : land animals

3.1. *Animal sounds*

The sounds that animals perceive are mechanical vibrations in air, water, or, more seldom, the solid substratum. As a means of communication they have excellent physical properties. They are comparatively easy to produce and to modulate in pitch, sequence, rhythm, duration and quality, and in man they have of course become the dominant medium of social converse through the use of spoken language. From the earliest times man has supplemented his vocal powers also with artificial—mostly louder—signal-making devices such as bells, gongs, whistles, horns, drums and detonators. Sounds can be made to catch the hearer's ear either by day or night, in dense cover, dark burrows or turbid water. No line of sight is necessary, so that caller and listener can both stay hidden. The virtual absence of sound shadows enables the auditory organs to be constantly alert for sounds coming from any direction, without needing to be specially orientated themselves or to scan their surroundings as eyes have generally to do. Most of the higher animals possess a hearing sense capable of analysing—and thus identifying and interpreting—sounds, and in some cases even of locating the direction and range of the source. One should not fail to notice in passing that although quite simple mechanical devices can often be used to produce sounds, the auditory systems that perceive them are frequently, like eyes, marvels of sensitivity and physical precision.

Audible and visible signals are often alternative or complementary methods of communicating, as has long been recognised in birds. Many of the most brilliant vocalists, like the nightingale or hermit-thrush, are secretive and cryptically coloured. Writing of the tropical rain-forest of British Guiana, Beebe (1917, p. 102) observes that ' of the . . . remaining groups [of birds with solitary habits] all were inhabitants of dense jungle and without exception possessed of remarkable vocal powers. These had an interesting generic resemblance in that the tones of the songs or calls were uniformly loud and, in the majority of cases, staccato, or with an insistent rhythm. To anyone familiar with these birds in life it is sufficient to mention tinamou, jungle pigeons, owls, goatsuckers, trogons, motmots, cuckoos, barbets, jacamars, puffbirds, goldbirds, cotingas and woodhewers, to recall memories that first are aural and then optical.'

In the higher animals sounds are for the most part purposefully produced, and mere noise, incidental to other activities, tends to be suppressed. The woodpecker excavating dry wood cannot entirely avoid being heard, and in some large animals the swish of wings or thud of hoofs may be loud enough to carry to a fair distance. According to Fitzsimons (1920, 3, p. 260), 'on a calm night the [African] elephants can often be located nearly a quarter of a mile away by the rumbling of the gas in their bowels'.

Being three-dimensional objects living for the most part in an illuminated world, animals have a relatively difficult task in evolving invisibility; the evolution of inaudibility, however, is in general very much easier, and most animals have virtually achieved it. Sound signals therefore tend to be very positive, intentional and unmistakable things. They generally appear to have been evolved through the development of what were in the first place accidental noises. Thus the woodpeckers have perfected the mechanical tattoo of the bill on a resonant surface of wood (or metal) as a territorial song-like display; the snipe (*Capella* spp.) 'drums' in flight with outer tail-feathers extended; and, when it is roding, the American woodcock (*Philohela minor*) precedes the vocal part of its flight-song with vibrant trilling sounds produced by three specially attenuated outer primary feathers in each wing. The male of the broad-tailed hummingbird (*Selasphorus platycercus*) and of other species in the same genus also produces a trilling sound with its wings as it flies, and lacks the vocal 'song' characteristic of the males of most other hummingbirds. Many similar examples could be found among birds. Among mammals the rabbit (*Oryctolagus cuniculus*) and various ruminants stamp their feet, and the beaver (*Castor* spp.) smacks the water with its tail. In these simple cases there is not much difficulty in connecting the specialised signal with its accidental forerunner; but even where the sound-producing structures are much more complex, as in the pneumatic vocal organs of vertebrates and the elaborate stridulating devices found in many of the higher animals, they have all doubtless originated in a similar way for the purpose of developing and putting to use sounds initially appearing as accidental noise.

Many of the voluntary sounds produced by animals can readily be assigned to functions differing little from those previously described for visual signals (p. 24). Of the interspecific categories, which include sounds particularly directed at animals of other kinds, only a few representative examples need be given. (a) Warning sounds probably include the rattle of the rattlesnakes (*see* p. 53) and the drone of angry wasps and bees. (b) Intimidating sounds, for the purpose of momentarily startling an assailant while the intended victim escapes, or driving it off, may include (in addition to the examples given earlier) the loud snap of some species of the click-beetles (Elateridae), the unexpected 'explosion' of many game-birds when disturbed off the ground, or the hiss of the cat. No better instance can be given, however, than that of the bombadier beetles (*Brachinus*); in the words of Kirby and Spence (1817, II, p. 246), 'the most common species (*B. crepitans*), . . . when pursued by its great enemy, *Calosoma inquisitor*, seems at first to have no mode of escape; when suddenly a loud explosion is heard and a blue smoke, attended by a very disagreeable scent, is seen to proceed from its anus, which immediately stops the progress of its assailant: when it has recovered from the effect of it, and the pursuit is renewed, a second discharge again arrests its course. The bombadier can fire its artillery twenty times in succession if necessary, and so gain time to effect its escape'.

A good many other animals can be provoked to produce sounds under the stress of danger, and, though it is not always clear that the sounds increase the chances of survival of the individual that makes them, they may in some cases serve to warn others to evade the danger. Many birds and mammals cry out when seized, and the normally silent rabbits and hares seem to reserve their principal vocal efforts for such occasions. The dor-beetle (*Geotrupes stercorarius*) and rock-lobster (*Palinurus vulgaris*) emit a creaking stridulation when taken in the hand, and the death's-head caterpillar (*Acherontia atropos*), in similar circumstances, makes 'a loud snapping noise, which has been compared to a series of electric sparks' (Newman, 1869, p. 6). The death's-head moth itself will also react when handled, producing a short shrill reedy buzz or squeak, at the same time rearing itself up by stretching out the legs, and exposing the brightly-coloured abdomen by raising or arching it between slightly parted wings; in this case the sound and posture together appear to be an intimidation display.

Decoying sounds (c) are produced by many birds, such as the common partridge (*Perdix perdix*), the ruffed grouse (*Bonasa umbellus*), and the snowy owl (*Nyctea scandiaca*) as part of their display to distract predators from their broods of young chicks. (d) Dissembling sounds, intended to deceive the hearer by giving a false signal, are perhaps uncommon. It might be very difficult to recognise with certainty an example of vocal Batesian mimicry; there is, however, the singular case of the death's-head hawk-moth (*Acherontia atropos*), which, as just described, emits a mouse-like squeak and is said to use it when, with unexpected impunity, it enters and feeds on the honey in the nests and skeps of bees. According to Barrett (1895, v.2, p. 20), 'the stridulous voice of the moth has been observed to arrest and control bees in a manner similar to that produced by the voice of their queen'. The queen honey-bee can, of course, quiet the workers by 'piping' (Allen, 1956, p. 19), and it has been discovered that a similar effect can be produced artificially by rubbing a wet finger on a pane of glass to make a creaking noise (*cf.* Ribbands, 1953, p. 287). Various mice, including *Apodemus sylvaticus*, similarly rob bees of their brood and honey, apparently without getting stung, possibly employing a similar vocal imitation.

When we turn to the intraspecific use of sound signals, we find in man, of course, a degree of development of vocal communication that enormously excels anything to be found elsewhere in the animal kingdom. No other single attribute has contributed more fundamentally to his dominant position in the natural world than the power of articulate speech; and, if we couple it with its modern derivatives of writing, printing, and the rest, none is so uniquely his own. He still retains also a certain power of expressing himself through inarticulate sounds such as laughter, crying, cheering and singing; and it is to be noticed that these, like speech, are used almost exclusively for intraspecific communication. If he is addressing himself to domestic animals, or on rare occasions to wild ones in an effort to decoy them or intimidate them, he does it by trying to adapt the ordinary skills and methods of social communication to serve what is clearly a secondary purpose. In other words, the evolution of voice in man and the shaping of his vocal powers, even in the case of the most primitive inarticulate sounds, have been primarily concerned with social integration and affairs; and the interspecific uses to which sound can be put have played little if any part in the process. This should no doubt be borne in mind in trying to assess the relative importance of intra- and interspecific communication, whether vocal or not, in animals generally.

3.2. *Intraspecific sound signals in birds*

There is no class of animals in which audible signals are more widely used for intraspecific communication than birds. Among members of flocks of birds on the move, social contact is not infrequently maintained by calls specially developed or reserved for the purpose; well-known examples of this may be found in the cardueline finches, titmice, larks, numerous Limicolae (especially at night), and geese. Birds use their voices also to express fear, alarm, warning, threat and other emotions and special states of mind. The loudest, most elaborate, diagnostic and persistently produced of these sounds are usually designated as song, particularly when they are more or less directly associated with intraspecific competition.

So close is the connection between song and the proclaiming of territory that the territorial function of song is generally regarded as being its primary use, and its other functions, such, for example, as enabling the sexes to meet for mating, are considered to be only secondary. This does not, of course, preclude the loss or atrophy of the primary function in special cases and its retention, chiefly at any rate, for the attraction and securing of a mate, as has been claimed in the case of the brown towhee (*Pipilo fuscus*) (Quaintance, 1938, p. 101).

In the great majority of birds it is only the male that sings. The explanation of this appears to be that territory is usually held only just before and during the breeding season, when dispersion is concerned with *pairs* as units, and it is normally sufficient for one member of the pair to act on behalf of both. To this extent there is a division of labour between the sexes, the female having on her side to produce the eggs and often to take the major part in nest-building and incubation. An optimum dispersion of males can be attained through territorial behaviour at the start of the breeding season, and, in so far as each then manages to secure a mate, the objective will have been efficiently achieved. The fact that the pattern of territories is very largely established by the males in many passerine species before the females begin to appear on the scene confirms the view that this is the actual procedure followed.

There are cases, relatively few in number, where territories are held by single birds under what may be described as asexual conditions, outside the breeding season. In these circumstances all territory holders may have to sing alike regardless of their real but dormant sex. The most widely known example is provided by the European robin (*Erithacus rubecula*) during the autumn and winter months. It begins to establish its winter dispersion in August and September, and is probably, round the year, the most persistent song-bird in Britain; but once sexuality is reawakened and the female becomes mated in spring she very rarely continues to sing (Lack, 1943, p. 26). Another similar example has been described in the mourning chat (*Oenanthe lugens*), in which the sexes are also alike. In their winter quarters in Egypt Hartley (1949) found that they established regular individual territories and sang in them; and since no birds were met with which did not sing it was probable that the hens were doing so as well as the cocks.

It will turn out later to be a general rule, not only in birds but in almost all animals, that the male tends to play the dominant role in the dispersion of the adult population and in all forms of nuptial competition and display; and his mate or mates and dependents accept and conform to the pattern and organisation that the males have established. This subject is considerably developed later, especially in Chapters 11 and 12.

Bird-song exemplifies particularly well the kind of signal evolved for social integration. In character and content it is generally quite specific, and often if not always identifies the individual singer. It is usually a seasonal phenomenon, reaching a peak at the time of maximum dispersive activity, just before the breeding season. It is normally produced by adult males, and is a declamation directed primarily at other adult males, supplying the characteristic cohesive-antagonistic bond between them. It is of a loudness consistent with the distance (related to the territory-size) over which it needs to carry. It frequently shows a characteristic diurnal pattern of production, as well as an annual one, beginning at a precisely determined hour at dawn and then, after a short intensive period during which it is usually the principal occupation of all the singers, gradually declining, its function being for the time sufficiently discharged. It commonly reaches a secondary peak at dusk. It is emulative, and can often be stimulated to greater intensity by the sound of rival voices. Thus, in the still hour before dawn, the farmyard cocks of our youth used one at a time to answer back to each other's crowing far across the parish; or again, in quite different surroundings, silent nightingales can often be stirred into action by human whistling. In many species the finest and most spirited singing of the day comes in the dawn chorus at the peak season of territorial competition, just about the time that nesting begins.

In view of what follows later it seems as well to mention, however, that the songs of birds need not always be related to the occupation of individual territories: communal singing, for instance, is by no means infrequent. It occurs in a number of groups, and is familiar to many people in the screaming (or in some species chattering) flight parties of swifts (Apodidae), or in the piping parties of oystercatchers (*Haematopus*). Hudson (1922, p. 256) describes 'the resounding, laughter-like, screaming concerts of *Homorus*', one of the Funariidae (now known as *Pseudoseisura*) found in the pampas of Argentina, 'which may be heard distinctly two miles away'. Starlings usually gather in groups and sing together before flying to a roost, and they join in a notable chorus when they arrive there. An observation is recounted on another page (p. 478) of hundreds of male white-winged crossbills (*Loxia leucoptera*), scattered for miles through the forests of Gaspé county, Quebec, in late July—most probably on the eve of one of their irruptions—their restless flocks constantly singing together in chorus.

It will be seen repeatedly in later chapters that there are fundamental similarities between forms of social organisation that at first sight appear very diverse. Vocalisation may in fact be just as appropriate at a 'lek' or prenuptial communal display, or at a collective roost, or in a close-packed breeding colony, as it is under a simple territorial system. What is required of phonetic signals in order to make them effective naturally tends to be different in these different circumstances, and so to some extent does the response that the sound is adapted to produce.

3.3. *In insects*

With these observations in mind it is interesting to turn to the terrestrial insects, some of which have specialised in the use of sound displays on similar lines to the birds. Few of them are more famous for their singing than the cicadas (Hemiptera-Homoptera, Fam. Cicadidae). The males of these large insects have elaborate sonic organs, consisting of a pair of drums vibrated by very powerful muscles, each adjacent to a large resonating chamber, and capable in some species of producing sounds that are among the loudest and most piercing in the animal kingdom: in one of the South American cicadas they are said to carry for a mile. The songs of the different species are very distinctive; that of the North American *Cicada tibicen*, for example, is a strident, vibrant or siren-like monotone lasting many seconds, by no means discordant in its proper place, emanating from the leafy depths of a tree where the solitary singer lies concealed through the quiet drowsy heat of a summer afternoon. The signals are purely intraspecific; like the songs of tree-frogs they are silenced by the slightest disturbance, and it may therefore require caution and patience actually to set eyes on the singer. This species of cicada frequently retains his station for long periods, and from the wide spacing-out of the singers there appears to be a mutual relationship akin to individual distance, if not actual territory. As with birds, singing is emulative, one stimulating another, and this sometimes results in antiphonic duets or trios (*cf.* Darwin, 1874, p. 434).

In some of the other species a number of males may be heard singing in a group together: but it will be noticed that distantly spaced and closely assembled male-displays are common alternatives, examples of both of which can be found in almost every class of animals in which any such display is recognised. Probably the best-known of the concerted singers is the American seventeen-year cicada (*Magicicada septendecim*). To the naturalist, at least, ' the music or song produced by the myriads of these insects in a warm day . . . is wonderful. It is not deafening, as many describe it; even at its height it does not interrupt ordinary conversation. It seems like an atmosphere of wild, monotonous sound, in which all other sounds float with perfect distinctness ' (Marlatt, 1907, p. 84, quoting G. B. Smith).

What makes us suspect that the cicada's song is again primarily of concern to the male sex is that the males are not only the sole possessors of the ' timbals ' and ' mirrors ' that produce and amplify the sound, but they also have much larger hearing organs than the females (Imms, 1931, p. 126). We find just the same thing in the American bull-frog (*Rana catesbiana*) and its nearest relatives, where the male's ear-drums are two or three times the area of the female's (*see* pl. II). It is no doubt for a similar reason that in many insects, especially in the Diptera and Hymenoptera, the males also have larger eyes than the females. Hitherto there has been no satisfactory explanation of these phenomena.

Stridulation, the production of sound by rubbing roughened surfaces together, is particularly characteristic of the Orthoptera (crickets, grasshoppers etc.), and it is clear from the variety of instruments used that it has developed independently a number of times within the order. The phonetic organs, consisting of a file and scraper in association with a vibratile surface (*e.g.* the forewings) or in a few cases a resonating chamber, are highly perfected, and in the great majority of species they are confined to the males. The songs are again specifically distinct, a fact reflected in the varied form of the stridulating surfaces. They differ in pitch, the commonest being between 4 and 13 kc/sec., but they are known to range from 1·2 to over 60 kc/sec.; actually the frequencies depend more on wing-resonance than on the fundamental pitch of the file and scraper, and they can be modulated in different species to produce a great variety of trills, rattles, chirps, and warbles (*cf.* Pierce, 1948).

Mention was made in the previous chapter (p. 29) of the song-flights of the cracker-locust (*Cercotettix verruculatus*), which performs a jerky, clicking flight in the same sort of way as the Carolina locust. In *C. carlingianus* of the arid western United States the ' obstreperous crackle ' is said to be audible for a quarter of a mile (Snodgrass, 1925, p. 412, quoting from Scudder). It has an obvious general resemblance to the song-flights of the males of many kinds of birds, and in fact it was observed long ago (Gosse,

1840, p. 269) that the male cracker-locust has a closely analogous attachment to a particular site, from which it refuses to be driven more than a few yards.

It is known in many species that the songs are emulative. A large katydid (*Corycus kraussi*) belonging to the Mecopodinae, with expanded forewings beautifully patterned like a pair of green leaves, which I kept for six months after its arrival in Aberdeen from the Cameroons in a crate of bananas, usually started its (nocturnal) melodious trill every time the clock struck, and sometimes responded when the telephone rang. Pierce (1948, p. 188) was able to induce a male of the northern true katydid (*Pterophylla camillifolia*) to sing in daylight while a phonograph record of its own song was played to it, and later to imitate to some extent the noises made by an assistant.

Snodgrass (1925, p. 435) found that with two male field-crickets (*Acheta assimilis*) placed in a cage, ' whenever one began to play his fiddle, the other started up '; and one of them would dash at the other with open jaws, chirping at the shrillest pitch, though ' neither male ever inflicted any actual damage on his rival '. (According to Darwin, 1874, p. 443, similarly confined males of *G. campestris* will fight till one kills the other.) Snodgrass says, ' *Gryllus* sings to express himself and not to " charm the females ". In fact, it is often hard to feel certain whether he is singing or swearing.' ' When we listen to insects singing, the question always arises of why they do it, and we might as well admit that we do not know what motive impels them. It is probably an instinct with the males to use their stridulating organs, but in many cases the tones emitted are clearly modified to the physical or emotional state of the player. The music seems in some way to be connected with the mating of the sexes, and the usual idea is that the sounds are attractive to females. With many of the crickets, however, the real attraction that the male has for the female is a liquid exuded on his back, the song being apparently a mere advertisement of his wares' (p. 425-426). This recalls the earlier conclusion of Sharp (1895, p. 200), who says that ' the musical powers of the saltatorial Orthoptera are . . . specially characteristic of the male sex. There is evidence that these powers are of great importance to the creatures, though in what way is far from clear'.

Stridulation by female orthopterans is occasionally reported, and it is interesting to discover that, while ' female stridulation forms a regular part of the behaviour of some species, . . . [it] seems to be much more prevalent in the absence of males ' (Ragge, 1955). It appears here that, when the normal epideictic sex is lacking, the females take over the dispersionary task, in a way which is in some ways reminiscent of the female robins that sing during the part of the year when they are unmated.

The singing of cicadas and Orthoptera appears to have the exactly the same kind of functions as that of birds, and there seems no reason to doubt its principal purpose as an epideictic display, concerned with dispersion and the regulation of numbers. Sometimes the song seems to have an immediate territorial function, as in some of the cicadas, or in the house-cricket (*Gryllus*

Male and female bronze frogs (*Rana clamitans*), showing the much larger relative size of the ear-drum in the male. (Photo by Dr. Robert S. Simmons.)

domesticus), which sings from the mouth of its hole; sometimes the singing is communal, as in the common American field-cricket (*Acheta assimilis*) or the seventeen-year cicada, when the total audible volume may supply an index of population-density, and lead to dispersive responses, either by immediate movement of individuals, or by delayed action on the recruitment rate. There is a considerable literature associated with cicadas and their songs; a general introduction to it may be found especially in Myers' *Insect singers* (1929), and in the work of Ossiannilsson (1949).

It is perhaps worth noting that a good many insects, presumably in order that they may be more easily heard, tend to sing at times of day when general ambient noise is low, and birds are silent, either in the stillness of night or in the hottest part of the day.

Dispersion presents a formidable problem to insect larvae that burrow deep in the ground or in wood. Several of them, including those of the dor- and stag-beetles (*Geotrupes* and *Lucanus*), emit sounds, which, since they are made by larvae, cannot in any case be held to have a sexual or epigamic function. Sounds are produced also by adults in some of the wood-boring beetles. The death-watch (*Xestobium tessellatum*), for instance, makes its persistent ticking by inclining its body forward and forcibly beating its sternum against the wood (Meixner, 1933, p. 1127); it can be heard most often at night in early summer. Stridulation in beetles is frequently confined to the males, but in *Phonapate*, a wood-boring genus distantly related to *Xestobium*, the special stridulating machinery appears only in the female (Imms, 1946, p. 100). Such sporadic reversals of the roles of the sexes are found in many other groups; they are discussed in greater detail in Chapters 11 and 12.

3.4. *In land-crabs*

A number of arthropods other than insects make sound signals, including various spiders. Among the Crustacea most forms are, of course, aquatic, and reference to those that stridulate under water is more appropriately deferred until the following chapter where the subject of underwater sound is reviewed.

The very interesting case may be considered here, however, of the circumtropical, swift-footed, air-breathing crabs of the genus *Ocypode*.* The adult males of ten of the eleven known species can make sounds, by rubbing a rasp on the ' hand ' of their bigger unequal claw (chela) against a ridge on the second joint of the same limb (Miers, 1883). The following account of it is taken from Anderson (1894): ' In *Ocypoda ceratophthalma* the stridulating organ consists of a ridge coarsely striated above, finely striated below, borne on the inner surface of the hand of the large chela. This ridge is rubbed across a smooth raised ridge on the ischium of the same chela, and by slowly rubbing the opposed ridges together, and placing the crab over the mouth of a wide-necked bottle to act, like the crab's burrow,

* This is the original orthography of Fabricius, 1798, but it is often emended to *Ocypoda*.

as a resonator, an exact reproduction of the sound emitted by the crab, during life, can be obtained. One bright hot sunshining morning in November, as I walked along the shore of Bingaroo, one of the Lakadive Islands, which is occasionally visited by the inhabitants of the other islands of the same atoll, I was surprised to hear a loud creaking noise, that appeared to proceed from the edge of the scrub jungle that covers the island. At first I imagined it must be caused by frogs, so perfectly did it resemble the croaking of these animals. However, on tracing the sound to its source, I discovered that it proceeded from the burrows of the Ocypode crabs which here fringed the beach at high-water mark. These burrows are frequently, in coral sand, very wide at their mouths (six to eight inches), and then taper gradually downwards, so that they act as excellent resonators.* The cause of the stridulation of the crabs was by no means apparent, the animals were all lying hidden in their burrows, and several were croaking at the same time, as if in concert.'

Alcock (1902, p. 174) tells us that this species, *O. ceratophthalma*, is protectively coloured, can look like a stone, and is called the grey ocypode. *O. macrocera*, on the other hand, is conspicuous, even at rest, and is known as the red ocypode. He is speaking of the latter, on the shores of the Bay of Bengal north of Madras, in the following account (pp. 214-216). ' In order to understand the matter, it should be known that these crabs, although they do not seem to have any social co-operation, are gregarious, and that each one has a burrow of its own. Though they may sometimes be seen marching in battalions across the sand, yet as a rule they stay close to their burrows. . . . It seems probable, therefore, that it would be advantageous to the species as a whole if the rights of property in burrows were rigidly respected, and if each member possessed some means of giving notice that its burrow was occupied—or, as Mr. Stebbing has expressed it, that it was " not at home " to callers; and I think that this consideration gives us a clue to the use of the stridulating mechanism.'

This behaviour has many of the typical features of epideictic display, in which the males advertise their possession of a token property—in the form of the burrow—by sounds emitted more or less in chorus with their rivals. The crab rests in his burrow during the day—or, in some regions, while the tide is in—emerging in due course to wander abroad over the beach, which is a common feeding ground, looking for sandhoppers, carrion and other food.

The related fiddler-crabs (*Uca*) also own burrows, but here the males appear on the sand, making a visual display of their single huge hand, which in *U. annulipes*, according to Alcock (loc. cit., p. 217) is ' often twice as big as the body ', and of a ' beautiful cherry-red colour '. They are notoriously pugnacious one towards another, one of the best-known American species being named *Uca pugilator*. However, they make no use of sound, and we should therefore defer consideration of them to a later page (p. 175), only

* An illustrated account of the digging of these burrows has been given by Cott (1929).

adding here that the observant Alcock concludes (loc. cit., p. 213) that *Ocypode* and *Uca* ' are about the most gifted members of the whole crustacean class, and among Arthropods are inferior to none but the social insects '.

3.5. *In amphibia*

Many of the frogs and toads have pneumatic vocal organs, which produce sounds by driving air through the orifices connecting the air-sacs, mouth-cavity and lungs. It is the males, as usual, that trill and croak; the females are generally either mute or merely make subdued sounds (though Gadow, 1901, p. 47, says that both sexes can croak in *Rana temporaria*). The parallel between the Anura and the Orthoptera or cicadas is in general very close. Singing is principally confined to the period just before and during the pairing season. Huge air-sacs under the throat and abdomen, placed ventrally or laterally, appear to serve both as reservoirs of air and as resonators, and even very small species like the North American tree-toad and spring peeper (*Hyla versicolor* and *H. crucifer*) can produce loud far-carrying calls, often in a deafening chorus which continues for hours at a time. The ear-drums, as noted earlier, are sometimes larger in the males, just as in the cicadas. Examples can be found either of species which trill in solitude, though within earshot of their rivals, or of those that join in a chorus, like the frogs of Aristophanes. Even the sounds produced are sometimes closely similar in quality to those of insects, and it may not be easy with an unknown species to be certain whether it is a frog or a locust (*cf.* the names ' cricket-frog ', ' *Acris gryllus* ', etc.).

Some of the singers are difficult to locate on account of a ventriloquial quality (*e.g.* the South American *Paludicola* spp., *see* Budgett, 1907, p. 62), and also because they are cryptically coloured to resemble the bark or leaves on which they rest; moreover they characteristically stop on the slightest hint of disturbance. Though *Hyla versicolor* joins in a noisy chorus at the spawning ponds in May at the height of the season, it also sings in isolation in bushes and trees during June and July, at least in southern Quebec. (My latest record was on 8th August 1942, near Lake St. John; see also Wright, 1914, p. 46). The closely related *H. regilla* of the Pacific slopes of North America may be heard at any season of the year in California, but the single prolonged note made by individuals in solitude is lower pitched than the short double spring chorus note (Storer, 1925, p. 222).

Frogs and toads that sing in chorus gather for the purpose at their breeding places. In many species it is known that the males arrive first. It might be imagined that the function of their far-carrying sound was to attract others to the spot, especially females. But Cummins (1920) showed, by trapping various species of frogs as they entered their breeding ponds, that their arrivals occurred in waves ' during periods of high relative humidity, coincident with temperature ranging between about 41° and 52° F.' Intense migration often followed periods during which there had been no croaking, and ' great vocal activity was not followed by increased migration. It is concluded *that voice does not direct* the movement of frogs into the pond ' (p. 342). This refers to four different early-spawning species, all common in the eastern United States, namely *Rana pipiens*, *R. sylvatica*, *Chorolophus nigrita* and *Hyla versicolor*.

When Cummins turned his attention to sex-recognition in the two *Rana* species, he found that it ' results from differential behaviour—*i.e.* submission, or struggling and croaking, when clasped '. Sight is believed to play no part.

From this it looks as though frogs of these species, both males and females, can locate their spawning ponds without the aid of the males' chorus, and, furthermore, that it is the males which find the females in the crowd, and not *vice versa*, by trying to clasp them. It must be noted in the case of some other Anura, however, including the American toad (*Bufo americanus*), that ' it has been shown that the voice of the male has a strong influence of attraction on the female ' (Noble, 1931, p. 406, quoting G. B. Wellman).

In the various species so far mentioned, whether the males sing all together or each on its own station, the spawning pond is common to them all. There are certainly some species, however, such as the well-known bull-frog and green frog of eastern North America (*Rana catesbiana* and *R. clamitans*), in which the males not only take up solitary stations, for example on the shallow margin of a lake at a long distance from one another, but also spawn in isolated pairs. These therefore somewhat resemble territory-holding birds; and, according to Noble (1931, p. 405), green frogs are able to home from a distance of several hundred feet. In these species the male's voice no doubt attracts him a mate, as it does in singing birds; he lies all but submerged to produce an occasional deep, loud, booming call, the sound of which is carried to a great distance by the water.

Though in most warm or continental countries the songs of frogs and toads are among the sounds especially associated with the spring, there are a number of species which come into voice again in September or October, including many in the southern United States (Noble, 1931, p. 409), and even subarctic species such as *Pseudacris septentrionalis* and *Rana cantabrigensis* in northern Alberta (Harper, 1931, p. 69).

It is as true of the Anura as it is of the birds and Orthoptera, that the songs are specifically distinct, and in regions where ten or a dozen species occur, as, for example, in eastern Canada, there is little difficulty in identifying them all by ear, when once each note has been correctly attributed to its source.

3.6. *In reptiles*

Most of the living reptiles are silent, though there are some notable exceptions such as the American alligator. Ditmars says (1907, p. 89): ' Among reptiles, the alligator is unique in giving voice to a *loud* noise, or bellow. In the Southern swamps the night air carries the call of a large individual for a mile or more. The " bellowing " of an alligator is hard to describe, as it varies greatly in cadence according to the size of the reptile, and from a sound like the gentle " mooing " of a cow with the small alligator of about five feet, ranges to a thundering and tremulous blast of the big male—ten feet or more in length. As the patriarch gives voice to his roars, the scent glands—on the under surface of the chin—are opened and fine, steamy jets of a powerful musky-smelling fluid float off into the heavy, miasmatic atmosphere of the bayou. The odour may be carried for miles and to the negroes it always signifies ," a big, ol' 'gater "'. S. F. Clarke (1891, p. 182) states that, ' during the breeding season, in late May, throughout June, and extending into July somewhat, the males are very active, wandering about to various ponds and rivers in search of females. Fierce battles are said to take place during this time between the excited males. . . . It is in the breeding season that their bellowing is mostly heard, and more in the night than during the day.' According to him the nest of the alligator is very large, and is built by the female. It is reported to be used for more than one season, and this was borne out by the presence of roots of a palmetto and a grape-vine extending nearly through the largest nest he found. Brehm (1925, p. 534) says of *Crocodilus niloticus* also that it always returns to nest at the same site; the places are known to the natives, who regularly destroy the eggs.

In some countries members of the Crocodilia (including alligators) become torpid in the cooler or drier months of the year, when they bury themselves in mud or occupy excavated underwater caverns. For this purpose also they must establish a claim to a particular site; and though considerable seasonal movements may be necessary on account of changing water-levels, it is well known that individuals become attached to particular localities. This is most often recorded of the large old males, and both Bates (1864, p. 307), referring to *Caiman niger* in the upper Amazon, and Hamilton (1947, p. 313), referring to *Crocodilus niloticus* in South Africa, observe that when some monster frequenting a particular spot is destroyed, another, perhaps of slightly smaller size, arrives some days or weeks afterwards to take the vacant station.

It seems certain that among these sociable animals there exists some form of hierarchy. All the Crocodilia, regardless of age or sex, possess the pair of large musk glands which open ventrally on either side of the broad throat, under the rami of the jaws; these are used not only in the breeding season, but at all times, and may be everted during a struggle (Brehm, 1925, p. 507). All species of Crocodilia are probably capable of roaring or making sounds, at least in moments of fear or anger, but the bellowing of the bulls seems to be the special feature of *Alligator mississippiensis*. Belt (1874, p. 9), however, refers to an ' alligator ' in the San Juan river, Nicaragua, which ' stretched up his head and gave a bellow like a bull ', and as this is apparently outside the range of *A. mississippiensis* it must refer to a species of *Caiman*, most probably *C. sclerops*.

Among Ophidia the rattlesnakes (*Crotalus* and *Sistrurus*) are exceptional, though there are other crotaline snakes that beat the ground with their tails. It is usually assumed that their characteristic whirring sound is directed at intruders, to warn them not to come closer, or probably at their prey, to decoy or hypnotise it. In the opinion of Ditmars (1907, p. 429) the evidence is not really sufficient to decide the true functions of the rattle. Snakes have neither a tympanum nor a middle-ear cavity, but they are of course by no means deaf; however, there is no evidence at all to link the rattle with any form of social or sexual function.

Crowing sounds have sometimes been credited to cobras and other snakes, though never in circumstances which put the statements entirely beyond doubt (*see e.g.* a letter by T. Phillips to the editor of *The Times*, London, 28th March 1956). There is a lizard, however,—a gecko living in the deserts of Persia and Turkestan called *Teratoscincus scincus*—which has a stridulating organ on the upper side of the tail, consisting of ' a series of large, transverse, nail-like plates. By rubbing these plates upon each other, this gecko produces a shrill, cricket-like noise, sitting at night in front of his house, perhaps to attract grasshoppers. The noise is made by both sexes ' (Gadow, 1901, p. 507). It may well be doubted whether an animal sitting in front of its house stridulating is really out to attract grasshoppers: as we have seen already in the birds, insects and land-crabs, it is much more likely to be an epideictic display, directly concerned with ownership claims, or some other form of social integration. Gadow (loc. cit.) also goes on to mention another desert gecko, *Ptenopus*, of south-west Africa, that produces a similar chirping noise by voice.

The American musk-turtles (*Sternotherus*) produce what have been described as stridulating sounds, but Risley (1933, p. 701) states that he had ' never been able to ascertain with any degree of certainty how they were produced, nor . . . that they were made only by the males'. He observed movements of the throat at the time of the sounds. Most tortoises make more or less feeble piping sounds in the pairing season, according to Gadow (loc. cit., p. 350).

3.7. *In mammals*

Scattered through the great class of mammals there are a number of very notable vocalists; on the other hand, there are still many species that are on the whole silent, and resort to sound-production, if at all, either in moments of peril, or in infancy, or in conversing with their young. We have to deal here with a group of animals having an outstanding capacity for expressing their highly-developed and complex emotions; for example, the purring of cats appears to convey a social signal indicative of a kind of pleasure, good temper or well-being, which probably exists in the other classes of animals, but is seldom as clearly manifest as it is here. Nevertheless, the uses to which mammals put their vocal powers bear a close general resemblance to those already considered in the preceding groups.

The selected examples given below are arranged for the most part systematically; consideration of the Cetacea is deferred until the following chapter on underwater sound (p. 66).

To begin with the Insectivora, the shrews (Soricidae) are among the more vocal of the small mammals. According to Crowcroft (1955, p. 74-75) the *Sorex* species utter sounds of at least two types, namely staccato squeaks and soft twittering or whispering sounds. ' Staccato squeaks constitute the shrews' principal means of defense and attack. The notes are uttered in rapid succession punctuated by harsh snarls during which the snout contracts. The individual squeaks and the posture during their delivery are very reminiscent of the barking of dogs. It has apparently escaped notice that the staccato squeaks identify the male sex in *Sorex araneus*.' ' The noisy nature of the fighting of shrews is its most prominent feature, and many disputes are decided by screaming alone.' On the other hand ' when first placed in a cage, common shrews utter an almost continuous twitter of soft notes as they explore their surroundings. . . . Such sounds . . . may act as a warning to other members of the species and thus aid mutual avoidance.'

The bats, however, by far transcend the vocal accomplishments of the insectivores: indeed, as an order, their only rivals among the mammals as sonic specialists appear to be the Primates and the Cetacea. The Microchiroptera have, of course, evolved marvellous adaptations of the larynx and ear (and probably of the nose as a sound-channel), to provide an echolocation system, by means of which they perceive solid objects as they fly.

Their voices are also used for various forms of social communication. Allen (1939, p. 220) describes a roosting cave of pig-nosed bats (*Erophylla planifrons*) in the Bahamas, from which there emerged a continual squeaking; a flashlight revealed several hundred bats covering the limestone walls, ' but all separated from one another by a greater or less distance, so that none of them were in contact. Perhaps their bickering cries were incident to the adjustment and proper spacing of the individuals. . . . ' In the roosting of flying foxes, also, ' the sharp sounds emitted have perhaps a threat value and may therefore play a part in bringing about a proper spacing of the animals while they are at rest ' (*idem*, p. 233). This closely parallels the corresponding behaviour of some birds, to which reference will be made later in the chapter on roosts (p. 286).

A particularly interesting case, reported in 1917 by Lang and Chapin (and quoted here from Allen, loc. cit., p. 234) is that of the large hammer-headed fruit-bat (*Hyspignathus monstrosus*) of the Congo region, which has, in the males only, an enormous enlargement of the larynx, and also a pair of immense pharyngeal air-sacs, ' attaining nearly a third the volume of the entire body-cavity '. According to Lang and Chapin, ' in no other mammal is everything so entirely subordinated to the organs of voice'. The larynx alone fills most of the thorax, and the posterior tip of the cricoid cartilage is practically in contact with the diaphragm, so that the heart and lungs are crowded back and sideways. Every evening throughout February and March the old males, to the number of about thirty, were seen to assemble one by one, in a particular spot across the Ituri river, and at a set time, ' about 6.15 to be exact '; each of them began to give a loud *pwŏk*! or *kwŏk*!, repeated twice or thrice a second. Without serious interruption the concert continued till 10 or 11 o'clock. ' During this performance their utter lack of fear was amazing. Neither talking, rapping on the trees, lighting a lantern, nor even firing a gun could induce them to cease their calling.'

This has almost all the characteristics of the kind of epideictic display known as a ' lek ' (*see* p. 194). It primarily concerns the males (nine shot were all males), which indulge in a symbolic and often openly competitive communal display, with vigour and persistence, resorting at a set daily hour to a customary place. In this case, however, it was impossible to know whether the site was a traditional one, as it generally is in an avian lek. As we have noticed before in various authors' descriptions of epideictic displays, ' its precise meaning was not determined '; but it falls easily into line with a number of others assembled in this chapter, and with many more still to be described in Chapter 11.

In the well-known African fruit-bats called ' epaulet-bats ' (*Epomophorus*) their English (and also Latin) name refers to a pair of sacs or depressions, one on each shoulder, which can be everted in visible display to reveal a conspicuous patch of long white hairs. This is an analogous development to the pouch of the springbuck described in the previous chapter (p. 33). Though they roost by day in companies, the males take up solitary stations at night and call to each other. ' The note is very high pitched, and sounds very much like the metallic ringing noise produced by sharply striking a blacksmith's anvil with a metal hammer ' (Fitzsimons, 1919, I, p. 90). In South Africa, where they occur even in city parks, they are often called ' anvil bats '. This case is similar to the last, except that the males are dispersed while singing; and it is already becoming clear that in most if not all social activities of animals concerned with dispersion, these two dispositions are common alternatives, depending chiefly on the ' individual distance ' associated with the particular performance. Not realising this, Allen (1939, p. 236) concluded: ' since the voice is largely helpful as indicative of location, it may be that its constant use by the males apprises each of the presence of another male, and this helps in spacing the individuals, but this can hardly be the case in the hammer-headed bat, where apparently the calling males come together to a particular place for their choruses. Nor is it as yet obvious what effect, if any, the notes may have upon the actions of females. Altogether, the sounds produced by these African fruit bats appear to be without parallel elsewhere in the group.' Viewed as epideictic phenomena, however, they differ scarcely at all from scores of displays of similar function in other animal groups.

The Primates, as noted already, are another loquacious order; their number includes the South American howler monkeys (*Alouatta* spp.) which are capable of emitting some of the loudest or farthest carrying air-borne sounds to be found in the animal kingdom. In this genus the rami of the mandible are exceptionally deep, and between them lies a heart-shaped air-filled resonator—a shell of thin hard bone developed from the basihyal, about five cm. in diameter in a large male; the posterior side is unossified, and the opening is partly closed by a membrane. In the female a similar sound-box is found, but it is about one-third smaller. The resonator is but a part of their extraordinary vocal equipment. The larynx itself is immense; the cricoid is ossified, and the thyroid cartilage is three times the size of that of man; the epiglottis is ten cm. in length and five cm. in breadth, which gives an indication of the size of the laryngeal chamber wherein the vocal cords are found (*cf.* Owen, 1868, III, p. 598).

Howler monkeys indulge in typical choral displays, every male of the group or clan participating simultaneously, and these have a most impressive peak at dawn. ‘ Invariably, as the light comes some clan or clans will break forth into howls, and these will be followed by similar roars from groups in adjacent areas. The vocalizations then spread from clan to clan until the majority of groups in the island have howled.’ ‘ I believe the roars are important . . . in regulating the territorial ranges ’ (Carpenter, 1934, pp. 99 and 115). He says also: ‘ Of the cues . . . which co-ordinate the movements of clans and hence their spacing, vocalizations are without doubt the most important. The howls at dawn and in the early forenoon seem to signalize the location of a clan to other nearby ones ’ (loc. cit., p. 118). According to Bates (1863, I, p. 72), ‘ morning and evening the howling monkeys make the most fearful and harrowing noise, under which it is difficult to keep up one’s buoyancy of spirit’. The deep volume of sound, he writes in another place (p. 294), ‘ is produced by a drum-shaped expansion of the larynx. It was curious to watch the [captive] animal whilst venting its hollow cavernous roar, and observe how small was the muscular exertion employed. When howlers are seen in the forest there are generally three or four of them mounted on the topmost branches of a tree.’ Bates’s description of the ‘ uproar of life at sunset ’ in the Amazonian forests, to which the howlers’ ‘ dismal roaring ’ contributes so conspicuously, is quoted later, in Chapter 15 (p. 336). As to its carrying power, Salvin says: ‘ it would certainly not be over estimating the distance to say two miles. When the sound came over the lake of Yzabal [in Guatemala], unhindered by trees, a league would be more like the distance at which the Mono’s cry may be heard ’ (quoted from Forbes, 1896, I, p. 201).

Less spectacular modifications and enlargements of the larynx occur in other platyrrhines, such as the spider-monkeys and marmosets. In the Old World monkeys also, the basihyal is often expanded to receive an inflated sac from the larynx, though of much more modest dimensions than that of the howlers. In the tailless apes, namely the gorilla, chimpanzee, orang-utan and gibbons, there are laryngeal air-sacs, which, in the male, may extend down the neck to the upper part of the chest. The roar of the male gorilla is, of course, very loud, and the cries of the vociferous gibbons, if not so loud as those of the Amazonian howlers, are far more varied in their range of notes. The hoolock, a species of gibbon found in Assam and Upper Burma, receives its name from the powerful double call, which at a distance ‘ much resembles the human voice; it is a peculiar wailing note, audible afar, and . . . one of the most familiar forest sounds. The calls commence at daybreak, . . . several of the flock joining in the cry, like hounds giving tongue. . . . [They] remain silent throughout the middle of the day, but recommence calling towards evening, though to a less extent than in the earlier part of the day' (Blanford, 1888-1891, p. 7). We should note the diurnal rhythm that so frequently characterises song-displays.

In man the vocal equipment lacks any elaborate accessories, and indeed the voice is not remarkably loud. It is important to observe, however, that there is a great enlargement of the larynx at maturity, especially in the male, accompanied by a marked change in the pitch and volume of the voice, which results in one of his most conspicuous secondary sexual characters. The voice is not only man's principal vehicle of sociable communication, but (as in the shrew) his chief weapon in many of the more primitive forms of competition with his fellows that stop short of bloodshed. The war-whoops and battle cries of primitive tribes as a prelude to combat are well authenticated; and it is interesting to notice, for the sake of the light it may throw on the analogous displays of the lower animals, that the fighter himself takes heart from the cries at the same time as he frightens the foe. The drums and other loud instruments devised long ago to enhance these effects survive of course in the martial music of our own day.

In comparison with the two preceding orders, the Rodentia contribute little to the volume of animal sound. The families in the suborder Sciuromorpha are perhaps the most vocal of them, and the squirrels themselves and some of their close relatives may even be described as noisy. Seton (1909, p. 345-346) refers to some of the sustained performances of the eastern chipmunk (*Tamias striatus*) as 'singing', and describes the morning chorus in which they take part in the spring, and again in the autumn, shortly before they retire into hibernation. 'When the morning is bright and warm some lusty fellow gets up on a perch and begins to "chuck-chuck-chuck". If psychologically well tuned, his invitation at once provokes abundant and rapturous response. Every chipmunk mounts his perch, and they make the woods ring for several minutes with their united voices' (loc. cit., p. 361).

Marmots and woodchucks are well known for their shrill whistles; they are generally excitable and make their loud calls whenever they are in the least alarmed; but, although some of the marmots live in communities, they do not appear to 'sing' spontaneously in any kind of epideictic display.

Some of the familiar species of mice and voles (*e.g. Clethrionomys glareolus*) use their voices freely. C. H. D. Clarke (1944, p. 23) found in the Yukon Territory, 'in the alplands of the St. Elias area, and in many forests near timberline . . . a small mouse which had the peculiar habit of coming frequently to an entrance of its runway system and singing, in a voice similar to that known for shrews and house mice'. This was an undescribed *Microtus* related to *M. abbreviatus* (*see* Rand, 1945, p. 100). Its behaviour reminds us of the house-cricket, the ocypode crabs and the geckos, which have all appeared earlier in this chapter.

Vocally the most notable member of the order is, perhaps, the South American vizcacha (*Lagostomus trichodactylus*), a very interesting social rodent belonging to the Chinchillidae. A short but valuable biography of it has been given by W. H. Hudson (first published 1892; *see* 1922, pp. 289-303). The vizcacha lives in permanent villages or warrens, which in the course of many years become raised above the level of the surrounding pampas through the accumulation of portable objects dragged to the site, and later covered under the excavated soil. This helps to keep them dry in the sudden heavy rainstorms; and since the surroundings are carefully cleared of obstructions, the residents can also see and be seen afar. In the long-ago days of which Hudson wrote, so common were the ' vizcacheras ' ' that, in some directions, a person might ride five hundred miles and never advance half a mile without seeing one or more of them ; ' moreover they are interconnected by paths, and visiting is a regular occupation. The vizcachas have a striking horizontally-banded pattern on the face, which must serve as a recognition-character; but, being largely nocturnal, their chief method of communication is by voice; and, says Hudson (loc. cit., p. 307), ' I doubt if there is in the world any other four-footed thing so loquacious, or with a dialect so extensive '. They hold ' a perpetual discussion all night long,' and in addition to many other notes and cries, at times the male ' bursts into piercing tones that may be heard a mile off.' Their cries are emulative or infectious, and any ' sudden sound, as the report of a gun, or a clap of unexpected thunder, will produce a most extraordinary effect. No sooner has the report broken on the stillness of night than a perfect storm of cries bursts forth over the surrounding country. After eight or nine seconds there is in the storm a momentary lull or pause; and then it breaks forth again, apparently louder than before. . . . It sounds as if thousands and tens of thousands of them were striving to express every emotion at the highest pitch of their voices; so that the effect is indescribable, and fills a stranger with astonishment.'

Two of the best known families of the Carnivora, the Felidae and Canidae, are much given to vocalising. ' The roar of the lion heard at close quarters at night impresses more by its volume than by its loudness. It is like the rumbling of an immense bass organ. . . . When a lion really decides to do his best vocally he can be heard over an astonishing distance. At night or early morning when it is still, or the breeze favourable, five or six miles is not an ambitious estimate. . . . With a little practice the deeper and gruffer tones of the male may easily be distinguished from those of the female ' (Hamilton, 1947, pp. 134-135). For carrying-power even the howler monkey must evidently give way to the lion. Though they roar nightly, and must be accounted very noisy, there is no standard explanation of their purpose in doing so. When hunting they are silent; but they roar especially ' when mating, after a kill, and sometimes when travelling to and from water.

Detached members are constantly calling up their mates ' (*idem*, p. 134). The late Col. Stevenson-Hamilton did not ' think it correct to say that lions roar to drive game in a certain direction. . . . It is possible that one of their reasons for roaring on leaving a certain spot, as they do, is to mislead the game as to their position.'

Where the game is sufficiently abundant the lion is a sociable carnivore, but the ' prides ' into which it is often rather loosely associated, do not each have the exclusive run of a particular territory. There is nothing at all improbable, therefore, in the idea that roaring serves chiefly for social integration and for giving an indication of population-density. Much the same occurs in the other well-known sociable carnivores, namely the hyaenas, wolves, jackals and wild dogs. It has been stated that lions roar far less frequently when their numbers have been thinned and they have become scarce (Shortridge, 1934, I, p. 82, quoting T. Roosevelt and E. Heller, 1914). A century ago, when they were already becoming rare in India, Blanford—the well-known authority on Indian mammals—was struck by the noisiness of the abundant African lions he found when on a visit to Abyssinia (Lyddeker, 1893, I, p. 364): possibly this is connected with the same thing.

By comparison, the tiger and many other large cats are relatively silent, and far less sociable; the tiger, for example, has a roar ' very similar to that of the lion, a prolonged moaning, thrilling sound, repeated twice or thrice, becoming louder and quicker, and ending with three or four repetitions of the last portion of it ' (Blanford, 1888, p. 62). As will be seen later (p. 184), tigers have a persistent spatial organisation, but they may be so sparsely distributed that roaring is of little value in maintaining contact, except perhaps between a mated pair. In the same way, the barking of foxes (*Vulpes* spp.) is largely confined to domestic signalling between the dog and vixen, or the vixen and cubs.

In the Canada lynx, according to Seton (1909, p. 682), there is a ' frightful caterwauling ' or ' yowling song ', similar to that of the domestic cat, heard both in the breeding season and at other times, and this ' is its chief means of communicating with its distant fellows'. In the domestic cat, as is all too well known, the males frequently join together in a hideous chorus. These cases are rather closely paralleled by the wolves and their immediate relatives. Wolves (*Canis lupus* and *C. occidentalis*) are said to howl sometimes when actually hunting in company, rather like dogs that give tongue in the chase; but they also ' howl at the moon ', which is very possibly a kind of song. In the much noisier coyote (*C. latrans*) this is certainly the case. Seton says (loc. cit., pp. 814-815): ' Most of the many calls of the coyote are signals to its companions, but some of them seem to be the outcome of the pleasure it finds in making a noise. The most popular of its noises is the evening song, uttered soon after sunset, close to camp. This is a series of short barks, increasing in power and pitch till it changes into a long squall. One coyote begins and immediately two or more join in. . . . It is kept up for a minute or two, then ceases till some new impulse seizes them. . . . Two or three coyotes will meet each night in a certain elevated place to sing. They have several of these recognised choir-lofts, but they never use the same on two nights in succession. Sometimes . . ., in dead calm moonlight nights, each coyote gets up on his singing perch and pours out his loudest and finest notes. This is passed on from one point to another, till the whole mountain seems ringing with the weird music, and, from its very wildness and the vast stretch of the country that is concerned, the effect is truly impressive.'

Three of the most typical recurrent features of epideictic display can be recognised here, namely the emulation of one coyote by another, the vespertine hour, and the use of established sites. The sexes are not readily distinguishable and whether or not both take part, or only the males, is unknown. Somewhat similar concerts of ' the most unearthly and startling music ' are given by the palaearctic jackals (*C. aureus*); Jerdon (1874, p. 143) tells us that ' the natives assert that they cry after every watch of the night'. Indian and Eskimo sledge dogs also show some tendency to join in rounds of howling at dusk and dawn.

Matthews (1939, p. 48) describes the corresponding pandemonium of spotted hyaenas (*Crocuta crocuta*) in Tanganyika. ' The noise they made was indescribably hideous. Shrill shrieks and yells, accompanied by deep emetic gurgling and groans, made a background for wild peals of laughter. . . . They appeared to be chasing each other round the tents and tree-trunks.'

The Artiodactyls, especially the ruminants, contain many species that are more or less sociable, polygamous, and vocal, at least in the rutting season. One species that differs greatly from the majority, the oriental muntjac or barking deer (*Cervulus muntjac*), may be mentioned first; it is a solitary creature, living singly or in pairs in hill forests, and has a cry that is ' very like a single bark from a dog, and is very loud for the size of the animal. It is often repeated at intervals, usually in the morning and evening, sometimes after dark ' (Blanford, 1889-1891, p. 533). We may be interested, though scarcely surprised, to find that a deer living in solitude should have developed a call considerably louder than those of the more typical sociable deer, which is apparently used at all seasons, and also evinces the dawn-and-dusk emphasis so frequent in epideictic phenomena.

Among these and other polygamous mammals the furnishings and accomplishments of the males sometimes reach a very high pitch of development, and an ability to roar or bellow is a common secondary sexual character. The sound is usually directed at other males as a conventional form of challenge and may be largely or entirely confined to the rutting season. A number of examples could be drawn from the Carnivora Pinnipedia, such as the sea-elephants (*Mirounga*) or fur-seals (*Callorhinus*); in both of these the bulls attain a massive strength, and weigh several times as much as the cows. They have to contend not only with other successful males enjoying a social rank similar to their own, but also with a possibly much larger number of unsuccessful competitors, the foremost of which would usurp their place on the slightest admission of fear or weakness. Their bellowing is being used to uphold their right to the stance and the females they have appropriated, as a challenge defying any rival to dislodge them; and this is one of the normal functions of what we have referred to throughout this chapter as ‘ song ’.

The roaring of the stags of the red deer (*Cervus elaphus*) and other species of *Cervus*, of the moose (*Alces* spp.), and many other deer is of the same character. In *C. elaphus* there is an enlargement of the larynx in the stag at rutting time, and they begin to roar (in September in Scotland) as the season draws near, and the sparring and competition that has been going on in the stag herds in a desultory way all summer mounts to a climax. At almost any time stags will measure up to their full height on their hind legs, briefly milling at each other with their fore feet; or they may wrestle with their antlers once these are full grown. They roar sometimes out of season, and often if not always in the autumn before they seek the hinds. Here again the roaring appears to be largely concerned with establishing and maintaining a male hierarchy. As Darwin (1874, p. 803) pointed out, the voice of the stag does not serve to call up the hinds, since it is actually the stag that goes after the hinds; and in the red deer and moose it is rather the female's voice that draws the male—a fact long made use of in North America in hunting the bull moose, which can be called up towards the hunter by the skilful use of a birch-bark ‘ moose-call ’ imitating the voice of the cow.

Compared with most animals, mammals tend to be rather large in size and to have correspondingly low overall population-densities. Consequently, in species with uniformly dispersed populations consisting of solitary members or pairs, sound signals can be used to maintain social contacts only if they are very loud. Many of the gregarious species, on the other hand, living in herds or colonies, can very easily maintain their social integration by sight rather than ear, and it is again in the dense environments of the forest and jungle that audible recognition is most widely resorted to.

Nevertheless, the mammals provide some of the best examples we have yet seen of the dispersionary uses of sonic signals, and at the same time they parallel extremely closely the functional pattern we have found in other groups, using their voices to reinforce personal claims to property and social status, especially in the male sex, and joining in concerted displays that reveal the density of numbers.

3.8. *Summary of Chapter* 3

1. For signalling, sounds have excellent qualities. They are easily produced (though elaborate organs are required for their reception), endlessly variable in cadence, quality and pitch, and capable of being broadcast and heard by individuals that are visually hidden. They frequently supplant visible signals in dense or dark surroundings. The signals originate from accidental noises, though the latter tend in general to be suppressed in

animals. A brief analysis is given of the uses made of sound, following the pattern adopted for visual signals in Chapter 2; it is noteworthy that, just as in man where spoken language is the principal medium of social converse, and interspecific uses of sound are almost non-existent, in most animals the social uses of sound also predominate.

2. Birds as a class are the most vocal of animals, and use sound signals for many intraspecific purposes—among others for the expression of emotions, for the conduct of conventional competition, and for epideictic displays. The loudest, most persistent and distinctive signals—used for the last two purposes especially—are usually designated as song, and the primary function of this in many species is the proclamation of territorial ownership.

Song is typically associated with the reproductive cycle, and is normally produced only by adult males, since they assume at that time the task of maintaining the social organisation and dispersion of the whole population, the females passively conforming to the pattern they establish. Dispersion at that season, in fact, is based on mated pairs as units, and the male takes practically the whole responsibility for it; the female reciprocates by producing the eggs, and often the nest, and sometimes undertaking most of the incubation as well. In the non-breeding season many birds become virtually asexual, and in some that are solitary and territorial at that time all individuals sing regardless of the latent sex (*e.g.* in the European robin).

Bird-song features prominently in later chapters (especially 11 and 12) and is not exhaustively discussed here. It should be remarked, however, that it is generally quite specific, even identifying the individual singer in some species; that it is characteristically directed at other adult males; that it is emulative; and that it has a marked diurnal cycle of intensity, with a peak at dawn and a smaller one at dusk. It is not confined to territorially-dispersed individuals, but may also be produced by assembled parties engaging in epideictic displays.

3. In insects, we find the cicadas for instance have rather similar vocal adaptations. Again singing is exclusively the adult male's job, and the prediction that it is primarily directed at other adult males is confirmed by their having much larger auditory organs than the females. (The same is seen in some frogs, such as *Rana catesbiana.*) The singers may in some cases be widely separated from one another, but in other species they are gathered into choral parties, sometimes of thousands.

Turning to the grasshoppers, we again find species in which the males are territorial, and engage in song-flights or sing from a fixed site. Threat displays, combat, and antiphonal singing by rivals are all well known. Several earlier observers have concluded that the songs have little to do with courtship or mating, and in ascribing an epideictic and dispersionary function to them we are therefore rationalising them for the first time.

Passing note is taken of social sound-signals produced by insect larvae, by adults of both sexes and by females instead of males where the epideictic relations of the sexes are reversed. Similar phenomena recur in many classes of animals.

4. Most soniferous Crustacea are aquatic and therefore deferred to Chapter 4, but land-crabs of the genus *Ocypode* provide parallel development in the use of stridulation in defence of occupied burrows.

5. There is a close parallel between the Amphibia and Orthoptera; even the sounds they produce can be extremely alike. Again we find both communal and solitary singers in different species. Singing is largely prenuptial, done by the males, but not apparently concerned with courtship or attracting additional individuals to the mating grounds. The epideictic explanation is again the first one to accord perfectly with the facts.

6. Reptiles are mostly rather silent, but the American alligator is a notable exception. It has a loud bellow, in the nature of a territorial challenge, confined to adult males in the breeding season. Alligators have some sort of social hierarchy and a traditional pattern of nest-sites and lying-up places.

In a few lizards both sexes stridulate, and some turtles produce sounds. (The rattlesnakes' rattling is probably not used as a social signal.)

7. In mammals the now-familiar dispersionary uses of social communication by sound have developed over and over again in different groups—shrews, bats, howler-monkeys, anthropoid apes, rodents, carnivores (especially Felidae and Canidae), seals, and deer. All the common characteristics are found, as to sex, season, conventional hour, conventional assembly place, dispersed territory-system, and emulative or competitive character—as well as the common exceptions. An impressive amount of parallel evolution is in fact revealed by comparing these cases, both together and with the earlier groups.

Chapter 4

Social integration by underwater sound and low-frequency vibrations

4.1. *Sound and hearing under water*

Though it has long been known that some aquatic animals such as porpoises, seals, fish and shrimps produce sounds of considerable power and variety, it was not until the great development of sensitive underwater listening devices in the 1939-1945 war that the full scale of underwater animal sound was revealed. In the succeeding fifteen years the subject has been pursued considerably further. As is well known, sound waves are transmitted more than four times as fast in water as in air at ordinary temperatures and pressures (which means that the wave-length at each frequency is increased by the same factor), and in general they carry farther; that is to say that water transmits sound with less absorption of energy than air. There is no special difficulty in producing sound under water, for example by stridulating devices; but any pneumatic organs, such as those of the fish and Cetacea, are liable to contract as the depth and pressure are increased, since the volume of contained gas varies inversely with the pressure; and except when the animal is fairly close to the surface the pneumatic organs consequently tend to be reduced in size, if not practically collapsed. This can, of course, be prevented by keeping them pumped up to match the external pressure, which is known to be possible to some extent in fishes; it might also be done in the complex air-chambers, of unknown function, which some porpoises have in the blubber of their heads, in more or less the same way as it does in their ear cavities (*see* p. 69).

Hearing is rendered difficult by the fact that the bodies of most aquatic animals have a density and compressibility about equal to that of water; they are consequently ' transparent ' to water-borne sound and do not stop or absorb its energy (Pumphrey, 1950, p. 11). What is required is some structure within the body, either much denser or much more compressible than water, to serve as a detector, with a sensory organ anchored to it. A gas-chamber is satisfactory for the purpose, except for the complication of having to allow for its change in volume whenever the animal changes its position in depth; but far more frequent is the use of a heavy body or stone. According to Pumphrey (loc. cit., p. 13), it can be shown that by using a mass only two or three times as dense as water a sensitivity can be acquired approaching the maximum theoretically attainable with an infinitely dense mass. The substance used is often a concretion, chiefly of calcium carbonate, which in the form of calcite has a density of 2·6-2·7; it occurs in the so-called statoliths or otoliths of many marine organisms, from medusae to fishes. Alternatively, many species use grains of sand taken in from outside, of almost the same density (for quartz, *c.* 2·6). It is very interesting to notice the development in one order of the Crustacea, the Mysidacea, of statocysts incorporating calcium fluoride, which necessitates the extraction of a relatively uncommon element, but has the superior density of 3·2.

4.2 *Sonic signalling in Cetacea*

In the Cetacea, when they dive to a considerable depth, the lungs and other non-osseous parts of the respiratory passages progressively collapse under the increasing external pressure, until finally the tympanic cavities of the middle ears are virtually the only air-filled spaces remaining. These are apparently kept 'pumped up' as far as possible by the air displaced from other parts of the pneumatic system. Each is further guarded from collapse, evidently, by a bony tympanic bulla of exceptional thickness, strength and density; but even so there are two cases known in the blue whale (*Balaenoptera musculus*) where the bulla has been actually crushed or imploded under excessive pressure, and suffered extensive comminuted fractures, which have afterwards been more or less repaired during the life-time of the whale (Fraser and Purves, 1953). An ability to hear appears to be of paramount importance to whales, at whatever depth they swim, as witness these elaborate adaptations of the middle ear, which keep its cavity inflated when all other parts of the respiratory passages have collapsed. The pre-eminence of the auditory sense is not surprising when we consider that the toothed whales are wholly devoid of a sense of smell, having no olfactory nerves, and in the whalebone whales these nerves are only rudimentary; and that vision in water is at best indifferent, and deteriorates rapidly with the increasing darkness as the whale descends from the surface. (Those river-dolphins belonging to the Platanistidae that live in the waters of huge muddy rivers such as the Amazon, Ganges and Yangtse-Kiang, are variously nearly or quite blind.) Because the Cetacea have only a minute external auditory meatus blocked with wax, it has frequently been concluded that their sense of hearing is also defective; but this is obviously due to a misconception, since an external ear and tympanum are not required in the reception of sound under water, as stated already. In fact the whales have both a compressible gas-chamber (the tympanic cavity), and a dense surrounding bone as well, possibly making an accessory contribution to the reception of water-borne sound. Their auditory nerves are always large, and there can be little doubt that hearing is their principal source of knowledge of the outer world.

Sound-production in the Cetacea has been studied in captive bottle-nosed dolphins (*Tursiops truncatus*) in the Marineland aquarium, Florida. McBride and Hebb (1948, p. 112) tell us that 'the porpoise's auditory sensitivity is clearly evident at all times, together with a tendency to vocalisation whenever the animals are excited.' 'There are three noises with "language" value. . . . One is a snapping noise made with the jaws, as a form of intimidation by a dominant animal towards subordinates. The other two, whistling and barking, are forms of vocalisation. Both . . . are quite loud; they can be heard readily outside the tank. . . . The whistling is accompanied by a stream of bubbles from the blowhole.' The authors also state (p. 117) that in reaction to a strange object of large extent in the water, such as a seine net, 'they will remain schooled and swim excitedly in a tight pack. . . . The whistling is frequent and loud during such a period.'

Kellogg, Kohler and Morris (1953), studying the same species, found that of the various noises produced by *Tursiops*, two occurred almost continuously. The first was a bird-like whistle, approximately 0·5 sec. in duration appearing in several 'melodies' or pitch-patterns, the most common of which had a rising inflection and resembled the cheep of a canary. The frequency began generally in the neighbourhood of 7 kc/sec. and finished at about 15. The second noise was a series of 'clicks' or 'clacks', described as the 'rusty-hinge' or 'creaking-door' sound; they resembled the drumming of a woodpecker, but the tempo was often even faster, varying from 5 to 100 per second. In the latter case the sound became the groan or bark, which has already been referred to in the quotation from the previous authors.

Kellogg, Kohler and Morris were seeking to discover whether porpoise sounds could serve as 'sonar' signals, permitting the animals to perceive nearby objects, and possibly the sea floor, by echo-location, in the same sort of way as bats do. This is a very interesting notion, and an adaptation one might particularly expect in the whales, occupying as they do a perceptible world dominated by sound and hearing. The authors conclude that 'although the facts . . . do not establish conclusively that the bottle-nose dolphin actually uses echo-ranging, they offer good circumstantial evidence to that effect.' However, the authors regarded as 'something of a paradox' (p. 240) the relatively low, readily audible pitch of the sounds emitted, and the somewhat disappointing lack of overtones. Bats emit ultra-sonic signals, at frequencies which their ears have become specially modified to receive, with a wave-length of the order of 0·7 cm. The shorter the wave-length the greater is the resolving power, and the better the definition of the 'sound-shapes' perceived; but also, unfortunately, the greater are the losses and attenuation with distance. Water is not such a suitable medium as air for sonar-location, as far as the animal is concerned, because to get as high a resolving power (that is to say, to use the same wave-length) as bats, the frequency would, of course, have to be over four times as high, and in the neighbourhood of 250 kc/sec. The principal output of the whistle of *Tursiops* is at about 15 kc/sec., with a wave-length of 10 cm.*

It is probable that the bold black-and-white patterns and patches found in many of the whales, such as the lesser rorqual (*Balaenoptera acutorostrata*), the killer whale (*Orcinus*), common porpoise (*Phocaena*) and various dolphins (*Delphinus*, *Lagenorhynchus*, etc.), are used for visual recognition at close range; but there are even more species that are almost uniformly black or grey. In either case at distances greater than a few metres contact can only be maintained sonically; and in fact the contact function of sound accords well with what we know about their vocal powers and habits. Another sidelight is thrown on the subordination of sight by the uncommonness of visible secondary sexual characters, so usual in other mammals; the narwhal (*Monodon*), with its huge tusk in the male, and the killer whale, with its greatly enlarged flippers and fin, are the principal exceptions. There are sexual differences in size; these are very great in the sperm-whale (*Physeter catodon*), the large bulls being almost twice the length of the mature females, and not insignificant in the whalebone whales (Mystacoceti), where the female is the larger sex.

It is to be noted that McBride and Hebb observed the use of sound signals in maintaining social dominance, a common function of animal ' song '.

There are reports of sounds heard from the common dolphin (*Delphinus delphis*) (Kullenberg, 1947), and a few other species, especially the beluga or white whale (*Delphinapterus leucas*), which has been called the ' sea-canary '. Schevill and Lawrence (1949) give a most interesting description of what they could hear under water in the vicinity of a herd of belugas in the Lower St. Lawrence off the mouth of the Saguenay. ' The noises which we thus ascribed to *Delphinapterus* were heard as high-pitched resonant whistles and squeals, varied with ticking and clucking sounds slightly reminiscent of a string orchestra tuning up, as well as mewing and occasional chirps. Some of the sounds were bell-like, and a few rather resembled an echo sounder. Occasionally the calls would suggest a crowd of children shouting in the distance. At times there were sharp reports. . . . On two occasions we heard trilling, which quite justified the name of " sea canary ".'

It is well known (*cf.* Owen 1868, III, p. 588) that the majority of whales, if not all, have a labyrinth of accessory air-ducts and saccules in the adipose tissue of the head, which variously communicate with the common narial passage just inside the blow-hole, dorsal and external to the great nostril valves. They are easily disclosed by dissection in the common porpoise

* Since this was written, Kellogg (1958) has clearly demonstrated by a series of experiments the ability of *Tursiops* to locate and differentiate between objects (*e.g.* fish of different sizes) by means of reflected sound. Frequencies up to 80 kc/sec. have been recorded.

and dolphin, the superficial cavities being larger in the former (*see* fig. 3). No function has been ascribed to these elaborate organs, the walls of which are provided with a layer of muscular fibres; but in view of the extreme modification of the larynx, the absence of vocal cords, and the difficulty of making laryngeal sounds when denied the use of the mouth, it seems very possible that they have something to do with the voice.

4.3 *In fishes*

The sensory equipment of fishes is far more comprehensive than that of whales: in fact its compass and variety are possibly unequalled by that of

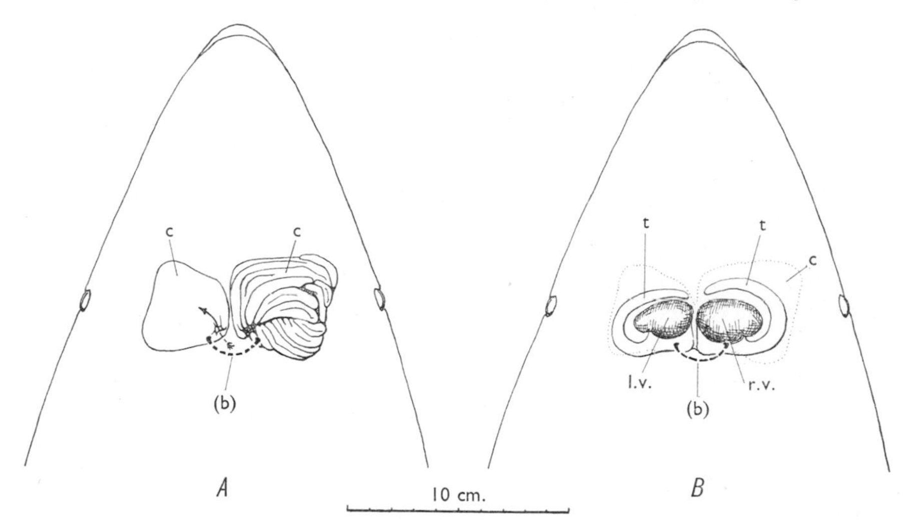

FIG. 3. Frontal air-sacs of the porpoise (*Phocaena communis*). In the blubber of the forehead, underneath the crescentic blowhole (*b*), the nasal chamber communicates upwards (in *A*) with an asymmetrical pair of flattened air-sacs (*c*) with thin roofs and leathery, muscular, convoluted floors. When these are dissected away (in *B*) another pair of blind curved finger-like air tubes (*t*) are revealed, also embedded in the blubber. At this level we see the entrances to the huge nasal passages each firmly closed by a thick muscular valve (*l.v.*, *r.v*) The frontal air-sacs may be connected with sound-production.

animals of any other kind. Many, and perhaps most, of the bony fishes have colour-vision. Their olfactory organs are well developed, and those of the cartilaginous fishes are very elaborate. The sense of hearing is acute and discrimination of pitch excellent in some, for example the minnow *Phoxinus* and the catfish *Ameiurus*, where this has been investigated; and all fishes have also the lateral-line system which is known to be sensitive to low-frequency vibrations (*see* p. 78). In the sharks and rays there are ampullary organs—possibly responsive to temperature change, or possibly electrical detectors. Cutaneous sensory buds are usually distributed all over the body-surface and mouth-cavity; and, in addition, more or less elongated tactile barbels and filaments are not at all uncommon. Finally, a small number of fishes (so far as present knowledge goes) make electrical discharges, which appear capable of use as signals; in a few cases they have been shown to elicit aggressive responses in the fish perceiving them (*see* p. 85-6).

Fishes need not therefore rely on sound to the same extent as whales for the purpose of attracting mutual attention and signalling. There are, nevertheless, many scores of species that are known to make sounds, either when taken in hand or else even loud enough to be heard by people in boats—a fact already known to Aristotle. Indeed the field is by no means a new one: it was the subject of an investigation by Johannes Müller (1857) a century ago; Sørensen wrote and published a 245-page thesis on it in 1884, and Bridge (1904) required almost ten pages in the *Cambridge Natural History* for a concise summary of the knowledge of it existing at that time. The latter is still possibly the best general account in English.

In their hearing apparatus, fishes employ both methods of receiving underwater sound-waves, namely by the use of heavy ear-stones or otoliths alone, and by the use of the gas-filled swim-bladder as an initial pick-up (*see* especially von Frisch, 1936). In the most celebrated family of noise-makers, the drums or Sciaenidae, where the bladder is used to create sound and apparently not to receive it, the otoliths are so exceptionally large that they were worn in mediaeval times as amulets. But in the Ostariophysi (Cyprinidae, Siluridae, Characinidae and Gymnotidae) pressure deformations of the air-bladder are transmitted to the auditory capsule by an articulated chain of little bones called Weber's ossicles, formed from parts of three vertebrae, in a manner analogous to the working of the mammalian ear, and equally elaborate. In the cod family (Gadidae) and others, a tubular extension of the swim-bladder leads forward on either side to a window in the periotic capsule, at the back of the skull, where pressure changes are transferred to the perilymph; and in some of the herrings (including *Clupea* spp.), each air-tube ends anteriorly in a sac, abutting directly on a protruding outgrowth of the utriculus, which is part of the membranous labyrinth itself. Though these adaptations are almost certainly concerned with hearing, one should not fail to observe that they must equally provide sensitive gauges of water-pressure, capable of indicating even very small changes of depth (*cf.* Schreiber, 1935); but these two functions are not of course mutually exclusive, any more than are the functions of hearing and balance in the labyrinth.

Fishes produce sounds either by some kind of stridulating mechanism, usually in conjunction with a resonator which may be bony but is very frequently the swim-bladder, or by vibrating the walls of the swim-bladder directly with muscles. The variety of stridulating devices is as great or greater than in any single order of insects, which implies that sound-production has developed over and over again in the Teleostei and is not a 'primitive character'. The devices involve grating or striking the teeth, either in the mouth (*Mola*) or the pharynx (*Scomber*), or rasping of special surfaces, associated with the anterior dorsal fin (*e.g. Balistes, Capros,*

Gasterosteus), the pectoral fin (*Doras*), pelvic fin (*Capros* and *Gasterosteus* also), the operculum (*Cottus*), or internal parts of the pectoral girdle (*Balistes* again). Even where similar structures are concerned, the anatomical relations vary in different genera.

Sounds derived from muscular deformations of the swim-bladder are made by fish of several families. The muscles may originate on the skull, and be inserted on the walls of the bladder or on special bony springs to which the bladder is attached; or they can originate on the abdominal wall; or they may be 'intrinsic muscles' wholly applied to the bladder itself. Unlike the smooth muscle-fibres in the lungs of the higher vertebrates, muscles in the swim-bladder of fishes are apparently always striated, and innervated by 'voluntary' nerve-fibres; this has no doubt aided the specialisation of the bladder as a vocal organ. It is interesting that the Sciaenidae, which produce their powerful signals by the muscular drumming of the bladder, appear to detect sound by means of their otoliths, which attain a larger size in the species that drum than in those that do not (Evans, 1940, p. 71). It could well be that the simultaneous adaptation of the bladder for transmitting and receiving sounds presents serious difficulties, and that combining the two functions is scarcely practicable.

In the Sciaenidae, and probably in some of the others, there is a marked sexual difference in vocal development, the sounds of the males being much the louder; in some the females are said to be silent, or not at any rate given to singing. One species in this family, the croaker (*Micropogon undulatus*), has been studied off the Atlantic coast of the United States by Dobrin (1947) and Knudsen, Alford and Emling (1948). The latter authors state (p. 424): 'The sounds produced by an individual croaker consist of a series of "taps" that continue for about one and one-half seconds at a rate of about seven taps per second. The series is repeated at intervals of 3 to 7 seconds. The sounds resemble the tapping of a woodpecker on a dry pole. In an area with a high concentration of croakers, the sound during a period of great activity is a continuous roar, and the sounds of individuals may be heard only infrequently over the chorus of the whole population.' The species is a gregarious one, caught commercially in gill-nets in the same way as herring. The authors continue: 'Croaker noise occurs principally during the feeding period, which starts in the evening as the bottom begins to darken. This, together with the migratory habits of the fish, accounts for the very pronounced diurnal and seasonal variation in croaker noise. . . .' Their graphs are reproduced in fig. 29 (*see* p. 338).

What is described here has the now familiar appearance of a dusk chorus, and it may well occasion initial surprise to find such a thing in the sea. However, diurnal cycles are very common in marine animals living in the photic zone, and synchronisation of activities timed by means of dawn and dusk changes in light-intensity are normal, especially in habitats not directly influenced by the tides.

Many years ago Couch (1863, II, p. 27) described how the gurnard

(*Trigla gurnardus*), a bottom-living fish, 'sometimes in the fine weather of summer . . . will assemble together in large numbers, and mount to the surface over deep water . . . and when thus aloft they move along at a slow pace, and rising and sinking in the water for short distances, and uttering a short grunt, as if in self-gratification.' Gurnards are among the best-known vocalists, and their common name originally means 'groaners' or 'grunters'. Moulton (1956) has discovered that in the related American sea-robins (*Prionotus* spp.) there is a particular staccato call, first produced at the onset of the breeding season. By transmitting imitative pulses of sound into the sea at ten-minute intervals he found he could on many occasions evoke responding calls from one or more sea-robins, though they had remained silent in the intervals, thus showing that the signals have for them the emulous character typical of epideictic songs.

Knudsen, Alford and Emling (loc. cit., pp. 424-426) describe another different case, that of the toadfish (*Opsanus tau*), a small inshore blenny-like species of the family Batrachidae. These have hiding-places where they lurk, and they fight amongst themselves (Bigelow and Welsh, 1925, p. 358); no doubt they have a solitary or territorial type of dispersion. 'Toadfish, individually, seem to produce higher noise levels than any other form of marine life thus far identified and reported, with the possible exception of the porpoise. The sound produced is an intermittent, low-pitched "boop" of about one-half second duration, similar to a boat whistle or sometimes like the cooing of a dove. . . . These fish are shallow water bottom dwellers which nest under rocks, tin cans and similar debris. Toadfish are not gregarious and apparently do not occur in sufficient concentration to produce the continuous roar generated by croakers during peak activity. Toadfish noise is subject to little diurnal variation.'

A very similar case is that of the California singing fish (*Porichthys notatus*), which belongs to the same small family as the toadfish. Its habits have been carefully described by Greene (1924); the swim-bladder is U-shaped, with the horns forwards, each bearing on the outer side a powerful striated muscle innervated by the vagus. Posteriorly the common sac is divided by a transverse diaphragm, perforated by a pupil-like hole, and it is the motion of this diaphragm which initiates the sound. The singing fish spawns in June and July in tide-pools, making a nest in a cavity under a rock, and depositing the eggs on the roof of the chamber, where they are guarded by the male. Any disturbance of this duty calls forth a protesting sound.

'The sound produced is that of a low croaking or grunting noise. The tone pitch and quality produced by the vibrations vary in different specimens ,' and in an aquarium these could be easily recognised by voice individually. They were pugnacious, and when the largest of them 'aggressively attacked other specimens, they would produce the noise but while swimming away from the aggressor. At other times, when the aquarium was quiet and only the most gentle movements occurred, a more soothing tone of low intensity was used by various members of the aquarium family as they swam back and forth in each other's proximity. Take it all in all, the behavior of the specimens was not unlike the group behavior of the more familiar land animals. One could scarcely refrain from the conclusion that these fishes used their noise-producing air-bladders under conditions of colonial activity—fright, combat, defense and friendly association.' Viewed in the perspective already acquired from the study of the numerous very similar cases described in the previous chapter, there can be little doubt that the loud croak is a normal song, clearly associated with establishing and retaining both a locus and a status in the community.

Loud phonetic signals are made by many freshwater fish as well as marine ones, for example the loaches (Cobitidae), which have a partly ossified air-bladder, the North American freshwater drum, *Aplodinotus grunniens*, the ferocious South American piranha (*Serra salmus*), or the Brazilian Tetragonopterid, mentioned by von Ihering (1930), in a paper devoted to the voices of freshwater fish.

4.4. *In Crustacea*

Examples were given in the previous chapter (p. 49) of the stridulation of land-crabs. The more typical aquatic Crustacea include other soniferous species, and these are likewise confined, so far as is known, to the Decapoda. Possibly the largest are the spiny-lobsters (*Palinurus*, etc.) which have a pair of stridulating organs, each comprising 20 to 30 fine parallel ridges (in *P. vulgaris*) lining a hollowed surface, on the base of the second antennae; by raising both the antennae at once these can be rubbed along the smooth edges of the rostrum, and a brief but quite loud creaking sound produced. It can readily be stimulated by picking up the animal, and is then evidently used in response to the situation of emergency. Dijkgraaf (1955) found that he could imitate this sound by rubbing with a moistened finger on glass; when he did so the *Palinurus* (a male) in his aquarium at once responded by a silent ' defence-movement ' of the antennae. The next time he made the sound its reaction was repeated, but the third time it answered equally promptly with an audible creak. Moulton (1957) studied the West Indian species *Panulirus argus*, and found that it had two forms of stridulation, one a ' rasp ' exactly corresponding to the creak just described, and the other a slow rattle of longer duration. He heard the rattle only from groups of lobsters confined together in a cage or ' live-car ' under water; it consisted of 5-6 pulses lasting about one-quarter second. It was detected (with a hydrophone) only during the daytime, when the lobsters remained heaped in the corners of the live-car, and seemed ' to be more in the nature of a conversational sound ' (loc. cit., p. 291). Moulton was unsuccessful in obtaining any response by playing recordings into aquaria containing *Panulirus*. In the spiny lobsters the stridulating organs are found equally in both sexes.

Much the most notable noise-makers in the Crustacea, however, are the snapping or pistol-shrimps, belonging to the genera *Alphaeus* (the

Crangon of American authors, but not closely related to the common shrimp *Crangon vulgaris*) and *Synalpheus*. Both these are enormous genera, said to contain 215 and 150 species respectively, of tropical or warm-temperate distribution. The claw structure indicates that all the species snap to some extent (Johnson, Everest and Young, 1947, p. 127). The best-known British species is *Alphaeus ruber*, not uncommon at Plymouth.

The snap is made with the larger of the two asymmetrical chelae or great claws, left- and right-handed snappers being equally common; both sexes possess the device, but it is much larger in mature males, and in them the ponderous 'hand' and claw may approach in mass the whole of the rest of

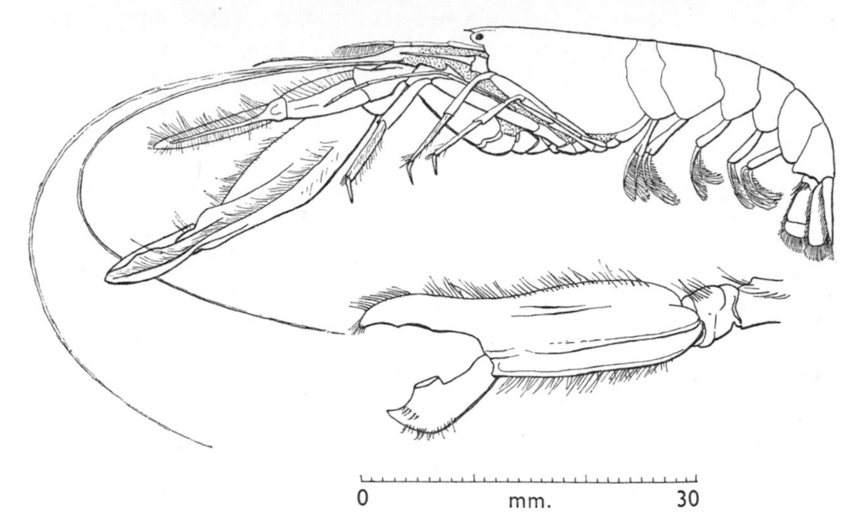

FIG. 4. Snapping-shrimp (*Alphaeus ruber*). Large male, from Plymouth. The snapping claw is as usual on the shrimp's right side. It is shown below in ventral view on the same scale.

the body. In *A. ruber*, which is here described (figs. 4 and 5) there is a projection like a hammer-face on the inner edge of the claw near its tip, which closes into a deep, rimmed socket on the hand when the pincers shut. The hammer-face has also a concave surface, so that, in closing the pincers, the hollows meet like a pair of opposed shallow cups. In order to make a snap, the claw is first abducted to the full extent, till it stands out at right angles to the hand; then suddenly the jaws are snapped shut, and there is a loud crack, which sounds in an aquarium tank as if the glass had snapped. Johnson, Everest and Young state that, in closing, the plunger-like mechanism ejects a vigorous jet of water; the writer missed this in observing *A. ruber*. I found that snapping could be stimulated by provoking the shrimps with a glass rod; but they also snapped at long intervals spontaneously. MacGinitie (1949, p. 278) refers to this in *A. californiensis*, which he says 'accounts for some of the "pistol shots" that one hears so frequently while collecting in certain areas'. The former authors say that the habit of snapping is associated with defensive and offensive activities, that the sudden gush of water may frighten away enemies approaching too near, and that an antagonist may also be driven away or sometimes killed by a direct blow of the small hammer.

Under natural conditions *A. californiensis* lives in shallow water, from low tide mark down, inhabiting burrows or crevices, often in pairs. 'The snapping shrimp are gregarious to the extent that large numbers of a given species are found either in pairs or as solitary specimens in isolated retreats within the area. The occupants of these retreats make a noisy protest when

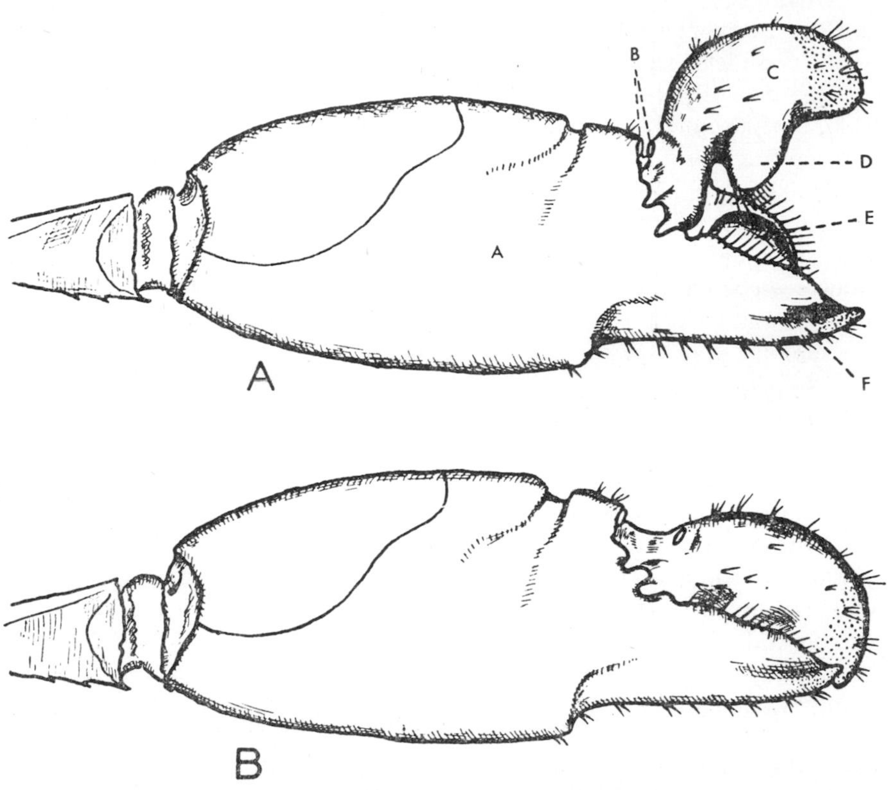

FIG. 5. Snapping mechanism of the male *Alphaeus californiensis*. *A*, snapper open; *B*, snapper closed—A, palm; B, sucker; C, movable finger; D, plunger; E, socket; F, thumb. (From Johnson, Everest and Young, 1947.)

an intruder approaches. Specimens have been observed to leave their burrows in search of food, especially at night, but those under observation always returned to their respective retreats. These retreats in some instances were occupied steadily by the same specimen for several months' (loc. cit., p. 134). 'The sound emitted by an individual shrimp is a single sharp "snap" or "crack" produced only occasionally. It is the combined snapping of members of a large population that results in a continuous loud underwater crackle over or near the shrimp beds' (loc. cit., p. 124).

The chorus can reach a high intensity. The authors liken it to the 'brisk burning of twigs', which may 'merge into a sizzle and finally a hiss'.

It has its strongest components at frequencies higher than those characteristic of fish-noise and water-noise; and when the sea is quiet and ambient noise low, 'the transmitted shrimp sound is appreciable [through a hydrophone] to a distance of over a mile from the boundary of the bed'.

This case has much the character of a lek, in that the performers hold particular stances, from which they contribute to the communal chorus. Johson, Everest and Young (loc. cit., pp. 133-134) say that the bedlam is never-ending. 'There is a small diurnal variation. . . . At night the levels are 2 to 5 db. higher than in the daytime. In addition, there is a slight peak in the noise level shortly before sunrise and after sunset'.

The two other kinds of stridulating shrimps are mentioned by Alcock (1902, p. 106 and 118) from Indian waters. The first is *Gonodactylus chiragra*, a 'locust shrimp', in which the terminal claw of the first pereiopods folds down like a penknife blade into a groove cut in the whole length of the joint proximal to it. The 'knife-blade' flies open and makes a sharp click as it snaps past an erectile spine on the 'case'. The other is an unnamed musical prawn, which has a burred rasp on the inner edge of the inner ramus of the uropod, and when this rubs against the telson 'a soft, thrilling sound, like the subdued note of a grasshopper, results'.

4.5. *In insects*

A great number of the aquatic Hemiptera can stridulate. In the Corixidae the anterior limbs bear the sound-making organs—on the tarsus—the mechanism being less perfectly developed in the female; in some there are different structures on the dorsal surface of the body. In *Ranatra* there are coxal stridulating surfaces, both nymph and adult being capable of sound-production (Imms, 1946, p. 354 and 371). The males of some Notonectidae are also said to stridulate (Skaife, 1954, p. 104), but corresponding sonic adaptations do not appear to have been developed in other groups of aquatic insects.

4.6. *Vibratory sense in Gyrinus*

There is, however, a development of different and remarkable character in the whirligig beetles (Gyrinidae), which gives them a faculty or means of perception somewhat similar in effect to the lateral-line sense of vertebrates.

Whirligig beetles are, as is well known, very sociable. When undisturbed, and under suitable conditions, they consort together in dense swarms, swimming at the surface with their backs out of the water, in a state of continuous motion. The purpose of this behaviour has never been understood; it shows, however, some of the normal features of epideictic display, to be noted in the following account. According to de Wilde (1941, p. 385), 'In calm weather, when the water-surface is smooth, the whirling-beetles mostly remain in smaller or larger flocks, in open places, where they make their whirling gestures. They choose distinct, fixed places, where they may always be found; we do not know whether they were always the same

Evening mass-manoeuvres of starlings at a roost (*see* p. 286)
(Photo by Dr. Robert Carrick.)

A swarm of 3-4000 whirligig beetles, on the Rivière des Prairies at Ste. Dorothée, near Montreal; September 1931. (Author's photo.)

animals, but their number was fairly constant. When by a change of the wind, such a surface was covered with any floating material, the beetles would disappear'. They whirl only by day, and were never seen in moonlight; their food is obtained entirely beneath the surface.

Eggers (1927) observed that, in an aquarium, the beetles, moving at high speed, were able to avoid crashing into the wall of the tank, or colliding with other moving beetles. They could not detect other beetles at rest, or other light floating objects, and lost the sense altogether when they dived. When they are at the surface, he observed that the large second joint of the antenna with its fan of sensory hairs rides on the surface film, the distal part of the antenna being carried high and dry above it (fig. 6). The same

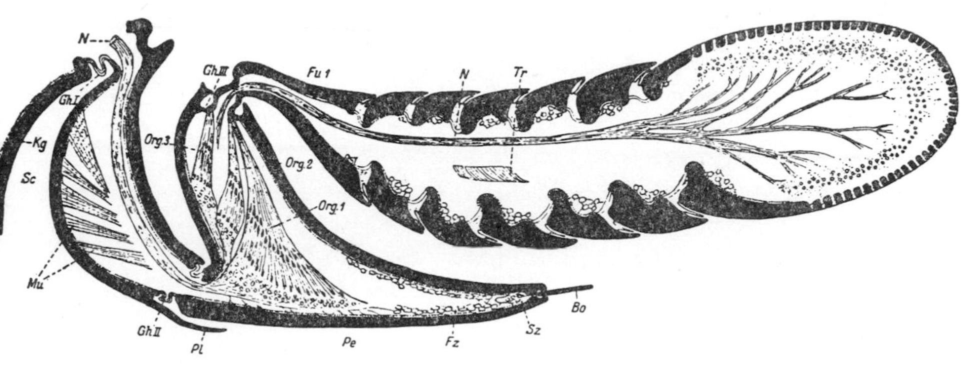

FIG. 6. Antenna of *Gyrinus marinus*, sagittal section. *Bo*, hair, probably sensory; *Fu* 1, 1st segment of flagellum; *Fz*, fatty tissue; *Gh I*, *Gh II*, *Gh III*, cuticle forming joints; *Kg*, recess in head, housing base of antenna; *Mu*, muscles of scape; *N*, antennary nerve; *Org*. 1, *Org*. 2, *Org*. 3, separate sections of Johnston's organ; *Pe*, pedicel; *Pl*, plate-like projection covering second joint; *Tr*, trachea; *Sc*, scape; *Sz*, sensory cells. (From F. Eggers, 1927.)

joint contains, as usual in insects, Johnston's organ, which is a chordotonal organ responding to sounds and other mechanical vibrations. He found that either removal of the sensory hairs or extirpation of Johnston's organs destroyed the faculty. Under normal conditions the whirling beetles could avoid a glass tank wall at a distance of one centimetre or more; but he thought he could increase their sensitivity to over 5 cm by scattering Lycopodium powder on the water-surface (loc. cit., p. 145).

His work has been in large part repeated and confirmed by de Wilde, who concludes (loc. cit., p. 399): 'When swimming on the surface in swarms, animals never crash against each other; an animal at rest cannot be avoided. The only possible conclusion is, that the "bow-wave" must be perceived'.

There is here a sensitivity to the low-frequency vibrations of the surface film set up by their own swimming movements, so that the beetles can detect one another, and catch the rebound of their own motion from ponderous objects in much the same manner as fish, though apparently at much smaller ranges, and only on the surface. The faculty is very suitable for social integration; the combined movements of the whole swarm must produce a sensible disturbance very comparable with a chorus of voices. The mutual participation, and the characteristic conservatism in the site of the display should not escape notice. Whirling is neither strictly seasonal nor immediately concerned with mating, and is enjoyed by adults of both sexes alike.

4.7. *Lateral line sense in aquatic vertebrates*

The majority of lower vertebrates, including the cyclostomes, fishes and aquatic amphibia, possess the ' lateral line system ' of sensory organs, generally taken to be employed primarily for the perception of low-frequency mechanical vibrations. The system lies on or near the body-surface, and consists typically of rows of sensory neuromast organs, most often either spaced out along a series of open grooves, or enclosed in fine canals which communicate at intervals through pores to the outside. The canals run in the leathery skin in the cartilaginous fish, but in the bony fish they are frequently protected by passing into the substance of special investing scales or dermal bones. They are distributed especially over the head, but the name of the system comes from the lateral line itself, which when normally developed runs back along each side of the body from the temporal region to the root of the tail, and sometimes to the end of the caudal fin itself (as in the Sciaenidae). Though it comes to extend therefore over as long a ' base ' as possible, the whole system is innervated by cranial nerves, and probably originated on the head. In addition to the canal system there are individual receptors, chiefly on the head also, each with a similar type of neuromast organ; the best known of these are the numerous ampullae of Lorenzini in the Elasmobranchs, which Sand (1938) showed to be thermoreceptors. The membranous labyrinth of the ear is another derivative of the same family of sense organs, specialised to yet other functions, namely hearing and balance.

Though the pores or canals on the head are usually retained, some fishes have more or less completely lost the lateral line itself, for example, the gobies (Gobiidae) and cyprinodonts (*e.g. Fundulus*), and some of the darters (Etheostomidae) and minnows (Cyprinidae). These are perhaps always small fishes living in shallow water, and in many of them the development of brilliant colour-patterns points to a high development of the sense of sight. In a few others, on the contrary, including the Hexagrammidae (the name *Hexagrammus* refers to the presence of ' six lines '), the lateral line is reduplicated up to five times at different levels along the flanks of the body.

There are great developments of the hind-brain in fishes, associated with the lateral line and other cutaneous senses. The lateralis roots of the facial, glossopharyngeal and vagus nerves frequently lead into special facial and vagus lobes of the medulla oblongata, which in exceptional cases (*e.g. Barbus* or *Carpiodes*) may be much larger than either the optic lobes, or the cerebrum, or cerebellum. The lobes in turn have strong connections with the cerebellum, which may also undergo corresponding enlargement or hypertrophy;

so that in the African freshwater elephant-fish *Mormyrus* the part of the cerebellum associated with the lateral-line sense, known as the 'valvula', is relatively larger in relation to the rest of the brain than is the cerebral cortex to the rest of the brain in man (*cf.* Herrick, 1924, p. 192. This may, however, be more concerned with the fish's electrical sense: *see* below, p. 85). In some deeper water species also, such as the scabbard-fish to be mentioned below, and the hake (*Merluccius*), the cerebellum has a pair of large acustico-lateralis lobes, which are the most notable feature of their neural anatomy (Evans, 1940, pp. 113-117). There can be no doubt of the importance of the system, nor that in some fish it has become the dominant sensory organ.

Actual dermal canals are confined to the true fishes. The cyclostomes have rows of pit-organs without connecting ducts. In the amphibia the system consists of shallow depressions, often in pairs, containing neuromast organs, but again there are no canals; the distribution and innervation of the sense organs follow patterns similar to those of fishes. The organs are present in free-living tadpoles and in the adults of many Urodela while they are aquatic; in the Anura they always disappear at metamorphosis, except in *Xenopus*.

The actual sensory cells are in compact groups, and each cell bears a hair-like process; the hairs are often united by a covering membrane, and project into the canals, or into the pits in which the organs lie; they are stimulated when they are bent over by fluid moving in the canal. The nerve-fibres that supply the neuromast organs, like those coming from the labyrinth of the ear, discharge spontaneously with a certain frequency at all times; and Sand (1937) showed that fluid moving in one direction along the canal excites the organs and increases the rate of discharge, and in the opposite direction it inhibits the organs and stops the nerve impulses. He confirmed the earlier work of Parker by finding that the system is extremely sensitive to low-frequency vibrations.

This function, as previously stated, is generally accepted as the primary one. There must, however, be other functions; for example, except in the more rigid parts of the head, the bodily movements of the fish itself must deform the canals and stimulate the organs; they also respond to touch; and Regnart (1931) showed that the response of fishes to the make and break of weak electric currents in the water was inhibited by cutting the lateralis nerves.

By means of the low-frequency vibrational sense fish are able to maintain their position in a stream in relation to nearby objects without the aid of vision, and it seems certain that they can similarly perceive and locate other fish in their neighbourhood, whether these are fellow members of their own school or other species, including potential predators and prey. In a discussion of the physical basis of the sense, Pumphrey (1950) has pointed out that at the frequencies of oscillation to which the lateral line is likely to respond, the corresponding wave-lengths are long; at 10 cycles/sec. the wave-length is 140 m. At distances greater than a wave-length, the displacement-amplitude of particles of the medium is inversely proportional to the distance of the source; but when the fish is near the source of the vibration, at a distance much less than a wave-length, the amplitude of displacement decreases with the *cube* of the distance. Thus a fish receiving such vibrations broadside from a source at a distance of the same order as its own body-length may feel an appreciably stronger signal in the middle of its body than it does at its head or tail, which are slightly further from the source; if it were to turn until it faced directly towards or away from the source, the difference felt between head and tail would rise to a maximum value; and in fact whatever its position, detectable differences in the strength of the vibrations at different parts of its body can be applied to locating the source. But, as he says, it is essentially a short-range sense, and as the distance increases it must ultimately become inadequate.

It seems clear, however, that the longer the ' base line ', the greater the effective range of the system; and as Dr. P. A. Orkin has pointed out to me, this may partly explain the remarkable length and attenuation independently acquired by a number of pelagic and deep-sea fish. These include the scabbard-fishes (*Lepidopsis* and *Trichiurus*); all of the eels (Anguillidae), and various other members of the Apodes, especially the deep-sea gulpers (*Saccopharynx*) and snipe eels (*Nematichthys*) with their flagelliform tails; the deal fish (*Trachypterus*) which may exceed 2 m in length, and the oar-fish (*Regalecus*) which may exceed 6 m; the grenadiers or rat-tails (Macruridae); the very rare frilled shark (*Chlamydoselachus anguineus*); and the rat-fish (*Chimaera*), also named from their long thin tails, on which the lateral line canals appear to be carried (ventrally) right to the terminal filament. In *Hariotta*, another rat-fish allied to the last, the snout or rostrum is also drawn out to a remarkable extent, and carries lateral line canals to the anterior tip as well. Several of the fish with long-attenuated tails have the caudal parts of the canals placed ventrally and close together, on either side of the ventral fin, as, for example, in the rat-fish, and in the deep-sea Heteromi, such as *Halosaurus*; though what the significance of this may be is not known. Of those mentioned which I have been able to examine, namely *Trachypterus*, *Trichiurus* and *Chimera*, all carry the canals to the tip of the tail, and it certainly suggests that the tail has been extended primarily to support the lateral line, and increase its range of usefulness (*see* figs. 10 and 11, pp. 120 and 122).

Attention may be drawn to the fact that few of these elongated deep-sea fish are provided with luminous organs; conversely, the luminescent species are for the most part small, short fish. This could in part be explained if these two totally different adaptations were providing alternative methods of overcoming the same problem, namely the maintenance of social contacts.

4.8. *Conclusion*

Pumphrey's (1950, p. 3) definition of sound ' as any mechanical disturbance whatever which is potentially referable to an external and localised source ' covers all kinds of underwater transmission with which we have been concerned in this chapter. We know much less about water-borne animal sounds than about air-borne ones, largely because conditions make it difficult for a human observer to hear them directly. Sound-production by aquatic animals may be more common than we know at present; but enough has been said to show that it is evidently a frequent means of underwater signalling.

The phenomenon of ' singing ', sometimes by males only, either in chorus or singly, at particular places and times, reappears here, in the Crustacea, aquatic insects, fishes and Cetacea; and it is apparent that sound-production and hearing can be used by aquatic species in their epideictic behaviour very much as they are employed in the terrestrial world. Sociable fishes and gyrinid beetles possess other different forms of mechanical perception, which appear to provide alternative methods of maintaining contact between the members of a school, and indicating its numerical size and density of crowding.

The emphasis in this chapter has again been specially placed on the social significance of sound-production and hearing, and above all on the integration of the population group as a social unit. It should perhaps be repeated, therefore, that, just as the ears can receive important information from the outside world that is not social at all or even in the nature of signals, so sound can sometimes be intentionally produced for non-social purposes, and directed interspecifically at other animals. Taking the animal kingdom as a whole, however, this appears to be a very minor aspect of sonification, compared with its social functions.

The requirement for social integration, as a prerequisite to the homeostatic control of dispersion, offers a simple explanation for the production of many sounds by animals, the purpose of which has not previously been understood; and this is perhaps nowhere more obvious than in those made by animals under the water.

A summary of this chapter and the next together will be found on p. 87.

Chapter 5

Social integration by electrical signals

5.1. *Electric organs in fish*

The possibility that electrical signals are employed socially by fish follows naturally upon the subject of the preceding pages; it is one that has received little consideration hitherto, although, having seen already what diversity there is in methods of animal communication, an adaptation of this kind would not surprise us.

There are several genera of fish, belonging to three or four families, capable of producing powerful electric shocks, and a similar number of others with electric organs of the same physiological character, but in varying degrees weaker. According to Lissmann (1958, p. 17) electrogenic organs have independently evolved at least seven times. The most powerful are found in the South American freshwater gymnotid *Electrophorus electricus*, usually called the electric ' eel ' on account of its shape rather than its affinities; it has two pairs of electric organs running along the sides of the tail. Like those of most if not all the other electric fish, except *Malapterurus*, the organs are modified striped muscles, supplied with very large ramifying nerve-fibres, the whole organ consisting of a number of longitudinal columns or piles of ' plates '. The electric charges on each plate are of the same order of size as the action potentials of muscle-fibres, but they are all polarised in the same direction to give a summation of thousands of small charges in series. *Electrophorus electricus* grows to a length of over 2 m, and according to Humboldt its shock can fell a horse wading in the water, if not kill it; the EMF built up is known to attain about 600 volts (Coates, Altamirano and Grundfest, 1954). A number of other gymnotids are able to produce much feebler discharges.

Malapterurus electricus is a catfish (Siluridae) occurring in the Nile and rivers of tropical Africa. In length it commonly attains 50 cm and occasionally 1 m. The electric organ in this case is integumental, developed from the smooth muscles of the skin of the trunk and tail, and the whole is controlled by a single huge nerve-cell on each side. The shock is evidently less powerful than that of *Electrophorus*.

In the electric rays, belonging to the chiefly tropical genera *Torpedo*, *Narcine*, *Discopyge*, *Astrape* and *Temara*, the electric organs are developed from the branchial musculature, and the columns are arranged dorso-ventrally. They are bottom-living sea-fish. So are the American blennioid fishes known as star-gazers, *Astroscopus* spp., which attain a length of 20-30 cm, and have electric organs developed in the orbits behind the eyes, from the same rudiments as the eye-muscles; these are supplied by the oculomotor nerve, and grow to the relatively large size of $3 \times 2 \times 2$ cm (Dahlgren and Silvester, 1906). The shock they can give is apparently much less than that of the *Torpedo*. The skin overlying the electric organs is smooth and unossified in *Astroscopus*, and all the other strongly electric fish are entirely smooth-skinned and notably devoid of denticles or scales.

The remainder are not nearly so well known (for a general review and bibliography, *see* Garten, 1914). They include the skates and rays belonging to the large genus *Raia*, which are distributed predominantly in the North Atlantic and North Pacific regions. Several, and perhaps all, of these have a pair of large electric organs, lying one on each side of the tail, and developed from part of the axial muscles (Ewart, 1888). The shock is said to be barely appreciable if a large ray is grasped by the tail. Another group known to have weak electric organs are the Mormyridae, a family of tropical African freshwater fishes with numerous species, many if not all of which are thus equipped. They show some remarkable convergences with the South American Gymnotidae: both families have electric organs, both contain some members with long decurved trunk-like snouts, and both contain eel-like species. The mormyrid ' eel ' is *Gymnarchus niloticus*; like *Electrophorus* and *Gymnotus* it has lost both pelvic and caudal fins. It swims with the body held more or less straight, by the undulations of a very long median fin, which carry it forwards or backwards with equal facility (Budgett, 1901). The gymnotids have a similar fin, but it is the anal, whereas in *Gymnarchus* it is the dorsal.

It has also been reported recently (Kleerekoper and Sibakin, 1956a and b) that the lamprey *Petromyzon marinus* can set up a weak electric field around the head anterior to the eyes, which takes the form of rhythmic spike-potentials synchronised with the respiratory movements. The authors point out that the low frequency of the spikes (about 2 per sec.) excludes their use for ' radar ' orientation, but that they might perhaps enable the lamprey to sense the proximity of objects, such as its prey, which distorted the electrical field within a few inches of its head.

There are two classical explanations of the development of typical electric organs, namely that they are used either (*a*) to numb or kill prey, which can then be seized and swallowed with greater ease, or (*b*) to repel attacks by aggressors. In the case of *Torpedo* there is good evidence for both functions. Although it is rather a sluggish creature, the food found in its stomach includes quite large numbers of strong and active species, even dogfish and salmon for example, which as far as one can see it would be almost impossible for it to catch without the aid of the electric shock. Wilson (1953) succeeded in getting captive specimens of *T. nobiliana* to pounce on a dead fish moved by means of wires; they enfolded the bait beneath them in their ' wings ', and, when electrodes were planted in the bait and attached to waterproof leads, it was found that there was a simultaneous discharge sufficient to deflect a galvanometer (on the lowest sensitivity range) more than full scale. On one occasion convulsive contractions of the electric-organ region were seen when a *Torpedo* and a turbot met as both tried to seize the bait at once. A number of other observations suggested also that defensive shocks were given to various intruders in the tank, such as congers, dogfish and large lobsters, which were seen suddenly to start back on touching the *Torpedo*, and make a hurried escape.

When *Electrophorus* and *Malapterurus* are placed in a tank, they are liable to kill everything else confined with them, but this may be a consequence of crowded confinement and not a demonstration of their natural feeding methods, as will shortly appear. According to Bates (1863, II, p. 290) *Electrophorus* in the dry season inhabits small isolated ponds in the soft muddy beds of dried-up creeks along the Amazon, ponds which, he believed, the fish had excavated themselves, from two to eight feet in diameter and of a similar depth. These they shared with an abundance of small armoured catfish (*Loricaria*), which suggests that the eels were not in the habit of using their electric weapons, otherwise the catfish would no doubt all have been killed in such a confined space. Three captive electric eels belonging to Roule (1935, p. 164) were fed once a week on small roach (*Rutilus*), and darted after them, churning up the water in their excitement. In the process of catching the roach they discharged their electric organs, with the result that all the roach still at large were soon stunned or killed; but the eels began the chase by dashing after the fish and catching them alive; and the possibility cannot be excluded that the discharges were an expression of rivalry among individuals competing together in a confined space.

Brehm (1925, p. 232) supplies an interesting account of *Malapterurus electricus*. Kammerer, he says, had twelve of these fish, which he put in three one-metre-long tanks, four to each tank; as a result of nightly battles, the numbers were quickly reduced to two in each, and finally the weaker of these also succumbed and a sole survivor was left, which waxed strong thereafter. Their food consisted entirely of bottom organisms, especially worms of all kinds, and even when hungry they would never take fish. (Köhler, on the other hand, found they would take small Crucian carp). The electric catfish, he continues, is a decidedly nocturnal creature, which stays all day in a favourite place on the bottom where it has dug out a hollow with its tail; but by night it gently pokes about the tank discovering anything edible with its sensory barbels; its sight is very poor.

Thus there seems to be no very convincing evidence that either *Electrophorus* or *Malapterurus* deliberately use their electricity to immobilise their prey, though *Torpedo* undoubtedly does so. Too little appears to be known about *Astroscopus* to allow any conclusion to be drawn in that case; but in the remainder, including *Raia* and the Mormyridae, the capacity is evidently far too weak to serve effectively for such a purpose.

As to the second explanation, that the electric shock has been developed to repel attacks by predators, this seems a very likely function in the fish with powerful electric organs. Both *Electrophorus* and *Malapterurus* share their habitats with members of the Crocodilia, for example, against which their discharges might be very effective; *Torpedo* has been observed, as already related, to startle intruders and cause them to take flight. With the less powerful rays, gymnotids or mormyrids, however, the effectiveness of the shock as a repellent seems much more dubious, and it is the existence of these electrically-feebler species that prompts one to ask whether there may not be another primary or primitive function, such as signalling, to be served by electric organs.

Some confirmation that this is so can be drawn from the discovery of Lissmann (1951), relating to *Gymnarchus niloticus*. It appeared a possibility to him that this fish might use its electrical discharges to locate nearby unseen objects, and if so this would account for its remarkable skill in swimming quite rapidly backwards, guiding itself easily between the reeds among which it lives. Lissmann found that even when at rest, *Gymnarchus* emitted pulses of electrical discharges, at a frequency of between 258 and 318 per second, in the water-temperature range of 21-31·5° C. (Marey (1879) had much earlier found that both *Electrophorus* and *Torpedo* emit similar pulses on slight stimulation, with frequencies of about 150 and 165 per second respectively, and that these frequencies change with temperature.) Lissmann made a very significant further discovery; he says: '. . . if, by means of a beat frequency oscillator, electrical impulses comparable in intensity and frequency to those emitted by the fish are applied to the water, very marked responses are elicited. Similarly, if the fish's own pulses are fed back into the water some distance away from the animal, the fish is able to locate the source of stimulation and attacks the stimulating electrodes. Regnart's finding [*see* p. 79] that the lateral line of other fish is responsible for the lower limits of the perception of electric current is of interest in this connection '.

More recently Möhres (1957) has indicated quite specifically that the electric discharge-rhythms of mormyrids may have in fact a territorial function, ' analogous to the singing of birds '. He found that when a second individual of *Gnathonaemus* was introduced into the presence of one that was already an established resident, a ' tumultuous ' change ensued in the rhythm and strength of the discharges. This was mutually synchronised, and followed a well-marked concerted ritual in which both contestants took part—the resident usually contributing the dominant share. If one was inhibited from discharging for a few moments, the other one also stopped. A non-electrogenic fish of similar size elicited no comparable display; but if, after the non-electric inmate had settled down and become accepted, another *Gnathonaemus*—enclosed this time in a dense gauze cage (' in einem dichten Gazekäfig ')—was dipped into the surface of the water, the prevailing peace was instantly disrupted by an electrical outburst.

Lissmann (1958), in an important paper, has greatly extended his earlier work on the nature and functions of the discharges, both in the mormyrids such as *Gymnarchus*, and also in *Gymnotus carapo* and other gymnotids. His observations of *Gymnarchus* reveal that they can detect each other's presence at a considerable distance (p. 165). Not only would they identify and attack the electrodes that reproduced their own discharge patterns in an experimental aquarium, but, despite their small mouths, they were liable in some of their natural habitats in West Africa to inflict many casualties on one another, attacking the tail end especially. Lissmann made one of his experiments with an aquarium divided in half by an opaque cloth partition, and demonstrated that fish placed on either side of it were aware of each other's presence electrically, very much as Möhres had done; one could be made to respond to discharges elicited by mechanically disturbing the other. Burst of discharges from both fish coincided when they came close together.

His main purpose was, as already stated, to investigate the properties of the electric field surrounding the fish, as a possible method for locating objects in the water, such as other living things, capable of distorting the field. He showed that both *Gymnarchus* and *Gymnotus* could detect a stationary magnet, and discriminate between conductors and non-conductors in the water. A locating sense of the kind postulated, for objects whose conductivity is different from that of water, is in fact demonstrably present, although apparently effective only at a relatively short range. When electric fish detect one another's presence at a distance, therefore, we must conclude it is due to recognition of the particular discharge pattern emitted by the other individual—that is to say by the perception of a signal. That the frequencies and patterns of discharge are specifically distinct has been noted by a number of authors, including Lissmann (1958, p. 159), Möhres (loc. cit.), and Coates, Altamirano and Grundfest (1954); and there is indeed no reason to question the fact, also recognised by Keenleyside (1955, p. 199), that electrogenic organs are used at least in some species for the purpose of social integration.

From an evolutionary standpoint, the more distant recognition of other members of the population, especially in turbid water, may very probably have been the original advantage to be gained by developing special electrogenic muscle, which summated the small random discharges associated with normal muscular activity. If this were so, then the proximate electrostatic location of external objects, of the kind demonstrated by Lissmann, and similarly the numbing of prey, would be in the nature of secondary—although no doubt very valuable—by-products of increasing electrogenic power. The first of these two developments would in that case rather closely parallel what is found in bats, oil-birds (*Steatornis*) and Cetacea, where sound-production, originally evolved for making recognition signals, has been re-adapted for echo-location.

This explanation actually provides the best answer at present available to what Darwin (*Origin*, chap. 6) described as a ' case of special difficulty ', namely how the earlier steps in the evolution of electric organs could have been favoured by natural selection. Lissmann has shown us that electric signals carry to a distance considerably beyond the limits of the effective field of electrostatic object-detection; and even their feeblest precursors could have had a potential value for this sort of purpose—for instance in maintaining individual-distance relationships—long before they had become strong enough to be put to any known non-social use.

5.2. *Summary of Chapters* 4 *and* 5

4.1. Sound travels over four times as fast in water as in air, and with less attenuation. Most animal tissues, having about the same density and compressibility as water, are ' transparent ' to underwater sound, and in order to perceive it it is necessary to have in the body some structure either denser (*e.g.* a statolith or otolith) or more compressible (*e.g.* a gas-bladder) than water, with a sensory organ anchored to it. Ear-stones of calcium carbonate or quartz, with a density of about 2·6, give good results, but the mysid Crustacea go one better by substituting calcium fluoride (density 3·2). Gas-bladders unfortunately change in volume with changing depth, unless they are kept pumped up to match the external pressure.

4.2. The dolphins and porpoises are great producers of sound. They rely for its detection on the paired gas-filled middle-ear chambers and their surrounding massive petrous bones, evidently keeping the chambers under pressure by allowing the lungs and respiratory passages to collapse progressively as the whale descends, and driving more and more air into the middle ears. The olfactory sense is almost (in baleen whales) or wholly (in toothed whales) wanting, and eyes are of limited use except near the surface where it is light enough to see. The whale's is essentially an auditory world.

Investigators have described the various clicks, trills and warbles produced, and have sought evidence that the sounds subserve the function of echo-location; most of these are of a somewhat unsuitably low frequency, however, of say 15 kc/sec. or less; and it seems possible that echo-location has developed as a secondary function, the primary use being for social communication and integration.

4.3. In complete contrast to the Cetacea, fishes are probably provided with a greater variety of sensory apparatus than any other animals. Many produce sound signals, either by stridulation or by using the swim-bladder as a drum. In the Sciaenidae there is a sexual difference in sound-production, the males being generally the louder drummers, and having larger otoliths in their ears. In an American marine species, the croaker (*Micropogon undulatus*), the shoaling fish produce a massed chorus at dusk; this is, however, a strongly seasonal phenomenon, reaching its height in the pre-nuptial period. It has, therefore, a number of characteristics that identify it as an epideictic display. In various other fish, including gurnards (*Trigla*), toad-fish (*Opsanus tau*), and singing-fish (*Porichthys*), the ' songs ' have been variously shown to be emulative, aggressive and territorial, or otherwise socially employed.

4.4, 4.5. Closely analogous phenomena can be found in some of the aquatic Crustacea, such as rock-lobsters (Palinuridae) and pistol-shrimps (Alphaeidae). The latter in some places put on immense choral epideictic displays. A few of the aquatic insects, mostly Hemiptera, are known to stridulate.

4.6. The whirligig beetles (Gyrinidae) are particularly interesting because they have perfected a special antennal sense-organ that can detect the waves propagated along the surface of the water, arising from the motion of their companions. Their characteristic gregarious behaviour, never hitherto satisfactorily explained, appears to be an epideictic display rather like a lek, held each day in the same customary place although not confined to special hours.

4.7. The majority of lower vertebrates, and the fishes especially, have a lateral-line system sensitive to low-frequency mechanical vibrations, analogous to that of the whirligig beetles except that they perceive vibrations coming from any direction through the water and not merely along its surface. (In some species the lateral-line system may provide an electrical sense as well.) There is a physical advantage in having long lateral-line organs, and there several notable fish, especially among those living in dark abyssal waters, with greatly attenuated tails carrying long extensions of the sensory canals. Characteristically these long-tailed fish are not luminescent; and, conversely, luminescent fish that are presumably integrated socially by visual signals are rather generally short-bodied.

4.8. It appears in general that sound-production has been developed for social communication and display in aquatic animals in very much the same was as it has in terrestrial ones.

5.1. Special electrogenic organs, capable of producing electric discharges, have evolved independently in at least seven groups of fish, though unknown in any other class of animals. The organs are almost always derived from modified striped muscle. A few species, such as the *Torpedo*, the electric 'eel' (*Electrophorus*) and the electric catfish (*Malapterurus*), can produce numbing or killing discharges. Many others, especially in the rays and skates (Raiidae), the South American Gymnotidae, and the African Mormyridae, produce only feeble discharges. Lissmann has shown that these can be used to detect and locate near-by external objects whose conductivity differs from that of water (*e.g.* other fish)—but only at short distances. Only the *Torpedo* and its relatives are known for certain to use their electrogenic powers to paralyse their prey.

It has also been shown that the discharge patterns and frequencies of a number of the mormyrids and gymnotids are specifically distinct, and have a demonstrable social function, in some cases akin to the song-displays of sound-producing animals. Specific discharge patterns elicit a response in the percipient fish, which may reply with its own discharge sequence; it may also approach and attack the source of the signals—whether these are experimental electrodes or intruding rivals. Signal detection is effective at considerably greater distances than electrostatic object-location or prey-numbing.

In the evolution of electrogenic organs it seems on the whole probable that the signalling function was the original adaptation to have been selected, because it is the one that requires the least power. The other functions, though they may have come to predominate in certain cases, would then be secondary readaptations. There is an analogous situation in the re-adaptation of sound-signals, originally social, for echo-location, for example in bats.

Chapter 6

Social integration by olfactory signals

6.1. *Chemo-reception and scent-production*

For an understanding of the uses that animals make of scents and odours we have to rely, so far as direct observation is concerned, very largely on our own limited olfactory sense. There are no instruments by which odours can be detected, amplified, analysed or recorded. Their properties are not definable or measurable in terms of physical units, and the human brain itself is far less competent to resolve them quantitatively or describe them intelligibly than it is with things that are heard and seen. Nevertheless most of us can distinguish and identify hundreds of odours each separately learnt and retained unclassified in the mind—often with extraordinary tenacity. Some of them strongly affect our emotions. There is no doubt that certain other animals—mammals especially—have much greater olfactory powers than our own, and in some if not all the wet-nosed, or macrosmatic, mammals olfaction is unquestionably the dominant sense.

A good many indirect inferences can be made, however, from observation of the animals' behaviour, physiology and anatomy. The toothed whales, for example, having no olfactory organs or nerves at all, can be presumed to be anosmatic—lacking the sense entirely. In most birds the organs are very small, compared with those of the majority of other vertebrates and also compared with the birds' sensory provision for sight and hearing; but there are marked exceptions in the petrels, American vultures and certain other groups, which justify the inference that these birds have a fairly well-developed sense of smell (*cf.* Bang, 1960). In sharks the organs and corresponding bulbs in the brain are particularly large, and so on. The bony fishes generally possess, in addition to their olfactory apparatus, a cutaneous chemical sense (*cf.* Young, 1950, p. 210) which can be described as gustatory since what has been salvaged of it in the evolution of the land vertebrates survives in the taste-buds in the mouth. Furthermore there are large numbers of animals, notably among the insects and mammals, that have special glands and structures for the production and dissemination of scents, and in many cases these can be observed in use, fulfilling the same kinds of function, in social signalling for example, as do the visual patterns and audible sounds considered earlier.

Chemo-reception is naturally used for many purposes besides that of social communication—most commonly of all, perhaps in the search for food. Many aquatic animals such as fishes are inclined to be highly sensitive to chemical changes in their environment. Hasler and Wisby (1951), in a series of important experiments on one of the American freshwater minnows, *Hyborhynchus notatus*, found that the fish could be conditioned to distinguish between natural fresh waters from two different Wisconsin streams. When the fishes' nasal epithelium was cauterised, however, they became incapable of responding correctly, as they had previously been trained to do, which showed that the discriminating sense must be olfactory. The authors infer that it is probably the same faculty in the salmon and their relatives (Salmonidae) that enables them to home, on their return migration from the sea, to the particular river and tributary in which they were born. Such a degree of refinement in perception suggests that their osmetic organs may possibly be able to detect single molecules of particular aromatic substances.

Aquatic animals in particular tend to set up chemical changes in their environment as a result of their metabolism. Dissolved substances are removed from the water, and the products of respiration, excretion and secretion added to it. These additives are collectively known as external metabolites, and in some cases they have been shown to exert important effects both on members of the same population and on those of other species (*cf.* Lucas, 1938, 1947, 1956). The rate of accumulation of metabolites in the habitat normally depends on the degree of crowding; and it has been shown in a number of animals including flour-beetles, *Daphnia*, water-snails, fish, and amphibian larvae, that they can profoundly influence either the growth-rate of individual members of the population or the reproductive rate, or the life-expectancy, or combinations of these. ' Conditioning ' the medium has assumed a major importance in the population homeostasis of such animals as these, and merits much fuller discussion at a later stage in the book, especially in Chapter 23 (p. 557).

Most functional odours have presumably been derived through natural selection from metabolites excreted or secreted in the first place for some different purpose. Integumental glands originally supplying mucus or wax to the skin have probably often been elaborated in this way to produce the relatively powerful odours commonly employed as signals, for recognition and other purposes. It is interesting to notice in passing, as we have done in the previous chapters, that such organs—in this case chemical—frequently become converted into protective repellent devices, or used as interspecific warning signals or even as weapons. Thus a great number of insect species have developed repugnant odours, and so have very many other land-animals, such as the slug *Hyalina alliaria*, many snakes, nestling hoopoes (*Upupa epops*) (*cf.* Sutter, 1946), and skunks (*Mephitis*, *Conepatus* and *Spilogale*.).

The borderline between metabolites that are primarily odorous and those that are definitely poisonous or venomous is not always a clear one; certain materials like formic acid, used by some ants as a repellent, partake of both qualities. An odour is defined as such solely by its ability to excite the olfactory senses, and as far as this sensation alone is concerned the effects, however offensive, can scarcely be regarded as toxic. Odours purposefully produced can be, and most frequently are, purely abstract and symbolic signals—conveying an enticement, perhaps, or a threat; whereas poisons—substances that can inflict somatic injury—when similarly produced are real direct weapons. Such distinctions as these, however, tend to break down in particular cases, and to become meaningless among the lower animal phyla where there may be no chemo-receptive sense that depends on discrete olfactory organs.

6.2. *Olfactory signals in insects and myriapods*

Among the insects the classical example of the use of scent for intraspecific communication is the so-called 'assembling' of moths, in which males are attracted from a distance by an odour emanating from the virgin female. This phenomenon has been recorded in at least eleven different families of Lepidoptera (Kettlewell, 1946, and general reference works). In many of them there is a marked sexual dimorphism in the antennae, in which the olfactory sense resides, those of the males being large and ciliate, bipectinate or plumose. The phenomenon can be very spectacular, especially in the large saturniids such as the emperor and peacock moths (*Saturnia pavonia* and *S. pyri*) in Europe and the *cecropia* and *polyphemus* in North America. The virgin female emits a scent, in most cases imperceptible to us (Eltringham, 1933, p. 85), often at a particular time of day, from the posterior segments of the abdomen which are protruded and retracted rhythmically. The males assemble to her from surprising distances (as much as a mile in *Lymantria dispar*), coming up-wind. As soon as one of them has paired with her, the emission ceases and the attraction is over. It has been suggested that the male's olfactory sense may be entirely specialised or confined to the detection of the female's scent; in some cases the latter has been extracted, and it has been shown that identical aromatic substances may be used by different species of moths. Habitat, season, and time-of-day factors are thought to be effective in preventing confusion (*cf.* Kettlewell, loc. cit.).

The most striking feature of this phenomenon is that it appears to be exclusively epigamic, and differs quite radically from any instance of intraspecific communication considered hitherto. Here the female emits her scent, like a beacon light, to attract a mate, and as soon as this function has been successfully discharged, signalling stops altogether. If the standard function of secondary-sexual display signals were to bring the sexes together and facilitate fertilisation—as many authors on this subject are inclined to think—then this would seem to be a highly effective means of achieving it. The more sedentary of the two sexes broadcasts a strong signal, being equipped with a special emitter for the purpose, while the more mobile partner evolves the complementary structure—a supersensitive detector—that guides him to the signal source. Except where the female is unnaturally confined by a human experimenter, so that the suitors cannot actually reach and satisfy her, there need be no spectacular concourse of males at all.

In the previous cases of social communication that we have had to consider, however, the typical situation has been to find the special equipment for both the production and the reception of signals developed in the same sex, normally the male. This reflects the essential difference in character between epideictic and epigamic displays—the former generally containing some element of rivalry or challenge, potentially capable of exerting a competitive stress on those taking part, usually members of the same sex, and the latter being on the contrary in the nature of an overture, soliciting a submission or acceptance between members of opposite sexes.

Even in epigamic displays, however, it is most often the male that takes the initiative, both in attracting the female and in any subsequent courtship—courtship being the sequence of marital behaviour which culminates in the fertilisation by the male of the female's eggs. The 'survival value' of courtship is a subject to be discussed in a later chapter (p. 249), where the unusualness of having females adapted to solicit the males in this direct and powerful manner will be more fully appreciated; but at first sight it looks like a simple and practical arrangement for securing the meeting of the sexes.

To emphasise the difference, it is worth describing a contrasting example from the Lepidoptera, well known to European entomologists, namely the display of the males of the ghost-moth (*Hepialus humuli* and other species of the genus). This is a visual rather than olfactory display, although the males do in fact exude a scent. 'The moth flies, like its congeners, at early dusk till nearly dark. The male has a habit . . . of swinging like a pendulum over a space of about a yard of ground, backwards and forwards; and not infrequently two or three males swing to and fro over the same spot, now and then two, or even three, coming into contact and swinging off again, just like flies when at play in a room. They are extremely conspicuous from their white appearance, and sometimes dozens may be seen at a time swinging, either singly or in little companies, over different parts of a field. The female [dull yellow] never flies in this manner, but takes a tolerably straight course with rather heavy flight, and, evidently with intention, strikes in her course against one of the swinging males, which instantly leaves its station and follows her to where she alights a short distance off. This has been repeatedly observed, and is open to no doubt' (Barrett, 1895, II, p. 168); pairing ensues. This is a typical communal display or lek, like several others we have met already, and it is a standard example in so far as the males are the more conspicuously coloured sex, take the initiative, and engage primarily in displaying to one another.

Neither the epideictic nor the epigamic roles of the sexes are freely interchangeable. This is another point on which it is desirable to defer the main discussion until later (in the section on Reversed Sexual Roles in Chapter 12, and the one on Polygamy in Chapter 21). It turns in the first place on the fact that, in any one episode of reproduction, the female requires to be fertilised only once, and she cannot effectively receive more than one male, whereas the male is frequently capable of serving a series of females. There are of course many instances, among the birds and mammals especially, where the mated male becomes committed to one female for a whole breeding cycle, or even for life; but where no such enduring monogamous status exists and the relationship between mated individuals is evanescent and quickly consummated, there tend to be many males continuously available to fertilise every female. The displaying of males in a swarm, so common in insects, is made possible largely because of this circumstance; the same individuals can keep it going day after day as long as the mating season lasts, temporarily forsaking it only long enough to pair with and dismiss a succession of visiting females. Females could not easily maintain a corresponding swarm because in most species, as soon as they became successively impregnated by visiting males, they would have to forsake the swarm to deposit all their eggs, and would have no reason thereafter to return to it again; for this reason it would soon be disbanded.

Ultimately, as we shall see later on, it is the differences in size and motility between the ova and sperms that result in the male sex being normally the one that is freer to assume the active epideictic role at mating time, and to take responsibility for acquiring territorial concessions and establishing a place in the social hierarchy. That the male sex is also the one which normally attracts the mate and takes the lead in courtship turns out to be another closely related consequence of the same primordial differences.

Olfactory social signals are probably very important in the insects, especially perhaps in the Lepidoptera; it is common for the males in this order to have special scent scales or androconia. Some of the odours are readily perceptible to us, for example in the death's-head hawk-moth (*Acherontia atropos*), which has erectile tufts of long ochre-coloured scent-hairs on the underside of the abdomen just behind the thorax and near the bases of the hind legs. These are erected by the males at the moment of the threat display already described, and produce a clean fragrance resembling jasmine or lavender; but since they are lacking in the female whose threat posture and squeak otherwise exactly resemble the male's, this is probably not their primary function. It has been supposed that the male scents of Lepidoptera have an aphrodisiac effect (*cf.* Eltringham, 1933, p. 82), which means simply that they serve to release the appropriate mating response in the female; but this is inherently difficult to make sure of, and Eltringham actually quoted only one isolated observation suggesting their use by the male for this purpose (loc. cit., p. 83). In the light of our experience in the use of other means of social communication, it would not be surprising if alternatively or additionally they served, especially in drab-coloured nocturnal species, as distance-signals directed epideictically at other males.

Very interesting discoveries have been made regarding the use of scents by male bumble-bees, for the purpose of marking a series of points on a circuit, round which the male systematically patrols, hour after hour and day after day (Frank, 1941, and Haas, 1946 and 1952, quoted from Free and Butler, 1959, pp. 38-40). Frank found that one such circuit measured 275 m, and the marked male *Bombus terrestris* that made it completed 35 tours in 90 minutes. Haas discovered that the scent-marking took place once a day, first thing in the morning, and that the male gripped the leaves, twigs or pieces of bark he was marking in his mandibles and made gnawing movements, often whirring his wings in an excited manner as he did so. Thereafter the places were merely visited each time. Some of the points were held in common with other males whose circuits overlapped, 'so that in a given area there was in effect a network of interwoven routes along which the males flew in all directions, and, during favourable weather, scarcely a minute went by without at least one male arriving at each of the established visiting places' (Free and Butler, p. 39). The individual routes were found to vary slightly from day to day. Many of the species of *Bombus* share this habit, their scents being apparently specifically different; *Psithyrus* males do the same patrolling but do not appear to mark their visiting points osmetically.

FIG. 7. Male *Bombus hortorum* in the act of marking an apple leaf with scent, as he patrols his circuit in the direction of the arrow. (From A. Haas, 1946.)

Free and Butler (loc. cit., p. 40) conclude that the males behave in this way to ensure that any queen entering the patrolled network will quickly find a mate. This is no doubt a valuable consequence of having such a system; but it is difficult to resist the conclusion that the primary purpose of such a tireless and remarkable marathon flight is epideictic, directly concerned with providing an index of population-density. It parallels in a remarkable degree the network of olfactory 'sign-posts' established in common by wolves and other Canidae, to be described later in this chapter (p. 103); these are also visited only by males, although far less assiduously. We shall find that there are a number of similar cases among the mammals, implying that the basic dispersionary systems are probably in all cases the same. This epideictic explanation is reinforced by the fact that only male bumble-bees visit the scent marks, and females are evidently not attracted by them. The males have extremely large eyes, as is so common in Hymenoptera and Diptera, and appear to 'spot' the females visually on the wing (Free and Butler, p. 38).

An example of an odour produced in entirely different circumstances has been described, not in an insect but in the millipede *Zinaria butleri*, by F. N. Young (1958) in Indiana. Early one June morning an immense aggregation of larvae of this species was found and reported to him, occupying a strip about 12 feet wide and 25 feet long and piled in places to a depth of 4-5 inches. By afternoon they had dispersed, but they could still be exposed by hundreds everywhere under the mat of dead leaves where they were now feeding. Not only could they easily be heard, but the whole area had a distinct odour of crushed cherry leaves due to the HCN gas released from their repugnatorial glands. Apparently large aggregations of this species are not uncommon. They seem likely enough to have an epideictic function, and resemble in some respects the aggregations formed by processionary caterpillars (*Thaumetopoea*), to be mentioned in the next chapter (p. 115); their HCN production is reminiscent also of the flour beetle *Tribolium*, which similarly releases a gas from special glands into the flour in which it burrows, capable of exerting profound effects on growth and reproduction of the beetles where they become overcrowded (*see* p. 496). Some comparable conditioning of the millipede larvae, at a stage when their gonads are beginning to develop perhaps, may be involved in this case.

6.3 *In fish and reptiles*

It is regrettable that, in the nature of things, our knowledge and understanding of the application of odours to social communication is so limited. Vertebrates that employ scents, being generally bigger animals than insects, tend to produce them in larger and more detectable quantities. It is nevertheless remarkable that there should be fish living in the water that can produce odours readily appreciable to man. This occurs in the smelts (*Osmerus*), small salmonoid fishes that habitually gather in immense prenuptial and spawning swarms. The generic (but apparently not the English) name refers to the odour well known to continental fishermen who catch the European smelt (*O. eperlanus*) by thousands in these gatherings; it is usually said to resemble cucumber, but by some it is likened to violets. According to Seeley (1886, p. 334), 'some French writers have regarded the odour as a protection from enemies, and in certain localities the smell is said to be so penetrating that anything dipped in the water where the fishes abound becomes impregnated with it'. In Scandinavia, however, smelts have been much used for fish-bait, and this casts substantial doubt on the idea that the smell is a repellent. It seems, indeed, to be a case where an external metabolite, emanating from the cutaneous slime-glands, could quite possibly serve as a population index in the quiet shallows where the swarms gather, in a manner analogous, for instance, to a frogs' chorus. In certain respects, indeed, it is reminiscent of the millipedes just mentioned in the previous section. Experiments with certain other fish and amphibian larvae have shown that comparable metabolites—in at least one case also emanating, like the smelt's odour, from mucous glands in the skin—can have important regulative effects on growth and other vital statistics of the population concerned (*see* p. 557).

The Crocodilia, with their gular musk-glands, have been mentioned in Chapter 3 (p. 53). The bull alligator, it will be recalled, bellows during the mating season in the stillness of the night, in the southern swamps of the United States, and at the same time the glands open and ' fine steamy jets of a powerful musky-smelling fluid float off into the heavy, miasmatic atmosphere of the bayou. The odour may be carried for miles. . . . ' (Ditmars, 1907, p. 89). There is no reason to doubt that the alligator is in fact emitting two kinds of signal simultaneously, both for the same purpose, and that the roar is a ' song ', conveying an assertion of dominance and a challenge to other males within earshot. Females and young have similar smaller glands, which they evert whenever there is any kind of struggle—an action that recalls the analogous use of the scent apparatus in the male death's-head hawk-moth just described (p. 94). African crocodiles (*C. niloticus*) are also characterised by their strong musky smell, which often betrays their actual or recent presence (Hamilton, 1947, p. 314).

Scent-glands are found in many other reptiles, for example the fresh-water turtles; in parts of the south-eastern United States the little musk-turtles (*Sternotherus odoratus* and other species) are popularly known as ' stink-jims ', and produce an ' utterly revolting stench '; the musk of the related mud-turtles (*Kinosternon*) is merely ' rather nauseating ' (Carr, 1953, p. 104). There are two pairs of glands in *Sternotherus*, opening by ducts through the bone and integument on the ventral side near the borders of the carapace, one pair just behind and the other some distance in front of the bridge between the carapace and plastron (Risley, 1933, p. 687). The secretion is stimulated by handling the turtle, and the smell is therefore presumably defensive; but most probably, as in several other similar cases mentioned in this chapter, this is only a secondary development. Another species with somewhat the same kind of defensive odour ('like concentrated essence of fish ') is *Clemmys leprosa*, a freshwater turtle of southern Spain and north-west Africa (Gadow, 1901, p. 357). Many other members of the Chelonia have differently situated odoriferous glands, for example on the mentum, the base of the neck, or near the cloaca, but these are usually inoffensive.

In the snakes there is invariably a pair of anal glands or sacs, which produce in some if not all an odorous secretion. Two of the best-known species, closely related to one another and with near allies in many parts of the world, namely the European ringed snake (*Natrix natrix*) and the North American garter-snake (*Eutaenia sirtalis*), are among those which, when caught or maltreated, discharge a powerful and disgusting stench from the sacs, which probably serves them as a protection against predators. This is not true of the majority of snakes, however. Smith (1951, p. 213) informs us that the glands are present in both sexes at all ages, and are active at all times of the year. It has been shown that the male adder (*Vipera berus*) can follow the trail of the female, and that wherever the secretion has been discharged, it forms a point of attraction. The ringed snake is known to discharge its scent on herbage. Its use does not seem to be exclusively sexual, since the glands are developed even in the young at birth, and in some species it has been shown that the male is attracted rather by the odour of the female's skin. The chemo-receptive powers of snakes are evidently good: in addition to the common nasal sense they have developed, to a high degree, an active tongue in association with the organs of Jacobson, which are a pair of accessory olfactory sacs opening on to the palate behind the premaxillae. It therefore seems not unlikely that their scent deposits are capable of providing information about the presence or recent transit of other members of their own kind, in the same way as they do in many mammals.

6.4. *The persistence of scents*

Olfactory signals have the very characteristic quality of persistence, lingering more or less long at the place where they were initially deposited. Some of course are emitted into the air very much like auditory signals to produce an immediate if not instantaneous effect; but the majority, in all probability, are deposited on the ground or other objects where they have a chance of surviving for a longer time. This characteristic tends to be much exploited, in particular by the terrestrial mammals, which appear to specialise in olfactory communication more than any other group. Among those that live more or less solitarily, like some of the carnivores, and at the same time at the low population-densities that are generally incumbent on large animals, it is undoubtedly possible to maintain the necessary social integration by scent, even though any given individual may rather rarely come face to face with others, at least of his own sex. We shall see that a network of marked trails or signal-stations can be established, and regularly traversed or visited, in many species by the males especially.

6.5. *A digression on land-rights systems*

In the Introduction a brief indication has already been given of the fact that many alternative types of dispersionary pattern have evolved in different animals; the food-territory claimed and defended by the male in certain species of birds was chosen there as the main illustration only because it happens to be the simplest and most readily understood. These patterns and their maintenance form a large and complicated subject, naturally one of the most important to our theme, and Chapters 9 and 10 in particular are devoted to surveying them more fully. It is necessary here, however, before embarking on the uses of scent by mammals, to anticipate a little and at least differentiate the four main types of land-rights system that can exist.

It is one of the standard premises of homeostasis that populations of animals should be capable of self-perpetuation, without losing their identity too easily through free mixing and interbreeding with other populations. In most cases this implies the more or less permanent settlement of each population within its own area or habitat, on the conservation of which perpetuation so largely depends. We shall see later that free nomadism is quite exceptional, on land at least, and in most species each individual member becomes attached at an early age to a particular home area, in which it remains or to which it regularly returns at the same season for the rest of its life.

The home area or ' home-range ' may be larger or smaller, depending in part on the habits and powers of movement of the species concerned. Within it, individuals may, in the first place, be either (*A*) solitary (but admitting a pair when necessary as the basic unit, or a temporary family group), or (*B*) gregarious. Quite often their social organisation is not permanently committed to one or the other, but can switch from the solitary to the gregarious from season to season or from one time of day to another; moreover a continuous transition can frequently be found from onc extreme to the other. But the distinction nevertheless is very helpful in clearing one's mind when trying to analyse the complexities of dispersion. As a cross-classification with *A* and *B* there runs another, second pair of alternative possibilities, for the home area can also be either (*C*) exclusive to the solitary individual or flock that occupies it, or (*D*) extensively overlapping with the home areas of other individuals or flocks, so that the habitat is to a certain extent communal and not completely subdivided into separately held parcels. A transitional borderline can be discovered between these two also, but again it does not detract from their usefulness as a means of classification and comparison among the countless individual variants that exist.

An individual or pair with exclusive title to the possession of a parcel of ground (type *AC*) is usually said to hold a food-territory, or simply a territory, and this is quite common in birds in the breeding season. Type *AD* is probably even commoner, especially if we include reptiles and mammals also, and describes the case where the individual is solitary, but his home range overlaps that of a number of other solitary members of the population —as for instance in foxes or bears; typically in this category each individual or pair have some defended point within their home range, such as a den or nest or roosting site, which is exclusively theirs and qualifies them as property-owners and established members of the community, at the same time forming the focus of their activities. An alternative sort of overlapping system of rights among solitary individuals was seen earlier in male bumblebees (p. 95). *Mutatis mutandis* the same conditions apply in types *BC* and *BD*: in *BC* the social unit has an exclusive home range that its members hold in common with one another, as do rooks (*Corvus frugilegus*) in the breeding season (*see* p. 108), or such sociable mammals as rabbits (*Oryctolagus cuniculus*) or vizcachas (*Lagostomus*) that similarly have warrens or 'towns' to serve as foci; and as red deer (*Cervus elaphus*) tend to do—especially the females and their followers—where they are sufficiently abundant. In *BD*, where herds or schools have working ranges that overlap

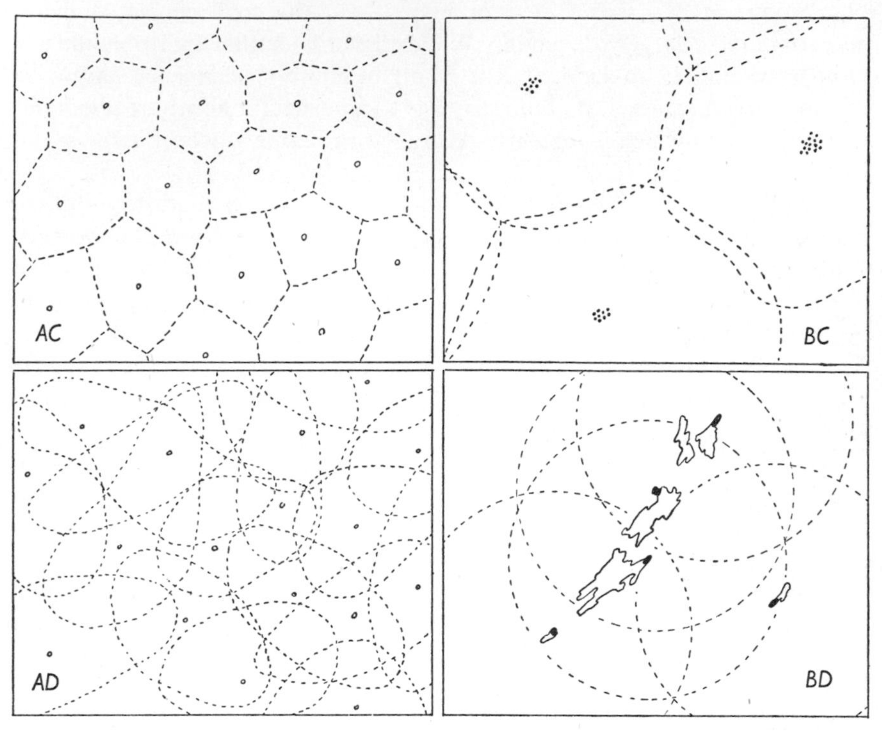

FIG. 8. The four basic types of 'home range', each implying established rights to seek food. *AC*, solitary and exclusive, frequently all defended. *AD*, solitary and overlapping; the base, if any, usually defended. *BC*, gregarious and exclusive (the diagram actually shows a fractional overlap, as in rooks). *BD*, gregarious and overlapping (shows a group of islands with five seabird colonies from which residents fan out in all directions up to a maximum radius).

with those of other similar groups, it is not possible as yet to give a well-proved example; but such a dispersionary pattern is theoretically practicable, and there seems little doubt that it is to be found—in colonial sea-birds, for instance, and perhaps quite commonly—in nature. These four types are shown diagrammatically in fig. 8.

We are constantly obliged to speak about 'populations' as if they were always discrete, finite entities, although it is perfectly clear that no one satisfactory definition can be given to cover every possible case. Because the terrain is generally varied and the different habitats broken up into islands or mosaics, populations of no great size are in practice often more or less isolated from one another. But even in the theoretical case of a habitat of unlimited extent, which permits a continuous dispersion of a particular species over a whole country or sub-continent, the effect of these land-tenure systems is to immobilise the stock occupying each section and give it the chance of self-perpetuation. Though any given local stock is uniformly contiguous on every side with neighbours of its own kind and has no natural boundaries, it need not violate any of our basic premises to refer to it as the ' population ' of that locality. Its members can still collectively possess a unique combination of group-characteristics—the local customs and speech-dialects found among continuous human populations are good cases in point—although these vary continuously outwards in every direction over an immensely wider region. Its essential quality is the opportunity for continued survival, through conservation of the habitat and its resources and through present and future adaptation by natural selection. We can safely vary the geographical dimensions of the term ' population ', therefore, to suit our convenience, or use it as a concept without any dimensional definition, just as we do, on a somewhat larger geographical scale, with the race or subspecies.

As to the opportunities for interbreeding that neighbouring members of a local population may be expected to enjoy, it is clear that there is often likely to be some kind of ' inverse square law ' effect, such that the probability of any two individuals meeting and mating diminishes very roughly as the square of the distance between their birthplaces. Such a generalisation naturally requires to be accepted with numerous reservations and exceptions in particular cases. As we shall see later, there is always likely to be some provision in every population, albeit usually on a modest scale, for sending out pioneers to settle and establish themselves at a distance.

6.6. *The use of scents in mammals*

As a class the mammals rely extensively on the use of scent to make their presence mutually known. Many of them are mainly nocturnal in habit, and ordinary visual communication is naturally impossible in the dark; but even in some of the most diurnal species scent provides an important medium for mutual recognition and advertisement. There are of course plenty of exceptions, some of which we have considered already in the foregoing chapters, where social integration is achieved mainly if not entirely by visual and audible methods.

The overriding importance of social integration is forcibly illustrated by the fact that wherever they go herbivores such as mice, rabbits and most ungulates leave persistent trails of scent behind them, notwithstanding the information this gives away to every keen-nosed carnivore as to their route and present whereabouts. In their turn the carnivores—especially the foxes and mustelids like the polecat—while hunting often trail even stronger and more unmistakable odours, which must alert every small mammal that scents them. When we begin to understand their great social and epideictic importance, however, the existence of such telltale signals in animals otherwise of secretive habits, often procryptically coloured, no longer appears so completely incongruous and baffling.

Many kinds of mammals habitually make use of established pathways and game-trails. These are often dictated by the lie of the land, following the most practicable route between two points, and their long-continued regular use may greatly improve the going and keep them clear of obstructing vegetation. They may be of considerable aid to their users in learning and remembering the topography of the home area, and undoubtedly they serve to concentrate the scent in places where other members of the same population are likely to come across it. In many parts of Europe the intricate network of paths of the rabbit (*Oryctolagus cuniculus*) has until recent years been a familiar feature of the pastoral landscape. In other countries similar trails are in use, for instance by lemmings (*Lemmus* and *Dicrostonyx*) on the arctic tundra, by porcupines (*Erethizon dorsatum*) in the boreal forests of the New World, by black and brown bears (*Ursus* spp.) over a still wider area, and by many kinds of deer (including *C. elaphus*) and antelopes. Some of these trails are adhered to year after year with great conservatism. Dugmore (1913, p. 81) tells us that ' in some places the rocks are worn away to a depth of one or two feet by the thousands and thousands of hoofs ' of the Newfoundland caribou (*Rangifer caribou terranovae*) that have passed over them for countless years on migration. It may be of interest to note in passing that in artiodactyls such as these and many of the carnivores including bears—though not in all carnivores, and not apparently in any rodents—when walking at normal pace their hindfeet fall exactly in the print of the forefeet, so that one complete movement of all four legs leaves virtually two footprints. The hindfoot has the advantage of using a foothold already tested by the forefoot, and where interdigital scent-glands are present, as in the ruminants, the ground-scent is concentrated at a minimum number of points. European rabbits, in following the surface trails already mentioned, tend to put their feet always in the same places, where the fore- and hindfeet make characteristic ' beats '; this is well known to rabbit snarers, who place their nooses in front of or on the marks of the forefeet (*cf.* Darling, 1937, p. 61; Thompson and Worden, 1956, p. 185).

Another well-known habit, with an apparently similar kind of epideictic function, is the establishment of special points to which animals go to deposit their scent and seek that of their fellows. There is no more familiar instance of this than the domestic dog. Any stone or tuft of vegetation, preferably rising above the ground surface, will serve his needs of the moment, though in a populous street there is a strong tendency to investigate and add to the scent-deposits of others. When two male dogs meet, the greeting ceremony usually includes an opportunity for sampling each other's individual odour, and it is variously accompanied by signs of goodwill, assertiveness or rivalry. Social urination, like the three-legged stance that directs it sideways, is entirely confined to males; the ' lamp-post ritual ' is strictly their affair, and can easily be distinguished from the epigamic interest they show in the odours left by females (especially if the latter are in heat), or from the female's reciprocal investigation of male odours. As an epideictic phenomenon it is readily understood—the dogs are making known their presence and keeping check on status relations with each other, acquiring in the process an indication of the field of male competition and local density of numbers.

In the wild, wolves do much the same, but there is apparently a more firmly established organisation. Seton (1909, p. 772) states that, ' incredible as it may seem at first sight, there is abundant proof that the whole of a region inhabited by wolves is laid out in signal stations or intelligence posts. Usually there is one at each mile or less, varying much with the nature of the ground. The marks of these depots, or odour-posts, are various; a stone, a tree, a bush, a buffalo skull, a post, a mound, or any similar object serves, providing only that it is conspicuous on account of its colour or position; usually it is more or less isolated, or else prominent by being at the crossing of two trails '. Farther on he continues: ' there can be no doubt that a newly arrived wolf is quickly aware of the visit that has recently been paid to the signal post—by a personal friend or foe, by a female in search of a mate, a young or old, sick or well, hungry, hunted, or gorged beast. From the trail he learns further the direction whence it came and whither it went. Thus the main items of news essential to his life are obtained by the system of signal posts'.

Many naturalists have been mystified by the corresponding habit of bears, such as the black bear (*Ursus americanus*). Here the males rear up as high as they can stand in order to bite and salivate, usually on saplings and small trees along their trails. Some of the gnawing trees continue to do duty for many (perhaps 10 or 20) years, and one of those examined in Texas by Bailey (1905, p. 188) appeared to have been given a single bite by the bear at each visit, ' leaving the marks of an opposing pair of canines '. Seton (loc. cit., p. 1062), who had himself seen hundreds of these bear-trees, chiefly in the Rocky Mountains, believed them to be ' sign-boards '—methods of communicating certain information, serving the same purpose as the urinary signal-posts of dogs, wolves and foxes. An interesting feature here, paralleled as we shall see later in the claw-marking of trees by some of the big cats (*see* p. 183), is the indication of the size of the visitor given by the height of its bite and possibly the width and depth of the tooth-marks.

Graf (1956) has described the ritual use of rubbing posts by the Roosevelt elk or wapiti (*Cervus canadensis roosevelti*). These are found ' particularly on the loafing resting grounds of the herd '. Young animals do not appear to indulge in the activities described; in the cows the ritual is first to ' nose ' the chosen post, next to scrape it with deliberate upward strokes of the incisors, third to rub the sides of the muzzle and chin on the post, and fourth to rub them against her own flanks. Adult males, five years old or more, perform a similar ritual only during the rut; the chief difference is that, instead of the teeth, the base or burr of the antler is used for the act of scraping. Graf believes that ' such sign posts may serve by intimidating strangers entering such a marked territory '. ' When bulls meet on common ground at the beginning of the rut the first activity on the part of the bulls is to indulge in highly vigorous sign-post activity ' (loc. cit., p. 167). He is of the opinion that in fact all the post-rubbing done by deer is an activity of this character, and that there is no evidence for the widespread idea that stags use the posts to rub off their velvet.

Experience with the red deer (*C. elaphus*) in Scotland confirms his observation that stags often go about with strips of velvet hanging over their eyes without making any attempt to rub them off. They may use particular rubbing posts for a number of seasons, often dead boughs of fallen pines, or small standing trees; and in the latter case the tree is usually girdled in the course of time and dies, and may finally be worn through and remain as a pointed stump. During the rut, long after the antlers are clean, the posts are in very active use. Stags may be seen nosing or sniffing them, and low posts are frequently marked by urination. In the eastern Highlands of Scotland, where the red deer population is high, there is no reason to think that stags or hinds ever occupy solitary territories, though outside the rutting season the segregated herds of stags and hinds year after year adhere to their own grounds, often in different glens. It may be added that rutting frequently occurs on open hillsides, where there are no trees or posts, and in such places stags can sometimes be seen thrashing the rushes and other tall vegetation with their horns.

The habit of ' musking ' is particularly associated with the Mustelidae, and has given them their generic and family name. Writing of the mink (*Mustela vison*), Svihla (1931, p. 368) states: ' I have often observed minks of both sexes rub their perineal region on the edge of a water pan in their cage. . . . This suggests that rubbing of glands may be of recognitional or informational value, used in the same manner as the urinating of dogs '. Seton (1909, p. 910) observed much the same in two captive male pine-martens (*Martes martes*), which ' made use of the musk gland at least every five minutes, pressing the parts on some projection of the cage '. Weasels (*Mustela nivalis*) are said to have regular musking places (Neal, 1948, p. 88), and a few years ago I noted such a place belonging to otters (*Lutra lutra*) beside a small loch in the Isle of Lewis, and approached from the water's edge by a short well-beaten path that lead no farther.

Since the scent-producing organs are imperfectly or wrongly described in several modern English works, it seems desirable to mention that a male badger (*Meles taxus*) examined by me in November 1957 (*see* fig. 9) had a very conspicuous median blind pouch, a centimetre above the vent and immediately beneath the base of the broad tail, large enough to hold a walnut, incipiently bi-lobed and lined with short hair, secreting from innumerable dermal glands a copious red-brown musky sebum; this stained the hair and skin all round the lips of the pouch and the anus. Entirely distinct from the pouch, the short ducts of a pair of muscle-coated glandular sacs (the size of hazelnuts) opened laterally just inside the anus, and could easily be extruded from it. They were closed with sphincters, and produced a light-brown and slightly pasty fluid. These latter sacs are, of course, the well-known anal glands, which produce the typical stinking musk that is especially notorious in the polecat, skunk and civet. The subcaudal pouch already described has nothing to do with them; its secretion is quite different—

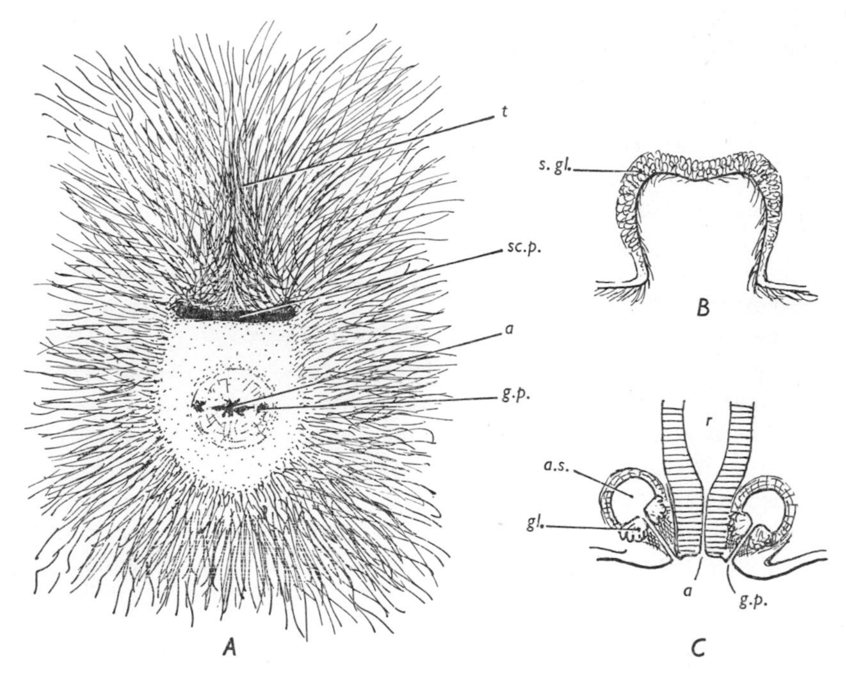

FIG. 9. Scent glands of the male badger (*Meles taxus*). *A*, external view of the anal triangle with the anus (*a*) extruded to show the pores of the anal glands (*g.p.*); above, under the root of the tail (*t*) is the large subcaudal pouch (*sc.p.*). *B*, horizontal section of the subcaudal pouch, showing the modified sebaceous glands (*s.gl.*). *C*, horizontal section through the anus (*a*), showing the rectum (*r*), the paired, muscular anal sacs (*a.s.*) each with a glandular collar (*gl.*), duct, and pore (*g.p.*). All $\times \frac{1}{2}$.

distinctly oily, clear, and rusty in colour. The pouch has no muscular capsule and no sphincter, being broadly open to the exterior; the badger is said to lick it and possibly in that case the secretion has some preening function, such as communicating an odour to the body.

One form or another of olfactory marking is no doubt very general in mammals. Ilse (1955) has described how young male lemurs, *Loris tardigradus*, marked the whole length of a horizontal pole in the cage where they were confined, by a form of rhythmic micturition in small drops; this he suggested was done in order to establish a territorial claim.

Pearson (1946) has provided a most valuable account of the scent-glands of the American short-tailed shrew (*Blarina brevicauda*). In addition to the pair of glands on the flanks possessed by most of all shrews (Soricidae), there is also a third gland on the midline of the abdomen; the last is composed of greatly enlarged sebaceous glands, whereas in the lateral glands these are mixed with enlarged sweat glands. The glands are better developed in males than females, and tend to be largest in males in breeding condition, and smallest in females which are oestrous, pregnant or lactating. The odour is offensive, and it is generally stated that in consequence few carnivorous mammals will eat shrews, though they may kill them; hawks and owls, however, take them very freely. As Pearson says, ' No one has explained satisfactorily why shrews possess scent glands. . . . If protection from predators is the function . . ., it is surprising that pregnant and nursing females, the most valuable segment of the population, have the weakest scent glands and are left relatively unprotected. It has also been suggested that since shrews have very poor vision the pairs locate each other during the breeding season by the odour produced in the scent glands, but this seems doubtful, at least in the case of *Blarina*, in view of the fact that the scent glands are reduced in oestrous females at the very time . . . they should be best developed. Perhaps instead the scent glands rub against the walls in the network of tunnels which make up the home territory of each shrew, stamping that network with the characteristic odour of its unsociable occupant and thus informing other foraging shrews that the tunnel is occupied. Song is used in a similar manner by many nesting birds to establish their territory '. This analogy can perhaps be carried even farther, when we compare the seasonal development of their scent to that of song for example in the robin (*Erithacus rubecula*), where the females sing in the neuter or asexual season, but are silent when they are paired and breeding, at the same time as the cocks are at the zenith of their performance. It may well be a matter of the seasonal change from dispersion-by-units-of-one to dispersion-by-pairs-or-families; in the latter period the male as usual would assume the whole responsibility and functional apparatus for representing his dependents.

In another insectivore, the common mole (*Talpa europaea*), the female actually undergoes a sort of endocrine sex-reversal in the long non-breeding season, as Harrison Matthews (1935) has shown, and develops large interstitial glands which are the homologue of the male's testes; this appears to be capable of explanation, in a very similar manner, as bringing the two sexes physiologically on to a common competitive footing.

In the beaver (*Castor* spp.) members of both sexes have modified preputial glands producing castor, a substance so attractive to beavers that it has long been used by trappers as a lure. It is normally deposited with the urine or faeces, and in specially-made ' mud-pies '; according to an old account (*cf.* Seton, 1909, p. 470), ' when two beaver lodges are in the vicinity of each other the animals proceed from one of them at night to a certain spot, deposit their castoreum, and then return to their lodge. The beavers in the other lodge, scenting this, repair to the same spot, cover it over with earth, and then make a similar deposit on top. This operation is repeated by each party alternately, until quite a mound is raised, sometimes to a height of four or five feet '. Whatever may be the precise details in this activity, it also appears likely to be epideictic in purpose.

The North American musk-rat (*Ondatra zibethica*) owes its name to the large inguinal glands possessed by both sexes. These aquatic rodents construct or adapt special landing places, on which they are in the habit of leaving their dung and musky odour. In Asia there is the equally well known musk-deer (*Moschus moschiferus*), a silent, solitary, retiring, mountain-dwelling species, of which the male develops a preputial musk-gland or ' pod ', which was formerly the source of the musk of commerce. Jerdon (1874, p. 268) records that ' the dung of the males smells of musk, but the body does not, and the females do not in the slightest degree '. The male has, ' besides, a peculiar organ or gland on the tail, which indeed is composed almost wholly of it. The tail of the male is triangular, nude above, thick, greasy, partially covered with short hair below, and with a tuft of hairs at the end, glued together by a viscid liquor. It has two large elliptical pores beneath, basal and lateral, the edges of which are somewhat mobile, and the fluid, which appears to be continually secreted, has a peculiar and rather offensive odour '. To complete the battery, a ring of glands surrounding the anus secrete ' an abominably offensive stuff '. One might surmise that, with all this communication equipment, the optimum population of musk-deer is probably sparse and scattered; its adaptations form a very interesting contrast and parallel to those developed for the same purpose by its compatriot, the barking deer or muntjac (p. 61).

Our theory of animal dispersion provides an explanation of yet another widespread habit of mammals, namely the adoption of regular dunging-places. Only a few of the more remarkable instances can be given here. The spotted hyaena (*Crocuta crocuta*) feeds very largely on bones, and its faeces turn white on drying, like those of dogs on a similar diet; the faeces were the source of the *album graecum* of medieval medicine, prescribed as a cure for ague, goitre, warts, and a large number of other ills. ' The animals' habit is to defaecate in regular latrine areas, spaces of limited extent covered with a plentiful deposit of album graecum. These places are conspicuous at a distance when the sun is shining on the chalky white faeces. Some of the latrine areas are very large: one seen near the Serronea river must have covered quite a quarter of an acre. A resident in Arusha was heard to describe such a deposit as " like a fall of snow " ' (Matthews, 1939, p. 47). This is a widespread African species whose behaviour is solitary or somewhat gregarious as circumstances dictate. The visual conspicuousness acquired by the site is quite possibly significant. The same habit is found in a number of antelopes, such as the Indian antelope (*Antilope cervicapra*), nilgai (*Boselaphus tragocamelus*) and Indian gazelle (*Gazella bennetti*). Blanford (1888-91, p. 528) says of the last that ' this species has the habit of dropping its dung in the same spot to a greater extent than the Indian antelope, but it not unfrequently resorts to heaps of nilgai dung for the sake of depositing its own '. The latter is at first sight a puzzling habit, but it falls into line with many other cases of shared social facilities between two or more species such as we find very often in the roosts of birds and bats, in the breeding colonies of birds, and in the mixed flocks of ungulates and other animals, to which fuller consideration is given in Chapter 17. The related impala of South Africa have a similar habit; in the Angolan species, *Aepyceros petersi*, Shortridge (1934, p. 599) refers to the traditional use of circular patches of ground year after year. ' These extraordinary aggregations, which average from 8 to 10 feet in diameter and a foot or more in depth, are conspicuously situated in open glades, the ground being bare and trodden down all round like a small circus ring '.

The last instance to be mentioned is that of the rhinoceroses. The great Indian rhinoceros (*Rh. unicornis*), now very rare, has the ' habit of depositing its dung in the same spot until a pile accumulates ' (Blanford, loc. cit., p. 474). This habit has recently been reviewed by Gee (1953b, p. 771), whose belief is that the dung heaps are used by any rhino which happens to be passing by, ' after the fashion of dogs at lamp posts ', and that they are not in the nature of individual monuments marking exclusively held territories. In that case we should conclude that this species—mostly solitary in habit—is basically dispersed on something like the *AD*-type pattern. Gee quotes the published observations of others to the effect that the dung has an overpowering odour, and that even when it is not their own it seems to attract the animals' attention and causes them to halt and defaecate at the same spot.

The African white rhinoceros (*Rh. simus*) also ' has a rather remarkable habit of always depositing its excrement at the same spot, and this in time forms great accumulations. When the mass assumes an inconvenient height, the animal scatters the dry mound with its horn ' (Fitzsimons, 1920, III, p. 210). The same author (loc. cit., p. 216) states that the black rhinoceros (*Rh. bicornis*) ' in localities where it is not hunted, usually deposits its dung in shallow hollows, which it scoops out under a tree or bush. These spots are visited regularly until a great pile of excrement has collected, which the rhino scatters at frequent intervals with its horn. Regular beaten tracks are often seen leading to these deposits, or from one to another '. He believes that ' this cleanly habit, which is common to other species of rhinoceroses and many other species of animals, tends strongly to keep infectious animal diseases in check '; but this is clearly a somewhat difficult suggestion to substantiate, and it might even be considered reasonable to claim the exact opposite. That it should really be explained as an epideictic phenomenon seems virtually certain when we view it in the light of all the other evidence.

Though exceptions are common, the tendency to predominance of the male sex in dispersionary activities shows up in the mammals almost or quite as strongly as in the birds. Much of the routine trail-marking falls to the males, and this may account for the apparently uneven sex-ratios not infrequently found by people trapping small mammals. In the European rabbit, for instance, a sociable burrowing species, whereas the sex-ratio figures obtained by ferreting and hole-trapping were found to be 0·73 and 0·79 males respectively to each female—the females being thus in the majority—snaring in the open gave an entirely different figure of 1·66:1 (Thompson and Worden, 1956, p. 62). For stoats (*Mustela erminea*) Deanesly (1944) found apparent sex-ratios of between 1·22 and 1·57 males per female, and for weasels (*M. nivalis*) 2·7:1. Trapping figures obtained at Aberdeen University (unpublished) were 1·6:1 and 2·4:1 for the same two species respectively. Here there is of course no check to show that the true sex-ratios are actually one to one, though the tentative assumption is reasonable. Voles, mice and shrews all tend to give similarly unbalanced trapping ratios, but in these there is evidence of a seasonal change in the apparent discrepancy, suggesting that it reflects a similar seasonal cycle in the intensity of male epideictic activity.

The mammals seem to provide us with particularly good material for revealing the kind of machinery required for the day to day maintenance of an orderly and regulated pattern of dispersion, through the use of specially evolved methods of social communication. We can see that it is necessary in most cases for the members of the local population to get to know one another individually, and that they have constantly to keep reinforcing their rights as residents—and presumably when under pressure making or yielding to adjustments—through a system of conventional observances.

The territory in Eliot Howard's generally accepted sense is clearly only one type of property holding, and there are others equally capable (as will become clearer in Chapters 9 and 10) of controlling the density of settlement and establishing individual claims to resident status. Man himself is a mammal, and there are, not unnaturally, examples of some of the same types of dispersionary patterns to be found in human—and especially the more primitive human—societies. Thus exclusive property held by the farmer or smallholder may be contrasted on the one hand with the status of the rural journeyman, who is allowed to extend his practice over a wide area recognised by custom, his acceptability in the neighbourhood being vouched by the fact that he is a known and respected member of the community and a householder in the village; or we may contrast it on the other hand with the property rights equally shared between the members of a club or syndicate to hunt game, or pasture their flocks perhaps, anywhere on commonly-held ground. Each of these three situations in a different way safeguards the vested interests established by membership in a localised society, and each is capable when necessary of restraining the admission of strangers or newcomers seeking similar privileges. To maintain his status of acceptability in his own small world and the rights that go with it, we may begin to suspect that the wild mammal can no more afford to neglect conventional observances—greetings, social assemblies and status-seeking exercises—than can his human counterpart. Indeed, in man and the mammals the social machinery is basically exactly the same, except that we belong to the mammalian minority that communicate socially by sight and hearing rather than scent; but in modern man the primordial function of social observance in regulating population-density has of course been largely displaced, obscured and forgotten.

6.7. *Summary of Chapter* 6

1. Our information on chemical signals comes from three sources, namely (i) direct observation aided by our own olfactory sense, (ii) comparative study of chemo-receptor and scent-producing organs and (iii) observations of animal behaviour. Chemical senses are of course valuable for many purposes other than communication, for instance in locating food. Fish and other aquatic animals obtain information about the water around them through their chemo-receptors, and in some cases they can be physiologically conditioned, and their growth and reproduction influenced, by the concentration of metabolites released into the water by their companions. Many terrestrial species have evolved protective stinking odours that are readily detected and avoided by predators.

2. Among the insects the best known case of communication by scent is the ' assembling ' of male moths to virgin females, which emit an attractive odour until they have been fertilised. This occurs in several families of Lepidoptera. Compared with the examples of intraspecific communication considered hitherto this is unique in being purely epigamic—the static sex emitting the signal and the active one carrying greatly enlarged receptors for detecting it; in most other cases of sexual dimorphism in communication equipment we have found that the male tends to have the best of both parts.

Although this seems a simple and effective method of securing the meeting of the sexes, the general principle of equipping the female to broadcast an epigamic signal is actually an unusual one, and possible reasons for this are discussed. Males tend to take the initiative not only in signalling but in courting the female after they have met. Largely on account of the fact that males can often fertilise many females whereas the female can receive only one male, the roles of the sexes in these matters are not readily interchangeable. Thus in many insects the males are free to join in epideictic swarms and dances day after day, whereas the females once fertilised go off to lay their eggs and thereafter have no reason to return.

Many male Lepidoptera produce scents, usually supposed to have evolved as stimulants to the female in courtship; but if Lepidoptera resemble other animals in this respect, at least some of the scents are likely to be primarily epideictic in function, directed at other males, and their effect if any on the females only secondary.

In the bumble-bees (*Bombus terrestris* and *B. hortorum* have been the most closely-studied species) each male spends hours every fine day flying round and round a circuit, visiting a series of scent-marked points en route. These marks may be shared with other males whose circuits overlap, so that the whole patrol system of the male population makes an interwoven network.

Females are not attracted by the marks, and the phenomenon appears to be directly concerned with territorial rights and the estimation of population-density; it resembles similar epideictic behaviour in various mammals, to be mentioned below.

A case is described of a huge aggregation of larval millipedes that produced a distinct odour of hydrocyanic gas from their repugnatorial glands—conceivably conditioning the onset of gonad development, in the same sort of way as is found with the gas secreted by flour-beetles.

3. Little is known about the social use of odours in fish, but there is at any rate the remarkable case of the smelts, which swarm before reproduction and produce a strong sweet odour resembling cucumber—possibly as part of a massed epideictic demonstration.

In reptiles, odours associated with aggression, and powerfully emitted by breeding bulls during the bellowing season, are produced by crocodiles and alligators. Their function in part at least is one of threat towards rivals and they fall therefore into the general category of epideictic signals. Various turtles also have scent glands, in a few cases foul-smelling and probably protective, but perhaps more generally epideictic. Apparently snakes usually leave a scent-trail behind them which has a social function, but some species may also be protected from predators by having powerful and unpleasant odours.

4. A distinctive general quality of odours used as signals is their ability to persist, especially when applied to solid objects, and to convey a message long after the communicator has left the site. They are especially practical therefore for social integration among stealthy, solitary animals that seldom meet to exchange visual signals, notably some of the mammals to be mentioned in §6.

5. An important digression is introduced at this point, explaining the general types of land-rights systems that are found in the dispersion of terrestrial and bottom-living animals. Thus space can be held or occupied (*A*) by animals of solitary habit (not excluding their mates and other dependents when necessary), or (*B*) by gregarious animals corporately. Running across this dual classification is another, since the ground can concurrently be occupied either (*C*) to the exclusion of other solitary individuals or flocks, or (*D*) with extensive overlaps between the territorial rights of several independent parties. This gives four main categories (*see* fig. 8, p. 100), namely type *AC*—the typical bird territory; *AD*—the very common situation found in, say, bumble-bees, or in foxes or bears, where each individual may defend privately his den or base but shares hunting grounds with neighbours; *BC*—where the social unit has a communal holding, as in rooks and probably European rabbits; and *BD*—where the flock probably has its own defended base as a rule, but shares feeding grounds with other flocks, as perhaps in some of the colonial sea-birds. These types all merge into each other in practice, and in the same population one type can often be quickly exchanged for another according to the season or even the time

of day. They must therefore be recognised as oversimplifications, but at the same time they have considerable analytical value.

The word ' population ' and what is meant by it is also discussed, and it is found justifiable to go on using it without necessarily giving it any dimensional definition, as being a practical general concept akin to such units as the race or subspecies. Localisation and integrity of stock are two of its most important characteristics.

6. Nearly half the chapter is devoted to reviewing the prolific development of scent as a means of communication in mammals. So great is its value for social integration here that herbivores usually leave telltale scent trails that freely give away their whereabouts to carnivores, and some of the latter sacrifice part of the surprise element in hunting by reason of their own strong smell.

The use of habitual paths and game-trails is widespread, and along these scent is deposited, sometimes by glands on the feet (*e.g.* in ruminants), but also by other methods. The dogs, wolves and their relatives (Canidae), among others, tend to have a pattern of common urination points to which routine visits are made by all the local inhabitants, though only the males participate in social urination. This is a perfectly typical epideictic activity, associated with establishing the status and property rights of males (acting on behalf of the whole population), and providing them with an index of rivalry. When two male dogs meet face to face, it is accompanied by other related forms of display. Male bears use traditional trail-side trees in an analogous way, rearing up as high as they can to slaver and bite, leaving the imprint of the canines as well as an odour. Various deer also mark trees as common sign-posts.

In the shrew *Blarina* the scent glands of males and females are equally developed in the ' neuter ' season, but in the breeding season those of the female regress, while the male's then attain their maximum size. This follows the same seasonal pattern as vocal song does in the European robin, and it is probably exactly analogous to it as a social signal. In moles, females develop testis-like organs in the neuter season, when, as with the shrews and the robin, all adults are presumably solitarily dispersed as co-equals.

In beaver, musk-rats and the various mustelids special musk-glands are developed in the anal region in both sexes (those of the badger are illustrated); in the musk-deer the analogous structures are hypertrophied in the male. All are used for ' marking ' as the animal moves about over its home range; and there is evidence from trapping records in several species of mammals that males tend to spend more travelling time abroad than females.

Anal glands may give strong specific odours to the faeces. There are various mammals that have customary communal dunging places, evidently serving the same epideictic function as before but in another way. Notable among them is the spotted hyaena (*Crocuta*), which builds up large patches of snow-white ' album graecum '—formed from the comminuted bone in its faeces. Various antelopes have traditional dunging places also, sometimes shared by more than one species; all the rhinoceroses appear to do the same, the dunging points being scattered at intervals and visited by all passers by.

The mammals give us considerable insight into the day to day tasks and duties entailed in maintaining an orderly dispersion. Their social behaviour often parallels that of man—especially primitive man—except that man uses mostly visual and auditory rather than olfactory methods for communication. Their land-rights systems are sometimes very similar to ours, and they show the same obligation to conventional rituals, *e.g.* in greeting acquaintances and strangers, in status-seeking, and in attending to communal ceremonies.

Chapter 7

Social integration through tactile perception

7.1. *Webs, silk lines and slime trails*

It would probably be impossible to find a definition, applicable to the whole animal kingdom, to distinguish the sense of touch from every other mechanically-stimulated sensation. This is of little importance, however, because we are concerned merely with grouping into convenient chapters and sections the media used by animals for receiving information from one another.

Touch is usually thought of as requiring actual contact between the percipient and some tangible object, and as residing especially in tactile organs or feelers, by means of which objects can be palpated, and information acquired on such qualities as their position, size, shape, number, hardness, texture, movement, inertia or viscosity. As far as we ourselves are concerned there is no sharp line between touch and feel—in the mechanical sense—though the latter extends to intangible qualities such as flow in air or water. As far as the material in this chapter is concerned, tactile perception must be accepted in the widest possible sense.

The range at which direct contact can be made with surrounding objects obviously depends on how far the animal can reach, and for this reason tactile organs, such as antennae, walking-legs, tentacles, barbels and fin-rays, are often specially elongated. In many of the spiders, however, where something resembling touch may possibly have become the dominant sense, perception is indirectly extended immensely further by lines of silk held like reins in the animal's feet. By means of these the struggles of trapped insects can be quickly detected in any part of a large orb web. The raft spiders (*Dolomedes*) use their mechanical sensitivity in another way ; they lie at the water's edge, with their front legs resting on the water, similarly alert to detect the movement of insects caught in the surface film, so that they can dart out and seize them. Examples like these give point to the difficulty of defining what is meant by a sense of touch, for *Dolomedes* in particular is using a method of perception mechanically similar to the vibrational sense of gyrinid beetles, which it seemed appropriate earlier on to include in the chapter on water-borne sound (p. 76).

Newly-hatched broods of spiders are frequently gregarious; in some cases they remain for a long time still living on their yolk and closely associated with their mothers (*see* p. 582). At the succeeding stage they often become airborne and scattered afar, riding on strands of gossamer, and after they alight they take up a solitary existence. When solitary, spiders are generally hostile to other members of their own species: Fabre (IX, p. 2 and 27), for instance, has described how two females of *Lycosa narbonnensis*, confined in a cage with their artificially-made burrows only 40 cm apart, fought to the death.

The orb webs of *Meta segmentata*, very abundant in gardens and elsewhere in Britain, disclose the individualistic, three-dimensional nature of the dispersionary pattern in a species such as this: even more impressive, perhaps, are the closely-packed horizontal nets of *Linyphia triangularis*, sometimes revealed in their thousands by the morning dew, extending for many acres through the whins and heather. However close they may be set, each web is clear of all the others, and where they are really crowded it would be difficult to find room for a single additional stance. Once in possession of a site, orb-spiders frequently spin one web after another from the same foundation or bridge-line. In the immature stages both sexes make webs on equal terms, but at maturity the male gives up web-spinning, and the large orbs of late summer belong exclusively to the females. Presumably spiders perceive pre-existing lines and webs in part at least by touching them with their feet, and, in avoiding them, they are in that case reacting to tangible signs of preoccupation by others.

Silk is used by various insects, especially the larvae of Lepidoptera, in order to deposit a persistent trail. Long (1955) has described how these silk strands are made use of in the social behaviour of the larvae of the large white butterfly (*Pieris brassicae*). For the first one or two instars after hatching the larvae are gregarious, leaving strands wherever they go, and while they feed they rest on a silk mat. Isolated individuals striking a silk trail follow it and are led to the area where the larvae have settled. At a later stage the aggregations split up into smaller groups, until finally the full-grown larvae are mostly found singly or in pairs. It appears likely that the tactile sense has an important part to play in their social integration and dispersion.

Examples of a rather more elaborate kind are found in the various lepidopterous larvae that weave large communal enveloping webs, including those known in North America, South Africa and Australia as 'tent-caterpillars' (*Malacosoma americana* and allied forms) and the lackey moths in Europe (*M. neustria*, etc.); and also the small ermines (Yponomeutidae). Among the most remarkable of all are the so-called 'processionary caterpillars', of the genus *Thaumetopoea* (*Cnethocampa*), which merit more extended attention.

The family of processionary moths is palaearctic in distribution, the best-known species being *T. processionea* and *T. pinivora*: the habits of the latter have been studied in the south of France, especially by Fabre (VI),

Edwards (1910) and Brindley (1910). The eggs are laid in clusters on the young shoots of *Pinus pinaster*, and hatch in September. The larvae are gregarious, and at once construct, and later progressively enlarge, a tent or nest of silk, in which they can take refuge and pass the inclement weather of winter. They feed at night, travelling to and from the nest in a single-file procession, each larva spinning a thread of silk wherever it goes. Fabre believed that the purpose of the silk was to lead them back again to the nest, but the observations and experiments of Edwards cast some doubt on this. In the procession there is a leader or ' primite ' in front, succeeded by any number of ' satellite ' individuals each following immediately on the tail of the next: a very long procession of about 260 full-grown larvae measured 26 feet (8 m) in length. The thread of silk left behind by a big procession became very thick, and in some places formed a band 2 or 3 mm wide: nevertheless, ' in spite of this it does not seem that it is the thread which guides a satellite in procession so much as the tail of the larva in front.' ' Two sections of a procession—each consisting of seven larvae—were progressing in the same direction, separated by about 18 inches: the second eventually caught up the first and joined it, though it did not at any time follow the silk thread laid down by the latter. On the contrary that thread was crossed several times, and joining up only occurred when the primite of the second in his wanderings accidentally struck the sixth larva of the first procession, when he stopped immediately, until the last larva had passed, and joined on in the right place. . . . It appeared . . . that touch rather than sight was the chief guide . . .' (Edwards, loc. cit., p. 433).

Periodically the procession is brought to a halt by the leader, and forms a ' circulating mass ' : ' the larvae do not come to rest, but continue to move slowly and in a very characteristic way, all the time moving their heads rapidly from side to side and depositing their threads. If the procession be a large one, the satellites entering into the newly-formed mass soon begin to crowd one another out in their efforts to reach it, until ranks of four or five deep are formed. . . . During the day the circulating mass formation may be maintained for an hour or two . . .' (loc. cit., p. 434). The particular processions here described were travelling over the sand in March, just before the larvae burrowed into the ground to pupate; and the circulating mass was succeeded either by mass-burrowing and pupation, or by a renewed procession; but massing occurs similarly at all stages of larval life and ' may result from a variety of causes, amongst which may be mentioned fatigue and external interference, and sometimes for no apparent reason at all. Fabre suggests cold and darkness as influences ' (*ibid.*). Though it does not concern us particularly, it may be noted that leaders are frequently changed, but there is a distinct tendency for the former leader to be the first to resume the procession when a group of larvae have been put into a heap (Brindley, loc. cit., p. 582).

No acceptable explanation of the processionary habit has apparently ever been offered, nor have any particular advantages been claimed. Brindley confirms Edwards' opinion that the silk threads are not followed or used as a guide. The larvae are to some extent protected by their poisonous or irritant hairs, but Brindley observed tachinid flies ovipositing in larvae during a procession, and comments that, '*prima facie*, such processions appear disadvantageous.' 'The meaning of processions which do not end by burying and the advantage obtained by continuous secretion of the thread remain puzzles. So also does the frequent formation of a "circulating mass" by a procession. It seems certain that contact between larvae is of much importance. . . .' (loc. cit., p. 587).

Remarkable as this behaviour is, most of its striking features can be paralleled elsewhere in the animal kingdom. The well-ordered pattern of the procession is common to the movement of flocks and schools in a good many species of vertebrates, and an equally military precision is shown, for instance, by the larvae of the sawfly *Neodiprion sertifer*. According to Husson (1952), colonies of these are found numbering about forty, all members of a single brood, and as they feed close together on the needles of *Pinus sylvestris* all in unison smartly raise and lower the head and forepart of the body every 3 to 5 seconds. The purpose of this extraordinary exercise is also unknown: conceivably it might be a deterrent to predators or ichneumon parasites, though like the processions of *Thaumetopoea* it seems rather likely to be an epideictic phenomenon. Mass-movements of an assembled population are among the most widespread forms of epideictic display—for instance at roosts or breeding stations—and the 'circulating mass' of the processionary caterpillars could readily serve in the same way to condition the individuals to the size of the group they are in: it is evident at any rate that an element of competition can be observed in their behaviour at this time.

Their trails of silk may have much the same significance as the scent-trails of mammals and other animals, discussed in the previous chapter. The progressive accumulation of webs over the bark and foliage, along the processionary routes to and from the refuge tents and elsewhere, must indicate exactly how heavily the ground has been travelled before, and this in turn could be weighed against the state of depletion already reached by the food-resources in different parts of the tree or forest. It would be quite possible for the silk lines to 'condition' the habitat in a manner analogous to the way the external metabolites of certain fish condition the water, or the gas-discharges of flour-beetles the medium in which they live. (These are phenomena already mentioned in passing in Chapter 6, but each of them will be found described separately at a later stage in the book.)

The processionary caterpillars appear in fact to provide an interesting example of the stylised conventions so frequently developed in epideictic behaviour, supplying the information on population-density required to operate the homeostatic machine. There are actually several other processionary insects, including the caterpillars of the communal moths (*Anaphe* spp.) of South Africa, which are members of a different family (Notodontidae) and have evolved the habit quite independently (Skaife, 1954, p. 171). The larvae of the fungus-gnat *Sciara militaris*, belonging to the dipteran family Mycetophilidae, provide another rather similar example. When fully grown and fed, a large number of these assemble and travel together in a long column: according to Sharp (1899, p. 464) ' millions of the larvae accumulate and form themselves by the aid of the viscous mucus into great strings or ribbons, and then slide along like serpents: these aggregates are said to be sometimes forty to a hundred feet long, five or six inches wide, and an inch in depth.' This refers to the common European species of ' army worm ' in which columns 12 to 15 feet long are often found: they are said to progress about an inch a minute; an American ' snake worm ' of similar habits is *S. fraterna* (Imms, 1957, p. 618). As usual with apparently epideictic phenomena, ' no satisfactory explanation has yet been advanced to account for the assemblage of these hordes of footless larvae ' (Imms, loc. cit.). However, it could again presumably give a direct indication of population-density, and it may be significant that in this case it particularly concerns larvae at what amounts to the prenuptial stage, when they are fully fed and about to be temporarily immobilised by pupation, in preparation for their final emergence as reproductive adults.

The possibility that slime can serve as a persistent recognisable mark in the same way as a silken thread or a trail of scent is at present purely conjectural. There are other mycetophilid larvae that also leave slimy tracks and strands, among them the so-called New Zealand glow-worm (*Boletophila luminosa*)—a photogenic larva that ' forms webs in dark ravines, along which it glides ' (Sharp, loc. cit., p. 463): the adult female is luminous like the larva, though the male is not.

By far the most familiar producers of slime trails are, of course, the snails and slugs. They secrete their mucus on the sole of the foot as an adhesive and lubricant essential to their special method of sliding along by peristaltic waves of contraction traversing the contact surface. Whether the trail is subsequently detectable by the same or another snail crossing it seems to be entirely unknown. Apart from the cephalopods, the molluscs—proverbially uncommunicative—are in general enigmatic as far as dispersionary adaptations are concerned. Snails and limpets (*Patella*) are known to have good homing abilities, but secondary sexual characters are scarcely at all developed and epideictic phenomena rather rarely detectable—perhaps partly because of the feeble locomotory powers and languid temperament of most species: many are also hermaphrodite, which likewise bespeaks an absence of, and perhaps at an earlier evolutionary stage a defection from, many of the commoner attributes of social convention.

7.2. *Feelers*

Long tactile appendages are to be found in many groups of animals. In the coelenterates, radiating tentacles appear to be the principal organs for the detection as well as the capture of food. In the many kinds of arthropods that have developed a long reach, on the other hand, it is generally the antennae or the walking-legs that serve as feelers; and in fish there are often special barbels developed from the surface of the head, or filamentous rays from the median and paired fins.

A highly developed sense of touch is particularly characteristic of animals living in surroundings where eyes are little use—for instance in the depths of the ocean, in the turbid waters of large rivers and estuaries, and in caves. Thus Alcock (1902, p. 241) refers to ' examples of deep-sea fishes whose defective eyesight is compensated by barbels or by a wonderful transformation of fin-rays into long streaming feelers ', and adds that the same thing is ' almost equally well illustrated by certain estuarine and shore fishes, such as the catfishes and Polynemidae, that inhabit turbid waters full of silt, in which it is difficult to see.' It has not always been recognised that elongated appendages of this kind are essentially tactile: Murray and Hjort (1912, p. 687), for example, state that ' several of these fishes, as for instance *Benthosaurus grallator*, are also provided with long filaments or whip-like appendages indicating pelagic habits ', apparently implying that they correspond to the supposed floatation or parachute outgrowths of plankton organisms. Similarly it has been suggested that the attenuated walking-legs of various deep-sea Crustacea and pycnogonids are adapted to walking on the soft oozes of the deep-sea floor; although, as Marshall (1954, p. 151) points out, the oozes now turn out not to be particularly soft. ' Underwater photographs have shown that the deep-sea sediments are scored and pitted with the tracks of animals: one photograph revealed heavy manganese nodules resting on the surface of the ooze. And the long-limbed creatures, all with fragile skeletons, are probably not much heavier than the surrounding water. Many cave-dwelling animals . . . have long, spindly limbs but the significance of this is not understood.' He goes on to suggest, nevertheless, that ' the longer a limb, the more tactile and olfactory hairs can be carried on it ', and, we should add, the farther it can reach.

A few examples are illustrated in fig. 10. In *Benthosaurus grallator*, just mentioned, a deep-sea scopelid fish with reduced eyes taken at 3000 m, the pectoral fins are only moderately long, but one anterior ray of each pelvic fin and one of the lower rays of the caudal are all longer than the body of the fish. In another scopelid, *Bathypterois*, also with reduced eyes, the longest feelers are a pair of special rays belonging to the pectoral fins, which arise detached from the rest, near the top of the opercular cleft: they are of a similar relative length to those of *Benthosaurus*. According to Murray (quoted from Bridge, 1904, p. 613) specimens of *Bathypterois longipes* ' when taken from the trawl were always dead, and the long pectoral rays were erected like an arch over the head, requiring considerable pressure to make them lie along the side of the body.' Notwithstanding this they have been illustrated by Murray and Hjort (loc. cit., p. 80) in a streamed position. The deep-sea macrourid *Gadomus magnifilis*, according to Marshall (1954, p. 242), lacks the elaborate appendages of the lateral line that are possessed

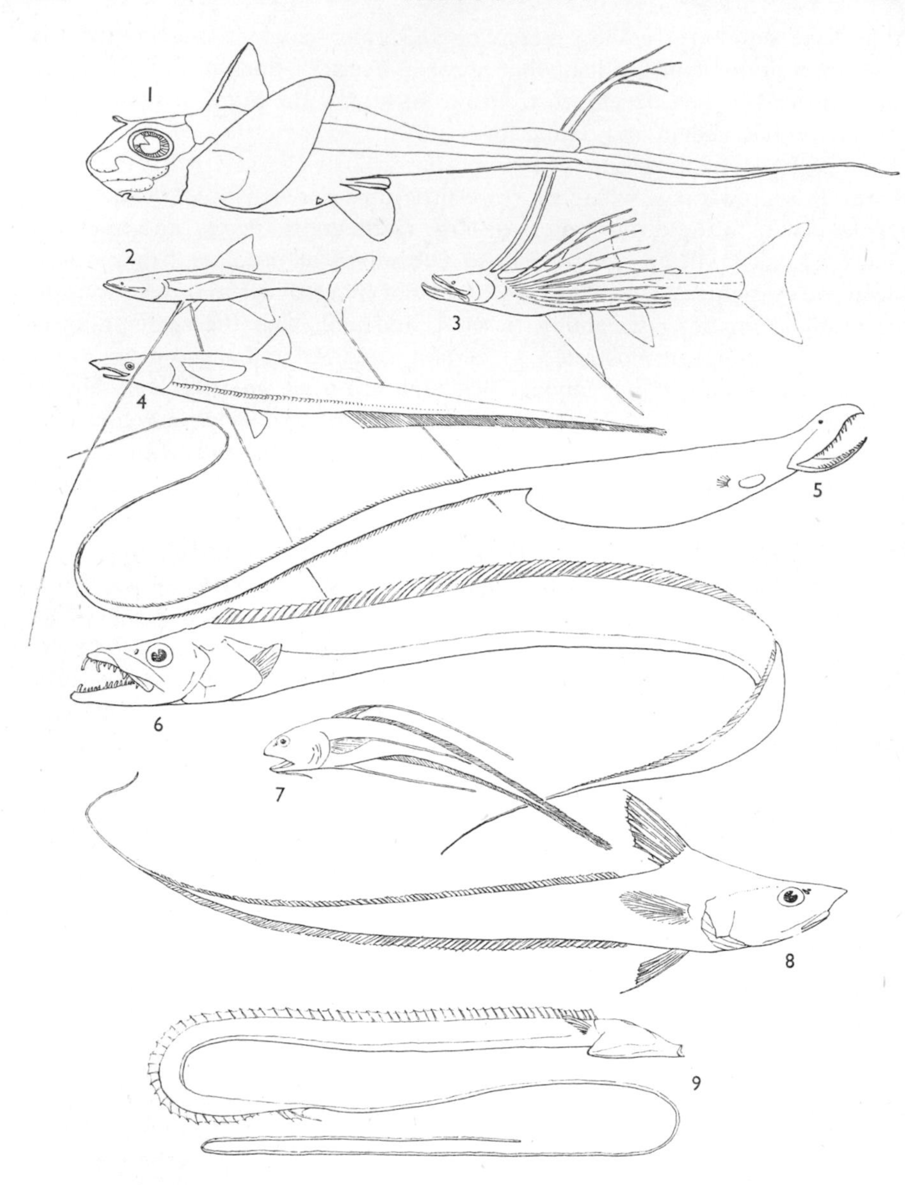

FIG. 10. Representative long-tailed and long-finned fish from the deep sea. (Redrawn, based chieflyon Goode and Bean's *Oceanic Ichthyology* (1896) and Murray and Hjort's *Depths of the Ocean* (1912).) 1. *Chimaera mirabilis* (Order Holocephali); 2. *Benthosaurus grallator* (O. Isospondyli); 3. *Bathypterois dubius* (O. Isospondyli); 4. *Aldrovandia macrochira* (O. Haplomi); 5. *Saccopharynx flagellum* (O. Apodes); 6. *Trichiurus lepturus* (O. Percomorphi); 7. *Bathygadus magnifilis* (O. Anacanthini); 8. *Macrurus* (*Gadomus*) *filicauda* (O. Anacanthini); 9. *Styelophorus chordatus* (O. Allotriognathi). Nos. 2 and 3 are both blind fish, or almost; No. 9 is known only from the type specimen, 33 in. in total length, caught in 1790 and preserved in the British Museum. In some at least of these fish the lateral-line canals extend to the very end of the caudal filament (*see* Fig. 11).

by a number of related 'rat-tails,' but has long sensory filaments growing from the dorsal, pectoral and pelvic fins, and a barbel from the chin. Perhaps the most remarkable barbels are those of the stomiatids, which may carry a photophore at the tip: that of *Grammatostomias flagellibarba* is three or four times the length of the body (*cf.* Marshall, 1954, fig. XI, 3). Good examples of similar organs in inshore fish living in the turbid shallows of river mouths and mangrove swamps are found in the genus *Polynemus*, which have a number of the anterior rays of the pectorals (seven in *P. paradiscus*) quite separate from the rest of the fin and not joined by a membrane, growing out as fine threads, the longest of which are double the body-length of the fish (Brehm, 1925, p. 378).

Though the long tails of deep-sea and pelagic fishes may be primarily adapted to carry the lateral-line canals and nerves (*see* p. 80), they are certainly capable of serving as feelers also. In the case of *Chimaera* it may be observed that the principal nerves near the end of the tail are the paired lateral line (vagus) nerves ventrally, and a pair of spinal nerves of slightly smaller size, running back beside the spinal cord dorsally: the spinal cord itself is thin-walled, with a very large internal canal, suggesting that it has little nervous function (fig. 11). Even at the extremity the tail is muscular, though so feebly that it can hardly contribute to propelling the fish, and the sensory canals and their nerves together with the supporting skeleton and integument occupy by far the greater part of the cross-section.

Some of the Crustacea living in the same habitats have equally long appendages. In the large genus of penaeid prawns, *Sergestes*, the antennae are usually very long, over six times as long as the body in *S. corniculum* (according to Plate 2 of Sund, 1920). They are curiously bent: 'there is a more rigid part up to the kink and then a much more flexible part like a lash' (Hardy, 1956, p. 230). Hardy thinks it possible that they could be used as fishing rods. In *Stenopus*, belonging to a closely related family, the antennules each have two flagella and the antennae one, making a total of six lashes, all of them two or three times as long as the body (in *S. spinosus*); in *S. hispidus* the pereiopods (walking-legs) are correspondingly long and slender, especially the third pair; hence the name *Stenopus*. It seems probable that all these structures can be spread out in different directions, as are the antennary filaments in the common shrimp (*Crangon*) and prawn (*Leander*), and that they can detect slight disturbances anywhere in the immediate neighbourhood.

Among the Brachyura, the spider-crabs have especially long pereiopods; *Inachus dorsettensis*, *Macropodia rostratus* and *Maia squinado* are well-known British species. The largest one known is *Macrocheira kampferi*, from Japan, whose tremendous legs may span up to 8 feet, and possibly 11 feet (2·5 to 3·3 m), according to reports. As might be expected, spider-crabs are inclined to be sluggish in their movements.

Mention must also be made of the pycnogonids—the anomalous marine spider-like animals belonging to the Arachnida, many of which are remarkably long-legged. They have no known sense organs other than eyes, of the simple ocellar type, and in some deep-sea species these are imperfect or lacking. They too are usually sluggish in habits; many of them live by sucking the juices of coelenterates and other animals. The four pairs of ambulatory legs attain six or seven times the body-length in *Nymphon* and *Colossendeis*: the latter when full grown spans about 15 inches (30-40 cm).

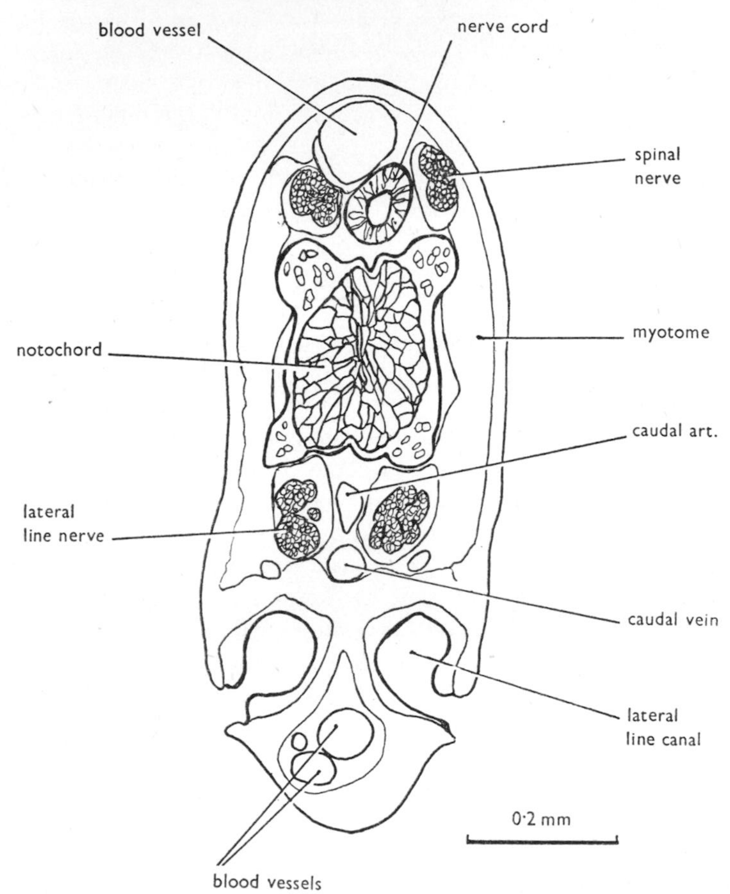

FIG. 11. Transverse section near the tip of the tail filament of the rat-fish (*Chimaera monstrosa*). The tail with its supporting structures carries the open lateral-line canals and their sensory nerves (branches of the vagus), and also a pair of spinal nerves running beside the nerve cord dorsally (the latter has a wide central canal and few axons). The large dorsal and ventral blood-vessels may give sufficient buoyancy to counteract the tendency of the filament to dangle when the fish is at rest. (×100.)

In the Mollusca, the long-armed bottom-living octopods can be seen to explore their surroundings by means of their appendages, the extremities of which are repeatedly coiled and uncoiled over the substratum as they move about. In rather different circumstances one of the bivalves, *Tellina*, which burrows on sandy seashores, has a long, mobile inhalent siphon, and when the tide is in and the animal is feeding, it ' curls over and explores the surface of the sand, sucking in loose particles on or near the surface ' (Yonge, 1949, p. 235). Holme (1950) showed that, in the habitat he studied on the Exe estuary in Devonshire, the rather dense populations of *T. tenuis* were ' over-dispersed ': that is to say, they were not scattered at random, but tended to avoid each other, so that the number of instances of individuals found less than 1 inch, or less than ½ inch, apart was significantly fewer than would be expected by chance. ' It is possible,' he concluded, ' that at moderate densities each individual occupies a territory delineated by the activities of the inhalent siphon on the soil surface ' (loc. cit., p. 279).

Even in the absence of direct observational evidence there is a strong possibility that most if not all the elongated appendages so far mentioned can be used for tactile communication between individuals. They frequently occur in groups of animals already known to possess well-developed social organisations, and endowed in the majority of species with other alternative organs for efficient signalling and perception: in most of these groups the tactile function seems to have become especially prominent in species living in surroundings where visual recognition at any rate would be impracticable, because it is too dark or muddy: this applies for instance among the highly mobile fish and prawns. Long appendages are present also in a variety of slow-moving, bottom-living creatures such as spider-crabs, pycnogonids, bivalves and brittle-starfish (to be mentioned below), in which they may be useful to their owners in securing a stance of sufficient size or establishing a minimum individual distance.

An instance of what appears to be a tactile communal display has been described by Newman (1917), in harvestmen belonging to the genus *Liobunum* (Opilones). He found in Texas a vast colony of these, estimated to contain between one and two thousand individuals; they were resting during the day, packed closely together in a somewhat circular area nearly five feet in diameter, on the underside of an overhanging shelf of rock. All were at first hanging from the ceiling motionless, but when he came within about six feet they began a curious rhythmic dance, raising their bodies up and down at the rate of about three times a second, without changing their footholds; ' and, curiously enough, the movement of the entire lot was in the most perfect unison '. After a minute or so it died down, but by poking a few of the nearest individuals he caused the dance to spread quickly again over the whole group. ' When the colony was first seen it was noted that the long legs of neighbouring individuals were closely interlocked and this mechanism was sufficient to account for the transmission of stimuli from one part of the colony to another'.

Another example may possibly be provided by the brittle-stars (Ophiuroidea). Vevers (1953) and McIntyre (1956) have shown by underwater photography that *Ophiothrix fragilis* sometimes occurs in British inshore waters in very dense beds on the sea-floor. One described by Vevers 5 miles off the Cornish coast near Looe Island was at least 2 miles long and half-a-mile broad, with an average density of brittle-stars of 97 per square metre, or nearly 400,000 per acre. There is some reason for thinking that the same bed was in existence in 1899. In another bed near the Eddystone a minimum figure was 340 per square metre (1·3 million per acre), but the animals were so thick on top of one another that the lower layer was partly obscured. Apparently individuals of all sizes are not found distributed at random, but in any one area they tend to be somewhat restricted in size-range; and in the Eddystone bed, with the higher density, the average size of individuals was 'considerably smaller' than on the Looe Island bed. It is possible therefore that an adjustment of the density takes place with growth in size of the individuals.

Though this is reminiscent of the huge beds of snapping-shrimps, for example, mentioned in an earlier chapter (p. 75), Allee (1931, p. 51) concluded, from his laboratory experiments on the aggregation of the brittle-star *Ophioderma brevispina*, that their grouping together is 'not the expression of a community spirit or of a social appetite. The only social trait necessarily present is that of toleration for the presence of other animals in a limited space, where they have collected as a result of tropistic reactions to environmental stimuli'. It seems fairly clear now, however, that even in such lowly orders of animals as these the populations must possess homeostatic adaptations of some kind, and in that case a form of social organisation would of course be prerequisite and necessarily present. Except for the podia or tentacles found on the arms, which are thought to be chiefly sensory in function, ophiuroids are devoid of obvious sense-organs: in some species, moreover, such as *Orchasterias columbiana* (Hyman, 1955, p. 592), *Asteronyx loveni*, *Astrophyton* and *Gorgonocephalus*, the arms are extraordinarily long. In view of this it seems most likely that whatever social integration they attain is achieved through tactile communication.

7.3. *Other contact aggregations*

There are various cases known in the higher vertebrates where a number of individuals pack in together, usually in a place of shelter, while roosting or hibernating. Thus on Mount Royal, Montreal, I have disturbed as many as ten grey squirrels (*Sciurus carolinensis*) from a single hole in a hollow red oak-tree in winter. Various kinds of bats, moreover, are known to roost in tight balls or clusters. After referring to this habit in *Pipistrellus subflavus* and *Scotomanes rufoniger*, Allen (1940, p. 220) continues: 'Some species of cave-dwelling bats will hang in close masses from the vault of their retreat, such as some members of the genus Tadarida, or they may be found packed away in crevices under hollow roofs so tightly that they lie more or less on or over one another and seem even to enjoy this close contact. I once had a number of the small Haitian Molossus, a related genus, brought to me alive from a large colony. When placed in a bag of netting, they all gradually worked their way upward to the top of the bag as it hung and finally quieted down in a close-packed mass, a position which seemed satisfactory to them notwithstanding the tropical heat.' Both of these are free-tailed bats

(Molossinae). It is recorded of the tropical American mastiff-bat, *M. obscurus*, that clusters of 100 and more have been found in hollow palm-trunks, consisting exclusively of either males or females (Flower and Lydekker, 1891, p. 670).

Rather similar habits are found in various birds, such as the Australian wood-swallows (*Artamus* spp., especially *A. cyanopterus*), which have been found roosting in dense clusters containing anything from a few birds up to 200: they ' hang together in clusters from the bole or branch of a tree ' and have been observed to return to the same site on successive nights (*cf.* Hindwood, 1956); much the same kind of aggregations have occasionally been found in various swifts, wrens, and other birds (*cf.* Lack, 1956, p. 127, 136). The functional purpose of the habit in warm blooded animals is usually, and no doubt in most cases rightly, assumed to be the retention of body-heat during repose, but in some instances clustering can be observed even in very warm conditions, as in the tropical bats mentioned above and in the wood-swallows (A. J. Marshall, 1957). There is thus a possibility, at least, that the mutual contact among the clustering individuals has in some species assumed a social and epideictic function; and although in such highly sensitive animals as these the nature of whatever communications pass between them is almost certain not to be exclusively tactile, nevertheless the subject is one that can appropriately be included in this chapter.

Many of the frogs and toads, and a variety of reptiles, including for instance the adders (*Vipera*), rattlesnakes (*Crotalus*) and garter-snakes (*Eutaenia*), gather as is well known into hibernacula, where they may sometimes be found passing the winter in a torpid state entwined or packed together in scores. Since they are all cold-blooded, the habit has probably little relevance to the retention of body-heat: an important part of its functional value appears likely to be social, but this is a matter that it seems better to defer until we come to Chapter 14 (*see* p. 316).

7.4. *Summary of Chapter* 7

1. The mechanical sense of touch cannot be rigorously defined but it normally requires actual contact with the objects perceived, and supplies information as to their position, shape, texture and other ' tangible ' qualities; its effective range can be increased by the development of elongated feelers. Many spiders succeed in extending tactile perception to a much greater distance by holding taut lines of silk in their feet, or, in *Dolomedes*, by using their feet to detect tremors on the surface of water. In many web-spinning spiders the property-claim of the individual appears to be co-extensive with the pitch in which the web is constructed.

Many insect larvae leave trails of silk which appear likely to play an important part in their social integration. The most remarkable case, perhaps, is that of the processionary caterpillars (*Thaumetopoea*), which move about in columns in single file, head to tail, leaving a cord of silk of a thickness proportional to the size and number of the individuals in the procession. When they cross old trails they do not follow them, but it is suggested that the pine trees in which they live and feed become ' conditioned ' by the gradual accumulation of silk, and that this indicates the extent to which the trees have already been worked over. From time to time every procession comes to a halt and forms a circulating mass: this appears to be an epideictic exercise, again providing an index of the size and number of those present in the procession. No acceptable explanation of the processionary behaviour-complex has been put forward previously. It is paralleled, in part at least, in several other insects, including another moth larva (*Anaphe*), and the larvae of the fungus-gnat (*Sciara*). Conventional activity appears to increase in intensity as the time for pupation (*i.e.* the prenuptial stage) approaches.

Slime trails, as produced by snails and a few other organisms, could presumably serve in the same way for mutual recognition and territorial marking, but so far no observations or experiments appear to have been directed to the subject.

2. A highly developed tactile sense is especially characteristic of animals living in habitats where vision is more or less precluded—deep in the sea, in muddy water, and in caves. They frequently possess long feelers derived from a variety of appendages, including walking-legs in the case of cephalopods, spider-crabs and pycnogonids; antennae in certain other Crustacea (*e.g.* penaeid prawns) and insects; barbels, fin-rays and filamentous tails in the case of different fishes. In the bivalve *Tellina*, living buried in the sand on seashores and in estuaries, the long and mobile inhalent siphon may be instrumental in keeping individuals spaced out and limiting population-density. A case is related of an assembly of over 1000 harvest-spiders, their legs overlapping, engaged in a communal dance-like display, the rhythm apparently being maintained by touch. Huge aggregations of brittle-starfish (Ophiuroidea) found in certain places, in so far as they have a social function, are probably integrated through the tactile sense.

3. A few examples are given of massing or clustering of animals, sometimes for roosting (*e.g.* in bats, swifts and wood-swallows), and sometimes for hibernation (*e.g.* in snakes and frogs): in all of these the sensation of contact is likely to provide a social bond.

Chapter 8

The social group and the status of the individual

8.1. *The possible functions of social organisation*

The general survey of the methods used for exchanging information among members of a population is now completed, and two immediate general observations can be made: first that, in one situation or another, animals have exploited practically every physical medium of communication open to them; and second that social integration in some form is evidently the general rule among animal populations and likely to be found virtually everywhere, except perhaps in some of the most primitive groups. There seems to be no question of dividing the animal kingdom neatly into two camps, the one containing animals that are social and the other those that are not: rather it emerges, as might have been expected, that socialisation is a general phenomenon, which from comparatively lowly and obscure beginnings has undergone progressive evolution, so that in the more advanced groups it has tended to become increasingly more conspicuous and complex. Although the extent to which social adaptations have been evolved in different animals, therefore, varies between wide extremes, the series is nevertheless a continuous one: the barnacle larva which at metamorphosis fixes itself or is inhibited from doing so, depending on the presence and number of other barnacle larvae already attached, is responding socially in a way that is not effectively different in principle from the one which animates the communal behaviour of the social insects or the most highly socialised birds and mammals. In each case the behaviour of the individual is conditioned by the presence and actions of other members of the population in the interests of regulating the dispersion of the group as a whole.

The conception of social integration, developed to promote common interests, not as being something sporadically evolved and thus confined to particular groups but rather as a normal characteristic of living animals, is not of course new: it was a conclusion reached long ago by Espinas (1878, pp. 7 and 356) and subsequently reiterated by Kropotkin (1902), Deegener (1918) and other authors (*cf.* Allee, 1931, p. 353).

That social behaviour can be of value in various ways to the common welfare of the society is generally assumed and accepted, although when it comes to defining what precisely the advantages are, the results are very apt to leave the enquirer unsatisfied. An aggregation of individuals, for example, may be an obvious prerequisite to successful breeding in certain species, but it seems quite impossible to maintain for this reason as a general proposition that ' social behaviour—the inter-relation of individuals within a group—is determined primarily by the mechanisms of reproductive physiology ' (Zuckerman, 1932, p. 29): for we have had many instances of social behaviour pertaining to the immature stages of the life-cycle or to asexual periods of the year, or involving segregated members of one sex, and we know it is common for animals to adopt different social patterns at different seasons or stages of development.

At first sight, at any rate, there seems no obvious reason for expecting that the benefits of sociality would all need to be in some fundamental way inter-connected. Thus Tinbergen (1953, p. 20) groups social co-operation into four quite distinct categories, according to the ends it serves, concerned respectively with (i) mating, (ii) care of the family, (iii) group life and (iv) fighting. Regarding the first, social gatherings as just stated can be seen to facilitate mating in some animals, though not in others: as an example we might cite the frogs and toads, many of which mass together at spawning time, though there are other species, in certain cases belonging actually to the same genera (*e.g. Rana catesbiana* and *R. clamitans*, *see* p. 52), in which solitary males attract single mates. In so far as social co-operation in this sense implies gregariousness, it is obviously not a fundamental and universal necessity for successful matings, but only a superficial adaptation—one among several possible alternatives—showing a pattern of presence or absence within single genera and even single species (as we shall see in Chapter 9, p. 151).

Similarly the second category includes only the rather scattered minority of animals, in which the parents actually succour their young, or the members of a brood are in some way mutually dependent.

Omitting the third category for the moment, and coming to the fourth—fighting—we find a commoner, more widespread form of social activity, with most important implications. That it should confer any kind of benefits at all, as Tinbergen points out, at first sight seems quite paradoxical, but the benefits turn out, as he discloses, to result from such things as the successful attainment of territories and of social standing in the hierarchy—matters that we have come to regard as playing an important part in the homeostatic machinery of dispersion.

It is into Tinbergen's third category, group behaviour, that the largest part of social activities seem to fall, including everything not concerned with fighting, mating and the family: it embraces the non-reproductive, and often feeding or travelling, swarms and flocks of insects, fishes, birds and mammals of many kinds, and various aggregations for sleeping and hibernating, besides a great number of the special instances of social activities detailed in the preceding pages. Abundant though the examples are, explanations are few and far from satisfying. The commonest one given is that of defence against predators: the merlin, for example, pursuing a tightly-bunched flock of snow-buntings, must generally succeed in detaching one from the rest before it can attempt a kill; similarly, wolves can be held at bay and defied by a ring of musk-oxen, standing stolidly together, flank to flank, with their calves behind them in the centre. But it is no less easy to find examples of the reverse—of predators attracted to a massed and defenceless prey: thus gannets and sharks pursue and harry—sometimes for days at a time—schools of herring and mackerel, and a wide variety of marine predators help themselves *ad libitum* to the spawning swarms of sub-arctic caplin (*Mallotus villosus*; *see* Templeman, 1948, p. 132) and sub-antarctic euphausians. Similarly flocks of lesser kestrels, pratincoles, rosy pastors and wattled starlings (*Creatophora carunculata*) attend and batten on swarming locusts; and, as earlier mentioned, parasitic Diptera take free advantage of contingents of processionary caterpillars (*see* p. 117). While communal defence may sometimes be one of the incidental benefits of social co-operation, therefore, it again cannot provide us with a fundamental cause: no more can the other equally favoured explanation, namely the greater collective vigilance that a herd can exercise in perceiving danger, for instance in geese, baboons or ungulates.

Allee (1931) gathered together and investigated a large number of cases of harmful and beneficial effects of animal aggregations. As with his predecessors in the study of social behaviour, there lies behind his work ' the idea that there is an essential unity within the phenomena to be discussed ' (Allee, 1931, p. 33), but the many and varied effects he describes appear to be too discrepant or isolated to lead to the formulation of common principles, other than a rather nebulous ' principle of co-operation ', which appears to him to lie at the basis of social evolution (loc. cit. p. 352).

The truth is that it has never yet been possible to offer any comprehensive explanation, or assign at all adequate advantages, to account for what is undoubtedly one of the commonest and therefore presumably one of the most important classes of animal endeavour and achievement. Our general hypothesis that sociality is the basis of conventional behaviour, and that together they provide an indispensable part of the machinery required for the homeostatic control of dispersion, therefore merits very serious attention, because it identifies for the first time a possible common underlying purpose. We remember too how often already it has suggested a lucid meaning for what had previously been puzzling and unaccountable facts.

8.2. *The role of threat*

In his thought-provoking book, *The meaning of animal colour and adornment*, Hingston (1933) forcefully expounds the view that threat is the commonest of all the purposes for which special animal adornments are developed; and in the foregoing chapters we have of course come across numerous instances in which methods and media of many kinds have been used for this particular type of communication. Though, often enough, exactly the same gestures of threat can be used either against conspecific rivals or against assailants of other species, it is at once clear that of the two the former is the more universal and fundamental function. This is shown, for example, by the common situation in which the equipment for threat display is not equally developed in all members of the species—as it would be if its primary use was against predators—but is stronger in or entirely confined to the adult males. Threat can be vigorously conveyed by the brandished claw of the fiddler-crab (*Uca*), the erected dark fins of the fighting-fish (*Betta*), the antlers and horns of ruminants, the roaring of bulls and lions, the songs of birds and crickets, the musk of alligators and death's-head moths (*Acherontia*); and in all these and innumerable other cases the apparatus employed is more or less hypertrophied in the males. Even in species where the females and immature stages share the equipment, threat is far more often employed intra- than inter-specifically. This does not mean to say that threatening devices primarily developed for use against marauders of other species are never found, although they are decidedly in the minority.

The organs used in threat are in some cases actual weapons, and their display then indicates directly the formidability of the owner and his immediate power to hurt: such are the fangs of snarling dogs, the tusks and horns of ungulates, or the pincers of crabs. More often, however, their significance is largely or entirely conventional. It may depend on a general aspect of fierce strength, genuinely muscular in bulls, for example, but often augmented by ' make-up ', such as the lion's mane, the inflated snout of the elephant-seal, the bristling hair of the dog's back, or the raised hackles of the fighting cock. Similarly, the aerial nose-dives of birds of prey simulate the movements of real attack; but often threat display has finally come to depend on purely formalised acts or postures, showing off some structure in itself harmless, such as a bright-coloured fin or crest (or in some birds even the expanded iris of the eye); or it may involve some other kind of signal such as a particular sound or odour, similarly made formidable solely by association.

All this is very well known, and so is the fact that the ordinary purpose of threats is to create an impression, and instil a respect for the power, always implied, to attack and hurt the opponent. Without an actual resort to painful and even deadly combat when the bluff is called, all threat would quickly lose its validity; but the wholesale wounding and killing of members by one another is generally damaging to the group and has consequently been suppressed by natural selection. (As a matter of interest there are special circumstances, detailed in Chapter 22, in which socially-induced mortality can temporarily benefit the population by reducing its numbers, but these are not the standard conditions of conventional competition with which we are dealing here.) As a result, therefore, a compromise is struck, and well-backed threats are generally all that are needed to win a decision and settle the immediate dispute: resort to violence and, still more, fights actually to the death are largely eliminated.

Direct competition between individuals, therefore, has generally come to assume conventional forms, and the individuals are adapted or conditioned to accept as final in the great majority of cases decisions reached by purely symbolic methods. Brute force and mutual savagery among members of the same species have at most a residual place in social competition. Even threat as a conventional substitute is itself relatively crude and direct, and not infrequently its harsher symbols have given place to specially evolved and often beautiful adornments, ceremonials and music, in the exercise of which individuals vie together more with pleasure than with bitterness. Out of the harmless enjoyment and companionship of such dual and social displays there can nevertheless clearly emerge a recognition of the merits and relative standing of the performers, and we can guess from our own experience that this is likely to be one of the main adaptive functions of sport or play.

As always with social attributes, the conventions governing social competition can only have been evolved by group-selection; and any immediate advantage accruing to the individual by killing and thus disposing of his rivals for ever must in the long run be overridden by the prejudicial effect of continuous bloodshed on the survival of the group as a whole.

In practice, conventions lay down codes of law, which have evolved to safeguard the general welfare and survival of the society, especially against the antisocial, subversive self-advancement of the individual. Individuals comply with them, as we shall see later, as a result of social pressure, and this compliance appears to be the exact counterpart of what we recognise in our own behaviour as moral conduct. The normal individual is predisposed, either through innate instinct or acquired conditioning, always to be governed by the conventional code, but conventions are essentially artificial things and can be reinforced only by equally conventional sanctions. Thus, should an individual break with them, no inherent disaster will automatically descend on his head. It is this situation which sets the stage for the moral choice between obeying what convention dictates, which is best for the welfare of the group, and promoting one's own self-interest when to do so is antisocial. We are perhaps rather generally inclined to think that the rational power of perceiving, however dimly, that such a choice exists, and being torn by a real temptation to pursue selfish ends on the chance of escaping the legal (*i.e.* conventional) penalties, is the exclusive predicament of man; but it seems likely enough to exist in an attenuated form, at least in the other mammals. For the non-rational creature, on the other hand, there appears to be no choice: conformity with the social code is unquestioned, and virtue automatic. This is a philosophically interesting topic to which we shall find ourselves returning at a later stage (p. 190).

8.3. *Social competition*

It is part of our general theory that free contest for food—the ultimate limiting resource—must in the long run end in over-exploitation and diminishing returns, and that this situation is avoided by substituting conventional rewards to take the place of actual food. Competition for these conventional rewards ideally operates in such a way that the population-density is brought to an equilibrium at the optimum level—this being the level at which food-resources are utilised to the fullest extent possible without depletion; and if the resource-level changes, or for any reason the density does depart from the optimum, forces are brought into play to restore the balance as early as possible.

The theory also implies that it is this substitution of conventional goals which lies at the root of social evolution. A raw natural resource such as food is something that theoretically can be sought and competed for by unconnected individuals in complete independence, all perfectly free to employ every means in their power to gain their individual ends. But as soon as artificial restrictions are imposed on the methods they may employ, and conventional goals are substituted in place of the actual resource itself, the situation becomes radically different. This may perhaps best be understood in the following way.

The underlying purpose of all the conventions that concern us here is to enable the food-resources to be conserved for the future, thus insuring one of the chief conditions on which the continued survival of the group depends. In practice, these conventions shield the resources against prodigal abuse from day to day by members of the contemporary population and thus safeguard the long-term prospects of group-survival. It has already been pointed out in Chapter 1 (p. 19) that conventions of this kind must, by their nature, always be properties of a concerted group, and can never be completely vested in or discharged by a lone individual in perpetual isolation: their observance has to be reinforced by the recognition and support of others who are bound by the same convention. In the absence of other parties they become meaningless.

It is this concerted group that appears to constitute the primordial germ of the society. Mutual integration and co-operation among individuals in fact play an obligatory part in the evolution of such conventions, which can only be regarded as essentially social. The social conventions themselves all ultimately spring from the need to develop substitute goals for mutual competition among rival members of the species—goals that are effective in preventing population-density from exceeding the optimum level. If we reduce this situation to its simplest terms it becomes possible, as we saw earlier, to define the elementary society as being *an organisation of individuals capable of providing conventional competition* among its members.

We have had abundant evidence already to show that this society-convention complex has undergone progressive evolution and has reached its greatest perfection in the insects and higher vertebrates. It is here that the homeostatic adaptations are generally easiest to see in operation, and consequently the present chapter is written with these particular groups, and above all the birds and mammals, especially in mind. We must not forget, however, that many typical features of social and epideictic behaviour—symbolic acts and signals, for instance, with mutually stimulating or emulative effects; threat, avoidance, individual distance, occupation of conventionally desirable sites, the use of traditional loci, adherence to conventional seasons and hours of the day, and so on—can be traced far down the scale of animal classification.

To fulfil its essential part in the machinery of dispersion, the society requires to become increasingly competitive as the density of population rises: the degree of competition provides the gauge of population-density, and, if the machinery functions properly, when it rises sufficiently it will evoke a response—generally either emigration or a lowered reproductive rate—to bring the pressure down. This response ought to occur before the safe limit of numbers is reached and permanent damage has been done to the resources of the habitat. At such times as these, or whenever any large-scale executive action requires to be taken, it may therefore be expected that social behaviour will be intensified and reach its highest pitch. There are two situations in which this may be regularly observed, namely, just before the onset of breeding, when the population is about to be increased or established for another year, and again in the face of adverse weather and shortage of food. Many birds and mammals vary much in the degree of sociability they show to one another at different times: wolves hunt alone or in small bands for the most part, but occasionally they form into packs of a score or more, apparently in winter only (Seton, 1909, p. 755). The similar flocking or packing of birds in winter or during very hard weather, especially in the buntings and finches, and game-birds like willow-grouse and ptarmigan (*Lagopus* spp.), or before and during weather-movements or regular migratory flights, provide other familiar examples.

Social groups, as has been stated previously, combine the two apparently opposite qualities of cohesion, which draws the individuals together, and mutual rivalry, which tends to keep each individual at a distance from its neighbours; these might properly be called synagonistic and antagonistic tendencies. It is only relatively recently that attention has been drawn to the general phenomenon of ' individual distance ', for instance by Hediger (1941, and 1955, p. 66) and Conder (1949), though it may well turn out to have been noticed by a number of earlier writers. For example, Allen (1940, p. 210), writing of flying foxes in their roosts, says that, ' though they seem desirous of companionship, hanging often in great multitudes in the selected treetops, there is also a negative factor which leads each member of the group to keep his fellows at arm's length, a sort of isolation in the crowd.' The essential feature of individual distance is the pre-emption of a minimum amount of space surrounding the individual, into which other individuals cannot intrude without rendering themselves liable to attack.

In the actual maintenance of this and other types of dispersion, avoidance is often effectively substituted for militance, at least so long as the competitive pressure is not too great.

Individual distance reveals in a simple form a further common characteristic of social behaviour. In accordance with the opposing synagonistic and antagonistic forces, the individual's status in the flock has two different but complementary elements. In the first place he has succeeded in joining the group and being admitted as a member of it; and, consequent upon this, he has secondly acquired the right to the sole occupation of a minimal conventional space or site within the group. The first element is abstract and subjective, the second concrete and objective. In flocks of birds the members often behave as though they all possessed equal status: this applies both to their membership and to their individual allowance of space.

It is well known, however, that equality does not necessarily prevail in these important relationships, and that, as far as subjective status is concerned, individuals in the group often stand at different levels on the social scale, some being dominant to others. In competition for the necessities of life the dominant individuals have the prior choice; and when it comes to the occupation of real property in the form of a stance or territory or home-range, submissive individuals lower in the scale may be driven to poorer marginal sites, or be obliged to do without. As a general rule, therefore, social competition is directed towards these dual and essential prizes—social status on the one hand, and a claim or tenure of property on the other. Depending on the particular type of social organisation, one of these goals may often appear to eclipse the other.

8.4. *The social hierarchy*

Competition for social status is naturally most conspicuous in gregarious animals, and it was in fact through observation of domestic hens and ducks in Norway that Schjelderup-Ebbe (1922) first discovered the existence of the peck-order. In any small flock of hens there soon develops a rather firmly fixed hierarchy, in which the top bird normally has the right to peck all the others without being pecked in return; and each of the others occupies a place subordinate to hers, usually in a linear series with respect to one another, down to the lowest bird, which all may peck without fear of retaliation. The series is not necessarily a straight line, but may include triangular loops and other irregularities. The relative positions of two birds may be decided once for all at their first meeting, either through actual fighting or because one yields to the other. The system naturally depends on individual acquaintance, and the peck-right, once established, can be reversed only ' in the event of a successful revolt which, with chickens, rarely occurs ' (Allee, 1938, p. 154). The pecks given vary in intensity all the way from slashing stabs that draw blood from the comb, or feathers from the neck, to mere preparatory gestures and threats. The same kind of hierarchy has been found to exist in cockerels, except that they are much more given to fighting than the hens, and the peck-order is less stable and more involved, with, for instance, no less than six triangles among eleven birds (Masure and Allee, 1934*a*, p. 310). In domestic pigeons the same authors showed that a permanent peck-right between individuals is not achieved at first meeting, but after a time there emerges instead a 'peck-dominance', such that although attacks are still delivered in both directions they are more frequent in one of them. In both species low-ranking birds tended to avoid social contacts and high-ranking birds to seek more than the average.

Though a sharp line probably cannot be drawn between them, hierarchies conforming more or less closely with either the peck-order or peck-dominance types have been found in several other species of birds in captivity. For example, peck-*orders* have been discovered by Noble, Wurm and Schmidt (1938) in American black-crowned night-herons (*Nycticorax nycticorax hoactli*), by Wessell and Leigh (1941) in white-throated sparrows (*Zonotrichia albicollis*), and by Jenkins (1944) in snow-geese (*Anser caerulescens*). Peck-*dominance*, on the other hand, more accurately represents the situation in budgerigars (*Melopsittacus undulatus*) (Masure and Allee, 1934*b*) and canaries (Shoemaker, 1939). It is far more difficult to investigate these relationships in the wild, but Odum (1942, p. 520) demonstrated a fairly enduring peck-order in a winter flock of colour-ringed black-capped chickadees (*Parus atricapillus*), as they visited a feeding-station over a period of months. Brian (1949) found the same to hold for great tits (*Parus major*): she says 'a straight line dominance order existed among the seven males, broken at one point by a triangular relationship between three of the birds. The females were closely associated with their mates and possessed a similar dominance order.' She found also that the birds, when not at the feeding place, tended to remain each in its own particular area, and 'the position of the bird in the dominance order corresponded with the distance of its territory from the feeding place; thus, the farther away a bird's territory, the lower its status.'

There is circumstantial evidence that dominance hierarchies are generally developed in all the classes of vertebrates. The linear peck-order is clearly a very simple kind of grading, dependent on small numbers and very intimate acquaintance of the members one with another. What can frequently be observed in nature is the emergence of one despotic individual, or an oligarchy, dominating the rest of the group. In the sage-grouse (*Centrocercus urophasianus*), for example, the lek or strutting display leads to the dominance of a top cock, and he, assisted by a 'sub-cock' and a few 'guard-cocks', fertilises hens by the score (*see* p. 214). Similarly in the fur-seals (*Callorhinus*) and elephant-seals (*Mirounga*), master bulls control the various beaches on which the cows come ashore. Associated with this kind of hierarchy there is in some species a progressive individual development from year to year in the signs of maleness, so that older males in the prime of life possess a formidable advantage in competition with younger ones on account of the obvious superiority of their horns, tusks, manes, colours, voices or bodily strength. Age does not invariably secure precedence, however, as has repeatedly been found in domestic fowls, and also in canaries (Shoemaker, loc. cit., p. 399) and golden-crowned and white-crowned sparrows (*Zonotrichia* spp.; Tompkins, 1933) but these are all species that mature rapidly and can breed at a year old.

The relation of the social hierarchy to sex varies in different cases. In general males dominate females: sometimes females appear to rank with their mates, as for example in Lorenz's (1931) jackdaws, Brian's great tits and Jenkins' snow geese, mentioned above; the converse may also occur, and Shoemaker (loc. cit., p. 391) found in breeding canaries that the females at a certain stage tended to usurp the dominance from their mates. Sometimes males and females are apparently graded without regard to sex, as in the black-crowned night-herons. The explanation of an organisation like the last one is perhaps that it is in effect asexual, and concerns a group of neuter individuals in which there are no established pairs and no interest in reproduction. The same explanation has been offered previously to account for the annual cycle of the scent glands in shrews, the interstitial glands in moles, and the winter singing of female robins and dippers (p. 106). The loss of sexually-distinctive plumages after the breeding season and the donning of neutral colours, especially by the males, in various families of birds is almost certainly another sign of the same essentially asexual condition (*see* Chapter 12).

An oligarchic or despotic system has been demonstrated in certain mammals in the laboratory, including mice (Uhrich, 1938; Scott and Fredericson, 1951, p. 273) and prairie-dogs (*Cynomys ludovicianus*; Anthony, 1955). In rabbits (*Oryctolagus cuniculus*) in the wild, Thompson and Worden (1956, p. 82) reported the existence of only a relatively small number of dominant, mainly old males, with the effect that ' remarkably few bucks took part in the affairs of the warren '. Short-tailed voles (*Microtus agrestis*; Clarke, 1955) and laboratory rats (*cf.* Grant and Chance, 1958) on the other hand have more continuously ranked hierarchies; and in the bottlenosed dolphin (*Tursiops*) McBride and Hebb (1948, p. 113) report that ' in all groups of males that have been observed there has been a well-developed and stable hierarchy of dominance, closely related to size. In the present group . . . the females do not appear to compete aggressively and do not form part of the hierarchy.'

Although the hierarchy can be regarded as an essentially abstract phenomenon, in practice the relative standing of individuals is greatly influenced by their concrete circumstances and situation at the moment. An individual guarding a nest-site, territory, established stance, roosting-place, or mate, normally dominates all rivals and intruders, including those that would stand higher in the scale if they met on neutral ground. Diffidence in trespassing on or contesting the property claims of others is very commonly seen, and has a strong influence on the competitive relationships between individuals. The same thing appears in the widespread practice of attacking and persecuting strangers and relegating newcomers to the lowest social rank. This has been reported among domestic fowls, black-capped chickadees (Odum, loc. cit., p. 521), spotted towhees (Tompkins, loc. cit.), vizachas (Hudson, 1922, p. 312), wild rats (Barnett, 1955), the lizard *Anolis* (Evans, 1938), and no doubt many other forms. The converse aspect of the same phenomenon is apparent in Shoemaker's finding in canaries (reported by Allee, 1938, p. 169) that, when they are 'confined in relatively small space, the social order becomes more simple and definite and there is no complication over the question of territorial rights.'

The possibility was suggested by Tompkins (1933, p. 100), as a result of her studies of three species of Emberizidae, that a clear-cut peck-order is perhaps more characteristic of solitary, individualistic birds, such as the spotted towhee ('*Pipilo maculatus*' = *P. erythrophthalmus* ssp.); and what she called 'supersedence', corresponding roughly to peck-dominance, is more characteristic of flocking species, such as the golden-crowned sparrow (*Zonotrichia atricapilla*). Odum (loc. cit., p. 522) points to a similar conclusion with reference to the chickadees, pigeons and budgerigars.

It is well known that many birds and fishes typically spend long periods of time in flocks or schools in a state of fraternity and apparent equality. Keenleyside (1955), as a result of laboratory studies of the rudd (*Scardinius erythrophthalmus*, a cyprinid fish), concluded that 'dominance hierarchies are not found in schools of rudd. All individuals in a group are of equal rank. This also seems to be characteristic of all fish schools.' Some kinds are gregarious throughout the year, but in many others flocking is especially characteristic of the non-breeding period, and as the reproductive season approaches the type of dispersion changes to an individually territorial one, for example in sticklebacks (*Gasterosteus*).

As already suggested, however, it would be incorrect to assume that in amicable and closely co-ordinated flocks the competitive element is non-existent. Not only is the regular spacing of individuals a rather general and prominent feature of flocks, alike in birds and in other groups, but there must also be a common vying for perfection among the performers in the execution of communal manoeuvres, of the kind that reach such polished precision in birds like plovers and sandpipers (Limicolae). Where the number of individuals is large (and in schools of herrings or flocks of finches and starlings it may reach the order of perhaps 10^5 or 10^6), individual meeting and acquaintance are out of the question, and establishing a detailed hierarchy therefore impracticable. Eager submission to marshalled discipline, however, is perhaps generally made in a competitive spirit, in the sense that the performer must strive to attain the standard set by his companions; and the emotional excitement aroused by a communal epideictic display, and the conditioning and stress that result from it, seem likely to spring largely from this source.

The hierarchy can make a very important contribution to the organisation of animal societies; and it seems probable that the size of social units is often partly determined by the consequent need for individual acquaintance, and the limitations imposed by a finite ability to remember 'faces', as it were. Many birds break up into small groups, or even proceed independently, in order to feed. When the status of each individual is known to the rest and accepted by them, actual combat for food and other resources is greatly diminished, and a corresponding amount of energy conserved. Though it has long been the generally accepted conclusion that animal numbers are in most cases ultimately limited by food supply, the amount of fighting over food actually observed in nature is remarkably small: what does take place seems to be largely confined to a minority of the higher animals, where it is probably for the most part incidental to the routine of sorting out social precedence among the individuals concerned.

8.5. *The functions of hierarchies*

Admission to the social group and advancement in social status within it appear to constitute one side of the dual prize to be sought in conventional competition: the other side, never perhaps wholly separable from it, takes the form of concrete possessions. In either event, the reward that comes from success is a matter of rights and privileges. Privilege is an exceedingly desirable prize, since without it the individual can be excluded from food, from reproduction, and from the habitat itself; or, as has already been demonstrated in voles and rats and as may well turn out to be common elsewhere, it can suffer such active persecution as to bring on pathological consequences (Chitty, 1952; Clarke, 1953; Barnett, 1956).

That food-privileges go with social status is well authenticated. The peck-orders of the chickadees and great tits mentioned earlier were in fact investigated by watching the displacement of one marked individual by another at their feeding places; and it was found that a subordinate individual would readily yield its place to a known superior. The peck-dominance of the domestic pigeons was similarly worked out largely by observing them at feeding times. With monkeys and apes much the same thing may be done: thus a dominant rhesus monkey, observing a subordinate stuffing itself with food in another part of the cage, may threaten it and, if the latter does not at once disgorge what it has crammed into its cheek-pouches, the former may proceed to raid these with its fingers (Zuckerman, 1932, p. 234).

As to reproduction, cases have already been cited of the fur-seals, rabbits and sage-grouse where a minority of dominant males virtually exclude all others from participation; it seems likewise extremely doubtful, to quote yet another example, whether the lowly cocks on the outskirts of a blackcock lek are ever approached by a female. In the majority of polygamous animals a proportion of subordinate males is similarly excluded from breeding. In every ape or monkey, according to Zuckerman (ibid.), 'the degree of its dominance determines how far its bodily appetites will be satisfied. Dominance determines the number of females that a male may possess.' In his study of flocks of canaries in captivity, Shoemaker (loc. cit., p. 388) observes: 'it is also pertinent to note that the three highest-ranking males in the flock represent the only pairs to raise any young to maturity.' The latter is interesting, because the attempts of other pairs came to grief in part through the disturbances caused by these 'invading' males. Such cases could be multiplied. The subject is one that we shall have to discuss again more fully in a later chapter (p. 502).

Machinery for grading the population when necessary into 'haves' and 'have-nots' has an obvious and important survival value. If there is a shortage of food, instead of this resulting in a general and uniform debilitation of all the members of the society alike, and perhaps their ultimate extinction, the dominant animals are given a preferred chance of sustaining life and vigour throughout the period of famine, and thereafter multiplying and replenishing the stock. Their dominance behaviour ensures that only as many as the remaining sources can sustain are allowed to partake of the food, and thus automatically the maximum number will survive; the excluded subordinates either perish quickly or emigrate to search for subsistence elsewhere. In a similar way, the successful individuals in reproduction are those that can hold their territorial status in the face of competition: those that cannot must again seek habitats elsewhere or refrain from breeding at all.

The function of the hierarchy, in fact, is always to identify the surplus individuals whenever the population-density requires to be thinned out, and it has thus an extremely high survival value for the society as a whole.

8.6. *The hierarchy in relation to 'social selection' and group-selection*

The hierarchy is a purely internal phenomenon arising among the members of a society, but it can nevertheless enormously affect their individual expectations of life and reproduction. Its establishment places in their own hands, therefore, a powerful selective force, which can conveniently be described as social selection. It is similar in character to the one Darwin believed to apply in the more restricted field of sexual selection (*Descent of Man*, p. 2), and is of such importance to our theme that at a later stage a whole chapter is devoted to it (*see* p. 223).

Dominant individuals, especially males, owe their social position largely to their ability to impress and win deference from others; and the recognition of their superiority presumably depends on the sum of many qualities, some of which can be classed as fiercer or more savage and others as arising from more abstract and refined conventions. Among the former are the possession of fighting weapons, and of skill and determination in offence and defence: among the latter are all the conventional social adornments and accomplishments that confer, in human society, elegance, grace and dignity, or, in the higher animals, the counterparts of these virtues.

Too extravagant a development of the weapons and ornaments by which social status is determined could obviously encumber and impair the chances of survival of the individual concerned, and ultimately even those of the race in which the development occurred: it is to be expected, therefore, that a strong external pressure of selection will always be brought to bear on what is socially admired, insuring that it remains consistent with the efficient discharge of other vital activities.

Attempts have in fact been made by a number of experimenters to try to show whether dominant individuals in fowls, pigeons, budgerigars and canaries were larger, heavier, differed in any of their internal organs, or showed greater ability in learning mazes, than the average of their group. In so far as adults are concerned, each of these attempts has drawn blank. This should not perhaps occasion much surprise, because, in animals already highly adapted to their way of life, if social selection tended to push the development of any character consistently away from the norm, this character would come under increasingly adverse selective pressure from other directions. It seems not improbable therefore that the experimental results confirm our expectation that in these cases social selection has stayed in line, or at least has not come into perceptible conflict, with other selective forces.

A hierarchy could not become established, however, in a society of identical individuals: it depends on the existence of variability—not necessarily genetic—among the members of the group. Variability provides not only the material basis of individual dominance and submission, but also—no less important—of individual recognition. In the characters concerned in social selection, variability may depend partly on age: older individuals may be larger, for instance, and thereby at an advantage over their juniors. But if genetic variability, especially, is to be maintained and not eliminated by social selection, then the status of the individual must be determined not by some single attribute, such as bodily size alone, but by a complex of factors all of which are taken into account. It is interesting to notice that in a small minority of animals, as we shall see later, including a number of spiders (*e.g. Theridion redimitum* and *Meta segmentata*) and the males of the ruff (*Philomachus pugnax*) in the breeding season, individual variability in visible colour and pattern seems to have run riot and, even to our eyes, no two individuals are alike.

It is not difficult to see that the hierarchy itself, and the conventional standards of mutual appraisal that play so great a part in determining the social status of the individual, are group-characters. Once again, no isolated individual can establish a hierarchy by itself, and, where a particular hierarchy exists, the whole of it cannot be manifested in any one of the several individuals that compose it, but only in the group of them collectively. Social characters all share this same collective quality; among them are many arbitrary conventional attributes, such as aesthetic taste and judgment, in which as we know from our own experience the average individual is strongly swayed by popular opinion, and readily adapts his standards to keep pace with contemporary fashion. The actual machinery by which the hierarchy is set up—for example by means of a lek display among the males in some species of grouse—is equally a group-attribute, demanding the acceptance by the competitors of a common conventional method for proceeding and reaching decisions, and always requiring two or more individuals acting in concert to put it in motion.

The reader may still feel sceptical or at least uncertain in mind about how selection could be brought into play at this level, and thus promote the evolution of social characteristics which belong only to a group collectively and not to the members separately. One of the keys to the puzzle is perhaps to be found in a very important principle, common to both animals and plants, namely, that any given stock tends to persist generation after generation in the same locality. The 'group' we have in mind is by no means a haphazard assembly of individuals drawn from far and wide and thrown casually and temporarily into a particular habitat: it is instead a population largely of common descent, self-perpetuating and potentially immortal. We shall find compelling evidence that this is the normal pattern of organisation, in the great majority of species, for reproduction and survival, just as we know it has been throughout human history right up till modern times. The local group is a persistent entity, with its own characteristic complement of genes distributed among its members—genes that are recombined in different permutations as one generation succeeds another, but that on the whole tend to change rather slowly in frequency and variety, considering the group as a whole. These local interbreeding stocks are of course the smallest subdivisions of the species that can accumulate the effects of 'ordinary' natural selection among individuals and undergo evolutionary change; and, although there is normally a trickle of gene-flow going on from place to place, one local stock can be expected to differ somewhat in its gene-complement from its neighbours.

In the case of a social group-character, what is passed from parents to offspring is the mechanism, in each individual, to respond correctly in the interests of the community—not in their own individual interests—in every one of a wide range of social situations. Thus individual members react when circumstances demand by collaboration in communal activities and can be conditioned thereby to respond in the appropriate way.

Under group-selection it is not a question of this individual or that being more successful in leaving progeny to posterity, but of whether the stock itself can survive at all. The homeostatic control of population-density and the conservation of resources are generally matters of supreme importance in deciding the long-term success or failure of local stocks of animals, and if ever the social machinery should break down or prove inadequate the consequences could be disastrous, and perhaps final. Does the epideictic display, for example, result in the expulsion of a sufficient surplus of the population, when food supplies are at a low ebb, to allow the remainder to survive until better days return? If not, it may spell extinction to the group.

All the homeostatic adaptations involved in a situation such as this have been elaborated and improved through the machinery of selection, and all are group-characters: they survive in those groups where they work efficiently, but disappear with the stocks that possess them whenever they prove inadequate. Where a group fails and goes under, its habitat may soon recover and fall into the possession of colonists from neighbouring, still successful groups, whose more effective homeostatic adaptations are thereby spread from place to place, and gradually disseminated through the whole range of the species.

8.7. *Summary of Chapter* 8

1. It is easily seen that animals cannot be divided simply into those that are social and those that are not: sociality is a general phenomenon, appearing more primitively in some groups and more elaborately evolved in others. This has been the conclusion of most previous authors on sociality from Espinas (1878) to Allee (1939).

Social behaviour would appear at first sight to be of survival value in a number of ways, for instance in securing protection from predators or in facilitating mating and reproduction. When these supposed advantages are examined, however, they seem by no means general or unequivocal, and very inadequate to account for so universal a phenomenon. It has to be admitted that no comprehensive or sufficient explanation of sociality has ever been given. Our general thesis, therefore, that it is the basis of conventional behaviour and as such an indispensable part of the machinery of dispersion, merits serious attention if only because it identifies a possible common purpose for the first time.

2. Threats constitute a very common type of communication. In most animals their prime use is intraspecific, as shown by the fact that the organs used to convey them are often not present in all individuals, but only perhaps in the adult males. Sometimes these organs are directly formidable, like the bared fangs of dogs, but often they are formidable only by association, as with the dog's bark or the raised hair on its back. However it may be conveyed, threat implies a power to hurt on sufficient provocation, but it has come to provide a conventional method of repelling rivals without resorting to wholesale bloodshed and fights to the death.

Direct competition can thus assume symbolic forms, and decisions between rivals can be reached by abstract conventional methods. In practice these need not be even as fierce or crude as what we commonly understand as threat, but can be refined into gentler, more pleasurable forms of competition, including play and sport.

Conventions lay down codes of law, especially preventing the subversive self-advancement of the individual: because they are artificial, breaking them incurs no automatic penalty, but only a social one. The possibility of choice, between personal but antisocial advantage on the one hand and conformity with convention on the other, lies at the root of morality and must extend in some degree to rational animals other than man. This is a subject reverted to later in the book.

3. Social competition is further examined in §3. The underlying purpose of the conventions with which we are concerned is to enable food-resources to

be efficiently exploited and conserved. These conventions are group-characters, not apparent in individuals separately but only when they form a concerted group; and such a group, it appears, is the primordium of the society. We can define the elementary society as an organisation capable of providing conventional competition among its members.

The society is required to provide forms of competition that will intensify as population-density rises and economic conditions fall, and thereby supply the indicator that regulates the adjustment of population-density and maintains the proper balance. Thus it may be predicted that social activity will tend to mount as a prelude to any major alteration in population-density, especially just before the onset of reproduction, but also for instance in hard weather or before emigration.

One of the characteristics of social relationships is their antithetic combination of cohesion and mutual rivalry. This is well seen in the typical flock of birds or school of fish, contained in a definite perimeter within which every individual strictly maintans its ' individual distance ' from its neighbours. Each such individual, furthermore, possesses the status of membership in the flock, which is an abstract property, and also the right of occupation of a certain physical space, which is concrete. All the members commonly behave as if they had equal status, but this is by no means true of all social groups, many of which especially in the higher animals reveal gradations of individual dominance and submission.

4. Such gradations or social hierarchies assume various forms, such as the clear-cut peck-order, the less rigid peck-dominance, the oligarchy and so on, and of these examples are given. In practice, hierarchies are often studied experimentally in relation to feeding, since individuals higher in the series have priority in taking food over those below them. Sometimes it is demonstrable that the right to breed is similarly graded, for instance in polygamous species. Not infrequently hierarchies apply primarily to males, the females and young being ungraded or assuming the rank of their male associate. They depend largely on individual acquaintance, and consequently do not develop or show up in large aggregations. But even in a large flock of sandpipers, for instance, where all the members appear to be equal, there is probably a strong competitive element latent in living up to the perfection of performance demanded in flight manoeuvres. Communal epideictic displays are thus able to stress and condition the performers *en masse*.

5. Hierarchies arise from competition for the abstract goal of social rank, but success in material situations (*e.g.* possession of a prized site) may itself often confer dominance and consequently the abstract and concrete goals are not always separable. The reward of success in every case is the securing of privilege, usually either to take food or to breed.

The hierarchy's function (not hitherto identified) is always to distinguish the ' haves ' from the ' have-nots ' whenever the population-density becomes excessive: thus where food is adequate for only a part of the population, the tail of the hierarchy is automatically cut off at the right level and excluded from feeding; individuals of lower rank can be similarly excluded from breeding. The hierarchy can thus be a most valuable tool in population homeostasis.

6. The hierarchy is an internal product of its own members, greatly affecting their individual survival and fertility, and thus placing in their hands a very powerful selective force—described here as ' social selection ' (*see* Chapter 12). This tends to enhance the symbols that best succeed in conventional competition, but it can always be overridden by ' ordinary ' natural selection. Dominant individuals arise in practice from around the genetic norm of the species and are not aberrant individuals. The hierarchy requires, however, that individuals be not all identical—phenotypic variability at least is essential—and the necessary mutual recognition among the members again depends on ' personal ' differences. A few species exist in which even to us no two individuals appear alike.

The hierarchy is a typical group-character, manifested in a collective group but evanescent when the members are isolated, and the same applies to the coventional processes by which it is established. To understand group-selection we ought first to recognise that normally local populations are largely of common descent, self-perpetuating and potentially immortal. They are the smallest subdivisions of the species of which this is true, and can be adapted to safeguard their own future. What is actually passed from parent to offspring is the mechanism for responding correctly in the interests of the group in a wide range of circumstances. What is at stake is whether the group itself can survive or will become extinct. If its social adaptations prove inadequate, the stock will decline or disappear and its ground be colonised by neighbouring stocks with more successful systems: it must be by this process that group-characters slowly evolve.

Chapter 9

Dispersion in the breeding season: birds

9.1. *Prenuptial social activity*

Social competition, as earlier noted (p. 133), may be expected to grow in intensity and reach a peak at or shortly before the onset of the breeding season. The reasons that lead to this prediction are, first that it would seem specially important to attain as nearly as possible an optimum dispersion of the breeding stock before reproduction begins; and second that, in most animals, the size of the breeding effort—that is to say the proportion of females mated, the number of eggs fertilised or young actually reared—can be varied to meet contemporary needs. Social competition, on our theory, is adapted to give a sensitive index of population-density in relation to current economic conditions, and thus to provide the basic information necessary for regulating both the dispersion of the population and the production of recruits.

At present we need to take this only as a tentative proposition, for a full examination of it must necessarily occupy the greater part of the pages between here and the end of the book. In this chapter and the next we are principally concerned with the first of the hypothetical requirements, namely the optimum deployment of the population as a prelude to breeding and the means by which it can be brought about. Chapter 11 is devoted to communal displays, the function of which seems to be most often related to the second requirement—that of providing a density-index that can in turn elicit a reproductive output of appropriate size. In practice, however, the two cannot be very rigidly separated, because the intensity of competition over matters of deployment and property-tenure as a rule reflects closely the existing pressure and density of population; and, conversely, in gregariously breeding species where deployment takes place by groups and not by individuals or pairs, communal display often figures largely in the process of getting the population dispersed.

9.2. *Territory and population-density*

The system of prenuptial deployment that has been most completely studied is the occupation of territories by birds, especially song-birds. In one very common type of case, represented by the corn-bunting, reed-bunting and yellowhammer (*Emberiza calandra, schoeniclus* and *citrinella*) in the British Isles (Howard, 1920) and by the song-sparrow (*Melospiza melodia*) over much of the United States (Nice, 1937), part of the population at least is resident all the year round in the vicinity of the breeding area. The claiming of territory begins intermittently on fine warm days even in the depth of winter. It concerns male birds exclusively, and these, when conditions are suitable, may temporarily desert the winter flock of which they are members, and disperse individually to their prospective claims, where they sing. According to the weather and the time of day, their behaviour can switch from gregarious sex-neutrality to the belligerent isolation of the overt male, with a correspondingly great difference in their demands of ' individual distance '. As the breeding season approaches, more and more time and energy are spent in the territorial activity, and before it actually begins, the deployment of the males is typically completed. They may have acquired mates at quite an early stage and these may aid to some extent in territorial defence, but the female's role is normally quite secondary, and territories are often established by males before they are mated.

Another numerous category includes many of the birds that are wholly migratory. In these as a rule the males return in spring to their breeding grounds in advance of the females, and on their arrival immediately deploy themselves, devoting great energy to the speedy fulfilment of their territorial demands, so that the pattern of tenure may be largely established by the time the females arrive. Again the initial process concerns only the males, and the point should perhaps be emphasised that it is just as distinct and separate from actual reproduction as is, for example, the migration which precedes it.

It is now well known that the scientific ' discovery ' of territory in birds has been made a number of times, but the author who first brought it into general prominence was Eliot Howard (1920). He conceived the term territory as including not only various kinds of dispersed claims like those of the buntings just mentioned, but also the nest-territories of such colonial birds as the guillemot (*Uria aalge*) and house-martin (*Delichon urbica*), and the stances of blackcocks (*Lyrurus tetrix*) and ruffs (*Philomachus pugnax*) at the lek. The nature of the territories occupied by different species, when reviewed, and the uses to which they appeared to be put, were apparently so diverse that he was led to the conclusion (loc. cit., Chapter 5) that territory ' serves more than one purpose, and not always the same purpose in the case of every species ': indeed, after inquiring ' whether there is not some way in which it has been serviceable alike to every species ', the answer seems inevitably to be that there is none.

Among its various functions he mentions that ' territory serves to restrict the movements of the males and to distribute them uniformly throughout all suitable localities ' (the guillemot and ruff possibly not being intended in this), and as a result ' each district has its allotted number of inhabitants, and the waste of energy and loss of time incurred in the process of mating is reduced to a minimum '. Moreover, in the case of the lapwing (*Vanellus vanellus*), ' the territory fulfils its purpose when once it limits the number of males, since, by doing so, it limits the number of families and prevents undue pressure upon the means of support '. Considering evidence from other species also, he concludes that territory ' roughly insures that the bird population of a given area is in proportion to the available means of subsistence, and it thus reduces the risk of prolonged exposure to which the young are always liable ' while their parents are away searching for food.

Howard left a good many loose ends to his theory, and many of his successors over the following twenty years were greatly exercised with trying to provide definitions of territory, but of all those they put forward none was ever completely acceptable (for a review *see* Nice, 1941). This impasse in the matter of definition, which has seemed to be such a serious drag on making headway with the subject right to the present time, arises of course from considering territory in isolation, and from not seeing it as part of a larger and much more general phenomenon—as just one among several categories into which we can conveniently group the systems of animal dispersion (*cf.* Tinbergen, 1957; *see* also fig. 8, p. 100). From our present standpoint, viewing it as an inseparable part of a larger whole, it appears never to have been capable of strict definition, and there consequently remains no further incentive for trying.

The dispersive function of territory in relation to food supply had already been expressed just as clearly by Altum (1868, translated by Mayr, 1935, p. 26). ' All the species of birds ', he says, ' which have specialised diets and which . . . limit their wanderings to small areas, can not and ought not to settle close to other pairs because of the danger of starvation. They need a territory of a definite size, which varies according to the productivity of any given locality.'

Howard's other very notable anticipator, Moffat (1903), approached the subject quite differently: he saw, in the spring rivalry of male birds, competition to secure a holding; success in this would enable them to breed, but defeat would prevent them from rearing a family in the neighbourhood at all. He postulated that this would have a powerful effect in stabilising the breeding population from year to year: whether there had been a large or a small mortality in the preceding winter the total number of birds in the country (*i.e.* habitat) would remain exactly the same, as a result of the exclusion of a surplus consisting of those that could not secure a ' parcel of land.' ' As long as the annual birth rate, or rather *number* of births, is constant, and has been so for a given number of years, it must be balanced by the annual death rate, and further increase of the species becomes impossible.' The germ of the idea at least is here that territory is connected not only with limiting population-density in the breeding-season, but also with limiting the output of young.

Moffat, in this all too little-known paper, must be credited with great insight. He understood the territorial and competitive function of song, and the fact that it was a signal directed primarily towards other males. 'The chief and primary use of song . . . is to advertise the presence in a certain area of an unvanquished cock-bird, who claims that area as his, and will allow no other cock-bird to enter without a battle' (loc. cit., p. 164). The same applies to plumage: 'have we not here some ground afforded us for suspecting that the bright plumage may have been originally evolved as "war paint"? In other words, as a sort of "warning coloration" to rival males, rather than attractive coloration to dazzle the females?' If so, he perceives, and as we now agree, Darwin's theory of sexual selection is wholly untenable. This was thirty years before Hingston (1933) expounded the same interpretation of animal colour and adornment.

In support of his idea of limitation of numbers, Moffat adduced evidence of the existence of a non-breeding surplus from the well-known facility with which birds that have suddenly lost their partners usually acquire another mate. There must therefore be numerous members of both sexes, mature and ready to mate, which are 'prevented from breeding simply by the fact that they have no suitable ground'. How else, he asks, is the existence of these surplus mates to be explained?

A spectacular demonstration of the reality of this phenomenon was obtained in a forested region in northern Maine, U.S.A., where continued attempts were made in 1949 and 1950 to shoot all the breeding birds on two small sample areas, for the purpose of trying to assess the effects of avian predation on a serious outbreak of spruce-budworm. What happened was that new individuals rapidly came into the depopulated areas, so that in the course of four weeks between two and three times as many males were shot as had originally held territories in them. More than forty species were present; and there was much diversity between them, from those which showed no apparent replacement, to the red-eyed vireo (*V. olivaceus*), of which five (1949) or six (1950) times as many were shot as had originally been present on the two areas (Stewart and Aldrich, 1951; Hensley and Cope, 1951).

Moffat's stimulating ideas, so strikingly in advance of their time, have never accorded with the climate of well-informed opinion on the subject: in Mrs. Nice's (1941) valuable review they appear merely 'a little far-fetched', and in Lack's (1954) book, on a theme so similar to Moffat's—the natural regulation of animal numbers—they actually find no mention.

Many other authors have, however, supported the notion that population-density, at least, is limited by territory. Huxley (1934, p. 277) discusses the limits of elasticity of territories, and the apparent fact that they cannot be indefinitely compressed. 'If this view is correct, territorial instinct . . . *will* be one of the more important of the factors determining the population of breeding pairs. . . . Whether it is ever the *final* limiting factor is a theoretical question which it is impossible at the moment to answer. What seems quite clear is that it does, in conjunction with other factors, play a part in determining the actual density of breeding population in those species in which it is manifested.' Of great horned owls (*Bubo virginianus*) according to Baumgartner (1939, p. 281), 'in areas where food and cover are optimum and human interference is negligible the chief limiting factor is, it appears, the definite territorial requirements of the males.' Among those who have expressed similar opinions in more recent times are Dasmann and Taber (1950, p. 161), referring to the Columbian black-tailed deer—'the existence of territorial behaviour has an obvious effect on population density'—and Cloudsley-Thompson (1952, p. 991). In a review, Burt (1949, p. 26) sums it up by saying that, 'whatever may be the biological implications of territoriality, and there are probably many, we can say with certainty that it is a potent factor which serves as a kind of governor in preventing over-population.'

9.3. *Territory size in relation to resources*

That the primary function of territory is to limit the density of population is strongly suggested by the work of Kluyver and Tinbergen (1953) on territory and the regulation of density in titmice. In each of two woodland districts of Holland—Hulshorst and Wageningen—there were more attractive mixed woods with high densities of tits, and less attractive pinewoods with lower densities. In the two types of habitat at Hulshorst the average figures over many years, in pairs per 10 hectares, were: great tit, 5·6 and 1·5; blue tit, 6·8 and 0·8; and coal tit, 5·0 and 2·2. The corresponding figures for Wageningen were: great tit (years 1934-42), 5·8 and 3·1; blue tit (years 1925-42), 2·0 and 0·8.

The authors found, however, that fluctuations in density from year to year were quite small in the mixed woods, but large in the pinewoods. In the case of the great tit at Hulshorst, for example, it can be shown that the coefficient of variation is nearly four times as big in the pinewoods as it is in the mixed. When the number of tits increased, density in the mixed woods rose much more slowly than density in the pinewoods, indicating that some mechanism was buffering the density in the mixed woods.

After considering what this buffer mechanism might be, they are led to assume the explanation that 'great tits are guided during habitat selection by two counteracting tendencies: a preference for certain habitats, and an aversion for densely populated localities.' 'The birds seem to prefer mixed woods to pinewoods, but as the mixed woods become more densely populated, the excess birds settle in the pinewoods. Thus the attractiveness of the mixed wood *per se* is counterbalanced by the repelling influence of the population already present. In one case this resulted in the mixed wood being always filled up to a constant level.'

These findings are exactly in line with our theoretical expectations, and the authors' inferences are almost the same as we should make. 'The biological significance of this principle of density regulation becomes clear, when we ask, what would happen if the birds were guided only by their preference for certain habitats and not by their aversion for crowding. The population then would be concentrated almost exclusively in the attractive habitats. When the population increased the densities of birds in these attractive habitats would become very great. We know that such concentrations of animals are in general unstable. There would follow an unfavourable trend in the rates of reproduction and mortality, eventually accompanied by an exhaustion of the resources of the habitat. On the other hand the surrounding unattractive habitats would be utilised to a very small extent. In the long run the species in question could only maintain a much smaller population than it does in reality. Thus the interaction of habitat preference and the aversion for concentrations prevents the development of top heavy populations in favourable habitats and ensures the utilisation of less attractive environments.' (This and the preceding quotations are from Kluyver and Tinbergen, pp. 285, 286 and 287.)

Some at least of the forces of dispersion can be clearly seen at work here, preventing the density in the good habitats from being forced beyond a self-imposed limit, and driving whatever varying surplus exists to seek its fortune in more marginal habitats. It would appear that Moffat's proposition regarding the exclusion of the surplus from the best country is substantially true of the titmice, though there is indeed no evidence that any individuals have been excluded from breeding entirely. For the great reed-warbler (*Acrocephalus arundinaceus*) in the Netherlands, Kluyver (1955, p. 40) has shown that the preferred habitat is in extensive reed-beds of *Phragmites*, but males sometimes locate themselves in beds of cat-tails (*Typha*) and club-rushes (*Scirpus lacustris*), where the females, from the heavy nature of the materials, cannot manage to construct a nest. He suggests that these males may be an overflow from the optimum habitat, and they are of course effectively excluded from breeding.

Another important correlation emerges from the work on titmice, which we shall be taking up again in a later chapter (p. 507), namely that clutch-size and the percentage of second broods are both depressed by crowding, so that the higher the population-density, the lower becomes the average fertility per pair (Kluyver, 1951, pp. 66 and 79). This of course is also in agreement with the dispersion theory, according to which there should exist generally a cause-and-effect relation between increased crowding and decreased fecundity, and the converse; and, as we shall see later, it agrees also with field studies on other animals and with the massive results of experiments on vertebrates and arthropods, which form the main substance of Chapter 21.

The productivity of a particular habitat generally varies from one year to another, and in extreme environments, on the arctic tundra for example, the range of variation may be very large. Pitelka, Tomich and Treichel (1955) made an important study of the relations between lemmings and their various avian predators, especially pomarine jaegers (*Stercorarius pomarinus*), near Barrow, Alaska. In 1951, when lemming numbers were low, the jaegers did not breed at all. In 1952, with moderate numbers of lemmings, the breeding density of the jaegers (strictly territorial birds) averaged 4 pairs per square mile, over the 7-9 square miles surveyed. In 1953, a high year for lemming numbers, the average breeding density over the area went up to 18 pairs per square mile. ' In the densely occupied area of 1953, minimal territory size was estimated at 15-20 acres, where in 1952, territory size on the most densely occupied area was approximately 110 acres ' (loc. cit., p. 115). Two other interesting points emerged: first, ' nesting success was lower in the dense population of 1953 than in the sparse one of 1952 '; and second, in 1953 but not in 1952 there were some breeding individuals with sub-adult plumage, an observation which has significance in relation to the social hierarchy and to the general question of deferred maturity (see Chapter 23, p. 565).

This example gives a rather striking illustration of the relation of territory-size to the resources of the habitat—a relation which, so far as birds with all-purpose territories are concerned, has in fact already been accepted or implied by most of the authors on the subject. It finds ready support in the commoner, converse experience that stable habitats tend to have constant populations year after year, that similar habitats have similar densities of territorial species, and that the ' carrying capacities ' of differing habitats are not by any means uniform. Lockie's study (1955) of the changes in population-density of short-eared owls in relation to a vole-plague points in very much the same direction. Notwithstanding the intense conservatism of distribution normally shown by animals, such results as these clearly suggest that dispersion can be a continuously dynamic process, capable of making rapid adjustments in a changing or unstable habitat.

It appears possible at this point to reach two conclusions: first, that in birds which behave like the tits the holding of territories can have the effect of imposing a ceiling on population-density in the breeding season; when the population rises, instead of the density becoming greater and greater, in the best habitats it soon approaches a limit, and any surplus is obliged to overflow into less favourable habitats, until these are likewise filled up. Secondly, it appears equally probable that the maximum degree of crowding and the minimum territory-size, which such birds as these are adapted to tolerate, are correlated with the presumptive productivity of the habitat (' presumptive ' because, at the time that territories are claimed in spring, the critical productivity later in the season when the young will have hatched often cannot be directly forecast): this second conclusion is implicit in the differences of population-density found in the two types of woodland in the case of the three species of tits, and in the correlation between territory-size and abundance of prey in the case of the jaeger.

9.4. *Territorial and colonial nesting as alternatives*

In a recent symposium on territory, Snow (1956, p. 446) wrote: ' It is not clear what part territory plays in breeding. The whole population shows essentially the same behaviour, and one can only guess what would happen if some members of the population behaved differently.' He was actually writing of the blackbird (*Turdus merula*), which, as he says, shows just such a uniformity of behaviour. There are many other species of birds, however, in which territory-holding is not an invariable rule, but can alternate in appropriate conditions with colony-nesting. In such cases, nature provides the desired demonstration, and valuable light is thrown on the functional relationship of these two common forms of breeding dispersion.

The types of birds that show such a duality of breeding habits include some species which we regard as ' normally ' colonial and ' exceptionally ' solitary or territorial, and some of which our common experience would be in general the opposite. An example of the former is found in the herring-gull (*Larus argentatus*), which, when it ranges into areas of oligotrophic fresh-water lakes and inland rivers, for example over immense areas of eastern and northern Canada, generally becomes a solitary breeder, intolerant of other gulls anywhere in the neighbourhood. Exactly the same applies to the glaucous gull (*L. hyperboreus*), which in many places occupies isolated eyries year after year like one of the raptors, and yet elsewhere occurs in colonies of considerable size. With these two species the colonies are generally by the open sea or on great lakes, and the solitary sites on narrower waters, usually either inland or on fiords. Many of the other species of *Larus* show the same duality, *L. marinus* being an example of one perhaps better known in the solitary phase than in the colonial. *L. canus* illustrates another slightly different variation in habit, since the nests may be either close together in tight colonies, or (most commonly) scattered at individual distances up to hundreds of yards in a diffuse but nevertheless distinct community, or also (rarely) quite solitary.

The terns provide closely parallel examples to the gulls; *Sterna hirundo* is colonial by the sea and often on great lakes, but usually solitary when it nests inland on river islands; solitary nesting also occurs in *S. macrura*, *Chlidonias niger*, *Hydroprogne caspia* and no doubt some others, as a variant of the commoner gregarious habit.

Many similar examples can be found among the herons (Ardeidae), all over the world, including the palaearctic *Ardea cinerea*, the North American *A. herodias*, *Butorides virescens* (generally solitary), *Nyctanassa violacea*, *Dichromanassa rufescens*, the Australian *Notophoyx pacifica* and *Nycticorax caledonicus*, and many others.

Among the ibises, the Indian black ibis (*Pseudibis papillosus*) is normally solitary but sometimes colonial; and in the common white stork (*Ciconia ciconia*), known to millions of Europeans as a solitary nester, the same thing is true: for example on Khorramabad, Persia, there are said to be over two hundred pairs in a colony (W. Brotherston, orally 1956). Another stork, *Dissoura episcopa*, shows the same inconsistency.

From the Anseres numerous illustrations could be given. The mute swan (*Cygnus olor*), generally a relentless opponent of all intruders on its well-defended territory, has an ancient and famous colony at the Abbotsbury swannery in Dorset, containing between 200 and 500 pairs. The eider-duck (*Somateria mollissima*) like *Larus canus* mentioned above, nests, according to conditions, in dense colonies (usually if not always on islands), or in diffuse communities, and these can be attenuated until they are not distinguishable from isolated nesting. Much the same is true of many other Anatidae, including other swans (*Cygnus* spp.), various geese (*Anser* spp.), scoters (*Melanitta* spp.), the tufted duck (*Aythya fuligula*), red-breasted merganser (*Mergus serrator*), etc.

Examples must also be given of the same phenomena in the Raptores, where their occurrence is very well known. The osprey (*Pandion haliaetus*), most frequently a solitary nester all over its extensive range, has or had colonies, among other places, in East Prussia, and at a number of places along the Atlantic coast of the United States from Florida to Massachusetts: the best known of these is on Gardiner's Island (and formerly on Plum Island) at the eastern end of Long Island, N.Y. In 1879 there were said to have been 500 nests on Plum Island, and 2000 ospreys nightly roosting there (*cf.* Bent, 1937, p. 355). The density of the colonies is very variable: Plum Island may have had something like 20-30 nests per 100 acres, whereas, according to Bent, in southern Massachusetts and eastern Rhode Island a scattered colony of about 60 occupied nests could be covered by a circle 8 miles in diameter—more like one nest in 500 acres, and therefore an extremely diffuse community.

A considerable number of the eagles and vultures associate loosely together at times in nesting communities, sometimes comprising several species. A particularly noteworthy example was the colony of white-bellied sea-eagles (*Haliaetus leucogaster*) which existed a century ago in Jerdon's day on Pigeon Island, off the west coast of India 30 miles south of Honavar: there were at least thirty or forty nests, and the ground beneath them was ' strewed and whitened with bones of sea snakes chiefly, and also of fish' (Jerdon, 1862, I, p. 85). This magnificent bird is commonly as solitary and exclusive as the other members of its genus. Tristram even ' found in one palm grove (in southern Algeria) no fewer than seven pairs of golden eagles breeding together ' (quoted from Jourdain, 1927, p. 72); and Gordon (1955, p. 60) mentions a ' Valley of the Eagles ' in Alberta, reported to him by Prof. William Rowan, where a similar colony apparently exists. Brown (1955, p. 41) describes an assortment of eight species of large raptors which frequented ' Eagle Hill ' in Kenya, and also (loc. cit., p. 119) the loosely-defined groupings of nests of fish-eagles (*Cuncuma vocifer*) on small islets in Lake Victoria. Mr. Eric Simms has informed me that near Aranda da Duero, in Burgos, Old Castile, there was in 1956 a colony of at least 60 pairs of griffon vultures (*Gyps fulvus*); and ' griffonries ' of over 100 pairs have been reported from Palestine (Tristram, 1884, p. 95). This species belongs to the typically sociable category, and solitary nesting is apparently uncommon.

Finally, among the falcons, the kestrel and lesser kestrel (*Falco tinnunculus* and *naumanni*) provide the most familiar examples. Of the former, Altum (1868) says it ' may have ten or even twenty nests close to each other ' in

Germany in the woods adjoining its hunting grounds in the open fields, but the latter species is actually far more inclined to be sociable, and, according to Witherby's *Handbook* (III, p. 32), colonies of 15 to 25 pairs are not uncommon, and have been recorded up to 100 pairs. Merlins (*F. aesalon*) and various other falcons and harriers (*Circus*) show the same tendency.

Such a list could be greatly extended, but there is really no need to confine it to particular species showing a duality of breeding habits: the same situation can be just as well illustrated by different species in a genus, such as the colonial fieldfare (*T. pilaris*) in a normally 'territorial' genus of thrushes (*Turdus*), or the colonial rook (*C. frugilegus*) in the chiefly solitary-nesting genus *Corvus*. Indeed, the fact that some birds hold typical territories, some breed gregariously in colonies, and some have yet other arrangements, is the essential problem which has to be explained.

It seems impossible to avoid the conclusion that, since territorial and colonial breeding are clearly alternatives, they must provide two ways of meeting some single common need, in respect of which their function is essentially the same. Which of the two methods is adopted may depend in part on the circumstances of the particular habitat, and in fact some species show a continuous range of behaviour between the extremes. In terms of the dispersion theory, they appear simply as alternative types of tenure and social structure, differing most markedly in the factor of individual distance, but each is as capable as the other of organising the population into a society and providing the competitive contacts between individuals on which the maintenance of dispersion depends.

It may be recalled that in the earlier chapters on methods of social integration, territorial and communal behaviour very frequently appeared in alternative roles in a great variety of animals, just as they do here. The essential condition in the present case is that the birds should seek to possess some conventional unit of real property, capable of being individually held against rivals, and conferring by its possession full membership of the society, including (in the situations considered in this chapter) a licence to breed.

9.5. *The nest-site as a form of property-tenure*

The precise nature of the property contested for is very variable, and this is one of the difficulties that makes the definition of 'territory' practically impossible. What ornithologists would generally regard as the most typical territory is a self-contained all-purpose area, the boundaries of which have been determined through contest with neighbours, and to which pairing, nesting and food-finding both for adults and nestlings are all confined. Patterns of territories of this kind are usually established *de novo* each year, but may in some cases tend to be more or less similar in successive years: sometimes the nest itself is built in practically or exactly the same place.

Many defended territories do not provide for all needs, and the owners may have to leave them temporarily on foraging excursions. One of the most remarkable known examples of this condition has been described by Swanberg (1951) in the thick-billed nutcracker (*Nucifraga c. caryocatactes*) in Sweden. In some areas nutcrackers depend for their staple diet for almost twelve months of the year on hazel-nuts: pairs mated for life hold perennial territories in the spruce forest, nesting there in spring and using the same area in September and October in which to bury vast stores of nuts, gathered from communal feeding grounds in neighbouring hazel-coppices—so many nuts indeed that they are still being dug up eight months later to provide the principal food for the next year's nestlings.

A very common type of non-selfsufficient holding is simply a nest-site, which requires the possession and defence of only a very restricted area, possibly only the nest itself. Nest-sites may be holes in trees, for example, and in such short supply that they are the limiting resource and put a ceiling on the breeding population anyway, without recourse to any self-imposed conventional restraints. This has been shown to be true of the pied flycatcher (*Muscicapa hypoleuca*) (von Haartman, 1956). But more often there is no such natural shortage of suitable situations, and in that case particular nest-sites often tend to become more or less permanent, and restricted in number by custom or tradition.

Nest-site holdings are frequently grouped together in a dense colony, but they can equally well be widely separated. There is no sharp line between what is a definite colony and what is rather a group of dispersed nest-sites or holdings, such as we find in lapwings (*Vanellus vanellus*), golden plovers (*Charadrius apricarius*), American goldfinches (*Carduelis tristis*) (Nice, 1939), common linnets (*C. cannabina*), or hawfinches (*Coccothraustes*) (Mountfort, 1956, p. 490). It may come as a surprise to find that eagles, such as the golden eagle, are essentially ' nest-site birds ', and that there is in this group a notable absence of boundary skirmishing or straightforward territory-defence. Instead their interest centres on the eyries (*cf.* Brown, 1955, p. 166; Gordon, 1955, p. 38); and, though it would be a mistake to exaggerate the average period of use of any particular site, eyries often remain in being for ten or twenty years and sometimes much longer. Usually, in Scotland at least, there are more recognised sites than there are breeding pairs, and the same ones are not necessarily occupied in successive years. Pairs are believed often to ' own ' more than one site. New eyries are probably built in most cases to replace ones that have become no longer tenable. There appears to be no lack of usable locations for nests, and adherence to established sites is consequently governed much less by necessity than by convention or custom. All the established nest-sites are probably known to each eagle over its wide home-range, and this local knowledge may be of service to them in keeping account of population-density in the area.

Exactly the same applies to many birds of prey other than eagles, including ospreys and various falcons: well-known nest-sites are sometimes tenanted year after year even though the individual occupiers may have been destroyed several times. It applies to white storks (*Ciconia ciconia*), which fight in spring to win or defend a traditional nest-site, but claim no wider realm (Haverschmidt, 1949, p. 34). It may be seen clearly in various geese (including *Branta bernicla*), which resort if possible to a previously established or known site to nest; so do eider-ducks (*Somateria*). More unexpectedly, it appears to apply to some of the Limicolae, including the redshank (*Tringa totanus*), which is ' non-territorial ' (Hale, 1956): in this species particular nest-locations may be reverted to, in some instances anyway, and used intermittently or regularly over a long period of years (E. Balfour, orally, 1956, with reference to Isbister, Orkney). There is evidently a similar tendency in the greenshank (*T. nebularia*) (Thompson, 1951, pp. 134, 135 142), and probably other related birds.

In colonial species nest-site tenure is the general rule. It applies almost invariably to those whose nest-foundations survive from year to year, such as the kittiwake (*Rissa tridactyla*) and some of the other gulls (*e.g. Larus hyperboreus*, mentioned earlier), various gannets and boobies (*Sula* spp.), most albatrosses and petrels (order Tubinares), and flamingos (*Phoenicopterus*) which, however, construct conical mounds of mud rather liable to be washed down again by the rain after the breeding season, so that many require to be built completely afresh each year (Maynard, 1888).

Actually the survival or obliteration of the remains of former nests need not make an important difference. Rooks and herons often repair those nests that remain, but commonly build additional ones. As a rule they are conservative about the location and also about the perimeter of the colony, and it is a matter of repeated observation that rooks sometimes destroy, in February and March especially, peripheral nests that have survived the winter, and even some just newly started (*e.g.* Yarrell, 1871-85, II, p. 296; Moffat, 1903, p. 161; I have frequently watched it myself in Old Aberdeen). This convention applies with particular force to most colonies of sea-birds: nest-sites tend to become crowded to the limit of tolerance, with the result that on the continuous surface of a flat-topped island their disposition may approach a regular pattern, almost like the cells of a honey-comb.

Among land-birds none form larger colonies than the notorious African weaver-birds known as queleas; *Q. quelea* in Senegal, for example, nests in dense colonies, exceptionally occupying up to 400 hectares (over 1,000 acres) and containing millions of nests. The latter are built exclusively by males, which commence their task almost simultaneously and complete it in the course of a few days, under conditions of tense rivalry; and it is only after the claims of the participating males have been consolidated by the completion of their nests that pair-formation ensues. Thereafter both the cock and hen defend the nest (Morel and Bourlière, 1956 pp. 100-103 and 109).

Colony nesting is certainly hemmed in with conventions. The habit naturally depends on the fact that the birds have developed a strong urge to nest in the colony, and not elsewhere, and it is quite evident, under conditions of moderate conservatism at least, that colonies generally have recognised boundaries. To acquire a site of sufficient size and status within the colony area is the surest passport to successful breeding. In birds like gannets or kittiwakes tradition can be very strong; and even the accretion of new sites on the fringes of the recognised breeding area usually proceeds extremely slowly: powerful inhibitions must be overcome before the individual can succeed in reproducing in a new place. This may be true even when there is a substantial excess of adult birds present, and when there is no real physical barrier to extending the colony area. At Cape St. Mary, Newfoundland, I estimated that there were some five to seven hundred 'bachelor' gannets standing apart in the special area reserved for the purpose, at a colony containing about 4000 nests (Wynne-Edwards, 1935, p. 589); and similar surpluses are well known at many other large gannetries, including the Bass Rock, Ailsa Craig and St. Kilda, as well as in New Zealand (*see also* Chapter 23, p. 571, and pl. XI).

If there were not this conventional limitation of the number of acceptable sites, it would be difficult to account for the individual competition that can normally be observed in securing one. Fulmars in Scotland contest for sites and spend long hours in occupation of them, as early as the beginning of November, though their eggs are never laid before May. Guillemots repair to and occupy their ledges on January and February mornings, soon to be followed by razorbills, kittiwakes and gannets, though none of them lay before May either. Even in the autumn, for several weeks after the young have all dispersed, adults of both fulmar and gannet continue to visit their breeding ledges, until late September in the case of the fulmar in north-east Scotland and until the end of October (*e.g.* 28th October 1957) in the case of the Bass Rock gannets.

Any idea that sea-birds are obliged to concentrate primarily in order to exploit the only existing usable locations for nesting, all of which are fully taken up, is quite untenable: there are well-attested cases of the colonisation of new breeding places, for instance by gannets at Noss, Shetland, since 1910; in 1956 there were some 2000 nests on cliffs that held none fifty years earlier. It is perfectly clear that numerous unoccupied but suitable places almost invariably exist at all times, and that habitable sites very seldom constitute a limiting resource: on the contrary, the limitations observed are primarily conventional and imposed by the birds' own code of behaviour.

Such a limiting convention appears to be just as necessary for colonial birds as the corresponding one for ordinarily territorial species, which imposes a minimum territory-size and thereby a maximum population-density, appropriate to the habitat. The one is the counterpart of the other: in the case of colonial birds it is on the number and size of colonies that the population-density on the feeding ground depends.

9.6. *Collective feeding rights in birds*

Colony-nesting sea-birds obtain most or all of their food from the water—that is to say, upon an area that could not by any practicable means be subdivided into separate holdings. The food-territory type of dispersion is consequently inapplicable, and it is an impressive confirmation of the indispensable need of some kind of conventional organisation that it appears always to be replaced by an alternative form of property-tenure, generally a nest-site. There are fourteen or fifteen families of birds, belonging to four orders, containing over two hundred species that regularly obtain their food from the sea during the breeding season, and, so long as they are actually living in a maritime habitat, there appears to be no exception among them to the rule of colonial nesting.

Practically the same holds good for the numerous species of swifts and swallows that obtain their food in mid-air, though the frequent requirement by members of both groups of a natural hole to nest in may interfere with the characteristic formation of close colonies.

Aquatic and aerial birds are not, as we have seen, the only ones with nest-site tenure and undivided feeding grounds. There are numerous more or less terrestrial species of which the same is true, and many of these are also colonial nesters. Some of them show reasonably clearly that population-density is subject to intrinsic control or limitation; that is to say, that the number of occupied nest-sites bears a relation to a definite area of the habitat, and also to its productivity.

As to the last, a general relation between population-density and productivity of the habitat is implied in the normal habitat-selection of all colonial birds. Lack (1954*b*) has called attention to the existence of such a relationship, for example, in the distribution and size of heronries in the Thames valley and the Netherlands. Referring to the latter he concludes: ' . . . the heron is dispersed far from evenly over northern Holland, the colonies being closer together where there are many dykes or rivers, and much larger where there are meres, marshes or shallow inlets.' ' The above type of distribution is, of course, characteristic of colonial birds in general ' (loc. cit., p. 118).

In the case of the rook (*Corvus frugilegus*), the Nicholsons (1930, p. 58) similarly came to the conclusion that food-supply (*i.e.* productivity) was the main limiting factor to be considered in accounting for the local range of the species in the Oxford district. Partition of the habitat among the population is based on the rookeries as collective units. ' A rookery may be considered to be established on a certain area of land or territory which extends around the nesting site. This area is claimed for the colony as a feeding ground from all other rooks. No rigid boundaries exist. There is much poaching from surrounding rookeries where they are not far apart ' (Roebuck, 1933, p. 20). ' Generally speaking, the feeding range of one rookery extends to meet those of the surrounding rookeries, with considerable overlapping at such vague boundaries ' (*the same*, p. 8). These areas are on the whole very stable, and may be maintained by established custom, and avoidance of neighbours, rather than boundary fighting. The dispersion-pattern of rooks in Britain can be followed through the winter also; then the site of one chosen rookery generally serves as a roost for a relatively huge area; these roosts are also associated with definite territories, comprising collectively those of all the member-rookeries (Philipson, 1933; Munro, 1948), and they are very persistent over the years (*cf*. Dewar, 1933; *see also* Chapter 14).

Collective territories have been studied in various other species of birds. In *Crotophaga ani*, a species living in groups of up to 15 or 25 individuals, Davis (1940*a*, p. 183) found in Cuba that the group-territory is ' vigorously defended against other individuals. Although any invasion is immediately repelled the colonies are on good terms with their known neighbours and settle boundary disputes quickly by " agreement ".' In *Guira guira*, a related species belonging to the same sub-family of cuckoos, the situation (Davis, 1940*b*) is more like that in the rook; the group have a much larger territory from which, however, members were on two occasions seen to drive intruders. Within the group-territory the pairs may have isolated nests and may ' defend the area around the nest in a desultory manner ' (loc. cit., p. 480).

These examples reveal with a sufficient degree of clarity that colony-nesters can have a territorial dispersion system which, in conjunction with a ceiling on colony membership, is capable of limiting population-density in a manner exactly parallel with that of typical territory-nesting species. The feeding area belongs collectively to the colony, and it might be expected that its boundaries would be much more difficult to change than in the case of an individual territory, because of the large number of members concerned and the part necessarily played by passive tradition and avoidance, rather than active militance, in deciding them. Within the group-territory, members do not usually nest at random, but only in recognised customary localities, the perimeter of which, though arbitrary, is usually very definite. Within the nesting area, owing to the minimum demands of individual nest-site tenants, there is room for only a finite number of breeding-pairs. As Moffat (1903, p. 160) says: ' The individuals belonging to the community cannot all nest in the space occupied. For example, we have a small rookery, confined to two trees, on the lawn at Ballyhyland. In a spring in which the number of nests in this rookery did not exceed thirty-five, and before any young birds of the year were fledged, I have several times put out of these two trees, by clapping my hands underneath, flocks of more than two hundred rooks. Within the rookery itself, then, there must be non-breeding birds, and there must be competition for space.' The competition for and stealing of nest-materials, so often witnessed in rookeries, are likely to be of conventional importance and to increase in stress, with discouraging effects, when numbers are high.

There seems to be no serious obstacle to accepting, as a working hypothesis, the idea suggested by the dispersion theory that colonies of birds, including of course sea-birds, are capable of discharging the important dispersive function of limiting the number of breeders, and of limiting it also to a maximum consistent with the conservation of the available resources of food. In some sea-birds there may strictly perhaps be no group-territory at all, but instead a general freedom to search for food wherever it may be accessible: if this is so, the balance between breeding numbers and food-resources may affect several neighbouring colonies mutually or collectively, and the need for steadfast tradition to preserve the *status quo* and strict control of recruitment in each one of them, will be all the more essential.

9.7. *Other forms of property-tenure in birds, especially bower-birds*

The all-purpose territory and the nest-site, either grouped into a colony or isolated, are not the only possible kinds of property tenure among breeding birds. Chapter 11 is devoted to communal displays, particularly associated with the breeding season, and in these the avian participants frequently occupy and defend display-stances, the possession of which may be pre-requisite to successful breeding in the same way as possession of a territory or a nest-site in the species so far considered.

Gould's manakin is one of the birds whose communal displays are described in Chapter 11: in this case the males occupy ' courts ', usually 30 to 40 feet apart, each of which consists of a small cleared patch of bare ground. These are the males' persistent property-holdings, occupied for as much as eight months of the year, and evidently discharging all dispersive functions: the males take no part in nesting, and the females visit the courts only for coition.

A development apparently of a parallel but usually more elaborate kind is the bower of the Australasian bower-birds (Ptilonorhynchidae, by some authors united with Paradiseidae). A comprehensive survey of the subject is given by Marshall (1954). In its simplest manifestations, apparently in the little-known black bower-bird (*Archboldia*) and in the toothbill or stagemaker (*Scenopooetes dentirostris*), the male possesses and defends a cleared area on the forest floor, in the latter species roughly circular or oval and between 3 and 8 feet in diameter. In the only instance so far described in *Archboldia*, this ' stage ' was merely strewn with dead fern fronds, but the toothbill cuts through the petioles of large leaves (up to 20 in. × 7 in.) of selected kinds, and arranges up to forty (exceptionally 100) of them face down, that is, pale side up, in a striking manner on the stage. The toothbill is a loud and persistent singer, perching much of the time on a ' singing-stick ' 2 to 10 feet above the stage, and often appearing to answer other males. It appears not to indulge much in visual antics, but guards the stage and replaces the leaves when they have wilted. Stages are used for up to six months, from August to January. Presumably, as in the bower-birds proper, the male takes no part in nidification.

In the more typical members of the family the males construct bowers, built of sticks and other vegetation, almost always on the ground and next to a display-stage, which is often ornamented with coloured objects, flowers, bones and the like. The bowers are highly specific in character: ' a glance at any bower and its decorations will reveal instantly the identity of builder and collector' (Marshall, loc. cit., p. 165). They can broadly be classified into two types, the avenue-bowers and the maypole-bowers.

It may be judged from Marshall's descriptions that bower-birds are not territorial, but that instead the bower has very much the same 'property' functions as a nest-site. In several species the sites of bowers tend to be fixed for a period of years, and, according to the species and nature of the bower, it may be built anew annually, or added to, in the latter case sometimes growing by degrees into a huge structure. It appears probable that there are commonly more mature males than there are bowers, since Marshall found in several species (*i.e.* the satin bower-bird and the toothbill) that when a male was removed, another invariably took its place within a week or so: bowers therefore appear capable of limiting the numbers of 'successful' males in the same way as nest-sites. Bowers are not necessarily very widely separated from one another, though dense colonies of them seem not to occur.

One of the principal forms of competition between bower-owners is stealing ornaments and wrecking each other's bowers. 'During the period of pre-nuptial gonad development neighbouring blue males (of the satin bower-bird, *Ptilonorhynchus violaceus*) are continually aware of each other's movements and raid each other's bowers and wreck and steal the coloured display-objects whenever they are left unguarded. . . . A wrecking rival will . . . hop in stealthily, rather than fly boldly through the open timber. . . . He comes by the most direct route from his position at the time he observes that the bower is undefended. At the bower the marauder works swiftly and silently and tears down beakfuls of the walls and strews them about in disorder. A wrecker rarely completes his task before he is disturbed by the swift swish of wings of the owner. Usually he snatches up a beakful of blue feathers or glass as he flees. He never stays to fight' (loc. cit., p. 44).

This is reminiscent of the plundering of sticks from rooks' nests, earlier mentioned, or of stones from the nests of Adelie penguins (*e.g.* Levick, 1914, pp. 48-49). It is very likely to impose a stress on the contestants, the severity of which is density-dependent, and possibly sufficient to discourage at least some attempts to establish new bowers in 'unauthorised' places, especially by individuals of inferior social status. It may be pointed out, in passing, that on the basis of the currently-held view of the bower, as being functionally connected only with displaying to the female, such persistent victimisation of one male by another, both of them very likely mated birds, is entirely unexplainable.

An interesting aspect of the bower-building habit has recently been pointed out by Gilliard (1956). He noticed, in New Guinea, 'that the species with the most elaborate bowers are often those with the least elaborate plumage. . . . Indeed, the bower in some maypole builders (and perhaps in certain of the avenue builders) may be of such transcendent importance in the behavioral pattern of the species that it nearly or completely replaces the visual morphological signals of the builder, and, through a transfer of the forces of sexual selection to inanimate objects, renders morphological ornamentation superfluous. The aptly named *Amblyornis inornatus*—a teepee builder—is the primary example of this hypothetical phenomenon. The male of this species, which constructs the most complicated and highly ornamented bower known, is crestless and virtually indistinguishable from the female; yet the males of all its close relatives wear elongated golden-orange crests and differ strongly from the females.'

Regarding the adaptive value of substituting a special stage or bower for other forms of property-tenure, it is possible only to offer very tentative suggestions. Birds that participate in leks may, of course, claim more or less continuous possession of individual perching or standing places, and there is no sharp break in the scale between this and the most elaborate developments in the bower-birds. It has an advantage over a nest-site as a place for display and contest, in that it allows the female to incubate eggs and feed young in seclusion—a precaution more likely to be important in a forest-dwelling solitary-nesting species than, for example, in a sea-bird. It may, secondly, have a special value in polygamous species, where a single male has connections with more than one nest. Thirdly, it is just possible that on account of the three-dimensional character of tropical rain-forests there is some advantage, in tree-nesting species, in having the process of dispersion referred to a common plane, the ground. Certainly in many marine organisms living in a three-dimensional medium, epideictic gatherings seem very often to take place close to the sea surface, perhaps for the same reason (*see* pp. 72, 382).

9.8. *Conclusion*

The various different forms of property-tenure we have considered in this chapter share three common characteristics. The first is that all provide in some form or another purely conventional objects of competition. The second is that they are contested for by adults (in many cases only the males) as a prelude to reproduction, and lead to a determinate pattern of local dispersion. The third is that the number of holdings in any given area is conventionally limited, and, since only successful contestants are entitled to breed, a ceiling can be imposed on breeding-density and the surplus (if any) excluded. There is a fourth characteristic, just as important but as yet not so widely demonstrated to be true as the others—though what evidence we have supports it—namely that the ceiling imposed by the different conventions (with a known and explainable exception in the case of the pied flycatcher) is related to the carrying capacity (or food-productivity) of the habitat.

9.9. *Summary of Chapter* 9

1. It can be predicted that social competition will tend to reach a peak shortly before breeding begins, because (i) it would seem to be prerequisite

to efficient reproduction to secure a prior optimal dispersion of the adults, and (ii) if the reproductive output is to be scaled to provide only what is justified by existing circumstances, the conditioning effects of the epideictic 'feed-back' machinery will have to be brought to bear in advance. This chapter relates to the first of these propositions, and only as it affects birds.

2. We begin with the study of 'territory'. Two of the earliest exponents of the territory theory, Altum (1868) and Moffat (1903), both conceived that a mosaic of feeding-territories exclusively occupied by single breeding pairs would limit the population to what was consistent with the available food-resources. Moffat went on to say that males which failed to secure a 'parcel of land' would be excluded from breeding in the neighbourhood. Howard (1920), concerned with the wider complex of property-tenure in general, including *e.g.* nest-sites in a colony, could find no universal common function served by defended possessions of all kinds, but he agreed with most modern authors that some forms of territoriality can limit density and prevent over-population. In some birds non-breeding surpluses have been shown to exist.

3. Kluyver and Tinbergen (1953) found that three species of titmice tended to hold much smaller territories when they nested in mixed woods and much larger when in pinewoods. The total number of tits varied considerably from year to year, but they tended to keep the mixed woods always filled up to a relatively constant capacity, whereas the density in the pinewoods evidently depended on how big an overflow there was from the mixed woods, and fluctuated much more strongly. This implies that tits have (i) a preference for some habitats more than others, and coupled with it (ii) an aversion to being over-crowded, which normally results in the overflow of a surplus.

A more conclusive illustration of the dependence of territory-size on the richness of the food-resources comes from a numerical study of the annual changes in breeding-density of pomarine jaegers, during a population cycle of their staple food, the lemming, at Point Barrow, Alaska.

4. There are species of birds belonging to many families that breed sometimes as solitary pairs and sometimes in colonies; they include various gulls, terns, herons, storks, geese, swans and raptors. Species in the same genus often show a similar duality, as between rooks and crows, or fieldfares and other thrushes. It is not the first time we have found territorial and communal behaviour appearing in alternative roles, and in this case it points to their offering the breeding population alternative methods of meeting a common requirement. This requirement appears to be a system of conventional units of property, capable of being held against rivals, possession of which automatically confers all the necessary privileges in the society including a licence to breed.

5. A very common kind of conventional token possession is simply a defended nest-site, either in a colony or isolated from others. The owners have to leave it in order to feed. Occasionally there is a natural shortage

of suitable sites, but generally their number is conventionally limited by tradition: isolated sites are used year after year, for instance by raptors, some waders and corvids, etc.; and in colonies, even if new nests are constructed annually, they must in some species at least (*e.g.* gannets) be within a recognised traditional perimeter. In some such cases a surplus of non-breeders in attendance but without nest-sites can be identified.

6. Birds that obtain their food from the sea are debarred from holding individual all-purpose territories because their feeding grounds are indivisible. It is very significant therefore that all the 200-odd species concerned, belonging to 14 or 15 families, are colonial breeders without exception. The same is broadly true of the aerial-feeding swifts and swallows, presumably for the same reason.

In colonial land-birds, colony-size is generally correlated with the productivity of the surrounding habitat. Moreover in the case of rookeries, each colony is typically situated within a recognised communal feeding area, not shared with birds of other colonies except in the belt of no-man's-land that may occur along the boundaries (*cf.* fig. 8, BC). In the ani in Cuba, the group-territory is vigorously defended.

Sea-birds perhaps cannot for practical reasons have even group-territories, and in that case the balance between population-density and food-resources calls for very strict control of recruitment, often backed by steadfast tradition.

7. Possession of token property holdings in a communal lek is discussed in Chapter 11, but the bowers of bower-birds are included here because they are individually isolated. The bower is a conventional property held by a male in competition with other males: it occupies a traditional site and the number of bowers is evidently limited by custom. Not all males have bowers, but only those with bowers can breed since females come to the bower for courtship and impregnation. Males often try to wreck their neighbours' bowers.

8. All these forms of property tenure (i) provide conventional objects of competition, which are (ii) contested for as a prelude to reproduction and lead to a determinate pattern of dispersion: all the objects are (iii) conventionally limited in number, so placing a ceiling on breeding-density and leading to the exclusion of the surplus adults if any; furthermore (iv) the ceiling imposed is, probably in all cases and certainly in some, related to local carrying-capacity (or food-productivity).

Chapter 10

Property-tenure in other groups

10.1. *General*

In the birds we considered in the last chapter there is often a sharp contrast in the pattern of dispersion between the breeding season and the remainder of the year: colonial breeders—many of the auks and petrels for instance—become broadly independent and solitary when they leave the nesting area, and individually territorial breeders on the other hand, such as the buntings or finches, may live very largely in flocks during their period of sexual dormancy. Whether the particular species is migratory or not, there generally tends to be some kind of annual cycle of dispersionary changes.

A close parallel to this situation is perhaps to be found only in a few groups of almost equally mobile animals—among the fishes, bats or seals, for example. More commonly dispersion is relatively fixed, and consequently no special re-deployment of the population may be required as a prelude to reproduction. In many of the short-lived invertebrates, however, marked seasonal changes in dispersion can occur for rather a different reason, since they depend on a developmental sequence in the growth of the individual, affecting its mode of life.

Illustrations of dispersionary systems found in animals other than birds have frequently appeared in the earlier chapters, but it is desirable at this point to provide a more general picture of the extent to which the kinds of social organisation found in birds, or others analogous to them, are spread through the animal kingdom. It will be recalled that in Chapter 6 (p. 99 and fig. 8) a simple scheme of the basic patterns of property-tenure likely to be found has already been given. One further general point may be reiterated here, namely that militance in the defence of property—often effected almost entirely without resort to actual force by the substitution of threat as a deterrent—is sometimes still further abated or made virtually unnecessary by an active avoidance, on the part of those concerned, of making any infringement upon the acknowledged property-rights of others.

Avoidance is probably especially important in the case of a communal territory such as is held by a rookery or a tribe of primitive men. It will ordinarily be backed by a fear of reprisals, and no doubt generally tends to produce a no-man's-land between neighbouring claims, and a lack of boundary definition. From this it is but a step to the claiming merely of restricted foci or sites, suitably dispersed, and thence to a common sharing of food-resources.

10.2. *Pelagic animals and others without fixed tenure*

Hitherto we have generally had in mind animals able to hold property-rights because they are living in close relation to the enduring solid ground. No animals are completely or for very prolonged periods aerial, but an immense number of aquatic species pass through either part or the whole of their lives without making contact with the solid substrate. These include the members of the marine and freshwater zooplankton and a large number of other pelagic and nektonic animals, some of them powerful swimmers such as cephalopods, fish and Cetacea. Their wanderings and dispersion need not for this reason be wholly fortuitous or haphazard: they can be and generally are confined within a particular geographical ambit which at least partly depends on their ability to seek out environmental conditions that conform with their specific preferences.

The general want of fixed objects to give a spatial framework to their social organisation may have a bearing on the well-known associations commonly established between small pelagic animals on the one hand and floating driftwood, jellyfish or any other similar large objects capable of serving as bases or rallying points. Some of these associations very likely confer other material benefits as well: the shoals of file-fishes mentioned by Alcock (1902, p. 75) as hovering round drifting logs in the Bay of Bengal appeared to him to be protected from sharks by their ability to match their background. Moseley (1879, p. 434) mentions enormous numbers of small fishes swarming under tangles of drift-wood, where they were harrassed by troops of dolphin-fish (*Coryphaena*) and sharks. In the case of a larger species, the rudder-fish (*Lirus perciformis*), the drift-wood supplies an important source of food in the form of barnacles. Murray and Hjort (1912, p. 98) noticed that under large loggerhead turtles (*Caretta caretta*), caught by hand south of the Azores, there were often quite a number of little silvery fish, which proved to be *Trachurus trachurus*; larger wreck-fish (*Polyprion americanum*) were sometimes also present, and beneath one or two of the turtles there were quantities of blue isopods.

In a very similar way, large jellyfish like *Cyanea* are fairly regularly accompanied by groups of small fishes, often including gadoids and carangids, and hyperiid amphipods. The little fish *Nomeus gronovii* is known to be constantly associated with the Portuguese man-of-war (*Physalia*)—just one or two to each ' host '. The idea formerly held that these little followers were immune from jellyfish stings appears to be unfounded; in fact they must avoid touching the 'live' tentacles, but they have generally been thought to obtain some security from third parties by living in such dangerous surroundings. Bainbridge (1952, p. 111), however, during an observation dive, watched for some time a small gadoid fish associated with a medusa (probably *Cyanea*), and actually it was generally *above* the bell, and made no attempt to conceal itself underneath it. In these cases there seems to be a distinct possibility of the existence of a form of tenure by the followers in their possession of a particular 'site', in the form of an oblivious and indifferent host.

The same thing may hold of the little pilot-fish, *Naucrates ductor*, which associates itself with much larger fish and with ships (provided they do not travel too fast), again usually in ones or twos. Yarrell (1859, p. 229) recounts a story of two that are said to have accompanied the ship *Peru* in 1831 from the eastern Mediterranean until after she came to anchor in Plymouth harbour, the whole period being 82 days. It is generally accepted that pilot-fish enjoy the scraps of food they pick up from their hosts, but 'the fact that some individuals have been caught with their stomachs full of small fishes suggests that they do at times pursue their prey on their own account' (Norman, 1936, p. 245). Similarly with the *Remora* or sucking-fish, which attaches itself especially to sharks, it is suggested that 'on some occasions the shark is used merely as a vehicle to carry the *Remora* to fresh feeding grounds' (*the same*, p. 246).

The question of what exact benefits are being derived from the association does not of course affect the general proposition that the followers quite possibly enjoy property-rights in their hosts, either individually or shared with others, that would otherwise be lacking in their moving and unstable world. Such a secondary dependence on the dispersion of another animal would put them in a relation more or less similar to that of a parasite to its host, a relation that may be seen particularly clearly if we compare the followers just mentioned with whale-barnacles (*Corunula* etc.) or whale-lice (*Cyamus mysticeti*), the last a genuine ectoparasite. Indeed, *Hyperia galba*, the amphipod best known for its constant association with jellyfish, has recently been found to contain in its digestive tract stinging cells derived from its host (*Cyanea capillata*) (Dahl, 1959): possibly it has reached a stage midway between primitive 'tenancy' and ultimate parasitic dependence.

It may be mentioned incidentally that some true parasites are apparently able to pre-empt an individual host or at least to forestall such a degree of infestation as might endanger the host's life or theirs; one may recall examples of crustacean parasites, for instance, such as the rhizopods *Sacculina* and *Peltogaster*, or the epicaridian *Bopyrus*, or the dinoflagellate *Blastodinium*, that are generally if not always found in solitary association with their hosts. Corresponding examples could no doubt be found in a number of marine commensals, as well as in other groups of true parasites.

But although some kind of system of established property-tenure appears to be so widespread and generally valuable in terrestrial and bottom-living animals, it seems clear that nothing of the kind is conceivably attainable by the great majority of free-swimming pelagic species: for them three-dimensional mobility and detachment are the keynotes of their way of life. The usual broad distinction still exists, however, between species normally solitary in habit and those that are gregarious and tend either to be clumped or co-ordinated into definite shoals.

It is in a continuous and practically featureless medium like the free waters of the sea that, in the absence of real property ties, the system of ' travel and encounter ' outlined in Chapter 2 (p. 27) is probably of most importance for social integration. Solitary individuals are presumed to keep in constant or intermittent contact with any near-by members of their own species by means of recognition marks and signals, and thus by observation in their day-to-day travel to obtain a frequency sample of the surrounding population-density. This together with a concurrent sample of the amount of food obtainable constitute all the essential information needed to invoke the dispersive machinery and stimulate whatever adjustment in the population level it is desirable and possible to make. Such a method would be consistent with the existence of the distinctive and elaborate specific recognition signs evolved by many pelagic animals—the luminescent signals and patterns, for example, that in some cases are perceived and responded to by other members of the species, although in circumstances where it may be impossible to relate them to any epigamic or sexual function. The swarming and mass displays that provide the subject of the next chapter, probably including even the vertical migrations of the plankton, will also appear equally capable of performing an epideictic function, and providing the participants with a special opportunity for assessing their density of numbers.

Terrestrial animals, as we know well (*see* p. 100), are generally organised into local, partly isolated populations, and this is believed to be an important factor in their evolution, not least through group-selection. A good many pelagic animals also, including some of the whales and porpoises, and fishes like mackerel (*Scomber*) and herring (*Clupea*) for example, normally live in schools; and, so far as the breeding season is concerned, life-membership in a particular self-perpetuating stock seems on the whole very likely. This might be inferred from what is known about the differentiation of local races in the herring, for example; and, by analogy with gregarious land animals or, among other migratory fish, with the salmon, such attachment to an ancestral breeding stock would certainly be expected. Man of course depends on the predictable constancy both of their breeding grounds and of their movements to and fro in his successful pursuit of special fisheries, and in some of his whaling enterprises.

It is more difficult to understand how a similar self-perpetuation of discrete ' local ' stocks could be achieved in the majority of permanent-plankton species, especially those living in oceanic waters. We are still very much in the dark on some important aspects of their population dynamics, but it seems logical enough to conclude that, since wholly planktonic organisms can have no permanent fixed source whence annual contingents of drifting pioneers set out, never to return, each system must somewhere provide a closed circuit. If so, it is quite conceivable that the nucleus of a particular stock might manage to persist, in spite of continual drift, and in a sufficient number of cases succeed in preserving its integrity for an indefinite succession of tours.

A special chapter is later devoted to the plankton (p. 366), but it may be noted here that one of the well-known characteristics of planktonic behaviour is the patchiness or clumping of species that occurs in almost all the groups of organisms concerned. From the point of view of the dispersion theory it would seem important that these patches should retain their identity, subdividing perhaps when economic conditions permit and even locally uniting, but always demanding continuous social effort to prevent their disintegration. We could indeed predict that the individual stock would have to remain perpetually active and socially alert in order to do this, even though it might be seasonally diminished in numbers, and that it could never rely on 'winter eggs' or similar resting stages to carry it through part of the annual cycle because these would be unable to maintain any social integration and the stock would too easily become scattered and mixed up with others. Certainly it is characteristic of permanent members of the marine plankton, no matter to what phylum they belong, to have rapidly developing eggs; and in one particular planktonic group, the Cladocera, in which winter eggs are widely produced, these are confined to freshwater species and lacking in those that inhabit the sea.

10.3. *The Atlantic freshwater eels*

There remains the anomalous genus *Anguilla* which, as far as the palaearctic *A. anguilla* is concerned at any rate, appears on the face of it to contradict any argument that species require to be subdivided into partly isolated, self-perpetuating units. The genus contains some sixteen species of 'freshwater' eels, all of which descend to enter the sea as they approach maturity late in life, and, embarking on their single spawning journey, become lost to sight. In the North Atlantic region there are two species—*A. anguilla*, inhabiting fresh waters from Morocco and south Russia (rarely) to the Baltic countries, Norway and Iceland; and the American *A. rostrata*, found in the eastern part of the continent from Greenland and Labrador south to the Gulf of Mexico, the West Indies and Brazil. According to present evidence, as is well known, all this vast freshwater range is stocked from a comparatively restricted spawning area in the subtropical western part of the North Atlantic. It is from there that the larvae emanate, to be borne away by the westerly and northerly surface currents: many of them are ultimately carried afar by the branches of the Gulf Stream. The usual surmise as to the adaptive value of the tremendous migrations the adults must undertake to reach this part of the ocean is that it is imperative to make use of the water currents for the transport and dispersal of the young.

The eels are first identifiable as small larvae under 10 mm long, and at that stage occur in a broadly oval region about 500 miles from north to south (30-22° N.) and 900 miles from east to west (48-65° W.) (*cf.* Schmidt, 1922, p. 194). *A. rostrata* can be distinguished from *A. anguilla* from a very early age by having more vertebrae; and Schmidt's analysis of his catches indicated a more south-westerly origin for the former, and a more north-easterly one for the latter.

Recently Tucker (1959) has questioned the validity of the specific distinctions, and suggested that they are purely phenotypic, produced by differences in the temperature of the water in which the eggs happen to have been spawned and to have undergone their initial development. He has postulated further that the whole European range of *A. anguilla* is continually repopulated by eels of American parentage that, by getting on to the right-hand side of the current, overshoot their goal and are carried across the ocean; and that no European eels ever succeed in returning to the spawning grounds.

There are, however, weighty objections to regarding the two eels as conspecific. It is true that they are very alike and that their vertebral counts overlap—but so slightly that, when large samples of each were compared, 99·4 per cent could be identified correctly on this single character alone (Schmidt, p. 182). When the details are plotted side by side in a graph (fig. 12), the bimodality is convincingly displayed. Next, although the larvae are often found mixed in the same hauls in the western North Atlantic, *rostrata* has a much faster growth rate, and metamorphoses into an elver at a different average size, after only one year compared with 2½-3 years in *anguilla.* Thirdly, the European eel, like the Japanese species (*A. japonica*) (*cf.* Bertin, 1956, p. 170), undergoes profound changes before it finally runs down to the sea: the nasal organs and eyes become greatly enlarged, the gut starts to degenerate, and the yellowish skin pigment is lost and replaced by black and silver. The American eel darkens somewhat in colour but shows neither silvering nor structural changes: presumably these are deferred until after it has disappeared into the sea, and this implies a rather deep-seated physiological difference that cannot readily be explained if *rostrata* and *anguilla* are genetically identical. Moreover, in the process of delivering the larvae to their separate destinations on opposite sides of the Atlantic it is difficult to believe that nothing is involved beyond passive drift: from the evidence we have of larval distribution it would appear practically certain, as stated by Schmidt (p. 204-5), that independent purposeful cross-current movement —eastward or westward as the case may be—by the larvae themselves is required to sort out the divergent streams. The evidence from total larval catches indicates that the European form outnumbers the American by something like six to one, and this approximately coincides with what is known about the range and abundance of the freshwater stages and with the returns from commercial catches in the two continents (Schmidt, p. 202).

On the spawning grounds the centres of distribution of the two forms appear to be distinctly different, although there is probably an overlap between their total spawning areas. One would be inclined to assume, in the absence of any evidence either for or against, that each species in its own centre formed a single promiscuously interbreeding mass, and to doubt the possibility that subsidiary self-perpetuating stocks could exist, forming a sort of mosaic within the major spawning areas. Any such segregation would entail that, after a long initial journey to some far-distant continent as a larva and ten years or so spent in fresh water, the individual could navigate back not merely to the general spawning area but to some point singled out within it where it would be reunited with its own close kin. This perhaps sounds improbable, but it is not impossible, having in mind the

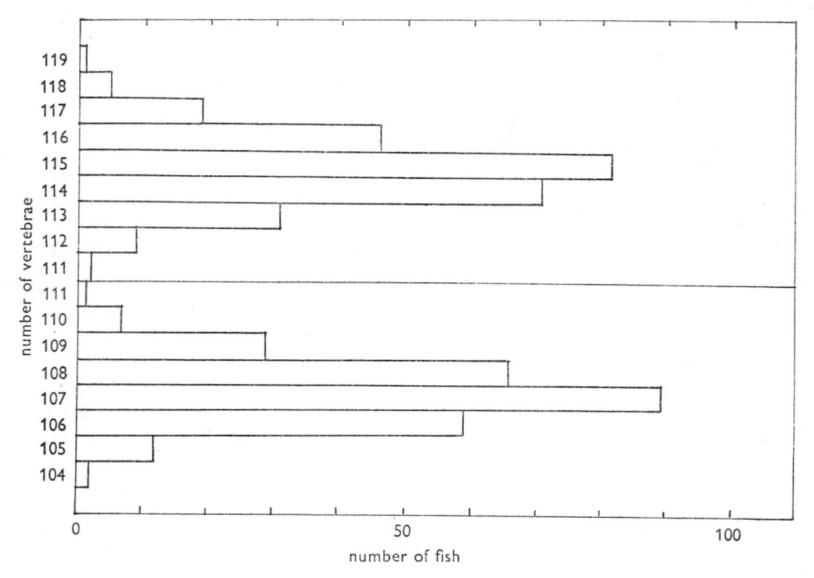

FIG. 12. Number of vertebrae in 266 eels from Denmark (*A. anguilla*) (top graph) and in 266 eels from Massachusetts (*A. rostrata*) (lower graph). (Redrawn from Schmidt, 1922, p. 182.)

ability of birds and salmon (not long ago just as hard to credit) to perform equivalent feats of navigation. Such a segregation would be highly desirable from many angles, one not mentioned hitherto being the possibility it offers of maintaining differently adapted ecotypes for environments as dissimilar as those of Labrador and Cuba, or Iceland and Spain. The observed segregation of the two species might be regarded as a straw in the wind, so far as this ' mosaic ' possibility is concerned.

If partial segregation of local stocks is achieved, it must be done within an area that would roughly accommodate the British Isles or the State of Texas. That the different populations, supposing them to exist, have not become more widely separated than this would allow, and each withdrawn to a spawning ground nearer ' home,' could be attributed to a need paralleled in many other aquatic animals to secure strictly defined conditions for the fertilisation and development of the eggs: none of the other *Anguilla* species breeds far outside the Tropics either. There is also the possibility of allowing a limited amount of genetic interchange—complementary to and no less important than local isolation—when such far-flung stocks as these all foregather in the same general area. Finally there is the presumed advantage of being in just the right region to make maximum use of the ocean currents for delivering the elvers.

Such speculations as these are justified only because the eels appear to be so puzzlingly different from the majority of animals in each seeking—in spite of enormous distances—a single reproductive concourse for the whole species. Perhaps as near a parallel as one could find would be in a bird like the greater shearwater (*Puffinus gravis*)—a pelagic species possibly mustering a million or two individuals at any one time and therefore not very numerous compared with either of the Atlantic eels: it nests only on the islands of Tristan da Cunha, though non-breeders frequent the whole Atlantic north to the Arctic Circle. This in the same sense also ' puts all its eggs in one basket ', but here it would be fully in keeping with what we know about birds (including a few other Tubinares) if there were a considerable degree of local inbreeding due to a tendency in virgin adults to return to the colony or area of the island of their birth, and, once established, to make it their permanent resort (*see* Chapter 19). Possibly too the greater shearwater was once more widely disseminated as a breeding bird, and the Tristan stronghold is all that now remains of its former range. The eel situation, on the other hand, has all the marks of an old-established evolutionary development that has long stood the test of time unchanged, and still remains as indispensable as ever to the thriving success of the species concerned. But presumably the riddle as to whether they enjoy the same kind of partial local isolation as other breeding animals must remain unsolved until (if ever) a deep-sea fishery can be developed for spawning eels.

10.4. *Territory and site-tenure in insects and other invertebrates*

At this point we must forsake the sea and return to the more easily studied terrestrial world. A number of examples appeared in Chapters 2 and 3 of insects whose dispersion involves some form of property-tenure, including the Carolina locust (p. 29) and cracker locust (p. 47), the males of which perform display flights over a particular small piece of ground; the domestic cricket, in which the male chirps from the mouth of its hole (p. 48); some species of cicadas in which the males sing from widely dispersed sites (p. 46), and the territorial pearly-eyed butterfly (p. 30). To those a few further examples may now be added.

In his work on the knapweed gall-fly (*Urophora jaceana*) Varley (1947, p. 153) found that the males ' are usually solitary, and even when they are active they tend to confine their movements within a small area. On 7 July 1936 seven males remained for 10 hr. within a foot of where they were first seen; they were observed at least eight times during this period. The males are pugnacious, and when two males meet they buffet one another with head and vibrating wings, or even grapple together, until one eventually retreats. . . . It is almost as if each male maintains a territory, as do the males of certain birds.'

There is next a remarkably interesting account by Pukowski (1933) of the habits and life-history of burying beetles (*Necrophorus* spp.). In the breeding season both sexes search for carrion lying on the ground, which they proceed to undermine and bury. As Fabre (VI, p. 131) had observed earlier, although several beetles may collaborate in the burial, more than one *breeding* pair are never found at a single corpse. Fraulein Pukowski observed (loc. cit., p. 530) that single unmated males are to be seen displaying in the neighbourhood of their claims, about sunset on warm summer evenings, each taking its stance on a grass-stem or stone nearby, with the head lowered and the abdomen raised and fully extended, a position she describes as 'up-ending' (*sterzeln*): in this posture the scent-glands near the tip of the abdomen are exposed. This in her opinion serves to attract a female. Males, and also the females in the pairs that are mated, will attack and fight furiously with any intruder of their own sex. This ensures that only one pair is finally left in possession, and that there will consequently be a sufficient supply of food for the brood. She points out that the contests also provide an effective form of intra-specific selection, which must not be confused with sexual selection.

She shows that these beetles have a number of other very 'advanced' habits. The adults actually prey and feed, not on the corpse itself, but on the larvae of carrion-flies, and sometimes of dor-beetles (*Geotrupes*), that live on the carrion. The female not only bravely guards her larval brood in their nest or 'crypt' below the burial, but feeds them at first on food regurgitated from her intestine; and for this the larvae may be seen to 'beg' (loc. cit., p. 570; one can see *Vespula* wasp larvae doing much the same).

Necrophorus thus shows a highly specialised form of tenure, connected with the sporadic occurrence of its peculiar sources of food. The connection of this with dispersion, and in particular with securing an optimum population-density in the breeding season, appears to be simple and direct.

Other cases of the defence of breeding-sites, nests or larvae can be found in a number of solitary as well as social Hymenoptera. Among the bees, for example, females of *Bombus* and *Halictus* species, emerging after hibernation, compete keenly for nest-sites; the queen bumble-bees sometimes kill one another (*cf.* Richards, 1953, p. 84; Butler, 1954, p. 26).

Quite a different case, but also exceptionally interesting, is presented by the dragonflies (Odonata-Anisoptera) whose spatial organisation has recently been studied by Moore (1952, 1953) and Jacobs (1955). Male dragonflies have been known for many years to show a kind of territorial behaviour, and to spend much of their time by the water in the vicinity of suitable spawning sites. The females come there relatively briefly for actual mating and egg-laying, and the 'teneral' or immature imagines, in which the sexes are alike, generally stay away altogether, at any rate in the earlier days of their pre-adult period of flight. Depending on the species, the distinctively coloured males either patrol a particular beat or egg-laying site, or occupy a convenient look-out perch, dashing at all intruders. If the intruder is another male of their own species a violent clash may occur, and the number of such interactions largely depends on the number of males present. Successful males tend to become localised, returning to the same site day after day. In the case of *Perithemis tenera*, Jacobs (loc. cit., p. 576) observed that, after the males arrive at the ponds, ' the initial spacing of the day requires little time, for the previous residents take their places quickly. As newcomers arrive, there is considerable wing-fluttering at sites, the appearance being strongly reminiscent of flocks of " dancing " male Diptera. During such periods of violent activity many males leave the pond, and the organization becomes more tranquil.' This tranquility may be disturbed throughout the day as newcomers arrive, but the number of sites or claims is strictly finite, so that many males are driven away to search for other suitable habitats. Vacancies caused by predation are likely to be filled quickly when excess males are present. Some males are driven to take up unsuitable sites, that is, where the females cannot spawn. Jacobs observed that ' males repeatedly unsuccessful in obtaining sites flee readily when dashed at by other males ' (loc. cit., p. 576), a situation full of significance in the light of what has been said about social dominance, and of the general phenomenon of lekking to be described in the next chapter.

Both Moore and Jacobs are satisfied that, as a result of the competition for space, ' once a certain approximate density is reached it is not exceeded ' (Moore, 1952, p. 96). Nevertheless, although the ' dashes of males at one another, pursuit displays, dual flights in *Plathemis*, and wing-fluttering and pursuit flights in *Perithemis*, seem to be forms of male competition,' ' the psychological motivation behind the competition remained obscure ' (Jacobs, loc. cit., p. 581).

It seems clear, however, that we have here a typical example of prenuptial epideictic behaviour, which in the first place imposes a limit on the number of males that have an opportunity to reproduce in each particular breeding habitat; and if there is an excessive number of male imagines emerging, the surplus is expelled as a result of social competition. The holding of territory has nothing to do with hunting or food for the occupying males, since they feed and even roost elsewhere, and all the circumstances suggest that it is concerned with controlling the output of reproduction.

It remains to be established, however, how the limitation effect could be extended to the females and their output of eggs. Moore observed that males in possession of beats dashed at all intruders, including females of their own species. In one of his experiments he dangled live insects one at a time, suspended on a light line from a fishing-rod. Sometimes the occupying male took little or no notice, but most often his reaction led either to a clash or to the assumption of the pre-tandem or tandem position. With males of the same species thus introduced, the result was 66 clashes and 50 pre-tandems or tandems: with females of the same species introduced, there were 8 clashes and 33 pre-tandems or tandems. The fact that in these dragonflies there is little sexual dimorphism such as would generally be expected in species where purely masculine competitions had evolved, and that the territorial males sometimes reacted to intruding females in exactly the same way as they did to intruding males, perhaps indicates that the same machinery can serve to expel excess females also, once a sufficient reproductive output has been secured. Jacobs, with her more dimorphic species, found that the females were attracted to the males, and appeared to be influenced by their courtship activities. Though no special attention was paid to the point, she incidentally records a male chasing females from the oviposition site (loc. cit., p. 577).

We noted in Chapter 7 how the orb-webs of spiders secure for their owners the exclusive use of the site in which they are pitched. Trapdoor spiders, in contrast, occupy burrows, and will not tolerate the presence of rivals in the immediate neighbourhood (*see* p. 115). The individual occupation of burrows, fixed tubes and other comparable structures is common, of course, in almost every major group of animals. It is especially prevalent in the annelid worms, both oligochaetes and polychaetes, and occurs in molluscs (*e.g. Pholas* and *Hiatella*), and Crustacea, for example in a number of amphipods such as *Jassa* and *Ampelisca*, or many of the decapods, including the ocypode crabs already described in Chapter 3 (p. 49). The large long-tailed Pacific beach-crab *Thalassina anomala*, living on grass-covered swampy flats, is remarkable for raising hillocks or towers $\frac{1}{2}$ to $\frac{3}{4}$ m in height containing a rounded retiring-cavity several centimetres wide. In the Moluccas where these facts were observed, the hillocks were in some places so close together as to make the going very difficult (Horst, 1893). This crab is known to stridulate when the tide is out (Pearse, 1914, p. 425). Two notable examples from the insects are the larval ant-lions (*Myrmeleo* spp.), which lie half-buried in their pitfalls, excavated in loose sandy ground, and the larval caddis-worms belonging to the genus *Hydropsyche* (Trichoptera), which spin underwater bag-nets in which to trap their food. In the vertebrates also, species that tenant burrows can be found in every class.

It is likely that in many of these cases the tenure of property is more or less closely interlocked with the process of dispersion and the limitation of population-density, but our knowledge of the behaviour of most of the species concerned is not yet full enough to show whether the site occupied by the individual is competitively acquired.

One such case, however, may be singled out because it has attracted considerable attention, namely that of the remarkable little fiddler-crabs, mostly belonging to the pantropical genus *Uca*. Some reference to their display equipment and behaviour has already been made in Chapter 3 (p. 50). The special feature of fiddler-crabs is, of course, the enlarged, shining, beautifully coloured chela or claw possessed by the full-grown males, which develops asymmetrically either on the left or the right side. The females and young animals carry a symmetrical pair of chelae, each of which weighs only about 8 per cent of the total weight of the owner, but in maturing and adult males the enlarging one undergoes rapid heterogonic growth, until it comes to exceed a third of the weight of the entire animal (Huxley, 1932, p. 9).

Excellent studies of the behaviour of fiddler-crabs have been published, especially by Pearse (1914) and Verwey (1930), and from these most of what

FIG. 13. Male fiddler-crab (*Uca pugilator*) standing over its burrow and signalling with the large coloured claw. (From Pearse, 1914; drawn by Hattee Wakeman.)

follows is taken. There are many species of *Uca*, living in the inter-tidal zone, some on sand and some on mud, some on open beaches and some in mangrove swamps. Both males and females excavate individual burrows, into which they retire very readily, and just before the tide submerges them they carefully close them with a plug of mud. The crabs' food consists of the microscopic living and organic matter in the mud and sand. Many species live in immense colonies. Fiddler-crabs are diurnal in their habits, and when the tide is out the males spend much of their time at the surface beckoning with their semaphore-like chelae. In some species at least (*e.g. Uca latimanus*) there is a readily-observable colour-change in the claw and carapace, which, though dull on the crab's first emergence from the burrow, soon becomes flushed with rich colour (Crane, 1941, p. 155). 'Each species proved to have a definite, individual display, differing so markedly from that of every other species observed, that closely related species could be recognised at a distance merely by the form of the display' (*the same*, p. 152). Not only are the particular shapes and colours of the claw different in each species, but so is the manner in which it is moved. The distal end is turned inwards across the front of the body, and in beckoning it is swung up and down in an arc in front of the crab's face, pivoting on the 'elbow' end, in a way

FIG. 14. Male fiddler-crabs wrestling, their large chelae locked together. (From Pearse 1914: drawn by Tom Jones.)

which, of course, is fancifully compared to the movements of a fiddler plying his bow. Some species wave it in a series of jerks, others with a more continuous movement, some more quickly upwards, others rapping it downwards, and so on. According to Verwey (loc. cit., p. 206), however, signalling is developed only in the densely gregarious species.

Uca stylifera and *beebei*, at least, often occupy their burrows for days or even weeks, without changing the location of the mouth by so much as half an inch (Crane, loc. cit., p. 150). Pearse tells us (loc. cit., p. 419) that 'a fiddler usually does not wander more than a meter or two from his hole, and is ever ready to dart into it at the slightest provocation.' Both males and females have this strong attachment to their individual sites. Pearse snared a number of *U. marionis*, a species found in the Philippines, and removed them to various distances. As he observed, 'it is by no means easy for a strange fiddler to make his way through a densely populated portion of a colony. He is set upon by every crab whose hole he approaches, and may lose his claw or even his life in such an engagement. An individual put down in strange surroundings acts shy and timid.'

They are inveterate fighters. Pearse says the females not infrequently fight with each other, and the males sometimes with females, but the principal combatants are the males. 'Each individual jealously guards the area about his own burrow, and immediately attacks any invader of this territory. His pugnacity is ever ready to show itself against his fellows that swarm about him and against numerous competitors of other kinds that also seek to eke out an existence from the area he has chosen for his own' (Pearse, p. 425). 'In fighting, males face each other, and often dance about excitedly, at the same time frantically waving the small chela. The large chelae are locked together, like two men shaking hands, and each contestant attempts to break off his opponent's claw by a sudden wrench. The strain is so great that when one of the fighters loosens his hold rather than lose his claw, he is often thrown backward into the air, sometimes as much as a meter' (*the same*, p. 421).

Various naturalists have expressed opinions as to the function of the male fiddler's display; most of the time it obviously cannot be immediately concerned with courting females. Alcock (1902, pp. 218-19) recounts that, 'though I spent many a precious hour on watch . . ., I could never see that these combats came to anything: the males always seemed to be in a state of passionate excitement, and the females to be always indifferent and unconcerned; and though the dismembered chelipeds of vanquished males could often be seen lying on the battlefield, I never had the satisfaction of beholding a good stand-up fight fought out to the sweet end, or a female rewarding a successful champion with her heartless person.' Hingston (1933, p. 269) concludes: 'Like other excessively developed weapons, the great nipper is an impediment to the crab. It is no light weight to carry about, also when feeding, the male can use only the little claw. . . . He would certainly be better off materially had he two small claws of equal length. Such a structure can only be an implement of threat.' 'The club is a psychological weapon, and for this reason brilliantly coloured, being a shining pink or a bright crimson that glitters, when brandished, like a fiery torch.' Crane (loc. cit., p. 151), considering the rival claims of courtship and proclamation of territory, rightly concludes that the display serves both of these functions.

Evidently the male's claw is an epideictic structure, employed as the fiddler-crab's chief competitive instrument in maintaining his foothold in the colony and his status in the social fabric. (Crane (loc. cit., p. 152) tells us that crippled crabs are definitely bullied by other crabs.) Like so many other epideictic structures, what serves to exalt the individual male in the eyes of rival males serves also to advance his suit in courtship at the mating season. As we shall see in the next chapter, there is an essential, or one might say foregone, connection between these two functions, since males that are successful or dominant among their rivals are the ones that manage to obtain the property qualification, without which they cannot be permitted to embark on reproduction. It is paramount, in fact, that males emerging successful among members of their own sex should be equally successful in winning mates.

Forms of tenure are likely to be discovered in numerous other groups of invertebrates living in contact with the substrate, but hitherto they have been little appreciated and consequently seldom or never sought. The instance was given earlier (p. 123) of the bivalve *Tellina tenuis* that burrows in estuarine or coastal mud and sand: it appears not to tolerate the presence of other members of its own kind within reach of its exhalent siphon, which it can extend and curl over so as to sweep the surrounding surface of the ground. The marine wood-boring isopod *Limnoria*, on the other hand, is very highly gregarious: once the first pioneers have colonised a new piece of timber, later comers tend, after searching some 4-8 square inches of the surface, to settle near them, causing clumps of infestation, until finally the burrows 'become nearly uniformly placed across the surface of the block' (Menzies & Widrig, 1955, p. 151).

10.5. *Territory and site-tenure in fishes, amphibia and reptiles*

Throughout the vertebrates territorial behaviour of one sort or another is a familiar phenomenon. It is common in fishes. The European brown trout and sea-trout (both forms of *Salmo trutta*) show a conspicuous 'individual distance' of at least three inches from the nearest neighbours, even in the post-larval stage: in contrast, salmon fry (*S. salar*), 'show all the signs of a gregarious nature. . . . In a group of fish at rest, brown trout and salmon fry can be distinguished by the degree of gregariousness exhibited' (Stuart, 1953, p. 36). By the time the parr stage is reached in trout there is 'an indisputable tendency . . . to hold and defend territories, but it was clear that not all parr behaved in this way' (Lindroth, 1955, p. 113). The latter author states later (p. 115) that 'as long as water is flowing over the bottom the fry, trout as well as salmon, stay in their territories.'

At the stage when salmon and sea-run trout are just coming in from the sea on the other hand, before ascending the streams where they breed, each species is well known to have the habit of lying in schools; but again when it comes to the actual spawning period itself, Stuart (loc. cit., p. 17) observes in the trout that 'during the time both sexes are together, the male, who appears to take no part in the selection of site or in the work of excavation, shows unmistakable reactions signifying the guarding of territory.' In most if not all typical salmonids the males evidently engage in prolonged mutual competition on their spawning grounds or redds, and the territories they occupy are probably in the nature of stances in a lek; to this further reference can therefore be appropriately deferred until Chapter 11 (p. 201).

Breeding territories, exclusive to single males or pairs, have been described in various fish, and in fact several interesting examples of territorial advertisement and defence, in the toadfish (p. 72), California singing-fish (p. 72) and electric eel (p. 85) among others, have already been mentioned in earlier chapters. Noble (1938, p. 141) in his review of sexual selection in fishes points out that 'a male stickleback [*Gasterosteus aculeatus*] drives all other males from a comparatively wide area. These other males fail to develop nuptial colours and build nests. In the case of the jewel fish [*Hemichromis bimaculatus*], rarely will more than one pair breed in a tank of 75-100 gallons. The reason for this is that the breeding male and later female are so active in their defense of territory that they keep all other sexually mature fish in a state of agitation, which prevents them from claiming any other suitable spot in the tank for egg-laying.'

A number of the Cyprinidae have also been studied, and the common North American species *Margariscus margarita*, *Notropis cornutus*, *Rhinichthys atratulus* and *Semotilus atromaculatus* are all territorial in the breeding season (Noble, loc. cit., p. 145). According to my own observations of these species, however, all but *R. atratulus* live gregariously, in typical minnow-schools, except when spawning. With the approach of the breeding season the males acquire so-called pearl-organs or nuptial tubercles on their heads, and, except in *Semotilus*, very gay nuptial colours. In yet another North American species of minnow, *Pimephales promelas*, where the eggs are laid on the underside of a lily-pad, they are fiercely guarded by a territory-holding male (Wynne-Edwards, 1932). Nevertheless, there is at least one cyprinid, *Chrosomus erythrogaster*, in which the males are brilliantly coloured, but are said neither to fight nor hold spawning territories (Noble, loc. cit.).

In the related family Catostomidae, the suckers, Reighard (1920, p. 31) found that 'no male occupies any particular locus of the spawning ground and attempts to defend it against other males. On the contrary each male is free to wander over the whole spawning ground. He may "pair" in any part of it, for he does not enter into combat with other males but cooperates with them. The female does not restrict her activities to any part of the spawning ground. Were she to do so she would finally be beset by so many males that spawning would be difficult or impossible. She does not actually reject any of the males, but when beset by so many that spawning is difficult, she seeks a new locus.' The relation between the sexes is here promiscuous. There is probably a collective or communal territory for the whole group, as in many other gregariously spawning fishes, in which all members share their property rights: the species investigated by Reighard, *Catostomus commersonnii*, *C. nigricans* and *Moxostoma aureolum*, are bottom-feeders, with a tendency to be gregarious at any season.

Attention has been drawn earlier to the same alternative types of breeding dispersion in Amphibia and reptiles. Most temperate-region frogs are strongly gregarious at spawning time, but a few, such as the North American bull-frog and green frog (*Rana catesbiana* and *R. clamitans*) have solitary, site-holding males (*see* p. 52). In the reptiles, instances of individual site-tenure, both in the form of excavated lairs in river banks, and in traditional nest-sites, have already been cited for various Crocodilia (p. 53). Evans

(1938) has shown that the lizard *Anolis sagrei*, which he studied in Cuba, exhibits territorial behaviour: non-breeding individuals of either sex may occupy exclusive areas, and during and after the breeding season they may often be dispersed in pairs and family groups. Each territory is on the average 400 sq. feet or so in area, and contains a suitable look-out or display perch, in the form of a tree, stump or fence-post.

On the other hand, various reptiles are gregarious at breeding time, usually resorting to traditional sites, as for instance the rattlesnakes (*Crotalus* and *Sistrurus*); these frequently breed in scores in special dens where they had gathered the previous autumn to hibernate, and where they mate after emerging in spring. Well-known examples of gregarious breeding are also found in turtles. Bates (1863, II, p. 247) describes how the hordes of the common neotropical freshwater species *Podocnemis expansa*, after pairing in the water, emerge at night on to certain traditional sand-islands of the Amazon and its tributaries to lay and bury their eggs. In his own words, 'the pregnant turtles descend from the interior pools to the main river in July and August, before the outlets dry up, and then seek in countless swarms their favourite sand-islands; for it is only a few praias that are selected by them out of the great number existing.'

Most if not all kinds of sea-turtles similarly breed and 'nest' at special traditional places, sometimes making long migrations to reach them. The 'most valuable reptile in the world,' *Chelonia mydas*, is the one about which most is known, but members of the other three genera of Chelonidae as well, namely the hawksbills, loggerheads and ridleys, apparently have similar gregarious breeding habits. Human exploitation has reduced these great reptiles to a fraction of their pristine numbers, but two or three centuries ago there were far-famed turtle-nesting beaches on such places as the coast of Nicaragua, the Florida Keys, Cayman Islands, Ascension Island, and many localities in the Indian Ocean, to which hundreds or thousands of turtles annually resorted to breed (for references, *see* Carr, 1952). A million eggs a year of *C. mydas* are still harvested on small islands off Sarawak, Borneo (London *Times*, 1 October 1957). There is some indication also that not rarely (*e.g.* in Florida) two or more species regularly frequented the same beaches, in a manner recalling the familiar situation in colonies of sea-birds or seals. It is interesting that some of these places are still resorted to year after year, but not enough is known to show whether anything in the nature of site-tenure exists, and probably where there is an epideictic display it takes place in the water. The males do not as a rule go ashore.

10.6. *Territory and site-tenure in the Pinnipedia*

The Pinnipedia (seals and their relatives), being sea-going mammals which come to land to breed, have habits in some important respects resembling those of sea-turtles and sea-birds. Their breeding places, if they are left unmolested, are resorted to annually for indefinitely long periods. The best-known seal 'rookeries' are those of the fur-seals, and especially of the northern species *Callorhinus alascanus* on the Pribilof Islands in the Bering Sea. Like the southern fur-seals (*Arctocephalus*), the sea-lions (*Otaria*, etc.), the elephant seal (*Mirounga*) and the grey seal (*Halichoerus*) the northern fur-seals are polygamous. The landing beaches are divided into territories won and held by combat among rival bulls, and as a result all the females are fertilised by a relatively very small number of successful or dominant males. The main reserve of males consists of bachelor bulls which have little if any share in breeding, although, it should be noted, they nevertheless assemble and make their presence felt at the rookery.

Professor R. Vaz Ferreira (1957, unpublished) has studied the longest established colonies of *Arctocephalus* and *Otaria* on the coast of Uruguay and drawn attention to the contrast in property-tenure between these two types, for instance at Lobos Island (*i.e.* ' seal island ') where both are found. In *Arctocephalus* site-tenure is exceedingly strong and the bulls will endure almost any exposure to storms and breakers in order to hold their ground, whereas harem maintenance is quite secondary. In *Otaria*, on the contrary, the bulls compete primarily for territory only at the beginning of the breeding season and thereafter each successful bull's first preoccupation becomes the herding of his females, in the face of a constant pressure of interference by sub-dominant or younger males. In spite of his vigilance, copulation by such non-territorial interlopers is frequent in this sea-lion. The sex-ratio among those present on land is estimated at 1:6 and about 1:7 respectively in *Arctocephalus* and *Otaria*. I am grateful to Professor Ferreira for allowing me to quote these observations.

Seals of the genus *Phoca*, and some others, besides the walruses (*Odobenus*), appear to be monogamous, and not to hold breeding territories. Most of them have well-developed communal activities, however, both in the water and ashore: harp-seals (*P. groenlandica*), for example, very characteristically rear their head and shoulders high up out of the water as they chase and gambol together by the hour in the arctic fiords, and often leap clear of the surface. Common seals (*P. vitulina*) spend much of their time ashore drying off and sleeping, especially when the tide is low, at all times of the year, and they usually assemble for the purpose, sometimes up to the number of scores, at certain customary spots. Some of the other pinnipedes have great traditional hauling-out places, sometimes quite separate and far removed from their breeding-haunts. The Atlantic walrus (*Odobenus rosmarus*) is a good example: in northern Hudson Bay there are well-known summer and autumn gathering places on suitable islands, where thousands assemble together, as at Seahorse Point on Southampton Island, Walrus Island, Bencas Island and Coats Island in the same region (Dunbar, 1956, p. 184). According to Freuchen (1935, p. 238) ' Walrus Island . . . owing to the swift currents is never frozen over, but had walruses all the year in very large numbers.' He mentions (p. 237) a similar assembly-point near Button Island at the east end of Hudson Strait, and observes that at Salisbury Island there is a resort exclusively used by cows and calves, 'in thousands upon thousands,' while the bulls are gathered on near-by Nottingham Island. The main purpose of these social gatherings, involving as they usually do a high proportion of immature and non-breeding animals, may quite possibly be epideictic. Unfortunately the use of modern powered boats and rifles makes it much easier to kill walruses at sea than to recover them from the water, because they quickly sink when shot, and the numbers in the Canadian Eastern Arctic have greatly declined in recent years.

Before leaving the pinnipedes, it is interesting to recall that, just as we would expect, recent individual marking experiments have shown a strongly-developed homing instinct or 'Ortstreue' in the Alaska fur-seal: the cows are known to come back to the same rookery and even the actual site where they were born (Dr. V. B. Scheffer, personal communication). Parallel adaptations, all conferring correspondingly marvellous powers of navigation, are probably found in all the vertebrate long-distance migrants: brief reference has already been made to the proven instances in fishes and birds, and in Chapter 19 much fuller consideration will be devoted to this important development. As stressed earlier in this chapter, it is a most valuable cog in the machinery of population-control, allowing each local group, notwithstanding the seasonal migration to far-away regions, to maintain its own integrity and govern its own recruitment, so that haphazard dispersion and fortuitous replenishment of the population becomes only the exception and not the rule.

10.7. *Property-tenure in land mammals*

A number of examples of property-tenure in land mammals have been given earlier, especially in the chapter on communication by olfactory signals (*see* p. 101). In the Carnivora it is common to find a form of dispersion in which independent and solitary families or individuals share a home range on terms of limited mutual toleration with other independent but known individuals or families. This is often associated with a network of trails or musking places on which scent is punctiliously deposited, the individually-recognised signal apparently serving alike as a passport and a registration or assertion of rights.

The signal-posts are not always olfactory ones. Sometimes they are howling places, where individuals gather, usually at dusk and dawn, for instance in certain bats (p. 55) and wolves (p. 60). Sometimes they are visually marked; and by way of a preface to the following example it should be pointed out that all the Felidae, large and small, appear to share the well-known habit of sharpening their murderous fore-claws by first embedding them, usually in soft wood or bark, and then drawing them out in such a way as to peel off any thin strips or laminae from the surface that have become loosened or desquamated and are ready to flake off, either on the top of the claw or along the sides and thickened margins: in domestic cats a trumpet or slipper-like sheath is frequently shed from the tip and more or less of the upper surface, all in one piece. By this very simple adaptation the claw is continually kept sharp in roughly the same manner as a pencil is repointed with a penknife—a fact that though doubtless known for centuries to observant people, has generally evaded zoologists. Even lions 'sharpen their claws on tree-trunks, and pieces of bark may be seen to have been thus detached at four or five feet from the ground' (Stevenson-Hamilton, 1947, p. 135). It is noticeable, incidentally, that the hind-claws of domestic cats are usually less sharp, and do not appear to be subjected to the same treatment.

The male lion makes his presence known to other lions, as we have seen (p. 60), by frequent roaring, but in the jaguar (*Felis concolor*) at least, to judge by the following experience related by Darwin (*Jour. Res.*, Chap. 7), a ritualised form of claw-sharpening appears to have been developed to serve a similar epideictic function. He says: 'One day, when hunting on the banks of the Uruguay, I was shown certain trees, to which these animals constantly recur for the purpose, as it is said, of sharpening their claws. I saw three well-known trees; in front the bark was worn smooth, as if by the breast of the animal, and on each side there were scratches, or rather grooves, extending in an oblique line, nearly a yard in length. The scars were of different ages. A common method of ascertaining whether a jaguar is in the neighbourhood is to examine these trees. I imagine this habit of the jaguar is exactly similar to one which may any day be seen in the common cat, as with outstretched legs and exserted claws it scrapes the leg of a chair; and I have heard of young fruit trees in England having been thus much injured. Some such habit must also be common to the puma, for on the bare hard soil of Patagonia I have frequently seen scores so deep that no other animal could have made them. The object of this practice is, I believe, to tear off the ragged points of their claws, and not, as the Gauchos think, to sharpen them.'

As to the last point, the gauchos may in fact have been just as near the truth, but when we compare the jaguar's habit with the earlier-recounted habit of bears (*see* p. 103), which not only claw but bite and slaver on their ceremonial trees, it suggests a nearly identical form of signal marking; and of course, in Darwin's narrative the gauchos themselves made use of it precisely as such, in order to discover whether there was a jaguar in the neighbourhood: it is reasonable to presume that another jaguar could do the same.

In the mammals it is also common to find individually held sites within the communal home range, just as we have seen in so many of the other animal groups, and these sites are, as usual, often conventional and very persistent over the years: two examples must suffice. According to Blanford (1889-91, p. 61), 'a well-known habit of wild animals, but especially marked in the case of the tiger, is the regularity with which haunts are selected in preference to others that appear equally well suited. Some one patch of high nul grass . . ., one dense thicket . . ., one especial pile of rocks amongst hundreds along the hill-side, will be the resort year after year of a tiger, and when the occupant is shot, another, after a brief interval, takes his place.' The second example is that of various species of hares, such as *Lepus europaeus* and *L. timidus*, that habitually lie by the hour in 'forms' or nest-like hollows in the vegetation. Those of the mountain hare in Scotland are often among the heather, and quite frequently they can be seen to be old from the way the thick woody heather stems have grown into shape around them. I have been informed by Mr. Edward Balfour that in Orkney, in permanent pasture land, he knows of forms of the common hare that have been constantly used over a period of twenty years since he was a boy: furthermore, when a hare has been bolted from a form and shot, others can afterwards be repeatedly shot from the same form at intervals, even during a single season. No doubt in a non-uniform habitat some forms and lairs are much preferred to others, and may even carry a higher social status, but undoubtedly the facts suggest once again that there tends to be an artificial restriction on the number of acceptable sites, which results in a density-dependent demand for the use of them and could therefore help to impose a conventional ceiling on the density of population which the hares were prepared to accept.

Turning to the more sociable mammals, some are of course much more sedentary than others, especially those that live in warrens, such as the European rabbit (*Oryctolagus cuniculus*), the vizcacha (*Lagostomus trichodactylus*), the prairie-dog (*Cynomys ludovicianus*) and various ground-squirrels such as *Citellus parryii* and *C. richardsonii*. Their possession of permanent underground dwellings, which continue on the same site from generation to generation, in itself provides a stable spatial framework on which their dispersion throughout the habitat can be based. There must as a rule be contact between neighbouring colonies, but the individual, once it has passed the adolescent stage, is no doubt normally identified with one particular community for the rest of its life. This has been indicated a number of times through individual marking in the European rabbit; and even within the warren, individuals come to be associated with particular parts of it. Thompson and Worden (1956, where the original references may be found) quote Southern's observation that the three or four dominant (and polygamous) males seemed each to have its own 'sphere of influence' in a different quadrant of the warren; and 'whereas the adult males were particularly conservative, the young males tended to move about through two of the quadrants. Young does were found to move about even more than the young males' (loc. cit., p. 106). Further, 'Southern noted that territorial conservatism was very marked among adult does, which tended to have their own particular stations, usually in the form of a patch of bare earth astride a burrow. This conservatism was combined with fiercely aggressive behaviour to other females, to young rabbits and sometimes to bucks' (loc. cit., p. 105). 'The hostility of these does to young rabbits generally, and to young does in particular, was clearly a factor in reducing the swollen summer populations to limits where the characteristic type of organisation could again operate ' (loc. cit., p. 83).

With captive rabbits in two-acre enclosures, under conditions closely approaching the natural ones, Myers and Mykytowycz (1958) similarly found that there was a strong tendency to subdivide into communal groups when numbers were high, each group occupying a well-defined area. ' A group's " territory," like its membership, is well recognised, and is also generally respected. During the breeding season the adult males are virtually confined to their group territory, and defend its boundaries. A dominant buck becomes subordinate in his relations with bucks in other groups if he enters their territory. High-ranking females seem to have the same proprietary feeling for the group territory as the bucks. . . Subordinate does and younger rabbits are less strictly confined to the group territory, and play no significant part in its defence. The urge to defend the group territory, as such, fades when breeding ceases, but intruders from other groups may be individually resented.'

The same authors go on to explain that ' within the group territory, individual rabbits possess home ranges within which they rest, feed, and reproduce ': these overlap to a considerable extent, and are not defended. ' The home range is merely a piece of land that the animal knows well, and to which its main daily activities are restricted. In males, the size of the home range depends upon status in the peck order, that of the dominant buck being virtually the same as the group territory.'

It needs but little exercise of one's imagination to conjure up a surprisingly anthropomorphic picture of these rabbit-warren societies in which individual temperaments and aspirations, rivalry, seniority, and the appropriation of place and privilege assume an importance, no doubt quite un-selfconsciously, not any less vital to success and survival than they do in our own.

Of the sociable ungulates some at least have communal territories within which they spend much or all of their lives, though they may be continually moving from place to place within them so as to take advantage of the best that the hour or season can offer in the way of cover or grazing. Segregation of the sexes, except in the rutting season, is not uncommon, either among the deer or the Bovidae (such as the yak, various wild sheep and goats, the ibex and tahr). In the red deer (*Cervus elaphus*) in Scotland, as a rule, the stags and hinds have their traditional grounds: this particular glen or shoulder is year after year the resort of a herd of hinds and calves, and that corrie or hillside is just as regularly known to harbour stags. In the rutting season solitary stags, probably for the most part younger mature beasts, often travel alone over the barren summits of the Cairngorms leaving their footprints in the autumn snow on their way from glen to glen: they have perhaps abandoned their own community in the face of competition and are on their way to try their luck elsewhere.

The dispersion of gregarious species such as these ungulates, in which the individual claims no private property but enjoys an unspecified share in the communal territory of the herd to which it belongs, resembles that of the schools of certain fishes such as *Catostomus*, considered earlier in this chapter (p. 180). So long as the group is capable of regulating its numbers through control of the birth rate or, by means of the social hierarchy and competitive pressure or persecution, expelling any unwanted surplus, communally-held territories constitute a system of dispersion as practical and efficient as any other. Writing of the Serengeti National Park in Tanganyika, Pearsall (1957, p. 112) comments that 'the absence of overgrazing effects in areas with large game concentrations is a matter of great interest and worthy of detailed study. It appears to be associated partly with the widespread and continued steady movement of the animals from one pasture to another.' In their study of the Columbian black-tailed deer (*Odocoileus columbianus*), Dasmann and Taber (1956, p. 161) concluded that 'the existence of territorial behavior in deer has an obvious effect on population density'; and the following account also throws light on an important aspect of the same situation in the case of the brindled gnu or blue wildebeest (*Gorgon taurinus*). 'Each herd appears to have its own well-marked feeding ground, and is seldom found far away from it while pasture remains, except on its journeys to and from water. Any infringement of grazing rights appears to be strongly resented, and I was once witness of a most interesting episode, when the herd bull of a certain troop chased a party of invaders back on to their own ground on the other side of a small stream, returning quietly to his own party so soon as his duty was done. Not the least remarkable phase of the incident was the sense of wrong-doing exhibited by the trespassers, which displayed not the smallest tendency to offer any resistance' (Stevenson-Hamilton, 1947, p. 83). We have come across one closely parallel case of this before in the cuculine bird known as the ani (*see* p. 159).

10.8. *Dispersion and property-tenure in primitive man*

In man the primitive systems of property-tenure are essentially the same as those already described. Territorial claims are, or were, vested in the nation, the tribe, the family or the individual, and not infrequently in various ways in all these categories at once. After a comprehensive review of the many customs and practices related to the control of numbers in every primitive hunting and agricultural people of whom we have knowledge, Carr-Saunders (1922, p. 203) reached the significant conclusion 'that these races are all, without exception, divided into groups which are strictly limited to definite areas—contrary to the still common notion that they wander where they please . . .' In the next paragraph he reiterates: 'it has been found that among all these races, without exception, groups of men are recognised as, if not owning, then as enjoying the usufruct of certain very clearly defined areas.'

The general types into which we have loosely grouped systems of tenure are all more or less recognisable in human societies. The collective holding, with communally shared rights, is of course represented by the tribal or national territory: it is simple and clear in the case of nomadic herdsmen or hunters, who move in a band together through the country that belongs to them, hunting their game or following the movements of the herds, wild or domestic, on which they depend. In most human societies, however, the tribal ground is subdivided into component holdings, though there may be certain specially localised resources of a tribal character, such as quarries for tool-making stone or metals, that are communally used. The tribal territory also serves to incorporate or unite a typical, partly-isolated interbreeding group, capable of competing as a tribe with and exerting pressure on other groups, both in peace and war, and in consequence conspicuously subject to group-selection, the effects of which are measurable in terms of tribal fortunes. The extreme facility with which local languages and fashions arise and create barriers to marriage and mutual understanding suggests that these may be an adaptation helping to promote the subdivision of the species into partially isolated groups exactly comparable with those we find elsewhere in the animal world. There are, as we shall see, many other characteristically human traits and customs apparently having an equally primitive ecological origin: for instance, the tendency for young adults to disperse the species, or, in human terms, to go out into the world to seek their fortunes, while their elders are not infrequently prepared to risk their lives rather than be dislodged from their established homes.

Primitive human nations like modern ones are, as just stated, usually subdivided into subsidiary territory-holding units. 'The fundamental unit of an Australian tribe is said to be a "local group": this may be identical with a family or it may include two or three families; it possesses exclusive rights over a well-defined hunting ground. A number of these groups occupying a definite territory may form a clan and a number of clans form a tribe' (Sollas, 1924, p. 290). Much the same kind of situation was formerly found in the Bushmen, Eskimos and American Indians. In the primitive Veddahs of Ceylon the whole country was not only elaborately divided into small hunting regions, of which each family possessed one, but 'the size of the tracts varied in accordance with the goodness of the land' (Carr-Saunders, loc. cit., p. 207), so that territory-size was related by purely conventional agreement to productivity and carrying-capacity, in a manner already familiar to us in birds.

Under normal environmental conditions such territorial systems lead not to over-population and starvation, but to a close control both of family-size and the exploitation of resources. Writing of the Australian aborigines, for example, Semon (1899, p. 233) reminds us that, 'if the hordes of a tribe are to live near each other in peace and amity, it is necessary that the number of the populace remain stationary. Provided the hordes increase, it would grow impossible for all to exist upon the yields of hunting and fishing and upon the produce of the wild-growing plants. As things are, the land is able to nourish only a scanty populace, so that we must regard it as fitting accommodation if the Australian tribe tries by artificial means to prevent the growth of the tribe, and thus render the population stationary. Some of the tribes attain this by exposing or killing a certain number of new-born infants. Others castrate a number of the youths as soon as they are grown up and before they enter the class of adults, or render them infertile by slitting the urethra. . . .' In comment on the last of these operations, apparently performed only during the elaborate ceremony initiating youths to manhood, it should be added that other authorities have doubted its ability in practice to affect reproductive power (*cf.* Carr-Saunders, loc. cit., p. 138).

The benefits of territory, so far as the exploitation of resources by hunting is concerned, have many modern illustrations, for example in the exercise of private rights to shooting and fishing for sport, in order to preserve the game and prevent the extermination so likely to ensue when the grounds or waters are freely thrown open to all comers. Not many years ago in Canada some provinces had already adopted in the fur-trade country systems of 'registered trap-lines,' each owned by one trapper and his family, while in other provinces there was no such system and, legally, crown land was completely free to all trappers. Under the registration scheme, successful Indian families conserve and husband their stock, deliberately avoiding over-exploitation; whereas in the absence of this protection of individual interests the tendency is to 'clean up' the game before someone else gets it, with devastating consequences, especially on susceptible species like the beaver (*Castor canadensis*).

These are clear illustrations of the important fundamental principle, which it is our main purpose to demonstrate, that in order to exploit the habitat to the full and yet continue to thrive, the consumers themselves must impose a limit on the demands they make on the resources.

It is essential to the welfare of the primitive human community that private territories, whether of the family or clan or tribe, should be properly respected. They are the local embodiment of a system of dispersion built up and slowly perfected over many generations, which 'works' and which must not be needlessly disturbed or overturned: the same applies to the elaborate conventions about reproduction, to which further reference will be made in Chapter 21. No single person conceived either system initially: they have evolved under natural selection, and it is far beyond the wit of primitive man completely to rationalise or 'explain' them. They are nevertheless intrinsic or man-made conventions, and not imposed from without. If they are broken serious consequences will follow, not at once and not necessarily affecting the actual law-breakers, but only ultimately, when the population mounts or the resources decline. In other words, breaking the conventions brings no immediate natural penalty to the offender, but is likely in the long run to have serious consequences for the whole group. To protect the group against this ultimate danger, a proximate expedient in the form of direct punishment of individual offenders at the hands of the community has been evolved, and this of course is essentially conventional or artificial, not necessarily bearing any natural relation to the ultimate danger, which is not even understood or appreciated. Thus the thief is put in jail, or loses his right hand or even his head, at the hands of his fellow tribesmen. Towards the same end a law-abiding conscience has been strongly selected for in the minds of members of the group, dictating an unquestioning submission to the elaborate but to a large extent adventitious and irrational code of taboos, customs, beliefs, fears and penalties that have secondarily arisen as inducements to good conduct. The manner in which the selfish advantage of the individual has thus been subordinated to the long-term welfare of the community can be noticed as a striking example of the overriding power of group-selection.

It is interesting to reflect once again that there is little new in this situation, as far as mankind is concerned, except in degree of complexity; all conventional behaviour is inherently social and moral in character; and, so far from being an exclusively human attribute, we find that the primary code of conventions evolved to prevent population-density from exceeding the optimum, stems not only from the lowest vertebrate classes, but appears well established among the invertebrate phyla as well.

The holding of ground that yields the means of subsistence is the simplest and most primitive form of property, whether the tenant lives alone in exclusive possession or shares the produce with other members of a group. Among the higher animals and especially in some of the birds and mammals, however, as we have seen earlier, forms of tenure have developed where the relationship between ' possessions ' and ' resources ' is much less direct and more symbolical: as for instance in the holding of a nest-site or a recognised lair which, like a ticket or licence, confers on the owner a right to participate in something quite different, namely the resources of the habitat or membership of a society. There is generally also a second equally important kind of possession, not ' real estate ' at all, but instead abstract and intangible: this is the status of the individual in the social hierarchy which is very closely and often inextricably bound up with the first kind.

Symbolic or token objects that by custom and use confer rights to the enjoyment of social status and real resources, with which, however, they have no direct or natural connection, are found everywhere in human societies: for example seals and signatures, crowns, badges and titles of rank and power, and ceremonial ritual and apparel. Closely connected with these are the jewels and precious metals which come to be valued primarily for aesthetic rather than utilitarian reasons, because of their beauty and rarity, and which are therefore sought by the wealthy, either through barter, in exchange for a surplus portion of the necessities of life, or through the still less ' real ' and, especially nowadays, extremely conventionalised payment of token money.

Recognition of the hereditary principle in the transfer both of individual property and of individual social status from ancestors in one generation to their descendants in the next, either through the male or rarely the female line, undergoes an important development among the more highly civilised races of man. Under stone-age conditions the bands of wandering hunters are necessarily small, and personal possessions few, simple and portable, but the introduction of crop husbandry necessitates a subdivision and permanent localisation of land-holdings, each in the possession of a small group, in order that those who till the soil and sow the seed may have the incentive before them of reaping the harvest. It can be surmised that, as the society increases in size and complexity, with the growth of personal and family wealth in servants, cattle, land, domestic equipment, robes, jewels and gold, and with the consequent widening of range in social standing between the richest and the poorest, the noblest and humblest, the principle of heritable possessions becomes firmly established. It follows, necessarily in a simplified and largely sex-limited manner, the natural course of inheritance of genetic factors from parent to offspring, and has grown out of the general custom in animal societies that property held by the social unit is retained in their possession by each succeeding generation. The emphasis on the male sex as the heirs of wealth and position reinforces their pre-existing epideictic responsibility and function, the womenfolk remaining in this as in other respects largely as obligate dependants.

10.9. *Summary of Chapter* 10

1. This chapter extends the review of breeding dispersion and associated phenomena of property-tenure to animals other than birds.

2. Pelagic animals present a special case because the fluid nature of their medium makes continuous tenure of immoveable property impossible. Nevertheless, semi-permanent attachments are often formed of smaller organisms to larger ones—especially if the latter are slow-moving creatures like jellyfish and siphonophores—or to inanimate floating objects. It seems not unlikely that members of the smaller satellite species, such as the pilot-fish, may ' own ' a particular host-individual and defend their rights in it against rivals. In these cases the dispersion of the satellites is tied in varying measure to the dispersion of their hosts. Parasites show an analogous dependence.

Although the great majority are debarred from acquiring property, pelagic animals can still be conditioned by and respond to population-density, either through ' travel and encounter ' between solitary individuals or through gathering into shoals or otherwise participating in social or epideictic activities: means of mutual recognition are always essential. The important principle of subdividing the species into partly isolated, self-perpetuating local stocks also meets with rather special difficulties in its application to permanent members of the marine plankton.

3. An even greater problem in the last respect is presented by the Atlantic freshwater eels (*Anguilla* spp.), because they concentrate to breed in a single area in the subtropical western Atlantic, and seem therefore to contradict the local stock principle entirely. One conceivable solution is that within

the spawning area (supposedly a 500×800-mile oval) there might be an established mosaic of subsidiary areas which are resorted to by different local stocks. This may sound improbable, but it is in line with existing evidence that the European and American species are at least partly segregated in that way. The case is anomalous and not yet capable of satisfactory explanation: few other animals show any close parallel.

4. Territorial phenomena are especially common in anthropods. Among the better known examples discussed at greater length are the burying-beetles, dragonflies and fiddler-crabs.

5, 6, 7. Among vertebrates they are widely known in every class. Examples are chosen from fish, amphibia and reptiles. A section is then devoted to the seals and their allies (Pinnipedia), which, like sea-birds and turtles, tend to have traditional breeding places frequented by more than one species. Among the land-mammals the principal attention is given to the cats, the hares and rabbits, and the gregarious ungulates.

8. The final section shows that in primitive man property-tenure follows the same lines as in lower groups. The normal subdivision of the species into local stocks is facilitated by the readiness with which tribal languages and customs diverge, as well as by strict observance of tribal boundaries. Population-density appears to be limited as usual by convention. The conventional discipline of individuals is reinforced both by conscience, which powerfully directs the individual to accept and comply with traditional authority, and by conventional punishment imposed by the social group.

Chapter 11

Communal nuptial displays

11.1. *Importance of the prenuptial period*

At the beginning of Chapter 9 we predicted that epideictic activities would be expected to reach a climax immediately before the onset of breeding for two reasons, both related to the fact that, as a general rule, the demands made by the population on their material resources are at that time about to climb more or less swiftly from the minimum to the maximum for the year. In the first place it is desirable to achieve as good a dispersion of the population as possible in advance of the rise in numbers, and thus insure that the most economical use will be made of whatever resources are available. In the last two chapters we have been chiefly concerned with this aspect of epideictic behaviour, which results in distributing the breeding stock over the habitat at appropriate densities.

The second circumstance that accentuates the need for epideictic display at this time is the fact that it generally rests in the hands of the breeding stock to vary their reproductive output according to whether a bigger or smaller replenishment is required to restore their population to the optimum. In a particular year an extra large breeding stock may have survived, or the habitat may be yielding a poorer food-crop than usual, and consequently the quota required to 'top up' the stock to match the carrying capacity of the habitat may be relatively small. On the other hand, when the stock has become depleted through the accidents of earlier events and survival is low, or when for some reason the available yield from the habitat promises to be higher than normal, a high or even maximum reproductive output may be desirable from all potential breeders. As we shall see in Chapter 21 which is specially devoted to the subject, there is now an unassailable body of experimental evidence, not only with laboratory vertebrates such as fish and mammals, but with Crustacea and insects also, all pointing to the same general result, namely that the reproductive effort and the resulting recruitment of the population are density-dependent in this way; field evidence supporting it from populations living in their natural environments is also constantly growing.

If recruitment is to be internally regulated, it is clear that the population needs to have opportunities for assessing its own density, and becoming appropriately conditioned by it, both before breeding starts and while it is going on, until no further possibility remains of controlling the quota of recruits.

Among individually territorial species of birds, or among those that occupy individual sites in a breeding colony, the intensity of competition in staking the initial claims of property automatically provides a direct index of population-density. But there are a great many other kinds of animals in which the demonstration of the strength of the breeding stock takes the rather different form of a special prenuptial forum or communal display. In the most typical cases this is held annually, during the same season and at the same hour of the day, in a traditional arena that may be, but often is not, the actual breeding place. As a rule all aspiring males take part in it together. In some cases it involves conventionalised contests among individual competitors, often for the use of symbolic sites or stances, and these contests serve both to test the pressure of numbers and at the same time to establish a social hierarchy. Wherever they have been mentioned on earlier pages they have been referred to by the general name of ' leks ', a word borrowed from Norwegian and Swedish (akin to the Yorkshire dialect word *lake*, and to the commoner English word *lark*) meaning a game or play, and originally introduced to British readers by Lloyd (1867) in his account of the blackcock in Scandinavia.

In other cases the prenuptial gatherings appear not to involve open rivalry between the individuals, but take the form of mass ceremonials, like the dancing flight of gnats. At their highest development these are sometimes conducted on a truly magnificent scale, as in the synchronous flashing of fireflies to be mentioned a little later (p. 198).

Displays of the latter kind are in general more common, and it is with them that much of this chapter is concerned. All such intensive epideictic activities, and most of all when they involve one sex more than the other, have a significance in the process commonly known as sexual selection, but this is a separate subject of sufficient importance to be deferred for the moment so that the following chapter (Chapter 12) can be entirely devoted to it.

11.2. *Communal displays in marine invertebrates*

Marine invertebrates rather frequently have some kind of mass spawning activity, but it is generally difficult to say whether it serves any purpose other than securing the fertilisation of eggs shed freely into the water. It is, however, worth calling attention to the fact that these gatherings usually have obvious resemblances to the communal epideictic displays of the higher animals. Such relatively lowly creatures as the common jellyfish *Aurelia aurita*, for example, sometimes appear at the sea-surface in sharply defined patches: one that I saw on 16th June 1957 from the high vantage point of Troup Head, Banffshire, in the water beneath the cliffs, contained many more than a thousand individuals, mostly of great size (up to perhaps 30 cm in diameter), densely packed in the space of an acre or two, and individually swimming at random. The most striking single feature of the aggregation was its sharp boundary, without any outlying stragglers, and the uniform density of the medusae within the perimeter.

Another example has already been given on an earlier page (p. 37) of one of the marine polychaete worms, the ' fire-worm ' of the West Indies (*Odontosyllis enopla*). This has a highly synchronised spawning display which has a lunar periodicity and is precisely set at the hour of nightfall. For twenty or thirty minutes the worms mass at the surface and emit flashes or continuous glows of light, depending on whether they are males or females, during the time they are discharging their sperm and ova into the water.

It would seem reasonable at first sight to attribute this phosphorescent display solely to promoting the assembly of ripe individuals, which, by their large numbers and synchronised spawning, can insure a high proportion of fertilisation in the floating eggs. Many other polychaetes have similar spawning excursions to the surface, though most of them lack the luminescent display. It will appear when the subject comes to be considered again later (p. 349), however, that the full explanation is probably not as simple as this: the existence of sexual dimorphism, of special ' castes ' of reproductive individuals, and of swarms exclusively of one sex, strongly suggest that there is an important epideictic function also, just as in the breeding swarms of many terrestrial species.

For free-swimming organisms, as these ' errant ' polychaetes are at the time, the surface is an obvious place to meet, the alternative being the sea-floor: these are the only horizons that can at all times be identified in the water, and thus readily sought and found. By repairing to one or the other, animals that at other times are scattered in depth can easily be concentrated tenfold or a hundredfold. In general the surface is greatly preferred, the most important reason being perhaps the general use of changing light-intensity as the pacemaker for diurnal synchronisation. On account of absorption, the transition between daylight and darkness is neither as rapid nor as profound on the sea-floor, even in moderately shallow water, as it is just below the surface; and beyond a few hundred fathoms the daylight cycle becomes extremely attenuated. This is the basis of the suggestion, presented in Chapter 16, that the well-known vertical diurnal migrations of plankton organisms are primarily epideictic in function. They are not all necessarily prenuptial or nuptial, however: some involve larval forms. For many reasons it is better to defer this important and interesting subject as well until a later stage, when another whole chapter can be given to it.

Some displays do of course take place on the sea-floor, and, if the water is shallow, they may show a diurnal periodicity. The case of the snapping shrimps has already been described at some length (*see* p. 73); it will be recalled that the species most thoroughly studied, *Alphaeus californiensis*, burrows in the sand, living in large colonies in shallow water, and the snaps made by thousands of individuals at times blend into a continuous crackling or hissing chorus, audible with a hydrophone for as much as a mile. Johnson, Everest and Young (1947, pp. 133-4) found a slight diurnal variation only, with noise levels 2-5 decibels higher at night than in the day, and with slight peaks shortly before sunrise and after sunset. By contrast, the mid-water chorus produced by the croaker (*Micropogon undulatus*), a gregarious fish, has a pronounced evening rise, reaching a peak just after sunset: it does not wholly die away, at least at the height of the season in June, until after midnight (*see* pp. 71 and 338, and fig. 29).

11.3. *Communal nuptial displays in insects and spiders*

Rather similar in character to the chorus of snapping-shrimps is the mass-stridulation of some of the Orthoptera, such as the field-crickets. *Acheta assimilis*, for example, common in eastern North America, begins its cheerful chorus as the day warms up and continues it far into the night. The insects are often thick on the ground, moving about as they feed, and not appearing to be restricted to individual territories. The singing is all done by the males, and, as already described in Chapter 3 (p. 48), in *Gryllus campestris* at least it is accompanied by unmistakeable signs of rivalry, including the emulation of one male by another, and, under certain circumstances, direct attacks with open jaws. The hundreds of singers within earshot at one time produce a steady volume of sound, hour after hour, the mere recollection of which after many years can quicken in the mind's eye the whole colourful world of a hot summer evening in southern Canada. The conclusion of the entomologists that have considered the purpose of the concert, two of whom were quoted earlier on p. 48, is that, though it is presumably something of great importance, it is not in a direct way connected with the attraction and wooing of females: they are consequently at a loss to explain it. As an epideictic demonstration, however, capable of indicating to the listener the density of numbers within earshot, and possibly influencing the individual distance between males, it at once assumes a comprehensible meaning.

More directly concerned with the immediately succeeding events of reproduction, perhaps, is the type of communal display shown by the cockchafers (*Melolontha melolontha* and related species), the males of which, soon after their emergence from pupation, swarm in the evening twilight on a number of successive evenings, ' flying round the tops of trees and shrubs with a loud humming sound ' (Imms, 1947, p. 123). In his essay on ' le hanneton des pins ', Fabre (VI, p. 157) describes how, in the related *M. fullo*, ' in silent, impetuous flight the males gyrate to and fro, spreading the great fans of their antennae '; the latter are many times the size of the female's and have seven sensory leaves in place of her six. As usual, twilight is a favourite time for social gatherings in the insects, though there are very many exceptions: in another beetle, *Hoplia philanthus*, ' the males assemble by myriads before noon in the meadows, when, in these infinite hosts, you will not find even one female ' (Kirby and Spence, 1858, p. 292).

In many of the mayflies (Ephemeroptera) and various Diptera Nematocera, such as the Chironomidae (non-biting midges), *Culicoides* (a genus of biting-midges), *Dicranomyia* (a genus of crane-flies) or *Trichocera* (the winter-gnat), groups of adults engage in varying numbers in the familiar aerial dances, each swarm often holding its station in calm weather for an hour or more at a time. In most if not all of these the dancers are exclusively males: a sample of sixteen *Dicranomyia* sp. taken at one sweep of the net in my garden at 8 p.m. on 20th August 1957 were all males. In *Culicoides*, Downes (1955) found that swarms of a hundred or more males formed over markers on the ground (such as cow-dung in a field): a female would occasionally enter the swarm and mating would follow at once. It is quite typical of insect leks, and bird leks also, as we shall see, that the female should appear and remain on the scene only long enough to be fertilised, whereas the males pour out their untiring energy for hours or days on end, whether there are any females present or not. Such a social organisation could, of course, lead to the dispersion of the species at a suitable density throughout the habitat, if, for example, the swarms had a maximum tolerable size, and the participants could see or otherwise know of the presence of any other nearby swarms. Small creatures such as these flies may well find it easier to identify another swarm of a hundred or more, with its characteristic location, shape and movement, than to recognise individuals of their own kind one at a time; and, as has been pointed out by others, swarming would also consequently aid solitary females in finding the males.

We may turn next to a case in the Hymemoptera. The Micheners (1951, p. 98) observe that ' a peculiar swarming or dancing can often be seen about the nests of various stingless bees [*e.g.* the tropical social bees *Trigona* and *Melipona*]. The drones hover or fly about in a fairly compact group usually near the nest entrance. Sometimes, although in our observations not often, workers join in these dances, which occur in any season and which are quite distinct from the dispersal swarms consisting of a queen, numerous workers, and perhaps some males which leave periodically to establish new colonies '.

There is virtually no end to the examples of this sort of male gathering in insects. Bates (1863, I, p. 249) describes how various kinds of butterflies, especially the beautiful *Callidryas*, congregated on moist sand along the banks of the Amazon; ' they assembled in densely-packed masses, sometimes two or three yards in circumference, their wings all held in an upright position, so that the beach looked as though variegated with crocuses '. All were of the male sex. Many others, including the tiger-swallowtail (*Papilio glaucus*), similarly gather into what Klots (1951, p. 39) descriptively calls ' mud puddle clubs ', which may persist for several days, and consist of bachelor males.

Gay though these sunlit gatherings are, they can scarcely be compared in splendour with the brilliant flash and sparkle of fireflies in the darkness.

Luminous organs are found in more than one family of beetles; although a minority of the well-known examples belong to the Elateridae, the typical glow-worms and fireflies—of which there are said to be about 2000 species—are all Lampyridae. The photophores are carried chiefly on one or more of the last three segments of the abdomen, on the ventral side, and are often present in both sexes, though usually better developed in one, not necessarily the male: in many the larvae and even the eggs are luminous also, and in one species of *Pyrocoelia*, the oriental *P. fumosa*, the larvae are luminous though the organs have been lost by the adults (Haneda, 1955, p. 361).

Some fireflies, like the species common in eastern North America (*e.g. Photinus pyralis*), fly most abundantly at and shortly after nightfall about trees and hedges and near woods, not going above tree-top height and generally keeping within a few yards of the ground, each steering an independent course. In the majority of species only the males engage in active flight, the females being either disinclined to move about, or actually flightless. Their light is produced in brief flashes, at intervals of a few seconds, and for each species the characteristic duration, interval, rate and colour of the flash can be recognised. So enchanting is the effect of scores of these insects together that in Japan some of the most famous places for fireflies (especially *Luciola cruciata*) have been placed under protection by the government as sanctuaries (Haneda, loc. cit.).

In certain Old World tropical species there has been evolved a most extraordinary development of the display: this is the assembly of thousands or even millions of fireflies, always on certain particular individual trees, night after night, where they flash in unison, with the regularity of a neon sign. Haneda speaks of it as an unforgettable and amazing spectacle, as he saw it in the Botanical Garden in Rabaul, where countless numbers alighted on a big silk-tree, flickering rhythmically about 70 times a minute, every day for a week while he was there, lasting from sunset to dawn notwithstanding the rain. In some species, like this one in New Britain, both males and females are present in equal numbers, but in others ' all the insects flashing in unison are males. The females presumably lie hidden in the jungle and the light seems to have nothing to do with mating ' (Harvey, 1940, p. 73).

The subject has accumulated a large literature (of which an excellent survey can be found in Harvey's *Bioluminescence*, 1952). There have been two common theories as to the function of the flashing: first, that it is a defence against aggressors—an old and now discredited view that originated with Kirby and Spence (1817, II, p. 427); and second, that it attracts the sexes together for mating. The latter is ' rather universally accepted at the present time ' (Harvey, 1952, p. 403) although, to judge from the reservations expressed by some writers (*e.g.* Skaife, 1953, p. 230), largely for want of anything better.

W. H. Hudson (1892, pp. 159-171) noted, of one of the South American fireflies, *Cratomorphus* sp., which was excessively abundant in the southern counties of La Plata: ' This insect is strictly diurnal in its habits—as much so, in fact, as diurnal butterflies. They are seen flying about, wooing their mates, and feeding on composite and umbelliferous flowers at all hours of the day, and are as active as wasps in the full glare of the sun. . . . I have never been able to detect it doing anything in the evening beyond flitting aimlessly about, like house flies in a room, hovering and revolving in company by the hour, apparently for amusement. Thus, the more closely we look at the facts, the more unsatisfactory does the explanation seem.' Together with this, however, should be considered the observations of Mast, McDermott and others (quoted from Harvey, 1952) that in *Luciola* and *Photinus* the flying males do locate and assemble about the feebler-flashing female, both when held captive and when resting freely on a grass-stem, as a result of an exchange of signals between them. Moreover, by responding to the male's flash after the right interval (2 seconds in *Photinus pyralis*) various writers have been able to attract fire-flies with flashlights (the same, p. 405).

The mutual attraction of mates is nevertheless clearly inadequate as a universal explanation for the whole range of phenomena involved. In the common European glow-worm (*Lampyris noctiluca*), the female has much brighter lights than the male, and she is flightless: each night she reappears at the same site, and it has been shown by Elmhirst (1912) that, after receiving the attentions of one or more males, her light soon disappears. The male's eyes actually are much the better developed, with 2500 ommatidia as opposed to 300 in the female (Imms, 1947: 93). Now all these facts are perfectly consistent with the view that the female is producing light in order that the male may find her, and in fact the circumstances very closely parallel the olfactory assembling of various moths (*see* p. 92). In many other species, on the contrary, it is the males that produce the brightest light, although they still seek the inactive females, and in *Phausis splendidula*, for instance, not only does the male produce more light, but his eyes are again very much the larger, with special glass-clear windows above and in front of them in the prothorax, where, as in many of these Lampyrids, it roofs over the head (Meixner, 1933-6, p. 1167). In *Pyrocoelia* (Harvey, 1952, p. 396) and *Luciola* (*cf. L. capensis*, Skaife, 1954, p. 229) the development is essentially similar. Instead of developing the transmitter in one sex and the receiver in the other, therefore, in these species the males have the best of both, and it is thus reasonable to conclude that their bright signals are much more readily perceived by others of their own sex than by the females.

It is clear that there are a good many circumstances in which flashing is in no way directly connected with mating: it may even occur in some species in immature larvae. The ' mating ' theory can in fact provide an adequate explanation only in a minority of fireflies. The view that in the great majority the primary function of flashing is epideictic, on the other hand, seems open to no similar objections, whether it concerns the males alone, or both sexes, or immatures. That it should often serve also the secondary function of drawing the sexes together is of course perfectly compatible with this: the same is true of the songs of territorial birds.

We can conclude this section with a case of prenuptial display in spiders. The Peckhams (quoted from Warburton, 1912, p. 84) discovered that in some of the Attidae, such as *Icius*, there was a very conspicuous prenuptial swarming, in which the males ' danced ' and were exceedingly quarrelsome, sparring frantically whenever they met, though their battles were entirely bloodless. ' Indeed,' said the observers, ' having watched hundreds of seemingly terrible battles between the males of this and other species, the conclusion has been forced upon us that they are all sham affairs, gotten up for the purpose of displaying before the females, who commonly stand by, interested spectators '. However, the contest does not actually appear to lead to mating, and in the light of comparable behaviour in other groups its interpretation as a form of mass courtship is almost certainly mistaken: rather it seems to be a perfectly typical lek, the primary purpose of which is to provide social competition among males.

11.4. *Communal nuptial displays in fishes, amphibia and reptiles*

There are two habits of the brown trout (*S. trutta*) described by Stuart (1953, pp. 10 and 16) that seem to fall within the category we are considering here. It will be recalled that this fish is at most times markedly anti-gregarious, and rarely tolerates the close presence of another individual. In autumn ripe fish that have been living in lakes during the summer are ready to enter tributary streams in order to spawn. On a number of occasions Stuart observed the formation of a large shoal of trout at the mouth of a spawning stream, which either remained in existence until the run started some days later, or else disappeared again in the interval. This transitory phase, during which the trout are strongly gregarious, immediately precedes the spawning migration and redispersion of the adult fish on the ' redds ': it contrasts so sharply with the behaviour of the species at other times that it presumably has some special significance. The habit is probably shared by other salmonids : one recalls the old accounts of the humpback salmon (*Oncorhynchus gorbuscha*) lying so thick in the entrances to the great rivers of Alaska that a man might be tempted to try to walk across on their backs. Of the even larger king salmon (*O. tschawytscha*) Turner (1886, p. 105) wrote that they arrived at the Yukon delta about the middle of June. ' The fish remain outside for several days before entering the freshwater so as to accustom themselves to the change of water. *The larger fish* usually enter first ' [my italics].

Examples are given in the next section of somewhat similar assemblies among such birds as herons and cranes, just before they proceed to their breeding colonies, and it is possible that these are all phenomena of the same kind, revealing to the prospective candidates what the ' field ' is like before they proceed to the next and bolder step of competing individually with their neighbours for personal status and property qualifications: it may already be sufficient to deter the novices and weaker spirits from further immediate participation.

Once the trout are on the spawning grounds, and during the spawning period, ' the males remain longer in the stream, while the females retreat rapidly after they have spawned ' (Stuart, loc. cit.). The males thus tend greatly to outnumber the females present at any one time on the redds, and the circumstances come to resemble those typical of a lek. In most of the many varieties of salmon and char the males become highly coloured, and one or both the jaws tend to grow out into a hook as the time for spawning approaches. This at its fullest development can lead to a more or less monstrous deformity of the whole anterior part of the head: on reaching full maturity in the late autumn the male Atlantic salmon (*Salmo salar*) temporarily

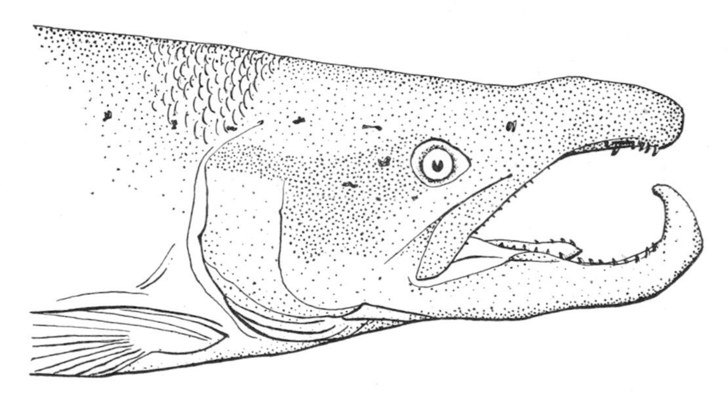

FIG. 15. Head of spawning male Atlantic salmon (*Salmo salar*) (Scotland, 26 Nov. 1957). The ' kip ' or hook on the lower jaw is an aggressive weapon or adornment, present only while the male is mature. The upper jaw is also temporarily overgrown and a socket is formed in the palate to receive the point of the kip. (Standard length of fish = 93 cm. Reduced to $\frac{1}{4}$.)

develops a vicious-looking ' kip ' on the lower jaw (fig. 15) which fits into a deep socket in the front of the palate and may occasionally perforate the forehead. The deformity is even more exaggerated in the various Pacific species. In *O. gorbuscha* the whole form of the male's body is grossly distorted in a manner suggested by the common name of ' hump-back '. The development of such notable secondary sexual characters as these takes place through the process commonly known as ' sexual selection ', the nature of which is to be examined in the following chapter, but there is no harm in anticipating the conclusion here that such signs of fierce virility as these result entirely from epideictic competition among rival males, and courtship and females have nothing to do with it. Their very existence in these salmonid fishes confirms the view that keen male competition exists on the spawning grounds, and that these partake in some measure of the character of a lek.

Earlier in this chapter reference was made once again to the submarine chorus of the sciaenid fish known as the croaker (*Micropogon undulatus*), a highly gregarious species common along the Atlantic coast of the United States. Both the males and females are provided with drumming muscles inserted on the air-bladder, and both are capable of making sounds, described as resembling the tattoo of a woodpecker on a dry pole (*see* Chapter 4, p. 71). The effect of a great shoal all sounding together is a continuous roar: as previously recounted there is a marked diurnal cycle, which builds up rapidly in the evening as soon as the bottom begins to darken (*see* Fig. 29, p. 338). There is also a seasonal cycle: during the period of the investigations by Knudsen and others (1948), the ' maximum daily overall pressure ' of sound rose from some 90 decibels in mid-May to a peak of 110 db. in early June, and then fell steadily to less than 80 db. in mid-July. According to Welsh and Breder (1924) the breeding season does not commence until August, so that the sound-display appears to be a strictly prenuptial one.

A corresponding build-up in social or communal activity before or during spawning is probably not uncommon in fishes. The swarming in large numbers at the sea-surface of the grey gurnard (*Trigla gurnardus*), an otherwise strictly bottom-living fish (except in its larval stages), has earlier been noted (p. 72): this has been seen in fine calm summer weather, and the gurnards (*i.e.* ' crooners ' or ' groaners ') observed to be uttering jubilant grunts as they swam (*cf.* Yarrell, 1859, and Couch, 1863). In the North Sea area where these observations were made they are said to spawn from April to August, laying pelagic eggs, which appear annually at the same places, where large spawning shoals collect (Meek, 1916, p. 363), and it seems likely that the surface-swarming display is closely related to reproduction.

Another rather different type of example is provided by the smelts (*Osmerus* spp.), which usually run into fresh water from the sea to spawn, or sometimes into streams from freshwater lakes: they swarm then in such dense shoals, and are covered with such rough nuptial tubercles in the case of the American *O. mordax* at least, that I have known it possible to plunge one's hand with fingers spread into the shallow water and at one grip lift four fish between the five fingers (at Lake Massawippi in the Eastern Townships of Quebec). In this species the males are the first to run: they remain longest on the spawning beds and most if not all the time outnumber the females (Greene, 1930, p. 116). As mentioned earlier, smelts secrete a sweet odour into the water, the concentration of which is presumably density-dependent (p. 96).

The epideictic behaviour of fish probably offers many other close parallels with that of land vertebrates, but in the marine species it is particularly difficult to observe. The general phenomenon of assembling, sometimes after a considerable migration, on a traditional spawning ground at a particular season, in itself reveals the development of social conventions wherever it is found.

Among the Amphibia it is sufficient to recall the choruses—sometimes almost deafening—of some of the frogs, in their most typical of communal nuptial gatherings. Their croaking is a characteristically male attribute, and in the few cases where there is any difference in size between the males' and females' ear-drums it is the male that has them larger and is better equipped to hear the sound. People who live in country districts in eastern Canada, at localities where in April the ditches and ponds are full of melt-water and three or four species of frogs may be in action at the same time, are sometimes driven almost to distraction by the insistent sound, and it needs no imagination to believe that it is capable of exerting a powerful density-dependent influence or stress on the participants themselves. This is most probably the real function of the chorus, since, as was shown in Chapter 3 (p. 51), it has nothing to do either with directing wandering frogs to the breeding sites or with sex-recognition and pairing itself: it may actually reach its fortissimo before amplexus and egg-laying begin.

In the reptiles we have noted the great nuptial gatherings of some of the aquatic turtles, both freshwater (like those of the Amazon) and, more especially in former times before their numbers were reduced, marine (*see* p. 181). Mention has also been made of the social gatherings of rattlesnakes, which first form in autumn in traditional ' dens ' where the snakes hibernate; before they disperse in the spring mating occurs, and in some of these places fifty or a hundred may be found together at that time (Ditmars, 1907, p. 428).

11.5. *Communal nuptial displays in birds*: *Corvidae and colony nesters*

There is such a wealth of material relating to nuptial displays in birds that the selection of representative examples calls for considerable restraint. We can start with what Darwin (1874, II, p. 113) referred to as a ' magpie marriage '. This concerns a species (*Pica pica*) that breeds in solitary pairs: nesting is generally in full swing in Britain in April, and frequently earlier. During the remainder of the year the birds often gather into roosts for the night, exceptionally comprising hundreds of individuals, but in the morning they generally disperse singly or in very small parties. During the first six weeks of the year, however, according to Stubbs (1910), flocks of magpies —from a few up to a hundred or more birds—may gather at any hour of the day and join in a ceremonial display. The same meeting places are in some cases resorted to year after year, and these may (though they do not necessarily) coincide with the roosting sites. The flocks are noisy, and the birds are engaged in displaying to one another by fanning out and expanding the white parts of their plumage—which Goodwin (1952, p. 116) says magpies do under most conditions of social and sexual excitement—and also by springing about and performing antics so astonishing that Stubbs regarded them as quite equal to anything he had seen among the birds of paradise at the London Zoo. Darwin's correspondent described the participants as sometimes fighting, although fighting was never seen by Stubbs and his collaborators.

It has generally been considered that these gatherings are concerned with pairing, as the name ' marriage ' implies, for, though magpie pairs are thought to remain together usually throughout the year and indeed for life, new matings have to occur some time. Writers in their descriptions have generally assumed that the individual performers were both males and females, but a shadow of doubt is cast on the marriage idea by Darwin's statement that a gamekeeper shot at one of these gatherings nineteen males on a single morning, with no mention of any females being taken.

Practices of this type are evidently widespread in the Corvidae. Hudson (1915) gives an account of a very similar spring display among jays (*Garrulus glandarius*). It is interesting that he, too, should conclude by remarking that ' again and again, when watching these gatherings at Savernake and other places where jays abound, I have been reminded of the descriptions given by Alfred Russel Wallace of the bird of paradise assemblies in the Malayan region '. It is clear from the standard works of ornithology that something of the sort occurs in various North American jays and magpies. In the numerous species of *Corvus*, also, some kind of ' crow's parliament ' is doubtless almost universal: they may then, like the European rook and jackdaw, mount upwards in magnificent aerial displays of hundreds of birds, sometimes towering to a great height. These assemblies are usually very noisy, and individuals may often be seen swooping and skirmishing in them. They are not necessarily prenuptial, being seen frequently in autumn, but it must be borne in mind that in this genus breeding generally starts early in the year, and also, of course, that epideictic displays frequently occur in birds at other seasons, especially in autumn.

In the common heron (*Ardea cinerea*), Lowe (1954, p. 72) records that before first occupying the nesting trees in early spring the birds ' assemble and stand on some piece of ground near by. These gathering grounds are at varying distances from the heronries, dependent upon local topography: once having settled the place, however, they return to it as unfailingly as to the heronry '. On this traditional site the birds engage not in combat but in a sort of desultory dance of feet and wings, which is stimulated principally by the arrival of each newcomer to the party. On one occasion birds were seen to give repeated demonstrations of ' that rapid zig-zagging descent which they often make when approaching the tree-tops '. No mating has been seen to result at the meetings, but they continue throughout February and well into March, ' until occupation of the colony is complete '. If some social machinery is really required to limit colony-size in herons, it appears very possible that this has an important part in it.

F. M. Chapman (1908, p. 86) describes another pre-nesting assembly in the brown pelican (*Pelecanus occidentalis*) in Florida. ' The records of Warden Kroegel show that, as a rule, pelicans, in flocks of from 500 to 1,000, arrive, apparently at night, in the vicinity of Pelican Island about November 1. At first they stay on the [Indian] river, their numbers rapidly increasing, and during this time they sail for hours over the island, possibly engaged in mating evolutions. The clans having gathered, at the end of a week the birds in a body take to the island. Nest building is begun at once, and the first eggs

Manx shearwaters assembled on the sea at dusk in late June, photographed off Skokholm Island, Pembrokeshire, with Skomer in the background. (From a colour transparency by Ben Feaver.)

are laid by December 1.' It is significant to notice in the present context that in 1907-8 7000 birds appeared, but finally only 1500 nests were built. 'What became of the remaining 4000 birds is a mystery' (loc. cit., p. 88).

Early in March Lockley (1942) found the Manx shearwaters (*Procellaria puffinus*) returning to Skokholm from the open sea, and gathering each evening on the water between there and Skomer (*see* pl. IV). 'As the sun slowly sank behind the lonely island of Grassholm the great rafts of shearwaters would swing as one being, their white breasts now a silver flash in the sun's last rays, and then . . . velvet black on the grey-blue sea.' They would settle, 'but soon ever restless they would rise, wheel, and perhaps split up for a while, making three or four separate flocks. Always these grew in size as more and more birds came gliding in from oceanic feeding grounds to the north and west and south'. Later in the spring the whole assembly contained probably more than a hundred thousand, evidently comprising the united population of Skokholm and Skomer. 'What are the birds doing in those great rafts all the evening,' he asks. 'Does their impatience and herd excitement force them to scream as on land? Assembling always in the same spot so many hours each evening, they surely must be doing something important?' This would indeed be true, if in fact they too were holding a mass demonstration of the whole breeding unit centred on these two islands, by which the magnitude of the year's breeding effort was being conditioned and regulated.

The various auks, as previously mentioned, pay intermittent early morning visits to their breeding colonies long before egg-laying time, appearing in their scores of thousands on the ledges at Fowlsheugh, in Kincardineshire, on occasional days in January and February: through these visits nest-site tenancies are presumably renewed and bachelor birds get their chance to establish themselves where a vacancy occurs. It should be possible for them to sense whether numbers are high, normal or low by the pressure of unsatisfied demands for a place on the ledges, and the stress on the individual bird resulting from this might have an effect, through the endocrine system, on the subsequent progress of the maturation of its gonads and its reproductive efficiency. In the puffin (*Fratercula arctica*) the Faeroese bird-fowlers reckon there are seven adults for every occupied nest-site, judging by the rapidity with which substitute birds repeatedly appear to incubate the egg when the occupant of the nest-burrow is removed (Williamson, 1948, p. 155), and it seems highly probable that some of these supernumeraries would have bred themselves had they been early enough possessed of a site.

It will appear clearly in the sequel that nuptial epideictic displays are frequently continued far into the actual nesting season, but this perhaps requires to be stated before mentioning the case of the common and arctic terns (*Sterna hirundo* and *macrura*). In spring they at first resort to their breeding grounds only at dawn, or at dawn and dusk, retiring during the rest of the twenty-four hours to some other temporary base or dispersing to feed (J. M. Cullen, 1957, unpublished). It is at these times that the well-known ' panics ' occur with the highest frequency. Most of the time the terns fly noisily in all directions above the colony site, but every so often there is a startling hush and all the birds swoop away in unison, usually towards the sea, as if in panic from some imaginary danger. In a few moments the incident is over, but nevertheless it is a very dramatic display. Cullen found that on 14th and 17th May, when the colony was first occupied, there was an average rate of 28 panics per hour, or in other words one every two minutes, whereas between 7th and 31st July, at the end of the season, it had fallen to 5 per hour.

These panics have attracted the notice of every student of the colonial terns, but no satisfactory explanation of their purpose has hitherto suggested itself. As an epideictic display, underlining by its silence the previous bedlam of noise, and breaking into the private affairs of individuals with a compulsory united manoeuvre, it is readily comprehensible, and finds a parallel in many other communal activities of sociable nesting birds, for example the great wheeling flights of puffins over the sea in front of their colonies, the evening manoeuvres of the Manx shearwaters just described, the piping parties of oystercatchers or the screaming parties of swifts.

11.6. *A second series in birds*: *the typical ' leks '*

The displays so far described have been in the nature of mass demonstrations in which, with the possible exception of that of the magpie, no one sex is known to have an exclusive or even predominant part. Though some of the birds mentioned or their close relatives have notable nuptial adornments and colours, these are possessed by both sexes in almost equal perfection, and sexual dimorphism is on the whole very slight. It is an interesting and rather general rule, though subject to some exceptions, that colony-nesting birds tend to have relatively indistinguishable sexes.

Very different, therefore, appear the nuptial displays of the birds that take part in a typical lek: in these the cocks, in most cases characterised by a striking sexual dimorphism, contest among themselves for the right to occupy one of a number of stances, by convention conferring the privilege of participation in breeding. A good many instances are known, but only in a relatively small number of these is there sufficient detail available to establish beyond doubt the more important facts.

One of those most fully described is the display of Gould's manakin (*Manacus vitellinus*), studied by F. M. Chapman (1935) on Barro Colorado island in the Panama Canal Zone. This small passerine bird belongs to a family, the Pipridae, noted for their ' fandango ' displays (*cf.* Sick, 1959): the fact that the same dancing-sites are used regularly for many years has long been known. The most extraordinary secondary sexual character of the highly coloured males, especially developed in the related genus *Machaeropterus*, is the modification of some of the secondary wing-feathers, in which the shafts are distally curved and greatly thickened and the vanes peculiarly shaped, so as to form a percussion instrument, producing a succession of loud whirrs and snaps: there is an associated hypertrophy of the wing-bones. In Chapman's opinion the primary feathers are also somewhat modified for sound-production; and examination of museum specimens reveals that in the males of *Manacus* the first four primaries are extremely attenuated (reminding one, on a miniature scale, of those of the American woodcock, *Philohela minor*, which has three outer primaries similarly narrowed and used for sound-production). The female's primaries are quite unmodified, and form an important difference between them and the similarly-coloured young males, particularly in those species of *Manacus* where these two have identically coloured plumages.

The males display on individual ' courts ', which they defend as a territorial focus. Four to seven males are commonly associated in a group, with courts spaced at intervals usually of 30 to 40 feet apart (extremes, 12-200 feet): each court is a cleared and cleaned patch of bare ground, ' usually irregularly elliptical in outline and averaging two and a half feet long by twenty inches in width. They are placed in forests with an undergrowth of small saplings, several of which grow at or near their borders ' (loc. cit., p. 489). In display, the male jumps about over the court from one perch to another one, two to four feet away, to the accompaniment of loud rattling whirrs and snaps, sometimes six to ten in succession and as rapidly as two to the second. On a still morning, over water, Chapman states that he had heard them at a distance of 300 yards.

The males in a group form a display unit, separated by a relatively long distance (minimum 300 yards) from any neighbouring group. They are evidently not all equals in social status: in fact a ' submissive ' male may be seen displaying to a ' dominant ' one, appearing to supplicate or try to win the attention of the latter, which is likely to remain impassive. Birds with the yellow areas of the plumage of deeper colour were in three out of four cases observed to be dominant. When Chapman placed a stuffed male in the court of a cock known as No. 1, a submissive bird, its presence was resented alike by the owner of the court and by its dominant neighbour No. 5, which suggests a corporate interest of the birds in the presence of interlopers or supernumeraries within their group.

He concludes that all the many tests ' appeared to show that the presence in or near a court of a stuffed male aroused more response than the presence of a stuffed female in similar situations. Sexual jealousy, therefore, was apparently stronger than sexual ardor ' (loc. cit., p. 518).

It was found that the court life of the males may last for eight months, and that they take no part whatever in nesting activities. The females present themselves, so far as determined, only for coition. On the appearance of a female the cocks do not go after her, but each occupies his court ' in an apparent attempt to induce the female to visit him there '; and she, in accepting one of the several available, ' seems to select her partner ' (loc. cit., p. 499).

There are several very significant points to be noticed in this ritual performance. As already observed on an earlier page (p. 160), no fundamental distinction can be drawn between the semi-permanent individual courts of the manakins and the bowers of the bower-birds, except that the latter are more widely separated. Both serve all the purposes of property, and it seems fairly safe to conclude that the number of courts tends to remain very constant, placing a ceiling on the number of males that are candidates for mating within a particular area. The number of females coming in for fertilisation must be known to all members of the display group whether they are performers or onlookers, and the regulation (*i.e.* the giving or withholding) of this essential office could thus presumably reside with them. One of the interesting features of polygamy, to be discussed in Chapter 21, is the potential control of the fertility of a whole group of females which it vests in one, or very few, males, and the possibility that this simplifies the process of controlling how many ova are fertilised.

The corresponding performance has been recently observed in the black-and-white manakin (*M. manacus*) in Trinidad by Darnton (1958). The lek began about sunrise, and had completely died away by about 9 a.m.

It is characteristic of a lek that the display of the males is primarily directed towards one another, and that its objective is to decide and maintain the social status of each. If there is any exclusion of surplus males from the courts, this must be decided and effected entirely among themselves: in other words, the males must determine which of their own number are eligible to breed, but it is interesting to have Chapman's opinion that, among the pre-selected group presented to her, the female appears to exercise her own choice.

Less elaborate traditional display grounds are used by groups of males of various hummingbirds. In the British Guiana forest E. M. Nicholson (1931*a*) found several groups of up to twenty birds of *Phaethornis superciliosus* (all those collected being males) occupying what he came to call leks: there, distributed on perches 5 to 30 feet from the ground, they were ' uttering, with great emphasis, a continuous flow of monosyllabic short-clipped notes, hard and insect-like in quality: " Jang-jang-jang, jang, jang . . .," ' and at the same time the tail, often fully fanned, twitched perpetually in rhythm. They frequently chased each other in lightning pursuits, using a different note, but the perched birds kept up the ' janging ' hour after hour with scarcely a break. The Indians said the same places were used year after year, and were haunted persistently except in the rainy season. T. A. W. Davis (1934) afterwards found the same thing in another species, *Pygmornis ruber*, and there are earlier-published accounts by Brewster and Chapman (1895) of similar male contests in the related species *Pygmornis longuemareus* and *Phaethornis guy*. Nicholson remarks, incidentally, that ' the sexual function of these assemblies is utterly obscure in the case of the hummingbirds ' (loc. cit., p. 82). In a more recent paper by Davis (1958) there are two very interesting further points relating to *Ph. superciliosus*: first, regarding the duration of the lekking season, it is probable that the leks are often in being throughout the breeding season, from July to April, with a break in December and January for the short rains—a period comparable in length with that of Gould's manakin; and second, that a lek he visited in 1937 and 1938 was seen from time to time in the same spot by Mr. D. B. Fanshawe up to 1949.

For perfection of colour, grace and splendour, however, none can surpass the males of the birds of paradise (Paradisaeidae), birds related to the Corvidae and of a similar or somewhat smaller size-range, whose geographical centre is in the region of New Guinea. The leks of several species were made known by Wallace (1869, II, p. 252), who elicited accounts of them from his native hunters though he never witnessed them himself. He describes the males' marvellous plumages: in the great bird of paradise (*Paradisaea apoda*), for example, in addition to the gorgeous metallic emerald and velvety yellow of the head, set off against a rich coffee-brown, ' from each side of the body, beneath the wings, springs a dense tuft of long and delicate plumes, sometimes two feet in length, of the most intense golden-orange colour and very glossy, but changing towards the tips into a pale brown. This tuft of plumage can be elevated and spread out at pleasure, so as almost to conceal the body of the bird ' (loc. cit., p. 390). In more recent times this and other species have thrived sufficiently in captivity to display from an aviary perch: it is known that at the climax of the acrobatic display-dance its head is thrust forward and far below its feet, the wings spread-eagled downwards into a semi-circle on either side, and the whole creation surmounted by a blaze of shimmering plumes. Every visible part of the bird is blended in the beautiful and dramatic effect; and in the ' king ' and ' magnificent ' birds of paradise (*Cicinnurus* and *Diphyllodes*), an incidental added touch of perfection is the delicate green lining revealed at a particular moment by opening the mouth.

The third example deserving passing mention is as different as possible from the previous two, and makes an exception to the rule that lek displays are associated with highly developed male adornments. This is the display of the great snipe (*Capella media*), a bird in which both males and females are coloured cryptically and indistinguishably. In Scandinavia from May till early July great snipes gather to their traditional marshy parade-grounds nightly at dusk, and take part in a choral ritual. ' From time to time encounters between males occur, the birds sparring with bills, rearing up at one another and striking with wings, but the fighting seems to be mainly formal ' (Witherby, 1940). It is not stated in this description how the males can be identified, but since the display continues through the incubation season it is perhaps unlikely that both members of a pair are always present. There is at least one other well-known species of bird with cryptic colouring and little sexual dimorphism that nevertheless has a full-blooded lek, and that is the North American sharp-tailed grouse (*Pedioecetes phasianellus*).

One of the most celebrated of all lekking birds is another palaearctic limicoline, the ruff (*Philomachus pugnax*). So great is the sexual dimorphism here in the nuptial season that the bird's name enjoys the distinction rare for a wild bird in English of having two genders, ruff and reeve, but in life the difference lasts only about four months and during the rest of the year both sexes are inconspicuously coloured and alike, except that the male is larger. In the concise, lucid phrases of Newton (1894, p. 798), ' The cock-bird, when out of his nuptial attire, or, to use the fenman's expression, when he has not " his show on ", and the hen at all seasons, offer no very remarkable deviation from ordinary sandpipers, and outwardly there is nothing, except the unequal size of the two sexes, to rouse suspicion of any abnormal peculiarity. But when spring comes all is changed. In a surprisingly short time the feathers clothing the face of the male are shed, and their place is taken by papillae or small caruncles of bright yellow or pale pink. From each side of his head sprouts a tuft of stiff curled feathers, giving the appearance of long ears, while the feathers of the throat change colour, and beneath and around it sprouts the frill or ruff. . . . The feathers which form this remarkable adornment, almost unique among birds, are, like those of the " ear-tufts ", stiff and incurved at the end, but much longer—measuring more than two inches. They are closely arrayed, capable of depression or elevation, and form a shield to the front of the breast impenetrable by the bill of a rival. More extraordinary than this, from one point of view, is the great variety of coloration that obtains in these temporary outgrowths. It has often been said that no one ever saw two ruffs alike. That is perhaps an over-statement, but, considering the really few colours that the birds exhibit, the variation is something marvellous, so that fifty examples or more may be compared without finding a very close resemblance between any two of them, while the individual variation is increased by the " ear-tufts ", which generally differ in colour from the frill, and thus produce a combination of diversity. The colours range from deep black to pure white, passing through chestnut or bay, and many tints of brown or ashy-grey, while often the feathers are more or less closely barred with some darker shade, and the black is very frequently glossed with violet, blue or green—or, in addition spangled with white, grey or gold colour.' He points out that first Montagu and later Bartlett demonstrated that each individual male in successive years assumes tufts and frill exactly the same in colour and markings as those he wore in preceding seasons: ' as an individual he is unchangeable in his wedding-garment '.

The best-known descriptions of the lek written in English are those of Selous (1906-7) and Turner (1920), both based on observations made in Holland. The display begins when the migrants arrive back in March or early April, and continues into June. The males gather on traditional ' hilling ' grounds, which are distinct from the adjacent breeding areas, in numbers from half-a-dozen or fewer up to forty or sixty: regular performers have individual ' hills ' or display stances. Each bird is readily recognisable, and Miss Turner observed that each one usually resorted to his own special patch. ' Each little hill was about eighteen inches in diameter, and absolutely devoid of any vegetation. The ruff is either as motionless as if he were carved in stone, or else he is vibrating like a toy on wires. It is the rapid, restless motion of the feet, and the dancing, which have worn down the grass and hardened the ground in these circular patches' (loc. cit., p. 146). Cocks tensely display to their neighbours, or chase and spar with them in rapid, excited movements. 'They rush round with the regularity of a clockwork mouse. When several are fighting together, they are an indistinguishable whirr and blurr of feathers. . . . They filled me with amazement. Why do they behave in this ridiculous manner, and what is the meaning of their extraordinary behaviour? How did it originate and what is the use of it?' (*the same*, p. 149).

Miss Turner found that their whole attitude struck her as a pose. Never in the time she watched them did she see a feather fly, nor any blood drawn. The noiselessness of the display in particular made it seem unreal. Females, by this time busy with their nests, took practically no part in the proceedings during the period of her observations in May, but one reeve paid apparently regular visits to the most dominant ruff in sight twice a day during a 3-day watch. All the six or seven males present 'bowed themselves to the ground as she approached, and remained in this devotional attitude for some seconds after she left'. Thereupon they either straightened up, or began to dance or sometimes rushed at one another in a characteristic whirl of wings.

'The arrival of an extra bird on the scene generally meant a display of energy on the part of all the ruffs, which had taken up their positions. They either rushed at him, or else gyrated on their own little hills. If all the hills were occupied the intruder flew away' (*the same*, p. 153).

This case has been recounted at some length, in order to emphasise the remarkable similarity of typical lekking displays one to another. We have the exaggerated adornments, the bouts of feverish excitement and energy, the absorbing, ritualised competition among males to have and to hold a fixed number of traditional and symbolic sites: there is once again an indication of an inequality in social status among those present, and of an unsatisfied surplus of males attempting to gain a place but being repulsed. The lek is in session until the season of egg-laying is over, so that the whole business of nesting and incubation appears to be entrusted to the female sex. Comparing it with the leks of Gould's manakin, the 'hermit' humming-birds, and the blackcock still to be described, we have a striking series of examples of convergent evolution in behaviour.

The special unique feature of the ruff is the individual polymorphism of colour and pattern in the males' nuptial ornaments. This has been a notorious puzzle to evolutionists, and no generally acceptable explanation of its advantage to the species has ever been suggested. On the present interpretation of the lek the possibility suggests itself that, like the banner of the jousting knight or the jockey's colours, it is an aid to individual recognition among a group of competitors. Individual recognition is one of the basic essentials in maintaining any sort of peck-order or hierarchy, and an

FIG. 16. Blackcock (*Lyrurus tetrix*) at a lek (Scotland 1961).

instant badge of identity may be especially needed in the lightning mêlées of the ruff, as well as in detecting the arrival of unknown intruders.

The classical examples of the lek are, of course, those of the grouse-like game-birds or Tetraonidae, especially the black grouse (*Lyrurus tetrix*). Here again the pronounced sexual dimorphism has resulted in the two sexes bearing separate names, blackcock and greyhen. The handsome cock is much the larger, with a glossy black plumage, erectile scarlet combs above the eyes, a broad white bar on the wing, and the famous black lyre-shaped tail, backed, when erect and spread, by snowy white under-tail coverts.

As usual in grouse, the display begins at the first light of dawn, when a number of cocks (in Scotland nowadays commonly numbering between five and twenty-five) gather on the customary ground, often itself called the lek, though the word more properly refers to the action that ensues. Some early strutting and display takes place on fine mornings as early as December and January; and from April until late June it is a daily event, at its height lasting for two or three hours in the morning, being resumed off and on during the day and more vigorously again at dusk. Like many other forms of epideictic display it frequently recurs after the moult in autumn.

The contest is essentially a ' territorial ' one, in which the participants defend their relative positions against the threats and sorties of their neighbours. There are no identifiable permanent sites, and sometimes part or the whole of the assembly shifts its ground as much as fifty yards, only to drift back to where it started. Lack (1939, p. 293) observed, however, that when the cocks first fly in to the lek, ' each seems to know its place, which suggests habitual use '. In full display the cock raises the combs, inflates the oesophageal pouches under the feathers of the neck, stiffly extends the half-opened wings downwards on either side so that their tips barely clear the ground, and raises and spreads the tail over the back to form a beautiful erect fan, overtopped in the centre by the white under-coverts and bordered below on either side by the down-turned, lyrate tips. The effect is magnificent. The most virile males are capable of retaining this tense state with no more than an occasional, barely-perceptible, relaxation for two hours at a stretch. At the same time they persistently keep up a soft chorus of notes often described as ' rookooing ', sounding to me like ' coo-coo-coorŏw ', with a resonant, vibrant quality that carries it to a considerable distance. The displaying males frequently run to meet, stop, face each other with lowered and extended heads tense and motionless, then suddenly hop or jump or patter their feet, only to be frozen into menacing immobility once more. They make short runs together with rapid mincing steps, meeting and jumping with a quick flutter of downspread wings; or one may advance while the other gives way. At these moments of formal individual bouts they have a special harsh sort of war-cry, ' coch-oich ', like tearing oilcloth.

On the outskirts one can sometimes see a low-caste or submissive cock, attracted to the lek but never offering more than an occasional brief, dispirited display, mostly ignored by the others but at intervals chased right off the place, frequently by two or three of the display-cocks in a gang. He appears to be one of the ' surplus ', not admitted by the others to the lek, and consequently having much diminished chances of breeding.

Hens are sometimes very conspicuous, especially in late April, but more often none at all are present. When a greyhen comes she does not remain stationary, but may be courted by a succession of cocks, parading round her mostly one at a time and sometimes chasing her about. Hens evidently fight each other at times, but I have not so far managed to see a lek where more than one hen was present at a time.

The best English descriptions of the blackcock lek are those of Millais (1892) and Lack (1939). Selous (1907) also published his observations of the black grouse, as he did of the ruff, but in both accounts his obsession with the Darwinian theory of sexual selection at times detracts from his objectivity. Other authors have been much readier to admit that courtship of females could not be the sole incentive to display, and some indeed have thought it a largely useless by-product of male ardour. Lack (loc. cit., p. 295) concludes that ' the fighting and threat display of the blackcock are so closely correlated with its territory that the latter presumably has a function. Most of the functions already ascribed to territory are clearly inapplicable, but it seems to have one obvious function, namely to reduce the amount of interference between neighbouring males in copulation '. Beyond this presumably quite minor side-effect, however, no adequate and reasonable objective has hitherto been offered.

The blackcock's nuptial rites rather closely resemble those of the North American prairie-chicken (*Tympanuchus cupido*), sharp-tailed grouse (*Pedioecetes phasianellus*) and sage-grouse (*Centrocercus urophasianus*). These have been well described, especially by J. W. Scott (1942, 1950). The sage-grouse has ' strutting grounds ' which as usual are resorted to year after year: ' indeed, the instinct to return to the same location is so strong that a public road passing through one area did not prevent the annual return ' (Scott, 1942, p. 479). At these places, however, the number of cocks present is relatively enormous, reaching 350. Hens are generally far out-numbered, as the following proportions of cocks to hens on various successive dates in April and May show:—171:1; 355:141; 72:3; 350:150; 302:11; 174:11; 238:8; 146:5. Only on two of these dates, April 12 and 20, were large numbers of females present. Demonstrations may occur sporadically during the day-time, usually reach a peak in the evening, may break out during the night even, and attain their culmination at and after dawn.

' Dominance in the cocks plays a tremendously important role in the sexual life of this species ' (*the same*, p. 483). On one particular display ground, Scott found four well-defined mating-spots, each with (*a*) a compact group of hens, (*b*) a ' master cock ' which undertook practically all the matings, (*c*) a chief rival or ' sub-cock ', and (*d*) three to six ' guard-cocks '. Of 154 matings analysed, 114 were done by master cocks, 20 by sub-cocks, 5 by guard-cocks and 15 by isolated cocks outside the main mating-spots.

(In another 20 matings not included in these figures it was impossible to determine the rank of the cock officiating.)

As a result of the incessant demands made on the master cocks on busy mornings, they appear to become temporarily incapacitated, and the sub-cocks and even guard-cocks and outsiders then have some chance. But they are all under a restraint from the master-cock, it appears: Scott has seen three or four hens piled in a heap while the temporary incapacity of the dominant bird ' and the restraint of the system ' prevented them from being mated, and some hens apparently left without being served (loc. cit., p. 495). This is very important, because it shows that such a restraint could easily place a ceiling on the number of hens fertilised.

Turning finally to the prairie-chicken (*Tympanuchus cupido*), whose special ' booming ' performance very closely resembles the two previous ones, there are still one or two valuable scraps of knowledge to be gleaned. The Hamerstroms (1955, p. 462) have found that, as usual, booming grounds tend to remain in use year after year. Marked birds have been shown to defend the same territories day by day, and to resort to the same booming ground in later years. Within this general pattern there is variation, however: new grounds have appeared in periods of population increase; and established grounds have disappeared during population declines. Occasionally they have shifted position for reasons that were not always clear; and they have commonly changed in size from year to year. Though the numbers of cocks attending varies very greatly from ground to ground, they showed in comparing two Wisconsin counties together that the largest lek of all, with 45-48 ' regulars ', was situated in the county with the larger acreage of suitable habitat and the denser population of grouse.

11.7. *The lek as a method of controlling density of breeding birds*

There emerges from the examples in section 11.6 the conception of a fairly stable established pattern of display-points, generally conservative yet capable of modification as the habitat undergoes secular change. The capacity of these display sites to accommodate the competitors cannot be indefinitely extended, and under natural undisturbed conditions the sites probably become ' sold out ' in normal years before the actual mating season starts. The result is that a surplus of males unable to win places is turned away.

The lek appears to provide a system of property-tenure essentially the same as nest-site tenure in a colony, so far as its dispersive function goes: in either system the number of sites or claims is limited by convention, and males that cannot secure a site are ' surplus ' and debarred from breeding. Both provide occasions for competition that is capable of indicating the relative population-density, and both can identify and turn away a surplus when one occurs. Both, it must be assumed, are meeting-places that serve a definite area of habitat in which the members find their living.

Beyond this, however, the similarity ceases. Lekking birds can disperse their actual nests, hiding them away from predators, and still further discourage nest-predation by, presumably, being completely free to choose new sites every year if they please. In none of the lekking birds is there any indication of the existence of ordinary nest-territories, because, no doubt, the lek itself is discharging the territorial function of controlling population-density.

There is another difference from the colony-nesting system to be noticed. The tenure of places in a lek may indeed limit the number of males permitted to breed, but this does not extend to the females. In the absence of more definite evidence it is at present little more than a surmise that when there is a surplus of females the number of matings is limited. That is to say, a constant number of males will undertake a fairly constant number of matings before their interest and willingness decline, and further matings are refused. At least this is a possible way in which a surplus of females could be dealt with, and consistent with occasional observations of established cocks being disinclined to mate, both in the black grouse and sage-hen. In a small promiscuous lek all the cocks are no doubt aware of, and conditioned by, all the matings that occur, and the summation and limitation of matings could conceivably be a collective operation. If the control is to be effective, the lek must remain in session throughout the season when females might come forward for mating, and after the quota for the year has been met its function would be to prevent subsequent clandestine matings. It is interesting and undoubtedly very significant that polygamy has evolved in several of the species concerned, with the result that the actual matings are in practice controlled by fewer and fewer males. In Chapter 21 the adaptive value of polygamy will be further discussed, but one of its apparent advantages is the tightening up that it makes possible on the number of females fertilised.

11.8. *Prenuptial communal behaviour in mammals*

Our knowledge of leks and related assemblies in mammals is extremely limited. An example of a gathering that in many ways appears to resemble a lek, in the hammer-headed fruit-bat (*Hypsignathus monstrosus*) of the Congo basin, was mentioned in Chapter 3 (p. 55). Chapin and Lang, it may be recalled, found a particular spot to which every evening at dusk about thirty of these bats assembled, all males to judge by the sample collected: there for the next three or four hours they produced an uninterrupted chorus of ‘ pwok’s ’, from which they were not dissuaded either by people talking, lantern-light or gunshots. This seems certain to be an epideictic demonstration of some description, and, if it involves only one sex, it probably takes place in the nuptial season: ‘ asexual ’ epideictic displays generally concern the whole population regardless of age and sex. However, the relations between the males and females in this case are entirely unknown.

It is possible that the choruses of epaulet bats (p. 56) are concerned with the tenure of established ‘ male-sites ’ rather than of breeding territories, and in that case they may have something of the character of leks also.

Some examples of what may be described as prenuptial mass-demonstrations rather than leks have already been mentioned in earlier chapters. There was the case (*see* p. 58) of the eastern chipmunk (*Tamias striatus*), recounted by Ernest Thompson Seton, which he observed to take part in a vocal chorus on fine mornings in spring. 'Every chipmunk mounts his perch, and they make the woods ring with their united voices.' The same displays sometimes recur in autumn before they retire into hibernation. The extraordinary signalling displays of the white-sided jack-rabbit (*Lepus alleni*) (*see* p. 33), which can hitch up its white belly-fur on to either flank so as to reach the top of its back where it shines in the sun, are given most frequently, Nelson (1909) was inclined to think, during the rutting time.

The European hare (*L. europaeus*) and no doubt other species of *Lepus* are promiscuous breeders, and do not establish pair-relationships. Early in the spring the males can be seen to gather into small parties, commonly of four or five but sometimes twice or three times as many, and often follow one another about at a leisurely pace, rather like a game of follow-my-leader, and generally making themselves far more conspicuous than at other seasons of the year. Of the proverbially mad March hare, Barrett-Hamilton (1911, II, p. 284) has this to say: 'His antics are often extraordinary, and include grunting, biting with ears thrown back like those of a vicious horse, kicking (as he jumps over his adversary, like a barn-door cock), bucking (strange, writhing, upright leaps into the air), and boxing with his rivals. It is true to say that such combats are rarely of a serious nature, and that they usually cease after March, although the sexual season does not; but for a time the fur flies freely. . . .' Some kind of hierarchy among the local males probably results, but all one can say in the present state of knowledge is that this type of demonstration and contest among males at such a season is common in many groups of animals, and no doubt in all cases serves the same kind of epideictic function.

We are told of the jaguar (*Felis concolor*) that, in the days when they were still sufficiently common at any rate, 'during the love season the males congregate, eight or sometimes more together' (Azara, 1838, p. 184). The same sort of thing was mentioned earlier (p. 60) of the males of the coyote (*Canis latrans*), which gathered nightly to one or other of their 'choir-lofts' to sing.

In the red deer (*Cervus elaphus*) and some other species of *Cervus*, the stags and hinds, the latter with their calves and yearling followers, keep largely separate except during the rut in autumn, each having their own customary feeding grounds. In Scotland most stags cast their antlers in April or early May, and during the next four months the new ones are growing. In the spring and early summer the male herds break up to some extent, and it is quite common then to find small parties of them frequenting the high corries. In July, and still more in August and September, where the herds have previously disbanded they tend to re-form, and by that time sometimes contain over a hundred stags (frequently with odd hinds and youngsters among them). This agrees with Darling's (1937, p. 79) observation that, 'as the velvet is shed from the antlers in August, the stags appear to gather more closely in companies, and there is daily movement which may involve the whole length of their summer territories. But as the rut approaches what previously has been play and high spirits becomes serious quarrelsomeness. . . .' Indeed there are continual signs of rivalry amongst them: these consist mostly of threat-displays, in which one turns to face a rival, once or twice lowering and tossing up the head, but two stags quite often rear up on their hind legs facing each other and thrash for a few seconds with their fore-feet. When the horns have stopped growing, and even before the rags of velvet are all off them, from late August to October, stags engage one another with their horns, pushing and wrestling against each other, but often in a quite brief and formalised manner. It seems almost certain that before the rut begins in October a male hierarchy is well established: in fact it is quite possible that the segregation of the sexes during the preceding months is primarily an adaptation to facilitate this, without interference with or by the females and calves. As early as August it is easy after a few minutes' observation to pick out which are the dominant stags in a herd; and on 1st September 1957, for example, I noticed a stag, identifiable by having only one mis-shapen and crumpled horn, hanging about the rear of the herd, a little behind the others and evidently keeping clear of them, whenever I caught sight of the group during a period of four hours.

Rivalry reaches its climax in October and November when the stags join the hinds at the usual rutting places. The stags spend much of the day ' roaring ' and demonstrating: the majority are spaced out at intervals, of at most a few hundred yards, but they are near enough to be constantly aware of their neighbours by sight, sound and, probably, scent (*see* p. 104). Their displays are clearly in the nature of challenges one to another, very much as in a lek. The hinds appear to be free to come and go—in spite of what the stag may do to try to prevent them—and to associate with any stag: they are presumably influenced by his display and, possibly, location, but this can only be secondary effect and is not the display's primary purpose.

In the related but larger wapiti (*Cervus canadensis*) there is an account quoted by Seton (1909, p. 52) of a ' dance ' by a group of bulls, observed in late August in Wyoming. His informant wrote: ' My attention was drawn by a column of dark-brown dust rising ahead of me . . . almost as if caused by a whirlwind. On reaching a point as close to it as I could get without crossing the creek, I found it was caused by a band of elk, numbering from twelve to twenty, who seemed to be trotting quite rapidly, with occasional awkward galloping plunges, in a circle perhaps thirty feet in diameter. . . . They were moving, not with heads up, but with noses only a foot or two from the ground. My impression is that they were all bulls '. This appears to be a more elaborate variant of communal male display. Analagous behaviour of some kind probably occurs in most or all the highly sociable ungulates, including antelopes of many kinds, wild cattle (such as the American bison, *cf.* Seton, 1909, p. 288), and horses.

11.9. *Communal ceremonies in primitive man*

The ' initiation ' ceremonies of certain primitive human tribes probably have a certain epideictic quality about them. The Australian aborigines, for example, were originally organised on a territorial basis into local groups, in the nature of things largely isolated from each other: the groups were associated into clans, and the clans into tribes. Notice of forthcoming initiation ceremony was given by messengers; and, summoned in this way, ' the scattered families of the tribe come marching from all quarters to the chosen spot, a sacred place, where they assemble in their hundreds. The change from a more or less solitary life to the crowded society of a great concourse acts upon all as a powerful stimulus, and by itself is calculated to bring about a state of mental exaltation ' (Sollas, 1924, p. 297). At nocturnal dances they appear besmeared with stripes of ochre, chalk or coal, and with birds' feathers stuck in their hair, especially the yellow tufts of the white cockatoo (Semon, 1899, p. 213). The central figures are the novices, who undergo prolonged religious rites and teaching, and are admonished to adhere to the tribal councils and interests and always to place them before their own. ' That the ceremony does indeed produce a deep and lasting effect ', Sollas says, ' is the testimony of those who have been best acquainted with the tribes '. We are ignorant of its full meaning, he concludes, though in general it celebrates the youth's assumption of a man's responsibilities. But it is to be observed that a frequent and central aspect of the ceremonies in different primitive peoples is the mark it confers of eligibility for marriage, a mark with which the males (and in some African tribes the females also) are often physically endued by tattooing or by submitting to circumcision or some analogous surgical operation.

Tribal gatherings of primitive peoples clearly fulfil many necessary functions. It is by sharing together in the common exercise of traditional ceremonies that the spiritual integrity of the tribe is maintained, to say nothing of the opportunities the meeting provides for settling differences, trading, communicating experiences and discoveries, and making political innovations. Ceremonial gatherings of the whole tribe, either annually or at not much greater intervals, are consequently very widespread. Many of the North American Indians, for example, including the Sarcee, Blackfoot and Assiniboine, and some of the Pacific coast tribes, used to hold a great midsummer festival lasting several days, known as the sun-dance: to it the whole people assembled, joining in a kind of spiritual revival meeting, associated with dancing, games and ceremonial hunting. These were the occasions for showing one's prowess and establishing one's rank and status in the tribe. In man, where there is no marked annual breeding season, it is not to be expected that annual prenuptial ceremonies will be developed, but gatherings like these appear to be the only possible source of what knowledge exists regarding changes in numbers in the population, and to be the basis, therefore, on which tribal customs relating to sexual abstinence, infanticide and other forms of birth-control are built up, reinforced or relaxed. Moreover the gatherings establish, or iron out the wrinkles in, public organisation and the social hierarchy, and provide the occasion for hearing and resolving disputes, on all of which the continued orderly dispersion or welfare of the tribe depends.

It has been pointed out earlier that social organisation in the animal kingdom appears always to originate from one prime necessity, which is to provide a self-regulating balance between population-density and resources: in order to secure a workable dispersion, in fact, the activities of individuals require to be co-ordinated. During the last three thousand years, and above all in the last three hundred years of agricultural and industrial development, man has lived in a world of enormously expanding productivity, in which population-growth has been able, unnoticed, to escape from all the formerly imposed restrictions, without ever seriously invoking the ultimate threat of ' over-fishing ', famine and disaster. The elaborate and highly perfected self-regulating machinery, which man used to possess like other species, has all gone by default, so utterly indeed that its former existence may seem now rather difficult to credit: though the evidence of controls everywhere exercised by primitive tribes (briefly reviewed in Chapter 21, p. 492) will be found to admit of no other conclusion. The ultimate threat still exists, and as things are going now it must inevitably be met face to face as soon as the ever-growing demands of world-population exceed the ceiling capacity of world-production. Planned restoration of the means of limiting population and of checking the extravagant rise in living standards, to halt the demand for and reckless exploitation of resources before this crisis comes, is possibly the most pressing biological challenge of our time.

11.10. *Summary of Chapter* 11

1. Communal displays provide an opportunity for assessing population-density, alternative to the other systems of breeding dispersion and property-tenure considered in the two preceding chapters.

2. In marine invertebrates spawning assemblies are common, but it is not easy to decide whether these have any function in addition to securing an efficient fertilisation of eggs shed into the water. They often resemble those communal displays in land animals, however, that much more definitely appear to have an epideictic function. Swarming in water is most likely to occur by concentration towards one of the two permanently-identifiable horizons, namely the bottom and surface, and the latter is preferred if the daylight cycle is used as a synchroniser.

3. Nuptial displays are common in insects, most often involving only the males in some kind of aerial dance or other assembly : to these the females in some cases come only briefly for fertilisation. Probably the most spectacular of all are the displays of tropical fireflies, in which the flashes of thousands of participants are given rhythmically and exactly in unison,

as they occupy night after night particular customary trees. Their displays do not lead to mating and have never been satisfactorily explained: the conclusion is that they are primarily epideictic.

4. Swarming displays of fish usually involve both sexes: in the shoaling sciaenid fish *Micropogon undulatus*, for example, sound-production rises to a seasonal prenuptial climax in June and also to a strong diurnal peak at sunset. In such a species as this that does not require to assemble specially for spawning since it is always gregarious, the drumming display is probably in the main epideictic also. The same applies to the nuptial concerts of the males in gregarious frogs, which are known to have no function either in aiding the assembly of males and females to the traditional breeding places, or in facilitating pair-formation. Nuptial assemblies are also briefly referred to in turtles and snakes.

5. The birds provide an immense wealth of material. In section 5 are mentioned special display assemblies of various Corvidae, and of representative colonial birds, such as Manx shearwaters, herons, pelicans, auks and terns: in the latter the well-known ' panics ' are shown to fit in very readily to the general pattern of epideictic displays. In some cases special pre-nesting assemblies are held that suggest a function of imposing a limit on the quota of breeders. In none of these birds does formal combat feature prominently, and characteristically both sexes, though often provided with nuptial adornments, are alike.

6. and 7. In various birds the males engage in leks. Here the sexes are usually highly dimorphic (though exceptions occur, *e.g.* in the great snipe and the sharp-tailed grouse). Typically the cocks occupy and hold by combat special stances in a special arena: the arena generally remains fixed in location from year to year and the stances are apparently limited in number by convention. The system appears to secure dispersion through the geographical placing of arenas or leks, and the limitation of the number of breeding males at each lek, whereby any surplus is relegated to the status of non-breeders. The lek remains in session as long as any females could come forward for fertilisation, that is to say for the whole breeding season; and it is inferred that the males may withhold coition once a sufficient quota of hens has been fertilised. Male display is primarily directed at other males. The females pay brief visits to the lek only to get served by a male; and the whole burden of nesting and incubation rests on them. Nest-territories do not exist, nor traditional nest-sites, since the lek supplies an alternative form of property-tenure. The males develop a hierarchy amongst themselves and polygamy is common: this may make control of the supposed quota of females fertilised simpler in operation.

All the particular species individually mentioned, in fact, depart from this generalised scheme in more or less minor respects and details: those considered most fully are Gould's manakin, the hermit hummingbirds, the ruff, black grouse and sage-hen: a number of others, including the birds of paradise, are more briefly mentioned.

8. In the mammals communal displays of this kind are relatively less common. The principal case described concerns the prenuptial and nuptial behaviour of the red deer.

9. In primitive man there was no well-defined annual breeding season: but 'initiation' and other public functions were widely held, for which all members of the tribe assembled, usually to a traditional or sacred place, and took part in religious rites, dances, games, or ceremonial hunts, as well as in the initiation of novices. Some of these ceremonies, at least, were deeply concerned with the inculcation and observance of tribal laws and customs, including the regulation of reproduction and property-tenure.

Unfortunately 'westernised' man, owing to the enormous rate of expansion of the resources of his world, has unwittingly discarded the machinery formerly existing for the self-regulation of human populations: if it is not somehow restored the ultimate consequences must inevitably be catastrophic.

Chapter 12

Display characters and natural selection

12.1. *The premises of ' sexual selection '*

This chapter turns aside to look more closely at the relation that clearly exists between dispersionary or epideictic behaviour on the one hand and what is rather loosely called ' sexual selection ', together with related phenomena, on the other. In Darwin's original words (1859, Chapter 4), sexual selection ' depends . . . on a struggle between the individuals of one sex, generally the males, for the possession of the other sex. The result is not death to the unsuccessful competitor, but few or no offspring '. Sexual selection appeared to him to be a process distinct from ordinary natural selection: in his view the latter arises from the struggle to survive all the manifold rigours of the environment, in competition with other organisms of the same and other kinds; whereas sexual selection depends only on two comparatively-speaking domestic factors, namely, success in competition with other males for the attentions of females, and female choice. Of these the second seemed to Darwin by far the more potent. He developed the subject in *The Descent of Man* (1871), devoting to it almost two-thirds of the book, or more than six hundred pages (this he referred to in his Introduction as being ' an inordinate length '). On the matter of female choice he wrote in Chapter 8, for example, ' it appears that female birds in a state of nature have, by long selection of the more attractive males, added to their beauty or other attractive qualities.'

There have been numerous critics of Darwin's theory of sexual selection, the majority of whom have cast doubt on the importance ascribed to female choice. Many so-called secondary sexual adornments and weapons are regularly employed in display by males in the absence of females, as we have noticed over and over again, and evidence of any sort of female choice is not very often revealed by actual observation. In the case best known to us, the human species, in which many quite typical secondary sexual characters have been developed, no one would venture to say that female choice was more important than male choice. Whatever free selection of partners does occur in animals, it is at all events a relatively inconspicuous process compared with the common and widespread exhibitions of rivalry among the males.

In many animals, moreover, as has often been pointed out, there are special nuptial ornaments equally or almost equally developed by both sexes, for example the delicate plumes of the egrets: as an expedient to account for this Darwin postulated a transfer of the characters concerned from the male, which first acquired them through sexual selection, to the female—a process by no means impossible, but on his theory merely accidental and without rationale or survival value.

It is chiefly in the principle of female choice, however, that sexual selection is supposed to differ from any of the forms of natural selection. Darwin thought of it as a deliberate process, rather closely resembling the artificial selection practised by man on his domestic animals and plants. If the importance of female choice has been misconceived or exaggerated, therefore, the question arises whether the whole paraphernalia of secondary sexual characters and nuptial ornaments are produced after all by nothing but 'ordinary natural selection', or whether there is still some difference in principle involved, even if it is not what Darwin thought. Moffat (1903, p. 166) was among the earliest writers to suggest that the phenomenon could be attributed to natural selection. J. S. Huxley (1938), who discussed it much more fully, concluded that, even if it is due to natural selection, and not to sexual selection as Darwin conceived it, nevertheless we can still distinguish 'epigamic selection' as such. 'Broadly speaking,' he wrote (loc. cit., p. 34), 'sexual selection is merely an aspect of natural selection, which owes its peculiarities to the fact that it is concerned with characters that subserve mating (epigamic characters), and are usually sex-limited. . . .' Later he refers to 'the existence of a sex-limited form of intra-specific selection (intra-sexual selection)'.

12.2. *Natural selection and competition*

It is helpful in this connection to review briefly the different kinds of forces or agents that contribute to natural selection. In the first place, a population of animals contains individuals adapted, in varying degrees, to cope successfully with a certain range of conditions in the physical environment—temperature, light-intensity, humidity, *p*H, salinity, and the like; and a change in any one of these conditions is capable of reducing the viability or fertility of individuals not properly adapted to meet it. The whole population, as a continuing unit, must in addition be so organised and situated as to withstand the mortality, partly selective but usually largely random or accidental, caused by the more violent and unpredictable manifestations of the elements; otherwise the *population* will not survive. There is a similarly varied group of biotic selective forces: the individual has to be able to make use of a range of available food-materials, and of types of vegetational cover; or, if it happens to be a parasite or some other kind of dependant, of the environment provided by its host. The stock as a whole must, in turn, be sufficiently resistant to parasitic diseases and sufficiently protected against predators to maintain itself, in spite of whatever mortality these agents may cause among its members.

There are potential selective forces inherent in all the situations mentioned, and together they form a major part of the whole complex of natural selection. It is generally true that individuals submit to them passively, without any element of mutual competition—though as will be seen shortly there is not always a sharp distinction between the competitive and non-competitive elements of selection pressure: true competition, it may be observed, can only arise over advantageous situations for which the demand exceeds the supply. Such forces as we are now considering may present recognised dangers, and individuals may consequently compete for places of safety from them; but the actual pattern of incidence of a disease, for example, depends on the details of how the population happens to be disposed at the time rather than on any competitive situation that the outbreak itself creates. The same is often true even in the case of predation. Thus a passing whale with its huge sieve of baleen engulfs any masses of swarming euphausians that happen to lie in its path, taking a random sample and offering little if any competitive chance of escape. On the other hand there are situations where social competition leads to the relegation of subordinate or surplus individuals to marginal, insecure positions where in consequence they fall easier victims to predators.

For the population-unit as a whole to survive in the face of this group of testing forces it has to be fortunately situated, endowed with the necessary resources of hereditary variation, and possessed of efficient homeostatic machinery by which the density can be made to respond to economic changes and to return towards the optimum when the balance is disturbed. The primary ingredients of successful homeostasis are, as a rule, a social organisation with an effective code of conventions, and a collective fecundity with a sufficient reserve of power to make good the mortality from uncontrollable causes even when this is heavy, yet capable of being automatically checked so as to keep numbers from rising unduly when chance mortality is light. Any population whose social organisation, recruitment-rate and death-rate prove incompetent and unbalanced will get into difficulties, either dwindling away, or being forced to ' overfish ' and consequently to starve or force itself out of its habitat: this is the normal way in which group-selection does its work. Other neighbouring populations may later take advantage of the room so left vacant—provided it has not been irreparably damaged; but open rivalry perhaps rather seldom takes a hand in selection at the inter-population level. Profiting thus by another's failures could only at a stretch be regarded as competitive.

At the interspecific level we can have two species competing for the same food, for instance. In Chapter 17 we shall see that in appropriate circumstances a mutual convention can arise between them, with the effect of a pact, that greatly diminishes the intensity of mutual rivalry over resources and may sometimes abolish it. But in other circumstances the competition can be indirect and concealed, and therefore present no possibility of co-operation. For instance in a lake, divers (*Colymbus* spp.) and pike (*Esox lucius*) may both depend on the same forage fish—perch (*Perca* spp.) perhaps. Their potential population optima are consequently linked, but their fishing methods are utterly different—the bird a dashing raider from above scattering the small fish in alarm, the pike a stealthy lurker in the shallows, waiting for its unwary prey to amble into reach; their paths probably never cross. In practice the relations between two such species are perhaps almost always buffered in various ways, but it is perfectly conceivable that some interplay of selective forces may arise between them, and in that case they are potential competitors. The same kind of bonds must exist between the southern whales and all the other predators, including penguins and fish, that exploit *Euphausia superba*. Indirect competition that never involves rivalry is apparently a common phenomenon in nature.

Another illustration may help to show what is competition and what is not. The great skua (*Catharacta skua*) obtains much of its food in the northern hemisphere by harrying gulls (especially *Larus argentatus*), selecting one that has just fed and chasing it about until in despair it vomits up the contents of its gullet; this the skua adroitly catches before it reaches the water. On land it is still the bully, and a group of herring-gulls disputing for scraps on a beach will give way at once to the skua, leaving it to consume the food; if they did not it would attack them. The last situation especially looks like direct competition, but in reality the skua is a semi-predator or parasite, largely dependent on the food-finding skill of the gulls for its own living. Its numbers are to that extent dependent on theirs, and to enter into competition with them would be its undoing. It may therefore be taken for granted that its depredations have no seriously depressing effect on their numbers or chances of survival. At the same time, strong selective forces may obviously be brought to bear on the individuals of both species through such a relationship.

What is of most interest in the present context is competition *within* the species. An important result of having social conventions is the diversion of competition between members of the society away from their limiting resource—food—into substituted and innocuous channels: the function of this being, of course, to limit population-density in an artificial manner at the optimum level, and to prevent over-exploiting the food-supply. The usual proximate objectives of competition are twofold, as we saw in Chapter 8, the first being concrete—to secure and hold certain property-rights; and the second abstract—to be and remain a member of the society, and often to hold a preferential status in its hierarchy. The two are not always completely separable, because the symbol of membership or dominance may actually be the possession of a particular kind of property.

The consequences of failure to do sufficiently well in this internal competition can be extremely serious: the unsuccessful individual may be prevented from breeding, or relegated as surplus to the establishment of the habitat and forced to get out. He may be luckier if he moves elsewhere, but at the extreme worst he will either find himself unable to obtain the bare necessities of life or succumb directly to the depressing effects of shock and persecution. If the self-regulating process works smoothly the proportion of adult members of the population unloaded in this way may seldom be large: nevertheless the power of social competition as a selective force is clearly enormous.

There can be no doubt that this is the power responsible for the evolution of the special adornments used by animals in all forms of social display. As we found in Chapter 8 (p. 131), the use of dangerous physical weapons for determining social status tends to be avoided as much as possible, because to have rivals freely maiming and killing each other, especially among the best elements in the stock, is against the interests of the population as a whole. Instead, the place of these is very largely taken by threat and other psychological weapons, which are wholly symbolic. At its most elementary level threat consists of exposing the real weapons that could be used to reinforce it—as in shaking one's fist. But it very easily becomes transferred from primary to secondary symbols: the dog not only snarls but growls, and the primitive human warrior paints his body with ochre or woad. The opponent becomes conditioned to respond to these secondary manifestations just as easily as if they were real weapons: the lion's roar can send chills even down human spines. The degree of respect or fear that secondary symbols command is purely a matter of interpretation, and depends on what is ' read into them ' by the opponent; and as the majority of encounters between rivals lead to nothing but bluff and sham fights, there must be a perpetual tendency, through the powerful social-selective machine, to go on embroidering the ' bark ' long after the ' bite ' has become sufficiently lethal for its purpose.

As a result of this transferred symbolism a pronounced element of unpracticality or extravagance tends to appear in the development of symbolic display characters, often contrasting very sharply with the marvellously functional frugality that we are accustomed to admire in the design of other bodily structures. It is this quality which suggests, as it did to Darwin, that some radically different process must be at work in their evolution. But actually the difference lies more in the type of goal attained than in the selective process itself. Where adaptations evolve in relation to ' *real* ' situations, they are as utilitarian and economical as possible, but where the situations are *conventional*, and thus abstract or ' unreal ', the resulting adaptations tend very naturally towards the bizarre. Selection in relation to these conventional situations, is, by definition, social selection, but fundamentally it can still be regarded as just a particular sector, lying within the general broad sweep of natural selection.

Thus it is quite clear that the degree of extravagance permitted in the development and use of display characters must be strictly curtailed when it comes into serious conflict with other selective pressures. A stag whose massive antlers begin to rob the bones of their strength or impair his fleetness of foot is obviously near the point at which the penalties outweigh the advantages. The ideal complement of adaptations in any individual or stock naturally calls for compromise on almost every detail. This is no exception; and there is no justification for the view, at one time popular with Darwinian writers, that sexual selection—supposedly a distinct process—could lead to an excessive development of secondary-sexual characters, as in the great Irish elk perhaps, and thus cause the downfall and extinction of the species.

12.3. *Sex- and age-differentiation*

What may be described as the epideictic approach to social selection provides an explanation for several phenomena in addition to the evolution of nuptial display characters. We appear, in the first place, perfectly able to understand why there should be such a thing as social competition at all, and what the contest is about. The rivalry may involve one sex or both, or the whole population, or, as will be made clearer later, it may, not uncommonly, be restricted to various immature and larval stages. Social competition is a normal feature of the dispersion process at all ages and has not any exclusive connection with sex and reproduction. As we have repeatedly stated, population-density can be adjusted by either of two processes, the first being the more direct method of promoting movement of individuals out of or into the area concerned, and the second being the slower method of adjusting the turn-over (birth-, recruitment- or death-rates) of the resident population. In the tremendous outburst of epideictic phenomena associated with sex and breeding both these processes are involved together; but social displays of an asexual kind are also exceedingly common in 'perennial' animals at seasons other than the reproductive one, as well as in both these and 'annual' species while they are still immature.

The function of the display characters possessed by the individual is to declare or assert their owner's status in relation to whatever social situation prevails at the time. In a species like the black-headed gull (*Larus ridibundus*), for example, plumage characters allow us (and undoubtedly allow the black-headed gulls themselves) to recognise, first, all members of the species regardless of age and sex; second, juveniles and yearlings; third, asexual adults; and fourth, mature adults, which wear the dark chocolate cap over the head from, roughly, February to July. In epideictic activities related to breeding, only birds in the last category need take one another seriously into account: in the winter roosts and resting places, on the other hand, all members of the community, having mouths to feed, are equally involved, except perhaps that the juveniles may be regarded according to prevailing conditions, either as deserving special tolerance or as being the most worthless and expendable of all the members of the group.

Characters that allow a ready distinction between young and adults are at least as common and widespread through the animal kingdom as those differentiating males from females. In part this is an inevitable consequence of growth in size, but very often there is a definite transformation at the onset of maturity: no group illustrates this better than the insects, where it is practically universal. In part it is due also to the specialisation of larva and adult for different ways of life—the one aquatic and the other terrestrial, or the one a slow-moving feeding phase and the other a mobile dispersal phase in the life-history. In many species the young stages are so completely cut off from the adult ones in time or place that there can be no occasion for cross-recognition between them, but there still remains a need for mutual recognition between members of any one self-contained age-group.

The development of what may loosely be called caste differences, together with enough individual variation for 'personal' recognition such as occurs in many vertebrates, may make it possible for an individual encountering another member of its own kind to size up at once the exact situation or relationship existing between them: it may recognise an acquaintance or a stranger, a rival superior or inferior to itself, a potential or actual mate, an undesirable intruder, a neutral or a welcome companion, a dependent or, if it is itself a juvenile, a protector or parent; or indeed in particular cases (*e.g.* the honey-bee) it may perceive other distinctions more subtle than these.

The society is held together by a fabric of individual relationships, depending on a ready and discriminating recognition that releases the appropriate response whenever the parties are in touch with each other. Personal recognition is no doubt a refined accomplishment, associated with hierarchies of the peck-order type. It is safe to conclude that it generally depends on minute details, as it does in man, though we have noticed a few cases, like that of the ruff (*Philomachus pugnax*), where individual differences are much more conspicuous. The same occurs in isolated instances in various other groups, for example the epeirid spiders, and some coccinellid and lucanid beetles, conceivably for the same purpose. Caste recognition, on the other hand, especially between males and females and sexual and asexual states, is evidently much more widely spread among animals, extending to all that breed in pairs and including some of the aquatic forms (*e.g.* among the fishes and polychaetes) that spawn collectively. There are, nevertheless, large groups in which even external sexual differences are hardly noticeable to us, especially among the molluscs, echinoderms and coelenterates.

12.4. *Epideictic characters in birds*

The principles of epideictic adaptation are particularly well illustrated by birds, in the characters of plumage, structure and voice that they have acquired for purposes of display. In analysing the types of distinctive plumages or vocal powers they possess it is convenient to adopt a simple scheme of classification, such as the following, to show the varying extent to which particular castes or categories can be distinguished from one another.

(Group I) *All ages and stages alike* (except for downy young or nestlings).

(Group II) *Adults of both sexes alike*, distinct from juvenile and immature stages, and either (*a*) without, or (*b*) with a special nuptial or high plumage, alternating seasonally with a different asexual plumage.

(Group III) *Adult males differentiated* by a 'higher' plumage (or song), (*a*) all the year round, or (*b*) in the nuptial season only. Females and immatures (and in some cases under III*b*, males in asexual dress also) tend to be alike.

(Group IV) *Adult females differentiated* by a 'higher' plumage (or song), (*a*) all the year round, or (*b*) in the nuptial season only. Males and immatures (and in some cases under IV*b*, females in asexual dress also) tend to be alike.

This scheme is deliberately over-simplified, and in fact transitions from Groups I to II, and from II to III and IV occur: occasionally, for instance, where both sexes have a special nuptial plumage as in II*b*, that of the male is appreciably higher than that of the female, showing a tendency towards III*b*, as in some of the egrets and herons; or it may be just the reverse, as in *Phalaropus* and to a smaller extent in *Charadrius morinellus* (the dotterel), where the females are the more brightly coloured though both sexes have a nuptial plumage, and the situation is therefore intermediate between II*b* and IV*b*.

A special reservation must be made regarding the omission from the scheme of the condition not uncommon, either in birds or other animals, of polymorphism existing among members of the same caste-group. This is possibly a complex phenomenon. In part it appears to concern the present subject, where it relates to the individual variations in nuptial plumage like those of the ruff, which have just been referred to in the preceding section (p. 229); and in part its character seems to be different and to lack plausible connection with social selection, though there is no alternative explanation to offer and its adaptive significance therefore remains obscure. This is the aspect of polymorphism shown by the colour-phases of the fulmar (*F. glacialis*) and various other petrels, the red-footed booby (*Sula sula*), the blue and snow geese (*Anser caerulescens*), the pratincoles, skuas (*Stercorarius* and *Catharacta*), some egrets, owls and wheatears (*Oenanthe*), the Gouldian finch (*Poephila gouldiae*), or the natal downy plumage of the arctic tern (*Sterna macrura*).

Into Group I, with no clear differentiation between immatures, males or females, fall the majority of the Order Tubinares (petrels, etc.), though in one family of this order, the albatrosses (Diomedeidae), there are commonly progressive changes in plumage and bill-colour during the long adolescent period, comparable with those of the large gulls: to this the royal albatross (*Diomedea epomophora*) provides a slight exception, since the visible differences between immatures and adults, though still present, are very much reduced (*cf.* Murphy, 1936, p. 579). Sexual differences are always slight throughout the order: the males tend to be rather larger, particularly for example in *Macronectes*; and the male Campbell-Island royal albatross (*D. e. epomophora*) has a white patch on the wing near the humeral flexure, which is absent in the female (and absent in both sexes in *D. e. sanfordi*) (Oliver, 1955). In *Fulmarus glacialis* the males have larger bills than the females, and in the Atlantic races of this species the sharp 'claw' which forms the terminal third of the upper mandible is sufficiently stronger in the males to allow reliable sexing of birds in the hand, and sometimes at close quarters in the field (Wynne-Edwards, 1952). Such a character would be quite sufficient for sex-identification between individuals displaying to one another, especially as the open bill is the principal object of display. Fulmars and some other Tubinares, therefore, show a tendency towards Group III*a*, in which the males have a distinctly higher development of display-characters than the females. In the wandering albatross (*D. exulans*), and perhaps others, there is a prenuptial change in colour of the bill and feet of prospective breeders, and in most cases the ear-covers turn pink (Matthews, 1929, p. 564), which therefore relates this species to Group II*b* as well. There are also evidently slight sexual differences as well as seasonal changes in the dark spotting of the crown in *D. epomophora* (Richdale, 1942, pp. 180-1).

In the geese (*Anser*, *Branta*, etc.) the differences between immature and adult birds are rather slight, but generally quite sufficient for age-determination at close quarters: usually the changes at maturity involve heightened colouring of the bill and legs or the appearance of small but significant details in the feathering, such as the diagnostic white forehead in *A. albifrons* and *A. erythropus* or the white patches on the sides of the neck in the brent, *B. bernicla*. The differences between juvenile and adult stages are least in the Canada geese (*B. canadensis*), and greatest in the snow geese (*Anser caerulescens*): in the latter the juvenile has a greyish plumage comparable with that of a young swan. On the whole it is probably more accurate to place the geese in Group II than in Group I.

Examples of birds in Group I turn out, in fact, to be decidedly uncommon. In the remaining non-passerine orders similar examples can easily be found, of species of families in which the differences between the juveniles and males and females are all rather slight, as in some of the nightjars (Caprimulgidae), swifts (Apodes), kingfishers (*e.g.* *Alcedo* but not *Megaceryle*), bee-eaters (*Merops*), and in the hoopoe (*Upupa*) and wryneck (*Jynx*), but in all of these the differences are probably sufficient at least for age-discrimination at close quarters. The differences are very small in many of the owls (Strigiformes); and in the sandpiper-like birds (*e.g.* *Tringa* and *Calidris*) they often show more prominently in the coloration of legs and bills than in the plumage.

Among the Passeres there are several families or genera, such as the larks (Alaudidae), pipits (*Anthus*), certain Old World warblers (Sylviidae), thrushes (*e.g.* *Hylocichla* and some species of *Turdus*), vireos (Vireonidae), and buntings (Emberizinae) where the plumages show little or no age- or sex-differences, but which nevertheless can be put straight into Group III because of the males' songs. This category can be extended to include many non-passerine birds also, in which the plumages are often similar and cryptic at all ages, but the males have a distinctive vocal display, for instance in some of the rails, Limicolines (*e.g.* curlew, snipe and woodcock) or quails. It might be thought that good examples of Group I could be found among the Corvidae, and especially in the genus *Corvus*, where there are no recognised sexual voice differences and the plumages in most species are wholly black, but in practice there is never much difficulty in distinguising the juveniles, if only by their browner or more sooty colouring or lack of glossiness and iridescence. The same is true of the magpies (*Pica*); and in the choughs (*Pyrrhocorax*) there are also colour differences in the bill and feet.

The conclusion is, regarding Group I, that on closer examination it is difficult to find examples in which there are not perfectly adequate visible or audible display signals serving to differentiate mature males from females or immatures or both. The most convincing cases are found in the Tubinares, in the families of shearwaters (Procellariidae), storm-petrels (Hydrobatidae) and diving-petrels (Pelecanoididae), but even here some species are known to show small sexual differences. Where such a close similarity prevails, however, between all free-flying members of a species, regardless of age and sex, it presumably means that the asexual social relationships that exist while they are at sea are of predominant importance in the evolution of their visible characters. When they come to land, there is likely to be active competition in obtaining or retaining nesting burrows: Lockley (1942, p. 87), for example, mentions that Manx shearwaters quarrel and fight for homes. But members of these three families typically make nests underground, and are active in display only at night, so that there is practically no chance for the development by social selection of visible nuptial characters. Where age- and sex-differences are better developed, as in the albatrosses, giant fulmar and fulmar, the birds display at exposed nest-sites in broad daylight, which is likely to be a significant factor.

The owls offer some confirmation of this, since most of them are also nocturnal and show practically no visible caste-differences, but here again in the snowy owl (*Nyctea scandiaca*), obliged to be diurnal in the continuous daylight of its arctic summer home, the immaculate mature males and lightly speckled females can easily be distinguished from each other, and the young birds start life with a much more darkly barred plumage.

Group II, on the other hand, is a very large one. In II*a*, where there is no seasonal change from non-breeding to breeding plumage, the only two distinguishable castes are adults and non-adults. Into this division come the swans (*Cygnus*), various geese already mentioned, whistling ducks (*Dendrocygna*), many of the Steganopodes such as the gannets and boobies (Sulidae), some pelicans (Pelecanidae) and tropic-birds (*Phaëton*); many herons (Ardeidae) and especially *Nycticorax* in which the immature plumage is entirely different; most vultures, eagles, falcons and hawks, pigeons and doves, and, among the Passeres, for example the swallows (Hirundinidae), dippers (*Cinclus*) and the Old World robin (*Erithacus*). The list could be greatly extended.

Several of these groups contain exceptions which fall into Group II*b* rather than II*a*, since nuptial adornments are developed—for instance in the American white pelican (*Pelecanus erythrorhynchos*), in which the adults grow a remarkable horny crest like an oversize rifle-sight in the middle of the upper mandible in the breeding season, and afterwards shed it. The Old World white pelican (*P. onocrotalus*) is one of the rather numerous birds that acquires a transient rosy flush on its plumage in the nuptial season, due in this case to carotenoid pigments that appear in the preen-oil at that time and are transferred to the feathers (Stegmann, 1956). Either rose, lilac or metallic grey transient blooms of this sort are common in the small gulls and terns, in tropic-birds and many others: the subject is a most interesting one and much neglected, though it is too involved to discuss fully here, but in most cases the transient colouring appears to be a nuptial adornment.

The Ardeidae also provide interesting exceptions: many of the egrets grow beautiful nuptial plumes called 'aigrettes'; and in the American yellow-crowned night-heron (*Nyctanassa violacea*) the handsome white crown and long crest-feathers become tinged with an ephemeral sulphur-yellow, at least in the 'highest' individuals, during the breeding season. This is another manifestation of the phenomenon just mentioned of a nuptial flush of carotenoid lipochromes appearing in the plumage. Naumann (1822, p. 119) says that the underparts of the black-crowned night-heron in Europe (*Nycticorax nycticorax*) may be similarly tinged with a delicate yellow in life; and various other herons possess a bluish bloom that quickly fades after death. In the common heron (*Ardea cinerea*), and a number of other species also, there are ephemeral changes in the colour of the legs and bill, at least in some individuals (Tucker, 1949).

The gulls (Laridae) are very typical representatives of Group II*b*. All species exhibit a graded series of distinctive juvenile and adolescent plumages speckled and washed with buffs, browns and greys. The adults have a pre-nuptial moult, which in almost all produces a more or less conspicuous change of the head and neck feathers: in the larger gulls the whole head becomes pure white, having been somewhat grey or speckled in winter plumage, and in many of the smaller species the moult produces a black hood. Analogous changes occur in the terns; and in both small gulls and terns, as mentioned already, roseate or metallic flushes often appear on the feathers.

There are few avian families with ornaments more weird and wonderful than those of the puffins and auklets, among the most modest of which is the parti-coloured 'neb' of the Atlantic puffin (*Fratercula arctica*). The sexes are of course alike at maturity. The Bering-Sea tufted puffin (*Lunda cirrhata*) has a huge bill at breeding time measuring up to 6 cm in length and 5 cm in height, salmon-red in colour with a light olive-green base: its black plumage is relieved only by a white triangle on each side of the face, from the hinder point of which, over the ear, springs a spray of sweeping, straw-coloured plumes. The rhinoceros auklet (*Cerorhinca monocerata*), and crested and whiskered auklets (*Aethia cristatella* and *Ae. pygmaea*) are if possible even more bizarre. All the twenty-odd living species of the Alcidae fall into Group II*b*.

Space does not permit us to consider more than a few examples in each group, but attention should perhaps be drawn here to the large number of Limicolae, including golden plovers (sub-species *Pluvialis*), godwits (*Limosa*), turnstones (*Arenaria*), knot, dunlin and other species of *Calidris* and of *Tringa*, in which the two sexes are alike in having a higher nuptial plumage, but the male is differentiated by possessing a song: they fall therefore into Group III*b* rather than II*b*.

A generalisation to be made about the birds in Groups I and II is that, with relatively few exceptions, they are colonial ' nest-site ' birds, rather than ' territory ' or ' lek ' birds. In the two latter categories there tends to be a division of labour between the male, who is much occupied in territorial and lekking displays, and the female, who undertakes nest-building and incubation. In colonial nest-site birds, on the contrary, very often the best indication that a site is claimed is the visible presence of an occupant: provided the sexes are similar in general appearance the presence of either member of a pair will be sufficient to guard the site. Thus in typical Group-II species all the mature adults, regardless of sex, have to take part in social competition, and the adornments associated with success in the breeding season come to be shared equally by both.

Sometimes, nevertheless, though both sexes develop adornments, the male's are appreciably nobler or stronger than the female's; and actually it is easily possible to find a transitional series leading into Group III. This may be due either to a somewhat unequal and predominant share of social competition falling upon the male, or perhaps to the actual epigamic function of the adornments in courtship, where the male and female necessarily have rather different roles—or to both these factors together. The point is referred to again later (p. 236).

Before leaving Group II, passing reference must be made once again to the anomalous European robin (*Erithacus rubecula*) and the dippers (*Cinclus* and *Cinclodes*), where the adults are alike in plumage. The robin is a territorial species (*cf.* Lack, 1943, p. 96); the palaearctic dipper (*C. cinclus*) would appear to be a nest-site species in the breeding season, judging by its strong preference for traditional sites, though like the hole-nesting tits (*Parus*) its form of property-holding may perhaps combine the characteristic features of both territory and nest-site tenure. In both cases the cocks and hens sing, but in the robin the hens sing only in the asexual season when the species is dispersed in a system of solitary territories, and the same is, somewhat less regularly, true of the dippers (personal observation; *see* also H. Richter, 1953; and Bakus, 1959, p. 198). In the breeding season, therefore, these birds can be assigned to Group III on the basis of sexual difference in song-display: at that time they differ little from, say, tits (*Paridae*) and nuthatches (Sittidae), where both sexes are also alike in possessing conspicuous plumage, and where in the nuptial season the males compete by voice.

In Group III the males are distinguished as a caste by visible colour-patterns or by song (or both), either (*a*) all the year round, or (*b*) in the nuptial season. This is, therefore, the category regarded as most typical in Darwin's theory of sexual selection, according to which the so-called secondary sexual characters are selected in the male largely as a result of female choice. In terms of social selection, on the other hand, Group III appears to be associated with a division of labour in the breeding season between the sexes, so that dispersion and epideictic competition are the affair of the males, and the domestic duties of nesting are largely or entirely performed by the females.

Straightforward cases in Group III are very numerous, particularly in the Passeres, where the males so frequently develop a song associated with competition for territories or widely-scattered nest-sites: they often have resplendent nuptial plumages also, for example in the tanagers (Tanagridae), the American goldfinch (*Carduelis tristis*), members of the genus *Passerina* such as the indigo and painted buntings, and the African bishop-birds (*Euplectes*) and wydahs (*Steganura* and *Vidua*); or they may remain in high plumage all the year round, as in numerous other finches (Fringillidae), many sunbirds (Nectariniidae), the cardinals (*Richmondena*) and golden orioles (*Oriolus*). There are, as we have seen in the previous chapter, examples of lekking birds among the Passeres also, including the birds of paradise (Paradisaeidae), manakins (Pipridae), and, stretching the term a little, bower-birds (Ptilorhynchidae), the males of most of which can compare in nuptial magnificence with anything to be found elsewhere in the avian class. In the non-passerine orders, the game-birds (Galli) are especially notable for the frequency and beauty of male ornamentation, as well as for their tendency to lek and fight, in some cases with sharp spurs. Most hummingbirds (Trochili) tend to have iridescent jewel-like patches of brilliant colour about the head and breast in the male, though a minority of the species have little or no sexual dimorphism. The ruff (*Philomachus*) need only be mentioned by name here, since it was fully considered earlier (p. 209).

In the diurnal birds of prey (Accipitres) the immature stages are almost always distinct: the female is often considerably larger than the male, but otherwise the sexes are alike. However, there are well-known exceptions, notably in the kestrel-like falcons (subgenus *Cerchneis*) and harriers (*Circus*), where in both cases the male is the more conspicuously coloured. Birds of prey are all or mostly nest-site birds. The extra size and strength of the female, most famous in falcons such as the peregrine and merlin, might be thought to qualify them for the following category, Group IV, but the size difference is associated in *Cerchneis* with a more brightly-plumaged male sex, and may therefore quite probably be neither epideictic nor epigamic in origin. It need not necessarily result from social selection at all and might, for instance, be in some way related to killing and carrying prey for the young, or providing the nutrient reserves demanded in producing a clutch of eggs. By way of analogy, in many species the body-temperature of the female is slightly higher than that of the male, a difference more likely to be correlated with, say, incubation than with any social activity. In March 1944 there was an intermittent but furious battle lasting several days in which two or sometimes three peregrines shared, around the Sun Life Building in Montreal, where a pair of peregrines had nested for six years previously: the sexes of the participants were not directly determined, but one bird, a male, was killed in the fight and later found partly eaten on the nest-ledge—after which a pair was left to breed. The inference was that there had been two males contesting for the possession of both the site and the female: that is, playing the active role in dispersion.

Reference has already been made to the fact that there is not a sharp distinction between Groups II and III, since there are many species in which both sexes obviously have nuptial adornments, although those of the male (much more rarely the female) are distinctly richer or more elaborate. A good example of this is found in the lapwing (*Vanellus*), a very beautiful plover in which both sexes have rounded wings and share a conspicuous flight pattern of black and white: many of the dark parts of the plumage have a handsome green or purple sheen, and around the base of the white-and-black tail there is a rich chestnut band. The bird's most characteristic ornament is an upsweeping erectile occipital crest, of which the longest feather may have a shaft-length of 100 mm. The wings are broader than in any other plover and provide an excellent recognition character. Their shape is due to the unusual length of the middle primaries: in most plovers the second primary is the longest, and the third and succeeding ones get rapidly shorter, but in the female lapwing the third to fifth primaries are actually longer than the second, and in the male the third to seventh are all longer than the second, and longer than their counterparts in the female also. The resulting difference in wing-shape between male and female permits easy identification of the sexes in flight. The male in spring has a tumbling display flight expressly to show off the breadth and bold pattern of the wings, and there is a characteristic drumming sound made when they beat vigorously against the air. The male has finally a joyous exuberant song which he repeats time after time during his flight displays.

Those who have studied the lapwing's breeding behaviour describe it as a territorial species (*cf.* Spencer, 1953, p. 31), but they find that the boundaries of individual territories are not firmly defined, and may even be subject to variation from day to day: these holdings are not feeding territories, since the owners usually go away to neutral ground to feed, and, after hatching, the young chicks do not necessarily remain at all on their parents' nest-ground. Lapwings show a considerable degree of sociability in their nesting habits, and often form diffuse colonies in suitable habitats, each pair claiming some two acres or so on which to place their nest. In some important respects, therefore, their breeding dispersion combines features of the colonial nest-site system and the territorial system, and it is possible that the social characters in plumage and voice which they have developed, combining as they do the attributes of Groups II and III, are correlated with this.

12.5. *Reversed sexual roles* (Group IV) *in birds and arthropods*

Group IV differs only in the reversal of the epideictic and epigamic roles of the sexes, the female being in this case the one with greater adornment and more aggressive temperament. The number of examples found in birds is not very large: they were reviewed at some length by Darwin (*Descent*, chap. 16), since when comparatively little has been added to our knowledge of them.

One of the most interesting cases is that of the tiny bustard-quail (*Turnix suscitator*) of India, Ceylon and the whole of south-east Asia, well described by Jerdon (1864, p. 109). The female is larger than the male, and has a conspicuous black area from the chin down the centre of the breast which the male lacks. According to Jerdon, ' the hen-birds are most pugnacious, especially about the breeding season, and this propensity is made use of, in the south of India, to effect their capture. For this purpose a small cage with a decoy bird is used, having a concealed spring compartment, made to fall by the snapping of a thread between the bars of the cage. It is set on the ground is some thick cover carefully protected. The decoy bird begins her loud purring call which can be heard a long way off, and any females within earshot run rapidly to the spot, and commence fighting with the caged bird, striking at the bars. This soon breaks the thread, the spring-cover falls, ringing a small bell at the same time by which the owner, who remains concealed near at hand, is warned of a capture; and he runs up and secures his prey, and sets the cage again in another locality. . . . The birds that are caught are all females, and in most cases are birds laying eggs at the time, for I have frequently known instances of some eight or ten of those captured so far advanced in the process as to lay their eggs in the bag in which they are carried, before the bird catcher had reached my house. . . . The females are said by the natives to desert their eggs, and to associate together in flocks, and the males are said to be employed in hatching the eggs, but I can neither confirm nor reject this from my own observation '.

This suggests that the species is territorial since the bird-catcher moved his decoy-trap after making a capture, and furthermore that the territorial claim is advertised by voice. ' It is the cock-bird that has to do all the hatching and looking after the young, for the hen, as soon as she has laid her first set of eggs, goes off to hunt up another male to look after her second laying, and so on, until matrimony palls for the season and she either indulges in lonely blessedness or joins one or two other ladies who are grass widows for the time being ' (Baker, 1928, V, p. 446).

Many other species of *Turnix* variously known as button-quail, painted quail, quail-plover, hemipodes, etc., are found both in the oriental region and in Africa and Australia: usually the female is larger than the male, but in most species the sexes are alike in plumage. In *T. tanki*, also found in India, the female has a bright rust-red collar which the male lacks. It may be mentioned that the Turnicidae have no close relatives, and constitute a very isolated group, of doubtful affinities.

Another case of polyandry rather closely parallel to that of the bustard-quail has recently been discovered by Hoffmann (1949) in the oriental 'water-pheasant' or pheasant-tailed jacana (*Hydrophasianus chirurgus*). This beautiful bird lays its eggs on floating lotus leaves, sometimes without any nest: a single female provides several males one after another with a clutch of four eggs, the interval between successive clutches being about 9-12 days; she can produce 3 or 4 clutches within a month, and from 7 to 10 in a whole breeding season. All the incubating as well as the tending and protecting of the young is done by the male: he may incubate a second clutch, or sit for a third time if the eggs are destroyed. The female is a little larger than the male and takes charge of territorial defence.

There are several examples among the true Limicolae, including the dotterel (*Charadrius morinellus*), where the dimorphism is at most slight and there is some overlap between the brighter males and the duller females: the male is thought to undertake most or all of the incubation. The dotterel is a bird strongly inclined to re-use the same nest-site from year to year. In the three species of phalarope (*Phalaropus fulicarius*, *lobatus* and *tricolor*) both sexes have a spring moult into a special nuptial dress, but that of the female is conspicuously the brighter of the two, and she is slightly the larger bird: phalaropes nest in colonies, and the females quarrel and display among themselves as well as courting the males. The males perform the whole duty of incubating the eggs and caring for the young, remaining on the breeding grounds long after the females have left. A similar situation obtains in the painted snipe (*Rostratula benghalensis*), a species distributed from Africa through south Asia to Australia: here the female's plumage is more richly coloured than the male's and she is a larger bird; in addition, her characteristic trachea is apparently more convoluted or kinked than that of male and immature birds (*cf.* Lowe, 1931, p. 529). 'The females fight for the males and challenge one another with a loud note, sounding as if someone was blowing into a bottle. Their display is a fan-like spread of the wings and tail over the head whilst the bird crouches on its breast. The display seems to be both a warning to other females or enemies and an invitation to the male, being always accompanied by a loud hissing' (Baker, loc. cit., VI, p. 47). There is a South American painted snipe (*Nycticryphes semi-collaris*) in which the sexes are alike, but no other near relatives.

There are many genera of birds besides these already mentioned, in which the females are larger than the males, including the kiwis (*Apteryx*), emus (*Dromaius*), cassowaries (*Casuarius*), and tinamous (Tinamidae), and in the last two the female is said to be the dominant sex in nuptial displays. There are additional examples too of females having adornments that are lacking in the male sex, though in some of these the significance of the secondary sexual differences is not immediately apparent: for instance, the colour pattern of the male of the American belted kingfisher (*Megaceryle alcyon*) is entirely grey-blue and white; the female has a very similar plumage (as is usual in kingfishers), with the notable addition of a colourful rusty band across the breast and down the flanks, below the grey ' belt ' that gives the species its common name. In juveniles of both sexes the feathers of the grey belt have rufous tips, which are lost with approaching maturity, so that all three ' castes ' are distinguishable. It seems probable that the adult male's should be regarded as the highest state of plumage, in spite of his having lost all trace of the conspicuous red colour.

Darwin (loc. cit., p. 733) explained the reversal of dominance between the sexes by postulating that it originates in species in which the females have become much more numerous than the males, and are thus led to court the males. ' Sexual selection,' he wrote, ' would then do its work, steadily adding to the attractions of the females.' A simple numerical unbalance will not provide the basis for an explanation on the present theory, however, since the males' role of dispersing the breeding population can continue quite independently of the state of the sex-ratio existing at the time. In a polygamous species such as the Alaska fur-seal (*Callorhinus*) it is possible to slaughter, in the Pribilof Islands alone, fifty thousand bachelor males annually without either upsetting the social system or lowering the fecundity-rate among the cows: epideictic competition continues to exist among the surviving bulls, greatly though they are outnumbered by the opposite sex.

We require an explanation that will indicate, first, how the inversion of the normal epideictic roles can arise, and second, why it is never a very common phenomenon, though it crops up sporadically in most groups of arthropods and vertebrates. As far as birds are concerned, a number of the cases we have noted refer to species belonging to families in which nest-site tenure is the common system of breeding dispersion and in which the majority of species have both sexes alike (*i.e.* Group II). The Group IV cases generally occur as more or less isolated exceptions to the rule in the families or orders to which they belong; and, though it seems rather a glimpse of the obvious to point it out, they all appear to have arisen from ancestral species with no secondary sexual differentiation in visible pattern. A species committed to a Group III regime, that is to say with dominant males appropriating the epideictic stage in the breeding season, could not easily switch over to Group IV without first reverting to the Group II condition of epideictic equality between the sexes.

Evidently there must be some kind of natural disadvantage or handicap about the Group IV regime: if this were not so we should expect Groups III and IV to be equally common and numerous, which they certainly are not. Only in the Phalaropidae do we find a complete family (actually consisting of three living species) belonging to Group IV. The remaining examples are all single genera or species, suggesting that the evolution of a dominant female sex is invariably something of quite recent origin, and, by inference, therefore a ' wrong turning ' or side-track likely to end in extinction. By contrast, the development of a dominant male sex runs right through large families, such as the Fringillidae, Parulidae, Sylviidae, Turdidae, Tetraonidae, and many others.

The essential clue to this problem has already been briefly touched upon in an earlier chapter (p. 93), where reference was made to the fundamental difference between the primary reproductive functions of the two sexes, and the effect of this in preventing a completely free interchange of their respective secondary sexual roles. The underlying factor is the great difference in size, motility and cost between spermatozoa and ova. The former can be produced in millions with relatively little drain on the resources of the body, whereas the latter tend to be loaded with valuable nutrients, and consequently to be far less expendable and greatly restricted in number. Copulation has been evolved over and over again in different animal groups for the primary purpose of securing a high fertilisation rate and reducing wastage of precious ova. The basic difference arises from the need for two gametes to unite their nuclei in sexual reproduction. Instead of both being alike, as they no doubt were in the beginning, the female gamete (in the Metazoa) always provides the nutrients required for embryonic development and as a consequence sacrifices its mobility; whereas the male gamete is stripped down to the essential nucleus and the locomotive apparatus required in searching for the passive partner. In copulation, therefore, the motile sperms are almost always passed into the female, rather than the eggs being passed into the male—though the latter is not impossible and does in effect occur in a few rare cases where the male sex possesses a brood-pouch, for example in the pipe-fishes (Syngnathidae) and sea-horses (Hippocampidae).

In both freshwater and marine animals a good many species spawn *en masse*, but it is on the whole more common even in aquatic habitats for animals to associate in pairs for the purpose of fertilisation, either with or without actual coition: in terrestrial animals copulation is universal. As we have seen, the male is constitutionally free to fecundate any number of females as long as the season of his maturity lasts, but the female on the other hand is potentially able to get all her ova fertilised through a single mating act: she is therefore likely to need the service of only one male once. Only where the habit has been developed of laying the eggs in separate batches can the female effectively use the services of several males. Pairing does not commonly entail the continued association of two individuals in monogamous marriage, except in the higher vertebrates, and the result of this is that, during the breeding season there is frequently a ' pool ' of males available at any time to pair with a female requiring it. Competition consequently arises readily among the males for an opportunity to participate in reproduction, and the epideictic role thus tends to fall to them. Polygamy by females (polyandry) is inevitably on a much more limited scale, because of the high cost of egg-production, and it can only be effective if the male, after serving the female, is immediately put out of commission, either by having to devote himself full-time to the care, protection or incubation of the eggs, or by dying or being actually slain; otherwise mating would become promiscuous and not polyandrous.

A Group IV regime thus presents considerable difficulties and limitations. Even in species with monogamous reproduction (at least in terms of a single breeding season), like the majority of birds, it loads on to the female the two extremely onerous duties of establishing the successful place in the social organisation that is pre-requisite to breeding, and then producing, frequently at almost the same time, a clutch of eggs. The magnitude of the latter task may be gauged from the fact that the female song-sparrow (*Melospiza melodia*), itself weighing some 22 g, usually produces 4-5 eggs weighing 2·3 g each, that is, a clutch amounting to about 50 per cent of its own body-weight (Nice, 1937, pp. 21 and 113): in a more extreme instance, the bob-white quail (*Colinus virginianus*), a 200-g bird, lays usually about a dozen eggs of 18 g each, so that the clutch-weight generally exceeds the layer's body-weight (Amadon, 1943, p. 225). *Turnix*, *Rostratula* and *Phalaropus* all normally lay clutches of four.

Before concluding this section brief reference may be made to the incidence of the Group-IV sexual relationship in animals other than birds. The same conclusion appears to be generally true, that a reversal in the epideictic and epigamic duties of the sexes is always a rare and exceptional development.

Some of the most remarkable examples are found in the copepod Crustacea, an order of small planktonic organisms, in which the male and female clasp in copulation: they are much given to miniature adornments and, as we shall find in Chapter 16, to what appear to be epideictic displays. Their decorations often consist of very long, brightly-coloured setae, which are sometimes plumose or feathery, though unfortunately in the most ornate species they are so fragile that perfect specimens are seldom captured. Many of the most handsome species have been illustrated in five beautiful colour-plates by Giesbrecht (1892), from whose classical work most of this information is taken (*see* fig. 17).

Frequently both male and female are alike, and both can be similarly and splendidly adorned, as in *Euaugaptilus filiger* (= *Augaptilus filigerus*) and related species of *Hemicalanus*, where the adults have as many as fifty long coloured plumes appended to their second antennae, thoracic appendages and tail. It is generally true of copepods that the females are somewhat larger and live longer than males, and there do not seem to be many conspicuous examples of the Group-III condition, with the males as distinctly the more ornate sex: in *Sapphirina*, however, the males have beautiful iridescent colours, whereas the females tend rather to have instead bright blue or red ovaries or egg-sacs and are therefore quite different though equally conspicuous in appearance.

In *Calocalanus pavo*, a typical Group-IV example, the tail-fork of the female bears a handsome fan of eight broad brick-red plumes, and others, both feathery and simple, appear on the large antennules. In *C. plumulosus* the left side of the caudal furca carries a single immensely elongated flexible orange-coloured plume, trailing like the tail of a kite and nearly six times as long as the body. In another family altogether, *Oithona plumifera* has

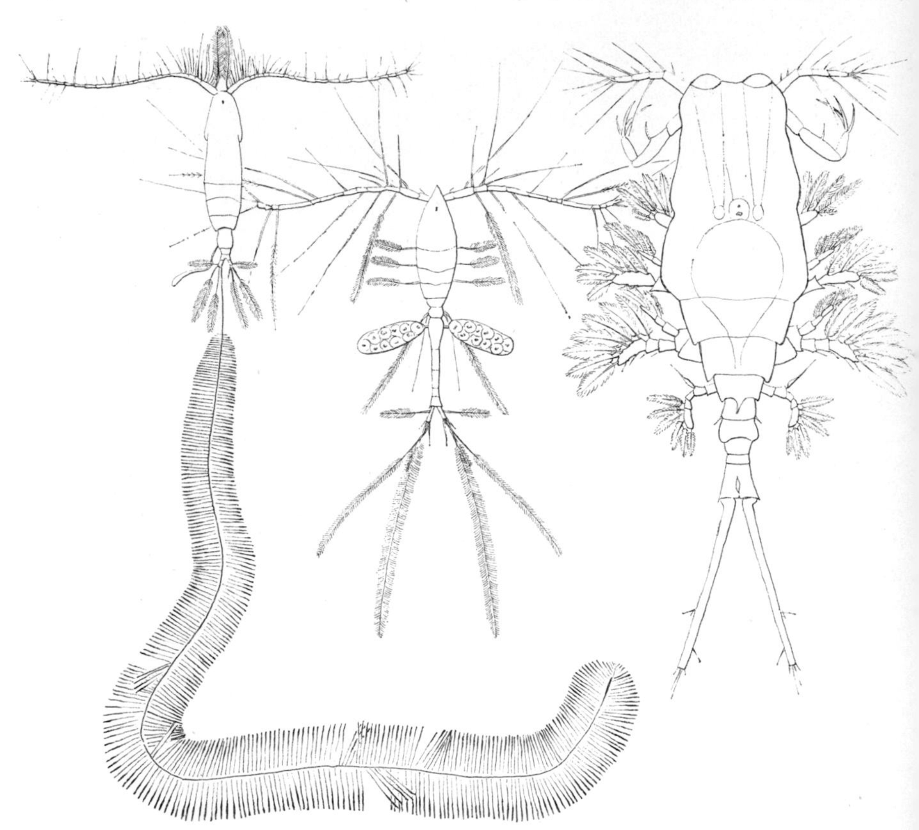

FIG. 17. Three resplendent female copepods belonging to different families. From left to right, *Calocalanus plumulosus* (×20), *Oithona plumifera* (×20), and *Copilia vitrea* (×16). The plumes are lacking in the males of *C. pavo* and *C. vitrea*, and in the male *O. plumifera* they are reduced. The plumes are brightly coloured, in *C. plumulosus* orange with a metallic iridescence, in *O. plumifera* rust-red, varying in intensity in different females, and in *C. vitrea* iridescent between orange-red and violet. (Redrawn from Giesbrecht, 1892.)

independently acquired a similar dimorphism, the female having long scarlet feathery setae on her appendages and tail. Just as remarkable as any of these is *Copilia vitrea*, belonging to a group with extraordinarily specialised eyes, which are paired, each having a lens and a long conical transmitting column leading back to the retinal pigment deep in the trunk of the body. In this species the trunk limbs all carry fans of brilliant orange plumes in the female, and, very significantly, her eye-lenses are much larger than those of the male.

This example perhaps more than any other serves to cast serious doubt on the function sometimes ascribed to all these showy adornments in copepods, namely that they are primarily floatation devices (*cf.* Murray and Hjort, 1912, p. 580; MacGinitie and MacGinitie, 1949, p. 255). When fanned out the expanded surfaces would naturally offer a drag against motion in any direction—up, down or horizontally—but many copepods if not all have an oil-sac that serves them as a float. At the same time many also make vertical migrations of scores of metres under their own power, both of which circumstances detract from the value to such active animals of any sort of parachute. It must be emphasised too that the adornments are always restricted to adult stages and sometimes to one sex.

Among the decapod Crustacea an example of the same reversed relation is reported by Alcock (1902, p. 269) in some species of *Aristaeus*, a genus of deep-sea prawns. In *A. crassipes* the rostrum of the female is very much larger than that of the male. Since ' the rostrum is generally regarded as the most formidable weapon that a prawn possesses . . . we are inclined to think that the character of the sexes is reversed '. The females are also far larger and more numerous than the males—a condition parallel to that obtaining in some of the copepods just mentioned, such as *Calocalanus*.

There are doubtless a number of similar examples in insects. As remarked earlier (p. 49), in the beetle *Phonapate* only the female stridulates. Among the empidid flies, many of which have ' dancing ' swarms much like those of chironomids, although the swarms usually consist of males, sometimes they contain both sexes (*Hilara* spp.), and in *Empis borealis* (Howlett, 1907) and apparently *E. livida* also (*cf.* Hamm, 1908) females only. In the two latter species the males come singly to the swarm, bearing their customary courtship gift of fresh or living prey, in order to entice a female away.

12.6. *Recognisable castes in relation to dispersion and the social hierarchy*

The two preceding sections have shown, as far as birds are concerned, that there are very often three recognisable castes in the population, namely the dominant sex when mature (usually the males), the other sex when mature, and the immature individuals: this is generally true throughout Groups III and IV of the classification on p. 230. The differences between the castes are in the main visible ones, though voice may play a very important part in distinguishing the dominant sex in the breeding season. There is a minority of species where no castes are apparent (Group I), another group in which distinction is possible only between adults and young (Group II), and yet others where, in addition to the three basic castes, it is possible to differentiate sexual adults from neuters, and sometimes older adolescents from younger ones as well.

It has been pointed out that all these caste differences appear to be connected with the social organisation of the population, and to play their part in dispersion. In the shearwaters (*Procellaria*), for example, which belong to Group I, the population lives continually at sea in the non-breeding season, where food-resources are communal and dispersion is naturally asexual. The entire population then consists of a single caste in which all individuals are on a basis of mutual equality. In the breeding season, whenever the breeding birds leave the vicinity of their colonies in order to feed, they normally go to a distance and remain away for several days at a time, no doubt mingling with non-breeders offshore on the same unaltered footing of equality. Huge epideictic gatherings may be held at dusk when they return to the breeding station (*see* p. 205), but as their truly sexual activities take place largely in the dark or underground, visual dimorphism is not to be expected, and nuptial display is by voice (and probably by touch). Sex distinction may possibly exist on a vocal basis, though if so it is not conspicuous and has not been recognised by ornithologists.

In Group II social selection has evoked a recognisable superiority in birds that are fully mature, aiding them in securing and holding the nest-site which they must have as a licence to breed. Individuals lacking the full nuptial adornments or with traces of immature plumage have much reduced chances of obtaining a site, although they are sometimes sexually potent. Usually, however, the younger year-groups tend to stay away from the breeding colonies, as we can observe in the gulls or gannets. The younger members that do appear there form part of the potential reserve, able to breed if there should be a vacant place, but usually there is already a surplus of adults present which precludes their chances. In Group II, briefly, social selection has led to a differentiation between individuals qualified to compete for nest-sites and those belonging to the asexual or juvenile castes.

In Groups II, III and IV we have noted examples of special adornments that last only the length of the nuptial season, for instance in black-headed gulls (*Larus* spp.), golden plovers (*Pluvialis*), tanagers (*Piranga*), or phalaropes (*Phalaropus*): during the post-nuptial moult in these groups the higher castes become ‘ reduced to the ranks ’, or de-differentiated. The significance of this seems evident enough, in the reduction of all competitors for winter resources to a uniform basis of sexless equality.

Equality in the asexual season, however, is not universal or necessarily desirable. One of our main postulates is that, when conditions of population stress develop, they can often be quickly relieved by the emigration of the surplus—the latter being identified by their lower standing in a social hierarchy. It may be desirable, therefore, that juveniles should persist as a caste during the non-breeding season also, holding ordinarily a lower status and being more expendable than the adults. It is to be observed in this connection among the so-called partial migrants—in which some individuals stay all the year round near their breeding quarters while others migrate away to varying distances for the winter—it is the youngest birds that on the average migrate farthest, whereas the adults, and especially the adult males, are the caste most strongly represented in the permanently resident population. This is generally true of all the considerable variety of species (including the song-thrush, lapwing, herring-gull, gannet and many others) for which sufficient data are available to justify analysis (*see* Lack, 1943-4, p. 143). It seems not improbable that in regulating the dispersion of these species for the winter season, it is largely a social pressure that tends to segregate them and impose longer journeys on the junior members.

From this viewpoint it is possible to look upon the marks of immaturity in birds, mammals and certain other animals as positive characters, of survival value to the species, leading through the social machine in the first place to a differential mortality between young and adults—a differential known to exist everywhere among the species of birds for which life-expectancy figures are available through the recovery of marked individuals. The evidence is that in some species young individuals are obliged to undertake longer migrations, and that in these and many others the young suffer a higher death-rate in their first year than at any later stage in life. To a very large extent this mortality may result from the heavier social pressure that falls on individuals of juvenile status, forcing them out to seek new habitats in which to establish themselves, or to make what living they can under marginal conditions (*see* Chapters 22 and 23). In the second place, this same differentiation is of equal importance in bringing about the dissemination of the recruit age-class, and by this means promoting a desirable amount of out-breeding, strengthening and building up depleted stocks, and populating vacant or newly-created areas of habitable ground. It may be noted in passing that such a valuable piece of machinery as this, like all the many other components of the social system, could be evolved only through a process of group-selection.

Thus it does not always occur in birds belonging to Groups III and IV that the dominant sex remains in ' eclipse ' during the non-breeding part of the year. Often the urge or ability to sing or display visually is inhibited then, although one may still be able to recognise the dominant sex by its unchanged plumage characters—for example in the chaffinch (*Fringilla coelebs*), a species that tends to be gregarious in the winter months. However, it may be noted that chaffinches occasionally start to exhibit territorial behaviour and sing for a few days almost as soon as their post-nuptial moult is completed in September, and resume it again in earnest, after a few months' gap, in February. For two-thirds of the year, therefore, the cocks are liable to be involved in territorial competition with others of their sex.

A group that deserves our brief attention is the sub-family of ducks (Anatinae). In most of these there is a very strong sexual dimorphism: the drakes of some species are among the most brilliantly adorned of birds, whereas the ducks in the majority wear an inconspicuous camouflage that conceals them especially on the nest. These drakes, once their nuptial duties are completed (often as soon as the eggs are laid), forsake domestic life and undergo an eclipse moult, which strips them of all their finery and even to a large extent of their sex-identity, for three months or rather more. During this time they tend to consort together, accompanied by non-breeders, undergoing at one stage a rapid wing-moult that briefly robs them of the power of flight. Some species, such as the king-eider (*Somateria spectabilis*, *cf.* Wynne-Edwards, 1952*b*, p. 362), scoters (*Melanitta* spp.) and shelduck (*Tadorna tadorna*, *cf.* Coombes, 1950) regularly undertake special migrations to foregather in enormous numbers on traditional moulting grounds; and no doubt there is an epideictic and dispersive purpose behind this habit, co-ordinated on a broad regional scale. By the month of October most drakes are once more out of eclipse, and have attained the height of their perfection. Even before the moult is complete, males of the surface-feeding species (especially *Anas*) are already consorting in pairs once again with members of the opposite sex, and from that time on, until the females begin to nest six or seven months later, their society is composed of pairs as units; and whether the units are scattered singly or aggregated into flocks, the pairs appear to remain unbroken. Display between drakes goes on all through the winter, and it therefore appears to be entirely on them that the burden of maintaining the dispersion of the population falls: the females' dispersion in this case being merely the reflection of the males'. In the diving ducks and mergansers, on the other hand, the winter society is less often built up of betrothed pairs, and in many species partial or complete segregation of the sexes is common, although this does not interfere with the continued mutual competition and epideictic display of the males at this season.

12.7. *Social selection in relation to other selective forces*

It is easy to appreciate, as we saw earlier, how great are the rewards of success in social competition, how severe the penalties of failure, and what an immensely powerful and rapidly-acting force social selection must be. In the vertebrates its intensity generally reaches a well-marked climax in the prenuptial and nuptial period, and, in a very large number of species, it is then chiefly if not wholly confined to the male sex. In territorial or lekking birds, for example, it is the cocks alone that decide by contest amongst themselves which shall attain the property and status that qualifies them for breeding: the result is that when the hens appear on the scene they are obliged to mate (if they can) with one of these pre-selected cocks. They may, of course, have a fairly large choice, and may exercise it too as Chapman concluded the female manakins did (*see* p. 207), but at best they can choose only from the range of males that have already been selected by their peers. Under such circumstances it would appear to be the male fraternity that wields the real selective power and effectively determines what secondary sexual characters are most to be admired and feared: indeed the action of natural selection on the tastes and preferences of the female sex in the choice of a mate would automatically favour a choice that adopted exactly the same criteria and values as those which influence the males themselves. The consequence is that there tends to be a close correspondence, often amounting to identity, between the adornments of aggressive and of amatory display; a correspondence that no doubt accounts for the confusion between these two very distinct functions which has generally prevailed among students of the subject.

All that female choice is likely to do in furthering the development of secondary sexual characters in the male is to conform with and reinforce the principles already accepted by the males themselves. But it is interesting to reflect that in sexually dimorphic animals there is no such need for conformity in the opposite direction. When males exercise their choice of female mates, there is no reason why it should not be freer and more unconditional, since among the female sex self-established hierarchies are generally either weak or non-existent, but this is a matter that can better be considered again a little later in relation to the functions of amatory display.

The symbols that confer prestige on the male (and also the characters that render the female acceptable as a mate) are always subject to the overriding requirements of adequate efficiency in carrying out all the many activities essential to everyday living, such as locomotion, feeding and finding shelter; and as we have already remarked this must curb their continual tendency to extravagant development. Above all they must not handicap the wearer in genuine combat, since he has to be able when challenged to back up his threats and hold his own in a real fight. If they do handicap him, then he is in a false position, and likely to be supplanted without delay by individuals genuinely superior. This may account for the fickle nature of sexual adornments in many groups of animals, in which closely related species have elaborated quite different display-characters, as if the particular symbols in vogue were not something of fixed and unalterable value, but could grow and decline in effectiveness and without very much difficulty be replaced by others. Among closely related species where sexual dimorphism prevails, whether in birds or in most other groups, the males are almost always far more distinctive and different from each other than are the females, and the characters in which they differ most are usually the very ones that have a social display-value. There may at times even be some leaning towards novelty for its own sake, as there certainly is in human fashions, tending to distract selective pressure from any particular display-character before there has been time for it to become hypertrophied.

As mentioned in an earlier chapter (p. 140), the various attempts that have been made to measure the physical characters of dominant individuals in the peck-orders of domestic fowls, pigeons or cage-birds have uniformly failed to reveal significant departures from the normal average; and among those scarabeid beetles that have extraordinary horns and excrescences on the head and thorax of the males, and in which the males show an unusual range of individual size (*e.g. Dynastes gideon* and *Chalcosoma atlas*), 'the small males are quite numerous and have not disappeared as a result of selection' (Richards, 1927, p. 303). The existence of a hierarchy or peck-order of course implies a range or diversity of attributes among its numbers, and the individual development of symbolic characters or weapons such as horns may extend beyond the optimum as well as falling short of it: in other words the most extravagant adornments need not necessarily confer the highest status.

This can be brought about through the over-riding effects, not of ' ordinary ' natural selection, acting on the individual, but of selection at the group level. An extremely interesting illustration of a parallel sort of group-effect is to be found in a rare and perhaps unique kind of sexual dimorphism and caste distinction that has evolved in man: a distinction of much social significance but devoid of any direct connection with reproduction. It must be pointed out that in the human race the strongest fighters and the handsomest and most eligible bridegrooms are young man just entering the prime of life: women likewise attain their fullest beauty at the same age. After marriage they devote together twenty-five years or so to bringing up a family, and in due time the woman first and later the man ceases to be fertile. Following this procreative period the physical powers of both man and woman notably decline, but their long experience and ripening wisdom are only now coming to a maximum value in the service of the community. The ' middle-aged ' undergo conspicuous physical changes, analogous to those at puberty, especially in the whitening of the hair of the head and (in the man) of the beard: in Nordic races especially, the men also tend to grow bald on the crown. There is no real reason for regarding these, as is usually done, as inevitable consequences of wasting senility: the eye-lashes, by contrast, retain their pigmentation indefinitely, and baldness is extremely localised and confined to the male, in a few human races only; some other parts of the hair, such as eyebrows, at this time of life grow more vigorously than ever before. The more logical conclusion is that these changes are in some way positive and functional adaptations. For one thing they visibly distinguish the possessor as belonging to the senescent rather than the virile caste, whose social status is not to be judged on the same criteria as those that are chiefly relied on with men and women in their prime; and at the same time they dignify him with an aura of venerability, which in its proper sphere commands the respect and deference of his juniors. Chiefs, generals, pontiffs, judges, ministers of state, presidents and matriarchs are all commonly drawn from this caste. It is therefore very interesting to discover that in the Australian aborigines there was a formal graduation ' of much importance ', usually at the time when the oldest child of a couple was declared marriageable, and comparable to some extent with the earlier initiation ceremony on first entering manhood, into membership of this third and final degree of seniority (Semon, 1899, p. 233).

Because of his gift of spoken language man is in a unique position to take advantage of the great and immensely valuable experience to be gained in the course of a long life, and undoubtedly this is connected with his equally unique characteristic of surviving regularly for thirty years (at least in women) beyond the reproductive period. Few other animals even briefly outlive their sexual potency, the extreme case I have found in the literature being that of the goldfish (*Carassius*), which is said to be normally sterile after the seventh year but capable of living 17 years and probably considerably longer (Comfort, 1956, p. 73). No species other than man has any comparable physically-distinct 'elder' caste in its social organisation, and we must therefore place it among his most notable and exclusive adaptations. Indeed it would have seemed an anomalous example of inefficiency, had the average healthy members of a population been adapted to outlive their potential usefulness to the community.

Anyone who is inclined to question the validity of group-selection must face yet another paradox here, namely, how selection at the individual level could lead to the development of characters, such as those of venerability and the prolongation of old age, which do not take effect until after the period of procreation is past. In terms of group-selection, on the other hand, there is no difficulty in understanding the ascendency of those human groups that are best able to benefit by the councils of their elder statesmen.

12.8. *The functions of amatory display*

It is necessary to revert once more to the duality of nuptial displays, to which brief reference has already been made in this chapter (p. 246). It was pointed out that exactly the same adornments are often used by the individual both in competitive display directed at rivals and in epigamic or amatory display directed at members of the opposite sex. The two events are very likely to be mingled together, and consequently the majority of writers on the subject of 'courtship' in animals have more or less completely failed to distinguish them.

In dealing with dispersion we have continually been emphasising the epideictic function. Animals need to engage in epideictic activities at all seasons and stages in the life-cycle, whether they are larvae, juveniles, asexual adults or breeders, because population-density generally needs to be adjusted continually either as the individuals grow in size, or the seasons change, or the environment proves unstable. But, as we have so constantly seen, it is almost always before and during the breeding season, and particularly just when mating is taking place, that they reach their greatest exuberance.

The epigamic type of display therefore usually appears at the same time. Though it has also been developed far and wide through the animal kingdom it is less universal than the epideictic. Whenever it occurs we can safely infer that an initial barrier against pairing exists, standing between the two sexes, which has to be broken down by amatory behaviour. In the more dimorphic species it is generally a demonstrative male that elicits acquiescence in a reluctant female, but in many cases some kind of mutual stimulation is required.

The realisation that courtship implies an initial obstacle, that must be surmounted through epigamic display before pairing can take place, originated with Richards (1927, p. 344), in relation to insects. 'The existence of elaborate displays in the males of many species,' he writes, 'and the frequent independent evolution of complex scent-organs, whose only function seems to be to excite the female, lead irresistably to the conclusion that there must be a further type of reluctance (besides mere immaturity), which it is advantageous for the male to overcome as quickly as he can'. There is, in other words, a 'coyness' on the part of the female; and this demands an explanation, since it is not universally found: indeed a large number of species of insects copulate immediately after emergence. Richards considers the possibility (loc. cit., p. 345) that it is no more than a question of obtaining the necessary co-operation between the male and female in the complex train of action that leads to copulation, in cases where the female might at the outset merely be otherwise preoccupied, but he points out that there are many insects in which 'the virgin females seem to have no occupation but waiting for the males, whose advances, nevertheless, they reject at first'. A second possibility, that coyness may be beneficial in hindering over-copulation, is also dismissed, and the author concludes that 'there is still much obscurity surrounding the problem of female coyness, especially its physiology'.

According to our theory one of the two primary methods of regulating population-density is by suitably varying the recruitment rate. As we shall find in the three final chapters of the book, there are endless ways of doing this, but one of the most obvious and probably most common is by controlling the number of matings allowed to take place in the population as a whole. It would obviously be helpful in this kind of control if there were some initial resistance to mating, so that a positive effort would need to be made to overcome it. Where the male is the epideictic sex it normally falls to him to take the initiative, and it is only his ardour and insistence that can succeed in breaking down the female's coyness. Circumstances may cause the female to vary her resistance, but the net effect in cases of this kind is always to place the ultimate control of mating in the males' hands. Where polygamy occurs, incidentally, it becomes further concentrated into those of only a fraction of their number (*see* p. 515).

The extent of the courtship, and the parts played by the two sexes in it, can vary greatly from species to species. Richards (loc. cit., p. 334) mentions insects, drawn from the Lepidoptera, Hymenoptera, and other orders, in which the males gather round the pupa in order to copulate with the female at the instant of her emergence. In these cases reluctance on her part must be non-existant, and limitation of matings, if it occurs at all, must therefore depend solely on the self-control of the males. If both sexes always entered instantly and automatically into copulation at first sight, there could naturally be no mating control at all. Courtship need not be a completely one-sided affair; perhaps it seldom is. Both partners can even play very similar and equally active parts in it, as they do in various colonial birds with little or no sexual dimorphism and where, as we have concluded earlier, the various other 'dispersionary' responsibilities are also divided between the sexes. In these cases of mutual courtship an emotional build-up sufficient to overcome the initial inhibitions of both partners simultaneously is required before coition can occur. But whether the courtship is one-sided or mutual, it appears very probable that the ease with which the barrier is penetrated can be greatly influenced by outside circumstances such as the economic situation and the intensity of social competition.

Making pairing conditional on successful courtship could thus very easily be an adaptation to enable restrictions to be automatically imposed on the number of matings—the formidability of the barrier to be overcome being effectively density-dependent, since it would vary with social stress. Individuals that were undernourished or depressed would presumably have greater difficulty in achieving mating than would the dominant and the well fed, because in the former resistance would be strong and ardour correspondingly weak. Birds that had failed to secure a territory or a nest-site, for instance, or had been driven away from the lek, could readily be inhibited from all further sexual activity by such a method. On this view, courtship would fit into its place in the homeostatic machine and provide one of the devices by which effect is given to the feed-back arising from the epideictic system in controlling the reproductive rate.

It has been mentioned more than once that the adornments used in epideictic and in amatory display are frequently the same. There appear to be a good many cases, however, where they have been evolved exclusively for the former purpose, and are not used in courtship at all. In studying the ruffed grouse (*Bonasa umbellus*), for example, Allen (1934) found that the handsome neck-ruffs and tail-fan of this species are used only in the ' intimidation display ', which is employed indiscriminately against others regardless of sex (at least in captivity). The ' mating display ' on the other hand, is almost the exact negative of this, being gentle and solicitous where the other is aggressive and harsh: the neck-ruffs, instead of being erected, are depressed and the tail is dragged on the ground. A similar and frequently quoted case is that of the South American sun-bittern (*Eurypyga helias*), in which the male has a most spectacular sudden display of the wings and tail, which are thrown into a great boldly-patterned arch with an arresting white ' eye ' at each side, but so far as is known this is used only for intimidation; and the actual courtship between cock and hen ' is quite simple and involved no use of the wing pattern ' (Stonor, 1940, p. 12).

The converse development of adornments used solely for courtship and never for aggressive display seems at best to be relatively uncommon: examples of it probably exist, though I have not succeeded in finding a clear case among birds. Where there are elaborate mutual displays of adornments in courtship, as amongst the grebes and herons, the same crests and plumes seem also to be employed when the owner is repelling rivals or asserting property rights. However, there exist in the females of various sexually dimorphic mammals in addition to man—and we can readily detect them—distinctive feminine attractions of a positive character. The origin and function of these is no doubt complex—part social, part maternal and part epigamic: the last is very likely the most important component, and to this extent they may be developed through their direct influence in attracting the attentions of males.

In this section we have been chiefly concerned with differentiating the competitive and amatory functions of the prenuptial displays, generally lumped together as courtship. Displays between members of opposite sexes are not—especially in the birds—necessarily confined to promoting pair-formation and copulation, but frequently continue throughout the whole period of association between the mated pair. Greeting ceremonies at times of nest-relief during the incubation period are common especially in some of the larger birds, for instance in the penguins, albatrosses, gannets and storks, and generally reveal obvious displays of affection. This appears to imply that the pair-bond, which insures the devotion of the mated birds to the duty of rearing a family, is potentially terminable and requires to be reinforced by frequent elaborate expressions of steadfast affection: were these to dwindle in enthusiasm, the bond would presumably weaken and collapse, and parental duties be forsaken. Such demonstrations of constancy, during the progress of nest-building, incubation and feeding the young, could have a supplementary function to that of prenuptial courtship, provided they were similarly responsive to economic conditions: a sufficiently adverse stress on either member of the mated pair would inhibit its ability to display or respond, and lead to the dissolution of the pair and the desertion of the nest.

These conclusions regarding the purpose of amatory forms of courtship, and of the functional distinctness of prenuptial competitive displays, have been based largely on what is known to occur in birds, but it appears probable that they can be applied to other groups. Noble (1938), for example, in a valuable review of ‘ sexual selection ’ in fishes, points out the general use made of displays of nuptial adornments in achieving territorial dispersion: in the case of the stickleback (*Gasterosteus*), in a sufficiently confined space, display and intimidation by the dominant male can even inhibit the development of nuptial coloration in any other males present. The same adornments are commonly used both for intimidation and for overcoming female reluctance. In the jewel fish (*Hemichromis bimaculatus*) there is an element of female choice in the selection of a mate. In trout and salmon the bright nuptial colouring is again primarily epideictic and used, like the overgrown jaws, in competitive display against other males. Here the nesting activity of the female—when she begins to ‘ cut ’ the gravel on the redd by the sweeping movement of the tail, and especially when she finally crouches into the hollow she has formed—appears to alert and attract the male, who darts towards her and ‘ quivers ’ in epigamic display (*cf.* Jones, 1959, p. 101). It appears, however, that very much the same principles hold good in these fishes as those we have seen in birds.

12.9. *Summary of Chapter* 12

1. Dispersionary activities, and especially social competition, are evidently connected with the phenomenon loosely termed ‘ sexual selection ’.

2. A brief survey of the various forms of natural selection shows that some of these are unconnected with any kind of competition whereas others

are more or less the direct outcome of it. True competition only arises over prizes for which the demand exceeds the supply: it need not involve an open contest between recognised rivals but can be indirect and unseen, as often happens at the inter-specific level between species sharing common food-organisms. Competition for resources between individuals of the same species, however, is all normally diverted into substitute, conventionalised forms of social competition, and this consequently comes to acquire an enormous selective power.

It is this power that is responsible for the evolution of the adornments of social display. On account of the conventionalised character of social competition, these adornments tend to become symbolical and to develop a quality of extravagance or unpracticality, when compared with the clean-cut, utilitarian efficiency of all other kinds of bodily adaptations. But social selection is nevertheless not fundamentally different from other forms of natural selection.

3-5. Social selection accounts not only for the recognisable differentiation of the sexes in dimorphic species, but also the differentiation of breeders from non-breeders, or adults from immatures. Outward differences label the various different categories in the population, and allow an individual to recognise which are the potential rivals belonging to its own competitive group or caste and which can be met on other, non-competitive terms. Using the birds as an illustration, four common combinations of caste-differentiation are recognised:—

Group I, all ages and stages alike (except nestlings);
Group II, adults of both sexes alike, distinct from immature stages;
Group III, adult males differentiated by a higher plumage (or song);
Group IV, adult females so differentiated.

Examples are given belonging to all these groups, as well as of various intermediate and subordinate conditions. One interesting generalisation that seems to emerge is that, in birds, species belonging to Groups I and II, where the breeding adults are alike, are predominantly 'nest-site' birds as distinct from territory-holders or lekking birds.

Group IV takes account of a phenomenon that is always rare and sporadic, and for this reason examples of it are given from the arthropods as well as the birds. The rarity of having the female as the epideictically dominant sex can be traced ultimately to the primordial differences between ova and sperms, the former being the passive and the latter the active gametes. It is a blind alley from the evolutionary standpoint. In birds it occurs in *Turnix suscitator*, *Hydrophasianus chirurgus* and *Phalaropus* (among others), but it is perhaps commoner in the copepod Crustacea than in any other group.

6. Recognisable castes are closely connected with the social hierarchy. Sometimes outside the breeding season all individuals are united in a common asexual class, but often the adult or sex insignia are worn throughout the

year. In birds that are partial migrants the individuals of highest status, usually adult cocks, commonly remain behind in the breeding quarters throughout the winter, and those of lower status are obliged through social pressure to emigrate.

7. Regarding the question of female choice in the selection of male adornments, this may (like male choice of females) play some subordinate part, but in most strongly dimorphic species (Group III) female choice is conditioned by the fact that she must mate with a male that has already been successful in competition with other males: selection will therefore tend to favour females that accept males having qualities which make them dominate other males.

Symbolic display-characters naturally tend to become hypertrophied, but are always subject to the over-riding interests of the individual's welfare in other directions: their over-exaggeration is therefore effectively prevented by other forms of selection. The interests of the group again may over-ride those of the individual.

A remarkable group-effect is the evolution in man, with his unique faculty of passing on the wisdom born of experience to others by means of language, of an ' elder ' or patriarchal caste, whose members are physically differentiated from adults in the prime of life: they command a different type of respect through their grey hairs and bald heads, and live some 20 to 30 years beyond the normal age of family responsibilities and child-bearing. This is without parallel in other animals, and, involving as it does a prolongation of infertile old age, it is a very good example of an adaptation which could only be evolved by group-selection.

8. The last section is devoted to the relationship between epideictic and epigamic (amatory) displays. Epideictic activities are commonly maintained throughout the year, and the particular form of social display and organisation in use very often undergoes seasonal changes. Amatory behaviour, that is to say courtship in its strict sense, is primarily a nuptial activity, and its function is evidently (i) to break down a definite initial barrier to mating, and (ii) thereafter to maintain a potentially ephemeral pair-bond. Fertilisation is in many cases conditional on a successfully concluded courtship, and the barrier appears to have a very important function in allowing restrictions to be imposed on mating. In dimorphic species where the male is the epideictic sex he normally takes the initiative in courtship and consequently possesses the ultimate control of mating: in cases of mutual courtship (*e.g.* Group-II species) this responsibility is divided. The male's ardour and, where she is subjected to similar pressures, the female's reluctance are assumed to be sensitive to social and economic stress; that is, to be density-dependent. In the same way an inherent inconstancy in the pair-bond, such that it needs to be actively fostered by displays of affection, permits it to be dissolved, and procreation summarily terminated, under sufficient pressure from the homeostatic machine.

Chapter 13

Further consideration of castes in animal societies

13.1. *Recognition of age and reproductive status in certain invertebrates*

Recognition marks which appear to us to label the individual as belonging to a particular caste or age-group are to be found in the majority of animals. Where there are specialised independent larval forms such as those of various insects, social integration through mutual recognition is probably not at all uncommon among them: some conspicuous examples of this have been given in earlier chapters, for instance in the processionary behaviour of the larvae of certain moths (p. 116) and flies (p. 118), the underground stridulation of the larvae of the beetles *Geotrupes* and *Lucanus* (p. 49), and the luminescence of larval glow-worms and fire-flies (p. 198). Amphibian tadpoles are, in some species at least, identifiable as to species by their typical colour and shape, and are very commonly congregated into schools of their own kind: the same applies to the planktonic post-larvae of various marine teleost fishes, the majority of which have their own diagnostic patterns and colours of chromatophores, at first not in the least resembling those of the adults. Among the planktonic larvae of Crustacea, to take a different group, there may frequently be close resembalnces between species of the same genus, for instance in crabs such as *Portunus* spp. (*cf.* Lebour, 1928); though the larvae of different genera are usually quite distinct in colour, shape and armament, and probably also in behaviour. The constancy of such characters as these throughout the members of a single population, larval or adult, is in itself evidence of their adaptive function. In animals with well developed vision, mutual recognition is always likely to provide the functional basis for keeping the patterns standardised, through the process of group-selection.

Colours obviously cannot have an epideictic function unless their owners have eyes with which to appreciate them. The invertebrate phyla contain great numbers of brilliantly-coloured species, for instance among the sponges, anemones, planarians, nemertines, polychaetes and starfish, which are of conspicuous and immediately recognisable appearance. Since most of these animals either possess no eyes or have them in too rudimentary a state for use in mutual recognition, social selection cannot have played a part in the evolution of their visible appearance. Many of them are known to be nauseous or stinging animals, in which case easy identification is presumably an advantage both to themselves and their potential predators. But they must be entirely dependent on predators as selecting agents in evolving whatever warning colours they possess, since an ability to discriminate between different patterns by eye is essential to the process of selection. Very much the same kind of thing has happened in the evolution of conspicuous flowers, only here the advantage comes from attracting the selecting agent (generally an insect), not from repelling it. In the evolution of cryptic coloration the same process works in reverse, because selective pressure from predators is directed against the more visible and favours the better concealed members of the species, but there is an interesting difference in that the predator, in improving the camouflage of its prey by continual selection, not only reaps no advantage for itself but on the contrary progressively defeats its own interests.

Our knowledge of the acuity of perception of the eyes in the various invertebrates just mentioned is largely conjectural, and there may be some species among them that because of their extreme conspicuousness are able to recognise their own kind in spite of possessing very indifferent visual powers. Examples of this probably exist in the polychaete worms, especially perhaps in some of the luminous ones, such as the fire-worms (*Odontosyllis*) and the planktonic *Tomopteris*. It is naturally hard to decide exactly where in the animal scale the social significance of coloration begins. With the other media of social integration, particularly the chemical one, the difficulties of making adequate observation and a correct interpretation on this point are likely to be even greater.

The development of secondary sexual characters is the most dependable guide to the existence of social selection in the lower groups, but social integration and epideictic phenomena undoubtedly occur in species that lack any obvious sexual dimorphism, and even in many that are hermaphrodite. It is quite possible that the swarming of marine invertebrates at spawning time often has an epideictic function, as well as promoting mass fertilisation. There are examples of this in various brittle-stars (ophiuroids), and in polychaetes belonging to a number of different families. The fire-worm just mentioned has a nuptial gathering, more fully described on p. 349, for which the males and females, normally bottom-dwellers, swim up to the surface just after dusk, especially around the last quarter of the moon: they participate in a brief luminescent display in which each sex emits signals, with a distinctive secondary sexual difference between the sharp flashes of the males and the more continuous glow of the females. The females are also very much larger: they glow as they discharge their eggs, circling round and round at the same spot, and emitting luminous material along with and adhering to the eggs. At this signal 'a number of males rush in, with short intermittent flashes, toward the light for fertilisation', attending partly to the female and 'partly darting rapidly to and fro through the luminous trail'

(Crawshay, 1935). This may be wholly epigamic, the female's luminescence serving to show the position of the eggs, and the male's flashes being a signal to her to continue egg-laying, but it may on the other hand be partly epideictic, as will appear in Chapter 15.

Sexual dimorphism occurs fairly frequently in nereid worms when they metamorphose into the breeding or ' epitokous ' phase known as the heteronereis, which usually swims with lateral undulations to the surface to spawn. The transformation involves only the posterior half or three-quarters of the body of the worm (the appendages of which become specially enlarged for swimming), and in the male it is often more extensive than in the female, leaving fewer unaltered segments in front. There is an extraordinary complication in *Platynereis dumerilii*, in which no less than four reproductive types occur, three of them being represented in both sexes and the fourth a hermaphrodite. The complex life-history has been largely worked out by Hempelmann (1911) (*see* fig. 18). The worms may become mature as nereids without metamorphosis (called Form α), in which case a male enters the female's dwelling-tube and fertilises the eggs she lays there. These eggs are very yolky, are housed in brood-compartments in the female's tube and guarded until they are ready to be set free as creeping larvae. Alternatively the worms may mature as small heteronereids (Form β), freely swimming, and, in the females, laying less-yolky pelagic eggs that hatch into a very distinct type of planktonic larva called a Nectochaeta. Both kinds of larvae, however, by the time they reach the 10-segment stage, have turned into almost identical immature nereids. Form γ is a larger heteronereid, probably older than Form β, with presumably a similar reproductive cycle.

The heteronereid condition is associated with swarming and in the swarms males sometimes greatly predominate. Fage and Legendre (1923, p. 1151) have described a nuptial dance occurring at these times, which at Concarneau, Brittany, coincide with the first and third quarters of the moon during the summer. The time is just after nightfall, and the worms are seen by the light of a lamp. They appear quite suddenly, ' like an avalanche '. ' Groups of dancers form on all sides, composed of a female and many males. The female turns round and round; the males describing circles and loops around her. The agitation grows in proportion to the arrival of newcomers. Spawning occurs: the female releases eggs through the anus and by rupture of her somites; the males emit sperm into the mass. Others follow similarly. Scarcely has one group finished than a number of others replace them at the same spot . . . .'

It may be surmised that there is an epideictic or dispersive significance in the differentiation of so many castes, and that production of the different types could well be related to the population-density and economic conditions prevailing at any given time and place, possibly in a manner somewhat analogous to the ' phases ' of certain insects to be considered later in this chapter. Hempelmann's own deductions were in accordance with such a view. ' We can conclude with certainty,' he wrote, ' that unfavourable conditions, and above all poor unsatisfying feeding, induce the animals to metamorphose as early as possible into Form β. Our *Nereis* is enabled, by repeated cycles of [reproduction from the tube-dwelling nereid stage], in a

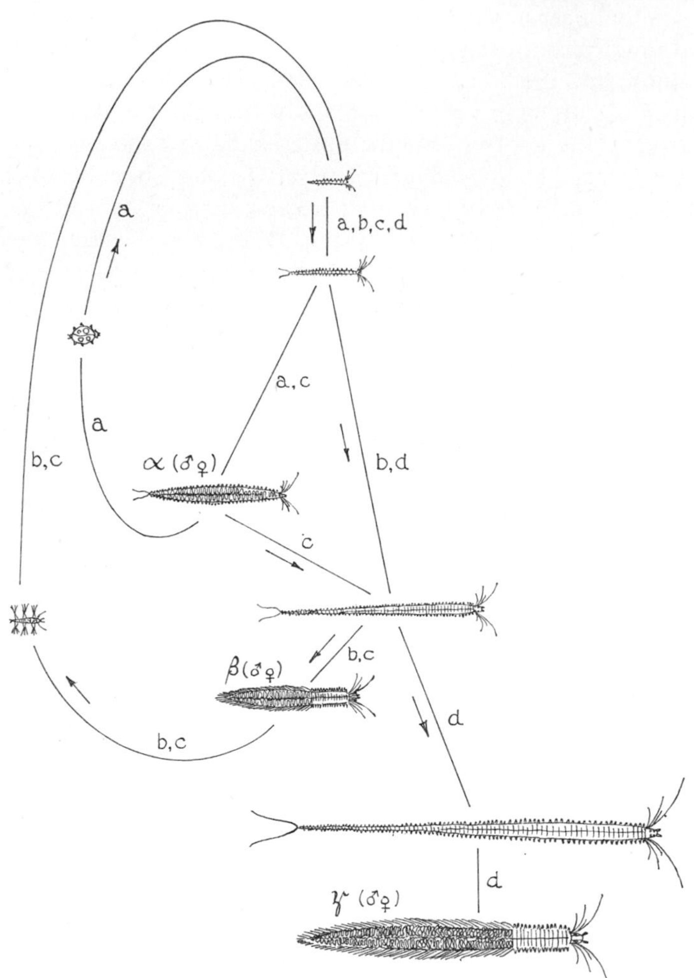

FIG. 18. Alternative growth and reproductive cycles in *Platynereis dumerilii*. The lines marked ' a ' show the life-cycle of individuals that culminate in reproductive Form α and ' b ' of those culminating in the heteronereid Form β. In cycle 'c' the worm matures twice, first as Form α and later, after retrogression and a further period of growth as the heteronereid β. Cycle ' d ' leads to the large heteronereid Form γ. (Redrawn after Hemplemann, 1931, p. 120.)

confined area offering perhaps very favourable conditions, to produce an enormous progeny, which serves to disseminate the stock only within that particular neighbourhood; whereas Form β makes it possible to spread the species more widely, and carry it far away from an unfavourable habitat' (loc. cit., p. 121). This is a subject to which we shall return in Chapter 15 (p. 350).

13.2. ' *High* ' *and* ' *low* ' *males*

The instances just described of the development of extra castes, over and above the basic three (*i.e.* male, female and immature) are not by any means unique, and the subject is worth pursuing further. A few of the Crustacea have two forms of males, which are conveniently described as high and low. A good case is that of the marine amphipods *Jassa falcata* and *J. marmorata*, the life-history and development of which I studied at Plymouth in 1927-8 (unpublished); they have since been more fully investigated by Mr. G. M. Spooner (unpublished). These two species live in pocket-like tubes (from which they make excursions), to be found in abundance on the submerged surfaces of buoys in Plymouth Sound. In *Jassa* as well as in many other amphipods and isopods the most important secondary sexual dimorphism is in the shape and size of the largest pair of appendages, the second gnathopods (fig. 19). The young stages resemble the females in the appearance of these limbs, but at the penultimate moult in the adolescent male the ' hand ' (which bears a terminal joint or claw) is already considerably enlarged and its shape differs somewhat from the corresponding joint of the female. At the last moult in the male the hand becomes relatively huge and acquires a strong prominent thumb on its palmar face, against which the claw closes. This is the common ' high ' type of male.

There is a much less numerous low type of male, usually considerably smaller in overall size. In this caste the final moult produces only a short feeble thumb compared with that of the high male, and the claw, which in other stages tends to have a cusp near the base, develops a tooth near the middle of its length. There is considerable individual variation in the extent to which these structures are developed in the low males, and it would probably be possible to find a complete series of intermediates between the low and high forms. Essentially the same situation occurs in both *J. falcata* and *J. marmorata*, and evidently also in several other species of this cosmopolitan genus. The high and low males are distinct enough to have led a number of authors to describe them under different specific names, which has resulted in great taxonomic confusion.

Once they are adult these amphipods never moult again, so that the male castes are end-products, and the presumption is, naturally, that some positive adaptive advantage results from their differentiation. Geoffrey Smith (1905, p. 313) described this kind of dimorphism as ' definitive ', there being no subsequent transformation by which the low could be converted into the high type.

Exactly the same phenomenon evidently occurs in another amphipod, *Orchestia darwinii* (Müller, 1869, pp. 24-26), even to the identity of the parts affected, namely the thumb and the position of the cusp on the claw of the second gnathopods. Müller described yet another apparently similar case in the tanaid isopod *Leptochelia dubia*. Here again the high male was found to be very much the commoner, and to outnumber the low form—which is smaller and has more or less female-type chelipeds (although male-type antennae)—by perhaps a hundred to one (loc. cit., p. 24). He observed this in Brazil. In the same or a very closely allied European species (typical *L. dubia*) Smith (loc. cit., p. 319) found a similar diversity in the size and ' maleness ' of the males, though he was not able to determine whether the high and low types were completely separate and thus bimodal in size and development, as they were in South America, or whether they merely constituted the ends of a continuous range of variation.

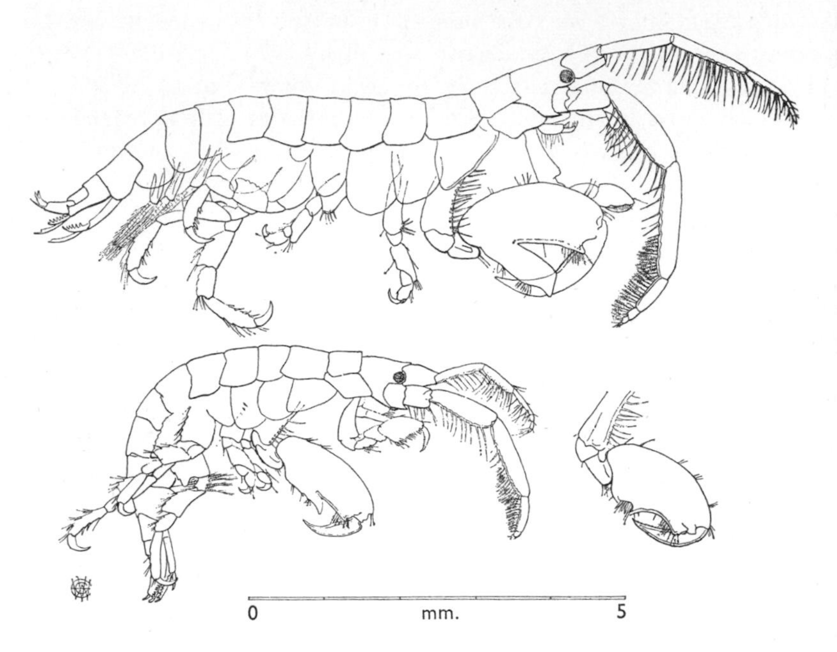

FIG. 19. *Jassa marmorata* Holmes. Above is a high male, large in size, with a massive, broad-thumbed second gnathopod (the principal secondary sexual character). Below is a low male, also in its final, adult instar, much smaller in size and with a weaker, small-thumbed ' hand ' joint. To the right of it is the thumbless second gnathopod of the adult female. (Plymouth Sound, Nov. 1927.)

In another isopod *Gnathia maxillaris* the larvae are external parasites on fish, though they may detach themselves when engorged, and the adults are free-living. Mature males and females are extremely dimorphic, the males developing a massive head with formidable nutcracker pincers, and the females being bag-like and distended with eggs. Smith (*cf.* 1909, p. 124) found an extraordinary length-variation in adult males and females of between 1 and 8 mm, the size depending on the duration of larval life. ' A large sample of males collected during full breeding-time fell into a distinctly bimodal curve in relation to size.' ' These facts appear . . . to show that . . . there is an antagonism between growth and maturity which tends to establish a division into high and low males ' (*the same*, 1905, p. 319).

In the crab *Inachus scorpio* Smith found another type of male dimorphism which he called ' facultative '. In this case the males have pre- and post-nuptial moults, and thus resemble many kinds of birds. At the former moult they acquire massive swollen chelae, and at the latter their chelae revert to the female type, which points of course to a seasonal alternation between sexual and asexual types of dispersion, in a manner already familiar to us. Small adult males after their first breeding season revert to the asexual type for the winter, and then at a later date, no doubt at the onset of the second breeding cycle, moult into a much larger or high type of adult male. The point of distinction in facultative male dimorphism is that the low can be altered into the high type by growth and moulting. Though Smith himself does not suggest it, it appears here that there may be two year-classes of mature adults in the breeding population, and two only: in addition, a number of ' middle males ' are found which are non-breeders remaining in the asexual caste. Whatever is the exact explanation of the presence of the high, middle and low males, the point of interest to us for the moment is that they are conspicuously distinct from one another in appearance.

The best-known case of all of high and low dimorphism is that of the earwig *Forficula auricularia*, an insect belonging to the Dermaptera. Here the terminal cerci are strongly developed in adults in the form of a pair of forceps showing a marked sexual dimorphism which permits instant recognition of males and females. In the nymphal stages the forceps of both sexes are exactly alike, and simple and weak compared with those of adults; the wings of course are lacking also. In adult females the blades of the forceps are relatively straight with incurved tips. In the males, on the other hand, they are strongly bowed like a blacksmith's tongs, and provided with heavy jaws at the base. Earwigs are very sociable insects, and in the adults of both sexes the forceps are constantly being displayed, raised and open, as the individual moves about and meets its companions. The males are more aggressive than the females, and in actual combat both sexes attempt with upturned abdomen to strike their opponents with the forceps: the owners never use them in seizing food, but they do employ them briefly as feelers (*organes sensoriels*) during coition (Lhoste, 1942).

As first clearly demonstrated by Bateson and Brindley (1892), there generally exist two types of male, having either long or short forceps, grouped in their sample around 7 and 3·5 mm in respective modal length. The high type averages a greater overall body-size, but the forceps-length is very strongly bimodal and it is possible to find numerous individuals of the same body-length that can be divided sharply into either the high or the low caste. The frequency of occurrence of the two types varies greatly from place to place. Bateson and Brindley's material was collected from the Farne Islands, in or near sea-bird nesting places, where earwigs were found to be extraordinarily abundant: ' under every stone or tussock ', they wrote, ' there was an almost continuous sheet of earwigs ' (loc. cit., p. 586). In a sample collected on 12th September 1892 there were 583 mature males: from the graph they give it can be ascertained that 244 (41·8 per cent) were of the low type, with forceps-length 2·5-4 mm (to the nearest half millimetre), and 331 (56·8 per cent) were of the high type (forceps-length 5-9 mm); eight (1 per cent) fell at the intermediate length of 4·5 mm. On the mainland nearby high males were in fair quantity but not so abundant as on the Farnes, whereas in a garden near Durham no high males were found at all. In an Aberdeen garden in 1938 the proportion was 9 (1·5 per cent) high to 581 (98·5 per cent) low (Morison, 1941).

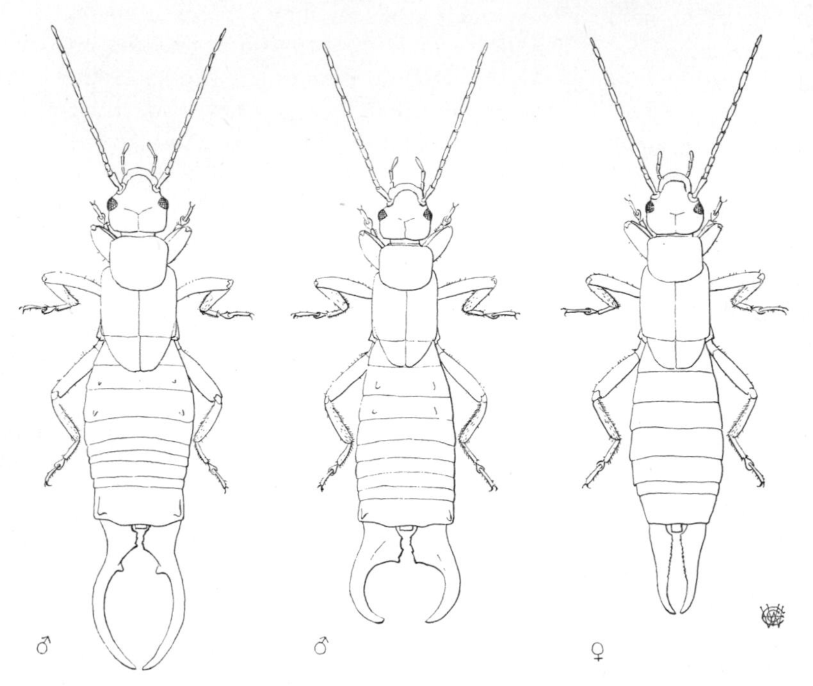

FIG. 20. The earwig, *Forficula auricularia*. Left, adult male with forceps 5·0 mm long; centre, adult male with forceps 3·3 mm long; right, adult female (all ×4). (Orkney, August 1961.)

Our knowledge of this case of male dimorphism was very greatly extended by Diakonov (1925), who studied *Forficula auricularia* and also *F. tomis* (which like most earwigs has the same duality of males) near Perm in Russia. He showed that the females are monomorphic, with a normal distribution over the whole range of their forceps length; that the characters for ' high ' and ' low ' in males are not genetically fixed but depend on environmental conditions during development; that the testes are equally developed and functional in both high and low castes; and that in the same locality the proportions of high and low males can vary from year to year, actually from a ratio of 70 : 30 in 1918 to one of 36 : 64 in 1920, other years being intermediate.

Diakonov's specimens were found in two localities, largely under the bark on dead tree-stumps, and by 1921, as a result of two years of collecting, he had in one section completely stripped off all the bark right down to the roots of the stumps. Thereafter the earwig nymphs and adults took to burrowing in the soft rotting wood, a habitat never exploited so long as there was bark remaining; and he showed that, whereas the undisturbed populations contained high and low males in approximately equal numbers (54 : 46), those from the wood-fibre habitat were very significantly different in proportion at 37·2 : 62·8. He concluded that the numerical interrelations 'may change under the influence of altered conditions of nutrition and other external factors' (loc. cit., p. 225).

In another part of the investigation he disproved the suggestion made by an earlier author that the dimorphism was due to some kind of parasitic castration, caused by intestinal gregarines, but he found the presence of a large parasitic larva belonging to a tachinid fly, which, if it did not result in killing the earwig, reduced its final body-size and also inhibited the development of large-type forceps. 'All severe exhaustion of the larva and unfavourable conditions as to nutrition lead to the formation of forcipes of the short type. This is observed on infection with the parasitic fly, causing general exhaustion, as well as on unilateral regeneration of the forcipes themselves resulting in local exhaustion, and on raising the larvae under conditions of starvation or under the generally unfavourable conditions of a laboratory' (loc. cit., p. 227).

Diakonov's work was soon followed by a very exhaustive study by Kuhl (1928), founded on four large samples of *F. auricularia*—two from Heligoland in the summers of 1924 and 25, and one each from Frankfurt and Holstein. Only in the last of these samples did there turn out to be a conspicuous bimodality of forceps-length, such as the previous authors had found; in the other three there were so few males with long forceps that the length-distribution curve showed simply one tall peak, with its upper limb extended a little asymmetrically in the direction of the upper length-groups without making any second hump.

Notwithstanding the unassailable evidence to the contrary, Kuhl persuaded himself that the bimodality previously established was nothing but an illusion, and that in fact there was a continuous range in forceps-length, which Bateson, Diakonov and the rest had never been mathematically justified in dividing into two sections. He made it much clearer than his predecessors that variation in length was only part of the story and that there were just as great individual differences in breadth and curvature, which wrought equally important consequences upon the shape and strength of the pincers. A truer analysis might be obtained in fact by grouping the male forceps into four categories instead of two, namely narrow short, narrow long, broad short and broad long. It became clear to him also that the form ultimately taken by the forceps in any given male adult was determined largely if not entirely by (i) the nutritional plane and consequent size and vigour of the antecedent larva and nymph, and (ii) the physiological state of the insect at the time of the last moult, when, in a matter of minutes, the pressure exerted by the haemolymph on the expanding cerci would govern whether they were to be partially or fully distended, and so determine their permanent shape.

From his study of their use by living males he confirmed the fact that the forceps were often employed, in the closed position, to administer downward blows at other males, though they were apparently not used for nipping; and that they had a minor but necessary role in facilitating copulation. But the variability in their length and strength had no detectable effect under experimental conditions either on individual copulatory success or on the longevity of their possessors.

The net outcome of the 233 pages of Kuhl's paper is in effect the negation of any adaptive function in the extraordinary variability that admittedly exists; it can all be attributed, according to him, to a fortuitous side-effect, that happens to show up very strongly in these highly labile male appendages. The cardinal issue seems to have been virtually ignored, namely that the variability, if not actually limited to adult males, is so much greater in them as to make what little occurs among females and nymphs seem quite insignificant by comparison; and furthermore, that the only characters affected are the male insignia themselves. Kuhl fully agrees with Diakonov's conclusion that forceps size-differences are in some way a reflection of what we have generally referred to as economic conditions, and that they are not to any important extent hereditary. This will appear, in fact, to be one of the common features of all the various types of intrasexual class distinctions with which we have to deal in this chapter.

When we consider all the circumstances, however, the likelihood that the variability of the male cerci is in some way positive and advantageous seems very strong. There is a possibility, for instance, that the visible strength of these weapons has a conventional significance, and tends to confer a commensurate social status on individual possessors; at high population-densities low males may tend to be debarred from copulating, just as the batchelors are for example in fur-seals. At present this can only be a conjecture; but if it were correct and led in practice to polygamy, or in some other way to the exclusion when necessary of part of the adult castes from reproduction, it could discharge a perfectly normal function in the homeostatic machine. At any rate, the kind of broad correlation we might have expected clearly exists, between the economic conditions on the one hand under which the individual grows up, and, on the other, the manifestation in the epideictic sex of his conventional status symbols: the richer the living, it appears, the more imposing they become.

Reference must be made next to what appears to be merely a variant of the same phenomenon, namely the existence of a wide, usually continuous or unimodal range in the development of male secondary sexual characters from a high extreme to a low extreme. This was mentioned on an earlier page (p. 247) in connection with some of the scarabaeid beetles, two of which, the East Indian *Dynastes* (= *Xylotrupes*) *gideon* and the European stag-beetle *Lucanus cervus*, were considered by Bateson and Brindley (1892) in the same paper that deals with the earwigs. Many additional examples among beetles have been assembled by Arrow (1951). The particular structure affected differs from family to family and genus to genus: in the stag-beetles (*Lucanus, Dorcus, Calcodes*) it is the huge mandibles; in *Dynastes, Golofa* and *Onthophagus* it is the horns on the head and thorax; in *Euchirus*, the excessively elongated antler-like forelegs; in *Chrysophora* and *Sagra*, the massive hind-legs; and in the weevil *Mecomastyx montraveli*, the non-sensory basal joint of the antennae (which may reach fifteen times the length of the female's corresponding part). There is very often an immense difference between the high and low extremes in the males of a single species (*see* fig. 21). The relationship between body-size and horn-size is typically of a 'heterogonic' character, such that a slight difference in body-size is likely to be correlated with a much greater relative difference in horn-size (*cf.* Huxley, 1932). More or less parallel situations occur in many other insects, and for example in the crustaceans *Gnathia* and *Jassa* already considered (p. 260).

It seems on the whole most probable that the function of this variability is to promote or permit some kind of social stratification when this is required. Male stag-beetles have long been known to emerge in spring about a week before the females, and to fight fiercely at that time, using their great jaws (*cf.* Darwin, *Descent*, p. 459). Beebe (1944) has described the methods of fighting between males in two species of dynastids, the 'middleweight' *Strategus aloeus* and the 'heavyweight' *Megasoma* (the actual species depicted by Beebe being *M. actaeon* according to Arrow, p. 61). The *Megasoma* attempt to set one another off balance by tripping with their fore tarsi, pushing and butting with their great horns ('exactly like two antlered deer'), and trying ultimately to get their horn underneath and then rear up, in order to throw their opponent over on his back. (Regarding the female, 'the only certainty was that she showed not the remotest interest in the encounter or in either of her suitors' (Beebe, loc. cit., p. 55).) *Strategus*, though quite differently armed, fought in almost exactly the same manner. Arrow (p. 111) mentioned that, in the South American dynastid *Diloboderus abderus* and in the African dung-beetle, *Heliocopris gigas*, pairs of males in each species have been found dead with horns interlocked, as happens from time to time with sparring deer.

Not a few of these beetles, belonging to several different families, have evolved a persistent pair-bond between monogamous mates: the male and female work together to construct a nest in which the female lays her eggs, in small clutches or singly; and in the dung-beetles (Coprinae), for instance, these are provisioned, tended and guarded by the parents (Fabre, V, p. 120). In the east-European dor-beetle known as the vine-cutter (*Lethrus apterus*) the female does most of the nest-digging and the male brings provisions for the brood and is seen guarding the entrance to the burrow. In *Diloboderus*, just mentioned as sometimes becoming interlocked with its opponents, the males 'are apparently active in the day time and energetically defend the nests against other males. After driving off his rival, the victor, according to Daguerre, expresses his satisfaction by a vigorous chirping' (Arrow, p. 56). The general situation is distinctly reminiscent of what can be found in birds, with a corresponding system of localised nesting pairs, and contests between males in defence of property and occasionally even the development of 'song' of a sort.

There seems no reason to doubt that the horns developed by the males in very many thousands of species of beetles are perfectly normal secondary sexual characters, in function resembling those of birds in Group III of the scheme outlined in the last chapter (p. 230). Some of the adornments have

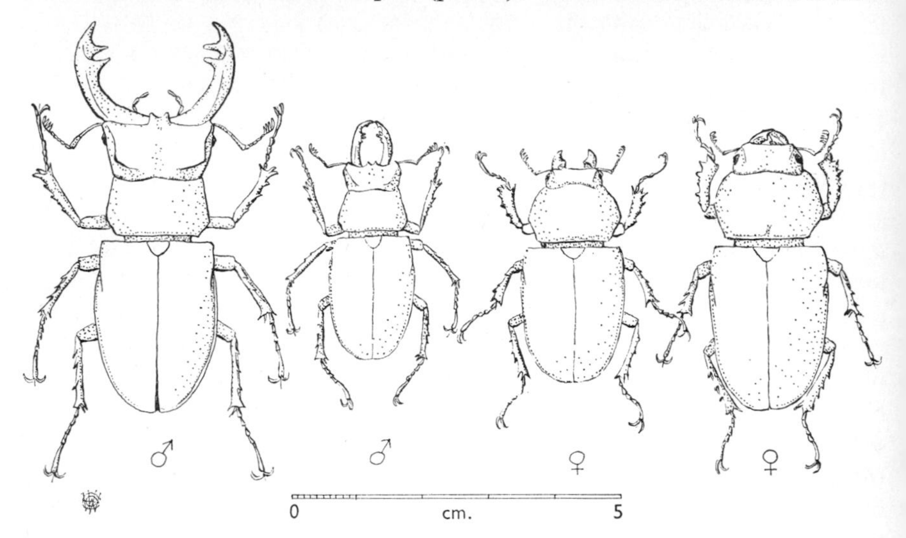

FIG. 21. Male and female stag-beetles (*Lucanus cervus*). The drawings indicate the size range of adults of each sex in British populations (slightly reduced).

obviously developed from actual weapons, such as jaws and feet, and, as very commonly happens with such structures, their use has subsequently been to a greater or less extent formalised. In the end they become transformed into symbols of threat and magnificence, and as such tend strongly to extravagant development. Their existence implies a corresponding tendency to develop a hierarchial system among their owners, with gradations of social status; and it is this that provides an intelligible explanation of the extraordinary ranges in body-size and horn-size occurring in the horned Coleoptera, paralleled only by those in the Crustacea mentioned above. There is, of course, as we have seen in other contexts, abundant evidence that social hierarchies can be developed in insects, and in Crustacea, in very much the same way as they can in vertebrates.

Arrow (pp. 120-3), though he had no alternative positive suggestion to offer, objected that the eye-sight, of lucanid and dynastid beetles especially, is not sufficiently good to admit visual recognition or discrimination of the finer gradations of horn-development (he was discussing Darwin's postulate of selection by female choice): he regarded the horns as mere useless by-products of increase in body-size. Certainly it is established that vision with compound eyes of the type these beetles possess cannot compare in acuity and resolution with that of many vertebrates, but colour-vision has been demonstrated in Coleoptera (*cf.* Wigglesworth, 1953, p. 147), and colour-dimorphism between the sexes, often with resplendent males, is also quite frequently established. ' One of the most lovely of all beetles, the glistening sky-blue and silver *Hoplia coerulea*, found in southern France, has a female of an earthen brown colour ' (Arrow, p. 140). The Hercules beetle (*Dynastes hercules*), one of the largest of horned species, has a dull-coloured hornless female, ' while the male is very glossy, his immense horns and anterior parts brilliant black and the hinder half shining green ' (ibid.). In view of this and of the general commonness in insects, and in other arthropods with compound eyes, of sex-differentiation on a visual basis (*cf.* locusts, earwigs, dragonflies, butterflies, Diptera, fiddler-crabs, copepods), and of the close resemblance between what has evolved here and in very different animals such as vertebrates, it seems not unreasonable to attribute great importance to visual symbols in the social selection of the Coleoptera. It is conceivable that the gross scale of size-variability found in these insects and Crustacea is itself a direct outcome of the limitations of compound-eye vision.

The individual size and development of horns does not in every case provide a simple unimodal distribution in beetles. In the oriental lucanids belonging to the genus *Dorcus* there may be two quite different alternative shapes of antler-like mandibles in the males of a single species: the same is true of the mandibles of *Calcodes* spp. (Arrow, pp. 88-91), and of the horns of the large Brazilian dynastid *Enema pan* (*the same*, p. 56). This is really a question of polymorphism among the males, showing most prominently in those of the highest grade, and it is a different phenomenon therefore from the segregation of high and low male types, with which we are at present concerned. Where we have met male-polymorphism before, for example, in the ruff (*Philomachus pugnax*), we have tentatively attributed it to promoting individual recognition. A genuine case of high-low segregation, however, appears to occur in another South American dynastid, *Megaceras jason*, where there is an unbridged gap, in the size and shape of the humps and horns, between ' huge-humped giants ' and ' more ordinary-looking males ' (*the same*, p. 106). In *Megasoma* likewise Beebe (loc. cit., p. 56) thought that the males seemed ' to fall into two general nodes ' to which he gave the names Major and Minor.

A situation rather different from any of these is found in the dragon-flies (Odonata), in many species of which the imago, after emergence, undergoes a considerable change of colour over a period of hours, days, or even one or two weeks, before it finally acquires the full adult characters. Commonly the immature or ' teneral ' male resembles the female, and in sexually dimorphic species usually undergoes a greater change in appearance after emergence than she does. Jacobs (1955, p. 572) has shown, in the case of *Plathemis lydia* in Indiana, that when mature territory-holding males (*see* Chapter 10, p. 174) were removed in sufficient numbers from the breeding place, immatures gradually took their places and became localised territory-holders. These so-called immatures, though actually completely fertile and able to copulate successfully, are under ordinary conditions easily driven off by the fully-developed males.*

This situation in the dragonflies is rather closely paralleled in the birds, in a few of which high and low caste males, differing in age, can also readily be distinguished. In the satin bower-bird (*Ptilonorhynchus violaceus*) of Australia the full-plumaged male is ' almost uniformly black except that the exposed edges of its feathers have a curious refractive property which transforms them to a beautiful lilac-blue in the sunlight ' (Marshall, 1954, p. 27). The ' sub-adult ' or low-caste display-plumage, worn by males approaching the age of, Marshall suspects, 4-6 years, is essentially cryptic in coloration, very much like that of the female, but the brilliant blue or lilac iris resembles the full male's. ' While displaying in the sunshine, the greenish plumage of the immature male seems to take on a mauve gloss, but this is of course quite different from the flashing spectacle provided by the completely adult birds ' (*the same*, p. 29). Immature, however, appears to be scarcely the proper term for these birds since, ' before they have grown a single blue feather, young males build full-scale bowers, annex females, achieve spermatogeneses, and sometimes breed ' (*the same*, p. 68).

A comparable case is that of the long-tailed manakin (*Chiroxiphia linearis*), whose song and dance in Costa Rica have been described by Slud (1957). One form of male display is a beautiful and exotic *pas de deux*. In another more communal type of display the males move between scattered perches with slow wing-beats in ' floating flight ', with the featherweight buoyancy of a Morpho butterfly, whose gorgeous blue colour they also resemble. The birds taking part in this are always full-plumaged males. ' Immature ' males lack the striking blue of the high plumage, and in the olive-green colour of the body they resemble the female, but they are easily distinguished by already possessing the male's red crown. Slud says (p. 338), ' I have seen immature males dancing with each other exactly like the adults, but I never happened to find a mixed partnership between a young male and an adult male '. He never saw the young males making the butterfly-flights. In both this species and the white-throated manakin (*Corapipo leucorrhoa*) ' the young male seems to display one part of his plumage and the adult another '. (It is interesting to note, with reference to the analogous situation found in so many other lekking birds, that during the floating flights of the males in no instance did Slud discover the presence

* It should perhaps be noted here in passing, not with the idea of confusing the issue but rather of revealing deeper complexities without attempting to explain them, that the females in various species of dragonflies occur in two, three or even five alternative colour varieties, for example in *Aeshna juncea*, *Pyrrhosoma nymphula*, *Coenagrion tenellum* and *Ischnura elegans*, among common species occurring in Britain (*cf.* Longfield, 1949).

of any female; and during the dances he saw a female only occasionally (loc. cit., p. 339).)

Even in these cases there is, of course, a gradual transition from the lower to the higher type of plumage, and indeed instances of such a progressive upgrading of display characters with advancing age are not at all uncommon in long-lived animals. The three species of waxwings (*Bombycilla*) show such a variation in the size and number of the red wax-like tips on their secondary wing-feathers, which may be a progressive character of this sort, but much clearer examples are found in the progressive growth of manes, tusks, horns and antlers in mammals, in the jaw-hook of the salmon, or the growth of the large chelae in decapod Crustacea.

Cold-blooded vertebrates such as reptiles and fishes, moreover, continue to grow throughout life, and there is consequently no fixed standard adult size such as we find in birds and most mammals. The mere age-spread of the breeding stock may consequently tend to confer on the older and physically larger individuals a dominant status in relation to their juniors. The virtual cessation of growth at or before maturity is a general condition associated with the evolution of constant body-temperature in birds and mammals—just as it is with those arthropods that never moult again after reaching the adult stage. It is very interesting, however, that the pinnipedes and Cetacea have independently re-acquired the capacity to grow throughout life: their epiphyses never fuse with the shafts of the bones—a condition no doubt rendered possible only by the fact that their aquatic habit frees them from having to support the body in locomotion as normal land mammals do. In the sperm-whale (*Physeter catodon*) and all the seals the adult growth-rate of the males is considerably higher than that of the females. In the elephant-seal (*Mirounga leonina*), for example, females can attain maturity at two years old, when they are about eight feet long, and may be expected to gain another four feet in length in the succeeding 10 or 12 years of life. The males become mature at age four and a length of about 10 feet, while still growing very rapidly, and on the average attain a length of some 18 feet if they live till they are 16 (Laws, 1953, p. 48). In pinnipedes generally and this species in particular it is very well known that size is correlated with social status, especially in the males, and it seems fairly safe to infer that in all the mammals that have evolved this condition of persistent growth it is primarily an adaptation which facilitates the development and maintenance of a graded social hierarchy.

Though we have been considering in this section animals belonging to very diverse groups, there appears to be a genuine element of unity in the common characteristic they present. All show a conspicuous diversity, among members of a single adult caste, in the size or perfection of what could well be described as the caste insignia. In point of fact the caste concerned is almost always the mature male sex, and the insignia are the male secondary sexual characters. The condition, wherever we have seen it, seems likely to be connected fundamentally in one way or another with the regulation of population-density. Either it varies in its manifestation according to economic conditions, as in Diakonov's earwigs, or it promotes the formation of a male hierarchy in the breeding season, thereby influencing breeding dispersion and the reproductive output, as in dragonflies or seals.

13.3. *Phase in insects*

The well-known phenomenon of phase in insects appears, even more clearly than that of high and low male castes, to be fundamentally related to the processes of dispersion. Bearing in mind the two basic methods by which population-density may be intrinsically controlled, namely by adjusting the reproductive or recruitment rate on the one hand, and by inducing emigration or re-settlement on the other, it can be seen that whereas the high and low caste system is essentially connected with the first method, the phase system is rather more closely tied to the second. In its most typical manifestations, for example in locusts, there are two principal phases—the one sedentary and the other migratory. The difference between them is not a genetic one: an individual egg can develop into either, the determining factor being the degree of crowding and competition in which the insect grows up.

In the Aphididae or plant-lice something of this kind occurs. The life-histories of these insects are usually very complicated, but one of the more conspicuously distinguishing features among the several types of adults produced is that some are wingless and others are winged. Increased crowding or a lowering of the nutrient or water content of the food-plant generally leads to a rise in the proportion of winged to wingless forms; the same is true in some of the related families such as the Adelgidae. There are often notable differences in fecundity, moreover, between the various winged or apterous generations of females: those that found new colonies on host plants where no population previously exists tend not only to be the most productive, but also to be parthenogenetic and viviparous. Sexual forms on the contrary usually develop in a period of declining population—at the end of the favourable season for instance—and in these the females are oviparous and lay a very small number of eggs, in some genera actually only one (*cf.* Imms, 1957, p. 449).

In the locusts (Acrididae) the differences both in appearance and way of life between the solitary and migratory phases can be so profound that they were not correctly associated together in people's minds until 1921, when Uvarov first advanced the ' working hypothesis ' that the three Linnean species hitherto known as *Locusta migratoria*, *L. danica* and *L. migratorioides* were actually forms of a single polymorphic stock. His phase theory has of course become the foundation stone of a vast amount of current research on the subject.

Locusts differ from grasshoppers only in the possession of a swarming phase. The distinction is not sharply defined because there are intermediate or annectant species in which the migratory specialisations are only adumbrated or imperfectly developed, a fact of considerable value, as will be seen later, in tracing the evolution of the swarming habit. The most highly evolved locusts are quite few in number; and, though they belong to several different genera in the tropical and warm-temperate zones, they are closely parallel with one another in their adaptations and habits. Those of greatest economic importance, to which most attention has been given, are the migratory locust (*L. migratoria*), with an immense distribution including Africa, southern Europe and Asia and part of Australasia, the desert locust (*Schistocerca gregaria*) of Africa and the Middle East, the red locust (*Nomadacris semifasciata*) and the brown locust (*Locustana pardalina*) of Africa, the Moroccan locust (*Dociostaurus maroccanus*) which extends from the Mediterranean countries to the steppes of Asia, and the Australian plague locust (*Chortoicetes terminifera*). A general account of them can be obtained from Uvarov's book (1928), and a valuable modern review of the phase theory from Key (1950); these are the principal sources from which the following factual material is drawn.

From time to time locust populations produce swarms that typically emigrate from the focus of production, which can be considered as semi-permanent and is known as an outbreak area. The swarms consist first of hoppers, intensely gregarious, which travel in bands on foot, and on meeting unite. When the hoppers moult into the adult state they become winged, and a period follows in which they indulge in ever-growing and uniting assembly-flights of short duration, still in the area of origin. After a very large swarm has been built up, and at a time when the weather is suitable, in most species the whole body takes off in a sustained emigration flight, which may continue on a more or less steady course for as long as 1-3 days and nights without stopping. Frequently the swarm flies downwind, and in favourable weather it can reach a point hundreds of miles from where it started, before alighting. Finally the hordes descend, having exhausted their food-reserves, and normally this is the end of the flight and the place where the swarm remains and breeds. Many months later, or in temperate latitudes the following spring, a new generation of voracious hoppers emerges, and this is when the greatest havoc is wrought on crops and other vegetation; but often before the end of that season the fortunes of the swarm begin to decline: sooner or later the predators and parasites take their heavy toll, and ultimately the great invasion dissolves and vanishes. The range of the population is once more restricted to its permanent home in the original outbreak area—until at some unpredictable future date, possibly years later, the whole chain of events repeats itself.

The migratory locusts differ from the non-migratory solitary ones in their extreme gregariousness, which causes them to seek and maintain at all times the closest contact with their companions. The two types differ also in form and colour, and strictly it is on the basis of morphology and not of behaviour that the different phase designations should be applied: this is not merely a question of hair-splitting, because there is not always a perfect correlation between the phase of a particular individual and its behaviour.

The extreme phases are known, regardless of species, as *gregaria* and *solitaria* respectively; intermediates when found constitute the phase *transiens*. Characteristically the wings are longest in *gregaria*, and this can be expressed quantitatively by the ratio of the length of the fore-wing to that of the hind femur (which is actually somewhat shorter in *gregaria*); *solitaria* tends to have a ratio less than 2 and *gregaria* more than 2. Other morphometric characters can also be used to differentiate between the phases. Correlated with them are pronounced colour differences. The *solitaria* phase, though rather variable, tends to be predominantly green, grey or drab, whereas the *gregaria* phase is conspicuously coloured, especially in the hopper stages, with a yellowish or orange background boldly marked and striped with black.

The outbreak centres are regions of unstable climate, generally with a variable or uncertain rainfall, where under prolonged favourable conditions the resident population (of *solitaria* status) can undergo tremendous increase. A subsequent return of arid conditions may lead to a crowding of locusts into places where adequate food and humidity still remain; and it is such conditions as these that lead to the production of *gregaria* types—a phenomenon known as gregarisation. *Gregaria* hoppers are much more active and excitable, with higher metabolic requirements, and mere passive concentration is soon superseded by deliberate aggregation. In the evocation of *gregaria* phase-types, as laboratory experiments have shown, high temperature, low humidity, sub-optimal food, and visual stimulation by the sight of others all play significant parts, and sight is also important in the integration of the social bands. Once mature, the *gregaria* females, at least in the brown and Moroccan locusts, lay several egg-pods, whereas the *solitaria* females normally lay only one. The whole sequence is therefore one of rapidly increasing instability: it has been referred to as a vicious circle, and is in fact a concerted and purposeful inflationary system building up towards its culminating eruption.

The emigratory flight is not an invariable consequence of gregarisation, because changes in the environment may put a stop to the process after the initial stages. If it does take place and if the swarm is not overtaken on the way by mass-destruction, perhaps by being forced down at sea or by meeting with storms, there will ultimately come a time, perhaps long afterwards, when the condition evoking swarm behaviour no longer obtains: then " degregarisation ' or dissociation ensues, leading to the production of *solitaria* types which may presumably survive for a time in their colonist home. Meanwhile the departure of the swarm has relieved the congestion in the original outbreak area, and the surviving population there likewise reverts to the *solitaria* status.

In other members of the Acrididae swarming is either undeveloped, or in various degrees less elaborate than in the typical locusts. In the Rocky Mountain locust (*Melanoplus mexicanus*) a *gregaria* phase was apparently produced formerly (though it is now rare), at least in part of its range, but the other American and Siberian species of *Melanoplus* do not show any phase differences, though outbreaks sometimes occur which are very destructive. In such species as these and the non-phasic Australian *Austroicetes cruciata* any exodus that does arise from a region where over-population occurs will in most cases be comparatively local: the 'invasion area' may frequently exceed the outbreak area in size, but it is more or less contiguous with it and there is no sustained flight to some unknown far-off destination.

In what may be called the locust-phenomenon there appear to be two evolutionary stages, the one antecedent to the other. At stage one, represented by the species mentioned in the last paragraph, the vagaries of the environment are liable to promote enormous fluctuations in numbers from year to year: whenever population pressure builds up to a sufficient threshhold, it is relieved by a straightforward process of emigration—a situation familiar in unstable habitats in many groups of animals, to which attention is particularly directed in Chapter 20. The major purpose of the emigration at this stage appears to be the relief of congestion, though, as a minor or secondary advantage, it no doubt helps the species to disseminate itself and colonise new habitats as these become available.

At stage two, attained by the truly migratory locusts and best exemplified by the most advanced types such as the brown, Moroccan and desert species, emigration appears to serve the same two functions but with the notable difference that the second is now far more important that the first. Here the environment is yet more fickle, to the point of being dangerously insecure so far as the long-term survival of the species is concerned. When conditions permit the population to build up, the opportunity is seized upon: there is a highly developed group-reaction which leads to the production of a definite swarm, of large size—probably the maximum that the available resources allow, morphologically and physiologically adapted for the purpose of travelling to a great distance. The emigrants commonly attain sexual maturity only during or after the actual flight, which points clearly to the function of the migration, namely to attempt to establish a new colony beyond the climatic influences prevailing at the point of departure. The commonplace adaptation of unloading an expendable surplus into no-man's-land, where at least it can do no further harm and may conceivably find a vacant home, has here apparently been modified and developed into a long-distance mass-dispersal device, enabling those species that have it to exploit the isolated and relatively transient habitats of the desert fringe, to which they are specially adapted. The fact that in most species the swarms do not seem to succeed in establishing new centres of population need occasion little surprise; so slow is the pace of long-term climatic changes that were they to do so once in a century or a millennium the adaptation might still suffice to ensure survival for the race. One of the species concerned, however, the desert locust, does apparently rely continually on the short-term establishment of swarms: it is said to have no semipermanent centres like the other migratory locusts but to breed sporadically and intermittently over a huge area, wherever a transitory foothold happens to be established.

Analogous phase differences are at any rate adumbrated in other groups of insects, including members of quite a different family of Orthoptera, the Tettigonidae (*cf.* Key, loc. cit., p. 369), and a variety of lepidopteran caterpillars, such as the widespread army-worms (*Laphygma* spp.), the African lawn caterpillar (*Spodoptera abyssinia*), and in Europe the larvae of the silver-Y (*Plusia gamma*), various quaker moths (*Orthosia* spp.), the emperor moth (*Saturnia pavonia*), and large-white butterfly (*Pieris brassicae*) (Long, 1953). In general, crowding causes a darkened coloration which may be very marked and lead to bolder contrasts and greater conspicuousness. The parallel with locusts is rather close in certain cases, and the phase-names *gregaria*, *transiens* and *solitaria* have been applied to the army-worms (Skaife, 1954, p. 174). In *Plusia gamma* Long (1953, 1955) found that crowding increased the rate of development and reduced the number of larval instars from six to five: the crowded larvae had a higher fat and lower water content, were much more active and spent on the average 25 per cent more time feeding than the solitaries. He showed that both colour and developmental rate depend on the frequency of tactile contacts between individuals, which provide them with what we could perhaps tentatively regard as an epideictic index.

In some of these species the crowding is not merely fortuitous, but results from a positive and organised gregariousness of the larvae, described by Long (1945) as sub-social behaviour. The aggregations may be very persistent, and could quite possibly have an epideictic effect on the individuals composing them. Whether the subsequent fecundity of the adults is affected by the phase of the larva, however, is not known, but it may be noted that *Plusia gamma* is one of the best-known emigrant or irrupting moths in Europe, spreading in some years over most of the continent from centres in the Mediterranean area, and that *Pieris brassicae* is almost as notable in this respect among the European butterflies. Some at least of the species of army worms are also subject to intermittent emigration, including *Laphygma exigua*, of wide distribution, which occasionally reaches Britain. One of the migratory butterflies of North America, the buckeye (*Precis lavinia* (= *coenia*)), is polymorphic in the adult imaginal stage, and is said to number among its variants 'static' and 'migratory' phases; 'sometimes one or the other, but not both, will occur in an area' (Klots, 1951, p. 109).

There is still much to be learnt about the life-histories of all these Lepidoptera, but what is known of the phase phenomenon in them suggests that it parallels the situation in the Orthoptera. Hitherto it has been studied almost exclusively in these two orders of insects, but Long (1955, p. 436) draws attention to what appear to be single similar instances in the Hemiptera and the Coleoptera, the latter being the cow-pea weevil (*Callosobruchus chinensis*).

The general correlation between phase and population-density is fully established; and there does not appear to be any obvious difficulty in going a stage farther and regarding phase as primarily an epideictic phenomenon.

Animal species in other groups, such as lemmings or crossbills, which, like the locusts, show the same phenomenon of mass-emigration from densely populated habitats, undergo similar profound changes in habits and social behaviour during what could be described as the fermenting period, preceding and leading up to the final exodus. The crossbills engage in day-long communal singing and excited manoeuvres (*see* p. 478), not altogether dissimilar from the assembly flights of the locusts. The evocation of a special dress, bearing a conspicuous visual pattern of particular epideictic significance, would presumably add still more to the revolutionary build-up: for a revolution it undoubtedly is that causes the whole accepted orderly system on which a stable society is based to be flung to the four winds, and the excited masses to set off on their hazardous adventure.

Some species of locusts, at any rate, appear to have developed further what was no doubt primarily a safety-valve mechanism, into a system for founding new colonies and disseminating the species, which is possibly of great importance in enabling them to exploit changeable and uncertain conditions in isolated, temporarily rich, but impermanent habitats. In this respect their swarming might be said to have a point in common with the swarming of honey-bees.

A characteristic of the evolution of phase-differences of which particular notice must be taken is that it leads to a type of polymorphism that is not controlled by a Mendelian system, but depends for its operation on contemporary environmental stimuli. The different phases, in fact, are different phenotypes built on a common genotype. Earlier in this chapter we saw that the same condition was found by Diakonov to be true of high and low dimorphism of the male *Forficula* (*see* p. 262), and in the following section it will be observed to obtain in the castes of all the social insects. These phenomena are all alike in showing a potency or plasticity of the developing egg, enabling it to be modified, sometimes profoundly, by the economic and social conditions into which it happens to be born; and to be modified, moreover, not in its own interests, but in those of the community. They are true adaptations of the living society itself, as an evolutionary unit, and once again it appears certain that they could have been evoked by natural selection only by its action at the group level.

13.4. *The social insects*

The Hymenoptera and termites provide us with material of extraordinary interest in the development and organisation of their colonial societies. The all-important proposition that societies exist for the purpose of controlling population, and the corollary that they owe their evolution to the process of group-selection, are both sufficiently unmistakeable here to have already been widely recognised.

Remarkable examples of parallel or convergent evolution can be seen in these groups. Highly organised societies, all of independent origin, have arisen in the termites, wasps, bees and ants, and each has led finally to the same result, in that reproduction is dominated by a single fecund female, the queen—all the other females being sterile as long as they remain subordinate to her. The other females, together with sterile males in the termites, form the worker castes, on which the construction, maintenance and defence of the nest and the care and feeding of the queen and larvae depend. As Richards (1953, p. 74) points out, the evolutionary advantage of having a female worker caste is, fairly certainly, 'a more accurate control of egg-production and the release of energy from that process for the other needs of the colony'. The highly specialised queen must automatically regulate her egg-laying rate in relation to complex and varied stimuli, such as the quantity and quality of the food she receives and the numerical strength of the colony, against a background consisting of 'remembered' information regarding antecedent conditions and rates of change of these various quantities, as well as innate or instinctive information, all of which must be integrated to give the best possible response for the immediate and future welfare of the colony. Actually her output may sometimes be afterwards diminished by egg-eating; and in this, it should be stated, not only the queen but the workers may take part (*cf.* Richards, loc. cit., p. 55; Ribbands, 1953, p. 235). In the termites moreover the rate of development of the immature insect can be speeded up or slowed down, depending on the colony's needs (Richards, loc. cit., p. 180), and this will affect the recruitment rate: also the workers may kill supernumerary members of any caste (*e.g.* soldiers) in order to preserve the optimum proportion in the colony (Skaife, 1954, p. 50). Honey-bee workers will similarly eject drones from the hive when these are not required (*cf.* Ribbands, 1953, p. 256).

It is the possession of sterile or sexually-inhibited castes that is the most characteristic distinctive adaptation of these insects. It carries with it the obligation of intimate dependence between the members of an established colony and thus the need to live together in a common nest. As already indicated, the evolution of sterility in a proportion of the individuals can only have been effected by selection at the group level, since it is self-evident that no agency can select in favour of sterility among organisms competing in status as individuals. Indeed in these closely-integrated societies it is, more than elsewhere, the group or colony that holds the spotlight as the vital evolutionary unit, undergoing intensive selection. According to the present theory, as already mentioned, all animal societies exist for the same primary purpose, which is to promote the optimum dispersion of the species. In the wasps and bees many progressive stages in the evolution of colonial life are known, easily bridging the gap between what are usually regarded as 'solitary' and 'social' habits; and there is consequently no difficulty in discerning the fundamental similarity of the higher insect societies to those in other animals. The difference is chiefly a matter of the relative degree of interdependence that exists between the individuals composing them.

Most of the characteristics we have observed in the societies of other animals can readily be found in those of one or other of the social insects.

Sometimes, for instance, the colonies are relatively permanent and the pattern of dispersion that they embody is therefore exceedingly stable. The tall weather-resistant mounds of some termites reach 10 or even 20 feet in height, and are no doubt sometimes of considerable antiquity: the same must apply for example to the huge ant-hills, occasionally 5 feet high and 20 feet in circumference, of the wood-ant (*Formica rufa*) so common in the coniferous woods of Europe and North America. A feral colony of honey-bees formerly known to me in a hollow maple tree in the streets of Montreal persisted to my knowledge for fifteen years—by comparison with the cases just mentioned no doubt a relatively short time. Along with these persistent establishments there is reason to think that some species hold fairly definite foraging territories. In the ants, ' as a rule, members of different colonies, even of the same species, are so hostile to one another that they cannot meet in number without a pitched battle. This hostility tends to restrict the feeding grounds of certain species within very narrow limits. It is generally admitted that the segregation of colonies is due to the presence of characteristic odors which vary with the species, colony and caste ' (Wheeler, 1910, p. 182). The habit of termites of living entirely in closed tunnels no doubt bespeaks a similar isolation in the internal geography of separate communities. On the other hand bees and wasps apparently enjoy undivided communal foraging rights with all other societies in the neighbourhood, once their own colony is established.

Many colonial bees and wasps build short-lived nests, for instance in deciduous vegetation, or under conditions of climate that preclude winter survival. The foundation of new colonies frequently depends on single fertilised queens, and between these strong rivalry is known to occur on occasion. Queens of the well-known species of *Vespula*, after emerging from hibernation, sometimes fight and kill each other over the possession of nest-sites (Richards, loc. cit., p. 56), a fact generally attributed to the shortage of sites: the same is equally well known to occur in bumble-bees (the same, p. 84). Even after she is established and has begun to build her nest, in these genera the queen is not secure, and may have to contest her claims with intending usurpers, which actually enter the nest. In such circumstances it is noteworthy that ' in humble bees and most kinds of social wasps the queen attacks any other individual attempting to lay eggs, and this behaviour is aroused much more by the act of oviposition than by the mere presence of another fertile female ' (the same, p. 38). It might reasonably be concluded from this that the antagonism is fundamentally more concerned with the regulation of numbers and population-density than with the simple possession of sites. Whatever dispersionary system does exist in wasps and bees it does not involve mutually exclusive foraging territories; in *Vespula*, moreover, the distribution of nests appears sometimes to be very uneven, and it is not uncommon to find two or more grouped in close proximity to one another (*cf.* Fox-Wilson, 1946).

The differentiation of castes in the social insects appears likely to have arisen in the first place through the development of a social hierarchy. The most characteristic division is between fertile and non-fertile castes, that is to say, in the Hymenoptera, between queens and workers. In some kinds of wasps—for example the tropical *Polybia* species—a number of co-equal egg-laying queens can be found in the same nest; and in the socially-primitive European *Polistes* wasp, *P. gallicus*, it has been found by Pardi in Italy that nearly every colony is begun by three or four queens. One of these, by a process described as 'terrorisation', after attacking the others with open jaws accompanied by a loud buzzing, establishes dominance over them, with the result that their ovaries ultimately degenerate, 'either because the wasps work harder or because they are not allowed to lay. Amongst these worker-like queens, which Pardi calls auxilliaries, there is a regular order of dominance, in which the queen always comes first. The position of a wasp in the scale is shown most clearly in the disposition of food, but it appears also in the frequency with which it shows aggressive behaviour and lays eggs, and in the amount of time it spends on the nest. A wasp high in the scale is more likely to be given food by one of its inferiors and less likely to give it away. Similar orders of precedence are known in many mammals and in birds . . .' (Richards, loc. cit., p. 62).

There is undoubtedly a very close resemblance between this 'peck-order' and the ones we have considered in previous chapters, especially in its influence on the participation of individuals in food-resources and in reproduction. The evolutionary step from this stage to a regular autocratic caste system does not appear to be either very profound or very difficult. Even where great physical differences of form and size have subsequently developed between one caste and another, it is invariably the case that workers and queens are genetically identical, and owe their differentiation to environmental influences during development; and when the queen meets with an accident and her dominance of the colony is thereby broken, workers freed from the normal repression both in *Polistes* and in various other wasps and bees, including *Apis mellifera*, can rapidly become mature and lay unfertilised (male-producing) eggs.

In the reproduction of the social insects the most primitive procedure is presumably the one in which a female, fertilised during a nuptial flight (or in the termites a male and female that have met and paired under similar circumstances) proceeds afterwards to found a new colony. The alternative is for an established colony to produce a swarm, containing representatives of all castes, which proceeds to found a new society elsewhere. The first system is particularly suitable for species with short-lived colonies, in which a force of workers is built up for the purpose of rearing a subsequent crop of fertile males and females; and after these have flown off on their nuptial flights, the workers' task being completed, they soon perish. This is the condition found in the *Vespula* wasps, whose workers die off in the late autumn after the drones and queens have been produced.

In many insect societies, however, more especially in warmer climates, the colony tends to become permanent, continuing from one generation to the next as the years go by. One advantage of this, perhaps, is the more stable pattern it establishes in the dispersion of the species, and the increased controlability of population-density consequent on this. Other advantages that may be presumed are the sustained efficient use of available nest-sites, in species such as honey-bees which cannot make suitable cavities for themselves, or the continued exploitation of the immense artefacts laboriously constructed by various termites and ants. All the termites and many of the species of ants still appear to found new colonies by the simple method of starting with a solitary royal pair or a queen, but in some of the ants and bees the fertilised queen that founds a new colony takes workers with her: in the case of the honey-bee usually some thousands of them. In this species, whose habits are generally so much better known than those of other social insects, it is established that the old queen accompanies the first emigrating swarm, leaving a daughter queen to assume her place in the original colony and renew its lease of life. As with other great emigrations, such as those of the locusts described in the last section, swarming is a revolution, an overthrowing of the whole pattern of previous orderly life; and it is an impressive experience to watch in an observation hive the mounting fever of excitement during the hour, and especially the last minutes, before the swarm bursts forth, jostling their queen among them (for an excellent description *see* M. D. Allen, 1956, pp. 19-22).

The immediate antecedent causes of swarming in honey-bees are still a subject of controversy. There is general agreement, however, that crowding is one important contributing factor. This may be taken to mean that one of the ends served by swarming is the relief of congestion in the nest. Moreover, in spite of the unusually tight control on the reproductive rate which the queen-system appears to have been particularly adapted to provide, over-population of the habitat is clearly an occasional possibility in an unpredictable environment. Whatever other ends may be served by swarming, however, there can be no doubt that the principal one is to reproduce the colony and disseminate the species, and this can therefore be looked for when economic conditions are sufficiently prosperous to warrant it, and at a moment when circumstances, such as the weather, are favourable for the attempt.

In a region where the species in question has built up to near the optimum density it is to be expected on theoretical grounds that swarms will tend to diminish in number, and that few of those that are released at such times will succeed in setting themselves up in new permanent homes: but the epideictic mechanism that prevents the establishment of a super-optimal number of colonies in any given area is at present unknown in the case of any of the swarming species. Something of the kind must certainly be anticipated, as a counterpart, for instance, to the initial competition found among solitary queens in species that found new colonies from scratch.

We saw earlier (p. 95) that in bumble-bees the males have a network of communal marked points which they visit systematically, comparable with the

' wolf telephones ' and musking places of carnivores described in Chapter 10 (p. 183). Thus Sladen (1912, p. 13) describes how, on a warm July day, one male after another of *Bombus pratorum* or *B. hortorum* can be seen to ' fly swiftly by and enter a dark hollow under a tree or shrub, where it pauses for a second, almost alighting, and then passes out and proceeds to another recess, where it again pauses and almost alights. Each succeeding bee flies in the same direction and visits the same spots. . . . A sweet fragrance, like the perfume of flowers, is perceptible about the pausing places '. Each species emits a different scent. There can be very little doubt that by this behaviour the males are performing an epideictic function, leading to the control of their own population-density, and perhaps on that basis to the placing of a ceiling on the number of females fertilised within a given area.

In the honey-bee no counterpart of this has yet been discovered. Queens and drones are produced especially in the spring and early summer. The virgin queens leave the hive singly at any time during this season and go off on a nuptial flight, returning home after fecundation. It is suspected, however, that the drones congregate at definite meeting-places, to which the females are attracted: but if this is true the places are presumably inaccessible, ' for instance around the tops of tall and large trees, which makes recognition and observation of the drones very difficult ' (Butler, 1954, p. 66). By inference from the difficulties experienced by queen breeders, who must attempt to get queens to pair with a selected strain of drones, Butler is ' inclined to believe that such places do exist and that drones from colonies from some miles around are attracted to them '. If this proves to be correct, our experience with other insects would suggest that some kind of epideictic display or lek would probably take place among the males; and this would as usual be able to provide the machinery for limiting the number of females fertilised and hence the number of potential swarms. In other words, it would be capable of supplying the missing epideictic phenomenon that controls the establishment of new colonies in a swarming species like the honey-bee.

The ants and termites have each developed the familiar and widespread adaptation of synchronised reproduction, a fuller description of which will be given in Chapter 15 (p. 352). In both groups the possession of wings and power of flight are confined to the young reproductive adults at the time of their nuptial flight. Usually both males and females are winged, but in some of the ants either the males or the females, but never both, are apterous. ' The nuptial flight for all the colonies of a particular species in the same neighbourhood usually takes place on the same day, or even at the same hour, so that the males of one colony have an opportunity of mating with the females from others. It is certain that the workers forcibly detain the impatient sexes in the nests till the propitious hour arrives ' (Wheeler, 1910, p. 183). This refers to the ants, which in some places and species produce such prolific nuptial hordes that they are ' visible from afar like clouds of smoke ' (*the same*, p. 184). The males and females usually mate in flight, often descending together to the ground. Thereafter they separate, and the females very soon actively divest themselves of their wings. The timing mechanism, both here and in the termites where the circumstances are on the whole generally similar, is generally conceded to depend on meteorological factors.

In passing, reference may be made here to the common use of the word ' swarm ' in describing these nuptial flights, though the phenomenon is very distinct biologically from the honey-bee type of swarm, which consists not of reproductives but of a unified task-force of workers derived from a single colony and accompanying a previously fecundated queen.

Both the ants and termites enter what is in effect a new element, the air, on their nuptial flights. There is no question that the retention of wings for this particular occasion—and for no other—helps both to promote cross-mating between members of different societies and to disseminate the species. The excursion of the winged adults is brief, partly perhaps because they are eagerly preyed upon by many predators and also because they have to be prepared to live on their reserves until they have established themselves in new colonies. The perfect synchronisation of emergence among those from many colonies must involve a very elaborate and accurate reponse to a complicated series of sensory stimuli, all duly integrated together; and the result no doubt is what makes possible the brevity of the nuptial flight.

As we shall find in Chapter 15, however, there is no very positive indication here of any epideictic function. There seem never to be any competitive displays or typical communal manoeuvres performed by those involved in the swarms. The alate insects themselves, which alone witness the flight, do not in most cases return to the perennial colonies from which they were released, and they cannot in that case have any effect on the continuing fecundity of the previously established queens living in them. Moreover, unlike the bees and wasps, most species of ants and termites appear to parcel the ground into exclusive colonial territories, each associated with a nest or nest-group, and these must provide a stable persistent framework as a basis for the homeostatic control of population-density. At the same time, the possibility that nuptial flights do have some secondary epideictic function, related especially perhaps to the foundation of new colonies, ought not to be wholly excluded.

13.5. *Summary of Chapter* 13

1. The possession of conspicuously labelled castes within any particular species, such as males, females and juveniles, is common to animals in many groups. Visually-distinguishable castes can be found, naturally, only where eyesight is adequately developed among their members, but the beginnings of sexual dimorphism go down the animal scale at least to the Annelida.

2. The second section is mainly devoted to the phenomenon of high and low castes of males, typically found in the Crustacea (especially amphipods and isopods) and earwigs (*Forficula*). Only size and secondary sexual

characters are affected by this dimorphism. A variant of the same thing is the wide range sometimes found among the males of a single species in the development of sexual adornments, for instance in stag-beetles (*Lucanus*). In the dragonflies (Odonata) the newly-emerged winged male is often coloured like a female, and is socially inferior to older males for some days until the breeding colours appear. A rather similar situation occurs in birds, such as the satin bower-bird (*Ptilonorhynchus violaceus*) and the long-tailed manakin (*Chiroxiphia linearis*), in which young males differ in display plumage from older males. Apparently in all these cases the ' low ' males are perfectly fertile, but subordinate to the ' high ' males; and the existence of these castes therefore promotes the establishment of a male hierarchy in the breeding season. Probably in all cases also, low and high maleness is not genetically determined.

3. This leads on to the subject of ' phase ' in insects, which is in many ways a similar development and even more clearly connected with dispersion. We have to do here with a primary phenomenon, namely the relief of congestion or over-population by emigration, and a secondary phenomenon, developed only in certain highly specialised locusts living in impermanent habitats—the dissemination of the species by swarming or mass-colonisation. The primary phenomenon has led to a purely epideictic physical differentiation of *solitaria* and *gregaria* phases, not only in locusts but in other insects such as lepidopteran larvae. As usual, the phases or castes are genetically identical; and the development of the individual into one phenotype or the other is controlled socially or economically.

4. The social insects—termites, wasps, bees and ants—show close parallels with one another in their social evolution. Each has evolved a method of living in compact co-operative colonies, dominated by a single reproductive queen and populated largely by sterile workers. The queen/worker organisation has almost certainly developed from a simple dominance relationship, resembling the peck-order system of precedence to be seen in the socially-primitive wasp *Polistes gallicus*. Their corporate way of life appears to have arisen primarily because of the accurate control of the reproductive output that it makes possible. The characteristic common feature of all social insects is the possession of sterile castes not differing genetically from the reproductive ones: the evolution of this sterility in certain individuals is another good example of the effects of group selection.

The dispersion of social-insect colonies and the control of colony-density is discussed in relation to various epideictic phenomena, including the rivalry of solitary queens in the social wasps and bumble-bees, and the activities of drones in bumble- and honey-bees. The synchronised nuptial flights of ants and termites, however, more fully discussed in Chapter 15, seem to have no obvious epideictic function.

Chapter 14

Communal roosts and similar gatherings

14.1. *Social roosting*

The habit of gathering to common roosts in which to sleep, generally by night but in some species by day, is very common in birds and occurs in other animals, especially the bats and insects which can also fly and are thus easily able to assemble from a wide area. Although it is such a familiar phenomenon to naturalists, no acceptable explanation has ever been offered to account for it: as Allee (1938, p. 32) says, the 'details as to how and why it operates are not known'.

We have already seen, especially in Chapter 11, that epideictic social gatherings tend to reach their peak just before and during the breeding season, the reason being that they allow the existing population-density to reveal itself and a system of social dominance to be built up, and these are the two main prerequisites for regulating the size of the reproductive output. Population-density, however, can be controlled not only through the regulation of the breeding effort, but also, as has been pointed out many times, by changing the rates of emigration and immigration. By this latter means the density can often be rapidly adjusted, and the process is one that can go on amongst active animals continuously, or can be invoked *ad hoc* at any season of the year. Especially in the supremely mobile groups of flying animals, adjustments are possible almost from day to day: with birds like starlings, ducks or plovers, for instance, a period of severe winter frost or deep snow may set up density-adjustments (known as 'weather movements') on a big scale, driving a hungry surplus to emigrate beyond the limits of frozen ground and water. Typically, for instance in the starlings, a proportion of the hardier spirits remain behind, to subsist at a lower density than before so long as food-supplies remain restricted; and when the milder weather returns there is likely to be a corresponding reflux.

In order to allow adjustments of this kind to take place it would be expected on theoretical grounds that some continuously-operating epideictic routine would be required; and it is this need that seems to be so perfectly met by the practice of assembling the local population daily for social exercises at a communal roost. During the day, when the individual bird is out feeding, it must be fully aware of—and presumably conditioned by—the prevailing level of subsistence available. At the end of the day it flies to the roost to take part in exciting social activities, and there is no difficulty in postulating that these reveal the other side of the picture, and condition the bird to the stress of competition that has to be faced within its present neighbourhood. These two essential indices—the availability of food relative to the number of mouths to be fed—supply all the information essential to elicit a response, either to be satisfied with the existing economy and remain, or to make a break and venture elsewhere. No doubt this is a somewhat over-simplified picture, but if it is in the main correct it offers a very simple explanation of the primary function and practically all the common features of communal roosting behaviour.

14.2. *The roosting habit in the starling*

The gathering of starlings (*Sturnus vulgaris*) to roost in spectacular numbers, sometimes in places such as Trafalgar Square in the heart of large cities, has attracted great attention. In the winter of 1928-29 I made a geographical survey of the roosts being used over most of the county of Devon and part of Cornwall, and the following winter another survey in Somerset and Gloucestershire, in the neighbourhood of Bristol (Wynne-Edwards, 1929 and 1931). The roosts were found by watching the directions of the birds' flight either at dusk or dawn, and the finished survey showed, therefore, not only the positions of the roosts themselves, but also of the areas of country served by each.

It emerged that each roost was associated with a recognised feeding area which could now better be described as a communal territory. The boundaries between contiguous territories were found not to be sharply defined, but to permit of a fair amount of overlapping or interdigitating along the marginal fringe: nevertheless there was no question that the system was essentially a territorial one. Roost-territories varied much in size, but in Devon the mean distance between neighbouring roosts (of which eleven were known, *see* fig. 22) was 16½ miles, and near Bristol three roosts in a triangle were separated by distances of 4½, 10¼ and 11¼ miles. In all this region the starling population was high, and it is estimated that the biggest roosts each housed between 100,000 and half-a-million birds nightly. Starling roosts are generally in young conifers, rhododendrons or laurels, sometimes in reeds, and often on buildings in cities.

There is sufficient evidence to show that the dispersion of roosts is, on a short-term basis, a stable system, and that the majority of birds are constant not only to their roost but to the particular feeding ground they frequent by day (Wynne-Edwards, 1929, p. 171). The accumulated droppings

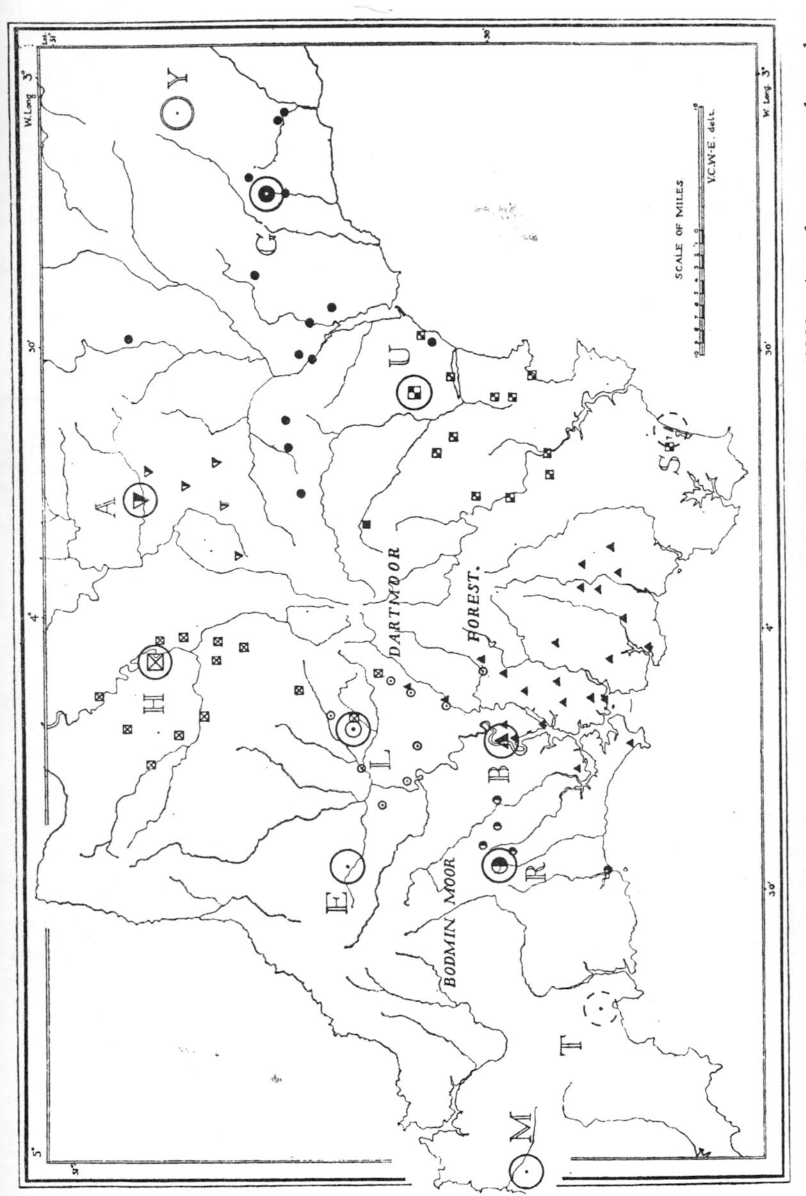

FIG. 22. Distribution of major starling roosts in Devon and Cornwall in January 1929. Actual roosts are shown by circles, each identified by a different mark at the centre. The feeding territories they served are indicated by the correspondingly marked observation points, from which starlings were watched at dawn and dusk coming from or going towards the different roosts. (From Wynne-Edwards, 1929, p. 145.)

usually defoliate and kill trees or shrubs if the starlings persist in using them for a number of successive years, but roosting-places on buildings, being indestructible, often continue in annual occupation for decades at least. Marples (1934, p. 193) recorded that the roost on Lundy Island was then known to have existed with only short breaks for 180 years, and that at Slapton Ley, Devon, for 135 years; * and of 247 roosts for which he had evidence, 107 had been in use for over ten years. The great Dublin roost in Phoenix Park has been going since 1845 (Kennedy *et al.*, 1954, p. 391) and perhaps much longer. The system is not necessarily so conservative, however, and the two roosts I studied most closely both shifted or divided during the period of observation.

A very significant discovery is the demonstration that each bird tends to have its own perch in the roost (*cf.* Kalmbach, 1932, p. 73; Jumber, 1956, p. 424-5). This is based on ' repeats ' of individually marked birds, and also on the fact that when starlings have been cleared from a particular ledge a gap is left which it takes several nights to fill up again. Mr. R. P. Bagnall-Oakeley has reported to me the observation of a very distinctive partial albino in a roost at Norton Creake, Norfolk, which had a regular roosting-site where it was seen on five successive visits in November-December 1957. (It was originally picked out in a feeding flock at a chicken farm at the same time.) The existence of a system of tenure within the roost is not at all surprising, but it is extraordinarily interesting and suggests at once that we are dealing with a social phenomenon closely analogous to a birds' breeding colony.

This observation may be coupled with the fact that going to roost is accompanied by excited community-singing, alternating with bickering and rivalry over perching sites. Moreover on fine evenings, especially early in the autumn, either a part or sometimes the whole of the noisy company not infrequently rises with a great roar of wings to engage in the most impressive aerial manoeuvres over the site: the massed flock may extend in a tight formation sometimes hundreds of yards in length, changing shape and direction like a giant amoeba silhouetted against the sky. On the grand scale these manoeuvres are by no means the least of the marvels of animal adaptation, so perfect is their co-ordination and so intense the urge to excel in their performance. It seems quite irrational to dismiss what is certainly the starling's most striking social accomplishment merely as a recreation devoid of purpose or survival value, and wiser to assume that a communal exercise so highly perfected is fulfilling an important function.

The roost thus resembles a breeding colony not only in the individual holdings it contains, in the exhibition of threat and rivalry, in the great vocal chorus (which gradually dies down as the birds become settled for the night, and rises to an equal pitch again before the morning exodus), but in the massed exhibitionary flights, comparable for instance with those of puffins circling in a great wheel over the sea in front of their breeding station or with

* This was still in being in the autumn of 1957.

the dashing air-borne parties of swifts. There tends also to be a similar conservatism in the preference for a traditional site, and a corresponding association of this as the focal point with an adjoining communal feeding-territory.

The social roosting habit in starlings continues twelve months in the year, but concerns for the most part non-breeding birds, so that from April to June the numbers gathering at nightfall are usually only a small fraction of the total population. With the fledging of the first broods of young in June the morning and evening flights begin to be conspicuous again. Many of the roosts occupied at that time are relatively small and temporary; and their combination into the great roosts of winter takes place gradually between August and November. The winter pattern is on the whole stable and may change relatively little in consecutive years.

Throughout this time, however, there are some individual birds that remain detached from the local roost and continue to sleep at night in their former nest-holes. Possibly those possessed of the most desirable nest-sites, in places where year-round feeding is to be had, retain continual possession in this way. In a starling nest-box at Old Aberdeen two birds customarily sleep at night all through the year, and may usually be heard softly singing within, in the darkness of a winter morning; but these birds participate in the social activities to the extent of sitting conspicuously on the chimney-pots in the late afternoon, singing with the other starlings as they watch for the evening flights to come past and move up to the roost.

In the British Isles great numbers of immigrant starlings arrive in the autumn from northern Europe; and in addition to these there are both native shorter-distance migrants and locally-bred first-winter birds that have never had a chance of acquiring a nest-site. Together these make up the bulk of the winter population and it is their dispersion especially that appears to be based on the roost-system. The solitary roosters form a small subsidiary, resident part of the population, evidently not completely integrated with the remainder. Their position is like that of the territorially interested males of finches, buntings or larks which detach themselves from their flocks early in the spring and begin to establish an independent dispersionary system: or again like that of the dominant individuals in species that are partial migrants which remain to see the winter through in their breeding quarters when all the rest of their company have migrated away. A dual machinery of this kind is probably not at all uncommon: Mrs. Rowan (1955), for example, found in the red-winged starling (*Onychognathus morio*) in South Africa that the population at all seasons similarly consisted of both resident pairs and roving flocks.

A further characteristic of the roosting habit must be mentioned, namely, the marked regularity in the times of arrival and departure of individual flocks and members of the roost. Sometimes a flight, undoubtedly consisting of the same individual birds, may be seen passing a particular point on their way out from their roost each morning, and on consecutive days with similar weather their passage (in relation to sunrise time) may be constant within one or two minutes (Wynne-Edwards, 1929, p. 172). When a large roost at Failand near Bristol was visited daily at dawn it was observed that the bulk of the birds actually left each morning in four great exoduses, of which the second and third were regularly the largest; and again it seemed very probable that each exodus consisted in the main of the same birds each day. The relation between the times of exodus and light-intensity was investigated, with interesting results to be described in the next chapter (p. 327). Towards the end of the period of observation (February-March 1930) the roost split: and it was possible to be certain from the photometer observations that the first and second exoduses had remained in the original place, while the third and fourth had moved to the new site not far away. A few days later the fourth exodus shifted again to another place, to be followed a week later by the third, and finally by the whole population (the same, 1931, p. 350). By this time it was 14th March, and the winter regime was nearly at an end, which may have had something to do with the birds' erratic behaviour; but the events served clearly to reveal the sectionalised character of the roosting community; and taking this in conjunction with observations of the constancy of smaller flocks in passage and on their feeding grounds we obtain a clear conception of the surprisingly orderly, in fact almost regimental, organisation that is required to provide for the efficient winter dispersion of such an abundant and mobile species as this.

14.3. *Communal roosting in other birds*

Some idea of the prevalence of sociable roosting in birds during the non-breeding periods can be obtained by going systematically through a selection of the species in which it occurs. Actually there is no order of birds in which the discovery of instances of sociable roosting would occasion any great surprise.

Although it is not quite a typical case, it is interesting to start with one of the penguins, partly because they are flightless and partly because their social roosting fulfils an important function, in the emperor penguin (*Aptenodytes forsteri*) at least, which is not epideictic at all. In this species dense shoulder-to-shoulder masses or ' huddles ' are very commonly formed, particularly in severe weather. Though huddles are more likely to occur at night, when the birds in them are mostly asleep (*cf.* Stonehouse, 1953, p. 8), the habit is not merely a nocturnal one: during a blizzard they may persist for 48 hours and perhaps longer. Moreover since the emperor penguins are presumably very largely aquatic except in the breeding season, it is at this time that the aggregations occur; and many of the participants are actually holding an egg. Prévost and Bourlière (1957) found that huddles were formed of anything from five or six thousand individuals down to a few hundreds, or even a few individuals; and by experiments they showed, first that birds sleeping within a huddle allowed their body temperature to drop 2° C. below that found in normal solitary individuals, and second, that in these circumstances the expenditure of fuel reserves, measured by the daily loss in body-weight, was actually halved. It must be borne in mind, therefore, with other species of birds and bats which roost in dense clusters, that they probably gain some corresponding benefit: although the temperature difference between the emperor penguin's body and that of the ambient atmosphere is of course very exceptional, no doubt occasionally exceeding 80° C. (*c.* 140° F.).

A roosting-site of the European cormorant (*Phalocrocorax carbo*), used all the year round by a score or two of birds in a region where the species does not breed, has been known to me for 15 years on a sea cliff at Findon, Kincardineshire; and according to Praed and Grant (1952, I, p. 34), the African pelican (*Pelecanus rufescens*) has regular roosting-places. There are no doubt many other examples in the pelecaniform birds.

Hudson (1892) has left us a graphic description of the formerly immense roosting flocks of the crested screamer (*Chauna torquata*) on the pampas of La Plata. These are huge birds, without near relatives, notable for their loud voice, double wing-spurs, and puffy (emphysematous) skin: at times they indulge in extraordinary communal song-flights, soaring upward to an immense height like, it is said, gigantic skylarks. They also gather in huge flocks by day in the swamps where they feed and do the same where they roost at night. Hudson says (loc. cit., p. 228): ' About nine o'clock we were eating supper in the rancho when suddenly the entire multitude of birds covering the marsh for miles around burst forth into a tremendous evening song. It is impossible to describe the effect of this mighty rush of sound; but let the reader try to imagine half-a-million voices, each far more powerful than that one which makes itself heard all over Regent's Park, bursting forth on the silent atmosphere of that dark lonely plain. One peculiarity was that in this mighty noise, which sounded louder than the sea thundering on a rocky coast, I seemed to be able to distinguish hundreds, even thousands, of individual voices. Forgetting my supper, I sat motionless, and overcome with astonishment, while the air, and even the frail rancho, seemed to be trembling in that tempest of sound '. Such is the spell of a great epideictic display.

To represent the Anseres the observations of Breckenridge (1953) may be cited, of the night-rafting of goldeneyes on the Mississippi River in the city of Minneapolis. The raft, containing at its maximum more than 600 birds, formed as an immediate response to the winter freeze-up in a place where a channel was always kept open by the heated effluent of a large power-plant. He found these birds gathered from as far as 10 miles down-stream and 27 miles up-stream, the former group having incidentally a distinctly different average arrival time from the latter. The roost lasted from early December till the end of March. Other similar rafts were established in St. Paul and, farther down, on the St. Croix River a short distance above its confluence with the Mississippi. The particularly favourable open-water situation for roosting at the Minneapolis site was no doubt a very important factor in bringing the ducks together. The counts showed that weather-movements occurred when severe cold diminished the number of open-water feeding places; and about 10th February the total had dropped to about 300: thereafter it mounted gradually to about 500 again by mid-March. The dispersionary machinery was thus evidently in action during the occupation of the roost.

Among the diurnal birds of prey, the California condor is well known to frequent traditional roosting areas, though the exact sites may be varied from night to night according to the weather (Kofoid, 1953, pp. 35-9). Kofoid located six such roosting areas, separated from each other by roughly fifty miles, in the principal surviving range of the species, and at one of these, Hopper Canyon in Ventura County, over twenty birds have been seen to collect on many occasions (loc. cit., p. 15). Turkey-vultures (*Cathartes aura*) have similar roosts on ledges and trees all over the southern United States. The particularly sociable red-footed falcon (*Falco vespertinus*) is another familiar example of a bird of prey that gathers to roost, evidently sometimes in great numbers (*cf.* Baker, 1928, V). Meinertzhagen (1956) has described a communal winter roost of Montagu's and marsh-harriers (*Circus pygargus* and *C. aeruginosus*) in Kenya, the resort of some 160 birds, evidently-assembled from a vast area; and Dresser (1878, V, p. 426) quotes an observation of Montagu's harriers collecting in thousands to roost near Loudun, in the Department of Vienne, France, after the breeding season.

Communal roosts are common among the limicoline birds such as plovers, knots and oyster-catchers. The knots (*Calidris canutus*) are among the most accomplished of all performers in massed aerial evolutions, not only at dusk and dawn but also during the day; their co-ordination is marvellous, and at times every bird appears to turn at once in a simultaneous flash. Their individual distance is small, and when hundreds are seen roosting (or resting by day) the surface of the ground they stand on is often completely hidden from view.

The morning and evening flights of gulls (*Laridae*) are among the most familiar signs of the roosting habit, in places either near the sea or inland, were gulls are at all plentiful. Depending in part presumably on the strength of river currents or tidal streams, gulls may choose to spend the night either on the water or on ice or dry land; and in the latter case especially on islands or cliffs where they are free from molestation. For the most part they have traditional gathering places, used year after year, but they do not go in much for noisy or showy demonstrations.

We have an interesting account by Beebe (1917, p. 95-98) of the roosting assembly of birds in the tropical jungle of British Guiana, and in particular of the golden-winged parakeet (*Brotogeris chrysopterus*). He mentions how the birds ' seem to lose all individuality and become imbued with a united flock spirit, which influenced all simultaneously, synchronously, as one bird '. ' To parakeets, going to roost was a rite, not to be performed singly as with caciques, nor lightly and with lack of dignity ani-fashion. Toward late afternoon the small companies of these birds . . . ceased from their two chief activities [feeding and screeching] and rising as if at the word of command, whirred swiftly toward some unusually high tree'. This tree was a subsidiary assembly or junction-place, from which, in due time, 'at an instant's signal, some reaction to a stimulus too delicate for our senses to detect, the whole company of several hundred birds was up and off like a whirlwind, all screaming their hardest. . . . They did not fly direct, but mounted high in the air and made several magnificent circles, a half-mile or a mile in diameter. . . . Finally, as if drawn in by a vortex, all banked sharply and spiralled downward and into a tall tree near the bamboos. This was the last resting place, and after a few moments, the mass of parakeets rose and pitched into the bamboos for the night'. The cacique (*Pionites melanocephala*) and the ani (*Crotophaga ani*), whose communal roosting habits he also describes, are respectively another kind of parrot and an aberrant cuckoo.

One of the most celebrated of communal roosters is the American chimney swift (*Chaetura pelagica*), a representative of the spine-tailed swifts, which do not scream like common swifts (*Apus* spp.) but chitter instead, and build bracket nests of short sticks cemented together with saliva. They usually choose for their night's resort a hollow tree or, since they frequent cities and towns in large numbers in the summer to breed, a chimney. The chimneys occupied are often unused furnace chimneys of churches and large buildings, two or three feet in diameter inside; and in these they may gather in thousands. Communal roosting is renewed during migration and on their arrival in summer quarters, but when nesting operations begin the swifts apparently disperse: I once witnessed their evening manoeuvres as early as 20th May, on an occasion when about 2000 disappeared for the night into a thirty-foot chimney at the back of the parish church in St. Jerome, Quebec. By late July the habit becomes strongly re-established, and at that date Audubon (1834, II, p. 333), in his often-quoted account, found in a sample of 115 birds he brought away only six females, the rest being either adult males (87) or recent fledglings (22): most of the females were presumably still on their nests. If so, the indication is that the males frequent the roost to a considerable extent even in the breeding season—a habit that we shall find again later in another species. As more and more young are fledged and parents are released from nesting the numbers of chimney-roosters mount to a peak in the latter half of September. Groskin (1945) counted over 12,000 at the maximum entering a ventilator flue at Ardmore, near Philadelphia, at that date.

He found unexpectedly that the numbers sometimes varied sharply from night to night, dropping occasionally by thousands, only to return in a day or two to about the previous ceiling. In addition to this it is apparent when his figures are plotted (*see* fig. 23) that the ceiling itself underwent a gradual secular change, rising up to a maximum and declining again finally to zero. Lowery (1939, p. 200) had previously pointed out that at this season there is a continual turn-over in the individual membership of the roosts, as migrants pass on to the south and new ones come in to take their places; and this, Groskin concluded, was probably happening at Ardmore (loc. cit., p. 364). These two independent findings put together seem to suggest the possibility that the whole vast passage-movement may be under orderly control, and that the arrival of new incomers from the north tends automatically to exert pressure on others to leave and continue on their way southward, thereby preventing over-saturation or bottlenecks arising at any particular stage of the journey. It is commonplace that, during migration, when populations are constantly shifting and the density is changing

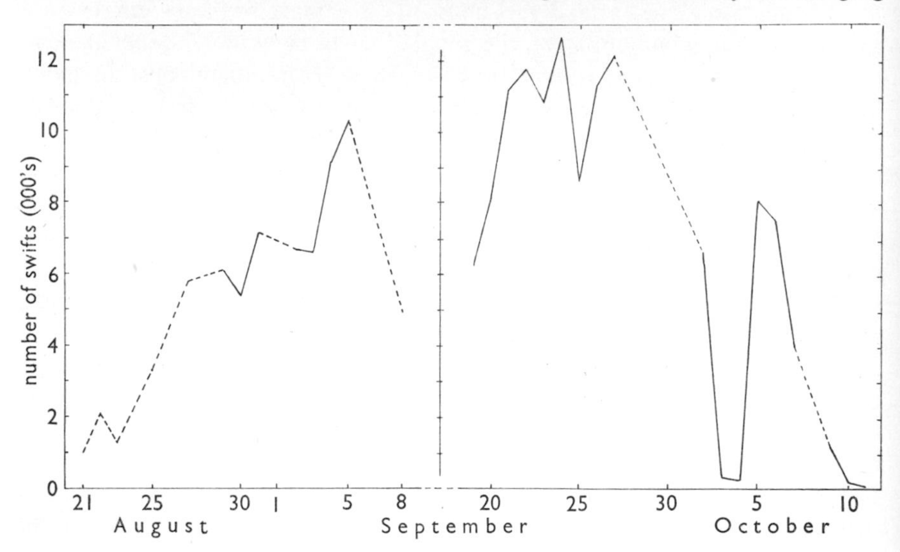

FIG. 23. Numbers of chimney swifts entering a roosting chimney at Ardmore, Pennsylvania, between 21 August and 11 October 1944 (figures from Groskin, 1945, p. 362). Note the rising trend from mid August to a peak in late September, and the decline thereafter; also the sharp fluctuations from night to night on some occasions during the southward migration, especially noticeable in the latter part. (Where the line is broken, observations were not made every night; there is a gap of 11 days in the middle.)

from day to day, most birds tend to be particularly gregarious; and there may be in this a general significance, implying that the flocking of migrants is in fact mainly epideictic in function, preventing or minimising the chaos that could easily arise if thousands or millions of uncoordinated and disorganised individuals of any one species were on the move together.

At the roost the swifts usually perform evening manoeuvres. Generally they make a long narrow stream well above the chimney top, circling to right or left in a wide wheel fifty yards across, and often joining up with the tail of the party to ' keep the pot boiling ' like boys on a slide. Frequently the direction is changed and sometimes after a straight run the band divides, some to each side, wheeling in opposite directions to make a figure of 8. Sometimes they scatter and re-assemble. At a certain stage a few birds start to dip as they pass the chimney, and next begin dropping from the passing stream, fluttering down into the shaft below. Soon the number disappearing on each circuit grows rapidly larger and, in the final minute, three-quarters of those coming over the opening are being drawn down, like smoke going in backwards, or 'like a thread from a rapidly revolving spool or reel' (Pickens, 1935, p. 150). Once inside, their individual distance may be practically minimal, consistent with each gaining a foothold on the wall. In the morning they re-appear unusually late: Groskin (p. 368) even found on one occasion that it was 11.40 a.m. by the time the last ones had left.

Swallows too are given to assembling for the night, and the habit is widely known in the common swallow (*Hirundo rustica*). The most spectacular swallow roost described in the literature is no doubt that found by Rudebeck (1955) and his companions in a reed-covered marsh near Lake Chrissie, Transvaal. The chief occupants were wintering swallows from Europe, but there were other species sharing the roost with them, including cattle egrets, long-tailed widow-birds and various weavers including queleas (*Q. quelea*). Some of these had arrived first and had attracted the observers' attention. Looking upward into the air they were 'amazed to find a cloud of swallows, so huge and dense as to be almost unbelievable. The birds were whirling around, fluttering up and down, and dashing in between each other. The size and density of the cloud and the activity of the birds were indeed extraordinary. Moreover an immense stream of swallows was flowing into the main flock from W.S.W. . . .' (loc. cit., p. 573). The flock was 400-500 m in diameter, and 175-200 m deep, the bottom of it being 50-75 m above the ground. By various methods they came to the conclusion that it contained about a million birds. 'The cloud of swallows was whirling around rather high up in the air until twilight. The birds even seemed to get more active before alighting. The same phenomenon was observed in swifts. . . . With incredible agility the birds were moving promiscuously around each other against the rapidly darkening sky. In the field-glasses the impression was really dizzying. Finally the birds alighted in the reeds. Some of them dropped suddenly and then it was a question of seconds until the others followed. Like a torrential shower of rain they hurled themselves down more or less vertically, and disappeared instantaneously. Even this sight was most spectacular. Within a short time . . . all the swallows descended. Afterwards none were to be seen, except for some belated flocks, which arrived at headlong speed and dived into the reed-bed. A few minutes later it was almost completely dark, and we left the observation place' (loc. cit., p. 574). Rudebeck gives an equally graphic description of the morning exodus, when the birds rose up in a series of 'explosions' and immediately departed to the west-south-west whence they had come the previous night, without any further display.

These observations were made on 13th-14th March 1954, which was presumably getting very near the time of the swallows' departure for the north. The main arrival in Britain, for example, is in the first two weeks in April (Witherby, I, p. 230), so that the advance parties were probably already on their way. It is to be expected that the evening display would reach a climax at this time, on the eve of such a great dispersionary change.

We may turn next to the Corvidae, the members of which are inclined to a similar habit in most parts of the world. It was a roost of the house-crow (*Corvus splendens*) that Willey (1904) described in the palm trees on the island of Barberyn, off the coast of Ceylon, in which there was a remarkable alternation of day-flying crows and night-flying fruit-bats, each passing the other in opposite directions at dusk and dawn (*see* below, p. 305). Apparently after 50 years it is still in use (Henry, 1955, p. 3). Probably all species of the genus *Corvus* gather, under some circumstances at least, into communal roosts. Roosts of ravens (*C. corax*) in this country sometimes contain up to one or two hundred birds, of carrion and hooded crows (*C. corone*) up perhaps to five or seven hundred, and in the case of the rook (*C. frugilegus*) the totals may run well into the thousands.

Reference has already been made in an earlier chapter (p. 158) to the territorial system obtaining in rooks. During the spring and summer, from the beginning of March to about August, their dispersion is based on individual rookeries, each with its established feeding area or communal territory: the breeding birds sleep at night in their own rookeries, and after the young are on the wing they too remain there, in company with their parents. For about six months, therefore, there is very little coming and going beyond the borders of the individual commune, or between one rookery and another. In early autumn the rooks begin going to their winter roost at night (*see* fig. 24). This is very generally, though perhaps not always, itself the site of a rookery, and it may serve a dozen or more surrounding rookeries as a focal point (*cf.* Munro, 1948). On their dispersal in the morning the birds usually go directly to their own rookeries, remaining there in noisy cawing and display flights for several hours on fine mornings, before they disperse to feed.

There is thus a subtle but important difference in the framework of the rooks' dispersion between summer and winter. In north-east Scotland, for example, there is no sign of any regular migratory movement either into or out of the region—the very dense native population being essentially self-contained, except in so far as a surplus may be produced, or alternatively a deficit exists to be made good from outside. Apart from this the same individuals occupy the same parishes and counties year in, year out, so far as is known. In winter, however, their dispersionary unit is based on the winter roost, and is a territory usually of the order of 100 square miles, whereas in summer it is based on the rookery, with a territorial unit most often of the order of only 10 square miles. In southern parts of Britain there is a large winter influx of rooks from continental Europe in autumn, and these are readily accepted into the winter-roost system. The latter appears likely to give much more flexibility of foraging, allowing the birds in an emergency, during periods of frost and snow for example, to concentrate wherever food can be got over the whole roost-territory, instead of centring them in relatively small local units based on individual rookeries, confined to a few square miles, probably of much less diversified ground. It also presumably permits the young birds to mix to some extent and be distributed as recruits over any of the component rookeries when they are ready to breed, thereby securing a certain measure of numerical adjustment and out-breeding.

Conservation of sites is of course characteristic of rooks both in their breeding colonies and winter roosts. Munro (1948, p. 25) cites a roost at Dunglass, East Lothian, known to have been used 70 years earlier, though two other famous roosts in the Edinburgh area recorded a century before had both been forsaken: at one of these, the Penicuik Crow Wood, many of the trees were blown down at one stage and the rooks had thereafter been systematically persecuted (W. Evans, 1922). Dewar (1933) and Philipson (1933) showed that there had been no change in the three main roosts in a

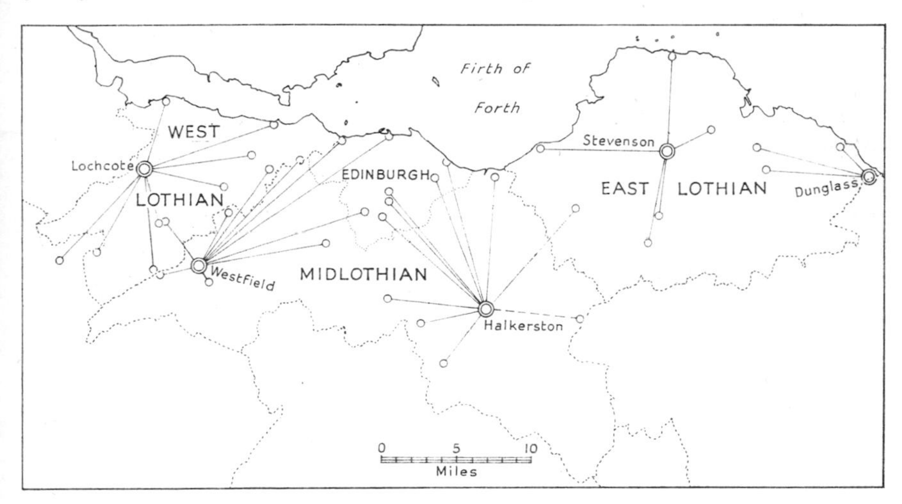

FIG. 24. Map of the Lothians, showing rook roosts in autumn and winter of 1946-7. The lines radiate out to rookeries lying near the perimeter of the large territory served by each roost. (Redrawn from Munro, 1948.)

large part of Northumberland and Durham between the years 1906 and 1932, and that they served practically the same areas at both these periods: a fourth roost had shifted in the interval.

Epideictic display flights, sometimes called crows' weddings or parliaments, are characteristic of rooks in autumn, especially in the afternoon as they work their way up towards the roost: in these they are often joined by jackdaws (*C. monedula*), which commonly share the same winter roost. As mentioned in the chapter on communal displays (p. 204), the birds often tower up, possibly a thousand feet or so, circling in an irregular column, and keeping up a continual chorus of ' caws ' and ' chacks ', with odd birds chasing, swooping and tumbling in their midst: usually this is alternated with massing together in a field. From the similar cases we have considered already we can tentatively connect these special displays with periods of large-scale adjustment in population-density. In the case of the rooks and jackdaws perhaps it is the lot of the young birds and migrants that requires to be settled in the autumn; and when the displays recur again, as they sometimes do, in February, it is on the eve of the break-up of the winter-roost system.

The Turdidae or thrushes are the next on our list. It is common for these to share their roosts with other species, in this country especially with starlings. Sometimes the starlings invade an existing roost of blackbirds (*Turdus merula*) or redwings (*T. musicus*).

Philipson (1937) studied a redwing roost in Northumberland in a winter when these northern visitors were particularly abundant, and found that they gathered nightly from a very large area (roughly 200 square miles on his published map), quite comparable with that of starlings or rooks. The next winter (1934-5) the number of redwings was down to perhaps one-tenth of their previous abundance, but they used the same roost and assembled from the same feeding territory—an interesting example of tenacity to an established dispersionary pattern.

The American robin (*T. migratorius*) shares the roosting habit with its congeners. One of the roosts observed by Brewster (1890, p. 36) reached a population conservatively estimated at 25,000 birds on 4th August 1875. Like other species, robins also use the same sites for many successive seasons. The most interesting feature of Brewster's observations was in finding cock birds assembling during the actual breeding season at a small roost in a lilac thicket in his own garden at Cambridge, Massachusetts. ' During the whole of May the roost was frequented by fifty or more birds, all apparently old males. By the middle of June these were joined by the first broods of young, and a month or so later by the old females with their second broods ' (Brewster, 1906, p. 394): by early August there were upwards of 700 robins coming in every night. We are reminded of the similar observation and inference by Audubon regarding the July roost of chimney-swifts. But the curious situation here is that American robins at nesting time are typical territory-holders, with a persistent song somewhat resembling that of the mistle-thrush (*T. viscivorus*): they have a particularly early waking-time and, during the militant territorial phase, must undoubtedly sleep in the actual place where they are going to take part in the dawn chorus. It looks as though, once the territorial plan is established and the robins have settled down to breed, some of the cocks are free to begin pioneering the succeeding pattern of dispersion; and consequently by the time the flood of fledglings is turned loose on the countryside a few weeks later, a roost-system is already established ready to receive them, and to provide them with the immediate spatial framework in which to begin their communal lives. It seems rather improbable that Brewster's May-roosting males were merely non-breeders, partly because there seemed not to be any non-breeding females among them, and partly because they were joined later, first by juveniles and then after an interval by old females with second broods of young.

Similar cases have been reported in the pied wagtail (*Motacilla alba*). Meiklejohn (1937) found a roost in North Wales on 28th June 1937, to which fully sixty wagtails had gathered, about two-thirds being juveniles. 'That the first broods of a species in one neighbourhood should flock and roost together whilst the second is still in the nest, is not extraordinary, but it is difficult to explain the presence of so many adult birds. The majority of them were males, so the possibility presents itself that some were the males of the breeding pairs of the neighbourhood whilst the rest were non-breeding birds.' This observation elicited a second report (loc. cit., p. 85) of a roost of about fifty pied wagtails occupied on 16th May, when all the birds seen were adults, but on the 30th May their number was found to have been increased by several birds of the year. The roost was still in being on 16th June, but had been forsaken by the 24th.

The most famous of pied wagtail roosts, in the plane trees standing in the centre of O'Connell Street in the city of Dublin, has also been tenanted through the breeding season (Moffat, 1931 and 1932). It appears to have been founded in the autumn of 1929. Within the first few years numbers up to 1000 or 1500 were reported, and by 1950 it was considered that 3600 were present during the winter months (Kennedy *et al.*, 1954, p. 380). In the spring of 1932 it was noticed that the numbers fell from about 200 to 70 in the course of the month of May, and to less than 50 in June, which suggests that many of those frequenting this roost in the nesting season may have been non-breeders.

In the evening at the Dublin roost 'as a rule the birds begin to arrive silently, but after the influx has gone on for about ten minutes the crowd bursts into a chorus of twitters, so loud as to be audible for nearly a hundred yards in spite of the noise of the city traffic. The twittering is kept up for about five minutes' (Moffat, 1931, p. 365). That is to say there is a massed demonstration, taking the form of a vocal chorus.

These mass-displays at the roost are, of course, presumed to have an epideictic function. A more openly competitive form of the same phenomenon has been observed in the Arabian babbler *Turdoides squamiceps*, a sociable passerine bird, which sometimes collects, before retiring to roost in a thick acacia, in flocks of up to twenty-five individuals. These birds then 'have a dance, which consists of a follow-my-leader opening, round and round a small bush, seeming to vie with each other in length of hop and efficiency in erecting the tail and wagging it when erect—all conducted in silence and lasting perhaps five minutes but with evident enjoyment and competition. If one good performer manages a particularly long hop or a good wag of erected tail, others will watch and repeat the performance in repetition; then quite suddenly one by one they fly off into the heart of a thick bush, where they settle down to roost, preceded by a good deal of preening, during which there is much chatter' (Meinertzhagen, 1954, p. 223).

14.4. *Conclusions regarding the epideictic function of bird roosts*

The description of the roosting behaviour of different kinds of birds could be very greatly extended. Many of the finches, buntings and sparrows have habits similar to those already described, besides members of a large number of other families such as the pigeons (Columbidae) and the American Icteridae (grackles, blackbirds, etc.), which have been skipped over in the preceding pages. Sufficient examples have been given, however, to bring out the fact that we are dealing with essentially the same phenomenon in every case—a phenomenon which though so widespread and familiar has never been in the least understood or accounted for. The essential features can be fairly easily picked out. First, though details about territory have not always been known to us, it is evidently characteristic that the roost should serve a particular area of country, and that it should normally be the only communal roost in that area. Secondly, the roost itself tends to persist at the same site, sometimes year after year, though if need be new roosts can generally be quickly established in replacement of old ones. Thirdly, the social behaviour connected with roosting is usually well synchronised and reaches its highest pitch at dusk and dawn—particularly at the former time. There is then, after the members are assembled, typically a collective display. This may consist of a chorus of voices or a massed flight or both, but sometimes there is an opportunity for the individual to show off to, or contest with, its immediate neighbours, and perhaps to claim its own special perch. Fourthly, communal roosting is typical of a way of life that comes into its own after the breeding season is over and is only suspended again when the breeding season approaches once more. There may in practice be an understandable overlap, but breeding behaviour and communal-roosting behaviour are essentially complementary or alternative activities.

It is quite clear that this general picture coincides very closely with our concept of a dispersionary adaptation. During the breeding season the birds concerned are dispersed according to one kind of territorial system, and during the asexual season according to another. In many respects the two systems are alike; and the similarities between a roost and a breeding colony are particularly striking, as we noted earlier. We observe also the characteristic synchronisation and the spirited participation of the members in what are frequently spectacular and vigorous displays. At any one roost the displays themselves fluctuate, depending on the season and often on the weather; and their compass and impressiveness are not merely a function of the number of birds taking part: in fact there is more than a hint that the intensity is highest at times when the population is working up to a major change, before migrating or, as migration proceeds, when the group is undergoing a progressive accumulation or turnover. At these latter times, in the face of the constant drift of transients, the roost-system appears capable of acting as a continuous regulator of population-density in its own neighbourhood and of feeding the migrants through the region in an orderly progression.

We may conclude, therefore, *that the primary function of the roost* is to bring the members of the population-unit together, so that whenever prevailing conditions demand it they can hold an epideictic demonstration.

The practical result of this is to stimulate the adjustment of population-density through emigration, when the economic and social pressure proves to be sufficiently high. Such demonstrations could theoretically take place at any time of the day, and constantly gregarious birds like the crested screamer (p. 289) and knot (p. 290), as we noted, may in fact hold them irregularly by day, and even by night in the former case. But we have repeatedly seen the advantage of using the hours of dusk and dawn for the purpose, because this is the time when the light-intensity is changing fastest and consequently it is the part of the twenty-four hours when the precise time is most easily gauged and synchronisation secured. Compared with the epideictic function, the fact that the birds afterwards remain together to sleep in the roost appears generally to be quite secondary in importance, and not to make any indispensable contribution towards safer or more comfortable repose. On the contrary, it quite commonly happens that predators are drawn to roosts and take a toll which they would find it difficult to equal in a scattered population; and in species like starlings and thrushes, with so many vociferous individuals present, minor commotions are liable to disturb the peace on and off throughout the night.

14.5. *Communal roosts in insectivorous bats*

The evening concerts and roosting behaviour of some of the African bats have already been mentioned in another context in Chapter 3 (p. 56). Being without exception crepuscular or nocturnal creatures, bats are essentially day-sleepers and tend to find places of retirement in dark holes, in trees, rock-faces and the burrows of other animals, and also in caves. Many of the insectivorous bats (Microchiroptera) seek seclusion also under the roofs of houses, in old towers and similar places.

There are, just as in birds, species that roost singly or in pairs, and others that at some or all seasons are more or less strongly gregarious. The most remarkable of the latter fall into two very different groups: first, there are the fruit-bats (family Pteropidae) of the tropical parts of the Old World, constituting the sub-order Megachiroptera, which for the most part hang up in trees in the day-time; and second, the various kinds of Microchiroptera, among them many horseshoe bats (Rhinolophidae), free-tailed bats (Molossidae), and members of the great genus *Myotis*, which frequent caves. So important is the latter specialisation that ' the existence of caves in a region no doubt has a marked influence on the bat fauna, conditioning their very presence and especially their abundance. . . If one were to map out the limestone regions of the world, one would undoubtedly find it an excellent guide to the localities for cave-haunting bats ' (Allen, 1939, pp. 75-6).

In the preceding section emphasis was laid on the resemblances that exist between breeding colonies and collective roosts in birds, and the supposed similarities in their dispersive functions; and in a later section of this chapter we shall be considering social hibernation as yet another adaptation of closely similar function. It is very significant to find that in bats it is sometimes difficult if not impossible to keep these three facets of gregarious behaviour separate, since the bats are liable to breed in their roosts when the appropriate season comes round and, in northern climates, to hibernate there also. In other words, one system of territorial foci may in some cases suffice for dispersionary purposes right through the year.

Actually it happens in rare instances that bats share their roosts with birds, changing places with them at dawn and dusk. A remarkable example of this was described by Pryer (1884) at the great birds'-nest caves of Gomanton, inland from Sandakan Bay in North Borneo, in the month of March. The caves are inhabited by vast numbers of swiftlets belonging to the two species of *Collocalia* which make, respectively, the ' white ' and ' black ' edible birds'-nests of that region, and also by myriads of free-tailed bats of the species *Myctinomus plicatus*. The caves are in a limestone formation and immensely lofty: in addition to cliff-entrances they have vertical pot-hole shafts emerging back from the cliff-top. Pryer took up his position where two of these emerged near together, one from the cave called ' Simud Itam ' and the other from ' Simud Putih '. ' The Malays told me to be sure and return to Simud Putih at 5 o'clock, as I should see the most wonderful sight in all Borneo—the departure of the bats and the return of the swifts. I accordingly took a seat on a block of limestone at the mouth of the cave. . . . Soon I heard a rushing sound, and, peering over the edge of the circular opening leading into Simud Itam, I saw columns of bats wheeling round the sides in regular order. Shortly after 5 o'clock, although the sun had not yet set, the columns began to rise above the edge, still in a circular flight: then they rose, wheeling round a high tree growing on the opposite side, and every few minutes a large flight would break off and, after rising high in the air, disappear in the distance: each flight contained many thousands. I counted ninteeen flocks go off in this way, and they continued to go off in a continual stream until it was too dark for me to see them any longer. . . .

' At a quarter to 6 the swifts began to come in to Simud Putih: a few had been flying in and out all day long, but now they began to pour in, at first in tens and then in hundreds, until the sound of their wings was like a strong gale of wind whistling through the rigging of a ship. They continued flying in until midnight, as I could still see them flashing by over my head as I went to sleep . . .' (loc. cit., p. 535).

' Arising before daybreak, I witnessed a reversal of the proceedings of the previous night, the swifts now going out of Simud Putih and the bats going into Simud Itam. The latter literally " rained " into their chasm for two hours after daylight. On looking up the air seemed filled with small specks, which flashed down perpendicularly with great rapidity and disappeared in the darkness below ' (loc. cit., p. 536). Incidentally, there is a peculiarly specialised hawk called *Machaeramphus alcinus*, the ' bat-eating buzzard ', found in the tropics of both Africa and south-east Asia: it has strictly crepuscular habits, and appears to make a good living at places like these. Pryer saw them ' working away on the bats in a very business-like manner ': the way they took them one after another strongly reminded him, he says, of a man eating oysters. They also caught swifts, and in every case swallowed their prey on the wing (loc. cit., p. 536).

Attention must be drawn in this case to the flight-manoeuvres of the bats before leaving the roost at night, so reminiscent of various birds.

It is another molossid, the Mexican free-tailed or guano bat (*Tadarida mexicana*), which frequents in such vast numbers the Carlsbad caverns of New Mexico and other caves of that region. ' The greatest bat-cave in the United States ' is said to be Ney Cave, north-west of San Antonio, Texas, and it is believed to have a population of 20 to 30 million bats—some three times as many as Carlsbad (Moore, 1948, p. 264). High above the entrance of Ney Cave when Moore and his party arrived (in August 1948) there soared a group of turkey-vultures, which were later joined by hawks (*Accipiter* and *Buteo*) and peregrine falcons. ' The reason for their presence shortly became apparent. Within a few minutes . . ., at 6.10 p.m., . . . a few bats fluttered out of the entrance. . . . Within five minutes, however, as if by pre-arranged signal, a great column of bats burst from the cave mouth. Flying in compact formation, the bats ascended at a sharp angle in order to clear a group of junipers on the cliff side; then then leveled off over the valley, the column maintaining its identity as far as the eye could follow. . . .

' In the meantime the hawks were busy. Diving into the flying column, it was seldom that they would emerge from the encounter without a bat in their claws ' (Moore, loc. cit., p. 263).

The flight continued for about twenty minutes and then suddenly dwindled and ceased. As the watchers had been told to expect, a second flight began about 7.10, and was still emerging with undiminished intensity when they left an hour later. They were informed that 5 to 6 hours were required for the outgoing flight, and that the incoming column began at about 3 a.m., and sometimes lasted as late as 11 a.m.

Few species of bats congregate in such enormous numbers as those in the two foregoing cases, and there are many old-established collective roosts that harbour only tens or hundreds of individuals at most. Venables (1943) made sustained observations on a roost of pipistrelles (*Pipistrellus pipistrellus*) in the roof of Binsey church, near Oxford, where the highest number recorded emerging on any night was 346. At least some of the bats evidently hibernated in the same place, and the number emerging depended partly on whether the temperature at the time was above a threshold of roughly 2° C., and on unfavourable weather factors such as gusty winds. They tended to come out earlier in relation to sunset-time in the summer months, when the days are longer and feeding-time shorter, than they did in spring and autumn.

In many vespertilionids there are seasonal changes in the use of particular roosts and in the mutual association of the sexes, with the result that four fairly distinct types of roost can often be distinguished. In their studies of the lump-nosed bat (*Corynorhinus rafinesquei*) in California, for example,

Dalquest (1947) and Pearson and his collaborators (1952) found (i) winter-roosts, (ii) nursery colonies occupied in spring and summer by pregnant and nursing females and their young, (iii) day-roosts of males and others, usually scattered and (iv) night-roosts, used in the midnight hours between the dusk and dawn feeding-periods. In some cases it appears that, for the individual male or female, the three types of roost that fulfil its various needs may all be at different places; and in other cases the same roost may combine two or more uses. The lump-nosed bat is not a true hibernator and males especially are likely to make sorties during the winter months: torpor is only temporary and may occur during sleep at all seasons. In the winter-roosts the sexes are mixed, though not very evenly (Pearson *et al.*, p. 277). Least is known about the night-roosts, which at any rate in some instances are not frequented at all in the daytime: at one mentioned by Dalquest (loc. cit., p. 26), in two small buildings, no less than six different species were trapped during one night.

It very often happens with bats, just as it does with birds or seals, that a number of species consort together in a breeding colony or roost. Guthrie (1933*a* and *b*) made a very interesting study of the cave-haunting species of *Myotis* in two limestone caves in Missouri—Rocheport and Hunter's caves—where six species were found. One of these, *M. grisescens*, was a summer resident, all but a remnant emigrating in autumn to winter-quarters elsewhere; and there were three others, *M. sodalis*, *M. lucifugus* and *Pipistrellus subflavus*, which were conspicuous winter residents, but absent in summer. The two remaining species (*M. keenii* and *Eptesicus fuscus*) were scarce and collected only in winter.

For *M. grisescens* one of the caves, namely Rocheport, served as a nursery colony. It was shown originally long ago by Rollinat and Trouessart (1896, p. 215), in their classic paper on the reproduction of bats, that the European *M. myotis* does not mature in its first autumn, but only in the following one, at the age of 17 months; and the same is now known to be true of *M. grisescens*, though not of all members of the genus. The French zoologists also found that the yearling virgin females accompany the female adults until the new generation of young are born in late May or June, after which they leave the nursery colony and go elsewhere. The authors (loc. cit., pp. 225-8) discovered, however, in the nursery colonies in the roof of the great church at Argenton, that their samples in May always included a few barren females —in sum about 10 per cent of the total catch—and they concluded that these must be a remnant of young females still virgin at two years old. In the circumstances this could not be more than a surmise, because the age of a female is difficult to determine after the first year of life, and older females may therefore have been included. The significant point is, however, that in this particular colony one-tenth of all the potentially adult females were non-breeders, a state of affairs very reminiscent of what may so commonly be found in a bird-colony.

The relationship of age-groups and sexes is evidently similar if not identical in *M. grisescens.* Very few non-breeding females and no males were found in summer in Rocheport cave by Miss Guthrie. In Hunter's cave, on the contrary, 15 miles away in a direct line, she found that before the end of June males and yearling females constituted the main population. It was not until 22nd September that old males were detected again in Rocheport, though the females were still greatly in the majority. ' Migration apparently occurred at somewhat different periods for old and young ' (Guthrie, 1933*a*). Mohr (1933, p. 51) visited a similar nursery colony of the same species in Tennessee in late June, which contained several thousand individuals: not one in ten of them was an adult male.

The segregation of the females during the pregnancy and lactation period, and the indication that there are customary nursery sites distinct from the haunts of the males and their followers, each presumably with its own collective feeding-territory, is common to many insectivorous bats, and on the whole rather similar to the system adopted by the red deer (*Cervus elaphus*) outlined on an earlier page (*see* p. 217).

There is a long history of discovery, going back now for a century, regarding the reproductive cycle of bats, which was very fully reviewed by Baker (in Baker and Bird, 1936). In the northern insectivorous bats it is generally true that copulation occurs in the autumn and the sperms then remain alive in the female's greatly distended uterus throughout the winter, until ovulation and conception occur in the spring. Guthrie (1933*b*) established that in *M. lucifugus* insemination took place either before the bats entered their winter cave or immediately thereafter; and most likely actually inside the cave itself. When she kept hibernating females in cages for a time, she found that the number of sperm in their uteri greatly diminished; and this appears to have been brought about both by the gradual evacuation of the uterine and vaginal contents and by the actual ingestion of sperms by the masses of leucocytes normally present in the uterus at this time: only rarely, however, were the sperms completely eliminated (loc. cit., p. 207).

These observations suggest that under improper or unfavourable hibernating conditions there might be a consequent reduction in fertility. There was for long a controversy as to whether copulation does or does not recur in the spring, but it is now certain that this happens more or less sporadically in some species. Miss Guthrie showed that it can take place in both *M. lucifugus* and *Pipistrellus subflavus*; and what is known in the case of *M. sodalis,* she concludes, lends itself to the same interpretation. It seems at first sight a completely baffling adaptation that many of the males should remain, as they do, in breeding condition throughout the hibernating period, but at least it is clear that rutting and social hibernation are closely bound up together. Since actual pregnancy never begins till the spring, no time appears to be gained for earlier gestation by achieving insemination six months in advance, and the advantage of the adaptation must therefore lie in a different direction. It seems possible that it is a homeostatic one—bringing the sexual phase of the cycle forward to a season when the weather is still warm enough for all the bats to be active and capable of joining together in a representative epideictic function. There appears to be no lack of opportunity in all the various circumstances for regulating the pregnancy rate.

The great majority of bats are uniparous, bearing a single young one at a time, and, no matter which ovary has produced the egg, carrying the embryo in the right horn of the uterus. A fair number appear to have twins very occasionally (including various species of *Myotis*); and some (including the European noctule and pipistrelle bats) to have them commonly, though only in certain parts of their range and not everywhere. A few species, among them the North American red bat (*Lasiurus borealis*), a long-distance migrant, are able at times to have three or even four young at a birth (*cf.* Allen, 1939, p. 248). It is, however, generally true that the reproductive rate of bats is low.

No great weight must be attached to what is only a casual observation, but it should be mentioned that Mohr (1953, p. 51) was told by the owner of the cave he visited in Tennessee (Indian Cave, north of Knoxville) ' that for more than two weeks he had been cleaning up young bats " by the shovelfull " from the floor of the cave '—' supposedly dropped from their mothers '. Though we know nothing of the circumstances—and there may have been overmuch human disturbance—this probably indicates that at least occasionally there is a wholesale infant mortality. Miss Guthrie (1933 *a*, p. 8) found that young *M. grisescens* born in captivity were completely neglected by their mothers, even when she had been careful not to handle them; Rollinat and Trouessart (loc. cit., p. 234), on the other hand, had almost no casualties with *M. myotis*. There may be here another important factor in regulating the rate of recruitment of the population: we shall find the same kind of thing occurs in a wide variety of animals to be discussed in Chapter 22 (p. 541).

Before leaving the cave bats, reference must be made to the probable use of roosts by passage-migrants, very much in the manner known to occur in birds such as chimney-swifts (*see* p. 292). Guthrie (1933*a*) found strong evidence of passage movements in *Myotis sodalis* in Rocheport Cave, especially in March, when the numbers present varied from a ' large mass ' to none, and back again to ' large clumps ', in successive weeks. Her interpretation was that this species was chiefly present during periods when migration was temporarily held up by the weather. Similarly, ' the only specimens of *M. keenii* were taken on February 6 and are considered to have been temporarily held in Rocheport Cave during the movements in which *M. lucifugus* and *M. sodalis* also took part ' (loc. cit., p. 17). It is well established from banding records that bats frequently move from cave to cave; and it is not difficult to see how a dispersionary system based on such known fixed points could provide a framework for controlling population-density over a wide area.

14.6. *Communal roosts in fruit-bats*

Though the members of the Pteropidae are spread over much of the warm regions of the Old World, those whose roosting habits are most familiar are the true flying-foxes, members of the genus *Pteropus* itself, which are found all through the southern parts of Asia and in Australasia. Many of these species customarily gather for the day into very large roosts that have long been referred to as 'camps' in Australia. The continent of Africa has no representative of *Pteropus*, though other kinds of fruit-bats occur there, including the anvil bat and hammer-headed bat, whose nightly concerts were mentioned in Chapter 3 (pp. 55-6). Curiously enough *Pteropus* species are present in Madagascar and the Comoro Islands—one of the well-known but unexplained anomalies in the zoogeography of that part of the world.

Roosting associations between bats and birds, analogous to the one already described in the case of a free-tailed bat, occur in this family also—flying-foxes in this case using the site by day and birds by night. Willey (1904) spent five days on the small island of Barberyn, just off the west coast of Ceylon, thirty-five miles south of Colombo, from 4th to 9th February 1904; and here he witnessed at sundown this most interesting sight, 'namely the passage in opposite directions across the strait, which divides the island from the mainland, of immense flocks of crows and flying foxes, the former bound for the island to rest for the night, the latter speeding their way to the mainland intent upon their nocturnal forage. The flying foxes flew on the average distinctly higher than the crows. . . .' Though the latter outnumbered the bats, Willey very characteristically estimated that 'weight for weight they probably represented an equivalent bulk of living matter'. The whole of the cross-movement occupied about half-an-hour. In the morning he observed the opposite transaction. The roosting site on the island was situated in the tree-tops of the coconut plantation.

A similar association has been described on the uninhabited South Sentinel Island in the Andaman group, where the birds that crossed the water each evening, changing places with the flying-foxes, were pied imperial pigeons (*Ducula bicolor*) (Ferrar, 1934). Both these accounts refer to the Indian flying-fox, *P. giganteus*, which on the mainland is, in a rather different way, often associated with human settlements. They 'frequently select for a permanent roost some large tree quite in the centre of a native village or at no great distance from its outskirts, gathering regularly in its top for their daytime sleep. Such trees may be in constant use for generations. . . .' (Allen, 1939, p. 210).

Living as they do in constantly warm climates, fruit-bats have no need to hibernate. In the New Hebrides the Bakers (1936, p. 126) found that the social habits of *P. geddiei* varied in a regular sequence through the year. 'Both sexes congregate together in large "camps" during the daytime from about September to about January. These camps are often in large Casuarina trees near the shore'. Despite the constancy of the climate there is a distinct breeding season: no pregnant specimens were found in November, December or January, and the evidence showed clearly that 'a large proportion of the conceptions take place in February and March. . . . Conception does occur, however, so late as June or July. . . .' When the females become pregnant they leave the camps near the shore and are difficult to find: the males on the other hand continue to lead a social life for a time. 'Later in the year, about June, when pregnancy is far advanced, the females may be found in inland camps which contain few or no males. At this season the males have given up their social life and live separately, and it is now difficult to obtain males'.

It seems evident from this that the social roosting habit is at its height in the period which in birds we should describe as prenuptial and nuptial. In these bats the five months between the admission of independent young into the population and the rut comprise the prenuptial phase, in which we have learnt to expect a peak development of epideictic behaviour.

The camp is the scene of all their main social activities; and the males apparently continue to occupy it as long as there is any possibility of females requiring insemination (presumably until June, at least exceptionally), just as male birds usually go on occupying a lek throughout the breeding season. It is clear from Ratcliffe's account (1932, p. 37) of the Australian species that social competition is a feature of the camps. Even in the daytime they 'are very rarely quiet—in fact they are most easily located by their sound. The animals are very quarrelsome, and much of the day is spent in noisy bickering'. More specifically MacGillivray (1860, p. 7136), in the notes appended to his original description of *P. geddiei* from the New Hebrides, stated: 'The males seem to be very pugnacious; I have seen them fighting among themselves on the trees. . . .' Semon (1899, p. 261) also observed that 'some of the animals are always awake, disengaging themselves from the mass, exchanging their place for another one, screaming and screeching, so that the presence of the flying-foxes makes itself heard at a considerable distance'. In one place Ratcliffe (loc. cit., p. 46) refers to a group-display by the bats when leaving their camp at nightfall: they first formed 'a packed wheeling mass above the trees', before moving off 'in a column about 100 yards wide and 100 feet deep (in the centre)'.

After the rut is over the males disperse, and the social emphasis appears to shift to the nursery camp, at least in *P. geddiei*. Inferences from these successive phases of behaviour must not be carried too far, but it may be noted as a general rule, that, up to the time of copulation, the initiative in multiplying the population rests very largely with the male sex: the moment the females have been fertilised, however, opportunity for further restriction in reproductive output becomes largely or entirely transferred to them, especially in mammals where they have sole charge of the foetus and young. If infantile mortality is of any importance in regulating the reproductive output of flying-foxes and other bats, this would help to explain the close epideictic association of females and their offspring into nursery groups. Ratcliffe (loc. cit., p. 42) states that his records indicated 'a considerable mortality among the young', especially perhaps at the hands of nocturnal predators at the stage when the young have to be left behind by their mothers at night because they are too heavy to carry. Incidentally, though Radcliffe recorded 100 per cent pregnancy among adult females in his flying-fox survey, it may be noted that out of 63 females shot in two camps, nine were classed as ' immature ' because of their very small mammary nipples, but in the absence of any sure method of distinguishing their age, it might be safer in the light of present knowledge simply to put down one-seventh (14 per cent) of this sample as being non-breeders: in fact the latter may well have consisted largely of recruit-class females.

As would normally be expected of all such social gathering-places, the flying-fox camps tend to be occupied regularly from year to year if undisturbed; in some cases, Ratcliffe states (loc. cit., p. 36), as long as the local settlers can remember, which is upwards of fifty years. He refers to the camp's ' territory ' as such, and the sites in Queensland he found to be of the order of 15 to 20 miles apart. *P. scapulatus* there formed the most populous camps, the biggest of which he estimated to cover 10 to 20 acres and in numbers to exceed—possibly considerably—quarter of a million (loc. cit., p. 47). The two largest ones he saw in New South Wales (both containing a mixture of species) were thought to hold as many as 100,000. Except for *P. conspicillatus*, a more tropical bat that was identified only in roosts exclusive to itself, mixed camps were the rule in the common Australian species, namely *P. poliocephalus*, *P. gouldi* and *P. scapulatus* (loc. cit., p. 41): Ratcliffe implies that each species generally kept to its own area of the camp (loc. cit., p. 37). *P. scapulatus*, the most numerous of all, has a breeding season opposite to that of the others, pairing about October and bearing its young six or seven months later in the Australian autumn; whereas *P. poliocephalus* and *gouldi* evidently pair about March and April, and have their young in spring—a difference that probably helps to prevent a perennial association of all three species. *P. gouldi* was found to make no obvious migrations (though the closely related *P. conspicilliatus* apparently did so, at least locally). *P. poliocephalus* is a regular summer migrant into the southern part of its range. *P. scapulatus*, also a migrant, is a great wanderer, somewhat uncertain in its movements, and dependent on the unpredictable blossoming of the eucalypts: Ratcliffe notes (loc. cit., p. 52) that ' when a district is already inhabited by another species of fruit-bat, *Pt. scapulatus* will almost always use their camping sites '.

One of the chief rewards of making a comparative study of a social habit like roosting is the unsuspected examples of convergent evolution that are brought to light by it. It is to be noticed in the last paragraph in particular that there is hardly a detail of roosting conduct developed in the flying-foxes that does not find its counterpart in the corresponding habits of birds, outlined in Sections 14.2 and 14.3.

14.7. *Communal roosting in insects*

Ability to travel from place to place without too much expenditure of effort seems to be a prerequisite for developing a system of dispersion based on daily or nightly visits to a roost, and it is fairly obvious that the collective roosting habit is best developed in those types of animals such as birds and bats that are able to fly. It need not occasion surprise, therefore, that similar adaptations should have appeared sporadically among the insects. In studying roosting habits we are rather strictly limited to terrestial species which are relatively easy to observe; and there is no doubt that this kind of behaviour in insects needs to take place on a fairly large scale if it is to attract attention without a deliberate search. We have noticed already certain forms of behaviour in insects that come within the fringes of this category, such as the nightly retirement of the various kinds of tent-caterpillars into their silken tents (*see* p. 116), or the withdrawal of social bees and wasps into their nests, or again the nightly gathering of tropical fire-flies to the trees or glades which provide the stage for their flashing displays.

There are, however, a number of examples that much more closely resemble those we have been considering in the preceding pages. One of these concerns the zebra butterfly, *Heliconius charitonius*, belonging to a conspicuously-coloured group of tropical American butterflies that are nauseous or distasteful to predators. Their social habits have been studied especially by F. M. Jones (1931) in Florida, in the period between January and April when the adults have lately emerged from pupation. The females are fertilised within a day or two of their appearance, but all members of the local group continue to use the roost for some weeks or months thereafter, presumably throughout the pairing and egg-laying season.

Jones found that the nocturnal roosts generally contained only 20 to 30 (occasionally 40) individuals of both sexes, resting in a compact group in a bush. By day the insects confined their activities to circumscribed areas each belonging to a single dormitory group, whose members seemed to keep within a few hundred feet of their sleeping-bush. This is an important observation, indicating that, as in the birds and bats, the roosts serve as foci in a territorial system. Individuals marked by snipping the wings were found sometimes to return to the same twig for a short succession of nights, though never for very many. In one year of observation (1930), three roosts were established within a hundred yards of each other; and it was found that ' there was a constant shift and interchange from one sleeping place to another. For example, a marked butterfly . . . was located, sleeping, on twenty-two out of twenty-six consecutive nights; and in this period it made eight changes, back and forth, between roost B and roost C. Its longest observed, unbroken stay on one roost was for six successive nights '. The three roosts together presumably comprised a single common territorial unit. Jones concluded from his study that the group-sleeping habit was unconnected with mating and had nothing to do with food. He thought that possibly a group of such nauseous insects might be better protected from predators by scent than a scattered population, while admitting that ' we have been unable to discover any reasonable alternative '. His work, as quoted here, is taken from an abstract presented to the Entomological Society of London by Poulton, and the latter appended the comment (loc.

cit., p. 10) that similar habits evidently occurred in other Heliconiidae and also in the equally distasteful Danaidae. Poulton thought that it might be 'to meet . . . the need for an enhanced aposematic advertisement during periods of rest', and the facts 'are otherwise inexplicable'. Incidentally, like many other insects, the zebra butterflies also gather at their roosting places during the daytime in dull weather.

The parallels in this case are surprisingly close to those we have considered earlier and it seems logical to consider this roosting behaviour as an analogous adaptation developed for a similar purpose, namely to provide a social territorial system which will serve to limit population density.

Collective roosting behaviour has also been discovered in various 'solitary' species of wasps and bees. The Raus (1916) in particular have given us an important account of their many observations, made in Kansas, on the sleep of insects, the greater part of them relating to hymenopterous species: their paper also contains as a valuable appendix a long annotated bibliography covering the earlier literature on the subject.

They describe first two congregations of the steel-blue wasp known as *Chalybion caeruleum*, one of which was on the underside of an overhanging rock and contained about thirty males and females: the other, found early one morning, consisted of two groups close together on the ceiling of a cow-shed. At the latter site the same evening the wasps were watched as they came in one by one, after being away all day, 'and lo and behold! they settled on the same spot and collected in two groups similar to those of the morning' (loc. cit., p. 229). Six weeks later, on returning to the farm, Dr. Rau found a much larger congregation on the same ceiling in the identical spot. He proceeded to mark a number of the wasps with paint, and ascertained that they returned night after night, some of them still bearing traces of paint two weeks later. He could offer no satisfactory explanation for their assemblies, at which he was convinced mating never occurred, though males and females were always present in almost equal numbers.

An equally striking case is that of another wasp *Elis quinquecincta*, of which several roosts were found: one of them contained as many as 105 individuals and another 'several hundreds'; but invariably in this species all were males. The Raus were entirely satisfied that only the males were gregarious in habit, 'and perhaps all in the region round about occupy one centrally located sleeping place' (loc. cit., p. 235). They rested on herbaceous vegetation, for example on asters two feet high, and when disturbed at dawn they dropped on to the grass underneath with spread wings, behaving in what seemed so stupid a manner that it was clear to the Raus that massing together in great numbers could not afford them any protection from birds. The most interesting observation on this wasp, however, was made at a very large roost in which a crowded clump of 27 plants of erigeron each contained from two to sixty males, already assembled as early as 5.30 p.m. 'By 7.05 o'clock when I returned to the place they were at rest on these plants in great numbers. Suddenly, without apparent provocation, they all arose and flew about in circles for several minutes before they settled again. Some cause other than myself was responsible for the disturbance, for I stood fifteen feet away. . . . They settled promptly to perfect quietude again, so that only a few minutes later I removed a whole plant with one vigorous stroke of the knife and carried it some distance without perceptably disturbing them '. The authors confess that ' we have no light upon the question of why the males congregate thus and where and how the females sleep ' (loc. cit., p. 237). The females in *Elis* are winged, incidentally, though in many of the Mutillidae, the family to which *Elis* belongs, they have become flightless.

In our previous experience of epideictic phenomena we have come across examples of all-male assemblies many times—most often in connection with some form of social activity that could be described as a lek. In the chimney-swifts (p. 291) and American robins (p. 296), and in some of the bats, we have, moreover, noted unisexual roosting under conditions not greatly dissimilar from those in the present instance. It all helps to emphasise the element of interchangeability, and unity of purpose, in the various common types of social gathering, in so far as their epideictic function is concerned. Presumably this all-male roost is discharging the same function as the all-male dances of gnats and midges, previously described (p. 197), only here the display is associated with the sleeping site instead of taking place at some other focus during the hours of day. The concerted flight of the whole group to join in social manoeuvres over the roost strikes a very familiar note, and equally supports the hypothesis that collective roosting behaviour in insects is an adaptation very closely parallel to that of birds and bats.

Space does not permit the recounting of all the Raus' interesting discoveries, but one more example must be included—that of the bee *Melissodes obliqua* (loc. cit., p. 242). On some charred stalks of burnt grass in a field a roost containing 28 bees in close proximity was found; and when it was revisited on succeeding nights the numbers counted were 29, 24 and 30 respectively. Bees marked with paint returned, though almost at once an uncontrollable change in the habitat diminished the site. Eighteen were collected as a sample, and in this case too there proved to be nothing but males present. At other roosts of the same species similar experiments were made with identical results. Sometimes, just as with birds and bats, two or more species of Hymenoptera are associated together in a roost. For instance, at two places the Raus found a small number of the bee *Melissodes obliqua* regularly coming to roost among larger numbers of the wasp *Prionony atratum* (loc. cit., p. 245). It is evident from papers they cite in their bibliography that *Prionony* spp. are especially given to roosting in mixed company: possibly they form the nucleus to which other species attach themselves. On a June evening on the scorched Kettleman plains in Fresno county, California, Bradley (1908, p. 128) found one roost after another of *Prionony* on the dry stems of a wild oat, to the number of ' scores, almost hundreds ', in the space of an hour before it got too dark to see. ' No less than seven species of aculeate Hymenoptera were represented in large numbers, belonging to three different families, four sub-families and five genera, *and each species so far as observed was always grouped separately.* Though a group of one species was in close juxtaposition to that of another on a neighbouring stem there was no intermingling of individuals ' (loc. cit., p. 129). Two other species belonged to *Priononyx* and one to *Melissodes.* Later he adds: ' There certainly was no protection received from their gregarious sleeping, since the groups . . . were very conspicuous objects in the failing light '. Though not universal, species-segregation is quite frequent too in bird and bat roosts where several are present, and in mixed breeding colonies of water-birds.

The oft-quoted discovery of Fabre (I, pp. 216, 225) is worth repeating, of a particularly large gathering of the hairy sand-wasp, *Ammophila hirsuta.* In August 1865 he and some companions had ascended Mont Ventoux (1912 m) in Provence; and close to the summit towards evening, while the party was busy rolling blocks of stone over the beetling crags into the north corrie, he uncovered a compact conglomeration of some hundreds of these wasps under a large flat slab—a discovery all the more unexpected since the species was until then known to him as a solitary insect living in the plains, thousands of feet below. It was neither the season nor the place for hibernation, and Fabre concluded that the wasps must be a party of migrants, but he did not know then as we do now of the communal roosting habits of many Hymenoptera, among them several other species of *Ammophila*; and it is generally thought that what he had uncovered was most likely a normal dormitory, notwithstanding the remarkable situation (*cf.* Rau and Rau, loc. cit., p. 266).

Fabre also describes (loc. cit., p. 236) two spectacular mountain-top swarms of the seven-spot ladybird (*Coccinella septempunctata*), one found on a warm October day on the outside of the flagstone roof and walls of the chapel on the summit of Mont Ventoux; in their myriads they covered the masonry like a pattern of coral beads. The other was in June on a lower neighbouring height, on the stone pedestal of a cross occupying the salient point of the hilltop. From the exceptionally conspicuous sites chosen for massing in the sunshine by these scarlet-and-black-coated beetles, we can suspect an epideictic purpose: there is of course no suggestion in this case of their frequenting a habitual roost. Most of the time the insects were immobile, but whenever the sun shone hot there was a continual exchange between those coming in to seek a place and the occupants themselves that were taking wing, only to return after a brief flight.

What appear to be epideictic swarms take, of course, an endless variety of forms in insects. Another example of a daytime resting assembly, which may be mentioned very briefly, relates to the empid fly *Hilara* (perhaps *H. litorea* Fall.), of which I found a motionless swarm containing not less than 20,000 individuals of both sexes on the underside of a boulder overhanging the water of the River Dee in Aberdeenshire, on the hot afternoon of 6th July 1957. In a dense layer they covered some six square feet of the rock surface: when disturbed they took wing readily but immediately returned to the resting swarm.

14.8. *Social hibernation, aestivation and similar seasonal gatherings*

More common in insects than daily gatherings are autumnal congregations of adults, usually in preparation for winter. In the course of a survey of hibernating arthropods (mostly insects) in forest environments in the Chicago region, Holmquist (1926, p. 398) found that 55 out of the 329 species he located were gregarious in winter (though this number included 15 species of perennially-social ants). Some of these gatherings recur at the same established sites from year to year, in a manner already familiar to us with other types of social arena.

A number of examples have been quoted by Imms (1947, p. 242). The two-spot ladybird (*Adalia bipunctata*) 'often hibernates in numbers of perhaps fifty or more under bark or in some other suitable place; the same individual tree may be used for this purpose for several years in succession'. He mentions the cluster-flies *Pollenia rudis* and *Musca autumnalis* which congregate, supposedly to overwinter, in buildings: my experience in eastern Canada was that they entered the house (in this case a summer cottage occupied from June to September) and buzzed on the window-panes in August, and were found dead on the floors and window-sills in masses the following spring. Imms refers to other kinds of Diptera of small size that sometimes appear in thousands on the window-panes and ceilings of houses from August to October. 'Such swarms are often found in the same house for several years in succession. The record seems to be held by the small greenish-brown fly *Limnophora septemnotata* which is known to have congregated annually for 16 years in the same house. All the individuals were impregnated females. . . . The tiny black and yellow fly *Chloropisca circumdata* behaves in a similar way, and is known to congregate on the same part of the ceiling in the same room of a house for several successive years.'

The last piece of information appears to be derived in part from a note by Scott (1916), which describes a vast swarm of small flies seen in Babraham House near Cambridge in October 1915. The swarm occurred in two rooms facing south-east—the drawing-room and the room above it. 'It was very much worse in the first-floor room, where it formed a truly astonishing sight. In the large bay-window every pane of glass was densely covered with countless myriads of small flies: on the upper sides of the projecting cross-pieces of wood between the panes the flies rested in masses, literally crawling over each other, while all the part of the ceiling near the window was almost as thickly covered as the window itself.' Several species were present, the two principal constituents being the *Chloropisca* (variously known as *circumdata*, *ornata* and *notata*, all of Miegen) and, it was thought, *Limnophora septemnotata*. The invasion had taken place for six years in succession: but 'apart from these immense hosts, which may definitely be called "swarms", Mrs. Adeane [the lady of the house] tells me that there has been a great quantity of flies in the sunny rooms every year, from at least as far back as 1888 onwards [*i.e.* 28 years], the largest number always being found in this same first-floor room '. Cambridge appears to have been especially liable to such swarms, for Sharp (1899, p. 505) tells of others in the Provost's Lodge at King's College in 1831 and again in 1870 (both perhaps in the same room), and also in the Museums (' always in the same apartments ').

In these cases each year's flies belong to a different generation from the last; and their tradition in using the same gathering places year after year presumably cannot depend on repeated individual experience as it does in longer-lived animals. Imms (loc. cit.,) says: ' When flies aggregate in the same room and at the same period in successive years, often in thousands of individuals, it is a complete mystery what guides them to so small an area.' Possibly, he suggests, individuals of previous years have left an odour behind which induces the newcomers to assemble in the identical place.

Actually it is by no means certain that the insects are seeking winter-quarters at all: indeed this is only the suggestion that is usually put forward for want of a better. Cluster-flies (*Pollenia rudis*), the ones that we used to find dead in our cottage in such quantities in the spring, actually lay their eggs in autumn (Colyer and Hammond, 1951, p. 286), and the over-wintering of some of the adults in the milder climate of England may be of no very vital importance to propagating the race. The frit-fly (*Oscinella frit*) is another species recorded as swarming in the same way, in which the summer adults lay their eggs in July-September: overwintering effectively depends on the larvae (Colyer and Hammond, loc. cit., p. 240), and very likely it does so in most of the species that have acquired this gregarious autumn habit. In some cases, as mentioned already for *Limnophora septemnotata*, the swarms consist solely of fertilised females ready to lay eggs. It appears not at all improbable that all these gatherings are purely epideictic in function, taking place as they do at the time of year when, after two or three spring and summer generations in some cases, the population has built up to a peak density and eggs are about to be laid for the final (overwintering) generation.

True collective hibernation does, of course, occur in insects; and the two-spot ladybird, mentioned at the beginning of this section, provides a well-known example of it, both in Europe and North America. Various Lepidoptera hibernate as adults, and in a few cases they mass together at suitable sites. As a boy I found such a gathering on 13th January 1919, when exploring a small dry cave near Austwick, Yorkshire: the candlelight suddenly revealed a patch of the low limestone ceiling covered with beautiful herald moths (*Scoliopteryx libatrix*) which seemed the more astonishing by contrast with the snow lying outside, and made a vivid and lasting impression. *Triphosa dubitata* is another moth with similar cave-haunting habits.

The possibility must not, of course, be dismissed that some of the smaller of these gatherings are completely non-social and result merely from a shortage of protected sites in which hibernation is possible. The facts do not suggest, however, that the swarms of some of the flies mentioned could be satisfactorily and completely accounted for in this way. Whether they last only into the autumn or whether they involve adults genuinely wintering, the assemblies occur in the period immediately preceding egg-laying and thus correspond with the numerous examples of prenuptial gatherings previously considered in different circumstances, especially in Chapters 10 and 11. As with those other gatherings, we are concerned here with a seasonal massing of particular castes of individuals at customary established sites; and we can at least tentatively conclude that in most cases, if not all, they present us with yet another variety of epideictic display whose purpose, as usual, is to supply the index of population-density by which the next step in the annual cycle is determined, namely, how to secure a reproductive output of the proper size to start off the next generation.

Strong support for this view comes from the exceptionally interesting case of the bogong moth (*Agrotis infusa*), an Australian noctuid, of which a thorough and valuable study has been made by Common (1952, 1954). In early summer (late October to mid-December) it migrates considerable distances from the plains into the Australian Alps of New South Wales; and there it forms immense aestivating aggregations in clefts and caves of granite outcrops on selected mountain-tops, in the cool alpine environment of 4000 to 6000 feet. Common found on Mount Gingera, A.C.T., that the moths were present throughout the summer until autumn (March-April), when they dispersed, presumably to their breeding-grounds; but that during the four months or so of their aestivation their numbers were not static. On a sample surface, formed by the roof of one of the larger caves, the area they covered decreased from a maximum of 96·2 sq. ft. on 1st January 1953 (estimated to contain 144,000) to 29·9 sq. ft. on 25th February—a reduction to less than a third; by 11th March it was up again to 63·3 sq. ft. and, after staying near this level for two weeks more, the cave was progressively emptied until every moth had departed.

Their behaviour was strongly gregarious: over large continuous areas the individuals were in physical contact, imbricated one over another, gripping the rock surface by their fore tarsi, with head and thorax burrowed under the wings and abdomens of those ahead of them (*see* Plate 3). They were not continuously quiescent, as might have been expected of aestivating insects, but had two pronounced diurnal activity periods—a stronger one just after sunset and a weaker one just before sunrise. At these times many of the moths vibrated their wings, producing a subdued buzzing sound; and a proportion of them became intensely active, flying first within the cave mouth but soon tending to join others already engaged in a dense milling flight over the summit of the mountain. 'Their flight appeared to be completely random in character, and, at its height, the noise was distinctly audible as a low, intense buzzing. The sound made by the frequent collisions of moths in flight was also audible as a constant clicking' (Common, 1954, p. 241). It continued for half-an-hour or more, but by an hour after sunset many of the

Bogong moths (*Agrotis infusa*) aestivating on the roof of a cave at Mt. Gingera, A.C.T., Australia, 13 February 1953. The time is dusk, and the closely-packed moths are beginning to crawl out into the bare spaces in preparation for the evening exodus flight. One has just taken wing in the bottom right-hand corner. (Photo by Dr. Robert Carrick.)

moths were returning to the caves. Another flight, more subdued no doubt because of the low temperature, occurred in the hour of dawn.

At no time during these periods of activity was copulation observed, though it was carefully watched for; and the dissection of samples of a hundred female moths taken from the caves at weekly intervals throughout the summer showed that spermatophores were never present in the bursa copulatrix, where they are normally deposited during copulation. Feeding did not normally take place either.

It is highly characteristic of the aestivating moths that their fat-body is greatly enlarged; and in fact their habits were first brought to light long ago by the revelation that the aborigines were in the habit of foregathering in days gone by for the special purpose of feasting on them, the interesting fact from our standpoint being that these people knew where to find them because they reappeared every year at exactly the same places. Indeed, the floors of some of the caves are covered with a fine silt-like material mixed with scales and particles of chitin—derived from the decomposed remains of dead moths—in places over a foot deep.

A study of the fat content of the moths throughout the season revealed significant changes, due in part to metabolism, but to a much greater extent to progressive changes in the populace. In the period 13th January to 4th February, for instance, there was a steady decline in numbers but a marked rise in average fat-content, suggesting that there had been a differential elimination of moths with depleted fat reserves. These may have perished or been compelled to emigrate; and at least in the subsequent period, from 4th to 25th February, most of the continuing decrease in numbers could be ascribed to emigration. During March, incidentally, large increases in population occurred, apparently due to the arrival of moths on passage-migration from ' camps ' farther south.

We have here what appears to be an almost perfect example of an epideictic phenomenon, associated in this case with the need to find a cool retreat for aestivation during the hot months when the moths' breeding grounds are dominated by harsh perennial grasses unpalatable to their larvae. We can recognise (1) the impressive gatherings, (2) regular and prenuptial as to season, (3) occurring in conspicuous traditional sites, (4) with a marked diurnal activity cycle culminating at dusk and dawn, when (5) there is a brief, crowded, exhibitionary flight of part of the swarm in the twilight, attended by an audible hum, and characterised by extraordinary clicking collisions. In addition, (6) the camps serve as way-stations for migrants, almost exactly as we have found the roosts of certain bats and birds to do, and probably for a closely similar reason; and there is finally (7) the interesting hint that some kind of differential elimination takes place during the summer, the individuals in poorer condition being forced to quit.

Compared with the insects, the epideictic developments of social hibernation in the vertebrates are generally much less imposing. We have already seen (p. 300) the association existing between rutting and hibernation in bats, where the females are for the most part inseminated after they join the hibernating group and before their winter-sleep begins. They do not ovulate and become actually pregnant, however, until shortly after they wake in the spring. The males go into hibernation while they are sexually potent and awake in the same condition. Courtship, pairing and conception thus all coincide in time with the period of assembly at the hibernating site.

Group-hibernation appears not to occur, except on a family scale, in mammals other than bats. Squirrels such as *Sciurus carolinensis* may gather to the number of ten or a dozen to sleep in the same hollow oak-tree during very cold weather, as I have found them doing in Mount Royal Park in Montreal, but their retirement seldom lasts more than a day or two at a time and is more in the nature of protracted roosting than hibernation. Some more typically social mammals such as beaver (*Castor* spp.) maintain their family communities, living in a more or less active state during the winter, but in none of these mammals do we know of any special gathering associated with the winter period. Typical rodent hibernators such as marmots and woodchucks (*Marmota*), ground-squirrels (*Tamias* and *Citellus*) or dormice (*Muscardinus*) most commonly occupy solitary winter nests, or at most they are associated in family groups. Exceptionally up to 13 and 14 have been found together in *Marmota bobac* and *M. marmotta* respectively (Barkow, 1846, p. 92): in the American woodchuck (*M. monax*), on the other hand, solitary hibernation seems to be the rule.

There are a few social hibernators among the reptiles of the north temperate zone. Woodbury and Hardy (1940) briefly described such behaviour in the desert tortoise (*Gopherus agassizii*) in Utah, the individuals of which tended to gather into dens during cold winter weather, dispersing again over the nearby areas during milder periods. Though the numbers congregating into a single den were not stated, marked individuals were found to have a home range of 10-40 acres, and there was some movement found from den to den. This all took place in the months of November to February. In hot weather in summer, in contrast, the tortoises descended short individual burrows. This seems to follow a fairly common type of dispersionary pattern, territorial rights being shared by a number of individuals that tend to become more gregarious (that is to say more demonstratively social) under the stress of hard weather.

Some of the commonest and best-known European and North American snakes are well known to hibernate in company. The previous reservation must be made again here, however, namely, that there may sometimes be a shortage of frost-free hibernacula and that this alone could be sufficient to induce aggregation. When we read (Service, 1902, p. 156) of the discovery in south-west Scotland, in a hole in a peat-moss, of forty adders (*Vipera berus*), an ' amazing number ' of common lizards (*Lacerta vivipara*) and ten toads (*Bufo bufo*), it suggests that the winter retreat they had chosen was an uncommonly favourable one. The individuals in these aggregations are often described as being coiled together. Smith (1951, p. 258) quotes the modern observations of Volsøe in Denmark, that a considerable number out of the total of several hundred adders sent to him in winter were found in shallow holes, alone: others, especially in dens two to five feet deep, were grouped in considerable numbers, 24 being the highest, together with various slow-worms (*Anguis fragilis*), lizards and toads. Volsøe showed that the males are not mature when they first emerge from hibernation and that consequently a short delay before mating occurs. The males emerge first and have often been found at that time taking part in apparently normal epideictic demonstrations, coiling up together and sparring with one another in a 'dance' (Smith, loc. cit., pp. 246-50). On the whole there seems to be rather little evidence in this case that social hibernation has any well-marked epideictic function.

In the timber rattlesnake (*Crotalus horridus*), in the north-eastern United States for example, the social habit is evidently far more strongly developed than it is in the adder. 'In the north this serpent shows a marked fondness for mountain ledges, cleft with many fissures and on which lie large shelving rocks. About such rugged situations large numbers of rattlesnakes gather in the fall, preparatory for the hibernating season. They appear to find the same places, year after year, making their way from the adjoining timber and lesser ledges as if led by some strange, instinctive power. On the main ledge they coil sociably in great clusters to enjoy the sun of "Indian summer" —but only for a limited number of days, when they retire into the deep fissures for the winter's sleep. For several weeks—in the spring—during the mating season, they linger on the main ledge in large numbers, but finally scatter to the timber for the warm months' (Ditmars, 1907, p. 443). There appears good reason to conclude, therefore, that these dens have developed into typical fixed points in a collective-territorial type of dispersion and that, since they are occupied also during the mating season, the gatherings that occur in them have an epideictic value.

The commonest of all snakes in the same region, the garter snake (*Eutaenia sirtalis*), is said to have similar hibernating dens, where the inmates linger in the spring (Ditmars, loc. cit., p. 236), but it appears not to be known whether mating takes place there or is delayed until after they disperse. Garter snakes are viviparous, incidentally, and sometimes produce prolific litters: Ditmars gives 51 as the highest number recorded in one brood. They share with other snakes the possession of scent glands and a very well-developed olfactory apparatus, and are thought to make use of these for social and dispersionary purposes (*see* p. 98); in fact, social integration is perhaps generally achieved in snakes through the medium of scent. Gregariousness such as we find in the present instances may well have been induced in the first place by the scarcity of suitable hibernating places: in some species it still perhaps amounts to little more than this unavoidable sort of assemblage, but in others, such as the rattlesnakes, it appears to have been developed as an auxilliary means of social integration.

It is probable that true social hibernation occurs in a number of the

Amphibia: some species of frogs, for example, are known to hibernate regularly in the mud at the bottom of particular ponds, not always those in which they afterwards spawn. Smith (1951, p. 136) mentions a case described from near Frankfurt in which, on the bottom of a clear, slowly flowing stream—the outlet of a lake—three separate collections of hibernating frogs (*R. temporaria*?) were found, two of them each containing between one and two hundred individuals and the third between four and five hundred. In frogs and toads the epideictic phase typically comes of course in the spring, after hibernation is over, when in many species the males join together in a prenuptial chorus.

Social hibernation appears also to take place in some species of fish. Carp (*Cyprinus carpio*) in particular have long been supposed to hibernate in company in Germany: they are said to seek the deepest part of the water for the purpose, excavating there a cauldron-shaped hollow in the soft mud and lying packed together in it ' like herrings in a barrel ' (Barkow, 1846, p. 92). There they remain completely inert, not reviving until May or even June, after which they soon begin to spawn (Seeley, 1886, p. 98).

To extend the phenomenon lastly to yet another group, it can be mentioned that terrestrial snails (*e.g. Helix* spp., *Cepaea*, etc.) are in some parts of Britain and elsewhere among the most conspicuous of sociable hibernators. *Helix aspersa*, for example, sometimes gathers in masses in the crevices of old walls, where scores of these large molluscs may frequently be found stuck together, with their shells sealed up by means of a hardened secretion that covers the mouth of the shell.

In so far as any of these gatherings have an epideictic function, the latter has no doubt become secondarily associated with the originally independent phenomenon of winter sleep, just as it has become associated in other circumstances with nocturnal or diurnal sleep in the parallel case of a roost. These two kinds of gathering are in one important respect very similar. True hibernation inhibits the social appetites of animals even more effectively than nocturnal sleep, and in both types, therefore, the associated social activities must be concentrated into the periods before the participants go to sleep and after they wake. In those animals whose adults hibernate and whose breeding season starts as soon as winter is over it is indeed not altogether surprising that hibernation should tend to become a social phenomenon, with prenuptial epideictic formalities tacked on to it, or that these should begin in autumn before animation is actually suspended.

14.9. *Moult-migrations in ducks*

In this chapter a variety of social activities have been described, some of them thought to relate to dispersion after the breeding season and during the asexual period of the annual cycle of behaviour. So numerous and diverse are these adaptations that any catalogue or survey of them must be very incomplete. There is still one whole category of social behaviour sufficiently important to deserve a chapter to itself, comprising the aggregations of plankton organisms produced by their daily vertical migrations (p. 366). All that is intended in the remaining sections of this chapter is to mention two additional types of communal activity, as different as possible from one another, and different also from any that have previously been considered.

The first of these arises from the so-called moult-migrations found in certain ducks. The case best known is that of the sheld-duck (*Tadorna tadorna*) in north-west Europe (*cf.* Coombes, 1950), which assembles in the month of July from a large part of the British Isles, the Low Countries and Scandinavia into the area of immense tidal flats at the mouths of the Weser and Elbe, especially near the island of Mellun, but extending westwards into the Waddensee inside the East Friesian islands. After their arrival in these quarters the birds become flightless for a time, as usual with ducks during the post-nuptial moult. The return movement to the breeding areas is a gradual affair and occupies much of the late autumn and winter.

The sheld-duck is one of the species of ducks in which the broods of ducklings gather up into ‘ creches ’ under the care of one or two nurses, so that the majority of breeding adults are released from domestic responsibilities almost two months before the young are fledged. They are at liberty to depart on the moult-migration therefore early in July.

In Morecambe Bay in north Lancashire, where he made his most intensive observations, Coombes (loc. cit., p. 411) found that their departure was attended with considerable social excitement. It was largely concentrated each day into two hours—the hours on either side of sunset—and was independent of the stage of the tide which so dominates this vast expanse of sand. There was an hour or so of preliminary flighting and flocking before the birds began to take off, and when they did so they followed a course inland and over the Pennine moors. ‘ The majority of flocks go through various flying evolutions before departing. Having flown in close order for several miles and risen to a height of 1000 feet or more, the flock often begins to split up, sections and individuals turning away from the east-south-east course, flying separately, circling, reforming, and again dividing. While this is going on, the sections or the whole flock often rush downwards for a distance, and gradually mount up again. When performing at lower altitudes the technique is often the “ falling leaf ”, but at higher altitudes the birds rush downwards together and the formation is not broken. Usually these changes of altitude are only a few hundred feet, but occasionally a flock will come back to alight in the bay, flying on a mile-long slant at considerable speed and making a rushing noise with their wings that can be heard at a distance. . . . There is a remarkable contrast in behaviour between the normally lethargic low-flying, slow-moving sheld-ducks at other seasons and these energetic migrants flying high and apparently at greater speed ’ (loc. cit., p. 410).

For the time being at any rate the juveniles and their attendants are left behind. Unfortunately we have no detailed account of the behaviour of the multitudes of adults once they have arrived at their destination.

Other ducks that perform similar migrations are the king-eider and common eider (*Somateria spectabilis* and *S. mollissima*) and the scoters (*Melanitta* spp.). In July and August 1950 my companions and I watched day after day a constant stream of flocks of king-eiders crossing the narrow waist of Baffin Island from west to east and passing out through Clyde Inlet to the sea (*see* Wynne-Edwards, 1952*b*, p. 362). On 4th August alone thousands were counted, and the total number involved must have been enormous: most of the flocks consisted entirely of drakes in full plumage, since the females do not migrate until September, and before that time we had left. There is little doubt that they all continued the 300 miles across Baffin Bay to the coast of West Greenland, where there is a well-recognised immense concentration reaching its maximum in August, and extending from the southern parts of Upernavik District and Disko Bay southwards beyond Egedesminde. Copper arrow-heads of a type locally used by Eskimos in King William Island, over a thousand miles to the west, have twice been found in eiders killed in this part of West Greenland (Salomonsen, 1950, p. 136).

In the other species mentioned, namely, the common eider and the scoters, similar concentrations on a much smaller scale occur in the Baltic, along the fiords of Labrador and Newfoundland and in the Gulf of St. Lawrence. Common scoters (*Melanitta nigra*), no doubt from the Baltic and Scandinavia, arrive in considerable numbers at this season along the east coast of Scotland. It is typical of all these moult-migrations that they are not developed on the same scale in every part of the breeding-range of the species concerned; and it is possible that neither the sheld-duck nor the king-eider, in spite of their wide distribution, have any other concentration-point to compare in magnitude with the big ones already known to us and mentioned above. Sheld-ducks for instance have a minor moulting ground in Bridgewater Bay in Somerset, and another at the mouth of the Scheldt in the Netherlands, in each of which the number assembling is apparently below 10,000.

No very acceptable explanation has hitherto suggested itself to account for the phenomenon, though almost certainly the assembly-areas are located at or near rich feeding grounds. It bears an undoubted general resemblance to the many kinds of social aggregations we have already considered; and it seems not improbable that like them it has evolved, at least locally, into a significant occasion for massing together and submitting to the pressure of accumulated numbers. If so, it presumably serves to promote the regulation of population-density in the species, in some cases over very wide areas.

14.10. *The acorn-woodpecker*

The other case remaining to be considered, that of the acorn-woodpecker (*Melanerpes formicivorus*), is indeed a proper ' curiosity of natural history '. Among other students, it attracted the interest of a distinguished former director of the Scripps Oceanographic Institution at La Jolla, the late Dr.

Acorn woodpecker: the ' bent tree settlement '. The two lower birds are males (white and red meet on the forehead) and the upper one on the tree a female (black frontal bar separates white and red). (From a painting by Allan Brooks in Ritter, 1938.)

W. E. Ritter, from whose book *The California woodpecker and I* (1938) the facts given here have largely been derived.

The specific name *formicivorus* is a misnomer, since the bird's economy is very closely tied to oaks and their acorns, in particular to some eight western species of *Quercus* and one of *Pasania*. Like the nutcrackers (*Nucifraga*) which in some parts of the northern world harvest the crop of hazel-nuts and store them to supply food for the whole year (*see* p. 155), these woodpeckers have developed the habit of storing acorns; but it is not so much the food-storing habit as the method by which it is done and attendant circumstances that compel our interest and indeed amazement.

It should be said that the acorn-woodpecker (*see* pl. VI) is exceptionally sociable for a member of this family, and is characteristically found in small parties of half-a-dozen or so together at all times. Such groups often roost in the same hole or holes (Ritter, p. 44): much more remarkably, it has been proved more than once that, 'for some birds at least, the entire breeding operation, from excavating a nest cavity to the end of feeding the young, is a communal affair participated in by several individuals—three, four, five or even more, males and females' (loc. cit., p. 47). It is also a very abundant species, possibly five times as numerous as the next commonest woodpecker in California, the red-shafted flicker (*Colaptes cafer*). The oak-woods that are its habitat do not everywhere form continuous forests, but over wide areas they are confined to 'arboreal islands' of no great extent: each of these may contain one or more 'settlements', to use Ritter's term, that is to say partially-isolated social groups. Typically the birds are permanently resident, though crop failure or drought may induce sporadic long-distance movements; and each group appears to be tied more or less to its own settlement and the territory that belongs to it. The crossing of territorial boundaries was, however, observed in the area of Ritter's most intensive observation, where there were two contiguous settlements. 'The general conclusion seems unavoidable that while these two groups are sufficiently defined in respect to the birds and to their territories to justify calling them settlements, still there is some going back and forth of individuals between them' (loc. cit., p. 88).

The focus of each settlement is the storage-tree (or sometimes trees). This may be an oak, but where the woodland is mixed, as it is at higher elevations in the mountains, it is likely to be a yellow pine (*Pinus ponderosa*) or one of the other conifers, which have a softer bark than oaks. At the harvest period, occupying the three months September, October and November, members of the settlement bring acorns to the storage-tree. Though sometimes trees are used that have a suitably furrowed or cracked bark in which the nuts can be securely wedged, the normal procedure is for the birds to make holes to receive them. 'The birds drill holes, each approximately the size of an acorn, typically in the bark of a tree. They then bring acorns, one by one, and insert them into the holes, the big end outward. As a rule, the acorns fit the hole so closely that after being vigorously pounded in by the birds they are flush with the surface of the bark, or may even be countersunk, and are held so tightly as to be removed with difficulty. Almost without exception the holes do not reach entirely through the bark ' (loc. cit., p. 33). The acorns thereafter can be dug out when required by the woodpeckers' specialised bills and tongues, but are moderately safe from the depredations of marauding squirrels and jays.

What interests us particularly is that this originally simple and straightforward food-storing habit has become extraordinarily conventionalised and secondarily overlaid—just as we found earlier could happen with other routine activities like roosting and hibernating—with an important social significance. The storage tree, once chosen, is permanent, in so far as the life of the tree makes this possible. As one of Ritter's informants expressed it, ' they do not store the acorns in trees indiscriminately. . . . In selecting a valley-oak as a granary they pick out one tree. It may be one of a group as is frequently the case. Then that is used for all time. Each colony has its own storage tree ' (loc. cit., p. 72). Holes are often punched in its bark in extraordinary numbers: a yellow pine in the San Jacinto Mountains, illustrated in Dawson's *Birds of California* and reproduced here (pl. VII), has its great bole riddled with holes to the number of 50,000, so that there is no room for any more within the chosen area. These require to be maintained or renewed; and it is interesting that the birds accurately vary the size of the hole to suit the species of acorn to be stored: those for the smaller live-oak acorns for example average 10 mm in diameter, while the big acorns of the black oak (*Q. kelloggii*), the most important of all the eight species in the birds' economy, require holes on the average 17 mm wide (loc. cit., p. 111).

The scale of the storing habit varies from settlement to settlement, and no doubt from year to year to some extent. Ritter (p. 25) contrasts two settlements in this respect, about ten miles apart with treeless hills in between: a more populous one at ' Saranap ', which stored on such an enormous scale that, ' at the oncoming of the new crop of nuts each year, it is hardly noticeable that any have been used from the storage tree during the preceding months '; and a less numerous one at ' Golf Course ', whose very moderate stores ' are entirely consumed some weeks or even months before the new crop is ready for use '. The same contrast was maintained through several years of observation. In addition to its use as a store the tree also functions as a regular feeding-place; and acorns or other nuts collected for immediate consumption are regularly carried there to be cracked and eaten (p. 85).

One of the generalisations made by Ritter (p. 81) is that ' the settlement idea is especially conspicuous at two seasons of the year—during the nesting and the harvesting activities '. This might easily have been predicted on general grounds, and thus observation gives confirmation of our premises.

The actual harvesting behaviour is particularly interesting. Acorns are typically collected from trees within the territory (loc. cit., p. 84), showing that, as usual, territorial dispersion is based on the procurement of food:

Acorn woodpecker: storage holes for acorns in a yellow pine in the San Jacinto Mountains, California, estimated to contain 50,000 holes. (From W. L. Dawson's 'Birds of California', II, p. 1027 (1923).)

the storage tree itself (if an oak) is not ordinarily harvested. The occasion is accompanied by varied forms of display. It should be interpolated here that, so conventionalised has the whole affair become, in the considered view of H. W. Henshaw, a life-long student of the species, the 'curious trait . . . seems to admit of no adequate explanation. With them, however, it is not by any means "all work and no play" but on the contrary the labor, if labor it be to them, is lightened by much gamboling and chasing each other in and out of the branches in circling sweeps, like boys playing at tag. Indeed, there is no reason why they should not make merry, for food is abundant and easily obtained, not only in the fall when the acorns are thus laid away, but during all the winter, a fact which serves to make their economy appear all the more inexplicable and useless' (Henshaw, 1876, pp. 260-1). In a later paper devoted exclusively to the storage phenomenon he added: 'In searching for the motives . . . we should not lose sight of the fact that the several acts in the process, the boring of the holes, the search for the acorns, the carrying them to the holes and the fitting them in, bear no resemblance to work in the ordinary sense of the term, but is play. I have seen the birds storing acorns many times, and always when thus engaged they fill the air with their cries and constantly play tag with each other as they fly back and forth' (Henshaw, 1921, p. 111).

It is evident from this that the process entails a definite competitive element, chases and cries being the normal accompaniments of social competition. A 'flapping, flopping and turnabout' form of flight-display is also much in evidence at the same time, simulating the flycatcher method of feeding on insects. 'Few kinds of wing action of the birds are more striking or more common than a shoot-off, usually upward, into the air from some high perch, followed by a sudden, short, about-face and return to the same perch. . . .' Ritter (loc. cit., pp. 139-40) says he has often seen them actually catch a flying insect and that this was undoubtedly the original purpose of the feat. 'But the performance is certainly not always to this end. I have watched it closely in many places at various times of year, and am certain it is often a performance, an action pattern, a special way of life—just that and nothing more.'

The conclusion is not difficult to reach that the acorn rite is something more than just harvesting: rather it resembles a harvest-festival. Food is ceremonially placed in a ceremonial tree, to an accompaniment of self-display and social competition, in a manner that has about it quite a reminiscent ring of the tribal stage in human evolution. In fact it seems to be the perfect example of an epideictic adaptation. On the one hand, competition is the means of applying stress, having potential consequences in relation to the social hierarchy, which can lead to the elimination of supernumeraries and determine the reproductive output of the community. On the other hand, the harvesting itself is a direct sampling of the food-resource on which the community must subsist until the following summer—the resource that actually determines the safety-limit of numbers. And the whole performance, finally, is focused on a traditional object, the tree, which symbolises the territorial system.

The ceremonial sample may afterwards of course be eaten: primitive man could in fact be expected to do the same with his sacrifices and offerings. Often, however, it is not, and remains in the tree till it rots or is eaten by grubs. Ceremonial it certainly is, as witness the communal performance, the established time-honoured site, and the illuminating observation that, not particularly rarely, pebbles are substituted for acorns on quite a large scale (Ritter, loc. cit., p. 128).

14.11. *Summary of Chapter* 14

1. Social roosting is a widespread phenomenon which has never been satisfactorily accounted for, but it clearly provides an opportunity for epideictic displays and may therefore be concerned with maintaining dispersion, especially, for instance, in birds in the non-breeding period of the year.

2. The roosting behaviour of the starling has been taken as a typical example. The roost-sites tend to be used year after year, and at night each receives birds from a surrounding communal territory. The number of birds attending can reach scores or even hundreds of thousands. Each tends to establish and claim an individual perch; and this frequently leads to rivalry and bickering. At times there are impressive massed manoeuvres over the site before the birds settle down for the night. In many of these characteristics the roost reveals striking parallels to a breeding colony, confirming the supposition that it has an epideictic function.

3. Communal roosting occurs in many orders and families of birds. The details vary from species to species, but it is clear that everywhere we are dealing with a single phenomenon not differing in any fundamental respect from what has been described in the starling. Many interesting and important points emerge: for instance, studies of the chimney-swift roosts, where huge catches of marked birds can be made, lead to the conclusion that the roost-system not only regulates the dispersion of birds resident in the area, but assumes a particularly important function in feeding the stream of passage-migrants through the region in a steady progression, preventing jams and bottlenecks developing as the waves of transients crowd each other along. Breeding-season roosts containing nothing but males sometimes develop in territory-holding birds; and these provide an established system into which the first-brood young can be received when they leave the nest.

4. This brief section sets out the main conclusion that social roosting must be a dispersionary adaptation, and that its primary function is to bring the members of a population-unit together, so that when conditions require it an epideictic demonstration and population-adjustment can take place.

5 and 6. Communal roosting is common among bats and appears to fulfil exactly the same purposes: except that here the diurnal roost and the breeding colony are inextricably mingled, still further confirming the likelihood that these two phenomena share a common dispersionary function.

Segregation of the sexes often occurs while the females are pregnant and nursing. The insectivorous bats and fruit-bats are dealt with in separate sections: the former seek caves and other dark places and the latter hang up in trees in ' camps ' during the day. Social roosting in flying-foxes is at its height in the prenuptial and mating period. There are few details of roosting behaviour in the bats that do not somewhere find a parallel in birds.

7. Some insects have also developed analogous behaviour, especially in the aculeate Hymenoptera. Though not a very common phenomenon in this class, cases are described where the roost is seen to be the focal point of a communal territory, where there is constancy of occupation for periods of weeks, where a flight-manoeuvre is held, and so on. Exclusively male roosts are also known; and at this point the roost shares obvious common features with a lek. Thus similar communal roosting habits are found in all three of the classes of highly mobile flying animals: the birds, bats and insects.

8. Annual gatherings, especially in preparation for hibernation, take place in many insects, at times on a very large scale. Whereas for lack of suitable refuges these may originally have been fortuitous assemblies, it is clear that they have often evolved important dispersionary functions, exactly like roosts. Virtually all the same characteristics occur, including the traditional sites (even in ' annual ' insects) and the dense swarming. The social aestivation of the bogong moth in the Australian Alps is a closely parallel phenomenon, providing a remarkable illustration of almost the whole known spectrum of epideictic adaptations.

Among other groups of land-animals that contain hibernators, social hibernation hardly appears in mammals other than bats, but occurs in various reptiles especially snakes. In the rattlesnakes, for instance, the hibernating place may be quickly transformed in the spring into the breeding colony, and in other snakes that disperse to breed it may become the arena for pre-nuptial display. Turning to aquatic animals, group-hibernation is described in the carp, and may occur in some Amphibia.

9. The moult-migrations of ducks, such as the sheld-duck, eiders and scoters, may similarly have taken on an epideictic role also.

10. Lastly, an account is given of the extraordinary social behaviour of the acorn-woodpecker in California. This bird lives in small communities each centred upon a storage-tree, in the bark of which hundreds or thousands of small holes are dug to receive acorns. The same tree is used permanently and the whole performance has become astonishingly conventionalised. In autumn there is a sort of harvest-festival, in which the acorn crop is sampled and the ceremonial tree stored, to the accompaniment of pursuits and displays of voice and flight. The acorns may or may not be subsequently consumed for food; and not very rarely stones are used in part to fill the holes. This case seems to present the perfect example of an epideictic rite, combining as it does a sampling of the food-supply, a territorial symbol (the tree), and social competition.

Chapter 15

Timing and Synchronisation

15.1. *Sychronisation of social activities*

The most common types of social function require the members of the group to be present together at a particular meeting-place and at a specific time. While this is the general rule, these two conditions are not entirely indispensable: some highly co-ordinated social engagements in animals, such as the dawn chorus of the males in a territory-holding species of song-bird, may involve a diffuse community that does not need to be specially aggregated for the purpose, but only specially alerted in order to achieve synchronisation. Other social acts, like the leaving of a sign at a marking-place in the case of various mammals or bumble-bees, though they require a visit to a particular site, may be free from any obligation as to timing. But in the foregoing chapters we have most often been concerned with social events that did demand the coincident attention of the participants both in time and place, and these conditions should therefore be regarded as normal.

Where the activity of the group is unmistakably social, as it is with the gathering and display of starlings at a roost, and where it is of relatively short duration so that it occupies only a small part of a 24-hour (or tidal or annual) period, the importance of co-ordinated timing is self-evident. It has been pointed out many times already that for an event of this kind that needs to recur daily it is particularly easy to attain synchronisation by visual means at dawn and dusk, during the period when the daylight intensity is changing fastest. There are, however, a good many incidents of animal behaviour in which the social aspects are at first sight much less obvious, which are just as strictly synchronised or periodic: the dawn and dusk choruses of birds just mentioned are an excellent example, for it is not until they have been recognised as essentially social and epideictic displays that the well-known synchronisation can be understood. Much attention has been given by comparative physiologists to the important subject of the timing mechanisms and life-processes by which synchronised activities are controlled (*cf.* Bünning, 1958), but the present chapter is concerned less with this aspect than with the possible functions and adaptive values of a number of representative examples of synchronised behaviour.

The faculty of time-keeping is developed to a remarkable degree in many animals which have come to possess an internal and to a certain extent autonomous clock or calendar-mechanism, requiring only at intervals to be checked and regulated against events in the world without. Periodically-recurrent activities may depend, that is to say, to a considerable extent on an intrinsic or physiological rhythm, but with very infrequent exceptions these rhythms are each tied to one or more of the periodic factors in the environment. The two most universal pace-making factors are, of course, the diurnal and annual solar cycles. Although tidal and lunar rhythms are also well known, they are far less common and will be considered only incidentally here when they appear in connection with diurnal and annual cycles. As an introduction, an example is set out in detail in the following section to show the type of interaction that commonly occurs between the internal physiological rhythm and the external pace-making events.

15.2. *Timing of the morning exodus of roosting starlings*

In February and March 1930 I made a daily series of light-intensity measurements at the starling-roost previously mentioned (p. 288), at Failand, Somerset, covering the half-hour or so in which the birds left the roost in the morning. Previously I had recorded (1929, p. 172) that there is a fairly high precision in the time at which the outgoing morning flight of starlings passes over a chosen observation-point. Although it is delayed on dull mornings as compared with fine ones by as much as 15 minutes, when consecutive fine mornings are compared the ' timing is accurate to within two minutes '. It was with the intention of investigating the nature of the timing-mechanism that the photometric observations reported here were undertaken the following year, but unfortunately the work was interrupted by my moving to Canada before it was quite completed and it has not hitherto been published.

Similar observations, on a restricted scale, were made by Nice (1935) at Columbus, Ohio, and appear to have given results consistent with mine: though no inference could easily be drawn from them except the general one ' that leaving the roost and returning to it were very closely correlated . . . with light '.

The data from Failand are presented here in the form of three graphs (figs. 25-27), covering about 12 days each. The observer was present on 27 out of the 36 consecutive mornings of the observation period; and each day a series of photometer readings was made to record the progressive rise in the intensity of the morning light: these daily series have been plotted to form a family of similar rising curves on each graph. All are reduced to a common time-zero, namely, the time of local sunrise on the day in question.

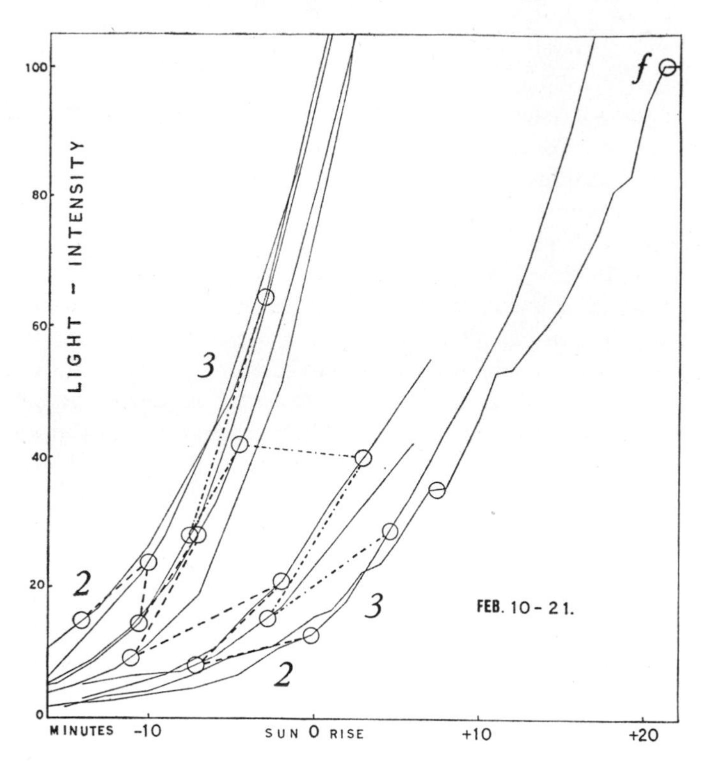

FIG. 25. Outward flights of starlings from the Failand roost, in relation to light-intensity and time, 10-21 February 1930. The graphs are explained in the text

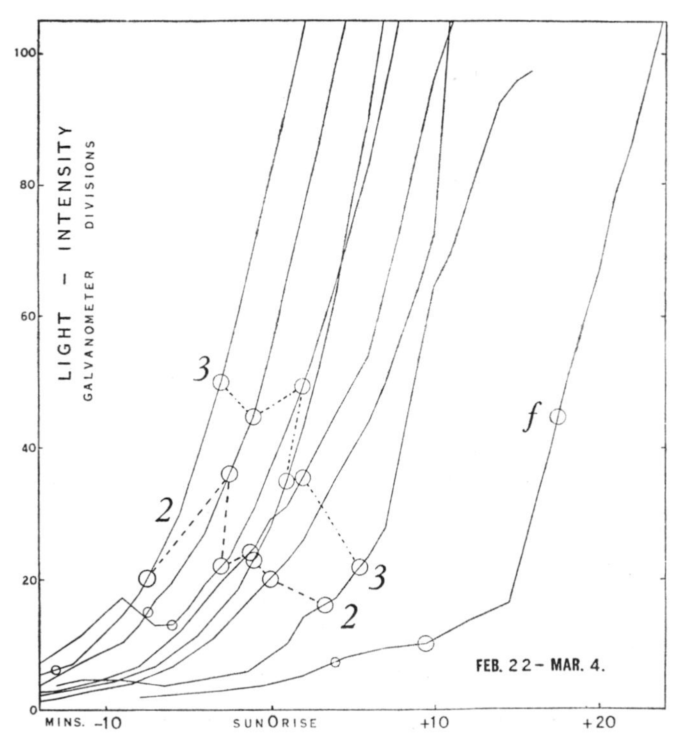

FIG. 26. Continuation of Fig. 25, from 22 February-4 March.

The consequence is that in each family of curves the brightest morning appears on the left side and the dullest morning on the right, regardless altogether of their calendar sequence.

It is to be seen that on the brightest mornings the light-intensity rises very smoothly, whereas on rainy or foggy mornings (the latter marked *f*) it is more irregular. The photometer was constructed in the Physics Department at Bristol University, and contained a potassium photoelectric cell opening towards the zenith: in the end it was never calibrated in terms of

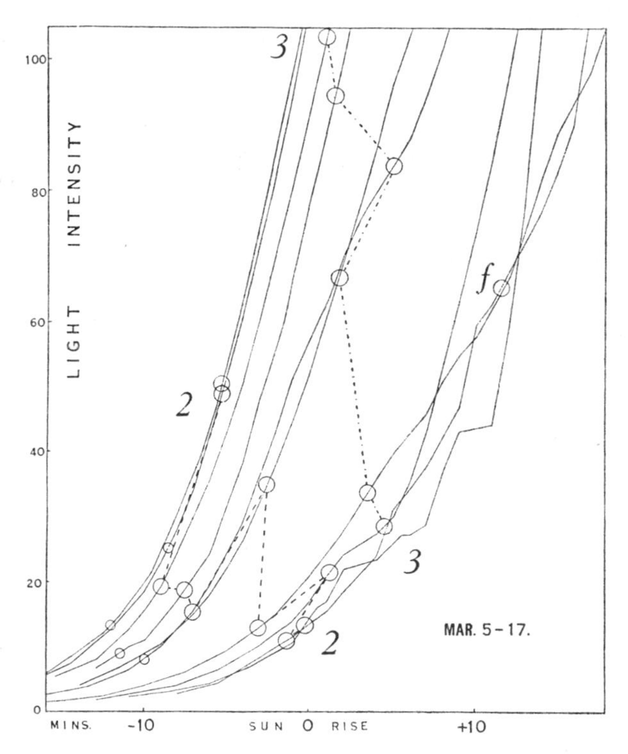

FIG. 27. Continuation of Figs. 25 and 26, from 5-17 March.

metre-candles or lux, but this is quite immaterial to the results. It is with pleasure that I record my belated thanks to Mr. M. G. Bennett for designing and building so sensitive and reliable an instrument.

On the curve for each day are marked first, by a smaller circle low down on the curve, the departure of the diminutive ‘ first exodus ’, and then in succession, by two larger circles, the times and light-intensities at the departure of the huge second and third exoduses respectively. As demonstrated in the last chapter (p. 288), these exoduses appear to have contained essentially the same body of birds each day: each was, in fact, a persistent social unit, whose behaviour could justifiably be compared from one day to the next. The second-exodus points, throughout each period, can therefore be joined by heavier broken lines, and the same can be done for the third exoduses.

When the trends of these broken lines in the three curves are compared, they appear in general to slope down towards the right, and this phenomenon is common to both exoduses in all three figures: it seems readily capable of explanation. Had the birds depended for their timing on the light-intensity alone, the broken lines would of course have been horizontal, centred upon the particular light-intensity which served as the trigger for each exodus. If, on the other hand, the birds were depending on an internal physiological rhythm alone, as we depend on an alarm clock, the broken lines would have been vertical, since the birds would be leaving at the same time relative to sunrise regardless of the weather or light-intensity. Neither of these conditions is revealed by the observations, however, but instead a compromise between the two. This is a common situation with periodically-recurring activities in animals: the rhythm, like a flywheel, is more or less easily influenced in its rate by the march of external events. The external event of chief importance here is the auroral rise in light-intensity, but clearly it is only partially successful in overriding the inertia of the established internal rhythm. The result is that on fine mornings the birds leave earlier but at a higher light-intensity, and on dull mornings later although in fact it is still considerably darker than average.

The slope or trend of the broken lines appears to depend to a great extent on how high up they cut across the light-intensity curves: the trend of the third-exodus line is thus invariably steeper than that of the second, but this is probably only a manifestation of the well-known 'logarithmic' type of response of the vertebrate eye in comparing lights of different intensities, and has no very important significance. There is, however, a progressive secular change taking place during the observation period, and as the weeks go by and the sun rises earlier and earlier the starlings' exoduses do not quite keep pace with it. In the last graph (March 5-17) the second exodus is leaving at roughly the same times and light-intensities as the third exodus did in the earliest graph (February 10-21). This phenomenon has been observed before and it illustrates how the relationship between the physiological rhythm and the pace-making external event can undergo modification, in this instance being actually influenced by changing day-length, and pursuing therefore part of an annual cycle.

On each graph there happens to be one curve marked *f*, which means that the morning in question was foggy; and in these conditions some factor, presumably the lack of normal visibility, had a marked retarding or inhibiting effect on the starlings' exodus. Something more is clearly involved on these days than a straightforward integration between light-intensity and the internal rhythm; and our simple explanation must be qualified accordingly, since it appears to apply only when visibility is unhindered. Another small point is the absence of a third exodus on some of the days in the third graph, due to the movement and break-up of the Failand roost, which began about 10th March. The observations were in fact continued until the 25th, but after the 17th even the residue of the second exodus was no longer certainly identifiable.

15.3. *The 24-hour cycle in birds*

The alternation of day and night influences the activities of almost all terrestrial and many aquatic animals; and the tendency for them to become adapted to exploit one of these alternative periods and remain dormant during the other leads us to classify most species as being either diurnal or nocturnal in habits. It is not with the general daily alternation of activity and rest that we are concerned here, however, but rather with particular activities that occur and recur at a fixed time in the 24-hour period.

Because of the tendency to specialise either in diurnal or in nocturnal habits there is a wholesale change in animals' activities at dawn and dusk; at these times one large part of the fauna is seeking shelter in which to lie up for the rest of the 24 hours, and another large part is rousing itself into full animation. They are times of transition, and for this reason very suitable for holding definite, relatively short, social and epideictic engagements, which will not then interrupt either the hours of rest or the main activity of the day. This advantage can be added to the more important one previously mentioned, that around sunrise and sunset it is far easier than at any other hour of the day to estimate the precise time by eye: moreover there is no natural 24-hour cycle apparent to any of the other sense-organs that is nearly so strong or regular as the daylight cycle. The outstanding importance of these transient periods of dawn and dusk in the social life of animals therefore does not seem particularly surprising. At the same time there is probably no obstacle to an animal developing a rhythm, keyed to the daylight cycle, which will prompt a particular recurrent activity at any other selected hour of day or night that happens to be suitable or convenient.

Starting our survey as usual with the birds, there are many examples that reveal association with early morning and evening in their social activities. Palmgren (1949, p. 574) showed that a caged mistle-thrush (*Turdus viscivorus*) had an absolute peak of activity just after sunset, after which the bird dropped at once into the deepest possible sleep. Before dawn its sleep again became especially deep; but on awakening in the morning it soon gave evidence of a second peak of activity, after which its activity-rate was greatly reduced for the rest of the day until evening returned.

In most song-birds the dawn chorus is far more intense and prolonged than the evening one. The time of first singing in the early morning has a constancy similar to that already noted in the exodus of starlings from a roost. It has long been known that some species are consistently earlier starters than others; and that in any particular locality there is consequently a customary sequence in the first entry of each performer into the chorus. There are frequently small differences between individuals of the same species, but when a particular male bird is timed from day to day its regularity is sometimes remarkable (*cf.* Schwan, 1920, p. 342). Allard (1930) has pointed out that the characteristic time for each species often bears a close relation to the onset of ' civil twilight '; but it appears from some of his graphs that various birds including the American robin (*Turdus migratorius*) undergo a secular change similar to that we have already observed in the starling, though in the opposite sense: so that they actually start singing at lower light-intensities at midsummer than they do in, say, April or August.

In very high latitudes, especially north of the Arctic Circle where daylight is continuous during the summer months, Palmgren (1935) has shown that certain passerine birds such as the chaffinch and brambling (*Fringilla coelebs* and *montifringilla*) tend to be quiescent in the period between about 6 and 11 p.m., usually rousing themselves and singing again before or soon after midnight: five hours seems to be their irreducible minimum period of sleep. Palmgren correlates this shift of the rest-period into the pre-midnight hours with the warmth and relative dryness of the afternoon, as the birds' time of sleepiness approaches; but it seems possible alternatively that it arises as a by-product of the overriding importance of dawn-song, which in this latitude gets earlier and earlier until it may begin at or even before midnight, with the consequence that the period of essential sleep must *pari passu* be displaced into the earlier hours of the ' night '. In my experience, however, for example in Baffin Island, there is no synchronised commencement of a morning chorus at latitude 70° N., such as can probably be observed everywhere south of 60°.

Allard (loc. cit., p. 458) draws attention to the fact that ' the early dawn singing of most birds is usually more or less distinctive in its character and manner of delivery '. The bluebird (*Sialia sialis*) and red-eyed vireo (*Vireo olivaceus*) sing hurriedly, and the kingbird (*Tyrannus tyrannus*) has a special ' mating song ', never heard except in the morning twilight hours, described by Hausman (1925, p. 324). Another member of the Tyrannidae, the wood-pewee (*Contopus virens*), has evolved a similar but more elaborate ' twilight composition '—a song ' so marvellous in its phrasing that Oldys regarded it as second to no form of bird-expression in technical rank. . . . It is a specialised song associated with a narrow range of illumination conditions occurring at this time ' (Allard, loc. cit.).

Allard also points out (p. 455) that the first morning song is not strictly the bird's time of awakening: sleeping birds with heads tucked under their wings he found to be very insensitive to illumination from a spotlight, and ' there is reason to believe that so long as this position is maintained the birds cannot become aware visually of the weak twilight changes with the advance of dawn '. In fact they must be awake and alert some time before the first song is delivered.

Thereafter singing is usually maintained at a high intensity for the first hour or so on a spring or early summer morning, after which it diminishes somewhat, though at the height of the season many species go on more or less regularly throughout the forenoon, and indeed all day long. The afternoon is the most silent daylight period, though there are species (including the wood-pewee and olive-sided flycatcher, *Nuttallornis borealis*) which make themselves particularly conspicuous in the heat of the day by continuing to sing while most other birds are quiet.

In the last hour of daylight bird-song reaches a secondary peak, almost always somewhat less vigorous and insistent that at dawn: as dusk falls some species habitually finish earlier and some later, but ' the same relations with the twilight intensities are not so nicely observed at the closing of the day, the birds often ceasing . . . relatively earlier with respect to the time of sunset than they begin them with respect to sunrise, although this is not an invariable rule ' (Allard, loc. cit.).

In most species the individual loudness of the voice is rather constant, and the total volume of song depends chiefly on the frequency of repetition or the persistency of individual singers, and on the proportion of potential males that are taking part. If the diurnal cycle of song-output for a typical passerine were represented graphically, the curve would remain at zero during the night hours, rise suddenly to its highest level at dawn, decline rapidly at first thereafter and then more slowly to a minimum daylight output-level between, say, one and four or five in the afternoon, and finally climb to a second maximum towards dusk. As the summer advances and the post-nuptial moult approaches, the total output diminishes; and the decline affects many species in a way that can be visualised by imagining the curve to sink bodily towards and finally below the base-line of the graph, that is, below the song-threshold; or alternatively by imagining the base-level gradually rising to submerge it. Thus song tends first to be given up altogether in the afternoon; and at the very end of the season, in late July or early August perhaps, its last vestiges can still be heard at dawn, often imperfectly rendered and brief in duration. To a certain extent the process is reversed in autumn or early spring when singing recommences, for early-season song is also most frequently heard in the early morning (*cf.* Howard, 1920, pp. 30-1), and has a tendency, particularly strong in the thrush family, to be resumed again in the late afternoon.

Outside the actual song-season, incidentally, many well-known birds, such as robins (*Erithacus*) and blackbirds (*Turdus merula*), greet the dawn just as regularly, and perhaps almost as aggressively, with chiding or alarm-notes: the same calls are repeated again with equal punctuality at dusk.

Among the non-passerine birds, the crowing of the farmyard cock no doubt provides the most familiar of all examples of dawn song; and it has of course long been metaphorically identified with the dawn itself. Daybreak crowing is equally typical of various other game-birds, including the red grouse (*Lagopus l. scoticus*), which has an unusually early starting time, determined by Peterle (1955, p. 61) near Aberdeen to be on the average 68 minutes before sunrise, during a period between 29th December and 5th May. The leks of the blackcock, sharp-tailed grouse, prairie-chicken and sage-grouse (*see* Chapter 11) also reach their climax at dawn, with a secondary peak at dusk at least in the case of the blackcock and sage-grouse. The manakins (p. 206) too hold their courts in the early morning hours, though at the height of the season they may occupy them much of the forenoon and afternoon (Chapman, 1935, p. 486).

The early-season visits of colonial sea-birds to their breeding ledges also usually take place in the early morning hours, for instance in terns (*Sterna hirundo*, etc.), kittiwakes (*Rissa tridactyla*), razorbills (*Alca torda*), guillemots (*Uria troile*) and puffins (*Fratercula arctica*). In the black guillemot (*Cepphus grylle*) Suomalainen (1939) and Koskimies (1949, p. 9) found a daily fluctuation throughout the breeding season in the number of birds present at a colony, which turned out always to be much higher between dawn and noon than in the latter half of the day: early in the season (May) it reached a well-marked peak about 4 or 5 a.m. Very much the same is true of rooks at a rookery in the winter (prenuptial period).

It is not always the dawn, however, that has been selected for the more intensive epideictic demonstrations. We noted earlier (p. 205) the evening massing of Manx shearwaters (*Procellaria puffinus*) into immense rafts on the sea adjacent to their breeding stations. The group-displays of the great snipe (*Capella media*) are said also be be held in the gathering darkness of night (*see* p. 209). Woodcocks (*Scolopax rusticola* and *Philohela minor*) are more lively and persistent in performing their roding flight and song at dusk than at dawn.

Among other birds with well-marked evening displays are the members of the nightjar family (Caprimulgidae). *Caprimulgus europaeus* has in the male a monotonous song—a reeling or churring note sometimes continued for minutes at a time, alternating between a slightly higher and a slightly lower pitch. When the time of first singing (the ' waking-time ') was recorded nightly throughout the breeding season from 15th May until 1st August, it showed not only the usual diurnal regularity but what appeared to be a superimposed lunar cycle (fig. 28); and during three successive lunations the waking-time was some 15 or 20 minutes later (with respect to local sunset) in the full-moon half of the month when the moon was between first and third quarters, than in the new-moon half (Wynne-Edwards, 1929*b*). Lehtonen (1951) more recently found that in autumn in Finland nightjars can remain flying, and therefore no doubt feeding, after dusk for as much as four or five hours when there is a moon in the sky to give additional light, whereas the average flight-duration is only 29 minutes on no-moon nights; and this suggests a possible reason for the difference in times of wakening between successive moonlit and moonless periods which I had observed in the breeding season: the more leisurely start of the evening's activities took place only when there would be moonlight to prolong either the evening or morning twilight.

15.4. *The 24-hour cycle in other vertebrates*

In the mammals ' diel ' behaviour cycles have been especially investigated in various mice and voles. Elton, Ford, Baker and Gardner (1931, p. 718) were the first to point out that these small herbivores generally appear to have a short-term cycle of feeding activity recurring every few hours, depending on their stomach-capacity and the time taken to digest a full meal, in addition to a day-night cycle of 24-hours. There is sometimes an indication, especially clear in the wood-mouse *Apodemus sylvaticus*, of two well-marked peaks of activity, found by Elton and his collaborators (loc. cit., p. 716) to occur shortly after dark and again about seven hours later, alike among wild *Apodemus* in Bagley Wood and in caged ones in the laboratory. Southern (1954*a*, p. 74) quotes the similar findings of Kalabukhov (1939), relating to the same species, of an activity peak shortly after dusk and another later in the night though in the related *A. flavicollis* only the first peak appeared. Under experimental conditions of continuous darkness the day-night rhythm

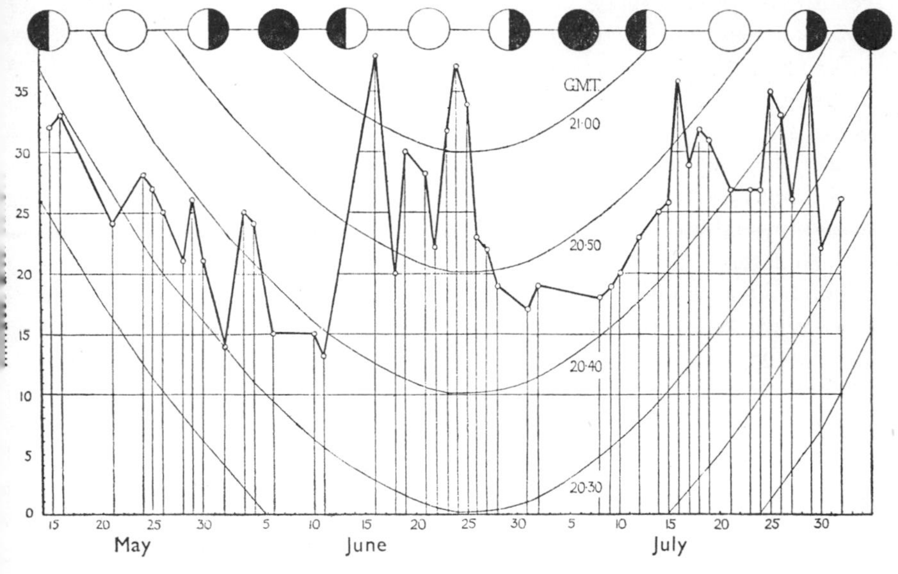

FIG. 28. Graph showing the times of evening wakening of a nightjar (a crepuscular bird) near Plymouth, Devon, in 1929. When there is a moon to prolong the twilight at dusk or dawn (shown along the top line) the bird delays its awakening roughly 15-20 min compared with the moonless periods. (From Wynne-Edwards, 1930*a*, p. 244.)

of *A. flavicollis* was found to persist in almost the original form for the remarkable span of 100 to 130 days, though that of *A. sylvaticus* in contrast soon weakened and broke down. R. S. Miller (1955, p. 518), whose work on *A. sylvaticus* confirmed and extended that of his predecessors and disclosed a similar activity peak directly after nightfall in captive bank-voles (*Clethrionomys glareolus*) while they were being subjected to a standard daylight-length of 16 hours, concludes 'that small mammal periodicities have extremely important ecological implications': however, there is as yet no indication in any of these cases that dawn and dusk activity peaks are correlated with epideictic behaviour.

In the still smaller shrews (Soricidae) Crowcroft (1954) found, as we might expect, that activity (of captive animals) was spread throughout the 24 hours; only the largest British species, the water-shrew (*Neomys fodiens*), showed a definite peak at about dawn; the common and pigmy shrews (*Sorex araneus* and *minutus*) each tended to have two prolonged high-activity periods, one in the night and the other in the day.

Scattered through the larger mammals, however, there are a number of vocal displays which closely resemble the dawn and dusk choruses of birds. Most of those known to the writer have already been mentioned in Chapter 3. A brief description was given there of community singing by hammer-headed fruit-bats in the Congo region: a number of males were observed to gather each evening at a set time (exactly 6.15 p.m.), and at a customary place, and there to engage for four hours or so in a more or less uninterrupted concert of rapidly-repeated and resounding 'pwoks'—produced with the aid of an enormously enlarged larynx and two huge pharyngeal air-sacs (*see* p. 55). A somewhat similar habit was cited in the related African anvil- or epaulet-bats, the males in this case dispersing from a collective roost before beginning their nightly concert of metallic ringing calls (p. 56). Mentioned there also were the loud and harrowing cries of howling monkeys in the rain-forests of South America: these Carpenter (1934) found in Panama to be especially concentrated at dawn, and again in the early afternoon—though Bates (1863) in Brazil referred to a similar concerted outburst at dusk. To Carpentner (loc. cit., p. 115) 'one of the most impressive periods of the day in the forest of Barro Colorado is that of the dawn, and one of the most conspicuous aspects of [it] is the roaring howls of the howling monkeys'.

Bates' description (loc. cit., II, p. 141) of an evening chorus he once heard in the jungles of the Tapajos river in Brazil is a classic of its kind. 'In this remote and solitary spot,' he wrote, 'I can say that I heard for the first and almost the only time the uproar of life at sunset, which Humboldt describes as having witnessed towards the sources of the Orinoco, but which is unknown on the banks of the larger rivers. The noises of animals began just as the sun sank behind the trees after a sweltering afternoon, leaving the sky above of the intensest shade of blue. Two flocks of howling monkeys, one close to our canoe, the other about a furlong distant, filled the echoing forest with their dismal roaring. Troops of parrots, including the hyacinthine macaw we were in search of, began then to pass over; the different styles of cawing and screaming of the various species making a terrible discord. Added to these noises were the songs of strange cicadas, one large kind perched high on the trees around our little haven setting up a most piercing chirp; it began with the usual harsh jarring note of its tribe, but this gradually and rapidly became shriller, until it ended in a long and loud note resembling the steam-whistle of a locomotive engine. Half-a-dozen of these wonderful performers made a considerable item in the evening concert. . . . The uproar of beasts, birds and insects lasted but a short time; the sky quickly lost its intense hue, and the night set in. Then began the tree-frogs—quack-quack, drum-drum, hoo-hoo; these accompanied by a melancholy night-jar, kept up their monotonous cries until very late'.

Another primate with crepuscular choruses is the gibbon (*Hylobates*), to which reference was also made in Chapter 3. Blanford (1888, p. 7) described the call of the species—known by the onomatopoeic name of 'hoolock'—as being a peculiar wailing note, audible afar: he said it commences at daybreak and continues until 9 or 10 a.m., several of the flock joining in together; after this the hoolocks are silent until, towards evening, calling recommences though to a less extent than in the early morning.

Other apparently similar cases earlier quoted are the spring-morning chorus of the chipmunk (p. 58), the evening choir of the coyotes on the prairies, uttered soon after sunset (p. 60), and the loud bark of the Asiatic muntjac or barking deer, which is repeated at intervals, usually in the morning and evening, but sometimes after dark (p. 61). Of the blue wildebeest (*Gorgon taurinus*) similarly, Hamilton (1947, p. 84) records that 'when undisturbed and calling each other [they] utter a peculiar and resonant "kwank", which may be heard most often in the early mornings'.

Since, taken all in all, they are a very vociferous group it may be expected that some of the frogs and toads will have dawn and dusk choruses. North America has an unusual number of vocal Anura, but in their nuptial concerts most of the common eastern species seem to show no very sharp daily maxima: *Rana pipiens*, *R. sylvatica* and *Hyla crucifer*, for example, can be heard incessantly day and night at the height of the season in late April or May, and the intensity of their piping appears, in a large measure at least, to be correlated with temperature. The 'spring-peeper' (*H. crucifer*), however, tends to be most vociferous in the late afternoon and evening; and the same is true to a more marked degree of the 'tree-toad' (*H. versicolor*) and the swamp tree-frog (*Pseudacris nigrita*), which are in my experience characteristically associated with evening, though they often sing also at other times of the 24 hours. Of the western *Hyla arenicolor* in California, Storer (1935, p. 208) recorded similarly: 'The chorus began about 4 p.m., soon after the sun had disappeared behind the San Jacinto Range, was strongest just after dark, from about 7.30 to 8.30 p.m., and continued into the night at least until 2 a.m. On other occasions males . . . were heard croaking up to about 6 a.m., but only a few notes were given as late as that hour in the morning.' Another western species *H. regilla* evidently has a similar song timetable (*the same*, p. 222).

In fishes the best case yet known of diurnally-synchronised social behaviour is probably that of the croaker (*Micropogon undulatus*), a pelagic marine fish that has been studied off the Atlantic coast of the United States (*see* p. 71). Croakers are gregarious and belong to the Sciaenidae or 'drums' which, as a family, have specialised in sound-production. Each individual croaker, it may be recalled, produces a roll of taps like the drumming of a woodpecker on a dry pole, lasting about 1·5 seconds and repeated at 3- to 7-second intervals. Where the schools are large and dense, the sound attains the volume of a continuous roar. Knudsen, Alford and Emling (1948) discovered, as may be seen from their measurements reproduced in fig. 29 (lower curve), a pronounced daily cycle in sound-production. The noise-level on a mid-June evening rises steeply from 5 p.m. to reach a high peak just about sunset (8.20-8.30 p.m.), after which the volume gradually subsides until midnight or soon after, and then very quickly dies away: the sound-pressure is measured in decibels which are, of course, logarithmic units; and the peak would therefore be much sharper if it were possible to sum on a linear scale the collective output-energy of all the individuals

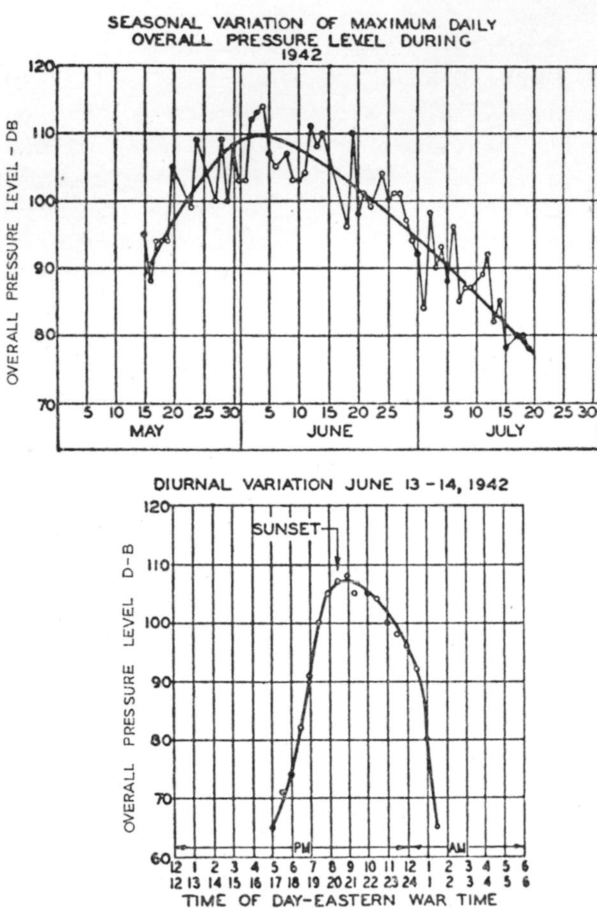

FIG. 29. Seasonal and diurnal variation of overall noise made by the croaker (*Micropogon undulatus*). (From Knudsen, Alford and Emling,1948.)

taking part. In addition to the 24-hour cycle there is a seasonal one, clearly evident in the upper curve, which shows the daily sound maxima over the period mid-May to mid-July, rising to a crest early in June. According to Hildebrand and Schroeder (1927, p. 284) the croaker has a long spawning season in the Chesapeake Bay area, beginning in August and lasting until December: the climax of the chorus consequently appears to antedate spawning by two months or more, but there can be no doubt of the close resemblance between this vocal demonstration and those we have concluded in many terrestrial animals to be prenuptial epideictic displays.

Another intermittent social activity sometimes showing a 24-hour cycle in fish is spawning: Gamulin and Hure (1956), for example, observed that, in the Adriatic, sardines (*Sardina pilchardus*) spawn in the evening, roughly between 6 and 8 p.m. A similar observation had earlier been made on the California sardine (*Sardinops caerulea*). It is not at all necessary to postulate any epideictic purpose in the synchronisation here, because the external fertilisation of eggs shed freely into the sea virtually necessitates well co-ordinated spawning to ensure a high enough rate of fertilisation; and such precise timing is very characteristic of marine animals in many groups which reproduce in this way. Sardines are highly gregarious fish at all times and are not known to have concerts or other special social displays like the croakers.

There remains another and to most people far more homely and familiar mass-phenomenon in fish that deserves to be examined in this context, namely the ' evening rise ' so well known in trout and some other kinds of freshwater fish. This was closely studied by van Someren (1940) over a period of several months in a small artificial lake near Guildford, in Surrey, which had earlier been stocked with brown trout (*Salmo trutta*). It is a perfectly definite synchronised performance on the part of the trout. ' Under favourable conditions the evening rise can be expected to start 20-30 min. after sundown. A few fish commence rising and in a few minutes the whole lake . . . is " ringed " with rising fish. The rise is at its height about 10-15 min. after starting, and ceases in about 30 min., with the exception of a few fish which rise sporadically until complete darkness falls. It is thus a twilight occurrence and starts almost coincidentally with the emergence of the bats and retiring of the swallows to roost.' Though trout may rise sporadically during the day ' there is never the concerted action characteristic of the evening rise '. ' It may be seen that rising fish cruise in small circles round a particular " beat ", subsequent rises by the same fish being to the right or left of the original rise ' (loc. cit., p. 93). On another page he states that ' the regularity of the evening rise is a very striking feature; it starts at almost the same time each evening after sundown and hence is later in the evening as the days lengthen. . . . It is a curious fact that the rise starts at the same time after sunset irrespective of whether the sky is clear or overcast ' (loc. cit., p. 103).

The phenomenon has, of course, repeatedly been mentioned by authors on angling, but there has been no final agreement as to its cause or explanation. It is generally uncritically accepted as a feeding-activity, though admittedly there is often no precise coincidence observable between a ' hatch ' of flies and the onset of the rise (*cf.* Skues, 1921, p. 44). Van Someren attempted to analyse it and found that it appeared to be affected by a number of environmental factors, though not in any very clear-cut or simple manner. The intensity of the rise varied greatly from night to night; and ' the majority of good rises took place when there was no wind or slight north-east or south-east wind ': other weather conditions, such as rain and temperature, may exercise some effect. ' An abundance of flies is certainly favourable for a good rise, though whether it is a factor important for the inducement of the rise is doubtful . . .' (loc. cit., p. 90): in other words, the same environmental conditions might suit both flies and fish independently. Contrary to the layman's notion, he concluded that the explanation ' that hunger is an important factor seems unlikely in view of the fact that the stomachs of rising fish may be already filled with underwater food '. Trout are not by any means the only fish that have an evening rise: in the same lake the roach (*Rutilis rutilus*) also rose in the evening; and their rise could be distinguished from that of the trout ' as a much gentler " dimpling " of the surface ' (loc. cit., p. 93). This is a species of different feeding habits from the trout, which rather rarely takes adult insects such as might be found floating on the surface (*cf.* Hartley, 1947, p. 152).

Where no alternative explanation offers itself for such a definite and widespread phenomenon we are justified in considering whether it is likely to be a social display with an epideictic function. It is to be seen especially in the summer months, that is to say in the non-breeding season, and could very easily have evolved into a communal demonstration of numbers, between members of a population sharing common property-rights rather than defending personal territories. It would in that case be comparable to the evening gregarious displays of birds before they go to roost. The dimpling of the water-surface to produce expanding rings of ripples may indeed be the primary purpose of the activity; and any feeding done be a secondary consequence of the fishes' rise and attentiveness to the surface. It is well known that the eyes of a fish or any animal under water, looking upward, see the outside world, right from zenith to horizon, in a distorted form through a round window immediately above them, the radius of which is a little more than the depth of the viewer: thus a freshwater fish one metre below the surface has a window about 2·7 m in diameter. The rest of the water-surface viewed from underneath reflects what lies below it like a mirror, and is generally therefore dark in colour, in strong contrast to the sky-lit window: furthermore the window is bordered by a rainbow fringe, in this case about 3 cm wide. Any disturbance of the window's calm surface, in the form of ripples for example, is consequently extremely conspicuous when viewed from below, each wave-train coming into view as a series of dark-and-light, colour-bordered stripes (*cf.* Ward, 1919, p. 142). In calm water it would probably be easy roughly to gauge the size and proximity of the origin of any incidental disturbance, such as the ' dimpling ' just described.

That the trout are participating mutually in a social activity is suggested not only by the synchronisation and conventional uniformity of their behaviour but in the observation that they are holding their individual stations relative to one another as they cruise in small circles during the performance, and are not dashing about at random all pursuing their own independent ends.

15.5. *The 24-hour cycle in insects*

The specialisation of diurnal or nocturnal habits is as characteristic of the insects as of any other group of higher animals: no illustration of this is more familiar than the broad distinction in the Lepidoptera between the typically diurnal butterflies and nocturnal moths. Consequently 24-hour activity rhythms are the rule throughout the group and indeed in the terrestrial arthropods as a whole. Gunn (1940) and Harker (1954) have studied in detail the well-known nocturnal proclivities of cockroaches, for example, and found that in *Blatta orientalis* and *Periplaneta americana* there is a special burst of activity in the early part of the night, mainly before midnight. The rhythm persists for at least three days under experimental conditions of constant light. Harker (1955) very interestingly discovered that, while the compound eyes alone enable a cockroach to behave normally with respect to the day-night cycle, the ocelli are necessary for the persistance of the rhythm: furthermore, the implantation of a sub-oesophageal ganglion into a rhythmless individual imposes the donor's rhythm upon it.

The rate of chemical processes in living cells is in general greatly affected by changes of temperature; and it is rather surprising that cold-blooded insects subject to considerable and rapid fluctuations in air temperature are capable of maintaining an internal rhythm at all, even for a few days. Many mammals and, to a considerably greater extent, birds, both of which, as we have seen, can develop highly persistent rhythms, are also subject to daily changes in body-temperature over a range of several centigrade degrees; and it is thus very interesting to find a growing body of evidence that certain mammalian activity-rhythms are independent of temperature (*cf.* Folk, Meltzer and Grindeland, 1958). Somehow the challenge has evidently been met of making the internal ' biochemical clock ' measure time accurately in spite of temperature fluctuations.

Such special activity peaks as the one just described are not uncommon, as we shall see. Notice has earlier been taken of the communal roosting habits of various solitary Hymenoptera and other insects, that involve nightly assemblies at a particular time, towards dusk; and there are, of course, great numbers of insects that rouse themselves into well-synchronised activity at the same time of day. Thus Rau (1932) found the time of rising of fireflies (*Photinus pyralis*) from the grass depended on light-intensity, varying with the time of sunset, the weather, and the shadiness of the habitat; and there are no doubt hosts of other insects similarly affected. Some species are almost exclusively crepuscular in their habits, at least in the adult stage: for example, Skaife (1954, p. 200) mentions a South African satyrid butterfly, the evening brown (*Melanitis leda*), which, contrary to the rule among butterflies, is not a diurnal flyer—nor indeed a strictly nocturnal one either, since its flight is chiefly confined to the evening and morning twilight. ' It becomes active after sunset and then seeks the open where it flies about until after dark.'

Somewhat the same cycle is followed in many other groups (*see* figs.

30 and 31). Studying the flight-activity of certain East African insects, Corbet and Tjønneland (1955) found that ' in the Ephemeroptera, Trichoptera and Chironomidae (considered as groups) flight activity was bimodal, the

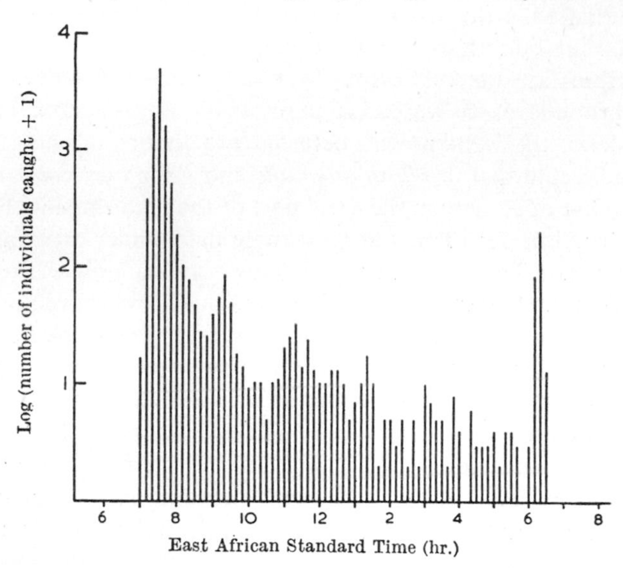

FIG. 30. Flight activity of an East African caddis-fly (*Cheumatopsyche* sp.) through the night of 5-6 January 1955. (From Corbet and Tjønneland, 1955.)

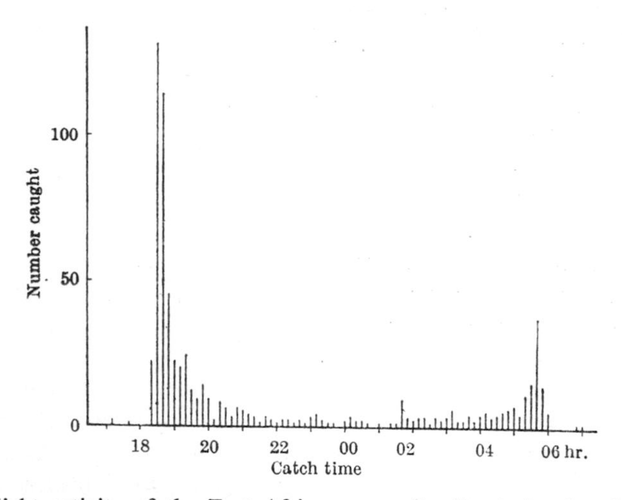

FIG. 31. Flight activity of the East African mosquito *Taeniorhynchus fuscopennatus*. Sunset is at 18.00 hr. (From Haddow, 1956.)

greatest numbers being caught shortly after sunset and again just before sunrise '. In the Trichoptera (caddis flies), with which the authors were predominantly concerned, six species were discovered to have a bimodal activity cycle with a maximum at dusk, and the seventh to have a maximum at dawn. No explanation of this perfectly definite cycle of behaviour has been readily forthcoming hitherto, and these authors comment accordingly that 'the significance of the two periods of activity in the life-histories of these insects is not clear. It seems likely that greater sexual activity occurs at dusk. . . .'

With the biting activity of certain East African mosquitoes such as *Taeniorhynchus fuscopennatus*, Haddow (1956) obtained similar results. The biters are the females and were caught on human bait in the forest canopy 65 feet above ground: their activity shows a very high peak at dusk and a much smaller one at dawn. 'The sudden nature of such activity-peaks is well shown by the figures for [this species], where 37 per cent. of the entire night catch was taken in the 20-min. period during 18.21-18.40 hr. The abruptness and short duration of the wave of activity suggest that a particular level of low light intensity may be important.'

Though there is evidently a high degree of synchronisation developed here, it would be unjustifiable to assume tacitly that the underlying cause is a social one. The mosquitoes were caught as they were alighting individually to feed and not while taking part in any obviously co-ordinated movement; and it seems not wholly impossible that such crepuscular and nocturnal insects, which had been resting during the preceding twelve or thirteen hours of daylight, would feel unanimously prompted by hunger to feed at the earliest renewal of opportunity. However, we must consider this case in relation to numerous others, in which no such facile co-ordinating motive is apparent, and to the fact that the same species shows a secondary peak about 20 minutes before sunrise. If this mosquito's habits at all resemble those of the familiar *Culex* species, only a minute proportion of the active individuals would be engaged in feeding at any given time, and in that case the representation of their collective behaviour in terms of a minor secondary activity such as biting could be entirely misleading. Indeed it seems not improbable that the number of biters is merely an index that reflects a general activity-peak.

This commonplace cycle of activity is followed also by many species of crane-flies (Tipulinae), as determined by the numbers caught in light-traps. Pinchin and Anderson (1936) found that the maxima for the two sexes came at different times, the females being most active at dusk and the males about an hour later; after this the numbers declined and became constant an hour before midnight, finally diminishing at dawn (but *see* below): twice as many males as females were trapped. No doubt significantly, in one of the eleven species concerned the females are in fact wingless.

Very important pioneer work on this subject was done by C. B. Williams (1935), who analysed the catches of a light-trap operated nightly at Rothamsted for a continuous period of 24 months (1933-5). The night was divided into eight equal periods, four before and four after midnight, varying in length according to the time of year: this has the desirable effect of putting the crepuscular insects always into the first or eighth period, regardless of season, and so on. The very interesting outcome of this laborious piece of work was the demonstration that many species have characteristic activity-peaks, not necessarily at dusk or dawn, but phased at various times in the night. In general, the largest catches are expected in the first and second periods, with a gradual decline to about half or two-thirds these numbers in the seventh and eighth periods. 'The Chironomidae (excluding *Ceratopogon*) are very consistent and show in each year a maximum in the first period and then rising again to a definite second maximum in the last period' (loc. cit., p. 536)—in other words, they follow the familiar dusk-dawn pattern; and 'the Tipulidae show maxima in the first period and second maxima at dawn in both years thereby resembling the Chironomidae' (p. 537). (The dawn sub-maximum did not appear in Pinchin and Anderson's results given above.)

Another group of Diptera, the Cecidomyidae, have their peak in the third (-fourth) period. Likewise the white ermine moth (*Spilosoma lubricipeda*) 'shows an almost similar distribution in each of the two years with a maximum in the fourth period, just before midnight' (loc. cit., p. 541). The buff ermine (*S. lutea*), on the contrary, had its peak in the sixth period; and similarly in the closely related muslin moth (*Cycnia mendica*) all the individuals captured came after midnight. The swift moth (*Hepialus lupulinus*) appeared, though only in small numbers, practically always in periods 1 and 2.

Some of the Lepidoptera are dawn or pre-dawn flyers, like the three following species of Tortricidae. '*Peronea variegana* Schiff., in both years, has a very definite maximum in period 8. *Argyroploce striana* Schiff. shows the same thing even more markedly, with, however, a sub-maximum in period 1 in 1934. In *Ancylis lundana* Fabr. the definiteness of the flight period is very striking, as in 1933 out of 83 individuals not one was captured before period 7. In 1934 periods 7 and 8 had most, but there were a few earlier captures' (loc. cit., pp. 547-8).

It should be noted—having in mind the possibility that these concentrated activity peaks are largely epideictic in purpose—that in the majority of the larger Lepidoptera (as with the tipulids mentioned earlier) it is chiefly, and in some species exclusively, the males that come to the light-traps (Williams, p. 548).

The synchronisation of activity at any selected hour of the day or night, and not merely at the obvious times of primary change-over at dusk and dawn, though relatively uncommon, is not met with here for the first time. A similar phenomenon by day occurs in the dragonflies, in which the peak density in the mating and egg-laying areas comes about noon (Moore, 1953, p. 348; *see also* above, p.173). There too the males greatly outnumber the females.

A case of synchronised oviposition, of particular interest here, has been reported in one of the blackflies, *Simulium damnosum*, in Liberia (Muirhead-Thomson, 1956). A dense swarm of two or three hundred fertilised females was observed to gather at dusk on vegetation dipping into water; the mass of small flies formed a solid line on certain selected leaves, advancing over the surface and leaving behind it a carpet of glistening eggs. Only two or three adjacent leaves on one twig were used and none on any of the eight other similar twigs trailing in the stream at that point. This assembly, of flies already fecundated and in the absence of males, taking place at dusk for the purpose of communal oviposition, seems very likely to have an epideictic significance, connected in some way with regulating the output or fate of the eggs.

Still another category of synchronised activity in insects which has received much attention relates to the emergence of imagines from the pupal stage. Its main features exactly parallel those of the general activity phenomena already considered. A review of the subject, together with new material on the Chironomidae, has been given by Palmen (1955), whose paper is the source of the material contained in this and the following paragraph. Cases are quoted there of different Lepidoptera, for instance, that emerge respectively at or near sunrise, about noon, and at sunset. In several species, including the clothes-moth *Ephestia kuhniella* (which has a maximum between 3 and 8 p.m.) and the dipterous flies *Drosophila* spp. and *Scopeuma stercoraria* (a dung-fly, with an emergence period from 9 a.m. to 2 p.m.), it has been found that an artificial shift in the timing of the day-night sequence will induce a corresponding shift in the daily emergence period: in other words the timing is (partly at least) photoperiodic; the daily temperature cycle may also have some (usually subsidiary) effect.

That this is another type of epideictic manifestation, analogous to previous ones, seems not improbable. We are concerned, in the first place, with the emergence of the adult stage in the insect's life-history, when prenuptial epideictic behaviour might be expected to develop without delay. It is not rare to find, as in the cecidomyid fly *Dasyneura alopecuri*, the chironomid *Allochironomus crassiforceps*, and the mosquito *Aëdes taeniorhynchus*, that the males emerge first; and in the moth *Parasemia plantaginis* only the males appear to have a definite emergence peak—between 6 and 9 p.m. in Finland where the observations were made—the females' emergences (rather few in total number) seeming to be randomly scattered. Sometimes, as in the midge *Pontomyia pacifica*, the emergence at sunset coincides with the swarming time, when the adults rise to take part in their display and mating flights. Synchronised behaviour, whether in the form of general activity-peaks or of specific events and duties, such as those reviewed in this section, appear likely to serve the purpose not only of aiding the estimation of total population-density, which is always prerequisite to any active control or regulation of numbers, but also in many instances of promoting reproduction itself.

Both these ends may be well served by the same adaptations which result in concentrating the activity of the whole adult population at a definite time of day and in a definite place or type of habitat. The same two purposes are very commonly combined in the social behaviour of animals, from perhaps as low down in the scale as the polychaetes right up to the birds and mammals;

and it is not always easy to separate them and judge the relative share of each in the evolution of social gatherings. In a dense population of mobile creatures such as is found in many of the chironomids or mosquitoes, however, it appears likely that copulation could be successfully achieved even if the activity of the adults were randomly distributed throughout the day or night—there being always enough on the wing to ensure the meeting of the sexes; and thus it is the epideictic function that usually appears to have most to gain from the simultaneous appearance of all the potential participants. It appears to be the epideictic aspect also that leads to the unisexual flights and sexual segregations, which are sometimes associated with particular activities. It seems not improbable that in many species of insects the peak or peaks in the diel cycle have been evolved largely to provide a form of social display, in general analogous to the dusk and dawn song-peaks found at the appropriate season in birds.

Twenty-four-hour cycles occur in invertebrates other than insects, but there seems little new to gain by pursuing the subject further at present. It will be reverted to in the next chapter, which is concerned with the vertical migrations of the plankton.

15.6. *Synchronised annual activities: the Palolo worm and some other Polychaeta*

The majority of accurately-timed annual zoological events are, as we have seen already, closely connected with reproduction. If adult individuals over a wide area mature simultaneously this is likely to ensure an adequate supply of mates of both sexes; and if maturation is further associated with congregating or mixing of individuals from neighbouring populations it can be expected to produce a suitable measure of out-breeding. Even in the flowering plants, which entirely lack any of the epideictic adaptations for controlling population-density such as we see in the higher animals, there is in most species a marked uniformity in the date of anthesis; with the result that in a given area the period of flowering often has a relatively sharp, brief peak. Reproduction itself, in other words, has often come to require the simultaneous participation of all the adult members of the population; and consequently, as pointed out in the previous section, it is sometimes difficult to make out whether synchronisation in the breeding season has any epideictic significance at all, or whether it is merely a condition necessary for securing cross-fertilisation, and wholly to be accounted for in this way.

Illustrations of this problem may be taken appropriately from one of the more lowly groups in the animal kingdom, the polychaete worms. One of the most celebrated and remarkable of all cases of synchronised breeding is found in the palolo worm (*Eunice viridis* (J. E. Gray)), which lives during its vegetative life in crevices of the coral rock in shallow water, among the reefs of Samoa, Tonga, Fiji, the Gilberts, New Hebrides, Seychelles (Hempelmann, 1931, p. 141), and Solomons. The maturing worm, as usual in this and several other families of polychaetes, has the body sharply divided into an anterior vegetative or ' atokous ' part, and a much narrower, longer, ' epitokous ' part that develops the reproductive organs. At the crucial hour on a particular spring night the epitokous part breaks off and swims freely to the sea-surface, leaving the anterior section behind, evidently to regenerate a new hinder end in due course. Undivided worms can only be obtained by breaking up the coral rock with a hammer and chisel, and they are liable to fragment in the process: Ehlers (1898, p. 401) records that an intact one measuring 312 mm had the atokous part 105 mm long (with 205 segments) and the epitokous 207 mm (with 339 segments): the atokous part was up to 4 mm wide, and the epitokous 1 mm or less. In preserved specimens I have examined from the Solomon Islands, through the kindness of Mr. Norman Tebble at the British Museum, the epitokous part was generally almost 3 mm wide in the male and 2 mm in the female. A piece showing the junction of the two parts of the body has been figured by Friedlaender (1898, p. 334), and the whole worm by Woodworth (*see* Hempelmann, loc. cit., p. 141).

There are several descriptions of the swarming, all in substantial agreement: the fullest and most dependable is that of Friedlander, but valuable observations were made earlier by the missionaries Stair (in Gray, 1847, and McIntosh, 1885), Whitmee (1875) and Powell (1882). There has been no recent review of the literature; and since it is a very interesting history it seems worth recounting in detail.*

The spawning date falls at the moon's last quarter in the months of October or November: at Apia, Samoa, the greatest swarm is likely to occur either on the day before or the day itself of the last quarter, depending on whether the almanac-time of the lunar phase falls early or late in the day (Friedlaender, p. 349). This is the ' great day ', but it is usually preceded by a much smaller rise the previous day and not rarely followed by one on the following day also. In some years spawnings occur in both months, but more often in only one: they appear not to fall earlier than 8th October in Samoa, but minor secondary swarms have been recorded on rare occasions as late as 22nd December (Whitmee, p. 496; Corney, 1922). Wide variation in the quantity of worms has been reported, as between good years and bad years.

Though the immediate dating factor is thus connected with the moon, in the long run the cycle is annual, that is to say solar, as Whitmee noted, and always culminates at the same calendar period: the principal spawning recurs after an interval of either 12 or 13 lunar months, whichever brings it nearest to the optimum period of spring in October-November. The immediate lunar stimulus perhaps works through the tides, though this is not necessarily so, since there are polychaetes such as *Odontosyllis enopla*, of which further mention is to be made shortly, which swarm only on moonless evenings, that is to say all through the second half of the lunar month, which covers a

* Since this section was written, some of the same ground has been covered in Korringa's interesting review of lunar periodicities in marine organisms (Korringa, 1957).

complete cycle of tides (Huntsman, 1948, p. 365). Here it is evidently a question of extinction of twilight below a very low intensity threshold.

The islanders are, or were a century ago, able to forecast the date exactly and seize the opportunity to catch enormous quantities of palolos, using for the purpose small, beautifully-made basket sieves: the worms were tied in leaves and baked, or, such is their delicacy of flavour, simply eaten ' undressed '. The boats put off in the darkness of the small hours; and Stair described how the worms could be felt by the hand, swimming at the surface, before there was yet light enough to see them. Friedlaender (p. 338) found the first ones to appear about 4 a.m., an hour and a half before dawn, but he relates that in easternmost Samoa they are said to start swarming about midnight. By sunrise (about 6 a.m.) they are to be seen ' in such countless myriads that the surface of the ocean is covered with them for a considerable extent ' (Stair): they tend to clump or aggregate strongly, awakening in Corney (1922) a ' recollection of vermicelli soup '. Friedlaender, who witnessed the event a number of times, says that often the palolos will gather into a particularly dense patch within a sharply-defined area of a few square metres, right at the surface. The worms ' move rapidly, and with considerable elegance, in a spiral manner, like a screw. The shortest, which were about 6 inches long, had generally two coils, while the longest, which were fully 18 inches long, had as many as six coils. . . . In places where the palolo were plentiful they seemed to be entangled in an inextricable mass ' (Whitmee, p. 498). Friedlaender (p. 339) says it is particularly the smaller individuals, described by him as only 6 cm long, that form these dense masses, while a relatively small number of the larger 20-40 cm worms can be seen snaking about between them.

When the epitokous part breaks away preparatory to swarming, the paired eyes are left behind on the head. Each epitokous segment, however, has developed on its ventral side a median ocellus, associated with the neural ganglion, and provided with a cuticular lens. These ' abdominal eyes ' (*Bauchaugen*) were first recognised by Ehlers (loc. cit., p. 403) and have been duly described and figured (*cf.* Hempelmann, 1931, p. 56). Powell had no doubt about the worms' ability to see. ' That the sight of these annelids is perfect,' he wrote, ' is evident from the way in which a single specimen will endeavour to escape the sieve. . . . No sooner has my sieve been put into the water than the animal has made off rapidly in an opposite direction; and when I have attempted to intercept its flight, it has immediately dodged again and escaped.'

There is a sexual dimorphism of the epitokous palolo, usually described as involving a colour-difference: the females are a ' dull dark blue-green ' and the males light brownish or whitish, this being due to the colour of their genital products, occupying almost the whole internal cavity and showing through the body-wall (Friedlaender, p. 340). In the preserved specimens I have seen, however, long after all colour has faded, there is a pronounced difference in the form of the segments, those of the male being shorter and fatter, scarcely more than a third as long as they are wide, whereas in the females they are almost quadrate, the length being much more than half the breadth.

Powell actually observed the discharge of both sperm and ova through segmental genital ducts, but most authors, including Friedlaender, speak of the gametes being released by the gaping of the body-walls. Whitmee observed this to take place about half-an-hour after sunrise; and Friedlaender (p. 339) stated that after a suitable short tour of swimming the worms break up, still right at the surface. The swarming could be regarded primarily as a sunrise phenomenon, or, alternatively, as one commencing in the latter hours of the night, but it cannot be decided with certainty whether the final timing depends on a 24-hour or a tidal cycle. Friedlaender (p. 346) reports the observations of Doctors Thilenius and Krämer, who placed a piece of coral rock in a bucket of sea-water on a Monday, went out to watch the palolo fishery between 4 and 7 a.m. on the Tuesday morning, and found when they got back that a mass of palolos (of about 10 cc) had emerged from the lump of rock while they had been away. But this indicates only that a physiological rhythm is involved and gives no clue as to its period.

The presence of different size-classes suggests that not all the worms bud off an epitokous section every year, but that in a small proportion of individuals they may be retained and released a year later, at a much greater size. The natives informed Friedlaender that on the day before the great swarm the palolos that appear are smaller, feebler, more watery and fragile than on the day itself; and he suggested that these might arise from worms budding off an epigamous body for the first time.

On one occasion at Apia a small species of *Aphrodite* (another polychaete of totally different appearance), only about a centimetre long, appeared at the same time and place, in numbers at least equal to those of the palolo itself: other polychaetes are also well known to mingle with the swarms (Friedlaender, p. 340).

There are resemblances between the palolo's performance and that of certain other swarming polychaetes, such as the 'fire-fish' or 'fire-worm' of Bermuda, *Odontosyllis enopla*, brief accounts of which have already been given (*see* pp. 37 and 195). In this species the spawning period is less restricted and may extend over a long succession of evenings after the full moon, for instance from the 6th to 17th day after the full moon in February one year and from the 1st to 8th day after it in the succeeding March (Huntsman, 1948, p. 364). Spawning can occur in many, perhaps all, months of the year. Here the display, for so it can be described, has a most precise relation to light. 'One could set one's watch within five minutes of the correct time by observing the first appearance of the worms in the evening and adding 50 to 55 minutes to the published time of sunset for the day,' according to a communication quoted by Huntsman. His own figures gave averages of $55\frac{3}{4}$ and $56\frac{1}{2}$ minutes after sunset during two consecutive spawning periods. Even when the worms are numerous the whole performance is over in another

13 to 25 minutes. Both sexes are luminescent at the swarming time, showing a notable sexual dimorphism in this respect. The female swims at the surface in small circles about two inches in diameter emitting a more or less continuous glow, and one or more males dart obliquely upwards towards her, with sharper intermittent flashes, evidently attracted by her light from distances as great as 10 or 15 feet (Galloway and Welch, 1911). On meeting ' they rotate together in somewhat wider circles, scattering eggs and sperm in the water '.

Huntsman stresses the frequent disparity in the ratio of the sexes present. ' To some extent in October, and to a marked extent in November, with full development of the spawning behaviour, males were more abundant than females, even to four or five times, and there might be a large number of males showing before females appeared. Early and late in the monthly period there might be no males. By February males were few or absent.' These were observations communicated to Huntsman by Edwin B. Damon: Huntsman himself saw only one certain male during the whole long February swarming period. In March no males were detected at first, but on the final days they appeared quickly after the females' display commenced, though never very numerous. In spite of the long-held assumptions regarding the swarming of this well-known marine worm, Huntsman (loc. cit., p. 367) concluded that, ' while the behaviour described has been considered to be associated with spawning, it is not certain that this is invariable. When females alone appear, the eggs will not be fertilised and may not be voided. . . . The matter requires investigation '.

This leads us on to consider another kind of polychaete, the interesting and much-studied European species *Platynereis dumerilii*. Sorby (1906, p. 436) described a remarkable swarm of this species that he saw from his yacht in Harwich harbour at 5 a.m. on 16th July 1898 (roughly an hour after sunrise): two hours later none could be seen. The swarm consisted of small-type heteronereids, the largest being less than 1½ inches long, to a number estimated at something like a million. All appeared to be males; and no ova were voided by any of the considerable numbers collected and kept alive in sea-water for several hours, until all had died. Similar swarms have been seen by other observers in this and related species; and quite commonly the males greatly outnumber the females, especially early in the lunar spawning period or ' run ' (*e.g.* Just, 1914, pp. 202 and 203; Aiyar and Panikkar, 1937, p. 252): on other, evidently less frequent, occasions the swarms consist solely of females (Just, loc. cit.). Unequal sex-ratios of a similar kind were found by Lillie and Just (1913, p. 147) in swarming *Nereis limbata* at Woods Hole, where on most evenings the males outnumbered the females by more than ten to one, although at the height of some runs great numbers of females did appear on a few nights.

Species of *Platynereis* have evolved some very specialised reproductive adaptations. For instance, in *P. megalops* there is an actual copulation—the sperms being transferred in a remarkable way through the mouth of the female, and carried to her body-cavity and ova through the lacerated walls of the gut (Just, loc. cit., p. 206). This does not occur in *P. dumerilii* (*cf.* Hempelmann, 1911) or the Madras species (Aiyar and Panikkar, loc. cit.), where the spiral swimming courtship of the male heteronereid about the female leads merely to normal sperm-shedding and ovulation. In *P. dumerilii* at Naples (the subject of Hempelmann's valuable monograph) there exists an extraordinary complication, already described in some detail on an earlier page (p. 257 and fig. 18): it involves sexually mature tube-dwelling nereids (Form α) which rear their larvae in the females' tubes, and two types of epitokous heteronereids (Forms β and γ) which spawn at the surface and produce a strikingly different, planktonic larva. As already stated in the earlier context (Chapter 13), Hempelmann showed that unfavourable conditions lead to the early production of small heteronereids (Form β), which are adapted through their pelagic reproduction to disseminate the population more widely: some larvae may thus be carried, before they settle, beyond the limits of the unfavourable habitat. Under locally rich conditions, on the other hand, the population may build up rapidly on the spot by repeated generations of tubicolous reproductives of Form α. Form γ is a larger and no doubt older heteronereid stage, suspected to have the same reproductive cycle as Form β; all three of these forms are dioecious, with the sexes separate. There exists in some parts of the range of *P. dumerilii*, though not at Naples, yet a fourth reproductive type (Form δ) that is hermaphrodite: its cycle is not known.

This is not the only polychaete with several reproductive forms. There is a boring cirratulid, *Dodecaceria concharum*, that makes its tubes in limestone, encrusting calcareous algae, or the shells of molluscs: it has three sexual breeding phases, one of which is parthenogenetic, appearing therefore only in females; the other two are dioecious with normal males and females, but one phase is sedentary and the other free-swimming (Caullery and Mesnil, 1898). Dehorne (1924, 1927) found, only in one habitat where the worms were living in a marly limestone, a fourth, completely different, form of reproduction in this species—by an asexual process, namely the proliferation of chains of daughter individuals within the original tube.

These complex phenomena are distinguished by a number of different terms. The existence of several types of reproductive cycle within a single species is referred to as ' heterogony '; and where one and the same individual may pursue different reproductive courses at different stages of its life, as in *P. dumerilii*, it is called ' dissogony '. Asexual or vegetative propagation through fission or budding is covered by the general term ' schizogony ', though the methods developed are so varied that they have received a series of distinguishing names. Schizogony is especially characteristic of the Syllidae, where it leads to the production of linear chains of zooids (*e.g. Autolytus*), lateral branching stolons (*Syllis ramosa*), or even rosettes of buds (*Trypanosyllis*) (*cf.* Potts, 1913).

Where there are different breeding-phases such as these, they may, of course, be only successive stages in a single reproductive sequence. Schizogony may first produce a host of daughter individuals, which then disperse and become sexually mature and collectively produce a vast number of fertilised eggs of larvae. Such 'compound reproduction' through the alternation of asexual and sexual generations is typical of many of the lower invertebrates and also of some Protozoa and many plants. But where there is no such temporal sequence, and where the different methods of reproduction are truly alternative and not merely alternating, it can be concluded with certainty that each method, in the conditions appropriate to it, has a special advantage over the others: if not, the alternatives could not have arisen through natural selection.

Particular environmental conditions in this latter case must evoke the appropriate type of cycle among the members of the population. Whenever conditions change sufficiently, or in passing to other different kinds of habitat, a threshold must be crossed and the cycle switched to the appropriate alternative type. The situation has a resemblance to the phase-systems found in a number of insects, not least in the differentiation, in *Platynereis dumerilii* and *Dodecaceria* for instance, of 'sedentary' and 'migratory' phases of reproduction. The determination of the best response in any given circumstances, and the evocation of the corresponding phase out of three or four alternatives, bespeaks a highly evolved mode of regulation. It appears just as likely in the various polychaetes as it did in the locusts that the phase-systems found have each independently developed in response to the exigencies of a fluctuating or impermanent type of habitat.

The position in the Polychaeta has been reviewed at some length in order to illustrate the degree of development of self-regulatory mechanisms found in animals belonging to such a relatively lowly group. It has been shown that adaptations modifying the dispersive effects of reproduction exist in some species: parallels have been found with the higher groups in the development of sexual dimorphism, nuptial displays, unisexual swarms or gatherings, and alternative reproductive phases. The original question, whether or not the swarming of the palolo has an epideictic function, still cannot be answered with certainty, but it is sufficiently clear that epideictic demonstrations are to be expected in a group that has developed these other closely related adaptations. It appears probable for the same reasons that the regulation of numbers and population-density is subject to a significant degree of intrinsic control in members of this group.

15.7. *The marriage-flights of termites and ants*

Examples of annual synchronisation of rather a different kind are provided by the so-called marriage-flights of termites and ants. These insects belong to two widely separate orders that developed their social organisations independently and at different times, the termites being very much the older group. They have evolved each in their turn a number of close parallels, one of the most notable of which is that they have all become flightless, with the sole exception of the sexual forms ready for the marriage-flight. These young winged adults, commonly of both sexes, emerge from their colonies and fly but once, as a rule for a matter of hours but sometimes only of minutes. There are exceptions among the ants, but generally the marriage-flight leads to pair-formation between the participants. In the ants copulation takes place and the female is thereby rendered fertile for the rest of her life: she immediately sheds her wings and, following the simplest pattern, is ready to found a new colony single-handed. The males become ' expendable ' when the flight is over and are no doubt mostly destroyed in a short time by predators. In the termites the male and female join forces, both being required for the foundation and building up of a new colony, in which they live as a royal couple. They too shed their wings soon after the flight is over.

In the ants the degree of conspicuousness and synchronisation of the marriage-flight varies very greatly from year to year and place to place. In western Europe and North America the great swarms of flying adults that attract general attention most commonly belong to one of the two species *Myrmica rubra* and *Lasius niger*; and in these, though the date of swarming is unpredictable and evidently depends on the weather—both before and on the day itself—there is often a high degree of synchronisation over a large area, comprising many square miles and sometimes a whole region of the country (*cf.* Donisthorpe, 1927, pp. 236-7). Hot days between July and September are usually selected. Instances have long been known of the workers restraining the ' alates ' from emerging until the time was ripe.

A marriage-flight of *L. niger* observed at Folkestone on 9th August 1911 was described in detail by Donisthorpe (loc. cit.) in order to illustrate the procedure, in a species where the males are considerably smaller than the females. ' A large colony . . . occurred in one of the pillars of a gateway to a house in a street in the town, the ants entering the masonry by a hole in the mortar at the base of the pillar. At five o'clock in the afternoon the workers were much excited, running all over the pavement and up and down the pillar, and a few winged ants were out, going in and out of the entrance to the nest. At six o'clock thousands of males and winged females appeared, emerging from the hole, swarming all over the pillar, and climbing to the top, and on the railings and shrubs in the garden. A few couples were observed in copula and these flew away together, but most of the winged ants flew off separately, rising straight into the air, and going up so high that they were lost to sight. The workers helped some of them to start, tapping them with their antennae and pushing them to the edge of the top of the pillar. More females than males occurred. By 6.25 p.m. nearly every winged ant had disappeared, and some few females were already on the ground without wings.' Wheeler (1910, p. 183) states of ants in general that ' it is rare to find colonies in the breeding season containing equal numbers of males and females. Usually one or the other sex greatly predominates and often only one is represented in the colony; ' and this condition, as he indicates, appears to favour cross-fertilisation between members of different colonies swarming at the same time, just as it does between dioecious plants.

Circumstances are very similar in the case of the termites. The following account, from Skaife (1954, p. 52), refers to the South African species *Amitermes atlanticus*. 'From about the middle of January onwards these winged adults can be found inside the nest. . . . They are kept prisoners for several weeks, for they are not allowed to leave the mound until the climatic conditions are just right for the wedding-flight.

'Sometime in March or April the first winter rains arrive at the Cape, and then, one day, when the weather is clearing after a good rain has soaked the ground, when the air is still and the temperature rising, the long-awaited signal is given for the great event of the year to begin. We do not know who or what decides that the time is auspicious. Nor do we know how the termites, hidden in the blackness of their air-conditioned home, learn what the weather conditions are like outside. But simultaneously in all the mounds in a given area, a great activity begins.

'The workers pierce a number of small round holes in the top of the mound, each hole being about one-eighth of an inch in diameter, so that a few square inches of the apex of the mound looks like the lid of a pepper-pot. This is the only time when the termites deliberately make openings in their bastion walls: under no other circumstances do they run the risk of allowing enemies access to their home.

'Soldiers appear at the holes, with jaws wide open and antennae waving, and then a number of workers come out into the open and these are followed by the winged males and females. There is a great bustle and excitement on top of the mound for a few minutes, with the insects milling around in seeming confusion. Then the princes and princesses take to flight, fluttering away from their home in a straggling swarm. When they have all gone the soldiers and workers retreat inside again, the holes are quickly closed and all is silent and still once more. Not all the winged individuals are allowed to leave at once; some are kept back to form a second, or even a third, swarm later in the season. Often the workers may be seen pulling some of [them] back into the mound and it seems obvious that the workers control the flight and decide how many may go and how many may be kept back.

'The flying termites do not, as a rule, travel very far. They are feeble fliers and an easy prey to all kinds of insect-eating animals. Birds, frogs, lizards, praying mantes, spiders and many other enemies have a great feast when the wedding-flight is on and very few indeed of the fliers that leave the nest survive for long.'

The two main purposes of marriage-flights both in termites and ants appear to be firstly the achievement of cross-breeding between members of different colonies, and secondly the dissemination of the species. Each colony is commonly made up of the progeny of one, and seldom of more than a very few, males and females; and the degree of out-breeding desirable is thus no doubt much higher than in ordinary populations of animals, in which there are no royal and sterile castes and all individuals are potential breeders. The system more resembles the cross-fertilisation commonly occurring between individual plants, each producing many flowers. This need for cross-fertilisation is possibly sufficient alone to account for the precision-timing of the marriage-flights, since the alates are highly vulnerable and this makes their expectation of life very short.

In the majority of ants and termites the colony itself has a permanently-fixed location; and in this also it resembles a sessile organism, such as a plant, and has a similar need for a dispersal stage in its life-history. As a consequence of the marriage-flight mated or fertilised females are widely disseminated, and those that find vacant places in suitable environments are able to establish new colonies. In the driver or legionary ants (Dorylinae), however, which are nomadic, the females are apterous and only the males are winged: presumably this allows for cross-fertilisation between different legions, the need for a dispersal-phase being obviated by the ants' continually wandering habits. As a reminder of the caution that must be exercised in generalising about ants, however, it may be added that in the Argentine ant (*Iridomyrex humilis*), an aggressive and successful species that has been accidentally spread by man into many parts of the world far distant from its original South American home, it is also true that only the males fly. Here the young females, though winged, are apparently never allowed out of the nest, but the males come out on warm sultry evenings when rain is threatening and are often attracted to lights (Skaife, loc. cit., p. 367). They are presumed to enter existing colonies in order to mate with young females there, and close synchronisation of their flight would seem therefore unnecessary. The interesting point is that the Argentine ant has huge compound colonies, usually containing many queens, and thus its communities are no doubt much more varied in genetic constitution than ones with only a single queen. Perhaps in order to ensure the perpetuation of this many-queened condition they have lost the practice of founding new colonies after a marriage-flight by means of single fertile females: instead the colony spreads by fission, budding off new communities containing a number of queens and workers from the start. Superficially, at least, it appears that the gain in genetic variability within the colony has been worth the sacrifice of the dispersal flight.

The development of these diverse practices in swarming and colony-foundation in different kinds of ants provides, incidentally, another interesting case of evolution at the ' group ' level.

In the marriage-flight phenomenon, whether in ants or termites, there does not appear to be any certainly-recognisable epideictic element. Once the participants are clear of the nest there is no indication of mutual social interest or of collective display—no mass-formations either in flight or on the ground. The conditions in which the flight takes place are, of course, quite different from those attending the nuptial displays of most animals: in the social insects with which we are concerned here there is a perennial territorial organisation based on established nests; and within each group-territory the population is continuously under regulation. Their intimate societies are presumably at all times highly sensitive to population-pressure; and the queen's egg-laying rate and the rearing of brood are closely geared to the economic conditions prevailing in the colony. A single queen confined in her nest may continue to lay eggs for many years; and consequently an epideictic display at the time of the marriage-flight could have very little practical bearing on her subsequent reproductive rate. The tentative conclusion from the evidence presented seems to be, therefore, that the marriage-flight does not form part of the epideictic machinery either of the termites or of the ants.

15.8. *Synchronised annual activities in birds*

In the birds, on the other hand, there are a number of examples of mutual synchronisation of particular events in the annual cycle, chiefly connected with migration or with some phase of nuptial or reproductive behaviour, that are not completely accounted for as being essential for securing mating and fertilisation or promoting reproductive success in some other direct way. In the popular mind the dates of return of migrants to their northern homes in spring are supposedly among the most regular of naturally-occurring phenomena. Observations over long periods of years show that as a rule the late-coming species, such as swifts, nightjars or hummingbirds, are more constant in their arrival dates than the earlier migrants: the latter are to a greater extent affected by the weather conditions they find on the journey and tend more to be early or late according to the state of advancement of the particular season. Migrants frequently arrive in waves, or batches, and it is a common experience to find the countryside overrun with a particular species, such as, in Scotland, the willow-warbler (*Phylloscopus trochilus*), in the course of a single night. It requires close study to obtain an accurate picture of the filling-up of a particular breeding habitat by migrants, but it probably takes longer as a rule than one would guess: even in the late-coming and relatively well-synchronised common swift, *Apus apus*, Lack (1956, p. 25) found that it took from the last days of April at least until the third week of May for the whole breeding colony to assemble.

Migration may be stimulated, as has been suggested earlier (p. 284), by mass-displays or other forms of epideictic behaviour, but it can have little social display value on its own account, especially among nocturnal migrants. In the spring it is not until the birds reach their destinations that prenuptial demonstrations normally begin. However, migrants stopping *en route* frequently hold communal displays, often singing freely in spring and sometimes engaging in remarkable collective concerts, but as has been pointed out before this very likely makes a significant contribution to the orderly progress of the migratory movement, by informing the birds of their transient density of numbers and pressing them to move on when an undesirable local congestion threatens.

A more or less closely synchronised arrival of the whole population in its summer quarters no doubt hastens the final attainment of the optimum breeding density, whether the species concerned nests in typical territories, or in scattered or colonial nest-sites. This process of dispersing and establishing the population for nesting is almost invariably attended by highly developed epideictic demonstrations and often by intense competition, and is, perhaps, the particular proceeding which has most to gain from the simultaneous attention of whole potential breeding populations—or, in dimorphic species where the male alone is responsible for dispersion, of the adult males. There is almost always an interval, frequently of two or three weeks, between the actual arrival of the main body of migrants at their breeding place and the commencement of laying, and it is during this time that the breeding density or quota is determined. Once this extremely important stage is attained, breeding itself can begin; and it is known that in some species the appearance of the first egg in the clutch may be practically simultaneous over a whole area of habitat or a whole colony. This synchrony is not necessarily confined to migratory species, and it is interesting to notice in passing that a closely similar adaptation has apparently developed in the American alligator (S. F. Clarke, 1891, p. 187).

A good example can be given from Dunnet's work (1955, p. 632) on a breeding population of starlings near Aberdeen. 'In each of three years, first clutches were remarkably synchronised. In 1950, 21 clutches were started in 10 days with a standard deviation of 2·3 days. The standard deviation for 48 clutches in 1951 was 2·6 days, and for 74 clutches in 1952, 2·7 days. Nest-building was much less synchronous than laying: while some nests are completed a few weeks before the first egg is laid, others are just complete by this stage, and in one case the clutch was started in a nest so incomplete that the egg lay on the bare floor of the box.' These findings confirmed the earlier work of Kluyver (1933) and others; and, as Dunnet observed (loc. cit., p. 652), such unison suggests that synchrony may have a high survival value. This seemed to him likely to arise in part from the need of keeping exactly in step with the life-cycle of the food-supply, on which the young are to depend when they are hatched. In this particular case one apparently critical condition was to leave sufficient time to rear second broods before the leatherjackets (tipulid larvae and pupae), which constituted fourth-fifths of the nesting starlings' food, had all emerged from the ground as winged adults, by about the middle of July.

At the same time as Dunnet was working at Craibstone, Carrick (unpublished) had another similar population under study at Torphins, 16 miles to the west and 400 feet higher above sea-level. There the same kind of synchrony of first laying occurred, but as compared with Craibstone the mean date of starting of some 30 clutches was 7·7 days later in 1950, 5·5 days later in 1951 (a late spring), and 7·0 days later in 1952. The quick-growing stage of the leatherjackets (in May) was also behind at Torphins, by about 12 days, but the final emergence of winged crane-flies in July was actually a few days earlier at this, the more inland station (loc. cit., p. 659, etc.).

A rather similar relation between nest-building and laying was reported by Gibb (1950, p. 512) in his study of the great and blue tits (*Parus major* and *caeruleus*). Nest-building began on the average 12 days before the first egg was laid in the mild spring of 1948, and 8 days in 1947; whereas in 1949 ' the first egg was already laid in many nests before the nest-lining was begun, and sometimes only as little as three days after the start of building '. ' A striking feature observed in the three seasons in Wytham was the simultaneous egg-laying both by individuals of the two species and by the two species of tits.' This appeared to be related to the crop of leaf-eating caterpillars of the moths *Cheimatobia brumata* and *Hibernia defolaria*, on which the nestlings were largely fed. When the quantity of caterpillars, measured by collecting their frass in trays, is plotted alongside the total number of nest-boxes containing broods of young, the correspondence is undoubtedly close: moreover the early and late seasons of 1948 and 1949 are compared, both the tits' breeding and the caterpillars' growth-cycle are seen to be displaced together by roughly a week. There appears no reason to doubt that a general correlation exists, both here and in the starling case, between the timing of the breeding season and the availability of a particular food-resource.

The question to be answered, however, is whether this correlation is sufficiently critical to explain the high degree of synchrony in the laying of the first egg. In the year 1951, for which Dunnet obtained complete figures, second broods were all fledged by 15th July, with about 10 days to spare before the biomass of leatherjackets had declined to one-fifth of its maximum value. In the case of the tits, in 1948 the peaks of ' broods in nest-boxes ' and ' abundance of caterpillars ' almost exactly coincided, though the ' broods ' curve thereafter tailed off much more slowly than did the caterpillars. In 1949 the tits were evidently behind the caterpillar cycle and were still close to their peak a week after the caterpillars had begun to decline, by which time the latter were already down to about one-quarter of their maximum abundance. The figure for the number of nest-boxes containing broods does not perfectly reflect the demands being made on the food-supply, because nestlings in their second week receive on the average about twice as many feeding-visits a day as those under a week old. Thus it seems probable that the heaviest demand in both years, and especially in 1949, was reached after the peak abundance of these particular larvae had passed. Dunnet found, incidentally, that the total takings by the starlings from the available resources of leatherjackets in each of two entire breeding-seasons amounted to only 1·9 and 7 per cent respectively; and similarly with the tits, Betts (1955, p. 318) estimated that, even in a year with a relatively low density of caterpillars, the tits did not take more than 5 per cent of the available supply.

Judgement may with advantage be reserved on the question of the underlying purpose of synchrony of egg-laying in these two cases, until certain others have been examined.

In one of the same species of tits, *Parus major*, Kluyver (1951, p. 44) found that, in five out of nine seasons, there was a significant difference in the start of laying between females two or more years old and those breeding for the first time; in the four remaining years the observed difference was not statistically significant. But when all nine years were put together the yearling females started on the average 2·1 days later than the older birds. When the histories of particular individual females were followed from year to year, there was again an average shift of 2·6 days earlier between their second breeding season and their first, and an average shift of 3·8 days earlier again between their third and second seasons. There may be a number of possible explanations for this phenomenon—one of them being that the social status of the individual is involved—but in any case it does not appear to tally with the opinion that rigorous synchrony of egg-laying is determined solely by the exigencies of the food-supply.

In some at least of the gulls it is thought that the same relation generally holds true, and that older birds on the average lay earlier than younger ones. In their study of the remarkable colony of kittiwakes (*Rissa tridactyla*), situated on the window-sills of a warehouse in North Shields, where the adults could be caught with a wire hook from inside the building and colour-ringed, Coulson and White (1958, p. 41) found that breeding birds returned earlier in spring than non-breeders, and that 'old' breeders returned earlier than 'young' ones. Thereafter the former were more assiduous at their nest-sites, and finished by starting to lay earlier. 'In 1955, birds breeding for at least the second time laid on average 8·8 days earlier than birds breeding for the first time. A similar difference of 7·2 days was found in 1956' (loc. cit., p. 43). These differences were highly significant (on the two years together $P<0{\cdot}001$).

Goethe (1937, p. 12) surmised from his observations that the older herring-gulls (*Larus argentatus*) were also the first to lay; and it seems possible that this factor entered into the differences found by Darling (1938, pp. 61-73) in four colonies of herring-gulls of varied sizes which he studied on Priest Island, Wester Ross (Scotland). In these colonies laying started earliest in the largest colony and latest in the smallest one. Fledging success was 48·6 per cent in the largest colony in 1936 and 41·4 per cent in 1937, whereas in the smallest colony it was zero and 12·5 per cent in the two years respectively. This is what might be expected to result from an age-difference, with the small colony containing young birds standing lower in the island hierarchy, and being less confident and attentive as parents. In two colonies of the lesser black-backed gull (*L. fuscus*) on the same island, though one was four times as populous as the other, there were no significant differences in fledging ratios in either year. Furthermore, a range of almost three weeks in the date of appearance of the earliest egg in the four herring-gull colonies in 1936 had dwindled to four days in 1937.

These gulls are not so closely synchronised in their egg-laying as the starlings and the tits and are cited here only to illustrate the influence of age on the laying-date. It may be observed with interest, however, that in the herring-gull the first visit to the nest-colony in spring is held up until the whole population has gathered in the neighbourhood: the colony itself remains strictly ' out of bounds ' until the very morning of synchronous occupation (Paludan, 1951, pp. 9 and 22).

As a final example of what is probably a widespread age-effect, the blackbird (*Turdus merula*) may appropriately be mentioned. Snow (1958, p. 3) found that in three successive years the average starting dates of yearling females were 25, 6 and 14 days later than those of older females; and he commented that ' if the date at which a female comes into breeding condition depends not only on the action of environmental factors . . . on her physiological cycle, but also on social status, in the wider sense, this would adequately account for the later breeding of young females '—a suggestion that concurs with our own conclusion.

Returning to the subject of synchronisation, it is not always the start of *laying* that must await the fulfilment of the pre-reproductive stage and the moment when the integrated society is perfectly in step. In the red-billed quelea (*Quelea quelea*), for example, the achievement of general co-ordination is marked by the beginning of nest-building. This highly sociable weaver-bird exists in vast hordes over much of the African continent and is sometimes a very serious pest of grain. Its nesting-places vie with those of the extinct passenger pigeon (*Ectopistes migratorius*) in population. One in Tanganyika covering nearly 100 acres was estimated to hold 2½ million nests; another covering 4 square miles might have held twice that number (Vesey-FitzGerald, 1958, p. 168). Morel and Bourlière (1956) studied one in Senegal occupying about 400 hectares (*c.* 1000 acres) and another smaller one of only some 20 hectares (50 acres). Nest-building is done entirely by the males—starting before they are mated—and the authors closely observed the synchronisation and speed with which it goes forward. They divided the building process into six descriptive stages, and noted that, whereas on the morning of 24th August most nests were in the third and fourth stages, by 6 p.m. 70 per cent in one sample tree had reached stage five: the next morning but one (26th August) all the nests were practically finished. It appears, they state, that a large colony can be entirely constructed in about a week (loc. cit., p. 101). The commencement is marked by intense activity, the males going and coming incessantly with the long green blades of grass used to weave the vertical ring-frame of the nest: building is also attended by an outburst of keen rivalry, evoking special ' intimidation displays ' between males and sometimes serious fighting. Their aggressiveness continues at least during incubation and perhaps longer.

The nest is barely finished before mating, courtship and laying succeed one another at the same lightning speed. The authors drew special attention to a distinct difference in timing, as it seemed to them, between the smaller colony and the larger one, situated 24 km away in a similar forest formation.

The difference appears, on examination, to have been scarcely more than 24 hours; and the importance attached to it emphasises the extraordinary pace of events and the striking appearance of unison existing in each colony taken alone.

15.9. *Synchronisation of breeding seasons of tropical birds*

The question still before us is whether exact co-incidence of timing of the breeding cycle throughout the community is an adaptation imposed solely by environmental factors, such as climate, cover and food-supply, in which case each pair of birds is primarily pursuing its own independent interests; or whether it has a social significance, related to establishment of a breeding quota and possibly a controlled level of egg-production for the community as a whole. The general close correlation existing between the timing of the breeding seasons and suitability of environmental conditions is not in question at all: the doubt arises solely as to whether this correlation is generally sufficiently critical to bring the whole population exactly into unison, down to a standard deviation of as little as a day on either side of the mean. The period concerned is invariably one of high social integration, characterised in birds by intense song and visual displays; and if this contributes to the achievement of co-ordinated timing, as it appears to do, it must be concluded that synchronisation has an adaptive value as a social phenomenon, benefiting the community as a whole. The existence of the age-effect discussed in the previous section is itself a strong indication that the precise timing of the breeding cycle is closely connected with the social organisation.

Evidence of a different kind comes from the well-known occurrence of definite and well-synchronised breeding seasons in tropical species, living in environments which in extreme cases are so equable and invariable throughout the year that it is a puzzle to know what external stimuli could serve to regulate an annual cycle. This may affect among others such large birds as African fish-eagles (*Cuncuma vocifer*), for instance, which, being presumably paired for life and seldom far separated from their mates, would appear to be free to nest at any time they chose. ' Like other eagles they seem, rather unexpectedly in tropical Africa, to have a regular breeding season, laying their eggs about August, hatching them in October, and fledging the young in December. Quite why they should have a regular breeding season in a place [Lake Victoria] where the supply of fish is fairly plentiful all the year round it is difficult to see . . .' (Brown, 1955, p. 121). Moreau (1950, p. 259) remarks that ' the time-limits of the breeding seasons [of African birds] are such that they must evidently be a consequence of strong natural selection; and it is a matter for surprise that selection should be able to operate so powerfully as it apparently does under some of the relatively uniform environmental conditions that have been described '. In an earlier paper (1936, p. 651) he had concluded similarly that ' the seasonal fluctuation in breeding activity at Amani . . . is far more marked and more applicable to the whole population than appears to be imposed by conditions'.

The same thing is true also of sea-birds on the coasts of Africa. Moreau (1950, p. 423) states that 'a continuous, or even an ill-defined, breeding season by any species, at any station, is very exceptional'. In other tropical countries, such as British Guiana (Beebe, 1925), the Galapagos Islands (Lack, 1950) and Indonesia (Voous, 1950), the situation appears to be similar. However, there are a good many exceptions: in equatorial Colombia, of ten species of birds studied by Miller (1955), eight were essentially non-cyclic and only two cyclic; and in the rain-forests of the Cameroons G. L. Bates (1908) found that most species had no distinct breeding season, though there was a marked exception in the weaver-finches (Ploceidae), including a bishop-bird (*Euplectes flammiceps*) and a widow-bird (*Coliuspasser serenus*); the males of both assumed their nuptial plumage in July. 'There is nothing in the changes of seasons . . . that seems to account for the habit these birds have of breeding at a distinct part of the year only' (loc. cit., p. 566).

There is one very remarkable case, in the sooty terns (*Sterna fuscata*) nesting on Ascension Island, of a regular breeding season not apparently linked to any solar or meteorological cycle, so that the dates of its commencement precess through the calendar. Chapin (1954) and Chapin and Wing (1959) pieced together a continuous sequence of nesting dates covering the period 1941-58 and found an average interval of 9·7 months between recurrent nestings. There are over a dozen colonies collectively known as the Wideawake Fair, in some three square miles of the south-western part of the island, and these are all in session together for the duration of each breeding period—about 6 months—and then all deserted for a non-breeding interval of three or four months. The close synchronisation and the number of terns involved is indicated by the fact that 40,000 or more eggs could be removed at one time from only a part of the area (Chapin, loc. cit., p. 1). The species has a circumtropical range and has a normal yearly breeding cycle everywhere else where its habits have been closely observed: from the rather scanty evidence available, however, Chapin concludes (loc. cit., p. 11) that in some of the islands of the central Pacific and especially Christmas Island there are two breeding seasons for sooty terns each year, about six months apart.

There are many problems raised by the 'wideawake birds' of Ascension —one being the question whether there could possibly be any kind of external pace-making stimulus with a period of 9·7 months or whether the rhythm is completely intrinsic and self-perpetuating. It seems highly probable that the integrity of the breeding period could be preserved only through a compelling social drive, which inhibits the birds from nesting except in unison with the rest of the population. According to our present theory the advantage and value of evolving a social organisation, such as the colonial form of breeding society and property-tenure found in these terns, lies in the means it provides of controlling the numbers and density of the population; and this particular society would break down in the absence of synchronisation. To maintain that colonial organisation, in short, the birds require to foregather not only at the same places but at the same times.

It seems safe to conclude that in almost every case where close annual synchronisation of activity among members of a population is found in animals, it confers a social advantage: that is to say, an advantage having a survival-value for the group as a unit, and if necessary overriding the individual interests of some of the members. The advantage may primitively be confined to promoting the union of the sexes or gametes: this still perhaps remains the principal purpose of the massed spawning of the palolo worm and, coupled with the dissemination of the species through the foundation of new colonies, it is so also in the marriage-flights of termites and ants considered earlier. Especially in the higher vertebrates, however, pair-formation may have taken place long before the commencement of the period of close synchronisation, and the pair-bond may not infrequently survive throughout life.

In seasonally changing climates there is usually a short-lived optimum period for reproduction which fixes the start of breeding within quite narrow limits; but conversely, in unvarying tropical climates it is likely that no such environmental influence will be felt at all. Regardless of any climatic considerations, however, social integration benefits from synchronisation; and it is social integration that supplies the means of controlling population-density homeostatically. It seems very probable, indeed, that it is the social advantage which is almost always responsible for the fine adjustment in timing and more or less entirely accounts for the definite annual breeding seasons frequently maintained in non-seasonal tropical climates.

There are, as we have repeatedly seen, a number of alternative types of social system for controlling population-density, and we know it is possible to find the means of achieving social integration and controlling dispersion without synchronised breeding. Some rodents, for instance, and primitive man, breed almost uniformly at all seasons and nevertheless regulate their reproductive rate (*see* pp. 492, 502). This clearly entails the establishment of a standing social regime that remains in efficient operation all the year round, or at any rate for most of it—a regime capable of taking a progressive account of recruitment and loss in the population. Closely synchronised breeding has no place in such a perennial system, where the recruitment-rate can be adjusted continuously to changes in economic conditions. It would seem possible to achieve a continuous balance on these lines only in a sedentary population of unvarying social behaviour, such as may perhaps be found in some of the non-synchronised tropical birds, or in various kinds of mammals that adhere to exactly the same system of property-tenure all through the year. On the other hand any birds that alternate between territorial and gregarious habits, or between breeding and non-breeding habitats or between nest-site tenure and roost-site tenure, no matter where they live, will have to adhere to a mutually synchronised cycle of behaviour.

It is probably for this reason that, among the long-lived animals, synchronised breeding is most common in those that are more mobile, such as the birds, fishes, bats, ungulates, seals and whales, whereas protracted breeding seasons are more common in the sedentary types.

15.10. *Summary of Chapter* 15

1. Members of a group engaging in conventional proceedings affecting the interests of the group as a whole commonly require to assemble together and this involves not only an appointed meeting-place but also a specified time. Synchronised activities in animals such as the dawn and dusk choruses of birds, or sharply-defined breeding seasons, may therefore often be suspected of fulfilling a social purpose.

2. Recurrent events in behaviour, whether they have a 24-hour, annual or other period, depend with few exceptions on the interaction between an internal physiological rhythm and an environmental pace-making stimulus. The interaction of these two factors is illustrated by an original study of the morning exodus of starlings from a large winter roost, in relation to sun-time and light-intensity: a compromise or balance is shown to be established between these two factors.

3. In the 24-hour cycle, telling the time is most easily done visually at dawn and dusk, when light-intensity changes fastest. These two periods are by far the commonest ones for holding social engagements of brief duration: at these times such engagements can moreover be conveniently sandwiched between the 'working' and resting parts of the daily cycle. The major peak of bird-song known as the dawn chorus, and the usually lesser peak at dusk, are among the best-known communal epideictic displays that follow this characteristic time-table. One or both of these hours generally sees the peak of other similar social demonstrations, such as leks or flight-manoeuvres.

4. Similar phenomena appear in mammals; and in some of the more vocal species, including certain fruit-bats, the howling monkeys and gibbons, dawn and dusk choruses are just as well developed. A diurnal cycle of sound output with morning or evening peaks is also found in certain frogs and fish. Other non-vocal social demonstrations with a similar periodicity appear to include the 'evening rise' of trout and other fish.

5. In insects 'diel' cycles of activity with well-marked dawn or dusk peaks are common; and there is no reason to doubt that in some cases the advantage conferred by this is a social one. Examples are also given of species with specialised activity-peaks at other hours of the day and night. The concentrations of insects produced by these sharp peaks may frequently have an epideictic purpose.

6. Turning to annual cycles, the first illustration given of precisely-timed behaviour is the mass-spawning of the Pacific palolo worm, *Eunice viridis*, which is concentrated into a few hours before and after dawn on two or three days in the year. The question arises whether this is merely an adaptation

to promote fertilisation or whether it has any epideictic function. For the sake of comparison a review is given of the reproductive behaviour of a number of other polychaetes; and it is shown that sexual dimorphism, nuptial displays, unisexual gatherings and even 'phase' (in the sense used of locusts) have been developed in the group. The last of these in particular is exemplified by *Platynereis dumerilii*. Epideictic phenomena are therefore to be expected.

7. In the marriage-flights of termites and ants—a parallel development in these two unrelated but often similar groups—we have another example of annual synchronisation. Here there seems to be no epideictic function; swarming promotes cross-fertilisation between different colonial stocks and serves to disseminate these sedentary animals. Reproductive rate is controlled independently, through the queens.

8. In birds, precise timing of certain events in the annual cycle is not so clearly related to promoting reproductive success. After the pairs have become established, the start of egg-laying, for example, is often markedly synchronised over a particular area or colony: examples are taken from the starling, and great and blue tits, where this timing is apparently correlated with exploiting a particular and abundant food-resource, at a slightly later stage, as food for the nestlings. There is some doubt, however, as to whether the correlation is critical enough to have evoked so marked a degree of synchronisation in egg-laying, particularly as it is known in the great tit that age influences the laying date. This age-effect, whereby the senior birds lay on the average earlier than the junior ones, is apparently widespread: other examples are given from the kittiwake and blackbird. It indicates that a social factor can influence the precise timing of egg-laying.

In the red-billed quelea, the sequence of highly synchronised events in breeding begins, not with egg-laying, but with nest-building, by males as yet unmated, and this kindles strong mutually-aggressive behaviour.

9. In tropical latitudes definite breeding seasons are fairly general (though not universal) among birds, even where it is difficult to detect any environmental change that would make one season more favourable for reproduction than another. Here the adaptive value of synchronisation is probably almost entirely social, arising from the epideictic effects of social organisation and the means thereby provided of controlling reproductive quotas and output. A very interesting and possibly unique case is that of the sooty terns on Ascension Island, which have well-synchronised colonial nestings, not annually but at average intervals (during the period 1941-58) of 9·7 months (10 lunar months).

Synchronisation of activities, however, is not indispensable to the conduct of an efficient, self-regulating society: it is especially characteristic of species that alternate seasonally between two or more habitats or types of social organisation. Where there is a single prolonged or perennial social regime, as probably in some tropical birds or in sedentary mammals, it has little or no place.

Chapter 16

Vertical migration of the Plankton

16.1. *Introduction*

A phenomenon that appears to share some of the characteristics of those considered in the last chapter is the very striking up-and-down diurnal movement that is made by a great variety of plankton animals. In the most common type of cycle the members of the particular species concerned swim up nearly or quite to the surface as dusk approaches, and sometimes, after dispersing to some extent during the dark hours, reassemble there at dawn: thereafter as the morning light increases they swim actively down, often for tens and more rarely for hundreds of metres, reaching their lowest level in the middle hours of the day, and mounting more or less rapidly again as the next evening approaches. Vertical migrations following some such diurnal pattern are found in pelagic organisms both in salt and fresh water in practically every group of animals adapted to this kind of habitat, and in larval and immature stages as well as in breeding adults. Nevertheless they are not universal; and the same species at different stages of its life-history, or at the same stage in different localities and circumstances, may sometimes be found to behave in totally different ways. The subject has received a great deal of attention, especially during the last fifty years; and it has been surmised by all those who have studied it closely that such a general and widespread phenomenon must have an equally general underlying cause or explanation. Yet of all the many problems presented by the natural history of pelagic animals this one 'seems perhaps more baffling than any other' (Hardy, 1956, p. 199). 'Vertical climbing uses up so much energy and has been developed so frequently in the animal kingdom, that it must clearly be of some very profound significance in the lives of these animals. What can be the meaning of it? We do not yet know for certain: we can only make some guesses. It is surely the planktonic puzzle No. 1' (*the same*, p. 200).

It is desirable to mention briefly at the outset the various kinds of organisms known to perform these movements. At the lower end of the scale something of the kind appears to have been demonstrated by Hasle (1950) in certain dinoflagellate Protozoa; this was in the enriched and sheltered waters of Oslo fiord, where the numbers of such green unicellular organisms are sometimes exceptionally high, and the water is correspondingly turbid. On account of their microscopic size and limited speed of locomotion (of the order of 4 to 15 mm/min) the extent of migration cannot be expected to be more than a few metres. Water samples were taken at frequent intervals at depths between the surface and 10 m, over two periods of 24 hours, and in these samples the numbers of organisms were counted. There were four common species of dinoflagellates, three of which showed rather divergent sequences of depth changes. In *Ceratium fuscus* on July 26-27 maximum numbers were found at the surface during the night and minimum numbers about midday: on the second occasion (July 28-29) this pattern repeated itself, whereas at a depth of 2 m the curve was inverted, pointing to a vertical interchange between the two levels. The weather at the time the observations were made was of course very calm. With *C. tripos*, though the numbers taken were smaller, the pattern appears to have been similar. *Goniaulax polyedra*, on the other hand, showed on the 26th-27th a maximum in the upper two metres during the late afternoon, and a maximum at a depth of 10 m during the night and early part of the morning: the figures for the 28th-29th were not published. The fourth species, *Prorocentrum micans*, showed higher surface numbers rather irregularly on each of the two observation days between 8 and 16 hrs and a maximum at 10 m around sunrise (4 a.m.) on the second day.

These results have been given in detail, because they have not as yet been replicated, and show considerable dissimilarities among themselves. It is not impossible therefore that they might relate to irregular movements rather than typical diurnal migrations: in any case they indicate the extent of movement possible in green flagellates, which constitute food for some of the smaller members of the macroplankton. It may be borne in mind that none of the other unicellular green organisms such as diatoms and desmids have any comparable powers of propulsion.

Diurnal variation in the numbers present at and ascending towards the surface has been demonstrated elsewhere in the Protozoa in the exceptionally fast-moving pelagic Tintinnidae, a family of ciliates characteristically invested in a lorica or test (Halme, 1937, p. 367): from these observations the occurrence of diurnal migrations can reasonably be inferred. In the Coelenterata there are well-documented examples in the Anthomedusae and Leptomedusae (Russell, 1925, 1928a, etc.); migrations are inferred also in Siphonophores. There seems to be no similar evidence for the scyphozoan jelly-fish but this may well be due to the difficulty of collecting and preserving these large fragile animals by the methods normally used in plankton research: the same is probably true of the ctenophores, though Esterly (1914, p. 34) did obtain indications of vertical migration in *Pleurobrachia*, and showed that ' during the summer the optimum conditions at the surface are found at sunrise or shortly after.' There is a well-known case of vertical migration in the beautiful transparent pelagic polychaete, *Tomopteris* (*e.g.* Russell, 1928, p. 88), and an equally familiar one in the various species of chaetognaths belonging to the genus *Sagitta* (the same, p. 88). Among the Mollusca, Cushing (1951, p. 158-9) cites night surfacing as circumstantial evidence of vertical migration in the pteropod *Hyalocylix striata*, and it has been observed that another pteropod, *Limacina retroversa*, can be taken in large shoals at the very surface when the sun is low, but may resort to deeper layers in the middle of the day (*cf.* Russell, 1927, p. 604). Vertical migration was demonstrated in *Limacina helicina* by Hardy and Gunther (1935, p. 240).

The Crustacea, however, are by far the most numerous exponents. Vertical migration is particularly characteristic of the copepods, the largest planktonic order, and only a little less so of the Cladocera. It is known to occur more or less commonly in ostracods, Cumacea, mysids, amphipods, euphausians and decapods (among species of pelagic or natant habits), both in the adults and during their developmental stages; and in some of the larger decapods which later settle down to live on the bottom there are planktonic larvae that are liable to diurnal migration also.

In the protochordate groups, the phenomenon has been recorded, for example, in the tornaria larva of balanoglossids (Russell, 1925, p. 807) and, among the tunicates, in salps (*cf.* Hardy and Gunther, loc. cit., p. 240) doliolids and Larvacea (Russell and Colman, 1934-5, pp. 210 and 230).

It appears, therefore, likely to affect in some degree every major constituent group of the zooplankton; and the result of further study will no doubt be to extend the number and range of species in which its occurrence is known. The examples given here represent a score or more of animal groups in which planktonic habits in general and this kind of behaviour in particular have been independently acquired, so that we are confronted with a very striking example of parallel evolution.

What explanation can be offered for so general (though not universal) a pattern of behaviour? To what extent are the participating species dependent or independent of one another? Are we dealing in some cases with the pursuit by predators of their vertically-migrating prey? The last question can undoubtedly be answered in the affirmative, at least as far as certain fishes and cephalopods are concerned. The drift-net fisheries for herrings (*Clupea harengus*), for example, have depended for centuries on the knowledge that the shoals approached the sea-surface between dusk and dawn, but descended lower and beyond the reach of floating nets during the daylight hours. Herrings feed to a substantial extent on copepods, such as *Calanus* and *Temora*, and on Cladocera, which themselves typically undergo vertical migrations, and there is no reason to doubt that the diurnal depth-changes of the herring reflect in the main those of their food-organisms. From their work on the diurnal migration of deep-water animals Welsh, Chace and Nunnemacher (1937, p. 193) concluded similarly that 'the striking correlation in numbers of copepods and fishes at the 400-meter level suggests that the fishes move with the copepods and thereby stay near their chief food supply.' Pelagic cephalopods are another group of which the diurnal movements are equally likely to be secondarily linked with those of their planktonic food species; and the same could perhaps be expected of some of the medusae, especially the larger species, which must be reckoned among the swifter and more powerful members of the pelagic fauna.

Members of the phytoplankton, as we have seen, are incapable of any diurnal movements other than those of relatively small amplitude. On account of their dependence on light for photosynthesis they multiply most abundantly in the uppermost layers of water; and animals which feed on them, for instance *Calanus*, when undergoing vertical movements, naturally do most of their feeding at night, while they are near the surface where the food is most abundant. In this instance, however, though a diel cycle may be imposed on their feeding activities, it is clear that their migrations do not stem from any simple pursuit of migratory prey. Rather frequently *Calanus* will remain in numbers at or near the surface throughout the 24 hours, and feeding then has been shown to go on both night and day (Gauld, 1953): it appears, therefore, that when *Calanus* does migrate, the daytime descent usually has the direct effect of reducing for the time being the availability of food.

16.2. *Hardy's theory*

It is unnecessary to review all the suggestions that have been made to account for the movements of organisms like *Calanus*: few are at all satisfactory, and it will suffice for our present purpose to accept the opinion of experts that we are in fact still faced with a major puzzle.

There is one general theory, however, deserving very careful attention, that was first proposed by Hardy in 1935 (Hardy and Gunther, 1935, p. 311; *see also* Hardy, 1938 and 1956). It is well known that the density of phytoplankton is ordinarily far from uniform from place to place over wide areas of the sea, and also that at certain seasons there are great 'blooms' or outbursts of unicellular plants that tend to accentuate this non-uniformity. It has long been noticed that where the phytoplankton is densest, animal organisms are frequently scarce or even absent. This may be in part due to the grazing effect of the zooplankton, which causes the stock of plant cells to diminish rapidly in the presence of herbivorous animals, with the result that the plants succeed in proliferating abundantly only where they are not being grazed. But it is now known that some of the plants (especially the dinoflagellates) produce substances, analagous to antibiotics, which they release into the water and which may be inimical or directly poisonous to other organisms such as copepods, euphausiids or even fish; with the result that these animals may actually be excluded from the dense phytoplankton patches. At any rate, feeding and living conditions for plankton animals certainly vary greatly from place to place within the surface layer; yet in terms of the geographical scale of this variation and the distances which may be involved, the feeble powers of locomotion of most plankton animals do not appear to afford them an effective means of searching in the horizontal plane in order to find an optimal habitat or to escape from an unfavourable one.

Hardy suggested that the vertical migrations could provide a means of overcoming this impasse by taking advantage of the relative differences in the speed and direction of water currents at different depths. 'The water masses are hardly ever moving at the same speed at different depths; the surface layers are nearly always travelling faster than the lower layers, and . . . often travelling in different directions. An animal which can swim to the right or left for only a few hundred feet will not get much change of environment in the sea, for all that effort; and in a diffuse light from above it will also be difficult for it to travel far in any one direction: it will most likely curve round towards its starting point again. But if the animal sinks downwards a hundred feet, and climbs up again the next evening, it may then find itself a mile or more away from where it was the night before, for there may well be a difference of a mile or two a day between the speed of the current at the surface and that at only 30 metres depth. A drop in the morning, and a climb of a hundred feet at evening, may well give it a horizontal movement (in relation to the surface) of some ten thousand feet!' (Hardy, 1956, p. 211).

This is an ingenious theory, suggesting a method of trial and error by which planktonic animals could locate good feeding grounds, but there are conflicting circumstances that we must take into account which appear to rule it out, at any rate as explaining the primary purpose of the habit.

In the first place, vertical migration is not confined to the marine plankton, but occurs just as characteristically in inland waters. In a pond or small lake it may be impossible for a plankton animal to get away from an unfavourable density of phytoplankton, because the bloom simultaneously covers the whole body of water; and, moreoever, unless there is a wind blowing, horizontal displacement of water-masses normally comes to a standstill. In a temperate lake in summer, however, there is frequently a discontinuity layer or thermocline developed within a few metres of the surface, with a warm water layer overlying a much cooler mass which fills the deeper parts of the lake; and when the wind causes horizontal drifting of the surface layer to leeward, an opposite return current commonly develops along the thermocline. This would in fact appear to produce conditions very suitable for Hardy's transport system, but in Lake Mendota, for instance, Birge (1897, p. 372) in his classic study found that 95 per cent of the planktonic Crustacea remained continuously above the thermocline, and they could not consequently have been availing themselves of the opportunity. Thermoclines also develop in the sea under certain conditions, and the limitation of particular plankton species to the layers above or alternatively the layers below has frequently been observed (*e.g.* Clarke, 1933, p. 428).

The next point is the relation of vertical migration to light intensity and to the 24-hour cycle. If the habit were evolved primarily for the purpose of permitting horizontal displacement, there would seem to be no particular need to make the transfer at one hour of the day rather than another. If the animal on ascending found itself entering an unfavourable environment, might it not be expected to descend at once and try again elsewhere, without any necessity of developing and conforming to a 24-hour periodicity? Changes in light intensity have been shown to be the one environmental factor with which vertical migrations quite definitely are correlated: the fact has been established, for instance, that in high arctic latitudes, where it is light throughout the 24 hours in summer and almost continuously dark in winter, *Calanus* does not perform regular migrations; they only develop to a marked extent with the return of a day-and-night rhythm in autumn (*cf.* Marshall and Orr, 1955, p. 128): so that, regardless apparently of feeding conditions, vertical migration in this case occurs only under the stimulus of regularly alternating light and darkness.

There is more substantial evidence than this that the habit must, at times anyway, be quite independent of feeding conditions. In part this is derived from direct observation. Clarke (1934, p. 445), for instance, studied the vertical distribution of *Calanus* at two stations in the Gulf of Maine, one over deep water and the other, two days afterwards, over Georges Bank. At the first site there was a marked diurnal movement, transcending the thermocline, but at the second there was a much less definite one of relatively small amplitude, the animals remaining above the thermocline all the time. After discussing the hydrographic conditions and the phytoplankton distribution, he concluded that 'in view of this situation not only does it appear that the differences in the behaviour at the two stations are not due to differences in the phytoplankton, but also it is difficult to understand how food could be an important factor in causing or controlling the diurnal migration which did occur at the deep station. . . . In their upward movement the animals passed right through the zone of abundance of phytoplankton.'

A rather weightier part of the evidence that migration may be independent of feeding conditions depends on the well-attested fact that, in a single species at the same locality and date, there may be differences of behaviour correlated with age and sex. In Lake Lucerne, for instance, Worthington (1931, p. 410) found that the young stages of *Daphnia* migrated much further than the adults; and the same occurred to a more pronounced degree in *Bosmina*, whose young behaved remarkably like those of *Daphnia*, though the adult's on the contrary remained near the surface all the time, rising a little if anything during daylight. In *Calanus* Clarke (1934, p. 437) found that at night the females rose much nearer the surface than the males or adolescent stages (copepodites IV and V): they tended to remain much more concentrated during the dark hours and not to descend in the morning as early as the males (loc. cit., p. 440). This pronounced tendency of the females to be more strongly migratory than the males and immatures in *Calanus* has been observed by other workers: referring to the Clyde sea area Nicholls (1933,

p. 153) for instance remarks that ' of all the different stages of *Calanus* none exhibits diurnal migration so well as the females in spite of the small numbers present.'

Calanus is known to feed largely on phytoplankton, and moreover ' on what is common in the microplankton at the time they are caught ' (Marshall and Orr, loc. cit., p. 102); ' copepodite stages III and IV eat the same as V and adults ': however, judged by the production of faecal pellets the females are considerably more voracious than the males (*the same*, pp. 112-113). Nevertheless, the feeding conditions required by the adults and older copepodid stages are likely to be very similar, and there is no indication at all of differences in diet that might be sufficient to account for the observed changes in the extent of vertical migration between, for instance, copepodid V and the adult female. The conclusion appears to be that caste differences like these in diurnal behaviour cannot be primarily connected with habitat selection.

The most direct evidence of the fundamental connection of vertical migration with changing light intensity comes from the fact that vertical migrations can be experimentally induced in *Daphnia*, for instance, under suitable conditions, by artificially varying the illumination in an aquarium, in an environment in which ' food ' (in the form of indian ink particles) is uniformly distributed from top to bottom (*cf.* Harris and Wolfe, 1955).

16.3. *Vertical migration as an epideictic exercise*

In previous chapters we have encountered a number of situations, often affecting terrestrial environments, that have an essential and surprising similarity to the vertical migrations of the plankton. Some of the planktonic species have reached a high level of social evolution, especially for instance the copepods, members of which, as mentioned in Chapter 12 (p. 241), have in some cases evolved splendid nuptial adornments, frequently in the female sex, and comparable with those of the most ornate decapod Crustacea or fishes. The euphausians are possibly even more spectacular, when they flash with brilliant luminescence in the darkness. It is certainly to be expected that such well endowed animals as these would assemble for epideictic displays. We must remember, however, that their powers of movement for purposes of aggregation are relatively much less than those of insects, birds or fishes; and though some pelagic organisms, it is true, are typically found in dense swarms, much of the time the majority are rather widely diffused through the mass of water.

The environment of a pelagic animal differs from that of a terrestrial or bottom-living form in a very important respect, already discussed on an earlier page (p. 166), namely that the water for practical purposes is a 3-dimensional continuum with no fixed points of reference. There is normally no solid substrate within reach on which to base a lasting pattern of territorial dispersion. Though the force of gravity can still be used to determine the vertical axis, plankton organisms usually find themselves in a habitat that is homogeneous and not steeply graded or stratified, at least in terms of the little world which bounds their individual motions: each is a moving point in space, seeking its food and encountering its neighbours more or less freely in all three dimensions.

For planktonic species living in the photic zone, however, there are two features of the environment that mitigate the endless physical uniformity of the space in which their lives are spent. The one of premier importance is the relatively rapid vertical gradient of illumination established during daylight, and the associated diurnal cycle by which it comes and goes. The second is the comparative nearness of the surface, which forms an accessible boundary to the aquatic world. Both of these are denied to bathypelagic species living in abyssal darkness.

The hypothesis to be examined here is that plankton animals have acquired the habit of vertical migration primarily as an epideictic device, capable of concentrating for purposes of display the population of a deep column of water into a thin layer, frequently just beneath the surface-film itself. The air-water interface is not necessarily the only level it would be possible to use for the purpose, although it appears to be the commonest and easiest: sometimes a thermocline may be made to do duty, or a layer defined by a particular light intensity; it is possible also that so-called bottom plankton in shallow water could concentrate equally well on the bottom. If this supposed dispersionary exercise is like others we have seen we should of course expect the concentration to be synchronised, most commonly either at dusk or dawn or both, and to be accompanied by swarming or some other form of social ritual or contest. The purpose of epideictic displays is to test the population density and stimulate responses which will hold it at or restore it to the optimum. Adjustments of density may in some cases be practicable only through regulating the reproductive output, and in this event only the adults, or even one representative sex, might require to undertake diurnal migrations; in other cases re-dispersion may be possible among the larvae or immature stages, possibly by Hardy's method; or the mortality rate may be capable of being influenced; and in these cases vertical migration may be necessary at different stages of the life-history. We need not expect complete uniformity of behaviour in all species, but can look for day-long leks or their counterpart at the height of the reproductive season, or for displays synchronised at other special hours of the day or night like those which we saw in some of the insects.

It seems evident enough that such a hypothesis must not be lightly dismissed, and that it is desirable to examine the relevant facts much more closely in the light of it, particularly those relating to synchronisation and social behaviour.

16.4. *Presence and absence of vertical migration*

It has been mentioned that diurnal vertical migration has been evolved independently in many groups of animals, probably in all or almost all those that have planktonic representatives. It will help to define the nature of the phenomenon to consider the types of environment where it does and does not occur, and the variety of behaviour patterns which are embraced by it.

The habit appears to be typically developed in animals that are truly planktonic, that is to say, living freely in water and completely cut off from the substrate. In shallow water it must be observed, however, that essentially benthic or bottom-living species not infrequently approach the surface in a rather similar manner, most often perhaps, like the palolo worm and other polychaetes considered in the last chapter, at spawning time. Russell (1928*a*, p. 93) records the appearance of the primarily bottom-living amphipod *Ampelisca* at various depths in night and dawn plankton-hauls. Such animals as these could no doubt avail themselves equally well of either the surface or the bottom for epideictic purposes. The whirligig-beetles (Gyrinidae), for example, live in extremely shallow fresh water and resort exclusively to the surface for their gregarious social displays (*see* p. 76). The amphipods mentioned, on the other hand, may merely be swimming up independently at a particular period of the 24 hours in order to feed, and their population-density may still be related basically to the sea-floor. Such situations as these tend to obscure the fringes of the migratory habit, but it is sufficiently clear that substantial long-sustained diurnal movements of the kind that concern us here are a general and characteristic feature of the plankton and of no other ecological category of aquatic animals.

It is self-evident that the habit must be confined to the so-called photic zone, since at depths below the farthest penetration of visible light there is no indication of the sequence of night and day, and probably no environmental factor capable of being used for any short-period or precise synchronisation at all. Social integration in any form undoubtedly poses especially formidable problems for bathypelagic animals. Welsh, Chace and Nunnemacher (1937) found that where the water is sufficiently clear the habit of vertical migration may be carried to an immense depth. The limit no doubt depends on the transparency of the water and the maximum depth to which a biologically significant amount of sunlight can penetrate: this appeared likely to be about 1,000 m in the Sargasso Sea, well known for its great clarity. In that area the blind acanthepyrid prawn *Hymenodora glacialis* was never taken above this depth (loc. cit., p. 191), though over the continental slope 300 miles off the coast of New Jersey, immature specimens were later found to be present up to the 800 m-level (Waterman, Nunnemacher, Chace and Clarke, 1939, p. 263). There was no convincing evidence of vertical migration in this species. Incidentally, it would be interesting to know how a blind bathypelagic species could determine the depth at which it was swimming and avoid ascending above, say, 1,000 m. Possibly ‘ vestigial eyes ’ may sometimes be retained just for this purpose, and be able to detect a limiting light-threshold near the lower limit of the photic zone. Some planktonic species may, however, have direct depth perception (*see below*, p. 376).

The investigation of diurnal movements of animals in deep water presents serious technical difficulties, but these same authors were able, by taking a series of hauls with 2-metre closing nets, at depths of 400 and 800 m simultaneously, spread at intervals of 3 to 4 hours throughout one diel period, to show clear indications of migratory movements in some of the sergestid and acanthepyrid prawns, and also in deep-water copepods and fishes. The fishes they concluded, as stated earlier, were probably only shadowing the movements of their copepod prey. Similar results were obtained in the later investigation over the continental slope, with a series of hauls down to 1,000 m.

Their findings had been partly anticipated by Murray and Hjort (1912, pp. 664-8), who compared the numbers of some of the same and similar species taken by day and by night on different dates and at different levels: their analysis showed that sometimes deep-water prawns could be caught (more often at night) in the uppermost nets within 50 m of the surface. This is a remarkable habit in animals otherwise closely identified in their adaptations and affinities with the dimly-lit mid-water depths of the ocean. It does not, however, seem to be particularly rare: perfectly co-ordinated and ' normal ' vertical migrations, from as deep as 500 m at midday right to the surface at dawn (2-6 hrs.), have for instance been demonstrated in *Euphausia superba* (Fraser, 1937).

It is evidently not necessary for the upward movement to culminate at the actual surface itself. To make direct visual observations of social behaviour under water at sea is extremely difficult and in fact, as we shall see, the few that have been made have necessarily been at or close to the surface; but as long as the participants are able to determine their depth accurately, aggregation or condensation can take place at any mutually pre-determined level. Two possible advantages in approaching the surface fairly closely for purposes of display might be, first, the increasingly precise synchronisation possible on rising towards the top, where the amount of light and speed of change at dawn and dusk are at a maximum; and second, the better conditions obtainable there for visual perception, when the ensuing demonstration depends on seeing the other participants by daylight.

It has been suggested of a number of different plankton species that they have each a particular characteristic or optimum light-intensity, which they attempt to follow up and down in the water as it undergoes its diurnal cycle. Several apparent examples have been collected by Cushing (1951, pp. 175-8), but it will be sufficient to mention only one of them here, namely the demonstration by Russell (1928b, p. 435), over a period of four months, of a close correlation between the day-depth of *Calanus* females and the varying penetration of light, affected as it is by the changing weather and transparency of the water. In their important experiments on *Daphnia*, in which the scale of vertical movement was reduced to the dimensions of an aquarium tank by the simple expedient of mixing a sufficient quantity of indian ink with the water, Harris and Wolfe (1955) showed that the animals are little affected by transient changes in light-intensity, such as might result from a cloud passing across the sun, but adjust their depth to slower gradual changes which more closely simulate the natural cycle of daylight. A *Daphnia* rising in response to a decreasing illumination from above is, of course, making a photopositive response, since it is moving towards the light, and one sinking as the light gets brighter is responding photonegatively. To these experimenters (loc. cit., p. 353) 'the particular value of the alternation of photopositive and photonegative phases appears . . . to lie in the possibility which it offers of the animal maintaining an equilibrium position which is related to an absolute light intensity. This revives the idea of a real "optimum" light level, for which there is much evidence in the field, and provides a mechanism for responding to the slow changes in light intensity which are the essence of the diurnal cycle.'

In relation to animal dispersion it is not difficult to suggest a possible advantage in the tendency to remain more or less permanently concentrated at a particular light-level. The situation so produced could be regarded as a special type of gregarious behaviour, by which the population broadly disseminated in two dimensions is aggregated in the third, with a corresponding gain in social integration. In the course of each day the migrating shoal sweeps up and down through the whole range of depth of their available habitat. It has recently been shown (Hardy and Bainbridge, 1951; Knight-Jones and Qasim, 1955) that some plankton animals are in fact directly sensitive to pressure-changes, although the means of perception used is still unknown, and they can thus no doubt determine their absolute depth quite independently of light-intensity; decapod larvae, ctenophores and certain hydromedusae have all been shown to possess this faculty in varying measure, but such typical vertical migrants as *Calanus*, *Tomopteris* and *Sagitta*, on the other hand, showed no response to experimental changes in pressure.

This tends to confirm the likelihood that in these animals position in depth is measurable only in terms of light-penetration, and that concentration of their numbers on a particular level or horizon must generally depend on seeking the appropriate light-intensity.

Some illustration needs to be given of a situation well known both in fresh water and the sea, in which the vertical movements terminate upwards or downwards at a well-marked thermocline or temperature-discontinuity. Most often the animals are found to ascend from their deep-water daytime position as far as the thermocline and no farther. Clarke (1933, p. 428), for instance, found that *Metridia lucens* continued to migrate towards the surface after dark, reaching their uppermost level just before dawn, but 'the zone of maximum abundance never rose above 18 meters and very few individuals were taken at the 6-meter level. The stratum of warm water (15° C.) present at the surface, sharply delineated between 10 and 20 meters from the cold water (5° to 8° C.) beneath it, undoubtedly played a large part in deterring the animals from moving much above the 18-meter level.' In this case the strong thermocline appears to be substituted for the actual water-surface as a 'ceiling' under which the population gathers.

16.5. *Synchronisation phenomena*

Vertical migrations do not always follow the accepted standard time-table. Not very rarely the highest numbers of some particular species or stage are taken in the surface hauls in the middle of the day, even in brilliant sunshine. This has often been observed in *Calanus*, especially between April and June, and sometimes, as we shall see below, it is associated with a visible epideictic display. Gardiner (1933, p. 598) found off the Northumbrian coast that the timing of the migratory cycle was practically reversed at this season of the year, not only in adults but in stages III, IV and V also. We shall return to a description of the displays in the following section (p. 382).

Something of value may be learnt from the apparently more trivial departures from the 'normal' routine, which are much more frequently observed. It has generally been broadly assumed as a normal rule that the migrating animals follow the declining light upwards in the evening, until the onset of darkness finally extinguishes the stimulus, after which purposeful upward movements cease and scattering or sinking results. At the first light of dawn, the animals are supposed to be attracted back again towards the surface in search of their optimum intensity once more; and, having attained it, to follow it down as the day progresses, completing the cycle. This hypothesis was first outlined by Michael (1911; *cf.* Russell, 1927a, p. 237) for '*Sagitta bipunctata*' in the sea off San Diego, California.

Certain other workers have found that their figures did not necessarily conform with this in detail, especially in the period between dusk and dawn. In Lake Lucerne, for instance, Worthington (1931, p. 429) showed that the main upward movement continued after dark, when the animals could no longer be simply following an optimum intensity, or indeed any photic stimulus. Studying the vertical movements of female *Metridia lucens* in the sea Clarke (1933, p. 428) found exactly the same thing: 'the observations,' he says, 'consistently indicate that the animals did not tend to become uniformly distributed after complete darkness but continued to migrate toward the surface, reaching their highest level just before dawn.'

This becomes more significant in the case of the copepod *Eurytemora velox*, studied in Lough Derg in Ireland by Southern and Gardiner (1932, p. 149): here not only did the maximum percentage appear in the top two metres before dawn, but there was no comparable concentration to be found at any other hour of the day. At all other times the population was, by comparison, widely scattered through the water column (*see* fig. 32). There seems to have been a definite, clearly-defined aggregation of the population at or before dawn, at this particular level. One is inevitably reminded of the many examples previously encountered of auroral demonstrations or activity-peaks in birds, fishes and insects, believed to have an epideictic function (*see* especially Chapter 15).

Just as in these other groups and environments, moreover, examples can be found in which surfacing is exclusively an evening phenomenon, and has no counterpart in the early morning. At certain times and places this is true of *Calanus finmarchicus*, and Nicholls (1933), in particular, found it to occur in the Clyde area. It was very well marked there in Stage IV copepodites in July: 'there was a distinct movement towards the surface before the sun

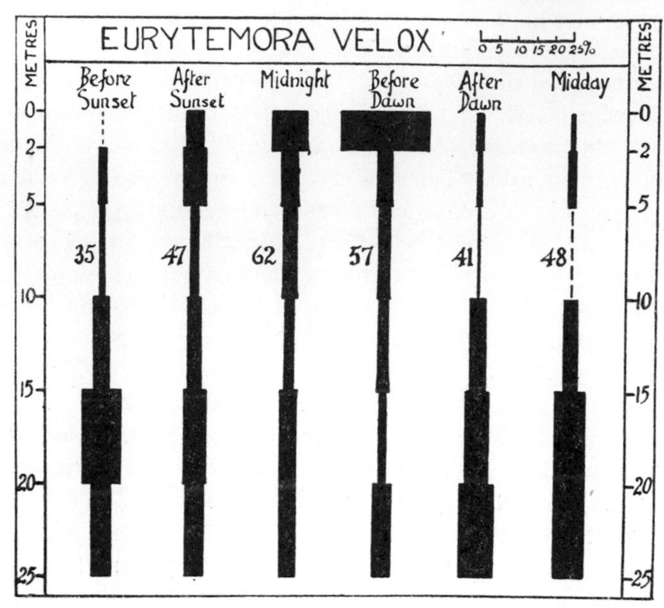

FIG. 32. Vertical migration of *Eurytemora velox* in Lough Derg. The histograms show the proportions of the total catch at each of six depths. Note the marked concentration at the surface before dawn. (From Southern and Gardiner, 1932.)

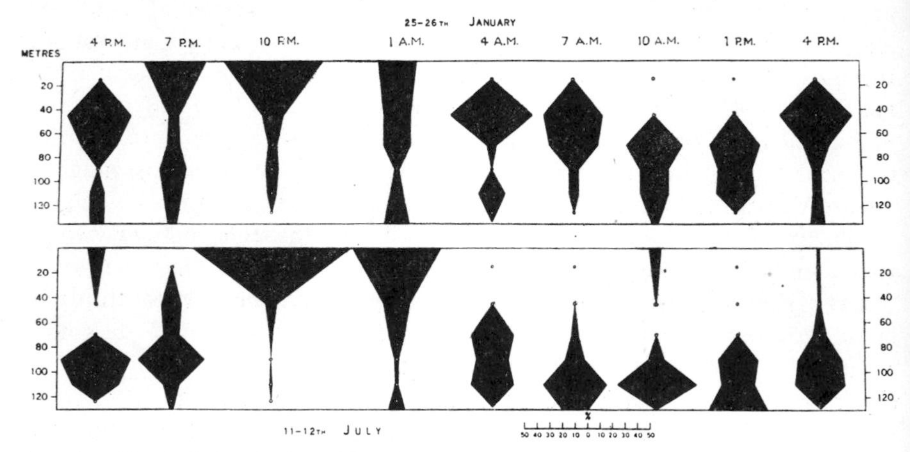

FIG. 33. Vertical distribution of female *Calanus* at 3-hourly intervals in Loch Fyne. Upper set, 25-26 January; lower set, 11-12 July. The massing in the upper layers occurs in the late evening, and soon after midnight they are leaving the surface. Sunset 16.27, sunrise 8.37 in January; sunset 20.25, sunrise 3.58 in July. (From Nicholls, 1933.)

had set and by 10 p.m. they were distributed in the upper layers. By 1 a.m. they were leaving the surface and at 4 a.m., by which time the sun had risen, very few were found in the 30-metre to surface layers' (loc. cit., p. 149).

Very much the same was true of the females, although the numbers taken were smaller: in January the massing near the surface was long after dark, with a maximum about 10 o'clock, but they had entirely dispersed by 4 a.m.; and in July a similar surface concentration occurred at the same hour, at this season just about nightfall, whereas by sunrise (4. a.m.) the gathering had disbanded, and all were found to be evenly distributed between about 50 and 100 m (loc. cit., p. 153) (*see* fig. 33). Practically the same has been observed off the coast of California, with surface concentrations only in the earlier part of the night (data from Farran, 1947 and Esterly, 1912, p. 283,

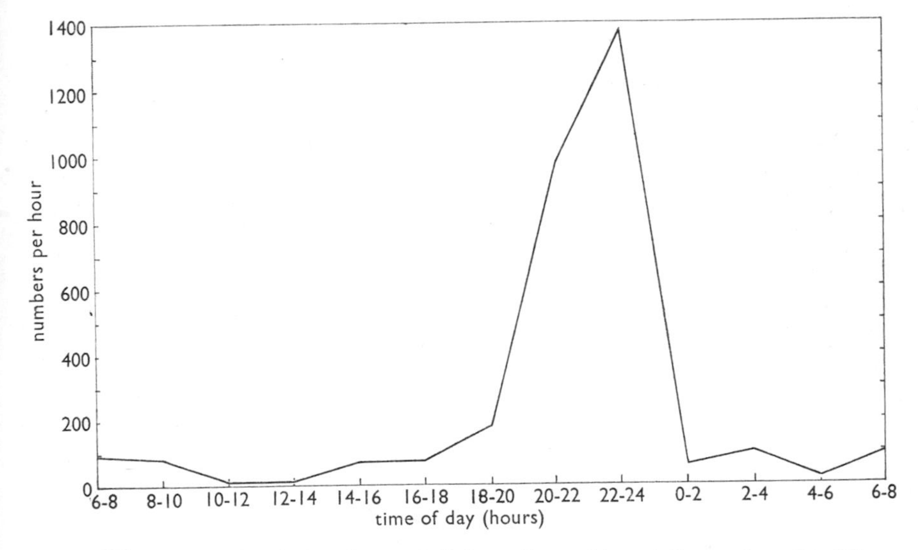

FIG. 34. 24-hour cycle of abundance of *Calanus finmarchicus* at the surface, San Diego region, California. The peak comes between 8 p.m. and midnight, and the species is virtually absent after 2 a.m. (*cf.* fig. 33). The curve is based on 117 successful net-hauls in June-August, from which >20,000 specimens were counted. (Redrawn from Esterly, 1912.)

respectively) (*see* fig. 34). Halme (1937, p. 363) similarly found a pronounced peak of numbers of ascending copepod nauplii in the surface layer, culminating at about 22 hrs and declining rapidly even before midnight; this was on the south coast of Finland in July.

Fig. 35 presents the results of Ruttner's (1905) work in the Grosser Plöner See near Kiel, and shows three differently-timed surface maxima: in *Hyalodaphnia karlbergensis* there appear the classical double peaks at dusk and dawn, though the former is the more pronounced: in the cyclopids (predominantly *Cyclops oithonoides*) there is a single dawn peak, and finally the rotifer *Conochilus*, and in the calanids *Diaptomus* and *Eurytemora* combined, have single midnight peaks. The resemblance between such curves as these and the figures obtained, for example by C. B. Williams, for the numbers of different species of insects caught at a light-trap (*see* p. 343) is very striking; and it is undoubtedly tempting to conclude that the parallel

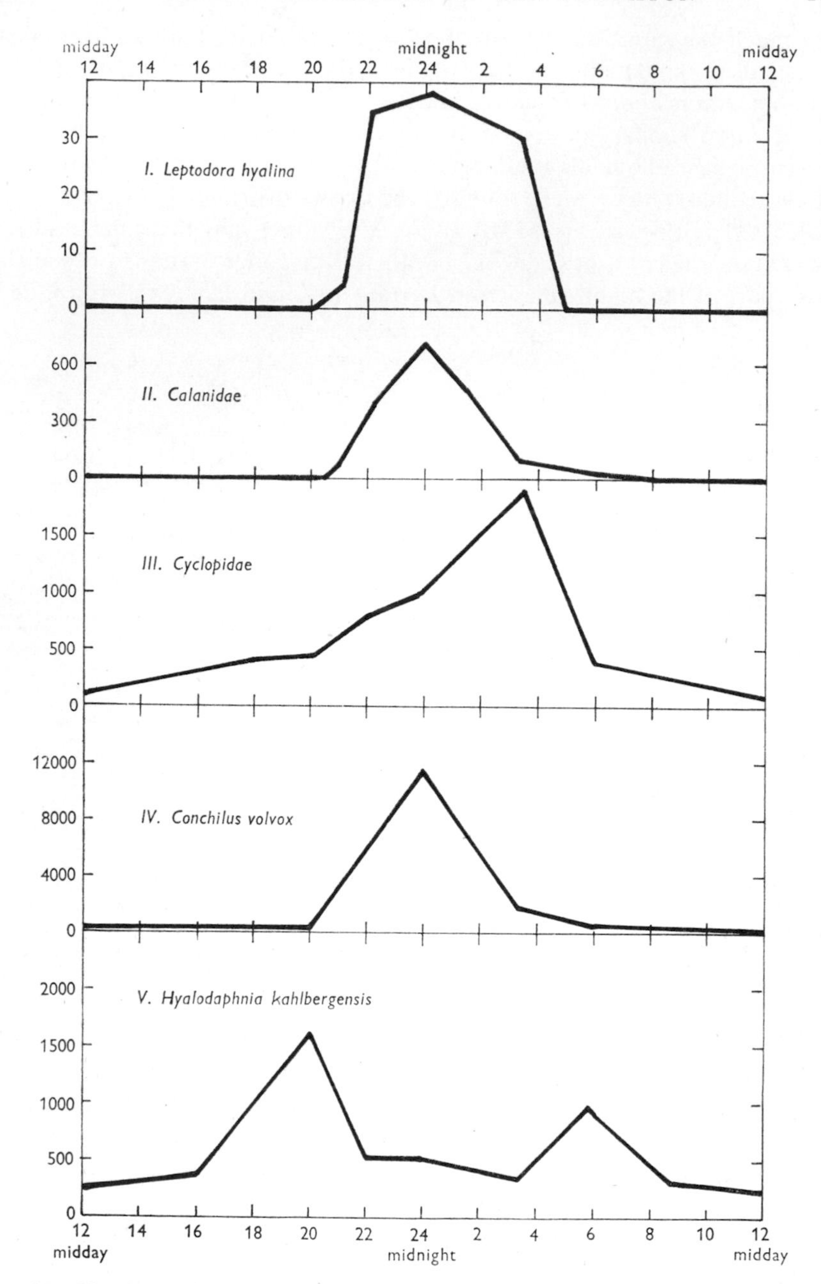

FIG. 35. Numbers of four freshwater planktonic Crustacea (I, II, III, and V) and one rotifer (IV) in surface layer in Grosser Plöner See near Kiel, plotted against clock time. Three of them show single midnight peaks, one a dawn peak, and one a classical double ' dusk and dawn ' peak. (Redrawn from Ruttner, 1905.)

series of synchronisation patterns found in the two have been evolved for the same purpose.

In the last chapter it was noted that often in the vertebrates and sometimes in the insects physiological rhythms are established in relation to diurnal cycles of behaviour, and that animals experimentally transferred into uninterrupted darkness or light will for a time continue to follow a customary time-table of activity in the absence of any external stimulus. The existence of such a rhythm of migration was demonstrated for *Calanus* by Esterly (1919, p. 37); in *Daphnia* on the other hand it appears that no rhythm is developed, since the animals can at any time be immediately induced to adapt their behaviour to an artificial day-length cycle, with a period even as short as 2 hours (Harris and Wolfe, 1955, p. 335). Internal rhythms may well be essential, however, to sustain the migratory movement after dark in species which culminate at the surface during the night or at dawn, or in those that reassemble at the surface after midnight sinking, in anticipation of the dawn; attention was particularly drawn to this last situation in copepods by Welsh and his collaborators (1937, p. 193)

16.6. *Evidence of social behaviour*.

It is a most characteristic feature of plankton populations, first clearly established by Haeckel, that their distribution is seldom uniform: a ' Hardy continuous plankton recorder,' for instance, towed horizontally at the same depth for a number of hours generally reveals a more or less pronounced patchiness from place to place for each species separately (*cf.* Hardy, 1936). Apparently all the main planktonic groups from the Protozoa upwards are affected in this way, and so also are the plant species. More than this, the patchiness of the zooplankton gives the impression of not being a completely random affair, and to Hardy (loc. cit., p. 537), who studied this phenomenon very closely in the Antarctic, there appeared to be an almost rhythmic repetition of high and low density as the continuous recorder progressed along its course. ' On reviewing the records as a whole,' he wrote ' it is difficult not to be impressed by the suggestion of an apparent rhythm in the fluctuations; a rhythm which is not constant for different groups, but one which appears to have a definite pattern on different occasions. . . . If [the fluctuations] were due to the action of the sea alone one would expect all organisms in the same area to fluctuate to the same " rhythm "; we see, however, different organisms fluctuating on the same records with their peaks at different intervals apart, suggesting that these differences are due to differences in the behaviour (vertical migrations ?) of the animals in relation to the water.'

Patchiness appears to be a positive and fundamental feature of plankton distribution: it suggests at once the possibility that, just as in terrestrial or benthic habitats, plankton species also tend to be organised into more or less discrete populations, the members of each one of which are socially integrated. So universal in animals is this tendency to subdivision and to the partial segregation of discrete groups or colonies, that it would not be in the least surprising to find it had somehow been carried over into the pelagic habitat, by the species which pioneered this way of life.

According to the interpretation which follows from our general theory of animal dispersion, subdivision into discrete populations marks a very important advance in the evolution of social organisation; and social organisation in its turn is valuable beyond estimation as being the key to attaining homeostatic control of population density. It is equally the basis of group-selection, on which further social evolution itself depends. If this is so, subdivision leading to ' patchiness ' has probably been evolved over and over again, therefore, by planktonic species, notwithstanding the formidability of achieving such a goal with relatively feeble creatures in a seemingly infinite medium like the sea.

The existence of patchiness indeed confirms the previous conclusion, to which we were led from the study of vertical migration, that social integration exists among plankton animals; and in consequence, under suitable circumstances, epideictic displays and social competition ought to be expected. They may be looked for as a prelude to reproduction, possibly in this case only involving the dominant sex, and also at other times: since they are density-sensitive in their occurrence they are likely to be most intense where populations are highest.

The most commonly recorded phenomenon that could be epideictic in nature is generally referred to as swarming. There have been many cases of this recorded, both in fresh water and in the sea, but as usual the much-studied *Calanus* provides the largest number of examples. Swarming of such diminutive animals would not generally be seen by an observer unless it occurred in very calm weather and sheltered waters, under conditions of optimal lighting, that is to say in the middle of the day—though it may actually be much more frequent, for instance, at dusk or dawn. *Calanus* sometimes reaches a very high density: in the words of Brook (1886, p. 48), ' in May and June they occur in myriads [in Loch Fyne]. At this time immense floating masses are to be seen in calm weather, which give the sea quite a red tint; ' the swarming individuals ' are so rich in brilliant red fat globules, that each shoal may be readily seen from a boat.' Marshall and Orr (1955, p. 127) state that ' the congregation of enormous numbers of *Calanus* right at the surface during the summer is a very striking phenomenon especially since it is opposed to their usual behaviour [*i.e.* of migrating to deep levels during the day]. It is a common sight in the Clyde on calm days during May and June to see the surface of the sea covered with small circles of expanding ripples like those caused by raindrops. These are usually caused by copepods, often *Calanus*, bumping against the surface film. The behaviour is not limited to dull days but may occur even in bright sunshine. It has been observed by many workers in many different places.' They mention the century-old identical observation made near the Isle of May in the entrance to the Firth of Forth by Harry Goodsir (a gifted naturalist who perished in arctic Canada a few years later with Franklin's last expedition): he noticed that 'the surface of the water presented a very curious appearance, as if a quantity of fine sand were constantly falling on it. I thought at first that this last circumstance proceeded from rain, but presently I found that both phenomena were caused by a great number of small red Entomostraca, which I had never before observed in such abundance' (Goodsir, 1843, p. 103): these he carefully described and identified as what we now call *Calanus*.

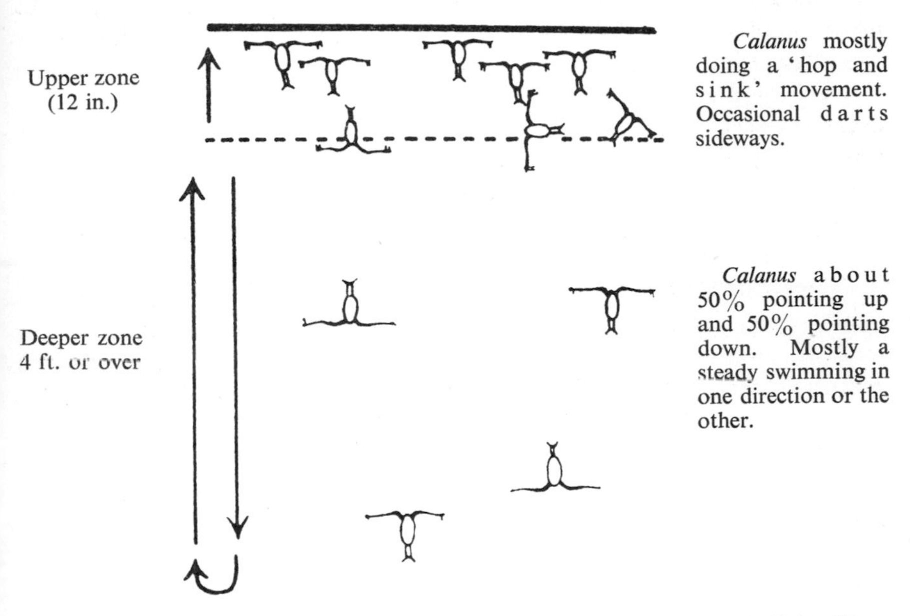

FIG. 36. Observed behaviour of *Calanus* in surface water layers during daylight. (From Bainbridge, 1952.)

These observations strongly suggest that during the surface swarming some kind of epideictic engagement or demonstration takes place. Fortunately it is unnecessary to leave the matter at this point, since Bainbridge (1952, p. 109) has actually watched the behaviour of *Calanus* under water and seen them bumping against the surface film from below. His observations were made in the spring months, in the Clyde waters off Millport, at a time when *Calanus* was abundant in the surface layers throughout the 24 hours, though not in particularly dense swarms. From his description it would appear that the display was considerably less vigorous than those described in the previous paragraph. 'Two zones of differing behaviour are clearly recognisable in the spring *Calanus*. In the upper 12-in. of water there is a high concentration of animals (up to as many as 10 per sq. ft. of surface) and there is a continuous gentle sinking and swimming up again vertically within this zone. Occasionally violent oblique or horizontal darts of several feet may occur, *especially when two or more animals come close to each other* [my italics]; and a good deal of horizontal movement may result from a sort of bouncing on the surface film. The Calanus in this zone are often aggregated into groups of a dozen or so which sometimes *swim round and round each other like a group of mayflies* ' [my italics]. Below this surface zone was another of unknown depth in which the individuals were scarcer and appeared to be swimming up or down in about equal numbers, as if travelling between the surface layer and another concentration deeper in the water—perhaps at 50-60 m, where tow-nettings showed that large numbers were present. ' An outstanding feature of all the populations observed was that, at any one time, as high a proportion as 50 per cent would be quite motionless, most of these remaining so for long periods and many so delicately balanced as not to be even sinking in the water. On some occasions, especially when the sky was overcast, the whole population would be hanging motionless or drifting passively ' (loc. cit., p. 110). Only once a clear horizontal movement was seen to take place by means of a series of oblique leaps and looping movements: there were few *Calanus* about at the time and the two seen made rapid progress.

There can be very little doubt that the first part of Bainbridge's description refers to an exhibition of social behaviour of just the same kind as we have encountered so many times before, and have repeatedly identified as an epideictic display. The sex and age of the participants in this case is not known: they need not necessarily have been adults at all, of course, since the dispersionary machinery is often set in motion in larval or adolescent populations. In so far as *Calanus* may have prenuptial displays of this general character, however, we may remember that in copepods the female appears not infrequently to be the dominant sex: it is the female *Calanus* for instance that performs the vertical migrations *par excellence*, as we have noted already; and in those remarkable cases, referred to in the chapter on sexual selection (*see* p. 242) of copepods in which the adults develop gorgeous coloured or irridescent plumes, it is most often the females which excel whenever the sexes are dimorphic. There is another possible link here, therefore, between the development of sexual dominance or adornments on the one hand and the relative constancy and perfection of diurnal vertical migration on the other.

A most interesting point is that copulation and the transfer of the spermatophores to the female apparently occur at an early stage, very soon after she has undergone her final moult. Males are at almost all times much scarcer than females, in part because their life is shorter. Egg-laying appears to start about a month after fecundation, and the females continue to live for some time after that (*e.g.* 74 days after laying commenced in the laboratory in one case, Marshall and Orr, loc. cit., p. 32). ' Egg production often takes place in a series of bursts, each lasting about a week. These . . . tend to recur at intervals of about two weeks . . . ' (*the same*, p. 33). The female, fecundated for life, fertilises the eggs from a sperm sac as they are laid. Interestingly, from our point of view, ' it seems as if females, though ready to lay eggs, do not necessarily do so but may hold them up until conditions are favourable. If this is so, it may explain why in the sea egg-production in *Calanus* often seems to coincide with diatom increases ' (*the same*, p. 36). The machinery is there, clearly enough, for regulating the reproductive output, and it appears to be used.

There has been reason to mention once or twice already the fact that differences in ' social behaviour ' are very commonly observed between different age-groups of planktonic animals, just as they may be between larvae and imagines in insects or between asexual and sexual stages in many other animals. In *Calanus* as a rule the first three post-larval or copepodite stages show little or no vertical migration, usually remaining near the surface all the time. Nicholls (1933) found that in the Clyde area copepodites of Stage IV, however, had a marked migration, with a late evening culmination in the upper layers: long before dawn, as we noted earlier, they were on their way down again. Stage V, in complete contrast, stayed massed in deep water the whole time, winter and summer: in July a few did come to the surface at night—possibly those, Nicholls suggested, about to moult and become adult. The adults, as we know, especially the females, are in most places outside the Arctic regular diurnal migrants.

Thus a considerable measure of independence in social regimen characterises the successive stages in the life-cycle, pointing to the probability that population-density can be regulated stage by stage.

The same no doubt occurs quite generally in planktonic Crustacea: Mackintosh and Wheeler (1929, p. 354) record that in one Antarctic season there was a high proportion of unusually small euphausians in the food of the blue and fin whales, and that the whales' stomachs consequently contained a more than usually mixed bag of sizes. These were not mingled indiscriminately but were assorted in different parts of the mass of stomach contents, ' suggesting that the whale had been feeding on separate shoals which differed in respect of the sizes of the individuals.'

The euphausians are among the most spectacular members of the plankton, and in Antarctic waters, where they are known as ' krill ': they often occur in patches containing prodigious numbers and attract a great variety of predators, including especially the whalebone whales. This is one of the dangers of aggregation: both Goodsir, and Marshall and Orr, incidentally, comment on the numbers of predatory fish lurking below the swarming *Calanus*.

To lend weight to this epideictic interpretation of plankton swarms we may give one further example by quoting the description of a day-time swarm of *Euphausia superba* (which takes two years to mature), seen by Hardy in Cumberland Bay, South Georgia, in early December, 1926: it is taken from his journal (*see* Hardy and Gunther, 1935, p. 210).

' For a whole day there was a dense swarm, like a red cloud, of closely packed Euphausians (*Euphausia superba*) against the jetty at our shore station. There must have been thousands and thousands in a close swarm some four feet across. They were all swimming hard and going round and round, sometimes in a circular course, sometimes in a " figure 8," but never breaking away from the one mass. The cloud would sometimes change shape, elongate this way or that. . . . At times they would form into two such moving parties and one would tend to separate from the other, so that the swarm became dumb-bell shaped; but as soon as the connecting link became of a certain thinness the one part would turn back and flow into the other to form one big swarm again. It was drawn into the whole like the pseudopodium of an amoeba; indeed the whole swarm appeared to behave as one large organism. . . .'

There is no mistaking the resemblance of this swarm, made up as it was of non-sexual individuals, to the now-familiar spectacle of an epideictic mass-display: even the comparison of the milling shoal to an amoeba has appeared before, though in totally different surroundings (*see* p. 286).

16.7. *Conclusion*

The evidence seems to justify the suspicion that we are dealing with another set of variations on the dispersionary theme: indeed, in concluding that animals living in the planktonic world hold epideictic gatherings and take part in regulating their own numbers, we are only extending to a new environment what we have found to be a general characteristic of the higher animals everywhere we have looked for it.

The subject this time is one that has attracted an unusual amount of speculation and theorising in the past without leading to any general agreement or conviction, and this is bound to instil caution and scepticism towards any new and sweeping explanation such as is offered here. It is characteristic of the dispersion theory, however, not to come up against generally agreed and well-established principles, but rather to suggest simple explanations for those very phenomena of group behaviour for which hitherto it has been particularly difficult to find adequate and acceptable functions. In the present case there seems to be nothing incommensurate between the vast effort spent—day in, day out—by the myriad millions of the zooplankton in seas and lakes all over the world, and the vast advantage supposedly gained, namely a general ecological integration and balance, with each species controlling its own population-density and protecting its vital resources from permanent harm. Here at any rate all four of the most familiar pieces of the plankton jigsaw puzzle—patchiness, swarming, vertical migration and the diurnal cycle—appear for once to fall naturally into place.

16.8. *Summary of Chapter* 16

1. Vertical migration which follows a diel cycle is found in practically all groups of plankton animals, and has thus evolved many times: above all it is characteristic of the small Crustacea on ' Entomostraca.' Great energy is expended on it, carrying the participants up and down through, in some

cases, tens or even hundreds of metres daily. There is no generally accepted current explanation.

2. Hardy has suggested that it enables the animals to sample different feeding grounds, by descending into a water-mass travelling at a different speed and direction from the surface layer; so that the following night they reascend to feed at the surface in a new place. This appears not to be the principle purpose of the habit, however, since it is clear that the primary connection is between migration and the daylight cycle, and sometimes it affects males, females, adolescent or young stages differently, and quite independently of their nutritional requirements.

3. It seems not at all unlikely at first sight that the migrations could have an epideictic function, by aggregating all the animals scattered in three dimensions through the water column into a thin layer at the surface, at a particular time each day; the surface being, for purposes of assembly, by far the most easily identified position in the habitat, as well as being the one at which light-changes are most pronounced and visibility (by reflected light) best. This hypothesis is therefore examined further.

4. Vertical migration is characteristic only of truly planktonic species: it does not occur below about 1,000 m, which is about the limit of effective penetration of daylight. The movement does not necessarily culminate always at the surface, but may stop at a thermocline: in some cases the species may apparently concentrate also at a particular light-level during the day.

5. The ' normal ' pattern of movement entails swimming up to gather at or near the surface in the evening, dispersing again during darkness, and reassembling near the surface at dawn: after this the animals descend to their ' day-depth.' The use of dawn and dusk as synchronisation times for social gatherings is already very familiar to us; and, very significantly, the available data here show that in fact some species have only an evening peak at the surface, others only a morning one, and still others a single peak during the night. The parallel with ' activity-cycles ' in insects, for example, is very striking.

6. It is suggested that the well-known ' patchiness ' of plankton distribution (in relation to area) is not an accident, but is the effect of the normal subdivision of animal species into more or less discrete populations. This subdivision is prerequisite to proper social organisation and consequently it promotes homeostatic control of numbers. Social behaviour is indicated in the plankton by the common phenomenon of swarming; and in *Calanus* and *Euphausia* details are given of what appear to be normal and typical epideictic displays. In *Calanus* as in some other copepods the female is the dominant sex and performs the daily migratory exercise more punctiliously than other castes in this genus. She is fecundated early in the season and can control the emission of eggs over the next few months. Segregation of age-classes is not uncommon in planktonic Crustacea.

7. The conclusion is drawn that the vertical migrations are epideictic in

origin, analogous, for example, to the daily roosting flights or collective 'dances' in various birds or insects; and that in fact the dispersion theory provides a workable interpretation of a large field of animal behaviour hitherto unexplained; that it unifies such characteristic features as patchiness and swarming with vertical migration; and that in so doing it is not in conflict with established principles.

Chapter 17

Associations between species

17.1. *Introduction*

The relationships that link the fortunes of two or more species together are very varied: in symbiosis, for instance, two partners may live together entirely to their mutual benefit. In the relation between the predator and its prey, on the other hand, or the parasite and its host, the advantage seems to be all on one side. The boundary, incidentally, between what is predation and what parasitism tends to become blurred when we are dealing with, say, the Ichneumonidae, or the wasps that ' provision ' their larvae (*cf.* Richards, 1953, p. 42). The burdens imposed by predators or parasites are normally capable of being borne without impairment to the prosperity of the group or, in the case of parasitism, of the individual upon which they fall: by the time the relationship has evolved to balanced maturity the host has often become adapted to tolerate the minimised injury caused by the parasite, just as the prey population becomes geared to cater for the regulated demands made upon it by its predators.

Compromise and adjustment in the interest of the other party are the essential ingredients of all such continuing relationships, for failure to reach a harmonious stable balance implies a situation damaging to one and possibly both of the parties involved, leading most likely to instability or deterioration which sooner or later brings the relationship to an end. Even in cases where one side gives and the other one takes, without any reciprocal benefit, there will generally have been selection acting on both sides to bring about an efficient and stable relationship.

On purely theoretical grounds Volterra (1928, p. 9) reached his well-known conclusion that two species, one of which feeds on the other, must undergo perpetual oscillations in numbers, because, he assumed, the predator population would inevitably increase until it overate its food supply, thereafter itself declining in consequence, thus giving the prey a chance to build up its numbers once more and so on in successive cycles. What he did not take into account is that it is immensely more efficient for the predator to conserve the stock of prey at a maximum all the time; and that consequently selection will quickly provide the group with a safeguard system of conventional tenure to prevent ' overfishing ' and eliminate the cause of Volterra's wasteful if not exceedingly dangerous oscillations.

In practice stability between predator and prey populations, as in other interspecific associations, seems to be the general rule, to be departed from only when the stabilising machinery is overtaxed. It depends on the demand by the user being adjusted in size to the expendable surplus of the giver; and also in some measure on the yield by the giver being stepped up to a maximum under the insistence of the user's demand. This situation indeed embodies, as has frequently been emphasised, one of the principal elements of the present theory.

There are certain associations between two species in which one party appears to receive some valuable concession virtually without cost to the donor: think, for instance, of the pilot-fish (*Naucrates ductor*) attending a cruising shark, or the small gadoids and amphipods that associate with pelagic jellyfish. These if we are not mistaken use the larger animal chiefly as a home or focus, tying their own community life, dispersion, and population-density to that of their impassive host, which may be as little affected by it as are the stones or trees that perform a similar function for benthic or terrestrial species (*see* p. 166).

Bearing in mind the possibility of evolving such harmonious associations as these, and allowing also the general development in all the higher animals of systems for limiting population-density through social conventions, the question next arises whether there are any associations of species that have evolved a kind of confederation or syndicate, under which they submit themselves to a common dispersionary regime: can there, in other words, ever be two (or even several) species that are jointly dispersed as one, under a single social system? The evidence is plain, as we shall see, that such a thing is not only possible but common.

The working hypothesis that every feature in the behaviour, physiology and structure of an organism is likely to have a positive use and value, could we but discover it—unless, of course, the contrary can be proved—is the product of long and illuminating experience; and it is with this in mind that we must turn our critical attention to the familiar mixed associations that are especially common, for example, among gregarious birds and mammals; and to those groups of sibling species earlier mentioned that live together and look alike, such as the common and arctic terns (*Sterna hirundo* and *S. macrura*), the numerous weaver-, bishop- and widow-birds (Ploceidae), the *Vespula* wasps and *Halictus* bees. Social behaviour *within* the species has so far proved to be primarily directed towards dispersionary ends; and it would not be wholly surprising therefore if some of these mixed social relationships turned out to have some cognate sort of function.

17.2 *Interspecific territoriality in birds*

One of the most obvious links between epideictic behaviour and the regulation of population-density—the one that we chose in Chapter 9 as an introduction to the various systems of property-tenure—is found in ' territorial ' birds in the breeding season. In the simplest cases the male claims a general-purpose territory, and defends both his possession and his social status by singing when his rivals sing and fighting them off if they dare to trespass. Territorial competition may often concern only the males, or, after pair-formation, the females may be more or less drawn into it: whatever the exact situation, it is not particularly difficult to discover by observation which castes in the community are involved and which are being ignored.

A substantial body of evidence has accumulated in recent years that in particular cases territorial rivalry can develop between members of different species. Howard (1920, pp. 220 *et seq.*) noticed that many Old World warblers display irritation when approached by quite different birds, such as hedge-sparrows (*Accentor modularis*) or titmice (Paridae), which happen to be their neighbours or to enter on their ground; and that this hostility becomes especially severe between species that are very closely related, such as the garden-warbler and blackcap (*Sylvia borin* and *atricapilla*). Here the reactions of neighbouring males to one another are ' similar in all respects ' whether the opponent belongs to their own species or to the other one. Blackcaps will often attack whitethroats (*S. communis*) and chiffchaffs as well. Chiffchaffs and willow-warblers (*Phylloscopus collybita* and *trochilus*), birds of unusually close visible resemblance, likewise show a mutual intolerance which he thought to be somewhat remarkable in view of their nearness of kin; and ' evidence of it can be found wherever the birds occupy the same ground ' (*the same*, p. 224).

Howard considered at great length the conditions under which this kind of hostility revealed itself—always between males, only in spring at the time of territorial establishment and only when at least one of the antagonists was occupying his intended claim; and in the end he was ' led to the only conclusion which seems consistent with the facts, namely, that there is a relationship between the " territory " and the hostility ' (p. 242): going a step farther, the true explanation of the hostility seemed to be that ' it roughly insures *that the number of pairs in any given area does not exceed the available means of support* ' (p. 246, my italics). This was clearly intended to be a plain statement of the principle we should define as interspecific dispersion.

Another example of interspecific territorialism of a rather unexpected sort was discovered by Hartley (1949) among three species of chats (*Oenanthe*) living as winter residents in Lower Egypt. Chats, like robins (*Erithacus rubecula*), take up individual territories in their winter quarters and may sing and fight in their defence; in the mourning chat (*O. lugens*), at least, it appears that both sexes sing at this season. This was the commonest species, the hooded chat (*O. monacha*) coming next; *O. leucopyga* was the rarest of winter residents and *O. deserti* occurred on passage during migration only. These birds, especially *O. lugens* and males of *O. monacha*, are all similar in appearance and equally so in feeding habits: in the area under study—an abandoned quarry—they occupied ground individually, and defended and fought off contestants in an apparently indiscriminate manner. In short, it seemed to Hartley ' that only an hypothesis of interspecific territorial practice—with the corollary of competition—will offer an explanation of the observed facts. Had the story of the occupation of the Wadi el Gibbu quarry been told of known individuals of a single species, it would have been regarded as a very normal story of territorial establishment, with the failure of the earlier arrivals to maintain themselves in the face of the later comers as the only unusual feature. The successful maintenance by one of the hooded chats of at least part of its territory throughout the winter suggests that there was something more than the mere replacement of a passage-migrant species—the hooded chats—by a winter resident species, the mourning chats ' (Hartley, p. 410).

Commenting on Hartley's observations, Stresemann (1950) recalled that previous observers had found a rather similar situation among members of the same genus during the breeding season, for example between the common wheatear, *O. oenanthe* and *O. leucomela* (= *pleschanka*) in Southern Ukraine. Next Simmons (1951, p. 408) confirmed the Egypt wintering observations, adding members of *O. leucomela* and *O. oenanthe* to Hartley's previous complex, as well as quite a different bird, a female blue rock-thrush (*Monticola solitarius*). He also found a similar but totally independent relationship obtaining among two species of shrikes, *Lanius collurio* and *L. nubicus*, during the time they passed through on migration—with the consequence that individual firmly-held territories proved to be of quite transient duration as the birds alternately halted and then moved off again on the next stage of their journey.

These last observations interest us particularly for two reasons: the first is the hint they give of the possible use of a pattern of transient individual territories in serving to feed migrants through the country in an orderly and uncongested manner just as we had reason earlier to suspect that a system of established communal roosts could do, for instance in chimney-swifts (*see* p. 292). The second point is that Simmons found some difficulty in seeing how this situation was to be reconciled with Gause's (1934) generally-accepted theory, namely, that no two species with a similar ecology can live in the same area.

A case with different features (and bearing closely on the last point) was studied by Lanyon (1956*a* and *b*) in the two North American meadowlarks, *Sturnella magna* and *S. neglecta*—a notorious pair of sibling species. The former or eastern meadowlark is rather darker and browner; and the western rather lighter and greyer: the sexes are alike, and in spite of having a striking pattern of yellow, black and white on the head, breast and underparts the two species are difficult to separate with certainty in the field, except that the males have completely different songs in the breeding season. They have an enormous area of overlap in the central part of the continent, which extends over the major part of the Mississippi basin. They are common ground-nesting birds that flourish in arable grasslands, pastures and prairies; and Lanyon found that, while there was a widespread tendency for *neglecta* to predominate in the drier, more upland areas and for *magna* to favour the river valleys and more humid vegetation, there were many habitats where they were completely mixed. Their territories, he found, were rather fluid during the nesting season, and subject to 'amoeboid changes in size and shape... as a result of changes in population-density, relocations of female centres, and changes in habitat availability' (1956*b*, p. 486); the males were polygamous, and egg-laying was protracted and unsynchronised. In such communities territorial behaviour was completely interspecific, the two species being dispersed as if they were one. 'A number of fixed behaviour patterns were recognised in the establishment and maintenance of meadowlark territories and these appeared to be identical in both species. Territorial defence and display between males of the two species was as frequent and intense as intra-specific display, resulting in the complete segregation of their territories. Such regularly occurring interspecific territorialism permitted the repeated decoying of resident males of one species with captive hand-raised males of the other' (the same, p. 488).

Simmons had thought he could see in interspecific territorialism a means of eliminating the effect of competition between closely allied species, and thus of aligning his observation with Gause's hypothesis; and to Lanyon also this seemed to be a reasonable view to take; but in fact, as Hartley realised, it cannot be so when individuals of the two species all compete freely together in claiming and holding their territories. It is only after it has won a territory in open competition that each individual can then enjoy the reward of getting what food it chooses without fighting over every meal—a reward earned by all successful contestants without specific distinction. Here indeed are two good species that do not readily hybridise, living in the same environment, holding the property which gives them their livelihood and for which they fight together on the same footing, to the exclusion of each other, under one common code of conventional behaviour: there can be no shadow of doubt that they contravene Gause's rule.

Another pair of sibling species which, in central Baffin Island if not elsewhere, are similarly intermingled on their breeding grounds, are the redpolls *Carduelis flammea* and *C. hornemanni* (Wynne-Edwards, 1952*b*, p. 383). Looking back on this situation in the light of present knowledge, and recalling that all eight occupied nests we found had been used in one or many earlier years, it appears probable that these redpolls are what in Chapter 9 we called nest-site birds (*see* p. 155) as far as property tenure is concerned, and that they were sharing a mutual pattern of establishing sites. Though they sang freely, we saw no fighting between males—perhaps because the population was below saturation: in fact one well-preserved but untenanted additional nest-site was located in our study area. The two species, though visibly quite distinguishable in either sex and not interbreeding in our region (as they have been reported to do locally elsewhere) are apparently identical in voice, habitat and food: outside the breeding season, and even a little way from the vicinity of the actual nest, *hornemanni* mingles or flocks apparently at random with *flammea* and the two behave socially as one. As not uncommonly occurs in these co-dispersed associations, the relative proportions of the two species change apparently fortuitously from year to year in the same place and from place to place in the same year (Wynne-Edwards, loc. cit., p. 381). Taking the broad view, *C. flammea* is the more versatile in its adaptability, extending its circumpolar breeding range far outside the arctic zone into the north temperate region, from which the strictly arctic *hornemanni* is completely excluded except as a winter visitor.

Passing reference can be made next to the bishop-birds *Euplectes hordeacea* and *E. nigroventris*, of which the males were found by Fuggles-Couchman (1943, p. 318) to hold rigid, mutually exclusive territories at one particular locality in eastern Tanganyika; actually two species belonging to the nearly related widow-birds, *Coliuspasser albonotatus* and *C. ardens*, were tolerated on the same ground by the *Euplectes* males. Lack (1953, p. 823), however, in his study of breeding *E. hordeacea*, also in Tanganyika, found that this species seemed to be completely tolerant to another bishop—*Euplectes capensis*—their territories being superimposed on one another's.

As a final illustration of territorial interaction between different species we may take the interesting and unusual situation found in the Anna and Allen hummingbirds (*Calypte anna* and *Selasphorus sasin*), studied in California by Pitelka and Legg. Pitelka (1942) confirmed initially that these hummingbirds do not have normal pair bonds, and found that males and females take up individual territories independently: males take no part in nesting and may chase out breeding females just as they chase any other hummingbirds trespassing on their domains. Later study showed (Legg and Pitelka, 1956, p. 402) that among Allen males there appeared actually to be two types of territories, one for feeding and the other for prenuptial display and mating: though the latter are relatively fixed, the former are subject to frequent changes. The system of tenure is thus rather more complex than usual. Between the two species there is considerable, though far from complete, ecological overlap: they are in vigorous and evident competition with one another in the overlap zone. *Anna* is a resident, non-migratory species in California, occupying territory before *sasin* arrives in the spring, and taking up a larger space individually. Though this excludes *sasin* from places it would readily occupy, it frequently happens however that *sasin* carves out its more modest holding in an outlying part of an *anna's* domain; and occasionally it actually displaces an established male of *anna* completely (Pitelka, 1951, p. 659). The conclusion has been established, as a result of several years of study of the same ground, that the interaction of the two species depresses the population-density of each below what it would be if the other were completely absent; this is true at least of the males, female interaction being more difficult to assess. The competition in Woolsey Canyon where this earlier study was made seemed to be distinctly biased in *anna's* favour (Pitelka, 1951, p. 660), but at Santa Cruz, on the contrary, the three instances satisfactorily observed of interspecific contests all ended in *sasin* males successfully displacing *anna* males (Legg and Pitelka, p. 403). It is a point of interest that hybrids very occasionally occur (Williamson, 1957).

There is nothing novel to the ecologist in the finding that related species compete with one another for resources and affect each other's success: what needs to be carefully noted is that in habitats where one does not completely dominate and eliminate the other but where both can live side by side more or less precariously balanced, selection appears to favour their incorporation into a single social dispersionary unit, in which they compete together for the same conventional possessions.

It is easy to see that this must be so if the food-resources of the habitat are to be conserved, for in order to protect them from over-exploitation one species must as usual transfer the main objective of competition to a conventional substitute, such as territory; and any second species coming along as a genuine competitor to exploit the same food-resource must make the identical transfer, so that the two are in direct competition for the same substitute objective, and thus are made collaborators in protecting their common food-supply.

It seems logical at first sight to conclude that the presence of the second species competing for not-unlimited resources is to the advantage of neither; that it would be better for both if the grounds of mutual competition were removed. If so, there must of course be a strong selective pressure favouring differentiation between the demands and requirements of closely allied species. This will not necessarily be wholly successful, however, since a complete partition of resources is likely to involve narrower specialisation, resulting in the end in tying each species to relatively few kinds of food-organisms, with a corresponding loss of resilience and stability should any of these fail. In practice, because of the undesirability of becoming too narrowly specialised in food and because of the limited range of possible foods available, partial overlap between species remains an exceedingly common phenomenon and is responsible for an elaborate development of social integration between species.

17.3. *Species associations in bird colonies*

As we saw in Chapter 9 there is no clear boundary between ‘ territorial ’ and colonial systems of property-tenure, but the distinction is nevertheless a convenient and practical one. A correspondingly wide range is to be seen in the degree of integration between species sharing the same colonies, from those at one end of the scale that show a pronounced similarity in ecology, behaviour and even appearance, to those that are not apparently competing for any common resources at all.

An example of what appears to be a rather highly integrated association is found between the common and arctic terns (*Sterna hirundo* and *S. macrura*) in colonies where they nest together. There is a substantial overlap in their geographical breeding ranges, including, amongst other regions, the whole of the British Isles. The birds are so much alike in appearance and voice that for the human observer sight-identification is not usually put beyond doubt until they alight on the ground at close range; and when, as often happens, one species greatly predominates in a colony, the other can very easily go undetected. Much of the published information on distribution must for this reason be accepted with reserve.

Early in the season Arctic terns take up pairing territories (J. M. Cullen, unpublished), and the same is likely to be true of the common tern. After pair-formation these territories are abandoned and a completely different system of nest-territories supercedes them; individual common and arctic terns are both known to return in successive years to nest on almost exactly the same spot, or within a matter of a few feet of it (Austin, 1949, and Cullen, 1957). There seems to be a fairly well-marked tendency for each species to form pure colonies on its own, though on the east coast of Scotland, for instance, mixed colonies are common. Only a careful and critical scrutiny will reveal the exact disposition of nests in a mixed colony (since the eggs are indistinguishable). On the Isle of May, where the observer can study the situation at leisure from a hide, arctic terns' nests (the minority here) are seen to be well mixed among those of the common tern. Dr. W. J. Eggeling, who has studied them very closely there, confirms that while the arctics' nests are not scattered at random among the others, there is no more than a rather slight tendency towards grouping and no definite partition of the breeding area between the two species. At Forvie, Aberdeenshire, the few scores of arctics' nests are mostly in two definite areas and little intermixed with the thousand and more nests of the common terns (Mr. Alastair Smith, orally).

As described on an earlier page, these two (and the Sandwich and roseate terns also) have a recurrent mass-display usually known as a ' panic '—in reality most probably an epideictic ritual. As a preliminary to it many sitting birds rise to join the cloud of others already poised in random flight high above the colony, and their harsh insistent cries swell to fill the whole air with noise. Suddenly there is a hush; and in dramatic silence all swoop in unison swiftly out towards the sea, but only for a matter of seconds before they turn back to renew the clamour. In colonies where arctics and commons are mixed they share these social commotions without noticeable specific distinction.

Tern colonies and populations are subject to apparently fickle changes; and the proportions of each species present in any one colony, such as the Isle of May, often vary substantially over the years (Eggeling, 1955, p. 79), just as they change from one colony to the next in any given year: for instance on three colony sites in the Heligoland Bay area the ratios of *macrura* to *hirundo* were estimated to be 1:5, 1:50 and 1:80 respectively (Grosskopf, 1947).

In the case of the herring-gull and lesser black-back (*Larus argentatus* and *fuscus*), which are perfectly distinct species in north-western Europe but are linked by a series of intergrades in other parts of the palaearctic region, there tends to be an ecological divergence in the respective breeding places, *fuscus* nesting more often inland or on flat moors and *argentatus* on sea-cliffs, islands and skerries: but they are nevertheless often associated at the same colonies and ecologically there remains a substantial overlap. Where they are closely associated each tends as a rule to be completely segregated into its own part of the colony; and only perhaps when one is in a small minority are its nest-sites likely to be mingled with those of the other.

At least as close as this are the associations between various members of the auk family. In more temperate latitudes of both sides of the North Atlantic the razorbill and common guillemot (*Alca torda* and *Uria aalge*) are very commonly found together; and somewhat farther north the most frequent combination is *U. aalge* and *U. lomvia.* These birds require very little space in which to incubate, or tend the single chick, but they differ specifically to some extent in their optimum requirements. The result is that the common guillemot tends to concentrate on broader and more extensive ledges and the tops of rock-stacks where a considerable number can be accommodated contiguously, without physical contact with other species: Brünnich's (*lomvia*) dominates on the steeper faces with smaller scattered ledges; and the razorbill is inclined to seek more protected hollows and crevices (*cf.* Sergeant, 1951, p. 582). Consequently the razorbill is usually in only a small minority on cliffs which it shares with Brünnich's (Salomonsen, 1950-1, p. 355). As so often happens with birds of overlapping ecology, in places where only one species is present alone it may extend its tenure into the type of site elsewhere largely pre-empted by its competitors.

The puffin (*Fratercula arctica*) is another frequent associate with these other three, preferring to lay its egg in concealment in a deep crevice or burrow. Most auks spend a considerable time in flight about their cliffs: and puffins in particular are given to circling in a huge 'merry-go-round', consisting of hundreds of birds, over the water in front of their nesting place: they may be joined in what certainly appears to be an epideictic exercise by other alcids—for instance razorbills. Somewhat similarly, guillemots and razorbills often mingle in forming the long rafts and strings on the sea below the nest-ledges.

Not only the gulls, terns and auks already mentioned, but practically all families of sea-birds, including cormorants, boobies, albatrosses, shear-waters and penguins, provide examples of near relatives sharing the same colonies: so also, of course, do the seals (Pinnipedia) and marine turtles (Chelonia) (*see* pp. 181-2). It is common, though not invariable, to find some segregation of the different species present, frequently as a result of a divergence in the exact type of nest-site preferred.

These sea-bird associations, as everyone knows, are not by any means necessarily confined to closely related birds. Large colonies more often than not contain a heterogeneous concentration of species. On Fowlsheugh—a typical example—the total population of perhaps a couple of hundred thousand birds is made up of common guillemots, razorbills, puffins (relatively few), kittiwakes (about half the total), herring-gulls, fulmars and shags (*Phalacrocorax aristotelis*); and to these might be added the land birds that share the same mile of cliffs, namely rock-doves (*Columba livia*), jackdaws (*Corvus monedula*) and, sometimes, a pair of peregrine falcons. Most parts of the world where seafowl nest could provide similar examples, many of them far more varied than this.

When we cast about for a parallel situation we find one scarcely less remarkable in the marsh birds. Colonies of herons notoriously tend to attract two or more species; and these may form the nucleus for an assembly of spoonbills (*Platal ea*), ibises (*e.g. Plegadis*), storks, cormorants, anhingas and many others, sometimes including birds of prey. One of the most astonishing assemblages on record is in the Keoladeo Ghana of Bharatpur (Rajasthan, India). This immense colony, described by Sálim Ali (1953), contains thousands of painted and open-billed storks (*Ibis leucocephalus* and *Anastomus oscitans*), four species of egrets (*Egretta alba, intermedia* and *garzetta*, and *Bubulcus ibis*), three of cormorants (*Phalacrocorax carbo, javanicus* and *fuscicollis*), a darter (*Anhinga melanogaster*), white ibises (*Threskiornis melanocephalus*), spoonbills (*Platalea leucorodia*) and four kinds of herons (*Ardeola grayii, Nycticorax, Ardea cinerea* and *purpurea*): these are all tree-nesters, and in addition there is a lower stratum in the marsh itself of various gallinules and, not least, Sarus cranes (*Grus antigone*). ‘ Although in many cases nests of a particular species predominate in a certain tree, there is no hard and fast segregation, and nests of four or five different species may often be found all in the same tree, perhaps several of each kind all cheek by jowl and touching one another. . . . It is usual to find nests of all three egrets along with those of cormorants, darters and white ibises packed in a single tree ’ (loc. cit., p. 533).

In the south of France the association of heron-like birds has been particularly studied by Valverde, paying special attention to the abundant and increasing little egret (*Egretta garzetta*). He observes (1955, p. 254) that, with only one known exception reported long ago in Portugal, this species always nests in company with other ardeids; the presence of other species already established in a heronry is consequently inferred to be an almost essential condition for its nesting. Though it has spread to such an extent that it now makes up two-thirds of the whole heron population of the Camargue, this dependence on other herons may largely depend on the fact that it is, in this part of the world, a species extending its range into a region where it has no established dispersionary framework of its own: we shall come across more or less parallel instances a little further on in connection with roosting associations, for example of flying-foxes (*see* p. 411). Its most frequent associate is the night-heron (*Nycticorax*), followed (except in the Camargue) by the common heron (*Ardea cinerea*). In southern France new breeding sites are usually pioneered by the night-heron, which thereafter serves in the role of ' call bird ' to attract other species: in southern Spain it is generally the cattle-egret (*Ardeola ibis*) that forms the nucleus of the heronries.

This attribute of colony-founding is correlated with strong manifestations of social behaviour in cattle-egrets and night-herons: the latter especially indulge in mass-flights and sudden group-manoeuvres when disturbed from the colony—very likely with an epideictic function. Nothing of this kind is to be seen in the little egrets; and in fact what Valverde describes as ' specific gregariousness ' seems to be much less important with them in this region than with the night-herons and cattle-egrets (loc. cit., p. 255).

Comparing the feeding habits of the several species Valverde (1956, p. 6) finds that, in the Camargue, though the little egrets are associated in some feeding habitats with purple herons and stilts (*Himantopus*), there is very little competition for food between these three; but between the little egret and the night-heron, on the other hand, there is most significant competition which exists alike in Italy and France. ' The two species eat the same things in the same proportions. Differences in the method, place and time of fishing seem to be the only valid ones, since even the differences in average prey-size . . . appear to be negligible ' (loc. cit., p. 22).

To a greater or less extent, therefore, the inhabitants of these colonies are drawing on the same food-resources, at least in so far as they depend on kinds of food that grow in or emerge from the water; and, because of the vast complexity of trophic inter-relationships, one species can seldom be entirely isolated or indifferent to the fortunes of the other members of the animal community. A simple if rather extreme illustration of this is the fact that all are dependent on the vagaries of climate; and in a bad monsoon year, for instance, the Keoladeo Ghana may contain no water at all and breeding there is entirely suspended (Sálim Ali, p. 534).

The same is broadly true of the associations of sea-birds; and a comparable illustration of their mutual dependence on the good behaviour of the physical environment may be found in Murphy's account (1936, p. 102 and especially p. 284) of an exceptional pendulum-like swing of the equatorial counter-current, El Nino—far southward along the coast of Peru—destroying in its course the plankton and abundant shoals of fish of the Humboldt Current and exterminating vast numbers of guano birds, such as cormorants, boobies and pelicans, that nest on islands in its path.

It has often been assumed that colonies of the kind we are considering are largely a fortuitous result of the common but independent requirements of the various birds using them—especially protection from ground-predators and proximity to adequate feeding-grounds—and that the different species are driven to associate together simply by the shortage of suitable places. It is true that the sites chosen, on cliffs and islands for instance, almost always meet these common needs; but in reality this can ordinarily be only a secondary cause of concentration. As pointed out earlier (p. 157), there are almost always suitable breeding places that are completely ignored by the birds—until the day that some final disaster befalls their old-established site, when in most cases they adopt a new one without delay, usually at no great distance. So deeply does convention enter into colony-formation that the little egret in southern Europe is largely dependent, as we have just seen, on the prior selection of sites by herons of other kinds.

That mutual conventions are often built up between the associates is revealed by the tendency for each species to confine its property demands to exactly the same area within the colony year after year. This applies with particular force to many of the gannets, auks and cormorants and to common kittiwakes, each of which usually return faithfully to their specific holdings; some others, however, including for instance the common and arctic terns, may not always be so constant. Species that are rapidly on the increase, like the herring-gull and fulmar in the West European region, evidently differ from the majority in being far less inhibited from adding to their traditional nesting estates, even though this may require the appropriation of sites formerly held by other species.

There appear to be two independent advantages that might be gained, as far as dispersion is concerned, by colony-associations. The first hangs on the presumption that species sharing common resources can benefit from being subjected to an interspecific or collective epideictic effect, so that the numbers of each species need not be regulated in complete independence, but instead all can be to some extent conditioned by the numerical strength of their competitors, and a mutual balance aimed at between combined population and feeding conditions. Independent tradition acts as a powerful stabiliser in the case of species living more or less in isolation; and mutual traditions such as we generally see in these colonies are potentially just as valuable to species bound together by a common interest in mutual food-resources.

The evidence suggests that the actual numerical proportions of the different species represented in a mixed colony may be to a very large extent arbitrary, but the total presumably requires regulation in relation to food-supply; and when the resources are of vast extent and indivisible, as they are in the sea, a moderate amount of overfishing could probably continue for a long time before the depletion had any serious effect on the economy of the colony. The process of adjusting the social conventions of a mixed colony through natural selection, involving as it must the elimination of populations with imperfect homeostatic adaptations, must therefore be exceedingly slow; and it is possible that there may be some connection here with the general observation that the breeding birds which regularly travel to the greatest distances and feed farthest offshore, such as the shearwaters, gannets and auks, tend to be the ones most strongly bound by inflexible traditions. The constancy that we can observe in the mutual dispositions of the different species in mixed colonies only serves to emphasise the homeostatic stability and conservatism that especially characterises deep-water birds.

The second possible advantage that might arise from the association of several species in the same colony could be the simplification of the dispersionary pattern. Thus Sergeant (1951, p. 584) has suggested that it may aid the homeward journey of the various auks, feeding far out to sea, to have their number swelled by others going the same way and converging on the same place, instead of having every species flying on a different course. It is also likely to be a necessary part of the education of birds like these to know the situation of all the breeding places over a wide area; or, if not, at least to be able to locate another one when, for example, a young recruit finds itself excluded from its native colony. We have in most cases no knowledge of the extent of the feeding area used by each colony, nor of whether any convention exists corresponding to a communal territory. On theoretical grounds this would seem desirable, but practical difficulties may make it impossible where adjacent colonies are close enough together to be within easy flight of the same grounds. With some species of offshore feeders the spacing of the colonies themselves is perhaps partly determined so as to prevent any material overlap. Much the same considerations apply to mixed roosts, to be considered later in the chapter (p. 411).

17.4. *The increase of the fulmar and herring-gull*

The two sea-birds that we have singled out, as far as western Europe is concerned, as being notable exceptions to the rule of strict conservatism that applies to most other species, are the herring-gull (*Larus argentatus*) and fulmar (*Fulmarus glacialis*); and it is necessary to examine these exceptions rather more carefully.

The herring-gull's numbers have built up remarkably in the last fifty years on both sides of the Atlantic, over practically the whole of its old-established coastal breeding range; and there has also been an important extension of range to Iceland. In many of its breeding colonies it has tended to encroach on ground formerly occupied by terns and by other species of gulls and to displace them, a fact that has been of concern to conservationists both on the Atlantic coast of North America (*cf.* Gross, 1951) and in Europe (*e.g.* in Scotland, Baxter and Rintoul, 1953, p. 646; Eggeling, 1955, p. 77; in Denmark, Jespersen, 1946, p. 42; etc.). It is not unique among the gulls in thus increasing, however, for the great black-back (*L. marinus*)—a larger and generally scarcer species that shares much of the same Atlantic range—has similarly increased and so, on a more limited scale, has the lesser black-back (*L. fuscus*). As far as the extension of range is concerned, Iceland has actually been colonised by no less than four species of gulls during the present century, namely, the black-headed gull (*L. ridibundus*), herring-gull, lesser black-back and common gull (*L. canus*) (Gudmundsson, 1951, p. 507).

The question of greatest interest here, from the standpoint of animal dispersion, is whether sudden increases of this kind are due to a breakdown of the homeostatic mechanism, such that population-density is no longer under control; or whether they can be attributed to a controlled and justifiable exploitation of increasing or newly-accessible resources. In the process of group-selection we must envisage the possibility that the homeostatic machinery can go wrong and that, if the population keeps on increasing far beyond the optimum ceiling, the food-resources will in most cases be damaged and the population itself will crash or even be exterminated in the ensuing catastrophe. On the relatively modest scale of most local groups this may be a very trifling incident to the species as a whole, but if such a sequence were to start up with a sea-bird, having the vast undivided resources of the sea to draw on, the increase might go a very long way indeed before any adverse effects were felt. This is essentially the situation that confronts civilised man: he has thrown off the primitive customs that prevented population increase—so long ago that their former existence is today scarcely even credible—and he draws more and more for his livelihood on the united resources of the world, many of which are rapidly diminishing. The sequence has now reached the point where, if the increase is not controlled and halted within the next two or three generations, a crash seems certain; and even if some fragments of the human species were to survive, civilisation as we know it would almost certainly have been lost for good in the resulting chaos. With the fossil fuels—coal and oil—largely gone, there might be no ' second chance ' (*cf.* Harrison Brown, 1954).

So far as the range-increase of the herring-gull is concerned, the probability is that the species is expanding under homeostatic control to exploit habitats in Iceland that have only lately become available to it. Iceland experienced a temperature amelioration during the first half of this century and during the same period there have been some notable accessions to the fauna. New colonisations involve not only the herring-gull and the three other species mentioned, but also a number of freshwater and land birds, some of them probably helped towards Iceland by the more frequent south-easterly winds that the climatic change appears to have brought (Gudmundsson, loc. cit., p. 512).

Regarding the general increase in herring-gull numbers within its traditional range, however, the issue is much less clear; but again the correlated and equally widespread increase of the greater black-back suggests that the economic situation for these birds has to an important degree genuinely improved, and that the changes in their population-density are legitimate and most likely still under homeostatic control.

The fulmar presents a rather different picture. Almost the whole of its increase has been achieved by the progressive colonisation of virgin ground. It is primarily a high-arctic breeder of circumpolar distribution, which in one narrow sector of its total range has thrust a great salient south-eastward more than 1000 miles into the temperate zone. Prior to 1800 there were already at least two colonies in Iceland and perhaps a few more, and a large one at St. Kilda—all of them presumed to be of long standing though records go back only to the seventeenth century. The subsequent spread has been chronicled in great detail by Fisher (1952); he traced its source to Iceland, where the birds appear to have started to take over new colony-sites along the south coast in the early years of the nineteenth century. They first began to colonise Faeroe not long before 1839 and were found nesting in Shetland (first on Foula) in 1878. Since these pioneer days breeding fulmars have completely encircled Iceland, overrun the Faeroes, and extended southward along almost the whole seaboard of the British Isles in ever-increasing numbers. The first colony appeared in Norway in 1921.

Two features of this episode should be specially emphasised; one is the fact that the fulmar is primarily a bird of the Arctic ice that has spread progressively into lower latitudes and warmer seas than it frequents as a breeding bird elsewhere. The other is that no detectable population increase or extension has occurred in any other part of the fulmar's vast range: in the Canadian Eastern Arctic it may even have declined by the loss of one or more large colonies during the same period.

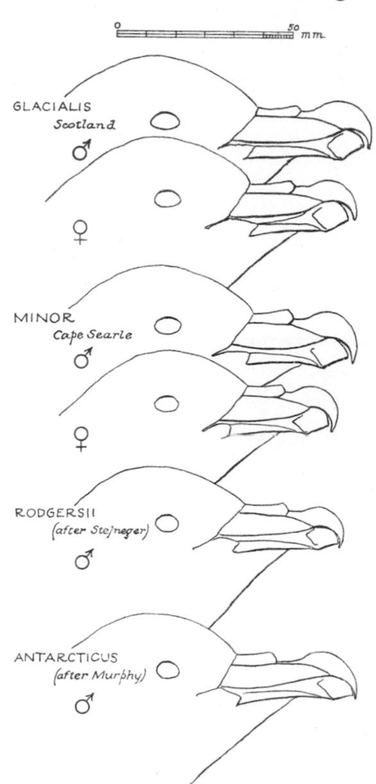

FIG. 37. Comparison of length, shape and sexual dimorphism of the bills in the named forms of *Fulmarus*. (From Wynne-Edwards, 1952*a*.)

The species is differentiated geographically in the size, shape and sexual dimorphism of the bill (figs. 37 and 38), and for this reason the Pacific and Canadian Arctic forms are accorded different subspecific names (*F. g. rodgersi* and *F. g. minor*). An analysis of this character among the Atlantic populations of *F. g. glacialis* has shown that, though the amounts are insufficient to warrant giving them separate names, the Spitzbergen stock is significantly different in bill-size from those found breeding in the area of increase—from Iceland to England; all the latter, however, are quite homogeneous among themselves. They have the largest bills of any of the forms of *F. glacialis*; and it is they alone that have participated in the explosive expansion (Wynne-Edwards, 1952*a*). The St. Kilda colony, incidentally, is thought to have remained fairly constant in population and to have contributed little if anything to the great tide that has been passing it by.

The expansion has been, in terms of evolutionary time, enormously rapid for such a slow-breeding bird. It has not been a case of first exploiting the possibilities for population-growth in Iceland and when that was full overflowing to the next available place, but rather of a succession of bold strides, of hundreds of miles at a time, taken first at intervals of many years

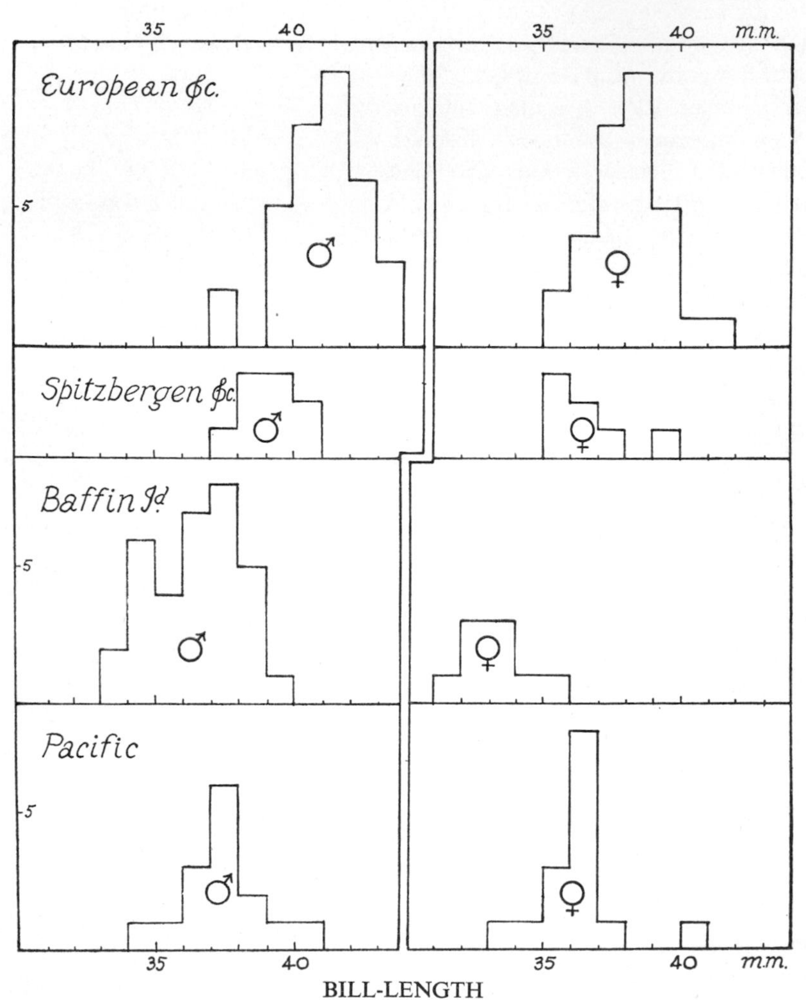

FIG. 38. Distribution of bill-length in the fulmar. The height of the column shows the number of birds in each millimetre group, and the mean for each population is shown by the location of the ♂ or ♀ sign. The contemporary geographical expansion of the fulmar's range (*see text*) concerns only the European stock, the others being unaffected. (From Wynne-Edwards, 1952*a*.)

but later becoming faster and leading to the establishment of a string of pioneer colonies—in succession at Faeroe, Shetland, the Hebrides, west of Ireland, and so on—from which subsequent occupation and filling in of the intervening coasts has been proceeding simultaneously all along the line. It is still going on.

Any such extension into completely new country must necessarily involve tapping new food-resources. Fisher (loc. cit., p. 433) has attempted to correlate the fulmar's increase with the growth of man's offshore fisheries—first whaling and later steam-trawling—which, he supposed, could in turn have provided new sources of food in the form of discarded carcases and fish-refuse. The interest taken by fulmars in these forms of nourishment should certainly not be underrated, but at the same time the geographical correlation between the postulated cause and effect does not stand up well to close examination. There is no evidence of any increase of fulmars having taken place in Spitsbergen, Greenland or the Davis Strait area—the chief northern whaling centres in the eighteenth and nineteenth centuries; all the colonists appear as we have seen to have sprung from the large-billed 'Icelandic' stock, in an area where by comparison very much less whaling was being done. In the present century the fulmar has spread continually south along the British coasts, at the same time as the trawler industry has expanded ever farther north, to Iceland, Bear Island and the Murman Coast. There has been no correlation in detail with the location of the principal trawl-fishing grounds: thus in Britain the fulmar struck south from Shetland along the outer Atlantic coasts, reaching Donegal in 1911 and Kerry in south-west Ireland in 1913, entering there a region where the amount of fishing was negligible compared with that on the North Sea coasts. No colony appeared on the 250-mile stretch between Aberdeen and Grimsby until the 1920s, notwithstanding the unparalleled intensity of fishing on the nearby Dogger Bank, at that time probably the most productive fishing ground in the world. To this day on the mainland of Europe from the Low Countries to the North Cape and round to the White Sea—an extent of 20° of latitude and 2000 miles by water, bordering some of the most highly exploited European fishing banks and former whaling grounds—the fulmar has established itself only tardily in Norway, near Ålesund, where there were colonies totalling about 500 pairs in 1955, and in the Lofoten Islands, where a new colony of a few pairs was starting up in 1957 (Wagner, 1958). Lastly, on the other side of the Atlantic, where whaling and fishing have been pursued on the great banks of Newfoundland and New England since the sixteenth century and countless numbers of fulmars are found as wintering and non-breeding birds, there has been no southward colonisation at all: the nearest New World colony is still in Baffin Island, some 1500 miles away.

It must be confessed, however, that if Fisher's hypothesis does not fit the facts there is no ready-made alternative to put in its place. The spontaneous 'flare up' of a single species is a phenomenon familiar to zoologists: another one going on at the present time is the lightning spread of the collared dove (*Streptopelia decaocto*) across Europe from the Balkans, as far as Scandinavia and Britain; other parts of the world are being invaded simultaneously by the cattle-egret (*Ardeola ibis*). By man's agency exotic animals and plants of immense variety have been introduced all over the world and in countless cases they have become firmly established—regardless of whether they happen to be birds, barnacles, mammals, or molluscs. They succeed in forcing their way into the existing economy, creating greater or less disturbance of the pre-existing fauna and flora in the process; other species have to give way and make room for them, but in most cases in a surprisingly short time everything settles down and equilibrium is restored with new and adjusted balances between the component species. This has happened with the European starling (*Sturnus vulgaris*) in North America, for example, and is proceeding rapidly with the Australian acorn-barnacle *Elminius modestus* in Europe.

There is no need to ascribe extensions like these to any failure of the homeostatic machinery, whether they occur spontaneously or otherwise. Every successful species must possess some initiative for exploiting new opportunities when they arise; these may come accidentally through an external agency—transport to new shores by wind or by man, for instance, or alternatively through some favourable change of climate; or they may come through an enhanced adaptation in the species itself, due to genetic recombination, perhaps, that allows it to advance beyond what had hitherto been environmental barriers.

In the fulmar's case the climatic change has—latterly at any rate—been going in the wrong direction for a polar bird and can almost certainly be ruled out as a causal agent. Fisher's studies (loc. cit., p. 80) have put it beyond reasonable doubt that the invasion originated in Iceland; and it would exactly fit the facts as he has presented them if we were to pin-point the Westmann Islands as the site of the first conquest made by this uniquely aggressive stock some time between 1713 and 1753 (loc. cit., p. 88). Where the Westmann pioneers came from is unknown, but quite likely it was North Iceland. The knowledge we have rather suggests that a special genotype may have arisen, originally in a single individual; an alternative though perhaps less likely possibility is that such an unorthodox turn in general behaviour might have been originated and spread as a new practice or fashion, purely by imitation, without involving any underlying alteration in the genetic mechanism. Whatever the explanation may be, it is certain that the propensity shown by this stock for striking out into lower latitudes, and being satisfied with small straggling colonies, is radically different from anything to be seen in fulmars elsewhere. Once the process had been set going with the bridgehead in the Westmann Islands, the advance continued without any major setback. Both of the older pre-existing colonies at Grimsey and St. Kilda appear to have been flowed around and by-passed, their inhabitants holding fast to their traditional conservatism and remaining unmoved by the new expansive tendencies.

If animal species are homeostatically equipped to adjust their population-densities, to cope with changes that occur in the amount of food available, it is clear that they must possess the basic behaviour mechanism for doing this in any new habitat they happen to colonise, as well as in those they have long traditionally occupied. Traditional practice may become an extremely important working tool in practical homeostasis, as we shall see in Chapter 19, but there must exist certain innate first principles to guide pioneers on which traditions can subsequently be founded. This probably prevents, in almost all cases where major extensions of range occur, any serious initial over-population and damage to resources: the probability is that there is usually a large margin of safety between the maximum density that the species will initially tolerate and the threshold at which chronic damage to the habitat could occur.

Our conclusion is, in brief, that the fulmar's spread is a wholly ' natural ' evolutionary phenomenon, basically unconnected with the activities of man. But it may be rather less justifiable to draw the same conclusion about the increasing population-density of the herring-gull and its relatives.

17.5 *Species-associations among non-colonial breeders*

As we know, the borderline between colonial and solitary breeding is quite indefinite and every intermediate gradation exists. The type of association that not infrequently is found among breeding birds of prey might well, in fact, have qualified for inclusion in the earlier section on mixed colonies.

A striking instance of it has been described by Leslie Brown (1952, 1955) at a place he named Eagle Hill in the Embu district of Kenya. One of a range of hills, rising some 1500 feet above the undulating bush-clad plains, it had a single sizeable rock-face: the upper part was covered with high forest, contrasting with the harsh prickly bush on the lower slopes. ' The total area of this hill above the 4200 ft. contour is approximately 4·2 sq. miles, and upon it there bred a pair each of *Aquila verreauxii*, *Polemaëtus bellicosus*, *Stephanoaëtus coronatus*, *Hieraaëtus spilogaster*, *Hieraaëtus ayersii*, and *Circaëtus cinereus* [all different kinds of eagles]. A pair of *Aquila wahlbergi* bred within half a mile of its foot and hunted over a part of it, and there were also a pair of peregrines, two pairs of augur buzzards, and a pair of gymnogenes. Between August and October 1950 six of the species of eagles listed had eggs or young in the nest, the only abstainer being *P. bellicosus*. This concentration of breeding pairs represented a density of less than 1 sq. mile per pair on the hill, though the birds hunted over a larger area. . . . The concentration seems more remarkable still when compared with the next hill in the range, with an area of 4·1 sq. miles above 4200 ft. and only one pair of *H. spilogaster* and one of *Buteo rufofuscus*, or with the whole of the rest of the range, totalling 11·6 sq. miles, and with only one pair of *Polemaëtus bellicosus*, a pair of augur buzzards and a pair of peregrines besides the two pairs noted above; Eagle Hill, in fact, supported nearly double the number of pairs found on the rest of the range on just over one-third of the area of the latter ' (Brown, 1952, p. 584). When food supply and availability of nesting sites are considered, however, Eagle Hill appears to be no different from its neighbours. ' The only explanation that seems to come anywhere near the mark—but again one for which I can adduce no positive shred of evidence—is that eagles of different species are inter-specifically gregarious, or in other, simpler words, they like each other's company. If so, this is the odder in that eagles are the last birds one would suspect of such a tendency . . .' (Brown, 1955, p. 42).

We are familiar with colonies or concentrations of breeding raptors belonging to single species (*see* p. 153), and the Eagle Hill case seems to be a normal sort of extension of their colonial-dispersion system, closely parallel to what is found in aquatic birds, whereby a mixed association or syndicate is formed. Another activity that much more often brings birds of prey together, either in single-species groups or mixed associations, is migration; and the extraordinary congregations that may be seen in autumn at specially favourable points, such as Point Pelee on Lake Ontario, Hawk Mountain in Pennsylvania, and Falsterbo at the southernmost tip of Sweden, where five or more species of raptors may sometimes be visible at the same time, have often been described.

A closely kindred subject, on which it would be desirable to have much more information, is the interchangeability of nest-sites and similar holdings between ecologically associated species. It is known to happen occasionally between different kinds of raptors, such as golden and white-tailed eagles, though Brown (1955, p. 206), with his wide experience, was inclined to belittle it. 'Among our twenty-six pairs of eagles in the Embu district there were two known cases in four years where eagles did actually use the nests known to have belonged to other eagles.' Evidently, however, it can occasionally occur. The smaller birds of prey show the same tendency: on rare occasions, for instance, the goshawk (*Astur gentilis*) will adopt a nest originally built by a common buzzard (*Buteo buteo*) or honey-buzzard (*Pernis apivorous*) (Holstein, 1942, p. 130); buzzards are more likely to make the reverse appropriation, and on one occasion in Finland I saw a pair of honey-buzzards circling over a nest that was known some years earlier to have been occupied by goshawks. Bent (1937, p. 148) mentions two cases of red-tailed hawks (*Buteo borealis*) appropriating former nests of the red-shouldered hawk (*B. lineatus*); and another case of a red-shoulder's nest found in 1920, known to have been occupied secondly by a pair of broad-wings (*B. platypterus*) in 1928, and thirdly by red-tails in 1929 (loc. cit., p. 150). Such instances could no doubt be multiplied.

There are indications of a similar tendency among limicoline birds. As we saw earlier (p. 156), some of these, such as redshanks (*Tringa totanus*), dotterel (*Charadrius morinellus*) and oystercatchers (*Haematopus ostralegus*), are not only 'nest-site birds' like the majority of the Limicolae, but are inclined to resort to the same sites in successive years, sometimes over a long period. Mr Edward Balfour, of Isbister, Orkney, has told me that some of these sites which he has known for many years may most often have been tenanted by redshanks, for instance, but have on various occasions been used by snipe (*Capella gallinago*). Thompson (1951, p. 142) quotes two examples from Norway, in one of which a greenshank (*Tringa nebularia*) was found sitting on a nest-site occupied three years earlier by a dusky red-shank (*T. erythropus*), and in the other a dusky redshank laid her eggs in a scrape which had held a greenshank's the previous year.

There is a suggestion in this evidence that a situation exists that is analogous to interspecific territorialism; and that any given area contains only a finite number of admissible tenancies, which can be distributed among the members of the local species-syndicate and are, to some extent at least, interchangeable between them. This would imply an unsuspected interest in and knowledge of the exact location of the nests, not only of neighbours belonging to the same species, but also to associated species. The relatively low rate of exchange among the eagles' nests rather suggets that tenants that survive to reclaim their holdings in subsequent years manage to dominate and drive off competitors of associated species almost or quite as easily as those of their own.

Another connection of rather the same sort, well known in Europe, occurs between the badger (*Meles taxus*) and the fox (*Vulpes vulpes*). Both these carnivores burrow in the earth to make their dens and both will continue, if sufficiently unmolested, to use the same sites for generations: for this reason such sites quite frequently became incorporated in medieval place-names, such as Brockley, Foxley, Broxton and Foxton. Of the two, badgers are the more persistent diggers and many tons of earth may be removed to the surface in the course of time at one of their old sets: the accommodation is usually sufficient for a number of badgers at a time, including two or three mated pairs. The possibility must be admitted that, by taking possession of part of the badger's ample dwelling, the more solitary fox is merely saving itself work; but, if so, this is condoned by the badgers to an extent that demands explanation—badgers being very formidable fighters and not frightened of tackling foxes; occasionally, in fact, they eject them (Neal, 1948, p. 127). The dispersion of both species is almost certainly related to a stable system of established earths, and between these there is a considerable amount of coming and going. Neal (p. 58) briefly summarises the situation by observing that ' foxes habitually live for short or long periods in badger sets . . .', and, in another context (loc. cit., p. 127), ' foxes are common inmates and in large sets a vixen may cub there at the same time as, but in a different part from, a badger family.'

These are two carnivores of rather similar size; and no doubt they overlap to a considerable extent in diet, in such items as rabbits, small rodents and beetles. Their association together in a mutual dispersion-syndicate would be exactly in line with that we have just seen among raptorial birds and can find similarly in the mixed herds of seals and ungulates, to be mentioned again below. The type of property-tenure represented by a burrow, with a tolerated household capacity that varies with economic conditions, is physically different in character from the systems we have been concerned with hitherto; but the resulting influence on conjoint dispersion is probably exactly the same.

17.6 *Species associations in roosts*

Owls tend to be very constant in using particular perches or trees for sleeping during the day-time; and at the proper season the same roosting site will sometimes be occupied for a number of years in succession. In the nineteen-thirties, when I lived in Montreal, there was a white pine (*Pinus strobus*) in a particular densely-wooded part of Mount Royal that came to be known as the Owl Tree. During the winter months of 1935-6 it was occupied by a great horned owl (*Bubo virginianus*). On 25th October 1936 it was found to be newly tenanted once more, but by a different individual of the same species, easily recognisable by its unusually grey plumage. On 8th November this owl's place was taken by a barred owl (*Strix varia*); and on the 11th by another great horned owl, not the grey bird. Here were three occupants of the same solitary perch, belonging to two species, within three weeks. At another place in the same park a barred owl was found in possession in December 1930; the following winter its exact place was adopted by a screech-owl (*Otus asio*). Southern (1954*b*, p. 386) has shown that tawny owls (*Strix aluco*) occupy territories throughout the winter, and that the pellet-stations, to which the owners resort to bring up their gastric rejectamenta of fur and bones, and the roosting-stations (which may be the same places, particularly in the late winter) tend to become permanent establishments. ' Occasionally the same perch would be used over a period during which I deduced from other evidence that the ownership of the territory had changed perhaps more than once. In other words the territory may be more permanent than the owners ' (loc. cit., p. 387). It seems likely, in the light of this, that what happened at the Owl Tree indicates a quick succession of ownership in a single established territory, perhaps during migration, involving two different species; and, if so, that some kind of interspecific territorialism may be inferred between species of owls of similar ecology.

A large part of Chapter 14 has already been devoted to the subject of communal roosts, and the conclusion there (p. 297) was that they serve as epideictic and dispersive centres for the highly mobile and gregarious animals that resort to them. Their development is practically confined to flying animals—the birds, bats and insects. A number of cases were mentioned of two or more species being associated at the same roost; and it should be made clear in advance that we are sometimes confronted here with associations between species that have no ecological similarity at all. We shall meet the same phenomenon again farther on in connection with animals feeding and travelling in mixed parties. Perhaps the most obvious examples of it are to be found in roosts in which, like Box and Cox, bats and birds alternate with one another—as described in a classic instance by Willey (1904), on a small island in Ceylon where the crows (*Corvus splendens*) came in at evening, passing the flying-foxes (*Pteropus giganteus*) which were just setting off for the night (*see* p. 305). In part such an association may depend on a real shortage of good safe sites, such as this small island might provide, or the birds'-nest caves in North Borneo (*see* p. 300) that are similarly shared between swiftlets and free-tailed bats. In part, however, it may again be a matter of simplifying the geographical pattern of dispersion-points, enhancing their permanency, minimising the knowledge of roost-site locations that each individual has to acquire, and assisting the migrant or newcomer in finding them.

It may be recalled that bat colonies and roosts—whether of insectivorous bats such as those in the caves of Missouri described by Guthrie (*see* p. 302), or of fruit-bats, whose camps in Australia have been particularly studied by Ratcliffe (*see* p. 306)—are frequently used in common by several species. It is probably significant in relation to what is stated in the previous paragraph that, in the fruit-bats, the most migratory, nomadic and unpredictable of the four species found in the eastern states of Australia—*Pteropus scapulatus*—is the one especially noted for its dependence on sites already occupied by other flying-foxes; it will almost always use their camps when it descends on a particular district (Ratcliffe, 1932, p. 52). We noted an apparently similar instance of a newcoming species adopting the established spatial framework of its near relatives in the range-extension of the little egret now going on in south-west Europe (*see* p. 398). Whether, in bat roosts, the several species segregate—each into its respective section—or whether they mingle together, seems to vary according to the species concerned, very much as it does in mixed breeding colonies of birds.

Some roosting associations in birds, such as that between rooks and jackdaws (*Corvus frugilegus* and *monedula*), are very stable, general and persistent; and others much more transient or casual. In his study of American-robin roosts in the neighbourhood of Cambridge, Massachusetts, Brewster (1890) found that most of them were also being used by other birds, including bronzed grackles, cowbirds, redwinged blackbirds, Baltimore orioles (all these four are icterids), kingbirds, cedar waxwings and brown thrashers. Of these, most are birds of the same order of size as the robin (*Turdus migratorius*), and probably all are liable to overlap at least slightly in food in late summer and autumn, either in taking invertebrates from the ground or soil, or in eating berries. Though attention has been called to the exceptions, it is the general rule to find roosts being shared by species of broadly similar feeding habits.

Not infrequently the impression is given that one species has invaded a roost previously established and occupied by another and has finally driven it out. This was my own impression when studying the starling roosts in Devonshire and Somerset in some of the instances where starlings invaded the roosts of redwings (*Turdus iliacus*); and at one of these I described the event as if it had been a battle (Wynne-Edwards, 1931, p. 347). This was probably a complete misunderstanding of what took place. The habit of planting themselves on other species is common in starlings, but the previous owners do not necessarily depart; and in two cases in the West of England, and another in Aberdeen, where blackbirds (*Turdus merula*) had been the first settlers, the two species continued in joint occupation. Driving out would result only from epideictic or economic pressure, one species being more influenced than the other by the increased density of population. What is probably a more common reason for the former occupants leaving is that their migratory habits are different from those of the starlings and the season merely arrives for them to move on.

Starlings, blackbirds and redwings are again all birds that take much of their food from the surface layer of soil beneath a sward of grass and share a common interest in such items as earthworms, tipulid and beetle larvae, and also in berries. There is, of course, a much stronger association, both in feeding and roosting flocks, between redwings and fieldfares (*T. pilaris*) while they are in Britain in autumn and winter; and they are no doubt closer still in their food requirements. In North America starlings commonly join up to roost with the various icterids, especially grackles, cowbirds, and rusty and redwinged blackbirds. Other examples of species of similar ecology sharing the same roost can be found in the chaffinch and brambling (*Fringilla coelebs* and *montifringilla*) (*cf.* Vleugel, 1941), and—most familiar of all in this country—among the five native species of *Larus* gulls, associated in varying combinations.

As pointed out earlier, roosting associations are also known in insects, in the Hymenoptera, having been recorded especially in wasps belonging to the genera *Priononyx* and *Melissodes* (*see* p. 310). Another probably analogous association occurs in the ' mud-puddle clubs ' of butterflies, in which several species frequently join forces (*see* p. 197; also Klots, 1951, p. 39, and especially Bates, 1863, I, p. 249).

There is not necessarily any economic connection between the species that join in a roost; and an example of this among birds can be found in Rudebeck's (1955) account of the remarkable Transvaal roost already mentioned in Chapter 14 (p. 293) where the principal occupants were roughly a million swallows (*Hirundo rustica*), with ancillary troops of queleas amounting to many thousands, ' Sakabula ' weaver-birds (*Diatropura progne*), pied starlings (*Spreo bicolor*) and even some 400 cattle-egrets (*Bubulcus ibis*). We can guess that here the great marsh full of tall reeds, offering a place of unusual safety, was one of the important factors in drawing together such a heterogeneous assembly.

In the earlier chapter we concluded that the roost is a very close counterpart of the breeding colony, so far as its epideictic and dispersionary functions are concerned. In some cases at least the sleeping perch takes the place of the nest-site as the object of individual ownership. It is interesting to find confirmation of the similarity here in the fact that roosts are evidently liable also to attract just the same kind of species-associations as breeding colonies do. It is appropriate to mention at the same time that analogous associations are sometimes found among species hibernating together, for example in insects (*see* p. 312), amphibia and reptiles (p. 316).

17.7 *Associations in herds of ungulates*

We can turn next to the interspecific associations of animals that are normally gregarious throughout the day and do not therefore need to assemble at a particular place in order to meet. Good examples of this occur in the ungulates, many of which have a characteristic habit of herding. As usual in these situations, however, not all members of the ungulate orders are gregarious: some are distinctively solitary and some given to associating, if at all, only in small parties. Thus the moose (*Alces*) and roe (*Capreolus*) are examples of deer that are solitary or never form herds, whereas the red deer (*Cervus*) and fallow deer (*Dama*) tend to do so wherever they are at all abundant. Among the Bovidae, the African duiker (*Cephalophus*) and its relatives, including the Indian four-horned antelope (*Tetraceros*), are solitary, although the majority of oxen, antelopes, sheep and goats are more or less closely sociable. The degree of sociability is related to the type of dispersion found; and it is probably a general rule that the herding species occupy communal territories.

Large herds are in general characteristic of the species that live in open country, feeding on grass and other low-growing plants that are frequently associated with rather arid climates. Under primeval conditions some of them were migratory—as in the American bison (*B. bison*) and barren-ground caribou (*Rangifer arcticus*) for example—travelling in temporarily enormous flocks, sometimes for hundreds of miles. Prairie-like country greatly enhances the visible epideictic effectiveness of animals in herds roaming together over their common feeding grounds. In bush and forest, on the other hand, permanent residence within a relatively small area held by one individual or a small band is on the whole more common. There are of course exceptions: for instance impala (*Aepyceros melampus*), which are among the most gregarious of all antelopes, ' favour wooded park-like country where they feed in the more open glades ' (Shortridge, 1934, p. 553).

Where they are found living in the same habitat, cursorial animals of different species sharing any tendency towards sociable habits are likely to be found mingled together in mixed herds. Antelope and other ungulates, for instance, ' habitually associate on open grass-flats or around favoured pans; and, in the more remote parts of South-West Africa, it is quite usual to find several species of antelope, together with giraffe, zebra, warthog, or ostrich, feeding in close proximity ' (Shortridge, p. 454). ' The tolerant if not definitely friendly terms most African ungulates so often extend to one another are particularly marked between the blue wildebeest and other species; where large game is still plentiful it is quite common to find zebra, and wildebeest, springbok and other antelope closely associated. On one occasion two warthog were observed feeding in the midst of a troop of wildebeest' (the same, p. 474). Such examples could be readily multiplied: Hamilton (1947, p. 79) mentions the herds of thousands of hartebeests (*Bubalis*) on the Athi Plains of East Africa, ' sometimes combined in considerable troops, at others mixed impartially with gazelles and zebras'.

The blue wildbeest (*Gorgon taurinus*) he particularises in the same respect as Shortridge: 'they are sociable creatures, and are not only often found in large herds where free from persecution but are frequently associated with other animals, notably with Burchell's zebra, for which they appear to have so strong a friendship that the individuals of the two species are often found grazing mixed up together; though on the alarm being given each draws away to the companions of his own type' (the same, p. 84). On one occasion he even noticed an orphan zebra foal being suckled by a cow wildbeest that had lost its calf (the same, p. 54).

The parallel with mixed roosts and breeding colonies is easily seen. There is a hint of a similar gradation in the intimacy with which the species mingle together; and it could be predicted that those with the closest associations, like the last two mentioned, are generally the ones with the largest overlap in food-plants and the greatest need consequently to form dispersionary syndicates. The herd itself in this case serves as the epideictic focus; and like the breeding colony and the roost it is liable to draw to itself stray individuals of other species or animals of rather different ecology: probably for this reason omnivorous ostriches, acacia-browsing giraffes, and even grazing or rooting warthogs, may occasionally be attracted. Single springbuck sometimes attach themselves similarly to other antelopes, zebras or ostriches (*cf.* Shortridge, p. 543).

At the risk of repetition the point should perhaps be explicitly made again here that there is no existing and generally-accepted theory regarding the purpose of the herding habit in mammals; or, for that matter, of comparable behaviour in any other major group of animals. Certainly these ungulates could scarcely do anything more certain to advertise their whereabouts to predators. We have been conditioned from childhood to regard herbivores as gentle and carnivores as fierce, but this is very largely an illusion as far as their relations with their own kind and with other animals are concerned. In fact bull African buffaloes, for example, are exceedingly formidable antagonists, 'far tougher than lions and far more solidly determined to get even with an enemy' (Hamilton, loc. cit., p. 73): the lion on the contrary, 'taking him altogether, . . . is a good-natured, lazy creature, seldom seeking trouble, and, so long as he is well fed, quite content to live and let live' (the same, p. 138). Indeed there is nothing axiomatic in the fact that many grazing ungulates live in herds: carnivores like lions, wolves and jackals are gregarious also, and woodland ungulates as we have noticed are frequently solitary.

In interpreting the herding habit as being primarily a dispersionary device, developed hand in hand with a communal territory system and alternative to the solitary type of individual land-tenure, we cut across these preconceptions about the world of animals being divided into predators and prey, or even into carnivores and herbivores; instead we can see each species as a unit requiring food which it must duly husband by having its populations regulated through one or other of the effective practical methods of density control. When several species band together into an association, it is logical to conclude that the habit is advantageous and that the advantage gained is similarly related to dispersion. There have been certain previous cases of interspecific territorialism where it can be shown that the numbers of one species present can influence the population-density of another in the same close association or syndicate; and it seems logical enough to look for some analogous interaction from mixed herding in ungulates.

17.8. *Associations in bird-flocks*

It is something very familiar to ornithologists to find mixed bands of small birds, chiefly outside the breeding season, foraging amicably together: less striking, admittedly, to observers in Britain or northern Europe than to those in the Tropics or temperate North America, where the numbers of species of insectivorous passerines are larger. In the jungle of British Guiana, according to Beebe (1917, p. 104), ' this was one of the most common phenomena and has been noticed by almost everyone who has spent any time there. The hosts of species of small and medium sized birds drifted together when their nesting season was past and roamed the jungle in small bands. It was remarkable how many different kinds were to be found in each little gathering. Sometimes when such a flock worked toward and across a glade it was possible to make an approximately complete census. I have counted twenty-eight birds in a flock of this kind, including twenty-three distinct species. The association reminded me strongly of birds migrating at night. There was the same steady drift in one direction and the same constant intercourse by means of short chirps and twitters, woodhewer calling to flycatcher, and manakin to antbird.' Winterbottom (1943) gives some lists of similar parties in Northern Rhodesia, the largest of which contained 42 birds of 15 species.

In North America it is a well-known dodge to follow up parties of chickadees (usually *Parus atricapillus*) in the woods at migration times, in order to have a look at the warblers and other birds that have gathered about them. The most regular members of such groups in eastern Canada, apart from the chickadees themselves, are the red-breasted nuthatch (*Sitta canadensis*), golden-crowned kinglet (*Regulus calendula*), brown creeper (*Certhia familiaris*) and downy woodpecker (*Dendrocopos pubescens*); these are likely to be found together at any time from August to May or June: the white-breasted nuthatch (*S. carolinensis*) and hairy woodpecker (*D. villosus*) are often with them. At migration times they are joined by a host of species, the most frequent perhaps being the ruby-crowned kinglet (*Regulus satrapa*) and myrtle warbler (*Dendroica coronata*), but with often a selection from the twenty-odd other warblers and vireos occurring in the region. On a specific occasion, 9th August 1931, I counted eleven species in one of these mixed parties at Brompton Lake, Quebec; and on 17th May 1935, at Mont Tremblant, nine species.

On a rather smaller scale the same may be observed in Britain, involving especially the nearest allies of some of the birds just mentioned, such as goldcrests (*R. regulus*) and tits (*Parus* spp.). ' In pine or fir woods,' wrote MacGillivray (1839, II, p. 411), goldcrests ' are generally seen in company with the coal tit, the long-tailed tit, frequently also the creeper, and occasionally the blue tit, and even the great tit. Their most common associate however is the first of these species. All these birds live together in perfect amity, none of them appearing on any occasion to interfere with its neighbours.'

Though it is particularly frequent in woodland insectivorous birds, the habit is by no means confined to them. Mixed companies of half a dozen species of buntings and finches, or North American sparrows, are common enough; and two or three kinds of thrushes or swallows or American blackbirds are just as likely to consort together. Knots (*Calidris canutus*) join forces with dunlins (*C. alpina*), and bar-tailed godwits (*Limosa lapponica*) with whimbrels and curlews (*Numenius phaeopus* and *arquata*). Examples of this kind could be extended indefinitely to all parts of the world. In virtually every case they concern birds behaving in the neuter or asexual manner and not holding individual territories.

It is easily noticed that certain particular types of birds, notably tits, are liable to be involved in mixed woodland parties wherever they occur; in addition to the palaearctic, nearctic, and ethiopian regions already mentioned, they appear in similar circumstances in Burma (Stanford, 1947). In South America it is the numerous Dendrocolaptidae that are correspondingly prominent (*cf.* D. E. Davis, 1946; Hudson, 1892, p. 255). Winterbottom (1943) described birds of this type as ' nucleus ' species, since they are generally more numerous and coherent than those joining in as minority groups and are more likely to form parties of their own: the other members, usually present in small numbers, he called ' circumference ' species. Davis (loc. cit.) made a rather different classification, into ' regulars ' and ' accidentals ', the former including all the species most commonly found in associations; this is evidently an equally practical division, since some of the most typical associates such as small woodpeckers or nuthatches are most often present singly and the question of numbers is of secondary importance. Combining this scheme with his own original one, Winterbottom (1949) devised a second classification, into ' nucleus species ', ' other regulars ', ' regular accidentals ' and ' accidentals '; with a fifth possible category of ' accidental nucleus species '; but the prospects of finally elucidating the problem through any such analytical treatment seem on the whole to recede with the multiplication of categories.

These authors and many others besides have discussed the function or purpose of mixed parties without finding any really satisfactory explanation. Usually it is concluded that somehow the association increases vigilance and gives more rapid warning of danger, or affords protection from predators through something resembling Allee's confusion effect (Allee, 1938, p. 117), or results in the readier discovery of concentrations of food. The objection that a flock of conversational birds is far more conspicuous to a predator than its members would be if alone and silent has been raised also.

Stanford (loc. cit., p. 508) made the following illuminating observation. 'In Burma it is noticeable that the areas in a square mile of climax forest, which attract passerine birds, are curiously limited. One can sit for hours in some places without seeing or hearing a bird, and might conclude the area was quite uninhabited. It will be found that these travelling flocks tend to take the same route day after day, and often several times a day, passing one particular bush, or tree, or patch of bamboo, and ignoring the jungle on either side of the route.' The keen-eyed and accurate Bates (1863, II, p. 335), who was one of the earliest naturalists to write on this subject, made an almost identical observation on the Upper Amazon. He noted the same apparent rarity of birds in the tropical forest, and the fact that they form these astonishingly varied 'associated bands of insect-hunters': he concluded, in addition, that 'there appeared to be only one of these flocks in each small district; and, as it traversed chiefly a limited tract of woods of second growth, I used to try different paths until I came up with it'. Lastly E. M. Nicholson (1931*b*, p. 74) actually cut special lines through the forest in British Guiana in order to study this point and found that 'on falling in with a mixed flock it became possible to follow it for hours . . . By this means it was shown that the first impression of these flocks being either migratory or engaged in wide wandering was quite mistaken; on the contrary they circulated in elliptical orbits or figures-of-eight within a remarkably confined area, occasionally doubling back on their tracks when a strong element of their highly sedentary membership found itself as far afield as it was prepared to go.'

Most casual observers have undoubtedly assumed just the contrary, especially when migrants are concerned, namely that the roving companies are true nomads, quite unlikely ever to pass over the same route a second time. But what Nicholson said of the tropical South American species is equally true of those in northern latitudes, so far as the 'nucleus species' and 'other regulars' are concerned—including the titmice, creepers, nuthatches and woodpeckers—namely that they tend to be firmly fixed residents during most or all of the mixed-party period of the year; and the creepers and woodpeckers, at least, are known to return habitually to their individual roosting sites at night. Our experience in other directions leads us to expect that the observations of naturalists in the Tropics will apply generally to roving bird parties and that, if they have not definite communal territories—and Davis (loc. cit., p. 179) was satisfied that they did not maintain a territory in the same active manner as his *Crotophaga* and *Guira* flocks did (*see* Chapter 9, p. 159)—then at least each is closely tied to a particular home range.

Reverting to the question of the function of mixed flocking in birds, there is nothing in the evidence in any way inconsistent with the view that this is yet another epideictic phenomenon, demonstrating to the members, in the case we have chosen for discussion, the density of population of insectivorous birds in woodland habitats; and leading as usual to adjustments of numbers when changes in economic conditions make these desirable. Into this community, as we can easily observe, the migrants of similar food-habits are admitted; and they are probably handed on through the countryside, on a system exactly parallel to those we have already recognised in communal roosts (p. 292) and neuter individual territories (p. 392), which automatically guard against the tendency of transients to pile up dangerously in any one place.

Though species of like food requirements may league together in this way, their own specific dispersion is not to be thought of as being completely submerged in that of the mixed group. It is very clear that some of the component species are more gregarious and tend to occur in larger numbers, whereas others are invariably represented only in ones and twos. Davis (loc. cit., p. 179), in commenting on this subject, points out that, though the number of flocks, the total of birds in each, and the total of species in each, may vary seasonally, ' nevertheless, when a species is present in a flock it is represented by about the same number of individuals at any season.' If the average flock is correlated with an average area of the habitat, the latter will support, let us say, a score of titmice or kinglets, but only two or three creepers and one nuthatch or woodpecker. Each species must clearly be governing its own population-density, through its own private social and epideictic system; the presumption, rather, is that in doing so it takes fully into account the densities of other birds with which its ecology overlaps.

Attendance in mixed bands need not actually occupy the whole day, as it would of course be expected to do it if conferred some continuous benefit: Hudson (loc. cit., p. 254) recorded that in La Plata birds began to assemble ' at about nine or ten o'clock in the morning, their number increasing through the day until it reaches its maximum between two and four o'clock in the afternoon, after which it begins to diminish, each bird going off to its customary shelter or dwelling-place '. All authors agree that the flocks are not in existence the whole time and that the nucleus-type species especially are liable to make up ' pure ' flocks of their own.

One final qualification must be made, that, as in all the other kinds of species-association we have considered up to now, the roving band of birds can also act as a general dispersionary focus and thus from time to time draw into itself species of unrelated habits, such as, in our example, fruit- or seed-eaters.

Much of what has been said in this section could be applied equally well, with minor modifications, to mixed parties of limicolines, gulls, ducks, pigeons, icterids, finches, and many other birds. That it leads us repeatedly to such simple and general explanations for common phenomena lends, of course, considerable confidence to the dispersion theory.

17.9. *Summary of Chapter* 17

1. In reviewing the kinds of association which link the fortunes of two or more species, attention is given in passing to that between predators and their prey: Volterra's theoretical oscillating relationship between the numbers of predators and prey is seen to leave out of account the normal safeguards against 'overfishing' evolved in part to forestall this very situation. Under natural conditions harmonious, stable associations between species can be evolved even where one party merely gives and the other just takes.

Chapters 17 and 18 are mostly concerned, however, with associations of a different sort, between species that enter into various kinds of social and epideictic union, apparently for the purpose of syndicating their dispersion and mutually conserving common resources.

2. The simplest, most elementary, illustration of this kind of dispersionary union is seen in the phenomenon known as interspecific territorialism in birds. Examples are recounted of ecologically similar (and usually closely-allied) species that compete for territory in the same habitats to the exclusion of each other—among migrant and wintering chats in North Africa, meadow-larks in the central United States, redpolls in the Arctic, hummingbirds in California, and various others. In each case the interaction of the two or more associated species has the effect of reducing the population-density of any one of them below what it could be in the absence of the other. They are thus in direct competition for the only contested commodity, namely territory; and yet they fail to comply with Gause's principle that species with identical ecological requirements cannot survive together in the same habitat at the same time.

The evidence does show a strong tendency for such competitors to separate ecologically, by narrower specialisation and partition of the habitat and its resources (*see also* Chapter 18, p. 433); but this road leads in the end to dependence on a smaller and smaller variety of food organisms, and, in the extreme case, having to submit and conform to every change in the economy of a single food-species. Wider roots and multiple resources, on the other hand, make for greater independence and more stability. Stability at the optimum level is the primary aim of dispersion, and as such is strongly favoured by group-selection.

Ecological exclusiveness has consequently to be weighed against the dangers of narrow specialisation and dependence on too few types of food; and, in a world of practicable resources of strictly finite variety, the resulting compromise leaves the extensive overlaps in food-resources that are to be seen among associated animals everywhere. In these circumstances selection favours social and epideictic co-ordination between associated species and an appropriate degree of dispersionary integration.

3. Another form of integration between species with overlapping resources is the sharing of breeding colonies. This is particularly common in

sea-birds and marsh birds, but occurs also in seals and turtles. In these mixed colonies the actual nesting or breeding sites belonging to the various species may be mixed together or more or less segregated; and ecological divergence in nest-site preference is common though not universal between closely related species occupying the same colonies. The degree of integration varies greatly, but two or more species among the terns or auks, for instance, may join together in epideictic displays.

The possibility is examined and dismissed that such colonial concentrations are always the result of a shortage of suitable locations for colonial species requiring nesting habitats. The adaptive value of mixed colonies comes primarily from the need, among species with overlapping food-resources, to take cognisance of the numbers of their competitors in regulating their own populations. There may be a minor advantage in simplifying the ' dispersionary map ' and the geographical education of the young, by combining species in fewer, more permanent, colonies.

4. Because their rapidly-expanding economies appear to be in strong contrast to the stable conservatism of most sea-birds, a digression is made to study the spread of the fulmar in Europe and the herring-gull on both sides of the North Atlantic. The question is whether these are examples of something that should on theoretical grounds be expected to happen occasionally, namely, a break-away from homeostatic conventions, leading to the unrestrained inflation and ultimately the collapse of the population concerned. Man's world population, for instance, is at present rushing headlong towards the brink along this very road.

In both cases, however, the balance of evidence is against such an interpretation. The increase of the herring-gull within its established range is paralleled by that of the great black-backed gull over much of the same area; and its extension to Iceland as a colonist is being shared by no less than three other gulls at the same time. In the fulmar, mainly a high-arctic bird, the increase has been confined to one narrow sector of its circumpolar range, where it has in the last 200 years thrown out a 1000-mile salient towards western Europe, far into the temperate zone. This event appears to have had a point of origin in Iceland in the 18th Century; it may or may not be the result of a genetical change. Fisher's theory, that the expansion was made possible by man's contemporary activities, first in whaling and later in the trawler fisheries, is rejected because there is insufficient geographical correlation between the spearhead advance of the fulmar on the one hand from Iceland to Faeroe, Shetland and the west of Ireland, and on the other the movement of the whaling (from Spitzbergen to Baffin Bay) and trawling industries (from the Dogger Bank north to Iceland and Arctic Norway) during the same period.

The spontaneous ' flare-up ' of a single species is a phenomenon well known to zoologists, as is also the conquest of vast areas by species artificially introduced to new continents. These events generally proceed rapidly, cause a varying amount of disturbance to the existing native fauna, but soon settle down to a new balance. This is probably a commonplace event in evolution; and both the herring-gull and fulmar may show examples of it.

The evolution of self-limiting adaptations in species drawing on the vast indivisible resources of the moving sea is difficult to comprehend; but it is in this same limitless sea that man himself has found his most convincing evidence of the reality of overfishing.

5. In this section another range of similar phenomena is reviewed, beginning with species-associations in nesting raptors and going on to the subject of traditional nest-sites used interchangeably by more than one species, both in raptors and Limicolae. In owls there is a similar interchangeability in traditional roosting perches. These phenomena have a distinct resemblance to simple interspecific territorialism. The association between European badgers and foxes in the use of common earths or dens seems likely to be yet another instance of dispersionary integration between species with overlapping resources.

6. Turning once more to the subject of roosts, mixed roosts generally serve species with more or less similar feeding habits, such as herring- and black-backed gulls, rooks and jackdaws, starlings and grackles; or, in bats or Hymenoptera, species similarly connected. But there are occasional instances where no obvious food-bond exists, as between crows and fruit-bats, or between swallows, queleas and cattle-egrets: here it may be admitted that the mutual attraction is probably largely provided by the good, safe site.

Analogous species-associations can occur in the hibernacula of insects, amphibia and reptiles.

7. The ungulates are seen to have dispersionary systems based as usual on the solitary (or family) holding or alternatively on the collective territory of the herd; as usual also there is no sharp line between the two types. Herding is more typical of open-country mammals, and family dispersion of those living in forests. There is no reason to doubt that mixed herds normally provide epideictic integration between species seeking largely the same food-plants, and that this is the main reason for their formation. The popular fallacy that herbivores are naturally gentle and gregarious whereas carnivores are fierce and solitary is exposed; and all species alike are seen to be units requiring food which they must duly husband.

8. The most familiar of all species-associations are those of wandering parties of small birds, such as titmice, goldcrests and warblers. These mixed flocks have attracted a considerable literature, but no explanation hitherto. They have been wrongly assumed to be nomadic, being essentially sedentary and associated with a definite home range. They attract passing migrants of similar habits and, like roosts, probably serve to pass them smoothly along through the countryside without congestion. Each species submerges only part of its social energies in the mixed flock which may assemble for only

part of the day; at other times each tends to maintain its independent epideictic and dispersive regime.

Though woodland insectivorous birds are used to provide the examples, the same conclusion could be applied to mixed parties of birds in many other categories and environments. The ' explanation ' offered by the theory of dispersion is in fact a general one.

Chapter 18

Siblings and mimics

18.1. *Differences in the extent and form of interspecific integration*

There are already sufficient indications that interspecific integration develops at an advanced evolutionary level of animal dispersion, as a special characteristic of the highest animal groups. Most of the examples in the preceding chapter refer to mammals and birds, and a few to insects; some additional ones relating to fishes, insects and Crustacea are still to be given. Judging by the evidence presented so far it appears that the degree of integration developed between different species can vary all the way from zero to almost unity. We are of course concerned only with species between which there is a potential overlap in useful resources; and we can neglect other kinds of inter-relationships, such as the direct bond between prey and predator, or the indirect one between two species serving as alternative preys for the same predator. The type of relationship we are considering at present is the kind that links two alternative predators feeding on the same prey.

For the moment we need to think only of sympatric species, that is to say, species co-existing in the same area and broadly the same kinds of habitat, between which overlap in desirable resources is a real possibility.

We should bear in mind that broadly speaking we have found two alternative ways of partitioning out the resources available to any single species when living in isolation from other competing animals. One is to allot a self-sufficient territory of adequate size to each individual; and the other is to admit a limited number of individuals to the free use of the pooled resources of a given area, the limitation being imposed by restricting the size of membership of a social group. These methods do not differ in principle, as can be demonstrated by imagining the number of users of a communal territory progressively reduced to one, and the territory consequently turned into an individual one; but the distinction is generally helpful in practice.

The clearest cases of integrated dispersion between two or more species are those in which there is interspecific territoriality at the individual or family level. Where they occur in the same habitat, the two meadowlarks *Sturnella magna* and *neglecta* (*see* p. 392) in the breeding season are dispersed as one:

the males compete indifferently for the same conventional status or property which they hold as individuals to the exclusion of each other. Together they make a homogeneous population which has a certain optimum and ceiling density; and if the complement supplied by one species should drop, that of the other could rise accordingly. Provided the species are so similar in ecology that the territory-size is the same for each, the optimum density will be unchanged whatever the proportions of the mixture.

In a case like this interspecific competition is not centred on the ultimate resources, but instead takes the form of social competition for conventional substitutes. By combining together in one integrated social unit, needless to say, the two species can be assured jointly of protection against ' overfishing '. Social competition, as we have seen in earlier chapters, depends very largely on display and threat, and is consequently much less lethal than bare-fisted competition would be for the real necessities of life. There tends automatically to be an upper limit to the intensity of social competition because whenever it mounts high enough it triggers the responses that reduce the population-density and relieve the stress: moreover it is conducted according to conventional rules that impose among other things a high degree of forbearance and urge submission to the established regime. It seems to be these formal or ethical attributes, arising from competition for artificial or symbolic goals instead of ultimate ones, which permit a situation like that between the meadowlarks to persist at all, even briefly and locally, and prevent a free test of strength that could only result in a sweeping victory for whichever one showed even the smallest advantage in each different section of the habitat.

Complete social integration between species, along these lines at any rate, seems to be the exception rather than the rule—although this does not mean it must always be extremely rare. More commonly, as has been stated already, selection forces a wedge between competing species and drives them apart in their ecology: each concedes where it proves inferior to the other, and each thus tends to become more specialised in its requirements as a result. Such a tendency is locally apparent in the meadowlarks—the western species over considerable areas occupying the higher and drier ground, and the eastern species the more humid river-valley lands (Lanyon, 1956*a*, p.102). It is more clearly seen in the wheatears *Oenanthe oenanthe* and *Oe. hispanica* on the islands of the Aegean Sea: where both occur together on an island they are said to separate by dividing the habitat between them. In this case *hispanica* occurs at low altitudes and *oenanthe* on the mountains; but on some of the smaller islands that are inhabited by only one species, this one has freely colonised the whole island regardless of altitude. On some islands, therefore, *oenanthe* descends to sea-level; and on one at least (the island of Kos), *hispanica* alone is present although the island contains a rather high mountain (*cf.* Stresemann, 1950). In a similar manner, where the common chaffinch (*Fringilla coelebs*) meets its close relative the blue chaffinch (*F. teydea*) on Tenerife and Gran Canaria (to which the latter is restricted or endemic), the former occupies the broad-leaved woodlands and tree-heath and the latter the conifers. Elsewhere throughout its wide European range *coelebs* is fairly common in conifers as well as in other woodlands (Lack and Southern, 1949, p. 614). Brian (1956) has shown that the ants *Myrmica rubra* and *M. scabrinodis* are similarly segregated into two varieties of habitat where they meet and compete in south-west Scotland.

A difference at least in the optimum habitat conditions in England has been shown for the two very similar and widespread snails *Cepaea hortensis* and *C. nemoralis*; but in northern Scotland, beyond the range of *nemoralis*, *hortensis* freely extends into the coastal sandhill habitat which in England *nemoralis* pre-empts. Such examples could easily be multiplied.

It may be asked whether there is any case in which two species, both normally dispersed in communal territories, divide a single uniform type of habitat between them in such a way that the ground is occupied by an inter-specific mosaic of pure colonies, instead of a mosaic of individual territories as in the meadowlarks. Possibly this happens with the ringed plovers *Charadrius hiaticula* and *C. semipalmatus* where they meet in Baffin Island (Wynne-Edwards, 1952*b*, p. 367) and it may enter into the tendency between the common and arctic terns to form pure colonies, or at least pure patches in mixed colonies.

Wherever we have to do with matters of dispersion over areas of consider-able size we are likely to meet with the effects of conservatism and tradition. The animal populations of such large areas tend to be self-perpetuating and the homeostatic control of their density tends accordingly to be largely self-contained and independent of other regions. As far as migratory birds are concerned it is well known that populations from different parts of the breeding range not infrequently remain geographically separate in their winter quarters also: this is a subject which has been discussed by Salomonsen (1955), who refers to the phenomenon of separated wintering areas as allohiemy. In maintaining this state of affairs custom certainly plays a most important part, for it is well established that individual birds in many species tend to return to exactly the same place in successive winters (*cf.* Whittle and Fletcher, 1924; Boyd, 1955). This is to be expected, because in their wintering area the birds maintain an active dispersionary system that usually depends on a traditional knowledge of roosts, feeding places and other key points.

As an illustration we can take the six fairly easily recognisable types of grey geese that winter in the British Isles, namely, the greylag (*Anser anser*), the European and Greenland whitefronts (*A. albifrons albifrons* and *A. albifrons flavirostris*), the lesser whitefront (*A. erythropus*), the bean (*A. fabalis fabalis*) and the pinkfoot (*A. fabalis brachyrhynchus*). Ornithologists know which ones to expect in any particular part of the country: in Aber-deenshire, for example, we are accustomed to large numbers of ‘ pinkfeet ’ mixed with smaller numbers of greylags. In central Scotland—Perthshire especially—the greylag is the dominant one of the two. At the New Grounds on the Severn Estuary, made famous by Peter Scott, more than 99 per cent of the visiting grey geese are European whitefronts. Two hundred miles west, in Wexford, Ireland, the densest wintering ground of the Greenland whitefront is centred, and so on. The lesser whitefront is a very rare bird and the bean goose now much less numerous than the greylag, common whitefront or pinkfoot; but in most areas good for wintering geese, occasional examples may turn up, either of these or the other species, in addition to the local dominants.

Though the geese show a high degree of conservatism in their use of winter-quarters, the latter are not immune from long-term changes. Boyd (1955, p. 122), in his discussion of the role of tradition in the winter distribution of the pinkfoot in Britain, shows from a study of recoveries of marked birds how easily changes could be brought about, especially by the scattering of second-winter birds. The only apparent explanation for the well-known adherence of each species to particular localities is the relative persistence or inertia of the established adults. With the passage of many generations of geese, changes do in fact occur. On the famous Wexford Slobs the Greenland whitefront has rapidly superseded the greylag over the last thirty years and had by 1950 replaced all but a small remnant (Kennedy and others, 1954, p. 79). Earlier than this the greylag had built up a large winter population centred on the Solway Firth, where it is now very common. During the second half of the last century the bean goose practically disappeared from eastern Scotland, where it had been the dominant type and where the pinkfoot now occupies its place (*cf.* Berry, 1939; Baxter and Rintoul, 1953): similar instances could be multiplied.

To some extent these changes may be correlated with alterations in the habitat which differentially favour one or another species: but when we remember the considerable range of variation in the habitats chosen by each particular type of goose, taking the country as a whole, and the commonness with which two or more species regularly occur together, it seems doubtful whether there is really much substantial difference in their ecological requirements. All are vegetarian, feeding at that time of year chiefly on grasses, fallen grain, clovers and plants of marshes and saltings; and no correlations have in fact been suggested between their distributional changes and contemporary alterations in the nature of the food-supply.

The possibility cannot be excluded, therefore, that tradition, established and gradually modified in a manner in which chance plays a great part, is largely responsible for the partition of the goose wintering grounds in the British Isles. With the various species localised and largely segregated as they are, the problems confronting each of them in controlling its own numbers must be correspondingly simplified. Were they all thrown into a mixed pool together and dispersed on one joint system—as their ecological similarity would probably allow them to be—any winter regulation of numbers would have to be done purely on a geographical basis and not on a racial basis at all; so that specific segregation is likely to be a considerable aid towards homeostasis. As it is, the separate species or races are for the most part separated into their own self-perpetuating groups, much like the local groups or demes within the range of a single species, and apparently for the same reasons. Interplay between them may result in a certain amount of interspecific selection, but adherence to the kind of geographical convention postulated here would, of course, reduce interspecific competition far below what would otherwise be expected.

It seems not impossible that geese are afraid to intrude themselves or at least to remain long in the presence of well-established and hostile related species; whereas they may be readily admitted and welcome at any of their own ' rightful ' specific haunts.

There is no need to stress that much of this is speculative; but the subject is one of great theoretical interest and may justifiably be pursued one step farther. If such conclusions were after all correct concerning the segregation of goose species in winter, could not the same be equally true of their summer breeding grounds? As in winter there are some considerable overlaps in specific ranges: the largest is between *albifrons* and *fabalis* throughout a great part of the tundra zone of the palaearctic region; *A. anser* occurs in Iceland along with the pinkfoot (a race of *fabalis*); but otherwise in the huge circum-polar breeding area of the grey geese there is virtually no mixing: the different forms are allopatric (replacing one another geographically) to an extent that is really no less conspicuous than it is in winter. Here also specific ecological differences almost certainly play their part: the greylag and lesser whitefront are less arctic than the others, the last in fact being more an alpine than a tundra breeder; but allopatry in such a case as this may also depend to an important extent on a history of social rather than ecological competition, and be preserved at least as much by tradition as by any physio-logical or adaptive differences that specially favour each species in its own respective area. It may be something of the same sort that prevents the great black-backed gull (*Larus marinus*) from nesting in east Scotland north of the Moray Firth, though hundreds and thousands of non-breeding individuals, mostly adolescents, frequent this coast, at least as far south as the Firth of Forth, throughout the summer: a century ago it is known to have nested on the Bass Rock (Baxter and Rintoul, 1953, p. 655).

Instances of this kind are probably to be found everywhere; and while we cannot at present explain them, it is desirable to notice the possibly strong influence that highly developed conservatism may have, in the absence of any outside opposition, in hindering the spread of animals beyond their tradi-tional borders into areas that could in fact provide a potential living. This might sometimes prove an evil consequence of tradition, to be weighed against the great positive contribution it makes to the regulation of numbers within the occupied area.

18.2. *Integration in the dispersion of sibling species*

Though the grey geese are physically and ecologically similar to one another, they are hardly to be considered as sibling species. This useful term, coined by Mayr (1942, p. 151), denotes any group of species that are all so much alike as to be troublesome to the field-naturalist or taxonomist in identification.

The resemblances between sibling species are partly a matter of close relationship and common features incidentally retained—as perhaps in species of the herring genus *Clupea*; but sometimes—and this is the aspect that specially interests us—the resemblances appear to be fostered by selection as if for an adaptive purpose. Visual similarity between two species amounting to virtual identity is everywhere the exception in the higher animals, as we noted in Chapter 2 (p. 34). Where it occurs, it may be a question of deceptive resemblance and in that case it comes within the definition of mimicry, to be considered in the next section. In a different connotation, however, we can find, at least in birds, species that have distinctive enough marks in the breeding season, though often only in the male sex, but proceed to obliterate them as soon as the breeding season is over.

This happens on a considerable scale in Africa among several dozen species of weavers (*Ploceus*), bishops (*Euplectes*) and widow-birds (*Coliuspasser*). The males have nuptial plumages that are not only distinctive but often brilliant in colour, and in some of the widows ornamented with extravagant tail-plumes. After breeding the males moult into an eclipse plumage in which they lose all their specific recognition marks and come to resemble the nondescript females and juveniles. Closely-related species become submerged in a sparrowy uniform which tends merely to be greener or more drab, plainer or more dappled, according to the particular genus. They are for the most part gregarious birds living in the open in grassland, and at that time tend to form large mixed flocks. I have not been fortunate enough to observe them in life, but it is evident from the literature that field-identification of species is for the most part impossible, and consequently not attempted, where several members of the same genus are liable to congregate together: probably for this reason there is very little information to be had regarding interspecific relations outside the breeding season. Marchant (1942, p. 187), for example, writes of the orange bishop (*E. orix franciscana*): ' only during the breeding season are these birds noticed, as at other times, when both sexes are in the sparrowy plumage, they are virtually impossible to detect among the hordes of drab weaver-birds which frequent the grass-lands '.

While there are several other groups of passerine birds notorious among ornithologists for their visual similarity, such as the American flycatchers belonging to *Empidonax*, and the phylloscopine and fan-tailed warblers (*Phylloscopus* and *Cisticola*), this lack of specific recognition marks, either of plumage or voice, is very unusual taking the avian class as a whole. It seems justifiable to conclude that there is a positive significance in the neutral similarity, which is in such marked contrast to the distinctiveness of the males when participating in nuptial dispersion with their bright colours or songs or both; and the conclusion to be drawn tentatively, at least in the case of the gregarious weaver-birds and their allies, is that in the asexual season they are jointly dispersed as one closely integrated complex, in which sibling species occurring in each neighbourhood are lumped together into one unit. Closer study of the composition of these asexual flocks seems very desirable in order to throw more light on the true position.

Another similar case in birds, involving only two species, has been commented on by Ticehurst (1938, p. 93) and concerns two laughing thrushes in the eastern Himalayan region, *Garrulax pectoralis* and *G. moniliger*. ' These two species ', he says, ' afford one of the most remarkable cases of parallelism in ornithology. Apart from two small characters, the difference between the two is that of size, the largest *moniliger* practically overlapping the smallest *pectoralis*. If the two were separated geographically they would at once be considered races of one species. Yet so far from being separated geographically they are not even, like some closely-allied species are, separated by difference in terrain; indeed, where one is found the other frequently is also, even in the same hunting party. Their nests, habits, eggs, courtship and terrain are identical; they share the same ecological niche.' Furthermore, the quite decided geographical variation in different parts of their range is exactly parallel in each. They are very gregarious, noisy birds, and Baker (1922, I, p. 151) informs us that ' they indulge in the same dances during the early part of the season and not infrequently at other times also, hopping about the ground, flirting and spreading their wings, bowing and performing like circus contortionists, all the time loudly applauding their own performances '.

This again suggests epideictic integration, at least outside the breeding season; and the same characteristic of apparently random complementary numbers, which we have seen in the terns and some other comparable examples, is suggested by Ticehurst's remark that ' in parts of the distribution of the two the status is by no means equal, the one or the other being much the commoner '.

It should be noted that in some pairs of sibling species, such as the two tree-creepers, *Certhia familiaris* and *C. brachydactyla*, in central and southern Europe, and the two titmice, *Parus montanus* and *P. palustris* (the willow- and marsh-tits), there is a high degree of visual similarity combined with a well-marked ecological divergence, so that the resemblances may perhaps have little significance in dispersion. It seems advisable to interpolate this comment, because the subject of sibling species appears likely to be complex; and it is certainly not intended to imply that a universal ' explanation ' can be given in terms of integrated dispersion.

The phenomenon is by no means confined to birds, and is probably very common in insects. For instance, in this country two closely-similar species of bluebottles, *Calliphora vomitoria* and *C. erythrocephala*, are usually found together and are distinguishable only by the colour of the cheeks which are situated low on the sides of the head and thus not visible except at close quarters: one has them red with black bristles and the other black with red bristles. The respective natural histories are not known to differ at all. Three gaily-striped hover-flies, *Helophilus pendulus*, *hybridus* and *trivittatus*, all found together along the east coast of Scotland—though the first is much the commonest—are difficult to distinguish without a hand-lens. It seems not unreasonable to suppose, however, in the case of relatively small insects which recognise their own kind visually that they can appreciate these details well enough if they come sufficiently close together—as these and many other flies can be observed to do when playing about on the wing: indeed, if such small characters are not intended for intraspecific recognition it is exceedingly difficult to suggest an alternative function.

Several of these sibling associations have been analysed by Diver (1940, p. 312), including another trio of common hover-fly species, *Syrphus torvus ribesii* and *vitripennis*, for which no gross differences in habitat preference could be worked out: however, all three were found together at only three out of 35 major stations at which records were made; two together were found at eleven stations and at the remaining 21 stations only one species was taken. Though the seasons at which each of the three was flying extended from April or May to September or October, the peak dates of the two most common species, *S. ribesii* and *S. vitripennis*, fell in June and September respectively: this would not of course affect the likelihood of their being epideictically integrated, since it is only the proportions in which they occur that change with the seasons.

With the grass-moths, *Crambus uliginosellus*, *C. sylvellus* and *C. pascuellus*, the differences are scarcely any larger, and the ecological overlap evidently extensive; Diver (loc. cit., p. 316) reported samples taken at one particular site where all three were established, two of them flying on the same dates in similar numbers; but discrete pure populations were found of *C. pascuellus*, the most common and widespread of the three. Differing numerical proportions of the species would as before be compatible with the possibility of mutual integration.

A very familiar instance is that of the common social wasps. We have six species in Britain, one of which, *Vespula austriaca*, is a ' cuckoo-wasp ' or nest-parasite on one of the others (*V. rufa*) and has no worker caste of its own. In the remaining five, a very little practice enables one to pick out the two tree-nesters, *V. sylvestris* and *V. norvegica*, quite easily from the three ground-nesters, *V. vulgaris*, *V. rufa* and *V. germanica*; but members of these two groups are superficially very alike among themselves (fig. 39). They have quite distinctive minute characters visible with a lens or at a visual range of a few millimetres perhaps—notably on the forehead or clypeus, where a simple black design or emblem usually appears on a background of yellow. Even in combination with general appearance, however, this does not serve to separate *vulgaris* and *rufa* workers with confidence. The workers are individually variable, and the specific characters are less pronounced in them than in the sexual forms. Workers of two or more species can very often be found visiting the same source of food at the same time.

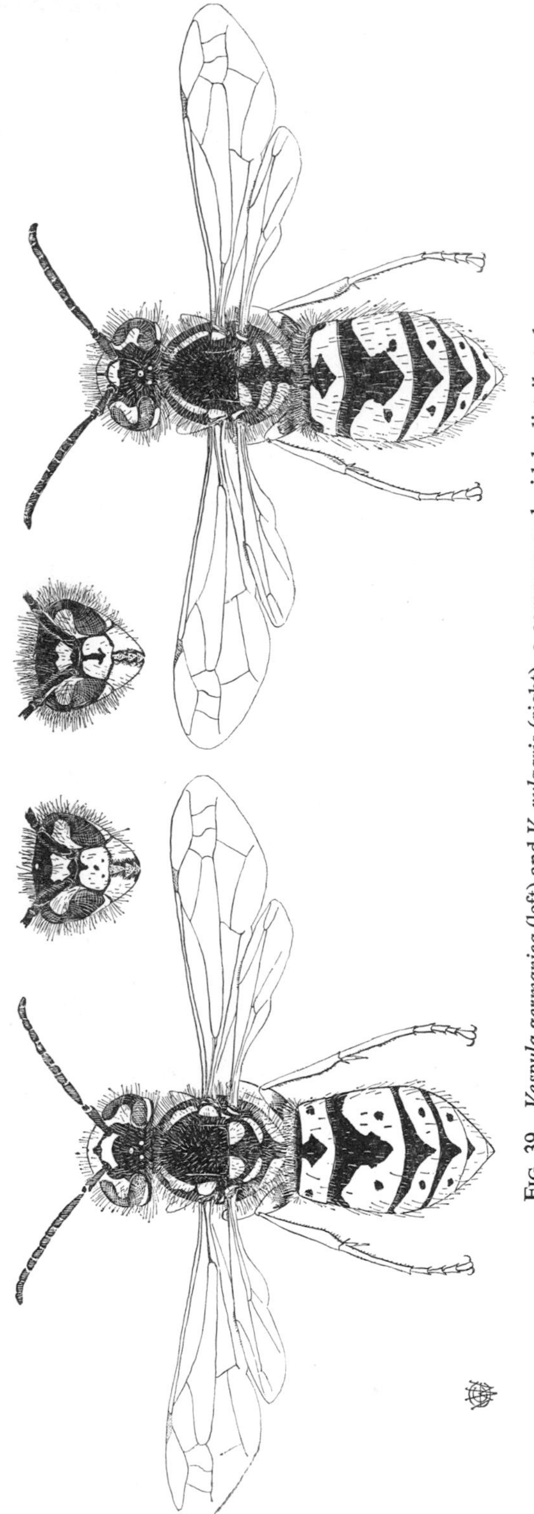

FIG. 39. *Vespula germanica* (left) and *V. vulgaris* (right), a common and widely distributed pair of sibling species, with no apparent ecological separation. Some individual workers bear an even closer resemblance in pattern than the two depicted here. Frontal views of their heads (above) show the specific characters of three black spots (left) or an anchor mark (right) on the yellow clypeus and also the difference in shape of the yellow corona, lying between the bases of the antennae. ($\times 2\frac{1}{2}$.)

Though the presence of several species living in the same area is characteristic of *Vespula* wasps in most temperate regions of the world, no comparative ecological study of the various members of the genus appears to have been made. There may be some persistent local and regional differences in the relative commonness of the species, but records show that great variations in the number of each occur from year to year, to a marked extent independently of one another (Barrington and Moffat, 1901). In ' wasp years ', when they are exceptionally abundant, it would seem that ' more than one species is commoner than usual ' (Beirne, 1944).

Such highly adapted insects as these are certain to have a well-developed homeostatic system of dispersion, notwithstanding their familiar fluctuations in numbers; they are largely carnivorous, feeding especially on small caterpillar-like larvae of other insects, and must protect their food-resources from over-exploitation in the normal way. Though further knowledge may reveal some non-visual and perfectly distinctive auditory or olfactory differences, the fact remains that they are much more alike than any of their supposed mimics are to them—such as the hornet-clearwing moth *Trochilium apiforme*, the wasp-beetle *Clytus arietis*, or the hover-fly *Volucella inanis*. Indeed the workers in each species-group are more like each other than they are to their own respective sexual forms, which can always be picked out at once—the queens by size and the males by their long antennae.

It is clear to the most casual observer that many species of bumble-bees and other bees share the honey and pollen produced by the same immense variety of flowers; and that, though there are well-marked ecological divergences within the large genus *Bombus*, considerable overlaps between species must exist. This is another familiar group very confusing to the beginner, in which, for example, the yellow-ochre-coloured ' carder ' bees, or the whitish- and buff-tailed banded types, or the red-tailed bees, form distinctive groups each frequently containing two or more very similar sympatric species, and each mimicked by cuckoo-bees of the genus *Psithyrus*.

This is less complicated, however, than the situation in the small partly-social bees belonging to *Halictus*. ' Thirty-six species are found in Britain and a much larger number in the United States . . . Many of the details of their organisation are still little understood. One of the chief difficulties in their study has been the large number of species, many of them very alike, especially in the female sex. Many species will often nest together in one bank of earth, and in any ordinary locality in southern England one may expect to find at least sixteen species. Another trouble is the variety of their behaviour, not only when different species are compared but within a single species either in different parts of Europe or in different years ' (Richards, 1953, p. 79).

Sibling associations are not confined to terrestrial animals. There is, for instance, the case of *Calanus finmarchicus* and *C. helgolandicus*, first differentiated in 1903 by Sars but later treated by most planktologists as being the same, until as recently as 1949 (*see* Marshall and Orr, 1955, p. 4).

These marine copepods do not become distinguishable until ' stage five ', after their penultimate moult; and their life-histories bear a close resemblance. The proportions vary from region to region: in the English Channel off Plymouth *helgolandicus* predominated in Russell's collections, never more than ten per cent of the population being *finmarchicus*; whereas at Millport *finmarchicus* predominated in Marshall and Orr's collections, though the proportions changed from time to time, so that *helgolandicus* progressively increased relative to *finmarchicus* during the period 1950-53. Both species occur also on the American side of the North Atlantic. It is possible that there is also a third sibling in this complex, comprising about one-tenth of the population at Tromsø in northern Norway in the spring of 1952, the remainder at that time being *finmarchicus* (Marshall, Orr and Rees, 1953).

Even more recently a close association of three species of the wood-boring marine isopod *Limnoria* has been found in Southampton Water (Eltringham and Hockley, 1958). This crustacean is the well-known gribble which destroys timber structures in the sea in somewhat the same way as furniture-beetles do on land. Until 1949 only one species, *L. lignorum*, was recognised in Europe, but in that year *L. quadripunctata* was described from the Netherlands and soon afterwards shown to be the more abundant of the two at Plymouth. Sometimes they are mixed together, sometimes their colonies are separate (Marine Biological Assoc., 1957, p. 196). In Southampton Water a third species, *L. tripunctata*, first described in 1957 from the shores of the Caribbean Sea, has now been identified. In a series of test-block samples taken from wharves and piles at Southampton and Calshot the proportions of the three were very inconstant, though *tripunctata* was the commonest in piling at both localities: the authors say that on several occasions members of different species were found in the same burrow, though they tended to occur in discrete colonies on the piling. The general characteristics of this association are not unlike those we have previously found among other closely-integrated gregarious species.

The examples presented in this and the preceeding chapter are at any rate sufficient to show that nearly-related species of very similar habits can be united ecologically, for a part of their annual cycles at least; and in some of the more extreme cases the evidence suggests that they can be united throughout the whole year. Such a union implies, in effect, that the members of an association are not in competition with one another as species, but only as individuals belonging to a mixed indivisible society. As individuals they enter into social competition for their stake and status in the community; but species-differences are submerged in a sort of *condominium*. This is clearly seen in mixed communities of meadowlarks (p. 392), in which the males of two species compete in spring for a single commodity, namely territory, without any detectable advantage on one side or the other.

Improbable as such a balance may seem at first sight, it serves to explain why Gause's principle evidently does not sever these highly-integrated associations. It must be remembered that our examples have been taken from a wide selection of the higher groups of animals; and there is no doubt that the phenomenon itself is sporadic in its occurrence, and everywhere the exception rather than the rule. The implication of this, as we have noticed earlier in several quite different connections, is that as a phenomenon it tends to be short-lived: the situation is probably one that frequently arises, but is usually ephemeral in terms of evolutionary time. However there may be exceptions, for we must not forget that a few of these generic and specific associations cover enormous areas of the earth's surface: the six British native *Vespulae*, for instance, inhabit virtually the whole of temperate Europe, Asia and North America; and their sympatric association has probably endured for an immensely long time, under the most varied ecological conditions. The fact that these particular wasps probably comprise a set of Müllerian mimics (*see* p. 436), and that this may have much to do with their close similarity of warning colour-patterns, does not greatly affect the issue that concerns us here—namely that they are nearly if not quite inseparable ecologically, and successfully evade the ' Gause effect '.

We have seen two levels or stages in the development of species-associations. At the lower level the member species enter into a loose epideictic consortium, either at the same colony or roost or in a mixed flock or herd; and at the higher level thay not only consort together but share a close physical resemblance as well. The inference to be drawn from this last development, very strikingly shown by the weaver-birds and in the other examples in this section, is that the resemblance has a definite function in promoting and facilitating epideictic integration. In other words, animals that form visually-integrated societies can most effectively be united by looking alike and wearing the same uniform.

The converse is likely to be true also, namely, that these sibling species may have difficulty in telling one another apart (*e.g.* in the weaver-birds again), and be unable therefore to make independent assessments of numbers of their own species alone. If the members of an association were just to form random proportions in a mixed population, without distinguishing and protecting the separate interests of each constituent party, the numbers of any of these might easily fall by chance to the vanishing point; and it is for this reason that we should expect that at some stage in its life-cycle each species would undergo a private internal homeostatic regulation, to ensure that its own numbers and share in the mixed population stay within safe limits. It is difficult to see, as has been indicated already, if and how this does occur in some of the cases given, either in the breeding season or at any other time of year; but in others—and again the weaver-birds provide a good illustration—the different constituents do sort themselves into species at the time of their nuptial dispersion and during the reproductive adjustment of population-density.

It seems probable that the inability of the individual members of species having a close physical resemblance to tell one another apart (except practically on contact) is often a real disadvantage: in the case of the grey geese it was thought to explain their adherence to traditional allopatry, at least in their winter-quarters, in order that epideictic confusion between the different races and populations would be minimised. Because of this difficulty it might be expected that natural selection would always favour the evolution of specific distinctiveness and clear recognition-marks.

Against this there must clearly be set some opposing advantage—something positive to be gained by associative integration. The answer seems to lie, as we saw earlier, in the desirability—almost the necessity—of sharing varied food-resources with other species. The only alternative is narrow specialisation on some restricted food that others do not want, and the linking of one's whole economy to the fortunes of one or two species of food-organisms. A varied diet undoubtedly makes for a stable economy; and even in the case of vegetarian insect larvae, many of which tend to specialise on particular food-plants, there generally prove to be passable substitutes when the first choice is lacking. The fact is that it is very difficult indeed to avoid overlap in desirable resources with other species; and in the case of food, if it is to be efficiently utilised and conserved and not over-exploited, the population-densities of all the species sharing it are bound to have a most important relevance to one another. It appears to be this factor that leads to multi-specific integration; and if two closely-related species come together and overlap in geographical range, there is some real advantage in total integration between them, permitting full use of every resource to both, and making ecological specialisation—always undesirable—unnecessary.

This whole subject is certainly among the most difficult we have encountered, and much more knowledge must be gained before we can expect to get completely to the bottom of it and provide fully satisfying solutions. Zoologists have long been accustomed to think of species with shared resources as being constantly and ruthlessly in competition with each other; but such a concept evidently needs to be reconsidered in the knowledge that among the higher animals with efficient adaptations for homeostatic control of numbers, in a habitat that is physically stable, there is no tendency whatever towards aggressive expansion of populations; and in particular the members of a coordinated league are not likely to initiate even the smallest changes at each other's expense. It is no doubt at times when the stability is upset, either by the physical factors in the environment, or by the breakdown of population-control in one or more of the members, that each species must defend and exploit its own interests as best it may in the heat of free competition, until a new equilibrium is reached. On a small scale something of this sort is no doubt frequently happening, particularly in marginal habitats, and may render total ecological integration between species very difficult to maintain.

18.3. *Mimicry and dispersion*

Mimicry in animals is usually taken to include only the kind of resemblance of one species to another whereby the mimic is mistaken for the model by ' third parties ', these being either predators or prey according to circumstances. In Batesian mimicry the model is generally a species distasteful to predators or else dangerous, like a wasp or bee, and the mimic is an innocuous animal that they deliberately avoid because of its conspicuous and confusing likeness to the model. The converse situation in which a predator is disguised as a harmless species in order to deceive its prey is perhaps more a theoretical possibility than a reality, and no account of it will be taken here.

In practice the great majority of models and mimics are insects, and the most numerous predators are birds, which is very much to be expected and needs no special explanation. Dangerous and unpalatable insects are often brightly marked with warning colours, and it is known that birds discover the significance of these by actual trial and not by hereditary instinct; the same is true for lizards and toads (for details *see* Cott, 1940). In Müllerian mimicry a number of unpalatable species share a common warning pattern whereby the experimental toll taken by young predators, in the course of learning to leave them alone, is proportionately reduced. It is still customary to regard the subject as somewhat controversial, perhaps largely because many details remain to be explained—though as to the reality of Batesian mimicry there can be no valid doubt, when we have for instance beetles and moths mimicking wasps, even to the possession of sham stings and going through the motions of using them when handled (Cott, p. 402). The experimental approach is rather difficult, but the practival effectiveness of mimicry has been demonstrated, for instance, in beetles that resemble the disagreeable Lycidae, in protecting them from predation by the West Indian lizard *Anolis sagrei* (Darlington, 1938).

Associations consisting of models and their mimics or of sets of Müllerian mimics are likely to encounter the same difficulties of mutual identification as the sibling associations reviewed in the last section; and it is therefore interesting to study the methods by which their effective dispersion is apparently maintained.

Protective mimicry is evolved through the selective action of predators. We have often had occasion to look at the relation between predators and their prey in connection with animal dispersion; and have had reason to emphasise the predator's interest in controlling its own population-density in order to prevent its demands on the prey from exceeding the optimum. Viewed from this standpoint it might at first appear that the prey-species scarcely need to protect themselves, since the predators can safely be trusted not to endanger their survival. In practice, perfect control of population-density is probably seldom held by any species for long: both the biotic and physical character of the environment is liable to prove unstable and to alter so that much of the time the optimum density becomes a moving target which the machinery for adjustment has a hard time to follow. It was pointed out earlier that the prey-species cannot rely on its predators always to be on hand when there is a population surplus to be removed, and it has instead to evolve its own intrinsic safeguards against overpopulation. Conversely, it must be adapted to resist the attentions of predators whenever their demands get out of hand and endanger its own survival.

A rather extreme view of the evolution of protective adaptations might be that it is an automatic and inevitable consequence of a very powerful form of selection, conferring as it does on the better-protected individual a greater chance of survival, and readily leading therefore to cryptic and mimetic resemblances of marvellous perfection; but that this does not really benefit the species as a whole or affect its fortunes in the slightest, since its numbers are not controlled by predation anyway. In other words, provided the predators take nothing but the expendable surplus, protective adaptations will develop just the same but will be adaptively quite neutral or valueless, since they give protection merely against something that cannot endanger the survival of the species. A theory on these lines in relation to the evolution of mimicry in insects was in fact put forward in 1927 by A. J. Nicholson. Insects so frequently have a very high rate of fecundity but lose all except a minute part of their output through non-selective agencies; and he concluded that 'the success of a species . . . does not depend on its relative freedom from attack by the selective agent [*i.e.*, predators], but on its power of avoiding attack by the non-selective eliminative agent' (Nicholson, 1927, p. 98).

This could perhaps be true under continuously stable conditions where the prey-species really never was endangered by its predators, but it can safely be assumed that even perfectly-regulated predators hunt within their own territorial claims for whatever they like and can catch; and that these adaptations against 'overfishing' give them only a blanket coverage for the resources of the habitat *en bloc*, without affording discriminating protection to every potential food-species separately. Moreover they have got to evolve their dispersive adaptations in the first place and maintain them through natural selection; and it would seem an important asset in these circumstances for the prey to be not less well protected than other species in the same plight, and especially to avoid attracting more than the minimum pressure of predation during emergency periods.

It is well recognised that Batesian mimics must not become too numerous relative to their models, otherwise their false warning (pseudaposematic) appearance will be disregarded by the predators. The highest permissible ratio of mimic to model can be expected to depend in part on the deterrent qualities of the model: provided the consequences of making a mistake are sufficiently formidable the predator may be very cautious, even in experimenting. It is a matter of repeated field-observation, however, that Batesian mimics do tend to be scarce in relation to their models.

Before a particular species of insect begins its evolution as a mimic, it may be presumed to have developed dispersionary adaptations of the usual type, which tend to produce population-densities optimal for the particular habitat. There seems to be much less chance of an abundant edible species being modified through selection to resemble a scarcer distasteful one than there would be if the numerical proportions were reversed and the edible species were the scarcer of the two. One reason for this is that if selection favours a certain variant of an abundant edible species, because it confers some degree of mimetic protection, the incidence of this variant in the population will go on increasing, up to the point where the advantage diminishes again because the mimic is becoming too numerous relative to the model. At this stage the drift of selection may even reverse if both the variant and the model suffer from severe predation. On the other hand if the incipient mimic has a relatively low optimum density to start with, it will still be sufficiently uncommon relative to its distasteful model even when the mimetic variety comes to predominate, and will retain a marked selective advantage over less perfectly mimetic forms.

A species that through evolution becomes a mimic may be expected initially, therefore, to be regulating its numbers homeostatically in relation to its own food-resources, its optimum density being in fact well below any danger-threshold in relation to that of its model. Independent changes may subsequently alter this relation; and it is possible therefore that in the course of time a ceiling on the mimic's numbers may come to be imposed by the abundance of its model: wherever the mimic exceeds a safe proportion it will forfeit the protection it has acquired against predators, and this will result in shifting the basis of its homeostatic control away from the resources of the habitat into a secondary dependence on the numerical density of its model. It will then have perfected a relationship to its model in some ways resembling the dependence of a parasite on its host.

There are many cases known in butterflies where the geographical range of the mimic is strictly confined within that of the model; and in some of the polymorphic mimics, to be discussed below, different polymorphs occur in different places in general accordance with the regional distribution of their various models. It might be concluded from this that the Batesian mimic is so dependent on its model that it cannot survive in its absence; and indeed this may sometimes be true: but it should not be forgotten that mimicry cannot easily *arise* by selection in the absence of a model, and mere conservatism and stability of distribution may for long keep the ranges of the associated pair almost or completely stationary, with the result that there may have been little or no tendency for the mimic to spread farther than its model. There are cases of apparently Batesian mimics flourishing in the absence of their models: the celebrated African butterfly *Hypolimnas misippus*, for instance, was accidentally introduced in the days of the slave-trade into tropical South America and the Antilles in the absence of its model, *Danaus chrysippus*, and has survived and spread there (Carpenter and Ford, 1933, p. 111; Klots, 1951, p. 278). However, in this case the New World monarch butterfly, *Danaus plexippus*, has a continent-wide distribution, and may possibly do duty as a protective ' model ' almost as well as the original Old World *chrysippus*.

In mimetic associations, whatever the type, overlapping of ecological resources such as we considered in the previous section, if it occurs at all, is purely a secondary factor. The great Müllerian association of orange-and-black beetles belonging especially to the Lycidae and related families (including some of the fireflies and soldier-beetles) and their varied mimics—drawn from the Hymenoptera, Diptera and Lepidoptera as well as from other types of beetles—show a great diversity of feeding-habits especially in their larval stages (*cf.* Cott, p. 403). In the same way the wasps are essentially predacious feeders and do not to any important extent seek the pollen or honey of flowers, as do many of their supposed mimics.

Both the model and the mimic require to assess their own population-densities independently of one another. One simple method of doing this is to use a different, usually non-visual medium for social recognition. The males of many butterflies, for instance, possess scent-glands on the wings; and though these are usually supposed to be employed only for courtship, they not improbably have an epideictic function as well, like most other forms of male adornment in different animal groups. The Danaidae, which often serve as models for other butterflies because they are distasteful, are known to possess such glands, and so also are members of the Papilionidae, a family notable for its adaptability in producing mimics. As just mentioned, some of the fireflies have ' lycoid ' warning colours and may belong to this Müllerian ring: but their own social integration is of course maintained by epideictic displays of flashing in the darkness, and probably has no connection with body-coloration at all. Auditory signals serve instantly to discriminate between the breeding males of the sibling phylloscopine warblers, such as the chiffchaff and willow-wren; and though these are not connected mimetically, the medium of sound would serve just as well if they were, and may possibly be used by some mimetic groups. There is thus no real difficulty about species that resemble one another in appearance being readily distinguishable by other means.

More attention has been paid to mimicry in the day-flying Lepidoptera than in any other group, partly because instances of it are particularly striking (*see* especially Punnett, 1915; Carpenter and Ford, 1933). As a premise to what follows it should be pointed out that in this group of animals sexual dimorphism of colour and pattern is rather common. Possibly in some cases it has evolved only for courtship purposes, but it would be anomalous in the light of our wide experience in other animals—including other insects—if it were not to have an epideictic function and if the dispersion of the breeding population were never consigned with it to the male sex. In fact there is circumstantial evidence of this in the frequent swarming of male butterflies at ' mud-puddle clubs ' (*see* p. 197), some of the participating species being sexually dimorphic in appearance (*e.g.* the common American sulphur, *Colias philodoce*).

Mimetic adaptations of unsurpassed perfection and intricacy have been evolved among the tropical swallowtails (*Papilio* spp.). There is one section of this large genus, the members of which are protected as adults by a nauseous quality acquired during their ' pharmacophagous ' larval stage, when they feed on *Aristolochia* plants which to most animals are poisonous. Innocuous species of swallowtails frequently mimic these, and mimic also the distasteful butterflies belonging to the Danaidae. In the most complex instances of mimicry—in *Papilio dardanus*, *P. polytes* or *P. memnon*, for example—there are two phenomena in particular that merit our attention. In the first of these, mimicry is confined to the females, the males being normal specifically-distinctive swallowtails; and in the second the females do not necessarily all resemble one model, but may be polymorphic, some mimicking one model, some another and some perhaps being non-mimetic.

Taking first the absence of mimicry in the male, we find this is not confined to species of *Papilio*. ' One of the most striking features of mimicry is its association with the female sex. It is very common to find species with non-mimetic males, having a pattern similar to their nearest allies, and mimetic females modified to resemble other and often widely different forms ' (Ford, in Carpenter and Ford, p. 121). Examples are to be found in the following species of *Papilio*: *dardanus*, *memnon*, *polytes*, *lysithous*, *cynorta*, *echerioides*; and, in the Nymphalidae, in the oriental *Euripus halitherses*, *Hypolimnas bolina*, *H. misippus* and *Argynnis hyperbius* (the last two being also found in Africa together with the same model, *Danaus chrysippus*); in the Satyridae in *Elyminias undularis*; also in a number of Pieridae including *Pareronia ceylonica* and several South American species; and, in the acraeine subfamily of the Danaidae themselves, in *Acraea alciope* (details in Punnett, 1915, except the last which is in Carpenter and Ford, p. 82).

This is clearly a situation that has developed a number of times in different butterfly groups. There are no known examples of the converse, in which mimicry would be confined to the males, though there are of course many in which both sexes are mimetic. There is no satisfactory ready-made explanation, the usual one put forward being that the male has less need of protection than the female. Ford, however, concludes (loc. cit., p. 123) that ' it seems more probable that selection has . . . definitely operated against the adoption of a mimetic pattern in the male while favouring it in the female.' He goes on to make a suggestion: ' An important function of the male is to stimulate the female to copulation. This seems to be performed by scent, touch and sight. It may be, therefore, that those males which conform to a certain uniform type are often more successful in stimulating the female than those which depart from it by reason of incipient mimicry.'

This seems by no means improbable; and it may be coupled with the advantage that the females are at least able to recognise the males visually in the first place even though their own identity is masked. But we are especially familiar in other sections of the animal kingdom with the existence of conspicuous, easily recognisable males coupled with concealingly-coloured females—for instance in the ducks or pheasants or possibly some of the ' blues ' (Lycaenidae) among butterflies. The functional similarity between cryptic and mimetic adaptations has long been recognised; and having this in mind another possibility commends itself even more strongly, namely, that in these mimetic species dispersion depends on visual recognition, and its discharge is vested exclusively in the male sex. We know this to be a very common situation in the higher animals and have already seen that it is likely to occur in butterflies. Non-mimetic males would in that case be able to recognise each other in the normal way and would respond by means of emigration and reproduction to regulate their population-density accordingly: this might induce the females to spread themselves out more widely where males were scarcer and seek the places where they were commoner, adjusting their own dispersion to that of the males. It seems difficult to avoid the conclusion that the absence of mimicry in the males, wherever this occurs, is simply a normal aid to social integration that has proved indispensable to the species concerned; and that this has as usual offset any lack of protection which the males suffer from their conspicuousness.

If this is the correct conclusion, then it is probable that in species where both sexes are mimetic and resemble their models closely, alternative non-visual methods are used for intra-specific recognition. Ford observes (loc. cit., p. 121) that in circumstances ' where both sexes are mimetic, the female is almost invariably the better mimic '; and it is possible that sometimes the males retain enough distinctiveness for intraspecific purposes while enjoying at the same time a considerable measure of mimetic protection. We may notice that, though mimics frequently imitate not only the shape and pattern but also the habits of their models, mimetic Papilios have retained the characteristic family habit of fluttering their wings when feeding, ' so that their true nature is often thus proclaimed in spite of their deceptive appearance ' (Carpenter and Ford, p. 91); and this can be coupled with Fryer's observation (1913, p. 230) of *P. polytes* that 'when attempting to discover each other the butterflies were extraordinarily blind to any individual which was not in motion '.

As a result of being non-mimetic in a number of the species concerned, the males give the impression of being to a large extent exempt also from the second phenomenon of special concern to us, namely mimetic polymorphism. Examples can, however, be found of Lepidoptera that are mimetic in both sexes where both are polymorphic together: for instance in *Hypolimnas dubius* (*see* pl. VIII), which has two forms mimicking two species of *Amauris* and belonging, incidentally, to two of the mimicry rings into which the highly polymorphic *Papilio dardanus* enters, as mentioned below; and equally in *Elymnias phegea*, which mimics *Planema macarista*, its two forms (common to both sexes) accurately representing the very different male and female sexes of the model. There are other mimics that have acquired a sexual dimorphism duplicating that of their model, but this is not quite the same situation as the last since only one form is found in each sex of the mimic, all the males belonging to one form and all the females to another.

Among the general Lepidoptera, and having nothing to do with mimicry so far as is known, there are cases of polymorphism confined to the female sex, for instance in the pallid varieties of the ' yellows ' or ' sulphurs ' (*Colias*), or in the *valesina* form of the silver-washed fritillary (*Argynnis paphia*). Fisher and Ford (1928) showed, in an investigation of thirty-five species of night-flying moths, that the females were significantly more variable in colour than the males. There is no obvious genetical reason for this restricted tendency in the male to vary and it may again be connected with his need for unmistakeable recognition as a standard member of his caste: the female, passive partner both in epigamic and epideictic rituals, is perhaps subject to a somewhat reduced pressure of social selection.

Mimetic polymorphism is very typical of the swallowtails, where it is always confined to the females. In *Papilio dardanus*—described by Poulton as the most wonderful butterfly in the world—three distinctive types of female (*see* pl. VIII) are found together, over immense areas of south and equatorial Africa: one (forma *trophonius*) mimics *Danaus chrysippus*; the second and most numerous (*hippocoon*) mimics *Amauris niavius*; and the third (*cenea*) mimics *Amauris echeria* (or in some parts of its range the rather similar *A. albimaculata*): all three, along with their models, are subject to some geographical variation and are completely different from the male. North of the equator there may occur a quite distinct form copying *Planema poggei*; and both in west Africa and Kenya additional types occur that are non-mimetic but easily distinguishable from the males by their lack of ' tails ' on the hindwings. Even this does not complete the tally, since in Abyssinia there are mimetic females that have not lost the tails, and in Madagascar and the Comoro Islands a very closely allied *Papilio* occurs, apparently retaining a pristine non-mimetic state, with only one normal kind of tailed female resembling the male.

South America has an equally elaborate example in *Papilio lysithous*; three female polymorphs each resembling a different pharmacophagous model can be found flying together in Brazil, and in all there are six named forms (Punnett, p. 44). In the oriental Tropics occur the celebrated *P. polytes* and *P. memnon*, each producing up to three sympatric female types. Sympatric polymorphism, whether in the female alone or in both sexes, is nevertheless on the whole a rare phenomenon: in addition to the examples so far mentioned it occurs also in the females of *Euripus halitherses*, belonging to a different family, the Nymphalidae; and in a remarkable set of three species, two of them belonging to the ' distasteful ' Danaidae—*Danaus chrysippus* and *Acraea encedon*—and one to the ' palatable ' Nymphalidae—*Hypolimnas misippus*—which have each three exactly parallel, named varieties associated into as many sympatric mimicry rings and varying in relative abundance from place to place (Punnett, p. 29; Carpenter and Ford, p. 109).

The particular interest of mimetic polymorphism to the student of

PLATE VIII

MIMICRY IN EAST AFRICAN BUTTERFLIES

Along the top row are three different species of danaids, all distasteful, which serve as models for three mimicry rings. The four butterflies in the second row are all *Papilio dardanus*, with the male (non-mimetic) at the left and three alternative mimetic forms of female. (A fourth, non-mimetic type of female which resembles the male is omitted.) The bottom row shows two species of *Hypolimnas*, namely *H. misippus* (left pair) which has a non-mimetic male and a *chrysippus*-like female, and *H. wahlbergi* (right pair) in which the sexes are alike but each appears in either of two mimetic forms.

1. *Danaus c. chrysippus* (L.).

2. *Amauris crawshayi* Butler.

3. *Amauris niavius* (Trimen).

4. *Papilio dardanus tibullus* Kirby, ♂.

5. *P. d. tibullus* ♀ f. *trophonius* Westwood.

6. *P. d. tibullus*, ♀ f. *cenea* Stoll.

7. *P. d. tibullus*, ♀ f. *hippocoonides* Haase.

8. *Hypolimnas misippus* (L.), ♂.

9. *H. misippus*, ♀.

10. *Hypolimnas dubius wahlbergi* f. *mima* (Trimen).

11. *H. d. wahlbergi* f. *wahlbergi* (Wallengren).)

1
2
3
4
5
6
7
8
9
10
11

animal dispersion is its implication that population-ceilings of mimics are in fact sometimes adapted to the abundance of their models. We began by imagining an initial stage before mimetic evolution started, in which the optimum density for a particular species was adapted in the normal way to the conservation of its food-resources; and the question arose whether, after the mimetic association had developed, this density-limiting factor might not be exchanged for another in which the mimic's limiting density became directly related to that of its model.

If we could reconstruct the history of the evolution of any one of the polymorphic species we might expect to find a preliminary stage in which there was as yet only one model, more or less faithfully copied. Such a stage is represented by *Papilio cynorta*, which imitates the female sex of *Planema epaea*. Now the *Planemae* are sexually dimorphic and serve as models for several other butterflies, among them *Elymnias phegea*, which has advanced a stage farther and succeeded in copying the male as well (its model is *Planema macarista*). To have brought this second adaptation about implies some added advantage, able to set the appropriate selective wheels in motion; and the only one apparent is the increase in numbers of the mimic relative to those of the model that it would allow. Assuming for simplicity that the population-densities of the respective *Planema* models, the pressure from predators, and so forth were the same in the two cases, the effective result of this double mimicry would be, presumably, that *Elymnias phegea* could afford a density of twice as many mimics as *Papilio cynorta* without any sacrifice of mimetic protection from predators. Where polymorphs have evolved which mimic different species of models a corresponding advantage is conferred, but there are subsidiary advantages also that have been variously discussed in the literature of the subject. Such a multi-specific mimic, released from dependence on the welfare of a single model, deploys its insurances more widely, stably and securely. Its geographical extent as a successful mimic is potentially as wide as the ranges of all its models put together and not merely of one of them alone; and perhaps by assuming several disguises it makes more difficult the predator's task of learning to detect its false pretences. The same elaborate adaptation has taken place, doubtless for the same reasons, in a number of other insects, such as the familiar hover-flies, *Volucella bombylans* and *Eristalis intricarius*: these feed on the pollen and honey of flowers as adults and closely resemble small workers of bumble-bees; each has a range of polymorphic forms whose furry abdomens mimic either the light-tailed *Bombus terrestris*, the all-buff *agrorum*, or the red-tailed *lapidarius* types.

The genetical basis of mimetic polymorphism has been worked out in some Lepidoptera, and is no less perfectly adapted than are the visible imitations attained outwardly by the insects. The many complex details of colour and pattern distinguishing the various polymorphs have evolved a simple dependence on a very few major genic differences, so that the progeny of crosses between different polymorphs develop into standard mimetic types and not into a swarm of intergrades (*cf.* Ford, loc. cit., p. 118).

Nevertheless, the ' population dynamics ' of a set of polymorphs, mimetic or otherwise, presents most difficult problems, which are still some way from solution. In the type of polymorphism we have studied here there can be no doubt that the selective agency responsible for bringing the situation into being is predation: but this does not automatically imply that predators are continuously responsible for maintaining the population-ceilings and the relative proportions of the various polymorph types to one another, as might casually be assumed.

We have no difficulty in imagining the populations of each different member of a Müllerian ring being independently regulated in terms of the food-resources available to the individual species concerned: there is no significance in this case attached to the relative numbers of each; and nothing but good is to be gained by the association of more and more species provided they are all genuinely and adequately distasteful. In the Batesian relationship, however, we have concluded that the population of a mimic must not exceed a ceiling that depends on the abundance of the model—a conclusion held by most authorities on mimicry.

It seems perfectly clear that the proportion of mimics to models is not a relationship that could be kept in balance and adjustment by predators alone, however adaptable the latter might be. A practical illustration of this may be taken from *Papilio polytes*, the genetics of which were worked out in Ceylon by Fryer (1913). In this species one of the three female polymorphs is a non-mimetic form very like the male, and the other two resemble pharmacophagous models. Two are thus protected from predators by mimicry and the third is not. If the numbers in the adult population were being controlled by predation, this would inevitably fall more heavily on the non-mimetic type and lead to its reduction in numbers relative to the others; but in fact it is sometimes the most numerous of the three and shows evidence of maintaining a stable relationship with the other two. Fryer (loc. cit., p. 250) summarised the position in Ceylon by pointing out that, if the statistics he had collected were reliable, ' the ratio between the mimetic and non-mimetic females is one which might be expected if it be assumed that there is no selection in favour of either of these forms of female; under these conditions the population is stable and may remain so indefinitely '.

It is scarcely to be expected that populations of mimetic species would have discarded the normal and relatively efficient adaptations for internal homeostasis possessed by other related insects in favour of a much more haphazard and undependable control by predators. Nicholson's (1927) essay on mimicry, quoted earlier, makes it abundantly clear that mimetic adaptations could evolve as a result of predation even if the predators were never at any time actually controlling the numbers of their prey. In this respect the situation is very little different from that of the cryptic insects, birds or mammals that retain the primary regulation of their own numbers, through territorial systems and the like, notwithstanding the pressure of predation upon them and its manifest selective effects. According to the dispersion theory it is a matter of fundamental principle that the higher animals are able to accommodate to these external pressures in exercising their homeostatic powers, and react spontaneously to restore the optimum whenever it has been departed from.

An interpretation along these lines accords with such otherwise puzzling features as the participation of mimetic insects in male swarms, the retention in some species of non-mimetic males presumably for the purpose of visual integration and the survival of non-mimetic female forms in some of the polymorphic *Papiliones*. If normal homeostatic processes are at work there is probably a ceiling imposed on the number of females fertilised per unit area, or the number of eggs laid, or both. Fryer (loc. cit., p. 234) was perplexed by the frequency of ineffective matings in his captive *polytes*; and after he had detected the fact initially he recorded only 15 fertile matings out of 34—a result that he could attribute neither to inbreeding nor the effects of captivity: in addition, female reluctance to pair was often very protracted. There may be a great surplus of adults normally produced, many of them doomed not to breed; and in these circumstances selective pressure from predators could be sufficiently heavy to perfect and maintain the mimetic types without ever reducing the breeding population below its optimum density. Even the survival of non-mimetic forms is not necessarily endangered although predation takes its heaviest toll of them. It cannot of course be imputed with certainty to predation, yet we may notice just the same that some of the polymorphic swallowtails have already lost their aboriginal stock of non-mimetic females, while other butterflies have become mimetic in both sexes, having lost their non-mimetic males as well.

Throughout this discussion of the possible methods of population-control in mimetic species we are hampered by the extreme complexity of the situation, the lack of sufficient knowledge and the fact that the phenomenon has not evolved in birds, whose behaviour is generally so much more easily and fully ascertained than that of tropical insects. We can therefore make no use of the analogy with birds which has so often helped in other situations. It is with some hesitation that this section has been retained in the book, but it would have left a regrettable gap had so interesting and relevant a topic been completely passed by.

18.4. *Summary of Chapter* 18

1. The extent of integration between associated species can vary from zero virtually to unity. As a result of (*a*) contesting for conventional objectives rather than the ultimate resources themselves, and (*b*) the homeostatic adaptations that prevent population-pressure from reaching acutely high levels, competition between associated species is essentially social and is conducted with considerable formality and forbearance; this allows situations like the one between the two sympatric species of meadowlarks in the

central United States to persist and tends to prevent the direct ' show-down ' between competing species which Gause found in the ciliate Protozoa.

Subdivision of the habitat, however, frequently occurs where the ranges of two closely-related species overlap, and a number of examples of this are given. Tradition can sometimes play an important part in deciding what is to be the domain of each contestant: as an illustration, the way in which six types of grey geese (*Anser* spp.) parcel out the British Isles among them in winter is cited. It is apparent here that tradition is more important than ecological specialisation in determining the pattern and that interspecific competition is largely if not entirely confined to the social sphere. It is not impossible that the same influence affects the limits of their breeding ranges also; and this may be equally true of other species.

2. In this section the problems of the dispersion of sympatric sibling species are surveyed. The African weaver-birds illustrate a sibling association in which, though the males at least are specifically distinctive in the breeding season, their recognition marks are afterwards very ' deliberately ' suppressed. Large mixed flocks are characteristic of the asexual season when field-identification of individual species becomes impossible. The inference is that interspecific differences have been submerged as an aid to consolidating a visually-integrated, multi-specific population, dispersed on one common unified system.

This case is not unique: others are given of birds that also look alike and share the same habitats. Examples are easily found in insects also and a good many are cited, drawn, *inter alia*, from the hover-flies, bees and wasps. Five British species of *Vespula* which are typically social are more like each other than any of their supposed mimics are to them, and have very extensive overlaps in their ecology. Such a situation is typical of the genus, many of the same species having a wide distribution in the north temperate zone. More confusing still are the *Halictus* bees, of which much larger groups of species can be found in many localities, often nesting in mixed associations in one bank of earth.

These sibling associations are not peculiar to terrestrial habitats; mention is made of a marine species-pair in *Calanus* and a trio in *Limnoria*.

Interspecific differences appear to be sublimated during the time spent in feeding on common resources, and individuals of associated species then compete only socially, as if they belonged to one homogeneous society—a sort of *condominium*. Gause's principle does not drive them asunder, or at any rate its action can be delayed.

Sibling associations are on the whole uncommon and sporadic in the higher animals, which implies that they generally have a rather short evolutionary life. The advantages and disadvantages of such a high level of integration between species are discussed; and this leads once again to consideration of the opposing selective tendencies of ecological differentiation *versus* the retention of broader assets and less specialisation.

Though this subject proves to be difficult and complex and not capable of complete resolution at the present time, there seems no doubt that the prevailing concept that species with shared resources must inevitably be in relentless competition needs to be modified and to take into account the non-expanding, non-aggressive character of the typical homeostatic animal population.

3. Mimicry raises further difficult though interesting questions in relation to dispersion. It is best developed in insects, which are the most numerous terrestrial animals and attract correspondingly heavy predation from birds and other vertebrates. In Batesian mimicry the benefit obtained depends on the predators not being confidently able to distinguish the mimic visually from its noxious or distasteful model; and in Müllerian mimicry a number of unpalatable species assume a common visual pattern which is thus more quickly learnt and avoided by predators.

It has been argued that these and other protective resemblances are nothing but the automatic and inevitable by-product of predatory selection; and that they do not benefit the prey species as a whole in the least, because populations are not limited by predators, which take only an expendable surplus, and predators cannot therefore threaten the species' survival.

In the ensuing discussion, dealing in general terms with predator-prey relations, it is seen that perfect homeostatic control of their own numbers by the predators cannot be realised all the time: they will be forced to over-exploit temporarily in emergencies. Even under stable controlled conditions their adaptations against 'overfishing' apply to particular *areas of ground*, with the contained resources *en bloc*, and do not afford insurance to each prey species separately. It is therefore to the advantage of each prey species to be at least as well protected or concealed as its fellows; and this suggestion that protective resemblances are without effect on the survival of the species concerned consequently appears to be invalid.

Selection is most likely to produce a lasting mimetic association of the Batesian type when the prospective mimic has a low optimum density relative to that of its model. Otherwise it may become too numerous, forfeit the mimetic protection, and even suffer a reversal of the selective trend.

Both mimic and model have their own respective optimum densities, and each must assess and regulate its own numerical strength independently. If they are very much alike this may not be feasible by visual means. Since the mimics often form a very small fraction of the combined whole, their numbers may be negligible as far as the models are concerned, and their existence ignored; but the converse can never be true: the mimics must always discriminate their own kind for epideictic purposes. This may fairly readily be done in some cases through an alternative medium, such as scent. It is significant, however, that in a number of mimetic butterflies, belonging to several families, only the females are mimics, the males being perfectly recognisable for what they are, and specifically distinctive. This is interpreted as indicating that in these species dispersion is done by visual means and is vested in the male sex—a very common situation among the higher animals, including many other insects, which have distinctive males and cryptic females.

Another development of equal interest is mimetic polymorphism, in which the mimics appear in several disguises, imitating different species of models. It is shown that this could develop as an adaptation to allow the population-density of the mimic to increase beyond the ceiling imposed by having only one model, without losing its mimetic protection.

It seems clear that the population-densities of mimics are not and could not effectively be controlled by predators; and that intrinsic homeostatic control is not only theoretically necessary, but in accordance with the known facts.

Chapter 19

The use of tradition

19.1. *Traditional conventions*

Convention supplies, as we have often seen, effective practical codes of conduct which can be substituted in the place of the cruder, more injurious forms of competition; and these codes are enforced by a system of purely social penalties. The greatest importance or survival value of this to the population lies, it will be recalled, in the formula it provides for preventing overpopulation and the overfishing thence resulting (*see* p. 10). Through the expedient of diverting competition away from food itself towards conventional prizes, such as the possession of property and social status, the power is acquired of stabilising the population-density at the optimum, and eliminating the uncontrolled scrambles for food that bring about overfishing and diminishing returns.

Conventional behaviour frequently turns out to be essentially inborn or instinctive—for example in the follow-my-leader reaction of the processionary caterpillars (*see* p. 115), or the defence by a male warbler or tit of a territory of the right size, in the right habitat. But a second very important constituent of conventions is found in tradition, the distinctive feature of which is the element of learning acquired by each new generation from its predecessors. Tradition and instinct are complementary to one another in conventional behaviour, tradition being especially resorted to in situations that do not lend themselves to regulation by universal instincts based, as it were, on 'first principles': where, for instance, conventional practice entails a knowledge of local geography. It is no doubt instinctive in acorn-woodpeckers to drill holes in the bark of trees of suitable consistency and to fit acorns into them at the appropriate season of the year, as an epideictic ritual (*see* p. 321); but the particular tree which is used for the purpose in any given locality necessarily becomes an object of tradition.

Very many of the acquired traditions of animals have, in fact, some such kind of topographical connotation. Perennial traditions are not easily developed in animals whose life-cycles are complete in a year or less, especially where successive generations are effectively isolated from one another in time; in these circumstances they are rare, though, rather surprisingly, not unknown, as we shall see. As would be expected they are especially common in long-lived animals with great powers of learning, such as man himself. In our own conduct we can observe that, while adherence to tradition can be strong and even binding, its hold on us has a kind of density-dependent quality. Wherever we are particularly subject to critical scrutiny—in facing a public, that is to say, to whom we owe it to uphold our social standing—we are most of all likely to conform with established custom: whereas when by ourselves, or among friends whose acceptance of us is not in doubt, traditional formalities are more easily set aside.

The reason for this lies, it seems, in the fact that tradition is reinforced by social pressure, and the penalties for non-conformity are social penalties; the result is that where there are no critics there is no compulsion to comply. The same really applies to all kinds of conventional behaviour. If the latter depends on instinct, and not on tradition, the individual is no doubt to a greater extent an automaton, compelled to conform with inborn rules; though it is actually possible to evolve even instinctive responses capable of being automatically varied in the same kind of way as traditional ones, in accordance with the varying stress of social conditions.

The dependence of conduct in these matters upon social pressure, as a general guiding principle, applies not only to man but to all animals bound by conventional obligations, either traditional or instinctive: these obligations tend to become more binding as social competition rises, and to be relaxed when it is diminished. Traditions themselves grow in power with use, the habit of obedience being progressively forged in the mind of the individual by repetition; and in a stable well-ordered community they usually become firmly set and difficult to disturb. They tend in fact to reinforce and perpetuate the population-stability from which their own strength is derived.

Conventions are attributes of the social group as a whole and not of the members individually; and they must consequently have been evolved and perfected through selection at the group level. The whole social machine, as we have frequently observed before, has been moulded by the same inter-group selective process; and the more perfectly it comes to function, the longer, presumably, tends to grow the average life-span of local groups as units. In some animals, like gannets, seals or termites, recognisable local clans, each possessing its own persistent geographical focus, are known to have continued in being at any rate for several centuries; this serves to emphasise how slow a process group-selection is liable to become, compared with selection amongst individuals.

As a form of conventional practice, tradition can of course effectively determine and impose a ceiling density on a population. When the local stock of one of these conservative species has been abnormally reduced by mortality—of the kind which affects the English stock of herons (*Ardea cinerea*), for instance, in hard winters—memory of antecedent traditions among the survivors can provide the pattern for restoring the population-density, as rapidly as possible, exactly to where it stood before the disaster.

It is but a step from this situation to contemplate another, where, after some greater catastrophe, the local population has been entirely wiped out, and its traditions lost beyond memory. To provide for such an event there has to be, necessarily, an inborn faculty in the colonists, arriving in due course to repopulate the vacant ground, to supply a practical formula for the forging of new traditions—a formula taking into account the potential carrying capacity of the ground. The growth and consolidation of tradition in such circumstances no doubt varies greatly in pace from one species or situation to another, and this may be reflected in the speed at which colonisation of new ground proceeds: in some circumstances an animal introduced into a new area may spread like a plague, and in others it may seem by comparison unaccountably cautious and hesitant. The European rabbit (*Oryctolagus cuniculus*) for instance has become a rampant scourge of huge areas of Australia. In Britain it was long extremely abundant also, until in 1953-54 the myxoma virus inflicted a death rate of roughly 99 per cent, over almost the whole country (Thompson and Worden, 1956, p. 158). Societies were often completely wiped out and their warrens thereafter deserted and forgotten; parts of the former dispersionary pattern were no doubt completely obliterated, and no inhabitants left with the traditional local knowledge needed to restore them. Taking into account the prolificity of the rabbit, which is evidently much more than enough to yield a hundredfold increase in four years, and the spectacular abundance of food resulting from the release of grazing pressure, the rabbit has made an astonishingly slow come-back in the succeeding years down to 1958. Then, in my home county of Aberdeenshire, myxomatosis sporadically reappeared, and the rabbit's progress was locally arrested once more; but I doubt if at that time the population had reached as much as a tenth of its pre-1953 density. One is tempted to conclude that rebuilding a traditional system of dispersion on new foundations can be a far slower and more deliberate affair than merely producing enough bodies to make good the original loss—especially perhaps in a very sociable or gregarious species like the rabbit, that depends on establishing a system of permanent burrows.

This case-history is probably not unique. The two right whales *Balaena glacialis* and *mysticetus* have both for some decades been reduced below the level of profitable commercial exploitation in the northern North Atlantic, but neither is extinct. They merely remain very scarce in their respective haunts, at population-densities which one may safely guess are less than one per cent of their original level. Neither species is subject to any mortality from whaling at the present time; yet the numbers remain in this depressed condition, for which it is difficult to account.

Possibly the same general phenomenon has contributed its share to the final collapse and extinction of some of the species of birds that have been over-exploited by man, such as the great auk (*Alca impennis*), eskimo curlew (*Numenius borealis*) and passenger pigeon (*Ectopistes migratorius*). The last in particular was once incredibly numerous, inhabiting much of the North American continent east of the Rockies. Its intense sociability, and habit of nesting in vast colonies sometimes many square miles in extent, and gathering in the non-breeding season to equally populous roosts for the night, made it unusually vulnerable to commercial exploitation. Indeed, a century ago hundreds (probably thousands) of men were regularly employed as pigeoners, and millions of carcases shipped in barrels to markets in the big cities like Chicago and New York. Persecution continued very systematically until the 1880s; but by 1889 or 1890 only a few small nestings could be discovered. The remnant managed to survive for another decade, however, until about 1900 when the species finally disappeared. (For amplification *see* Schorger, 1955.)

The pigeons fed when they could on beechnuts and chestnuts—both of which give very varying crops from year to year—though they could fall back when necessary on the less nutritious but much more abundant acorns. Prior to being seriously disturbed by man they appear to have had traditional nesting and roosting places (*e.g.* Trautman, 1940, p. 270); though, possibly in part because of the undependability of their food, they may sometimes have made rather intermittent use of them.

The older members of their great clans must presumably have carried with them on their wanderings an immense body of traditional local lore, relating to alternative places for feeding and assembly, distributed very likely over several states. Perhaps this dependence on elaborately organised traditions—so slowly built up but so easily extinguished under the ruthless mortality to which the pigeons were subjected—contributed not a little towards precluding any last-minute recovery.

The danger of destroying the framework of dispersion, when this depends largely on acquired tradition as it evidently did in the passenger pigeon, is one which has hitherto never been taken into account by man in his exploitation of natural resources of animals. The herring (*Clupea harengus*) comes at once to mind as a species by no means dissimilar in its highly gregarious nature and traditional migratory habits: it is possibly as vulnerable to collapse from this cause as the passenger pigeon—much more so, indeed, than any species composed of smaller and more localised population-units. We ought to be warned by experience and protect the southern North Sea stocks while there is still time, not assuming that they will be as resilient in recovering from the effects of overfishing as are the bottom-dwelling, less gregarious species like cod or plaice. The closer-knit the social fabric and the stronger the element of tradition, the slower and more difficult we may suppose is the task of re-creating the organisation once it has been dissolved.

It is perhaps unjustified to confine tradition by definition to its conventional uses—in the transmission from generation to generation, that is to say, of geographical customs, and of codes of display, communication and other forms of social conduct, with which we are particularly concerned. Novices can learn many things from initiates that are not in any way social or conventional: for instance, among tits, the technique of opening milk bottles in order to get the cream. This, once originated, was transmitted rapidly by imitation all over Britain and Denmark in the 1930s and '40s (Fisher and Hinde, 1949, 1951). Such an acquired technique might be called a tradition, though it is a type of imitative learning completely different in its biological application from the social traditions which concern us here. However, the boundaries of the field even of strictly social traditions appear to be rather difficult to define. For instance, the return of individual fish or birds to breed in their natal place is a typical social tradition; but the habit of various insects of laying their eggs on the particular species of food-plant on which they happen to have been reared, when alternatives are available, is much less clearly social although it still entails the element of arbitrary usage that strongly characterises convention.

19.2. *Constancy of tradition*

One of the ways in which long use of traditional sites is made known to us is through place-names referring to animals, given at a known period of history and still remaining appropriate at the present day. Most of these apply to sea-going animals, partly because their habitat till now has been largely immune from the inroads of human civilisation, whereas the notable assembly-places of inland animals have usually been molested by man; and partly because of their generally more striking habits of sociability. We have in Britain a number of ancient Viking names such as Lundy—' isle of puffins ', and Sulisgeir—' gannets' rock ', both dating from the 8th-10th century A.D., and still perfectly appropriate a thousand years later. Long-standing breeding places of seals and turtles are similarly known by old descriptive names in various parts of the world.

More tangible evidence comes from the stupendous deposits of guano which existed, until they were removed for use as fertiliser a century ago, on the arid bird-islands off the coast of Peru: deposits ' in glacier-like accumulations which had reached the maximum thickness possible under the laws of physics ', ' laid down during countless thousands of years ', and completely burying the more populous islands under a smooth lenticular dome (Murphy, 1936, p. 289). At Chincha Islands the maximum depth of guano was not less than 55 metres. In the exceptionally rich waters of this region, fed by the Humboldt current, so vast are the populations of sea-birds that almost every island off the Peruvian coast has been commandeered to hold nests, and numerous additional colonies occur on the mainland. It might be assumed that, where so much of the available ground appears to have been taken up, there would be little or no need for a special tradition to maintain the occupation of each particular site. But each island, in fact, has its own proper complement of cormorants, boobies and pelicans in varying proportions, and some are much more densely populated than others. Almost certainly each holds a typically autonomous and largely self-perpetuating community, obeying its traditional code as faithfully as in any bird-colony in other parts of the world.

The guano-bird *par excellence* is the cormorant *Phalacrocorax bougainvillii*. This belongs to a genus known elsewhere for its constancy. Thus Stuart Baker (1929, VI) tells of cliffs in Assam occupied by *Phalacrocorax carbo* and whitened by the droppings of centuries, where, the natives say, the birds have bred since the world began.

Another byword of constancy is the white stork (*Ciconia ciconia*); as Pliny remarked, it returns year by year to the same, usually solitary, nest. Nests commonly outlive their occupants but in the end generally grow top-heavy and tumble down, and have to be rebuilt. Haverschmidt (1949, p. 30) states that cases are known of sites being inhabited literally for centuries. So great, we may suppose, is the conventional obstacle to creating a new nest that, when it is undertaken, it is usually such an uphill business that the birds cannot rear a brood the same season (loc. cit.).

Constancy is exemplified no less by the smaller adjutant stork (*Leptotilos javanicus*); in Assam once more, Stuart Baker (1929, VI) relates that 'the nests . . . are used year after year for an immense period of time. A colony discovered by H. A. Hole in Sylhet in 1885 had been known to the hill tribes for as long as they had any traditions. When discovered it had fifteen nests and today, in 1929, it still has exactly the same number. When first seen it was in dense virgin forest; now it is surrounded by tea and cultivation, but the birds still breed there '.

This brings out a characteristic feature of traditional conduct in animals, namely the obstinate tenacity with which custom is adhered to, even in the face of destructive changes in the habitat. This has been pointed out also in the case of the carrion crow (*Corvus corone*) (Stewart, 1928); 'once chosen, it is remarkable how tenaciously some of [their] nesting trees are adhered to; for often a gamekeeper succeeds in destroying both nesting birds, and removing every vestige of nesting material, yet the following spring finds a fresh pair at the . . . site '. The same trait has led in the past to the decimation of many a gregarious species for commercial gain, including egrets, fur-seals and turtles, which have stubbornly returned each year to the scene of the slaughter until so few were left that the trade ceased to pay. Quite apart from any question of flagrant abuses, however, one begins to see increasingly clearly the immense and devastating effect that everyday human activities must have on the traditions and dispersionary stability of animals, notwithstanding their dogged resistance to change—through ploughing and planting, stock-rearing, burning and felling, building, noise, and pollution. A sure sign of this is the frequency with which in this book we have had to consult the past in order to find completely natural illustrations of dispersionary phenomena, still running with pristine perfection.

In looking to see where elaborate traditions have been evolved in the animal kingdom, we can obviously find them in profusion among the birds—above all in the larger, longer-lived species. Avian traditions are connected not only with breeding localities but, almost equally, with roosts, leks and similar places of assembly. In mammals traditions are also very common:

seals and bats, in particular, parallel the birds most closely in this respect; and in many tropical and subtropical bat-caves, centuries-old deposits of guano testify to the constancy of occupation. We noted in earlier chapters the commonness in mammals of traditional dunging-places (*see* p. 107), and the customary use of rubbing and marking posts, dens, warrens, trails, territories and singing-places (the last in coyotes and epaulet bats). The reptiles and amphibia could provide plenty of comparable examples, with their traditional hibernacula (rattlesnakes, etc.), communal breeding grounds (turtles, frogs), nest-sites, lairs and territories (alligators and crocodiles).

In the fish it becomes more difficult to find good instances of traditional behaviour, due in some measure perhaps to our much more limited opportunities for direct observation. Presumably it plays an important part in ensuring the repeated use of the same spawning beds year after year—a phenomenon well illustrated by the numerous species of whitefish (*Coregonus*), which inhabit boreal rivers and lakes and are frequently netted for food at their breeding sites. It is of course possible that suitable gravel or sand is in limited supply, and that the fish have no option but to exploit fully each suitable piece of bottom; but there are waters where two or more closely related species of whitefish are found together: for instance, in Squanga Lake in the Yukon (Wynne-Edwards, 1952*c*, p. 13), two species are stated each to preserve their separate ' traditional ' spawning areas. Brook trout (*Salvelinus fontinalis*) frequently breed in lakes when there are no suitable incoming streams for them to ascend, using the same beds year after year; and in the choice of these again tradition is equally likely to play a part. It is in members of this family, the Salmonidae, of course, that the ' homing ' faculty has been shown to be so highly perfected; but consideration of this comes more properly into the next section.

Among sea-fish the herring (*Clupea harengus*) in the North Sea area provides a good example of a migratory species which, as mentioned in the previous section, appears each year at the same places and the same dates, and spawns in selected areas, the general positions of many of which are known. The various fisheries depend very much on the regularity of the herrings' annual migrations. Herrings may live considerably more than ten years, spawning annually, and in general it appears very probable that an important element of learning (*i.e.* tradition) is involved in the conduct of their customary migratory cycles, which are naturally different for each of the major spawning groups or discrete populations inhabiting the region. In the case of the enigmatic freshwater eels (*Anguilla*), on the other hand, we must note that they spawn but once in their lives, and though they have of course been in the Sargasso Sea before during their earliest larval stages, the great journey they undertake in returning to it is presumably mediated very largely by inherited instinct; in the end, of course, the effect is the same either way, since the fish appears in the conventional spawning place at the appropriate time.

Tradition appears to be used, at any rate, in all the classes of vertebrates, as a valuable means of conditioning the individual to respond correctly, in accordance with the established local pattern of dispersion. It takes its place as the adjunct of hereditary instinct in governing conventional behaviour, and no doubt has the advantage of being much more readily adaptable, compared with instinct, when secular changes take place in the habitat.

The 'higher' adaptations evolved for regulating dispersion—and of these the ability to acquire traditions is certainly one—have repeatedly been found to occur, not only in the vertebrates but almost equally in the insects and Crustacea. Many of the latter group are long-lived, especially in the decapods, and as we have seen, they may have elaborate systems of property-tenure and epideictic displays. It would be surprising if they had not developed tradition as an aid to dispersionary control; but, in fact, it has so far proved impossible to find a good instance.

In the insects, the colonies of termites and of some of the Hymenoptera, such as honey-bees and ants, often occupy the same established sites for scores of years and sometimes much longer; these colonies are as a rule continuously occupied, and consequently the element of tradition involved in their use, though it probably plays a big part in stabilising dispersion, appears on the whole rather vegetative or elementary. Some much more curious cases have come to light, however, like that of the bogong moth in New South Wales, which assembles in the dry season at traditional caves among the mountains (*see* p. 314); or the ladybird beetles and various flies mentioned in Chapter 15, that likewise have recognised communal sites to which the current year-class is somehow guided for summering or wintering: in these the individual insects are believed never to survive the twelve months from one gathering-season to the next, and if this is true the 'tradition' involved must be different in nature from the ones previously considered. Usually it is assumed that an odour remains in the spot from the previous generation of occupants; even so, it would still be logical to regard the sites as traditional —marked by a special cachet that can be accepted, renewed and passed on from brood to brood. Such marking could of course define not only the general position but even the precise area that it was proper to occupy.

Since Chapter 15 was written C. B. Williams' *Insect Migration* (1958) has appeared,* bringing to notice an excellent additional case of the use of traditional sites by Lepidoptera. This concerns the famous monarch butterfly (*Danaus plexippus*) of North America, which performs two-way annual migrations of up to 1,000 miles each way. After having grown up and matured in any part of the United States or southern Canada during the summer months, the over-wintering generation flies south to Florida, the Gulf Coast, Mexico and California. 'Having reached the desired locality—usually close to the sea—they settle on trees and pass most of the winter in a state of semi-hibernation; usually sleeping but occasionally flying round in bright sunshine. There may be thousands of butterflies on a single tree and in some localities the same group of trees are used year after year. In Pacific

* And since then again, Urquhart's (1960) monograph *The Monarch Butterfly*.

Grove, about sixty miles south of San Francisco, the butterflies have come every autumn for over sixty years and " butterfly trees " are one of the sights of the neighbourhood ' (Williams, loc. cit., p. 15) (*see* pl. IX).

As far as is yet known there are no survivors from one winter to the next, so that the tradition may again be maintained by persistent olfactory signs; but we ought not at present to exclude the possibility that at least a small proportion of experienced individuals live through a second season, both in this species and in others with not less remarkable traditions, such as the bogong moth. Experiments now being made with individual-marking methods may in the end reveal whether some at least of these insect traditions are handed down by the imitation of elders, in what can safely be regarded as the usual method in vertebrates. This might quite well supply a selective advantage for prolonging adult survival beyond the single season which is normal in the great class Insecta: the most serious obstacle to be overcome, perhaps, being to keep the fragile fabric of the wings in a serviceable condition for so long. A good many kinds of insect larvae, of course, and some non-flying adults (like the royal castes of termites and ants) live for a number of years, and the same is frequently true of adults in the other arthropod classes such as the Crustacea and arachnids. Only in the Ephemeroptera (mayflies) can insect wings actually be molted and replaced: elsewhere their expansion is held back until the last instar of life, and this can probably be assumed to result from an inherent frailty of construction, and ephemeral airworthiness. If any monarch butterflies were to live two or more years as adults, they need not necessarily belong to the category of two-thousand-milers, but might be enlisted from stocks breeding nearer at hand to the southern wintering resorts.

The use of particular trees by swarms of fireflies in oriental countries (*see* p. 198) may be another example of perennial tradition. Morrison (1929) records that about Bangkok in Thailand the trees used are always *Sonneratia acida*, which grow along river-banks and in mangrove associations, often with their roots immersed; this had been noticed earlier by Bowering in 1855. It is not clear whether the same individual trees are selected season after season; but the general preference for *Sonneratia* is quite possibly traditional and not genetically fixed. It is here that the swarms consisting solely of males are found, and their synchronous flashing, in the opinion of Smith (1935), is ' the outstanding zoological phenomenon in a country that abounds in zoological features of great interest '.

The illustrations which have been given, though not intended to be exhaustive, are no doubt sufficient to demonstrate the potential constancy of tradition and its value in perpetuating an orderly, workable pattern of dispersion. Traditions directed towards this end are essentially geographical, and embody tested methods of applying general hereditary principles to particular local situations. The orderly pattern can always be built up afresh, by animals colonising vacant ground, on the basis of innate instinct alone; but this may be a gradual, hesitant process. Once it is satisfactorily established, the pattern persists by tradition, each new apprentice first learning his duty in the accepted scheme, and later duly discharging it, providing in his turn an example to his successors.

We have been concentrating our attention hitherto on long-term, perennial traditions; but it may be noticed that there is no sharp division between these and the transient customs of behaviour which last only for days or weeks. Roosts, like those of those of the solitary bees (p. 309), may be adhered to with regularity, but only for part of a single season; nevertheless the older initiates may be guiding the newcoming novices, so that the habit survives a steady turnover of individuals, just as with longer-lasting traditions.

Aside from the labelling of special sites by persistent olfactory marks, as appears sometimes to be done in insects, or by some other indirect method, an unbroken sequence of seniors and juniors, present at the same time in the same community, is normally required for the process of handing traditions down. For annually-used dispersion patterns, there require to be at least two year-classes present; and, other things being equal, the force of tradition is probably increased by raising the proportion of experienced individuals, which have rehearsed the drill once, and preferably several times, before. This may be one of the selective factors tending to promote longevity in animals.

Relatively short-lived passerine birds, such as the Turdidae or Sylvidae, often show little or no trace of a persistent pattern in the exact location of their breeding territories from year to year; in the migratory species especially, their disposition within the habitat tends to be worked out afresh each spring. The communal winter roosts of the thrushes, however, as we saw earlier (p. 296), may be perpetuated for years by tradition. A small number of senior birds would perhaps suffice to indicate the position of the local roost to a large company of newly-arrived fieldfares or redwings in autumn; whereas permanent breeding-territories might become practicable only if the majority of occupants were aged birds, with relatively few gaps to be filled in their ranks by newcomers: this is what evidently occurs in some of the larger species, and in the blackbird *Turdus merula* (Snow, 1959, p. 34). No hard and fast rules can be given, however: in the European dipper (*Cinclus cinclus*), for instance, nest-site locations are very persistent and static; one site is known to have been occupied at a particular pool on the Capel Burn, Dumfriesshire—it is believed continuously—from 1785 to 1936; and, after an interval, it had become tenanted again prior to 1953 (Baxter and Rintoul, 1953, p. 228).

19.3. ' Ortstreue ' *in migrants*

It could be claimed with justification that there are no attainments of animal behaviour more remarkable than the navigational powers of long-distance migrants, whether birds, bats, fishes or insects. In the popular mind birds have generally been regarded as pre-eminent in this sphere, and

Monarch butterflies (*Danaus plexippus*) overwintering on the 'butterfly trees' at Pacific Grove, California. They are sunning themselves on a bough of Monterey cypress. (From a colour transparency by Mabel Slack, December 1953.)

thought to be endowed with a sixth sense unknown to man. The truth that has begun to emerge in the last few years, while it disposes of this notion, seems to offer a solution scarcely less unexpected: it is that birds can determine their position by direct solar and astral navigation, observing by eye and appraising the positions of heavenly bodies, using their innate ' physiological clock ' as a timepiece, and carrying the requisite ephemeris and automatic computer preformed in their brains. What the ship's certificated first mate can do in five minutes, as the fruit of two centuries of human ingenuity and invention, with the aid of a sextant, chronometer, nautical almanac, chart, and the application of spherical trigonometry, the smallest warbler, in effect, without prior training or effort, can do in its head: or so it presently appears at least from the contemporary work of Sauer and Sauer (1959) and Sutter (1959).

When we turn to the fish, we find the belief that ' every salmon usually returns to the same river in which it was bred ' is at least as old as Isaak Walton's *The Compleat Angler* (1653, pp. 135-6); ' it has been observed ', he states there, ' by tying a Ribon in the tail of young Salmons . . . as they swimm'd towards the salt water, and then by taking a part of them again with the same mark, at the same place, at their return from the Sea '. Since about 1900 experiments of this kind have been made on a much larger scale, by means of fin-clipping or durable marking tags, and the accumulating evidence has continued to support the same conclusion. In 1905, for instance, 6,500 salmon smolts (*Salmo salar*) were marked in the Tay as they migrated downstream, each with a piece of silver wire through the dorsal fin, and, of the 110 recaptures by anglers and netsmen in subsequent years, all were again taken in the river or firth of Tay (Calderwood, 1907; Menzies, 1931, p. 60). In 1924, Hutton wrote: ' I do not think that there is now any doubt that, guided by this wonderful homing-instinct, salmon almost invariably return to their own rivers. I would go even further, for I believe they endeavour to get back to the very tributary in which they were born ' (1924, p. 11). The word ' almost ' shows that exceptions to the general rule were known by that time, but these turned out to be rather few in proportion to the number recaptured in the native river.

Notwithstanding such conclusions by many of those best qualified to judge, general scientific opinion long remained adamant and sceptical: that a smolt that went out to sea when, say, two years old would as a matter of normal course be able to locate its native river two or three years later, when evidence showed it might have travelled hundreds of miles in the interval, seemed sufficiently improbable to demand very rigorous proof—the more so since no compelling motive could be shown for the development of so remarkable a faculty. However, the fact that the great majority of salmon do tend to behave in this way must now be generally admitted. Hasler and his colleagues have proved experimentally that fishes have the ability to distinguish the odours of different natural fresh waters (Hasler and Wisby, 1951), as we noted in an earlier chapter (p. 91); and the tentative presumption is that, among other clues, the salmon becomes conditioned to remember the special odour of the water in which it passed its earliest years. Recently, Stuart (1957) has shown that brown trout (*Salmo trutta*) possess similar powers of homing and discrimination: young trout marked as fry or later, after migrating down to the loch from their home stream, were found to return to the latter in subsequent years with great fidelity; and trout transported experimentally from one spawning stream to another usually succeeded in regaining their home water with very little delay (in one case by the following morning).

How salmon navigate towards the coast from the open sea, or are able to swim in the right direction after reaching it so as to locate the parental river, is still unknown; but we would hardly expect the method used to resemble closely the one developed by birds. The two are parallel adaptations, each obviously of great complexity and precision. It is legitimate to infer that other migratory fishes use comparable methods in order to return to their wonted haunts; and the same must indeed also be true of all the great migrants, such as the flying-foxes and insectivorous bats, the seals and their kin, the whales and porpoises, the marine turtles, and such two-way migratory insects as the bogong moth and monarch butterfly, which reappear year by year in the same appointed places. Whatever the means of navigational perception evolved, it is in every case a marvel of practical efficiency, deserving our highest admiration.

In the birds the accumulated evidence of thousands of recoveries of marked individuals of many species, in every continent, combines to show that the return of the adult to the place of its birth is everywhere the normal rule; and that having once taken up residence as a breeder in a particular locality, the bird thereafter almost invariably returns to it annually for the rest of its life. This fidelity to the customary place is now often designated by adopting the German term *Ortstreue*.

There is a varying minority of birds, however, that, subsequent to their first visit to their winter quarters, do not establish themselves the following spring at or near their birthplace, but appear in some other locality at a distance: a minority, that is to say, that is not ' true ' to its place of origin. Three examples are given here to illustrate the kind of scattering that can usually be expected. Stoner (1941) found in the bank-swallow (or sand-martin, *Riparia riparia*) that, of 221 returns of birds in years subsequent to that in which they were originally banded (in New York State), all but seven were within four miles of the banding point. This does not necessarily carry great weight, since it can be assumed that the search for marked birds was much less intense at greater distances: the farthest was only nine miles away. However, 186 had been originally caught and marked as adults, and of these 144 recoveries (77 per cent) were in the same colonies as before: on the other hand, of 35 recoveries of individuals marked as nestlings, only 8 (23 per cent) were in their natal colonies, the remainder having found accommodation elsewhere in the neighbourhood.

In an analysis of the British ringing records of the blackbird (*Turdus merula*), Werth (1947) showed that 72 per cent returned to breed at their birthplaces, 21 per cent bred within five miles (combined total 93 per cent), and the remaining 7 per cent were recovered breeding more than five miles away; in the song-thrush (*T. philomelos*) the corresponding figures were 52, 31 (total 84), and 17 per cent. The total numbers of recoveries involved were 358 in the blackbird and 249 in the song-thrush; and the 'non-randomness of search' factor which always enters into these calculations can be taken as less, with two such observable birds in an English countryside, than with the bank-swallows. Figures for the fidelity of adults to their former breeding localities are not available for these two species, but would no doubt be considerably higher than those given here for young birds.

Delmée (1954) reported on a population of stock-doves (*Columba oenas*) breeding in the neighbourhood of Tournai, Belgium, in terracotta vases or pitchers slung as nest-boxes in the trees. About half the marked adults (52 per cent) returned to breed a second or later time—this depending evidently on a relatively short life-expectancy rather than on any infidelity; but only 7 per cent of the young ones, reported later, had managed to return and find a place in their native colony. Of all the 127 birds recovered in later seasons, comprising 82 originally ringed in the nest and 45 ringed as adults (and excluding 14 definitely shown to be transients), 100 were found either at the colony or within 10 km of it, 13 between 10 and 20 km, 12 between 20 and 50 km, and the remaining two at 71 and 75 km (loc. cit., p. 216). In Belgium the stock-dove is extensively shot by hunters, and this sample may therefore be taken as giving a more reliable indication of the true degree of *Ortstreue* than in the foregoing cases.

Migrant birds for the most part spend their lives in one of two localities, namely their breeding place and 'wintering' place; and there is enough evidence to indicate a similar tendency to be as faithful to the second as to the first. The wonderful gifts they possess as navigators have been evolved to enable them to undertake the journey successfully between the two. The need to perpetuate the societies into which their populations are loosely formed, in order to permit local traditions to grow and flourish and numbers to be controlled on a local basis by established custom, provides the real motive for the evolution of these extraordinary talents. It is this dependence on the basic principle that local populations must be self-perpetuating and self-regulating that gives selection the great power needed to evoke them. Furthermore, it is on this same condition of permanent attachment to ancestral ground that the whole phenomenon of group-selection itself depends. Seen in this light, it seems impossible to doubt that the habit of long-distance migration is only possible in animals which can evolve a sufficient faculty of precise navigation to enable them to return to the exact place whence they came.

The scale of the population units into which the species is organised no doubt differs from species to species; and young stock-doves appear content to secure a holding anywhere in a home area several kilometres in extent, without any strong compulsion to revert to a precise family site. In the returns of the blackbird and song-thrush a very similar condition is evident, with a fair proportion appearing at a little distance—up to five miles anyway—from the point of origin. It seems likely that this is perfectly consistent with local autonomy in dispersion, the homeostatic machine being geared to deal broadly with regulation of numbers over local areas of this order of size. In some cases the autonomous units may be much larger: for instance there is an indication that in the breeding season individual Leach's petrels (*Oceanodroma leucorrhoa*) may come and go between North Rona and St. Kilda, over a hundred miles apart: for an adult bird ringed in July 1958 on the former was recovered 14 days later on the latter (T. Bagenal, verbally, 1958). With the whole Atlantic Ocean as one's communal territory it might be necessary to establish some degree of inter-colonial co-ordination; and excursions to join in the nocturnal epideictic ceremonies of neighbouring colonies might be the means of attaining it.

In addition to the relatively minor effect of local scatter, there is an apparently different phenomenon, discovered to occur in a striking form in some of the ducks in the early days of bird-ringing, and called ' abmigration ' (A. L. Thomson, 1923, p. 276). It appears as an exception to the general rule of constancy to the native area, in that the migrants reappear in an entirely different region and establish themselves there. Different individual teal (*Anas crecca*) reared in Cumberland, for instance, were recovered in subsequent summers or autumns in Norway, Sweden, Denmark and Germany; and a wigeon (*A. penelope*) reared at the same place was found nearly six years later in Petchora, U.S.S.R. (Witherby, 1927). Other similar cases have since come to light, and it has been surmised that the phenomenon in the wigeon may be due to birds being led away in spring by foreign-bred mates they have picked up in their winter-quarters (Donker, 1959, p. 22).

An equally striking case is that of an adult female salmon (*S. salar*), stripped in a hatchery in Ross-shire, Scotland, then released, and caught a year later south of Sukkertoppen, West Greenland, 1,730 miles away (Menzies and Shearer, 1957). It is a point of some interest that, though salmon are at present known to spawn in only one Greenland river (Nielsen, 1961), yet they appear annually in considerable numbers along the west coast and are very well known to the Greenlanders.

It appears likely that young recruits may often find it difficult to win an immediate place in a prospering society composed largely of older, established individuals—as for instance in the stock-dove colony described near Tournai; the alternatives are to wait patiently until a vacancy occurs or to try to get in elsewhere. Usually at a sea-bird colony there are non-breeders present, and presumably in later years many of these do succeed in winning a nest-site and becoming fully established; but in shorter-lived species like the blackbird and song-thrush the indications are that most newly-matured adults persist in their search until they manage to secure a territory of some kind on which to establish themselves. The fact is well ascertained that by far the greatest part of the wandering and colonisation that takes place is done by young, unestablished birds.

It has frequently been pointed out on earlier pages that a compromise is required between strict conservatism to the ancestral soil on the one hand, and a tendency to pioneer into unoccupied ground on the other. Without the element of enterprise and opportunism the long-term prospects of survival of the stock would be poor indeed: ground lost to the species by local extermination from one cause or another would never be regained, and the machinery of group-selection itself would grind to a halt. The degree to which this tendency ought ideally to be developed is bound to vary according to circumstances. It ought to be greatest among animal species living precariously in harsh environments, like deserts and arctic barren lands, or in temporary bodies of water, where they are subject to repeated failures and catastrophes; and least among those living in the serener conditions of the Tropics. Attention will be particularly drawn in the next chapter to the fact that the phenomenon of mass-emigration is normally associated with animals in the former unstable types of habitat, whereas it is almost unknown in the latter.

The degree of isolation between neighbouring local stocks of the same species must be greatly influenced by the relative strength of this urge to colonise. Where it is strong there will be a constant and considerable interchange of genes between one population and another, leading to a relative uniformity in the genetic make-up of populations scattered over wide areas; and conversely where it is weak a more effective reproductive isolation will facilitate the differentiation of local races. The Atlantic salmon (*S. salar*) spawns in rivers from the Baltic to the Gulf of St. Lawrence—that is to say, over an immense area—but no subspecific differences have been detected between the European and Canadian stocks; and this may depend to no small extent on there being a relatively strong pioneering element in every population of sea-run salmon, tending to divert a substantial proportion of individuals from the normal parent-stream fidelity, and set them off on random explorations. Almost equally good examples are provided by the ducks just mentioned, namely the European teal and wigeon, of wide-ranging species showing little geographical variation; and it is interesting to recall that attention was originally drawn to the probable correlation of the two phenomena—abmigration and geographical invariability—at the time that abmigration was originally described (H. G. Alexander, 1923). Their stronger pioneering bent may have contributed something to the general tendency in arctic and antarctic animals to be wide-ranging and even circumpolar. Localised distributions of the ' endemic ' type are relatively uncommon among them; whereas in tropical countries, especially on archipelagos where physical barriers to interchange are added to instinctive ones, there tends to be a multiplicity of species and endemic races, to an extent that cannot be accounted for merely by the higher productivity and denser populations that exist there. Clearly there are other possibly much more weighty factors involved in bringing about this difference in geographical variability between polar and tropical faunas (and floras), and their relative importance cannot be seriously considered here; so that it must suffice to mention this aspect without laying emphasis on it.

19.4. *Summary of Chapter* 19

1. Conventional codes can be either instinctive or learnt; in the latter case they are called traditions or customs. The distinctive feature of tradition is its dependence on learning by new generations from their predecessors. Traditions are particularly useful where conventional behaviour involves a knowledge of local geography, or other detail too complex and changeable to be transmitted genetically. They are especially prevalent in long-lived animals (including man) and their great usefulness may constitute one of the factors tending to promote longevity (through group-selection).

Traditions, like innate conventions, are reinforced by social pressure, and non-compliance invokes only social penalties. They grow in power by repetition, and tend to cement the stability of the stable populations in which they flourish; but their initial growth and establishment may be slow. If for example harsh climatic conditions cause severe mortality in a population with well-established local traditions (*e.g.* herons in England), memory of these traditions by the survivors provides the pattern for restoring population-density, as quickly as possible, to its antecedent level. But if mortality were still more severe, and the species were locally exterminated, no traditions would survive; and the subsequent process of recolonisation would lack their guiding influence, and be either uncontrolled or relatively slow and hesitant. This loss of regulating traditions may have been responsible for the unaccountably slow come-back of European rabbits after myxomatosis in Britain, or of the northern right whales which remain scarce to this day although they have not been hunted for half a century. It may have contributed to the final collapse and extinction of the passenger pigeon and some other birds. Man has never considered this important factor in his interference with nature; we ought to take heed of it, for instance in the case of the North Sea herring, which shares many features with the passenger pigeon.

2. Local traditions can be extremely stable: they can survive for hundreds and even thousands of years, and be persisted in stubbornly in the face of injurious changes. All the vertebrate classes show them; surprisingly they occur also in some insects, even though the duration of social life is generally less than a year, and face-to-face contact may not then be known to take place between successive generations. A good example is provided by the aestivating conditions of the bogong moth, described in the last chapter; or the hibernating conventions of monarch butterflies, which annually resort to the same trees in Pacific Grove, California. Insect wings are frail and irreplaceable and seldom airworthy for more than one year; but those of danaids are well known to be exceptionally tough, and a sprinkling of second-year adults may turn out to provide the link which keeps this tradition alive.

Short-term conventional habits—loosely definable as customs, as opposed to perennial traditions—are of course very widespread, and may play an equally important part in maintaining temporary patterns of dispersion.

3. The navigational powers of long-distance migrants are among the most wonderful achievements of animal behaviour; by means of them birds can make recurrent journeys between individual summer and winter localities, salmon return to spawn in their native streams, and sea-turtles repair to nest on traditional beaches. The phenomenon of *Ortstreue*, which means the fidelity of the offspring to the land of its parents and of the individual migrant to its established haunts and ports of call, is a most important kind of tradition; it is a trait indispensable to true migrants, whether they are whales, ruminants, seals, bats, birds, turtles, fishes or insects.

Ortstreue allows the individual to remain a life-long member of a self-perpetuating local group and still enjoy the fruits of long-distance migration: this, as always, is the foundation of homeostasis. It is safe to infer that migratory habits could never develop apart from the navigational faculty that permits *Orststreue*.

As in all other animals, migrants must provide a pioneering minority that can cast the bonds of *Ortstreue* aside; abundant examples are known from bird-ringing, and some from tagged fish, including salmon. Pioneering tendencies must be stronger in harsher environments, where the need for recolonisation is more common; here therefore gene-flow is more likely to be rapid and geographical speciation to be diminished: in contrast, serene environments tend more to run to endemism.

Chapter 20

Fluctuations, irruptions and emigrations

20.1. *Fluctuations imposed by climate*

The populations of some animals are of course by no means stable, but tend to rise and fall, sometimes on a great scale. Those that during periods of exceptional temporary abundance interfere with man's economic or personal welfare, like voles, locusts, army-worms, wasps or jellyfish, are frequently regarded as plagues. Density-fluctuations can be of practical import to man for another reason when they affect species valued as fur-bearers, fish or game. In regions where fluctuations are recurrent and pronounced they constitute a striking phenomenon, interesting in its own right and because of the light it may be expected to throw on the general question of how animal numbers are regulated. Consequently they have attracted an unusual amount of attention from ecologists, and a tremendous literature.

According to our theory, animal population-densities are self-regulated, so far as this proves practicable, by the mediation of special homeostatic adaptations. It is particularly noticeable in temperate latitudes that the annual harvest from natural food-resources of many kinds is inconstant, varying greatly between good years and bad. In bumper years we see abundant crops of berries, nuts and seeds, adults and larvae of insects, or newborn prey for the larger raptors; figures obtained by the Swedish Institute of Experimental Forestry show, for instance, that spruce-seed production can vary something like a hundredfold between rich and poor years (*cf.* Svärdson, 1957, p. 337). Many of the less easily estimated sources of food fluctuate similarly, including the quantity and quality of pasture grasses and also of perennial forage plants such as heather (*Calluna vulgaris*). In the worst years the yield of some of these resources diminishes locally to nothing. We should expect the machinery regulating the population-densities of animals dependent on them to bring about continual adjustments seeking to match the population level to the changing carrying-capacity of the habitat.

We can expect that changes in food-availability will sometimes be so sudden and disorganising that the homeostatic machinery is unable to cope with the situation fast enough to maintain an optimum density; other things being equal, the more erratic the climate and environment, the more likely this would be to happen. Dispersionary control would be temporarily lost and starvation and resource-damage might occur.

The first hypothesis to examine as to the causes of fluctuations in animal numbers must clearly be that the instability is due primarily to fluctuations in the physical environment, and especially to those that are reflected in the availability of food. There is a marked correlation between the distribution of animals with unstable numbers on the one hand, and notoriously adverse climates on the other, including especially the high boreal and the steppe or desert regions (*cf.* Kalela, 1949, pp. 74-5; Lack, 1954, p. 238). Two of the most spectacular plague-animals known to the western world, the lemmings and the locusts, are each typical of one of these environments. From such extremes there is a general tendency for the incidence of fluctuations to diminish towards the stable, constant conditions found in tropical rain-forests, oceanic islands, or deep-sea habitats, where fluctuations appear to be relatively uncommon. The view that there is generally an underlying causal connection between climate and the rise and fall of density is, in fact, widely held: A. J. Nicholson, for instance, states that ' it is beyond dispute that the fluctuations in animal populations with which we are most familiar are dominated by climatic fluctuation ' (1947, p. 11). Parallel changes, and particularly years of peak abundance, are in some instances synchronised in well-separated populations of the same species covering a large area, and in the light of this it again appears that ' the only cause of the periodicity must . . . lie with the environment, and here the only possible factor which is acting in a similar way all over these regions is climate ' (Elton, 1924, p. 132).

Before going further it must be pointed out that there is a conspicuous third category of plague-animals, associated with human cultivation. Here one may guess it is usually the plough or the axe that provides the disruptive force. Wherever agricultural land-use is changed on a secular rotation the attainment of a permanent, stable and varied climax is systematically prevented. The pattern of homeostatic dispersion built up in a habitat which has hitherto long remained unchanged is likely to be totally destroyed by the felling of forests or the draining, ploughing or planting of land, and populations of the pre-existing fauna, if any survive, may be expected sometimes to go out of control: an example of this is perhaps to be found in the transient plagues of voles (*Microtus* or *Clethrionomys*) not rarely occurring at an early, favourable stage in the growth of young conifer plantations in Britain (*cf.* Chitty, 1952).

The common feature of these three plague-producing situations is the disruptive force that—recurrently in the arctic and desert, and in certain forms of cultivation—interferes with the achievement of a stable climax. The balance is either overturned, or its attainment repeatedly prevented. In the plant world there is a most interesting assemblage of opportunist species, commonly described as weeds, that exploit such disruptions, quickly colonising newly-broken ground by virtue of their powers of dispersal and rapid germination. There they flourish briefly, before there is any serious competition; but their existence is characteristically insecure and transient, and, where ecological succession is allowed to proceed naturally, by and by they are overwhelmed by the invading permanent climax of prairie or forest, in old pastures or long-abandoned land. It is recognised that weeds have a marked affinity in this respect to those arctic and alpine plants that occupy the unstable, pioneer habitats left by solifluction, frost-erosion, and the destructive action of wind and water; and the same holds equally good for much of the desert flora: all are perpetually at the colonist stage of the succession, gaining a precarious livelihood in the face of physical hardships, but enjoying an unusual freedom from competition for space.

Among animals there appear to be some which have developed an analogous kind of opportunism, and have become specialists in exploiting the life of pioneers. This is most obvious in mobile species such as locusts, or in the predators that follow and live on them like the rose-coloured starling (*Pastor roseus*); in the Arctic the snowy owl (*Nyctea nyctea*) and the jaegers (*Stercorarius*) similarly make their temporary habitations wherever a glut of lemmings can be found. At least part of the stock of these various species lives a true nomadic life, existing from hand to mouth with no permanent home—here one season and gone the next; and though some of them may have reservoirs of habitat under more permanent occupation, even these are generally subject to the dangers inherent in all erratic environments.

There is no sharp line to be drawn between these birds and those whose irregular winter wanderings are dictated by the varying local abundance or scarcity of the seeds and berries on which they feed, such as the Bohemian waxwing (*Bombycilla garrulus*), the redpolls and siskins (*Carduelis* spp.), the bramblings (*Fringilla montifringilla*) and the pine and evening grosbeaks (*Pinicola* and *Hesperiphona*), to choose a few examples among the many present in different parts of the holarctic region.

Then again there are species that recurrently advance as far as they can go in the face of the elements, pressing northwards, for instance, along the frontiers of their arctic range to the limit of habitability, only to be forced sooner or later to give ground or be locally exterminated. Their struggle for existence, as I have pointed out elsewhere (1952*b*, p. 384), is ' overwhelmingly against the physical world, now sufficiently benign, now below the threshold for successful reproduction, and now so violent that life is swept away, after which recolonisation alone can restore it '. In this world the best of homeostatic adaptations must yield to *force majeure*, and at the worst, permanent stability is never attainable by animals or plants of any kind.

The desert is no less fickle and ruthless towards living things than the Arctic. ' It may be stated as a general rule that the rainfall in desert places, even in places where it is relatively regular, is liable to great variation from the monthly and annual averages. This variation intensifies the struggle between animals and plants and their environment, and is an important factor in the production of a desert' (Buxton, 1923, p. 11). Moreover, 'every element of the desert climate is liable to rapid fluctuations, which are often quite irregular. . . .' (loc. cit., p. 64).

As far as animal numbers in these erratic surroundings are concerned, both the increases and decreases appear most often to be correlated with corresponding changes in the quantity and quality of the available food. Some of the best numerical data on fluctuations are derived from records of annual sales of fur-bearers taken in Canada, and one in particular, the lynx (*Lynx canadensis*), is especially well known for the long run of almost regular cycles of ten-year periods it has undergone since comprehensive records began in 1821 (*cf.* Elton and M. Nicholson, 1942). It is certain that these cycles closely reflect the fluctuations in abundance of its principle food, the snowshoe hare (*Lepus americanus*). Some writers have suggested that the cyclic relationship between the lynx and the hare is a delayed density-dependent interaction, of the kind theoretically postulated by Volterra (*see* p. 389); but evidence will be given in Chapter 22 (p. 552) suggesting that, under overpopulated conditions, hares and certain other animals may die readily of 'stress' diseases. Variation in the primary link of the chain—the vegetable food available to the hares—may therefore be one of the important factors promoting the fluctuations. It could be the sole effective external factor, and in that case its effects would presumably be subject to the modifying influence of intrinsic, essentially social factors affecting the hares' dispersion and their recruitment rate. Except for the last suggestion, these various alternative explanations have been reviewed by Lack (1954, p. 212, *q.v.*). The research in progress at Aberdeen on the red grouse (*Lagopus l. scoticus*), an essentially vegetarian, fluctuating species rather comparable with the snowshoe hare, points to the primary connection existing between grouse population-density and the condition of the staple food-resource, the heather *Calluna vulgaris* (unpublished). Siivonen (1957) concludes rather similarly that the quantity and quality of the food available in early spring, in the pre-laying period, initiates the fluctuations in numbers of grouse-like birds in Finland; and Errington (*cf.* 1954, p. 67) found that whenever heavy snowfalls in Wisconsin covered the grains and weed seeds on the ground for a couple of weeks or longer and thus deprived most of the bobwhites (*Colinus virginianus*) of nourishing food, drastic population declines were apt to occur, 'whether any cycle-like regularities showed up in the year to year population fluctuations or not'.

Some fluctuations are sufficiently regular in periodicity to be predictable and deserve the names of cycles; but these are in fact everywhere a minority. In the past the tendency has often been to overstate the degree of regularity. Cole (1954) has shown mathematically that cycles averaging between three and four years could arise from essentially random causes, particularly if account were taken of the influence of one year's population-size in that of the next, as seems to be appropriate in the circumstances. On the average, every third 'cycle' would have a higher peak than its predecessor or successor, and this would give a similar random basis for a periodicity somewhat exceeding nine years. The cause underlying any of the approximately cyclic fluctuations of animal numbers could thus be a completely random factor—for instance a climatic one—and there would still be some semblance of a rhythm.

The commonest average periods found in natural populations are of course 3-4 years and 9-10 years. The recovery-rate of the species concerned may influence the periodicity it shows; but on the whole it appears that among vegetarian species the shorter period is commoner in more arctic climates, and the longer period in subarctic and more temperate ones. The willow-grouse (*Lagopus lagopus*) has in the past followed a four-year cycle in the mountains of Norway and a ten-year one on the North Shore of the Gulf of St. Lawrence (*cf.* Lack, loc. cit., p. 219). Carnivores of course tend to reflect the fluctuations of their prey, but here again a single species, such as the Canadian red fox (*Vulpes fulva*), can show different periodicities in different regions, according to what kind of staple prey it feeds on. The inference seems to be that the fluctuating food-supply is in each case involved as a pace-making factor.

A certain minimum interval is likely to be required after a set-back to allow the habitat to recover its full carrying capacity and the population of the animal concerned to recruit its numerical strength; these could put a lower limit on the frequency of oscillation. This has been well brought out by Svärdson (1957), who quotes Swedish figures showing the changing annual index of seed and fruit production by various native trees important as food for birds. It is uncommon for two good seed-years to follow consecutively; the plant requires not only a favourable growing-season to produce a good crop, but also time to recover from the nutrient-exhaustion caused by the last effort. In his view, as in my own, the fluctuations in numbers and the irregular movements of northern vegetarian birds are closely correlated, like those of the carnivores, with the varying productivity of their staple foods; and it is on these that climatic fluctuations are imposing their primary effects.

The purpose of this introduction has been twofold: first to confirm the conclusion that, of the ultimate variables in the environment with sufficient influence to trigger recurrent natural fluctuations of animal numbers, climate is at any rate the principal one; consequently where climate is invariably well behaved, natural fluctuations are likely to be rare if they occur at all. And second, that there is nothing inconsistent or irreconcilable between the homeostatic control of numbers and the existence of fluctuations occurring in a variable environment.

20.2. *Irruptions*

It is true of most species that the swings of population-density are generally rather local events, and synchronisation over huge areas comprising tens or hundreds of thousands of square miles is infrequent rather than normal. In any confined area the build-up of numbers is typically more or less gradual, and the decline sudden: when the time comes, emigration, and mortality caused by stress, parasites and predators, combined in varying proportions, sometimes remove the surplus so fast that the event is described as a ' crash '.

Primitively, the crash is the sequel to a state of overpopulation. Generally the food supply has been fortuitously removed or fails to materialise, thus creating an emergency by lowering the carrying capacity of the habitat; or, what amounts to the same thing, the population is allowed to build up unchecked to a level which, in the event, turns out to exceed the optimum. Among highly mobile animals, especially birds and flying insects, the quickest way of alleviating a situation thus fraught with danger is to expel the surplus; consequently emigration, often on a tremendous scale, is very characteristic of these groups whenever they find themselves in this kind of difficulty.

Under such conditions discharging the surplus, to a distance at which its demands can no longer be felt, is the way to protect the habitat from damage and to safeguard the continued survival of the stock left behind. As a group-manoeuvre its survival value is obvious. The emigrants may all in fact perish after they leave the habitat—their fate being a matter of indifference so far as the immediate future of the local population is concerned. Students who have not recognised the overriding power of group-selection have sometimes wondered why, since so many of the emigrants usually succumb, selection has not long ago eliminated the inclination to emigrate, by constantly wiping out the individuals that responded to it. We can possibly obtain a truer perception of the problem by regarding the kind of emigration at present under discussion as being the result of an automatic social guillotine, that chops off any superfluous segment from the tail of the hierarchy whenever the occasion demands. The inclination to comply at all times with the rules of the social code, whatever these turn out to be for any given individual, will be neither stronger nor weaker in those chopped off than in those that remain behind, and will not therefore be under selection: but any stock in which this safety-valve device (depending on obedience to the appropriate code of behaviour) decayed and ceased to function would soon overtax its habitat and be exterminated.

In the limiting case, all individuals may have to go, if the habitat becomes completely untenable. To give a familiar illustration, there is an analogous, if somewhat different situation that faces many birds every autumn, when the summer food resources are near an end and a move to other quarters becomes imperative: in some species (including most boreal insectivorous birds) all finally depart on migration; and in others, known as partial migrants, a residue remains in the summer habitat throughout the winter as a rule consisting in the main of older birds, or, especially in sexually dimorphic species, of males—presumably those that occupy the highest ranks in the social hierarchy. Examples of this could be found in the robin (*Erithacus rubecula*) in the south of England (Lack, 1943, p. 110), or in the song-sparrow (*Melospiza melodia*) in Ohio (Nice, 1937, p. 40). In the latter more remain behind in mild winters than hard ones, indicating the part played in the phenomenon by carrying-capacity. Mrs. Nice found that the individual inclination to be either resident or migrant was not a matter of simple heredity, and this accords with our general experience in cases of social stratification, *e.g.* in phase-determination of locusts, caste-determination of social Hymenoptera, high and low maleness in Crustacea and insects, or peck-dominance in birds.

Where practically all individuals leave, this may prove the best or only road to survival, offering at least a chance of securing a foothold somewhere else or of returning successfully once the period of inclemency is passed. Attention was drawn some years ago to this kind of survival-value, arising from mass-emigration and leading to the dispersal of the species, by Kalela (1949, p. 75). In such a context interest centres on the emigrants, not on those remaining behind, since theirs is the better chance of posterity.

There is evidently a range of divergent effects that can result from emigration—starting at one end with the safety-valve type that successfully blows off an unwanted surplus; then passing through the state of balance where it is a toss-up whether emigrants or stay-at-homes fare better in the long run; and reaching at the other end the condition where the former territory becomes uninhabitable, and emigration offers the only hope of survival. These are but different phases of a single general adaptation, and one versatile social organisation can, according to circumstances, produce any one of the three effects.

In Chapter 13 attention was drawn to the development, among migratory locusts especially, of a kind of super-adaptation for survival by mass-emigration, in species occupying impermanent kinds of habitat. Here it is not just a question of passive submission to the recurrent failure of food resources, and being driven to look for another place to live, but of a systematic, constructive effort to create an expeditionary force, when conditions are good enough to permit it (*e.g.* after prolonged rains). It involves the production, in numbers as large as the existing resources can afford, of a distinctive caste, specially adapted in structure, metabolism and behaviour to travel and pioneer. As with other social phenomena, the evolution in locusts of this migratory or gregarious phase has been brought about by group-selection—it could not have been otherwise; and the presumed objective is to get one or more new ‘ permanent ’ colonies established before the lease on the existing habitat runs out—in terms of evolutionary time always an imminent event.

It is generally thought that the existing ‘ outbreak centres ’ of the typical migratory locusts enjoy their relative permanence because they are exceptionally favourable habitats. For any one species their number may be quite small. The swarms which intermittently emerge from them do not in ordinary experience become established, but sooner or later dwindle and die out: this indicates that the new habitats selected have proved unsuitable, because of insufficient resources, excessive interspecific competition, excessive predation, infestation with parasites, or some analogous cause. As with the weeds and arctic alpine plants mentioned earlier, the absence of interspecific competition might prove to be an important condition in enabling locusts to gain a livelihood in the arid, unstable climates to which they are so highly adapted. The apparent difficulty of establishing new centres of population, in spite of the elaborate swarming adaptations evolved for the purpose, is an indication of the precariousness of their way of life. The small number of existing species of migratory locusts may likewise bear witness to a rather limited measure of success.

20.3. *Other emigrants from arid regions*

The question arises, to what extent other kinds of animals have evolved the habit of mass-emigration to serve the same end. Several other types of insect, some of which are named on p. 274, have analogous phase-systems affecting their larval stages. Most of these species are well known as migrants when they become adult. The silver-Y moth (*Plusia gamma*) is one of the commonest immigrants to the countries of northern Europe, in some summers appearing in incalculable numbers and freely reproducing; but it does not overwinter in the higher latitudes. Its permanent home is in the arid belt extending from the shores of the Mediterranean eastward into Asia. Crowding of its larvae leads to a striking average darkening of their pigmentation, a quicker development and a fourfold rise in activity (Long, 1953, 1955). Again, under the right conditions the moth larvae known as army-worms (*Laphygma* spp.) behave, as their name suggests, in an intensely gregarious manner, and this is also associated with an appropriate phase-difference in colour: the adult moths are known to emigrate. The probability that these particular cases parallel the locusts, in producing whenever possible an expeditionary surplus to aid in the dispersal of the species, appears to be high. However, a phase-effect—that is to say a visible differentiation induced by crowding—could perhaps also be evolved as a releaser for the commonplace safety-valve type of emigratory response, and in most of the insect species that show it further research would be necessary to analyse its function.

The majority of insects that are specially notable for the emigrant swarms they intermittently produce have not evolved larval phase-differences. The most famous is the painted-lady butterfly (*Vanessa cardui*). It is known to over-winter regularly—that is to say, to have a permanent home—only in the arid regions of North Africa and western Mexico (Williams, 1958, p. 25); but from these, and possibly other centres in the drier parts of Asia and elsewhere, it peoples almost the whole of the five continents during its emigrations. Few other animal species are so completely cosmopolitan. Over almost all of its vast recorded range, however, it is but a transient visitor, often breeding for a generation or two, but not permanently sustaining itself. Like the locusts, its indigenous home is in the arid, unstable climates bordering the northern desert zone; and it seems not unlikely that its irruptions have been similarly evolved for the purpose of dispersing the species. So extraordinary are its powers of dispersal that it can reach almost any part of the earth's surface. The ' pioneering tendency ', to which reference was made earlier in this chapter, has become hypertrophied so greatly that no geographical races are recognisable; and the ordinary subdivision of a species into partly-isolated local self-perpetuating units, which appears to play such an indispensable role in evolution, may practically have ceased to exist. The uncommonness of such a situation, and of so cosmopolitan a range, points once again to its transitoriness in terms of evolutionary time: in the long run it may be an untenable kind of regime.

In some other Lepidoptera, as we have had occasion to note in the previous chapter, regular two-way migrations have developed, in some respects comparable to the annual migrations of birds or fishes. This type of behaviour has been fully reviewed by C. B. Williams (1958). In a good many cases (*e.g.* the clouded-yellow *Colias croceus*, or the red-admiral *Vanessa atalanta*) the outgoing early-summer flight is much the stronger of the two and the return flight in autumn often difficult to detect. It is thus not clear that the migration usually succeeds in contributing a significant number of recruits to the regular permanent strength of the population. In the clouded-yellow the swarms going out from the North African permanent centres vary year by year from virtually nothing to such hordes that the yellows become temporarily the commonest butterflies over parts of Europe; comparatively few can ever return to Africa, and biologically there seems to be a pronounced difference between this type of migration and the type displayed by birds. The clouded-yellow type of movement seems much more akin to that of the painted-lady or silver-Y, disseminating the ' seed crop ' of a species living in an arid, unstable, impermanent kind of habitat. The return southward in the fall of the year, when it can be observed, may in that case be the final effort to fulfil an expeditionary mission, on which the spring emigrants set out many months before, rather than an attempt to bring reinforcements back to the parent stock surviving in the original home.

There are two arid-climate vertebrates especially well known for their irruptive mass-emigrations, the first being Pallas's sand-grouse (*Syrrhaptes paradoxus*) from the steppes of Asia, and the second the springbuck (*Antidorcas euchore*) from the dry plains of South Africa. The events in both cases have now largely become things of the past, though the sand-grouse could presumably still irrupt into western Europe were natural conditions propitious: the springbuck, unfortunately, is never again likely to reach a sufficient strength of numbers to produce the vast treks of a century or more ago.

There were two great irruptions of Pallas's sand-grouse into Europe in the 19th century, the first in 1863 and the second and larger one in 1888. Both occurred in the spring, evidently taking their departure from an unlocated eastern centre in advance of the breeding season, and reaching even the shores of Britain by May. Smaller visitations were recorded in a number of other years between 1848 and 1909, the last that reached the British Isles in any strength being in 1908. As has been recognised by earlier authors (*cf.* La Touche, 1934, p. 221), these great emigrations appear to ' constitute attempts to increase the range ' of the species; breeding took place sporadically among the newly-arrived emigrants, and was even repeated the following year among the remnant that over-wintered, usually in sand-dune country, in Yorkshire and Morayshire, Jutland, the Netherlands, and no doubt elsewhere. The great wave of 1888 radiated at least 3,000 miles from its presumed origin east of the Caspian, chiefly, as far as we know, to the west and north-west. In several important respects, including the semi-desert origin, the intermittent exodus, the massed long-distance flights of strong individuals, and the futile attempts to breed abroad, the irruptions of Pallas's sand-grouse resemble those of the insects already described.

A hundred and twenty years ago the springbuck still existed in pristine abundance on the high open plains of South Africa. ' On the advent of the Dutch Voortrekkers, the high veld was found to be teeming with countless numbers of springbucks. Driven from one part of the country by drought, and consequent scarcity of food, they migrated in vast herds of as many as half a million. This vast army would pour from the dry, desert-like high veld of the north-west into the great Karoo-veld, devouring every particle of edible vegetation as they proceeded. So vast was the stream that various other species of animals, including sheep and goats, were caught up in the seething countless swarm of springbucks and irresistably borne away. . . . Lions, leopards, Cape hunting-dogs, hyaenas, jackals and other carnivorous animals and birds of prey, like an army of camp followers, kept in the wake of these migrating herds, preying on them whenever they felt so inclined. So prolific was the springbok that, in spite of their natural enemies, severe periodic droughts, and the pigmy Bushmen and Hottentots, these antelopes continued to increase in numbers until the advent of the European colonist . . . . ' (Fitzsimons, 1920, III, pp. 91-93).

On one great emigration in Namaqualand the herd reached the Atlantic Ocean, where they dashed into the waves, drank the salt water and died: their bodies lay in one continuous pile along the shore for over thirty miles (the same quoting, from W. C. Scully's *Between sun and sand*, 1898).

Among the significant pieces of information originating with Scully is the fact that ' the Boers divided the springboks into two sections: the *hou-bokken*, which usually remain on the same veld, and the *trek-bokken*, which were those which migrated. The latter were generally smaller, and not in such good condition as the former ' (Fitzsimons, loc. cit.). The possibility exists, therefore, that something incipiently analogous to a phase-system had developed; though it seems on the whole more probable that the difference in size and condition could have been produced by the social hierarchy.

Allusion has already been made to the springbuck in Chapter 2 on account of its remarkable semaphore of long white hair, normally concealed in a pocket along the mid-line of the back, and erected during the spectacular stiff-legged, bouncing display that gives the species its common name. It is intensely sociable, and Heape (1931, pp. 113-116) especially refers to the state of hysteria which precedes the trek, so similar to what is attributed below to the lemmings and crossbills, and to what we have seen to precede the exodus flight of gregarious locusts. Heape says that, in emigration, both lemmings and springbucks become 'fatalistic', losing all timidity, and courting danger. We may also note how, like the emigrant swarms of locusts and lemmings, they recklessly consume every blade of vegetation in their path.

Neither Pallas's sand-grouse not the springbuck appear to have reached, in their irruptive behaviour, the evolutionary development of the locusts. Both are perhaps still in a rather more primitive stage—compelled to evacuate their habitations more or less frequently by force of external circumstance and to look for others at short notice. They may not fully have attained the subtly more advanced level of specialisation to life in scattered centres in arid country, in which great expeditionary forces are systematically created as and when conditions permit, and then released as prospectors and colonists. It is difficult to make a comparison on the knowledge available, but at least it is not difficult to imagine how readily this kind of special adaptation might arise, given favourable circumstances.

20.4. *Irruptions from arctic and boreal regions*

The majority of birds whose recurrent but irregular invasions are familiar in temperate latitudes come from the north; some are of circumpolar distribution, like the snowy owl (*Nyctea nyctea*), the crossbills (especially *Loxia curvirostra*), the pine grosbeak (*Pinicola enucleator*) or the waxwing (*Bombycilla garrulus*); some are more restricted in distribution, like the little auk (*Plautus alle*) in the Atlantic sector, the great spotted woodpecker (*Dendrocopos major*) and Siberian nutcracker (*Nucifraga caryocatactes*) in the western palaearctic, the evening grosbeak (*Hesperiphona vespertina*) in North America. In the majority of these, emigration occurs after the breeding season or in autumn, and can be attributed directly to an emergent state of overpopulation that is being relieved by discharging the surplus to seek food elsewhere. Some of the species are largely dependent on one particular type of food—for instance, on lemmings in the case of the snowy owls, or on seeds of the ash-leaved maple (*Acer negundo*) in that of the wintering evening grosbeak—and these resources, as we have noted already, are themselves notably fickle and locally fluctuating in abundance.

A long list might be compiled of northern species that could be said to have irrupted, at least occasionally, coming beyond their normal range-limits. If aquatic species were excluded, it would consist largely of birds, with a few mammals and probably no insects: the latter of course make up a diminishing share of the fauna with the lowering summer temperatures and shortening growing-seasons of higher altitudes. It would be hard to draw a line between irrupting species that were genuinely arctic or high-boreal, and those of more middle latitudes that, like the European jay (*Garrulus glandarius*) for instance, at longer intervals put on a spectacular exodus. (The jay is a bird particularly associated with acorns and oaks (*Quercus* spp.).)

The whole subject of northern irrupting birds has been ably reviewed in a recent paper by Svärdson (1957). He concludes, as already mentioned, that their fluctuating numbers reflect the pronounced variations in food-supplies—whether these are of animal prey as they would be for the various owls or the goshawk (*Astur gentilis*), or, in more numerous herbivorous species, the buds, seeds and fruits of trees and other perennial plants. The variations in food-productivity are usually fairly local; and, in Svärdson's view, these erratic species of birds have become adapted to exploit this particular situation, and are prepared to travel every year after the breeding season as far as may be needful, in search of areas with adequate stocks of food: the population is to this extent a nomadic or floating one. In some years no movement is necessary; some years the flight continues as long and far as the birds can draw breath, because they fail to discover what they want and ultimately die. But the movement is not wholly an outward one from what may be regarded as the permanent range of the species: irregular return flights of emigrant survivors may also occur, just as shifts to and fro are continually being brought about in the centres of population of the birds remaining within the permanent range itself.

What is implied is that these birds are opportunists, able because of their higher mobility to seek out temporarily favourable habitats for wintering and breeding in a climatically capricious zone, automatically adjusting their population-density to the resources that they happen to find. Breeding is permissible wherever conditions permit, and may continue for one or several years before the local economy breaks down and the colony is abandoned. At its fullest expression this would be a completely nomadic way of life, lacking all the immensely valuable assets of permanent settlement and established social organisation.

Some degree of geographical localisation almost certainly remains in most of these species, however. For example, the common crossbill has many isolated populations far to the south of the boreal forests, usually in subalpine conifer zones in mountains, extending as far away as north-west Africa and Central America (Nicaragua). These may be post-glacial relict populations, or the result of more recent emigrant colonisations, but some have become subspecifically distinct from the typical *curvirostra*. The native Scottish race (*L. c. scotica*) has so large a bill as to approach the parrot-crossbill (*L. pytyopsittacus*); and it is significant to notice as the years go by how wave after wave of immigrant northern crossbills arrive in Scotland and spread over the country, without diluting this strongly distinctive local subspecies. The travellers no doubt avoid the pine forests already tenanted by local stocks; and one is inclined to suspect that even in the northern forests, as in Scotland, although the density of local crossbills rises and falls with the seed-crop and there is abundant evidence of large invading movements, a small garrison generally manages to hang on in the home area, and the local stock is not totally obliterated nor easily supplanted. Small passerine birds like the crossbills are capable of rapid increase in a favourable breeding season, and it seems probable on theoretical grounds that a great deal of the local fluctuations in breeding-population density can be attributed to autochthonous adjustment: selection is likely as usual to favour the self-perpetuation of local stocks, in so far as this is possible in extremely erratic environments. Strong support for this conclusion comes from the fact that notwithstanding the common crossbill's nomadic habits there is considerable geographical differentiation; even the more or less continuous boreal populations of North America, stretching from Newfoundland to Alaska, can conveniently be subdivided into three or four subspecies; and this is quite apart from any outlying endemic stocks established farther south.

The crossbill serves as a good illustration, because it is one of the birds most highly adapted to this irruptive way of life. Like the locusts and some of the other species discussed in the previous section of the chapter, its reproductive conventions may foster the creation of a large surplus in a bumper-crop year, to colonise and fill up the gaps, if any, in places where the species has actually been exterminated. It breeds very early as a rule, long before the seeds are shed from the cones in the first warm days of spring, and its wanderings generally commence in June and are over by the middle of August, except in years when they extend to enormous distances.

Many years ago in the spruce forest of Gaspé peninsular, Quebec, I found myself in the middle of a vast congregation of white-winged crossbills (*Loxia leucoptera*), through which I travelled on foot for ten miles on 23 and 24 July, 1934. All were males: presumably the females and young had already emigrated from this breeding area. The males were everywhere, in bunches and small excited flocks, singing in chorus almost incessantly, especially on the wing, as they descended with the ' slow-motion ' beats of their display flight. Reading in later years of the fever which mounts in the swarms of locusts and of other creatures on the eve of a great exodus, I have been led to think that this must have been the great epideictic display that preceded the crossbills' flight. Throughout the following winter the species was exceptionally abundant in the southern part of the province, four hundred miles or so to the south-west.

It is well authenticated that some of the crossbill's irruptions result in colonisations. Usually these do not survive indefinitely; but in the recently-established, extensive Scots pine plantations in Norfolk and Suffolk, a stock of *curvirostra* has now maintained itself since the irruption of 1910 (Witherby *et al.*, 1938); and it is possible that one in Tipperary has continued unbroken since 1838 (Kennedy *et al.*, 1954).

We have still something to learn here from a final example to be taken from the circumpolar fauna, namely the lemmings. In the first place, a series of fluctuations has been closely followed since 1945 (*see* Pitelka, 1957) by biologists at the Arctic Research Laboratory at Point Barrow, Alaska. Almost regularly-spaced peaks occurred in 1946, 1949, 1953, and 1956, with numbers so high that the vegetation cover was temporarily consumed or mown down. In the years between, the numbers fell to low minima. During each of the last three maxima exploration has shown that the high-density area was quite circumscribed; a survey from a small aeroplane in 1956 revealed that it stretched along the coast for about 200 miles and inland to distances of 15-20 miles; this was generally confirmed at a number of trapping stations on the ground. There is no indication that a similar periodic cycle has occurred anywhere else on the Alaskan Arctic slope: it has been a phenomenon confined, during this decade at least, to the restricted coastal fringe centring on the exposed northern foreland of Point Barrow.

In this habitat there are only two rodents—the brown and collared lemmings (*Lemmus* and *Dicrostonyx*), of which the second is scarcer and appears to have been little if at all involved in the cyclic changes; farther inland, however, there are three other voles—a *Clethrionomys* and two kinds of *Microtus*—and this richer fauna is certainly an indication of a somewhat more clement environment. The cycles were completely self-contained, involving a concentration and build-up of predators but no emigrations.

In Norway, the classical locality of lemming irruptions, the native species (*L. lemmus*) inhabits the alpine zone above the tree-line, and its southern populations are isolated from one another in detached mountain groups. When it has a prolific year in some particular district, other kinds of small rodents in the same area, living at lower elevations in the birch zone, or extending below into the conifer forest, may sometimes have a correlated increase (Collett, 1895, p. 20). These include the wood-lemming (*Myopus schisticolor*) and half a dozen other voles and mice, among them notably *Microtus ratticeps* and *M. agrestis*. In rather the same way as was found in Alaska, it is the arctic-alpine lemming, occupying the most rigorous habitat, that has the most frequent and violent fluctuations; and these culminate not infrequently in the famous mass-emigrations. Other species living lower down, and especially in the valleys far below, have considerably fewer surges, and very rarely attain numbers sufficient to produce an actual exodus from an overcrowded habitat.

In a prolific year the birth-rate soars and the young ones survive to an exceptional degree; one litter follows another throughout the entire summer (Collett, loc. cit., p. 19). Abnormal activity and excitement prevails, the food-supply becomes overtaxed, and, as the summer wears on, emigrations occur. 'It is as well to understand the really great magnitude of these movements, which attract much public attention and powerfully disturb the equilibrium of plant and animal communities below the Subalpine zone'. When the numbers are great, 'the lemmings go out over the lowlands, crossing fields and lakes and rivers and passing through busy towns, until those that survive reach the coast. Here they swim out and often reach islands several miles away. Though many drown, even these stay afloat and are cast ashore in drifts. There is usually a crescendo of migration pressure through the summer and autumn, and some very strong obsessive impulse seems to drive these fat but delicate little creatures onwards and downwards across powerful physical obstacles and without fear of man' (Elton, 1942

p. 213). Collett (loc. cit., p. 43) says that 'each individual appears to take its own road, even if proceeding in substantially the same direction as the others. Only at certain points may the wandering individuals be assembled in great numbers. . . . Generally they go their several ways, meet, bark at each other, and part again. If two old males meet, a fight often results, during which they may wound each other's throats so that death may be the consequence. Life quickly leaves them, and they die from the slightest injury'.

Finally, according to Elton again, 'among a great many obscure features of the phenomenon, one is thoroughly established by observation: practically none of the emigrants or their families return to the mountains. After a year or two, and mostly much sooner, all of them are dead' (loc. cit., p. 214).

20.5. *Conclusions about emigration*

It is possible by now to reach some general conclusions about emigration. We have recognised all along that it is, in the first place, an almost universally available means of rapidly reducing population-density in any area where the optimum has been exceeded. For this it is an *ad hoc* remedy capable of being invoked immediately by the animals themselves, without depending in the least on any outside agency like parasites or predators. For the purpose of density reduction its effect can be accurately and automatically adjusted to the size of the particular emergency: those members of the society standing higher in the hierarchy will continue to exert stress on those lower in the scale, until the population is thinned down to a threshold recognised to be conventionally consistent with the remaining food-resources. Emigration-pressure, that is to say, is density-dependent, and ceases to be felt at or below the optimum density level.

There is a second or parallel general function of emigration, which we have also recognised all along: it fulfils the need for extending or restocking the range where this would be advantageous, and for promoting a limited interchange of genes between neighbouring populations. The value of dividing the species into more or less isolated, self-perpetuating societies, each of which is a potentially immortal group in competition with its neighbours and subject to group-selection, must be offset against the need to safeguard the interests of the species as a whole—to spread abroad and fill up accidental voids that have arisen in the range, and enrich stocks that survive but are genetically impoverished. We have generally referred to this function of emigration as pioneering, and have seen again and again that pioneers tend to come from the junior section of the community, going out early in life to seek a 'foreign' home in some more or less distant locality, and establishing themselves permanently there.

Clearly this second function is entirely different from the first. It ought to be efficiently discharged above all by well-established, stable and successful populations; it would therefore be an expected refinement of the homeostatic machinery to budget for an appropriate surplus to send away as pioneers, even under the most serene economic conditions. On the other hand, in a population that has been accidentally decimated, the first priority will presumably be to restore its own numbers. Consequently the 'planned' production of pioneers for the common good of the species will come to be particularly associated with areas where favourable economic conditions prevail.

It is easy to see that both these functions must grow in urgency and magnitude with increasing instability in the physical environment. Where conditions are erratic, the homeostatic machinery will more often be taxed with unpredictable changes: sudden reductions of food-yields will make overpopulation a common accident, very likely requiring emigration to alleviate it. At the same time, major disasters will be more frequent, and the recolonisation of depopulated areas correspondingly a more common and pressing necessity: consequently the need for pioneers will also be greater. Pioneers can be afforded only in good years under these changeable conditions, so that selection will come to place increasing weight on producing them in quantity on every occasion when the infrequent opportunity offers.

Though pioneering is common to all animals, the examples in the last three sections have shown that the scale of adaptation extends without perceptible gaps from the normal state in a stable environment—where a small annual output of pioneers is regularly allowed for and dispatched, and safety-valve emigration to relieve congestion seldom exceeds a low level—right to the extreme state of specialisation to erratic conditions represented by the migratory locusts, in which the creation and release of a vast swarm of pioneers, on the rare occasions when it can be brought off, seems to have become the culminating objective of existence. Contrary to one's first impression, on reviewing the complex developments of phase-systems and the unbridled multiplication of gregarious locusts, there is no new principle involved. Lemmings, snowy owls and crossbills, painted-ladies and clouded-yellows, all in much the same manner appear to inflate their numbers 'deliberately' in favourable years, in order to release swarms of pioneers. What has primarily varied under natural selection is the effort devoted to pioneering, depending on the relative degree of stability and security, in populations living under different environmental conditions. Secondarily, of course, it has sometimes involved the development of special emigratory adaptations, most notably the phase-system itself.

Though the two functions of emigration are thus theoretically quite distinct, emigration in practice is usually one and the same process whatever the cause for which it is undertaken. The same social pressure can be exerted to disperse pioneers as to evict the surplus of overcrowding. The simple emigrants themselves are probably the same individuals no matter which of the two prime causes leads to their exclusion from the habitat: in neither case are they permitted to return home. Luck can turn an eviction into a timely act of colonisation; and condemned supernumeraries usually have some chance of useful survival elsewhere, just as pioneers can easily fail in their mission and perish. In practice, therefore, there is likely to be a mingling, both in motives and consequences, especially in what may be vaguely referred to as average environments. Where conditions are most stable and incidents hardly ever arise to disturb the even tenor of population-control, pioneering will be left as virtually the only cause that promotes emigration; and at the other extreme, in highly precarious and inconstant habitats, particularly those associated with desert and polar climates, pioneering will likewise dominate the picture of surplus-production, though demanding now relatively enormously greater specialisation and output of vital effort. It is in the great and heterogeneous range of environments in between these extremes that emigration is most often likely to be invoked for the simple adjustment of density.

It should not escape our purview that all the numerous redistributions or redispersions of individuals that normally take place as a result of seasonal changes in the environment, including especially the annual two-way migrations of birds, involve the same general social machinery in order to get the emigratory movements going and keep them moving. It is thus no coincidence that great, and sometimes excited, epideictic gatherings assemble on the eve of annual migrations, very much as they do in social groups working up to a 'safety-valve' exodus, or to the emission of pioneers.

20.6. *Summary of Chapter* 20

1. Departures from average in climatic factors are more frequent and have more effect on organic production in polar and arid climates than in more equable regions. Growth and food-production by plants can vary widely from year to year under these conditions. In general, the fluctuations in animal numbers tend to reflect the swings in available food-supplies so that optimum densities where possible shall not be exceeded and resource-damage be avoided. In a variable environment there is nothing irreconcilable between the homeostatic control of population-density and the occurrence of such fluctuations. Outbursts or plagues in animal populations are attributable to drastic changes in the environment, and these are not infrequently induced by man's activities, *e.g.* in agriculture or forestry.

2. When food-supplies are fortuitously removed or fail to materialise, a local emergency of overpopulation results; and, especially among highly mobile flying animals, this can be immediately relieved by emigration. The social hierarchy and code of behaviour to which all equally subscribe acts as a density-dependent device to force out whatever surplus exists. The exiles may be condemned to perish; or, in other circumstances, they may carry the torch of life for the population which must establish itself elsewhere or be exterminated. In very dangerous or impermanent habitats, such as those occupied by migratory locusts, special adaptations have developed for constructively building up an expeditionary force when conditions permit, in order if possible to establish new colonies elsewhere before the old ones are obliterated.

3. Migratory locusts are not alone in acquiring such adaptations. Some of the migratory Lepidoptera have larval phase-systems also; and these and various other butterflies and moths living in fickle, arid habitats build up surpluses for export in good years, which irregularly invade or irrupt into far-distant countries. The habit is paralleled by other animals of steppes and deserts, including Pallas's sand-grouse and the South African springbuck.

4. In the cold boreal and arctic regions most of the irrupting species are birds. Many are adapted to exploit intermittent or undependable crops of food, and are to a large extent nomadic in their search for regions where supplies are temporarily plentiful. There is reason, however, to think that even here a certain degree of localisation and permanency exists in the breeding stocks, and that local fluctuations in breeding density are to a considerable extent autochthonous. Special attention is paid to the crossbills and the lemmings, as avian and mammalian examples of boreal emigrants.

5. It can be concluded that two distinct biological functions are served by emigration. One is the 'safety-valve' function, to give immediate relief to overpopulation; and the other is the 'pioneering' function, to expand and replenish the range of the species as a whole and provide for gene-exchange. Emigration of the first kind is associated with quickly deteriorating economic conditions; whereas providing pioneers is something that can be afforded only when conditions are good. In more variable environments, both functions tend to become more important and to be exercised on a larger scale. Under extremely erratic conditions, producing pioneers develops into a dominating necessity whenever a good enough season occurs.

Though the functions are theoretically distinct, one social mechanism usually serves either to evict supernumeraries or to send off pioneers; the individuals expelled are in either case usually the same, being the junior fraction of the hierarchy. Supernumeraries have a chance of survival elsewhere, just as pioneers can fail in their mission and die; moreover in practice there may be a mingling of causes as well as consequences. Pioneering is the one cause of emigration operating in perfectly stable environments, and it becomes very much the dominating cause also in extremely variable ones: it is thus in the great range of intermediate conditions that safety-valve and pioneering emigrations are likely to be more or less equally developed.

The same general social machinery that controls safety-valve emigrations is involved in regulating seasonal redispersions of animals, such for example as the annual two-way migrations of birds.

Chapter 21

Recruitment through reproduction

21.1. *The equation of recruitment and loss*

It is self-evident that in every population remaining at a constant density, or returning each year to the same density at the same season, the rate of recruitment must be equal to the rate of loss of the constituent members. As most animals are seasonal breeders, this implies in the simplest imaginable case that a stable population requires each breeding season to make good the losses of the preceding twelve months: that is to say, the number of recruits must equal the number of members lost.

This simple equation, that recruitment and loss are numerically equal, is generally true wherever population-density (when averaged over a suitable period of time) remains constant, no matter whether the numerical value is great or small. Some animals reproduce faster than others and this is frequently true even of different populations of a single species living in different circumstances; yet their several population-densities can each remain constant over the years. This is possible because those that reproduce faster also die correspondingly quicker: the whole equation is given a higher numerical value, but both sides nevertheless remain balanced.

To preserve the balance, the two sides of the equation cannot be allowed to vary independently: a state of adjustment must exist between them; and the question immediately arises, which side is the independent and which the dependent variable? Are the losses somehow influenced so that they automatically match the number of recruits coming forward? Or is the reverse true, namely that however the losses vary, within manageable limits, recruitment can be adjusted to compensate for them?

This question has proved a controversial one, especially in recent years. In this country Lack has repeatedly put forward the hypothesis, so far as birds are concerned, that the reproductive rate seems in general as high as possible, 'each species breeding when conditions normally permit it to raise young, and laying a clutch corresponding to the largest number of young that it can successfully nourish' (Lack, 1956, p. 329). This means in effect that the reproductive throttle always stays wide open, and the rate is quite uninfluenced by antecedent losses, or population-densities existing at the time. One of the important general arguments evinced in support of this theory is that selection must tend strongly to favour those individuals that contribute the most numerous progeny to posterity. ' If one type of individual lays more eggs than another and the difference is hereditary, then the more fecund type must come to predominate over the other (even if there is over-population)—unless for some reason the individuals laying more eggs leave fewer, not more, eventual descendants' (Lack, 1954, p. 22).

This argument contemplates selection as acting at the level of the individual, relentlessly discriminating against those that leave fewer than the maximum possible number of progeny to posterity. It leaves entirely out of account the overriding effect of group-selection, that occurs between one population or society and another, and normally results in fixing the optimum breeding rate for the population as a whole. As we shall see, there is a most impressive body of evidence to show that, for any given individual, the annual production of offspring can vary between a potential maximum number and zero, but the average or normal figure lies somewhere in between, and is capable of being profoundly influenced by social conventions. The fallacy in Lack's hypothesis is the same as we met in the previous chapter over the question of why selection does not quickly eradicate the habit of emigration or irruption, when so many of the individuals embarking on it perish. What we are concerned with in both cases is compliance with a conventional code: breeding individuals that comply act in the common good to produce a share of the annual crop of recruits, large or small as the occasion demands. In the other context, those that are required by social convention to emigrate do so, in compliance with the relevant code—the same code that directs the remainder of the population to stay their ground. The codes in each case are social properties, reinforced by group-selection. Groups in which defection arises will be swiftly eliminated, when the breakdown of conventions leads either to dwindling numbers, or, more likely, to overpopulation and progressive ' overfishing ' with diminishing returns.

The apparent alternative to Lack's hypothesis is that the recruitment rate is the dependent variable, and can be continually modified as part of the homeostatic process by which an optimum population-density is maintained. Lack (1954, p. 22) dismissed this possibility, so far as birds were concerned. He pointed out that ' clutch-size could be adjusted to the mortality and achieve population balance only it if were much lower at high than at low population densities, which is not the case '; to underline his difficulty, one may recall that some birds have a fixed and practically unvarying clutch-size, like many of the plovers and sandpipers which always lay four eggs, whenever and wherever they breed. But the rate of egg-laying is not of course the same as the rate of recruitment to the adult population—this will become more obvious shortly; and as long as we make it clear that it is *recruitment* which is being postulated as the dependent variable, to be matched against the sum of mortality from all outside, uncontrollable sources, this hypothesis appears to come considerably nearer the truth.

Before going any further, however, it is desirable to break down the two sides of the equation each into its component parts. Recruitment, first of all, may be considered as enlisting newcomers to an adult population consisting of potential breeders: this is the ordinary meaning of the term, though in special circumstances we occasionally need to think of recruitment in other castes, such for instance as a non-breeding reserve, or a larval or juvenile society. Unless otherwise qualified, however, recruitment may be taken to mean admission to full membership of a population normally capable of sustaining itself by reproduction. We must notice, however, that acceptable recruits can arrive on the scene as immigrants from somewhere else, in addition to those that have grown up in their native place. Recruitment thus has two component fractions, respectively consisting of (i) immigrants and (ii) natives; in the vast majority of potentially immortal populations, the latter of course greatly preponderate.

Losses occurring in the membership of the population can likewise be analysed. First there are losses inflicted by independent agents, such as predators and parasites and the violence of the elements, and to these may be added deaths from old age; these may for present purposes be grouped together as ' uncontrollable '. In addition there are ' homeostatic ' losses, imposed from within by the action of the social machine, which can further be subdivided into (i) emigration losses and (ii) social mortality. The last category includes the effects of stress-disease, cannibalism, infanticide and certain other less common forms of intrinsic mortality.

The equation may therefore be amplified in the following way:

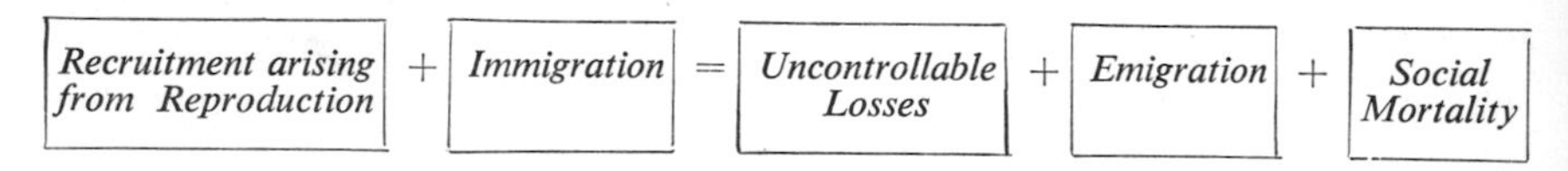

It is easy to see that of all the five components in the expanded equation, only one is incapable of being influenced from within through the physiology and behaviour of the members of the population themselves. This is the one labelled ' uncontrollable losses '. It therefore constitutes the true independent variable. All the other components can be homeostatically influenced by the operation of the social system, and to a greater or less extent all of them are in practice regulated in relation to one another in order that a balanced, optimum population-density may be maintained. It is very interesting and important to see that the homeostatic machine can go to work on either side of the equation, and that density can be raised or lowered by it with almost equal facility.

Our first formulation of the question ' Which is the independent variable? ' thus turns out to have been oversimplified, since both recruitment and losses are capable of being adjusted from within.

Emigration, and by implication immigration, formed the subject of the previous chapter. The present one is concerned with the remaining component on the left-hand side, namely recruitment arising from reproduction. Chapter 22 turns to the other right-hand side component—socially-induced mortality.

21.2. *Potential fecundity levels*

The view has long been widely held among zoologists, perhaps sometimes uncritically, that the power of reproduction in animals (fecundity) is broadly related to the sum of the dangers to which each species is exposed (*cf.* Carr-Saunders, 1922, p. 60). It is a commonplace observation that an evolutionary advance in the degree of parental care given to the eggs and young is correlated with a lowering of fecundity. Fish with free-floating pelagic eggs and larvae notoriously lay thousands—even hundreds of thousands—of eggs, whereas those like the 3-spined stickleback (*Gasterosteus aculeatus*) that build nests and guard their eggs and young, produce only a few score; those that are viviparous and retain their eggs within the body during a long development seldom have litters even as large as the stickleback's, and in a few species the number is down to single figures, for instance, in some of the sharks and dogfish (*e.g. Acanthias* spp.). The argument can of course be turned the other way: though fewer in number, the eggs laid by species that guard them jealously are generally larger and contain much more yolk; and perhaps the females are after all laying the maximum number of which they are physiologically capable, and the species is enabled to survive by giving them such assiduous care.

The general unlikelihood of the latter explanation is at once apparent. Some kinds of animals achieve an astonishing fecundity by supplementing sexual with asexual or parthenogenetic reproduction, which may start from infancy, in the early larval stages, so that a prolific multiplication can occur before sexually mature individuals appear at all. This is notably characteristic of species with specially hazardous life-histories, such as helminth endoparasites or the Cladocera inhabiting temporary ponds; their near relatives living in more secure circumstances are generally very much less fecund, although they may not differ much otherwise in size or general metabolism. It is safe to conclude, for instance, that prolificity is an adaptation to the way of life of the specialised parasite, just as surely as the loss of sensory, locomotor or digestive organs. An avian illustration could equally well be taken: as we saw earlier (p. 241) such a bird as the bobwhite quail may lay a clutch of say twelve eggs, collectively weighing about the same as the bird that laid them; a female song-sparrow with a clutch of four or five lays a weight of eggs about half as much as her own body-weight; whereas a gannet (*Sula bassana*), weighing between 3 and 4 kg produces only a single egg weighing 100 grams (ratio$>$30:1). If this egg be destroyed the gannet will lay a second time, and if that be removed, possibly a third (details from Gurney, 1913, supplemented by one original weight-measurement). The gannet is a very long-lived bird, and there is nothing in this evidence inconsistent with the hypothesis that clutch-size is broadly correlated with the sum of life's dangers, nor lending the least support to the opposite view, that it is determined by the physiological resources available to the female bird in producing yolk and egg-white.

The larger birds and mammals are sometimes very long-lived, and meet with few perils once they pass the stage of adolescence; their reproductive rates are—necessarily in non-expanding populations—correspondingly low. Many species of birds lay two or just one egg a year, and many mammals likewise produce only one or two young at a birth. There are reasons for concluding that this low rate is all that is ever required under natural conditions to maintain the stocks of the species concerned, and is directly correlated with longevity and the secure life, rather than with any physiological limitation upon nourishing a larger number. Some of the Cetacea, ruminants and primates (including man) that normally have one offspring at a time are capable of bearing two on occasion: but selection has undoubtedly been pressing in the opposite direction, since we can be certain their more primitive relatives and forebears had larger litters than they have.

Low fecundities, it may be noticed, are typical of the higher animals, though by no means confined to the vertebrates. The female minotaur beetle (*Geotrupes typhaeus*), for example, lays only some eight eggs in all (*cf.* Arrow, 1951, p. 44); in the related dung-beetle (*Copris lunaris*) the total is about the same, but in *Copris hispanus* it was not found to exceed three or four (Fabre, V, p. 134). In some of the pupiparous Diptera, such as the tsetse flies (*Glossina*), the female nourishes one embryo at a time, giving birth to a single full-grown larva ready to pupate; she may repeat the process twenty times or so, if she survives long enough (*cf.* Skaife, 1954, p. 295).

A few years ago I reviewed the evidence of low fecundity, so far as it applied to birds (Wynne-Edwards, 1955). Getting down to the one-egg (' uniparous ') species, especially common among the large raptors and sea-birds, one finds that as a very general rule they produce no second broods, in the way that so many more-fecund species commonly do. In a few highly specialised cases the remarkable condition has evolved, that only one egg is ever laid in one year, and even if it is removed or destroyed when newly-laid, there is no attempt to re-nest or replace it. This has arisen independently in several different families, namely the penguins (in *Aptenodytes forsteri*; Prevost, 1953), the albatrosses, shearwaters and petrels (where, among the eighty-odd species, egg-replacements have been exceptionally recorded only in two species of storm-petrels, *cf.* P. Davis, 1957, p. 98), and probably the auks (in the extinct great auk, *Alca impennis*; Martin, 1698). It is believed to be usual though not inevitable also in the California condor (*Gymnogyps*), in some completely unrelated vultures belonging to another suborder (*e.g.* *Aegypius monachus*; Kofoid, 1953, p. 85, and Mayaud, 1950, p. 602), and in other large raptors, including the golden eagle (*Aquila chrysaetos*) (Newton, 1864, p. 12).

These species have, therefore, largely or completely forfeited what is otherwise a normal attribute of birds, to replace lost eggs at least once. They are all, like the gannet, birds with relatively small eggs compared with their own large size and nutrient reserves, and the physiological cost of egg-replacement would presumably be trifling. What seems at first sight to be a lamentable and unexplainable defect, however, appears in a much more reasonable light as an adaptation lowering a little bit further still the reproductive potential of the population as a whole, in very long-lived species of birds.

Support for such a possibility comes from quite a different direction. A few highly-specialised species are not only uniparous, but are in one way or another prevented from breeding as often as once a year. For instance, most penguins have incubation and fledging periods amounting together to three or four months; in the emperor penguin (*A. forsteri*), the largest living species, it is five months: but its smaller congener, the king penguin (*A. patagonica*), produces a single chick that remains all the following winter in the nest, so that a complete reproductive cycle occupies rather longer than twelve months and the parents are thereby debarred from breeding more than twice in three years (*cf.* Stonehouse, 1956). In the great albatrosses *Diomedea epomophora* (the royal) and *D. exulans* (the wandering albatross), the same condition has independently arisen, though all other albatrosses can complete the cycle in less than a year. The chicks, moreover, in both the king penguin and the royal albatross reach adult body-weight three or more months before they finally go to sea, so that all that seems to be held up is the assumption of sea-going plumage (for references *see* Richdale, 1952, and Wynne-Edwards, 1955). The inference here is that these are cases where the reproductive potential has been reduced to approximately one-half an offspring per female per year. Many of the larger eagles likewise tend not to breed in the year following a successful nesting (*cf.* Brown, 1953, p. 110).

A few uniparous mammals are known to have a closely parallel adaptation, namely a period of gestation exceeding a year. Though most of them are large mammals, the condition is not closely correlated with size; moreover, as with the birds, the rule is that the smaller mammals produce proportionally the heavier litters: the neonatal two-ton calf of the blue whale (*Balaenoptera musculus*) is only about one-fortieth of its mother's weight, compared with litters exceeding a third of the mother's weight in various bats and small rodents (Leitch, Hytten and Billewicz, 1959). There is no reason to conclude, therefore, that a long pregnancy is inevitable in a big mammal because of the nutritive drain on the mother; it appears more plausible as an adaptation imposing a very low ceiling on potential fecundity in exceedingly long-lived animals.

As with the birds, this condition has arisen independently in different families. Both living species of elephants (*Elephas indicus* and *Loxodonta africana*) have gestation periods of nearly two years—17-23 months in the

Asiatic and about 22 months in the African. Neither of them has a sharply-seasonal breeding season; but pregnancies do not actually occur oftener than once in three or four years in *indicus*, or, on the average, four years in *africana* (Frade, 1955).

In the only known case of the great Indian rhinoceros (*Rh. unicornis*) reproducing in captivity, the calf was born prematurely after 18½ months, and the full period was presumed to be about 19 months (Gee, 1953*a*, p. 334).

In the Cetacea, the largest whales of all (*Balaenoptera* spp.) have gestation periods of only 9-10 months, but reliable estimates show that in the humpback (*Megaptera boops*), a somewhat smaller mystacocete, it is about 12 months; actual pregnancies appear normally to occur in *Balaenoptera* species in alternate years (Mackintosh, 1942, p. 222). Among the odontocetes, it has lately been confirmed that the sperm-whale (*Physeter catodon*) must have a foetal period of 15-16 months, and evidence shows that the relatively small and numerous pilot whale (*Globicephalus melaena*) also has a 16-month period (Fraser and Huggett, 1959). For other Cetacea we have no information. Among the pinnipede Carnivora, there are seals which, like the northern fur-seal (*Callorhinus*), carry their embryo for almost 12 months, but are nevertheless usually impregnated again within a matter of days, and thus bear an annual calf; but in the walrus (*Odobenus*) mating does not normally recur until the next year after parturition, so that the females calve only every second season (Freuchen, 1935, p. 256).

It is clear from this evidence that the maximum fecundity or breeding potential in some long-lived animals is considerably less than one offspring per year for each adult female.

21.3. *Fertility*

Following Carr-Saunders' differentiation (1922, p. 52), the term 'fertility' can be restricted to mean the actual output of reproduction being realised at any given epoch, in terms of living young.* An example will make this distinction clear. Lions and tigers (*Felis leo* and *F. tigris*) both have short gestation periods of about 15-16 weeks; both produce litters of several young at a birth, usually numbering between two and five or six. Allowing a reasonable minimum interval between pregnancies, their potential fecundity would appear rather high, as in the domestic cat. In practice, however, their actual fertility is much less. Stevenson-Hamilton (1947, p. 132) came to 'the definite conclusion that lionesses in a wild state do not produce cubs at intervals of less than two years' and, unless perhaps all the cubs of a litter have perished, he believed the period to be usually longer. Very much the same is true of the tiger (Pocock, 1939, p. 210). Moreover, according to these authorities the lion most commonly has only three cubs at a time, and

* 'Fecundity' and 'fertility' are not uniformly applied as scientific terms by different authors. In their widely-used textbook, Allee, Emerson, Park, Park and Schmidt (1949, p. 289) use the former for the actual egg-production realised, and the latter (as here) for the output of living young; others working with insects infesting grain and flour generally do the same (*e.g.*, Crombie, 1942, p. 311).

the tiger two or three—only about half the potential number. It should not be forgotten that the cubs are much smaller in proportion to the mother's size than are the kittens of the domestic cat, which normally bears between 50 and 150 kittens in a breeding life of say ten years.

According to the present theory of homeostatic control of population-density, recruitment through reproduction is one of the parameters that can be regulated in a density-dependent manner. It could be anticipated, therefore, that many species would have a certain potential fecundity, fully used only when a state of underpopulation existed and density needed to be rapidly raised; in stable conditions, when the optimum had not been much departed from, the coefficient of fertility required to make good current losses would be much smaller than this; and when overpopulation loomed up, as a result of failure of the food harvest for example, fertility in long-lived species might drop almost to nothing. We shall see that such a dependence as this between fertility and economic conditions is common in nature and in laboratory experiments; but it does not necessarily work just so, because there are alternative methods of reaching the same end.

There are numerous steps or stages in the process of raising recruits to the adult population, and a homeostatic brake can be applied at any of them. In many territorial species, when the carrying-capacity of the habitat changes, densities of breeding adults (and, inversely, numbers of non-breeders) rise or fall accordingly: this is the first point at which the recruitment rate can be controlled. The number of eggs and broods per female can be restricted. In birds, eggs may be neglected or forsaken, and in mammals, embryos aborted or resorbed within the uterus; thus the number of young brought into the world (*i.e.* the birth-rate, which is actually the measure of fertility) may be much less than the initial ovulation or fertilisation rates. Beyond this stage we trespass on the subject of the next chapter: but parental neglect, infanticide, cannibalism, and social pressures can bring about a variable but potentially enormous mortality among the young. It is possible, therefore, for a population to launch out on an apparently full-scale reproduction, the ultimate outcome of which, in terms of actual recruits to the breeding stock, tails off to a fractional residue, because of restraints imposed at subsequent stages of development by the action of the conventional guillotine. Although these particular illustrations refer to the higher vertebrates, the same variety of stages at which the brakes can be applied could be found in any of the higher animals.

In man, fecundity is relatively low: with a gestation period of 9 months few women can stand pregnancies at a rate of one a year except for brief periods. The most rapid birth-rates for whole countries barely attain 50 per year per 1,000 of the population. Assuming half the population are females, and, say, half of these are of an age when child-bearing is physiologically possible, only one in five of them would then actually give birth in any one year. The highest average *fertility*, assuming the correctness of these approximations, would be considerably lower than this (in Europe more like one birth in five years for married women of reproductive age). Normally there is a subsequent loss through infant and juvenile mortality; but assuming the residual annual rate of recruitment were only 20 per 1,000 instead of 50, the population could still double in 35 years. In fact, this is about the current position in the United States.

The populations of primitive men were not continually growing on any such scale, but were evidently under homeostatic control like those of other wild mammals. This can be deduced with certainty both by inference from the immensely long period in the Pleistocene through which mankind existed without attaining high densities, and also from the study of contemporary primitive tribes. Carr-Saunders (1922) showed with the utmost clarity that not only were all the primitive peoples of which we have knowledge dispersed on some definite territorial system, but all had adopted conventional customs of one form or another to restrict fertility and reduce family-size. The methods used fell into three main categories, each acting at a different stage in the process, namely abstention from intercourse, abortion of the unborn, and infanticide. There were also other ancillary homicidal traditions less universally developed such as head-hunting (the victims frequently being young people), sacrifice, and cannibalism; and in some Australian tribes, at least, mutilations performed as part of the ceremony of initiation may possibly also have impaired fertility.

The subject is one of absorbing interest, but can only be briefly summarised here. Abstention by women very frequently extended throughout the lactation of the previous child, during which time it was rigidly taboo to become pregnant. Lactations were quite commonly prolonged for two or three years, and in rare cases for five or longer. Abortion, the second method, was rather widespread among the American Indian tribes of both North and South America, and in the Eskimos; it was generally distributed in Africa and Oceania, though seldom reported from Australia. The third method, infanticide, has two distinct advantages over abortion, in that it does not endanger the mother, and the sex and congenital condition of the child are known: more girls than boys and all malformed or injured infants were commonly destroyed. This too was widespread in North America; it was prevalent among the African Bushmen, and practised on an enormous scale among the aboriginal tribes of Australia and Tasmania. Children were not necessarily killed deliberately, but could be left to starve or perish from exposure, in which case their deaths were often credulously attributed to magic, and feelings of compunction no doubt partly assuaged.

Primitive man, therefore, provides another good illustration of the principle that, under normal stable conditions, social conventions prevent the realisation of the full reproductive potential; the actual fertility is only sufficient to produce average families of three or four living children, and, taking account of subsequent losses, thereby to assure a practically constant population of adults in each generation.

21.4. *Carr-Saunders' principle of the Optimum Number*

It was from his comprehensive study of the territorial systems of primitive hunting and agricultural peoples, of the universal conventional restrictions they placed on fertility, of their seeming poise, enjoyment of satisfactory resources and normal freedom from want and disease, that Carr-Saunders was led to develop the principle of the 'optimum number' in human populations—a principle that, as he formulated it (although he assumed it could not be extended to man's pre-human ancestors let alone the rest of the animal kingdom) is in essentials nearly the same as my own.

Great interest consequently attaches here to his argument and deductions. It may not be irrelevant to mention that, on my tutor Mr. Elton's strong recommendation, I bought a copy of *The Population Problem* a few months before my final examinations at Oxford in 1927; but it was not until nineteen chapters of the present work had been finished, in 1959, that I re-read Carr-Saunders' chapter on 'The regulation of numbers among primitive races', and, not without astonishment and satisfaction, found so many of my own conclusions anticipated.

The principle rests in the first place on the premise that in any human society—depending on the contemporary state of culture, resources and the arts of production—as a consequence of mutual co-operation there will always be a particular optimum size for the population that will give the highest return of goods per head. 'If the population fails to reach that number or if it exceeds it, the return per head will not be so large as it would be if it attained that number' (loc. cit., p. 200). This particular concept, as it applies to industrial communities, was already current among economists at the time Carr-Saunders was writing. He realised, however, that mutual co-operation must have been an important factor even in the most primitive hunting races; and he could easily show that the principle of the optimum number held good for them also. 'There is that is to say, taking into account the abundance of game, the fertility of the land, the skilled methods in use, and all other factors, a density of population which, if attained, will enable the greatest possible income per head to be earned; if the density is greater or if it is less than this desirable density, the average income will be less than it might have been. Obviously it must be a very great advantage for any group to approximate to this desirable density' (loc. cit., p. 213). What is not entirely grasped, of course, is the supreme importance of not overexploiting renewable resources (overfishing); also it is not necessarily true of animals that the 'income per head' is reduced when population-density is too low: this depends on the extent and importance of collaboration in food-finding, which is commonly rather slight.

Starvation, he explains, is never acceptable as a means of limiting numbers, because it makes social conditions unstable, and all the accumulated skills in hunting and cultivating the ground become worthless. Instead, limitation must be achieved from within, either by decreasing fertility or by increasing elimination—the methods adopted being the three already alluded to. 'The view put forward here is that normally in every primitive race one or more of these customs are in use, and that the degree to which they are practised is such that there is an approach to the optimum number ' (loc. cit., p. 214).

When it comes to family limitation, ' the wishes of not only the parents but also of the relations and of the community in general have to be taken into account, and the practices are enforced by social pressure ' (loc. cit., p. 216). ' The problem we have to face is how these practices could come to be of the necessary intensity. Now men *and groups of men* [my italics] are naturally selected on account of their mental and physical characters. Those groups practising the most advantageous customs will have an advantage in the constant struggle between adjacent groups over those that practise less advantageous customs. Few customs can be more advantageous than those which limit the number of a group to the desirable number, and there is no difficulty in understanding how—once any of these three customs had originated—it would by a process of natural selection come to be so practised that it would produce an approximation to the desired number ' (loc. cit., p. 223). Notice especially the invocation of the necessary principle of group-selection to explain how selection comes to override individual advantage.

How can it be shown that primitive populations approximate to the optimum number? The answer is that it cannot be done conclusively, but only by the indirect evidence that such people usually appear to approach the highest standard of living within their reach. The former conception of savages always in a state of semi-starvation is utterly false; they are commonly people of graceful proportions and beautiful physique, able to endure hardship it is true, but wonderfully equipped for the lives they lead; they enjoy good health and live to an advanced age. The evidence suggests conditions which ' are not compatible with the state of existence on the bare means of subsistence ' (loc. cit., p. 234).

On the question how far these conclusions can be applied to prehistoric races, Carr-Saunders points out that ' they are clearly only applicable to races among whom social organisation has become established; for it is only when men begin to reap the advantages of co-operation that the conception of the optimum number arises ' (loc. cit., p. 239). This period may have begun, he surmises, as early as the Lower Palaeolithic era.

It is not a great step from this penetrating deduction to perceiving that the same principle of the optimum number must have applied long before actual co-operation in the sense he intended arose: for whenever a population depends on exploiting common resources or conserving them for posterity it holds equally good. The essential role of social organisation (and even of social pressure) in achieving the optimum is fully apprehended by Carr-Saunders. Social organisation he conceived as having originated along with ' the recognition of areas ' (*i.e.* land-tenure systems) in prehistoric man—though it now seems certain to us that the development of the inseparable partnership between social behaviour and property-holding goes back several hundred million years to palaeozoic times, and has taken a major part in shaping the course of animal evolution.

Carr-Saunders' theory came thirty or forty years before its time, and its unique contribution to the understanding of human population-dynamics (to say nothing of those of animals) has never been properly recognised, far less acclaimed.

21.5. *Experiments on the correlation of fertility with density*

So much experimental work has been done on the fertility *versus* density relationship in animal populations under laboratory conditions that anything like a complete review of it cannot easily be undertaken; nor is it really essential here since the broad results, whether from insect populations of *Drosophila* or *Tribolium*, or from vertebrates like guppies, rats or mice, all turn out to follow the same broad pattern. A fuller treatment and discussion, especially of the work on insects, may be found in the excellent textbook summary by T. Park in *Principles of animal ecology* (by Allee *et al.*, 1949, pp. 349-355).

The early work by Pearl on the fruit-fly *Drosophila melanogaster* showed, first of all, that the growth of a population within a ' limited and enclosed universe '—a milk-bottle containing a nutrient medium in which grow the yeasts that form the flies' actual food—is rapid as long as the density of adults is low, but slows down as density increases, until a definite population-ceiling is reached. This is effected by a marked drop in egg-production as the stocks become more crowded. He found that the same females, confined for alternate 24-hour periods at lower and higher densities, responded at once with an inverse alternation between higher and lower egg-production; and he concluded that the density-effect must be regulated primarily (though probably not solely) through a collision or interference action of the flies upon each other (Pearl, 1932, p. 83). Robertson and Sang (1944), later extending this work, proved that food-resources as well as actual crowding enter into the control of oviposition, and that crowding alone, when the flies are adequately fed, leads to only slight lowering of fecundity. This kind of homeostatic effect, depending on an integration between the frequency of individual contacts (with or without more complex epideictic phenomena) on the one hand, and available food-resources on the other, is of course the ordinary, expected type of density-regulation, and does not differ materially from what we have generally encountered before in comparable situations.

Turning to the ' confused flour-beetle ' (*Tribolium confusum*), the pioneer experiments of R. N. Chapman (1928) demonstrated very similarly that, in an enclosed universe in which the nutrient medium was a layer of flour 2 cm deep, a steady ceiling-population would ultimately be attained: an experimentally repeatable, almost constant density of individuals per gram of flour was finally arrived at, whether the culture was started with one pair of adults or many pairs, or whether the volume of flour was great or small. Chapman believed at first that the density-limiting effect was due entirely to cannibalism, and that adult beetles ate the eggs and immature stages in numbers proportional to the concentration of individuals per gram of flour (loc. cit., p. 122).

Many subsequent workers proved that this was only part of the explanation. MacLagan (1932), performing parallel experiments with *Tribolium* and *Sitophilus*, the grain-weevil, found in each case a drop in the number of eggs laid per female, associated with crowding. Food-supply enters into it, but not very directly: the female weevil, for instance, never uses all the grains available to her for oviposition however hard-pressed she may be; some effect, which MacLagan for want of a better term refers to as psychological, influences the regulating mechanism. He concludes that natural populations (in the same way as experimental ones) ' automatically check their own increase, by virtue of this density effect, and that the organism itself imposes the ultimate limit to its own abundance when all other factors (biotic and physical) normally inhibiting population increase have failed ' (1932, p. 452).

In *Sitophilus*, not only does the number of eggs laid drop, but eggs after they are laid may subsequently be punctured by females probing the same grains again for the purpose of oviposition. In a later paper MacLagan and Dunn (1936) referred to such controlling factors as these as ' autobiotic '—what we have often called intrinsic factors—operating directly through the members of the population themselves and resulting in inhibition of the rate of population growth.

Another autobiotic factor is the frequency of copulation. In *Sitophilus* at very low densities this is low because of the time spent in finding one another by beetles of opposite sexes; at somewhat higher densities it reaches its maximum frequency and above that declines, partly because of interference. The insects are sensitive to mechanical stimulation, and as far as copulation-frequency is concerned, crowding has a more intense depressing effect on males than females. In *Tribolium*, Park's early experiments (1933) clearly displayed a parallel effect: females lay more eggs when exposed to frequent recopulations than when impregnated only once, so that egg-production tends to rise at first with growing population-density; but the threshold is soon reached where increasing cannibalism of eggs and larvae overtakes this advantage, and the reproductive output thereafter progressively declines.

Later work on *Tribolium* has taken account of the ' conditioning ' of the medium by the accumulation of metabolites—waste-products and secretions. In the experiments so far recounted the cultures were at regular intervals sifted through a sheet of fine bolting silk, the various stages of the population being retained and counted, and the old flour replaced by fresh. Even a small proportion of heavily ' conditioned ' flour taken from run-down *Tribolium* populations, mixed with uncontaminated flour, will, it was found, depress the reproductive rate, affecting both sexes, but males apparently more than females. Their fecundity drops far below that found in uncontaminated cultures; length of larval life is prolonged also, with an accompanying rise in larval mortality (*cf.* Park, 1934).

Thus the output of fertilised eggs, and ultimate recruitment of new adults, is evidently far from being a simple and direct function of the intake of food by the parent generation. Crombie (1942), experimenting with *Tribolium* and other granary beetles (including *Rhizopertha*) showing similar effects, concluded for *Rhizopertha*, at least, that what the conditioned medium does is to restrain the laying of eggs, rather than alter the actual fecundity of the ovaries themselves.

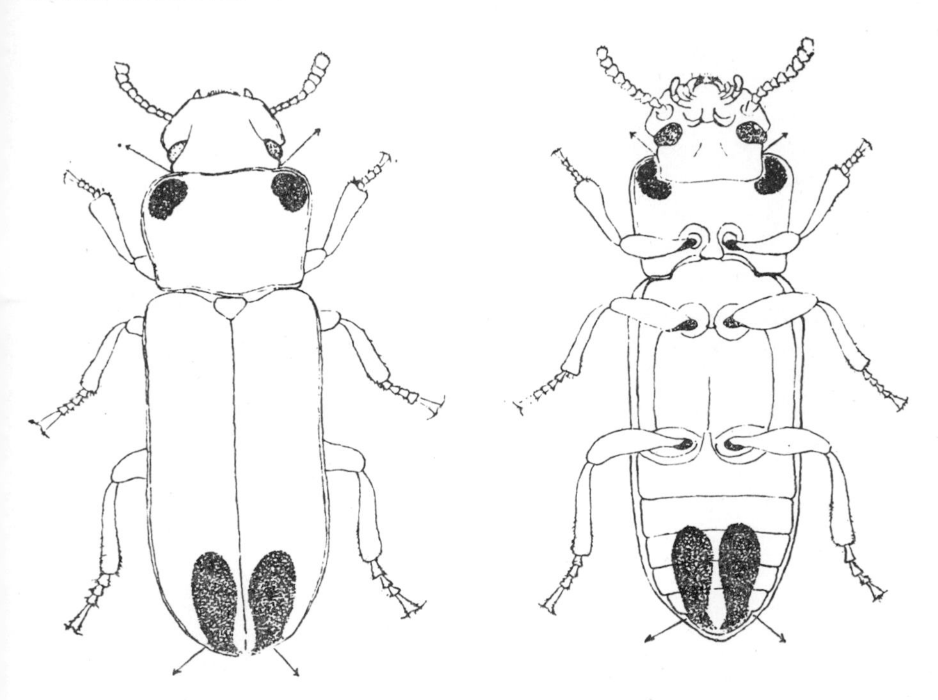

FIG. 40. Adult *Tribolium*, showing the positions of the turgid thoracic and abdominal reservoirs from which gas can be emitted at the four points marked by arrows. (From Roth, 1943.)

It is interesting at this point to disclose that both *Tribolium confusum* and its close relative *T. castaneum*, when adult, have paired odoriferous glands in the thorax and abdomen that produce an irritant gas (fig. 40). These are stimulated by disturbance and crowding, and the gas is liberated in maximal amounts from individuals massed (or chilled) in a confined space in the absence of any flour at all. Under these conditions it has been collected in quantity, crystallised by cooling, and identified as being mainly ethylquinone (Alexander and Barton, 1943). It is the same substance that causes the pinkish discoloration of conditioned flour, and it has been shown to be lethal in crystalline form when applied to first-instar larvae; and, as a gas, to induce developmental abnormalities in late larval and pupae (Roth and Howland, 1941; Roth, 1943). It probably has a depressing effect on the well-being of the adults (Park, 1949, p. 355) and seems reasonably likely to be the actual antaphrodisiac or fertility-depressing substance wholly or partly responsible for the experimental effects noted above. (Ethylquinone is well known as a spermicide in mammalian physiology.) It seems not illogical to suspect that its production is density-dependent, and leads to an appropriate restraint on the population's reproductive output.

Certainly there can remain no doubt in any of these insect experiments that the populations are effectively self-limiting; and the inference must inevitably be very strong that selection has perfected the adaptations so that population-densities always tend to balance themselves at the optimum level.

The next two series of experiments introduce a new and important parameter—that of exploiting a population, as a predator might do, and observing the response to varying levels of experimentally-imposed destruction. The first of these (A. J. Nicholson, 1955*a*) concerns the sheep blowfly (*Lucilia cuprina*) in Australia. The populations were initially left free to develop and maintain themselves under predetermined environmental conditions for long periods, generally about a year. In experiments numbered *A* to *D* there were four levels of exploitation, namely a control group (*A*) without any imposed destruction, and three others (*B*, *C*, and *D*) regularly subjected to the removal of 50, 75 and 90 per cent of newly emerged adults respectively. The actual numbers in the populations varied greatly from time to time during the long term of the experiments, because of a lag in homeostatic response resulting from the time it takes for eggs to develop into adult recruits; but ' the right kind of reaction to cause an approach to the equilibrium density is always engendered by density change '—that is to say, the response of the population always tended to compensate the oscillations that occurred, and restore its density towards the mean level.

Broadly, the result of destroying adults was to cause more adults to be recruited. In the controls, the average number of adults emerging daily was 573; but in populations *B*, *C*, and *D* the number emerging rose progressively to 712, 878 and 1,260. The adult population suffering 50 per cent imposed mortality (*B*) remained on the average almost as big as the control population (2,335 compared with 2,520); but stocks *C* and *D* averaged considerably less under their severer regime of destruction. As a compensatory reaction, in *C* and *D* average life-span of the adults was prolonged (from roughly $4\frac{1}{2}$-7 days), and the recruitment rate per adult per day much increased; so that what Nicholson calls the minimum coefficient of replacement, taken as 1 in the control population, rose progressively to 2, 4 and 10 in *B*, *C* and *D*. This indicates an extraordinary degree of resilience to destructive pressure: in terms of the fish experiments to be described next, it means that *Lucilia* populations could withstand the removal of a ' catch ' of 90 per cent of the adults without suffering by overfishing or yielding diminishing returns.

In these experiments there was one limiting factor still unmentioned, namely a restriction to 0.5 g per day of the amount of protein in the otherwise unlimited food supplied to the adult flies. The resulting intensity of competition among the adults to obtain the ground liver containing the protein was previously known to be correlated with egg-production. But here again, the relationship between egg-production and protein intake is not a directly quantitative one; thus population *D*, with about one-third the number of adults to share the protein, compared with the controls in *A*, actually averaged six times the production of eggs per day.

In the other *Lucilia* experiments the food-supply, not for the adults but for the developing larvae, was limited uniformly in each of four sets of conditions (*E* to *H*); the adults received a superabundance of all nutrients including protein. In big populations, like that produced by the controls (*E*), there was consequently intense competition for food among the larvae and relatively few of them survived to pupate, or emerge as adults. The stocks *F*, *G* and *H* were subjected this time to 75, 95 and 99 per cent destruction of the emerging adults respectively. Though their average adult populations were kept down by this high mortality, compared with the controls, all three stocks stood up to these astonishingly high tolls, and in every case the mean number of recruits continued to match the mean total number of deaths. In *H* the ' minimum coefficient of replacement ' was no less than 100 times as great as that of the controls.

These experiments reveal many points of great interest—for instance, an amazing ability to build up the recruitment rate of the particular life-history stage immediately preceding the one specially subject to destruction—either eggs or pupae or both in the cases reported here. Taking a more detached viewpoint, the results indicate how negligible an influence natural predation or parasitic mortality could have in dictating population-size and survival in a species like this, whose homeostatic control-mechanism can cope equally well with the regulation of numbers, whether independent mortality is zero or 99 per cent. In totally different circumstances from those of the *Drosophila* and *Tribolium* experiments, we see here the same powerful ability to seek and maintain a population balance adapted to the available resources of food, and a remarkable degree of independence from other kinds of ' environmental resistance '.

The experiments on fish-populations which particularly concern us were made by Silliman and Gutsell (1958), primarily in order to apply a practical test to the type of mathematical model erected by exponents of the overfishing theory. Four identical aquaria under uniform conditions were stocked with guppies (*Lebistes reticulatus*), two of them being kept as duplicate controls, and the other two as duplicate experimental populations. They were all allowed to develop in the presence of superabundant food until the 40th week, and all increased up to about the same ceiling level, so far as the total mass of fish was concerned, of approximately 32 grams in 17 litres of water (*see* fig. 41). This was to be expected: Breder and Coates (1932) had earlier found that two tanks when started with one gravid female and with assorted fish respectively, took only about 20 weeks to reach the same practically constant population of 9 or 10 fish each (very reminiscent of Chapman's flour-beetles).

After this equilibrating period, tanks *C* and *D* were allowed to continue without interference for 2½ years until the 174th week: during this period the biomass in each case underwent only minor oscillations. Tanks *A* and *B*, however, were subjected to 'cropping' at 3-week intervals—three weeks being the average turnover of each generation of guppies under the conditions of the experiment, equivalent to perhaps a year under the usual conditions of a natural fishery (*see* figs. 42 and 43).

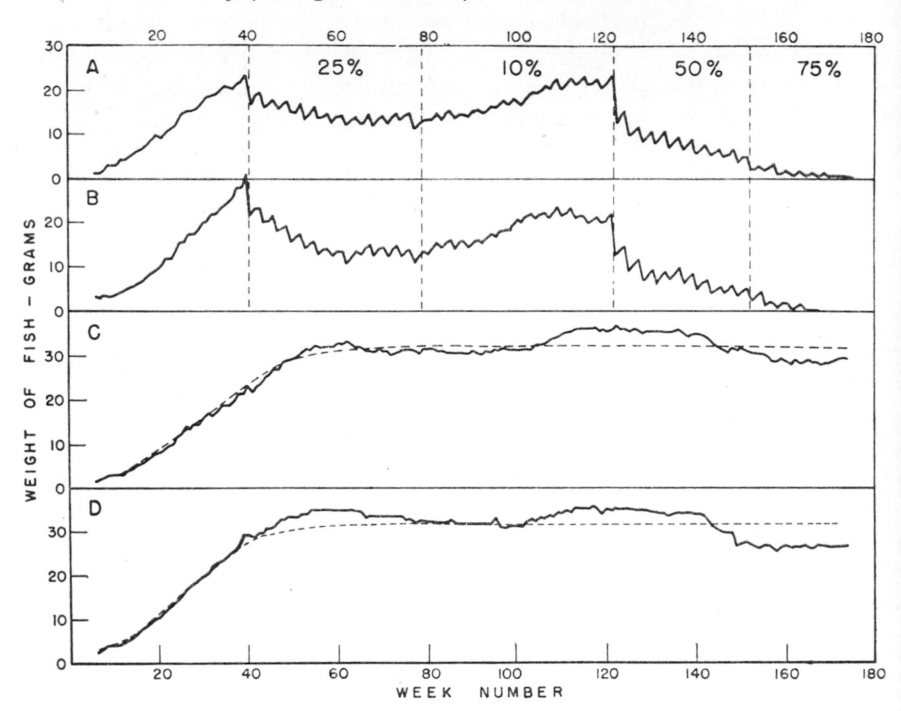

FIG. 41. Weekly population weights (biomass) during Silliman and Gutsell's fishing experiment with guppies. A and B are duplicate experimental populations exploited at four different rates in succession (25, 10, 50 and 75 per cent); C and D are duplicate controls. (From Silliman and Gutsell, 1958.)

At first, 25 per cent by weight of the fish were removed every three weeks from *A* and *B*. The number of adults and young fell rather sharply, but by week 60 there had been a great new influx of juveniles (guppies are of course viviparous), and a new stable ratio of young to adults was established; the biomass averaged considerably below that of the controls. After week 79 the exploitation rate was reduced to 10 per cent; the biomass now progressively increased, and, though *A* and *B* were not quite uniform in this respect, the proportion of juveniles to adults tended to decline again. From week 121-150, exploitation was at the high rate of 50 per cent; the result was a great increase in the proportion of juveniles, and a progressive decline of the whole biomass: rather severe overfishing, in other words, was taking place. Finally, from week 151, the exploitation rate was put up to 75 per cent,

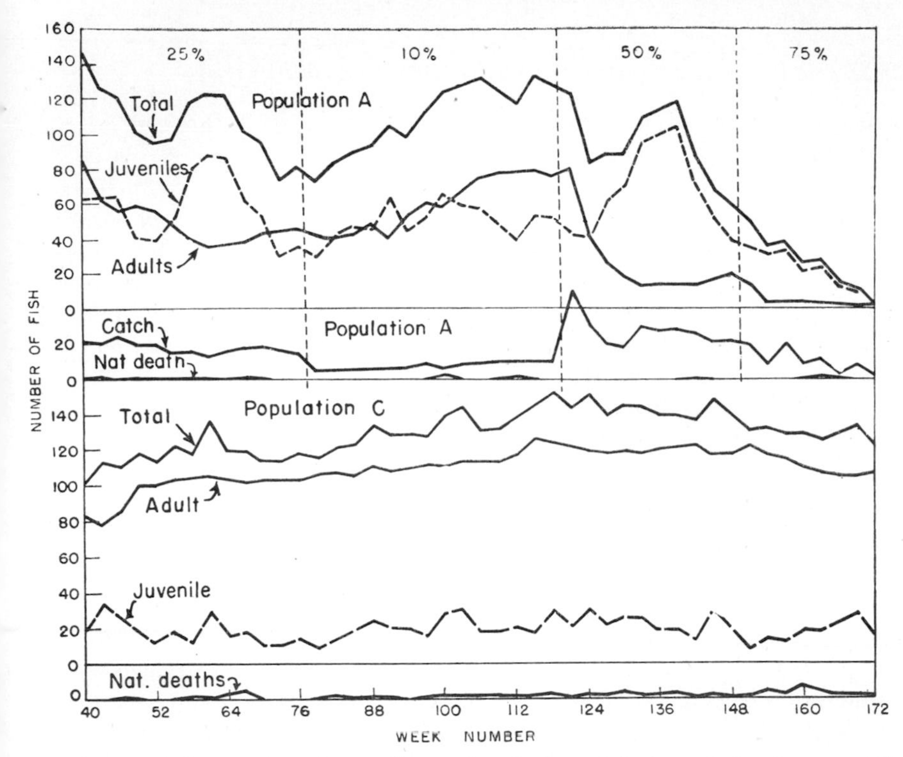

FIG. 42. Population, catch, and mortality numbers in the exploited population A, compared with the control C. Note rise in number of juveniles in A (*i.e.* increase in recruitment rate) following increases in exploitation rate at about 60th and 130th weeks. Compare with fig. 41 to see effect on total biomass. (From Silliman and Gutsell, 1958.)

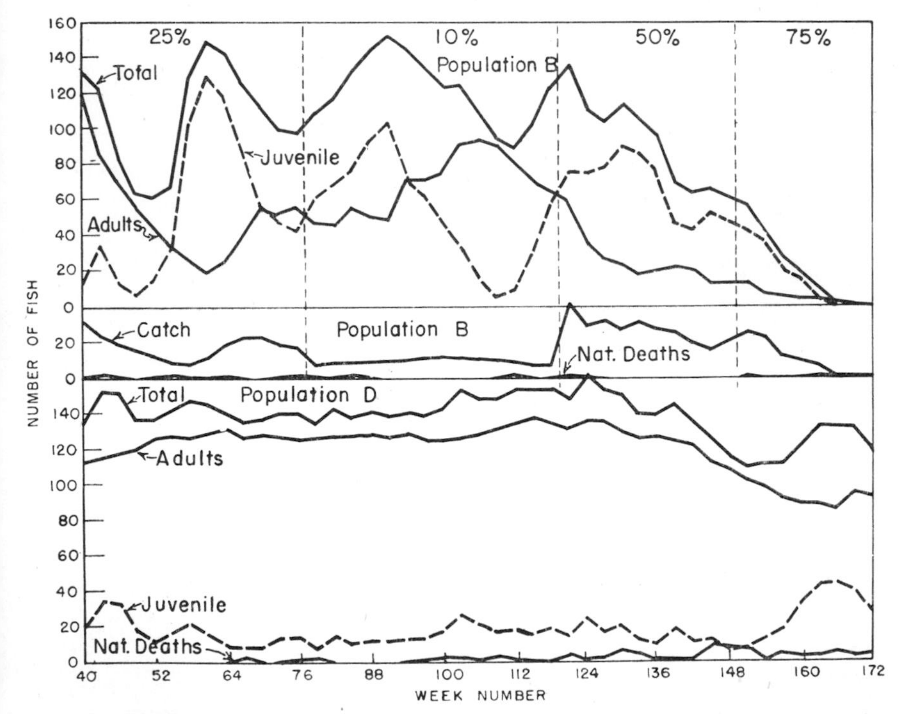

FIG. 43. Duplicate of the experiments shown in fig. 42 above, in this case comparing exploited population B with unexploited control D. Note close resemblance between these results and the ones simultaneously obtained for populations A and C. (From Silliman and Gutsell, 1958.)

and this led more or less swiftly to extinction, and the end of the experiment.

This fine demonstration of experimental overfishing is in some important respects nearly parallel to the experiments with *Lucilia*. Notice that in spite of superabundance of food, an individual space factor limited each population initially at the same level. At this equilibrium, the proportion of adults became high and the production and survival of young low; cannibalism was common (*see* also Breder and Coates, loc. cit., p. 149); and the effective recruitment rate just sufficient to replace the natural mortality of senile fish. Under exploitation, however, the life-history stage preceding the one exploited—that is, the juveniles—increased, roughly in proportion to the level of exploitation. Analysis by Silliman and Gutsell clearly demonstrates that exploitation at between 30 and 40 per cent would give the biggest sustained yield, in grams of fish caught: beyond that the stock cannot recruit itself fast enough to maintain its natural optimum population; and above about 60 per cent exploitation it is doomed to extinction (*see* fig. 44). *Lucilia* also showed a similar reduction in biomass and increased production of 'juveniles' under exploitation; but the 'sustained yield' went on increasing even up to 99 per cent, and the point of overfishing would not have been reached until some still higher exploitation rate, lying between 99 and 100 per cent. It is very instructive to see quantified in this way such great factor-differences in the rates of extrinsic mortality that different kinds of animals can withstand without the fundamental equation between recruitment and loss being violated.

It may be noted in passing that experiments to test the simple effect of crowding on fertility have been done with the domestic fowl. Pearl and Surface (1909) showed that mean annual egg-production over a period of three years was 10-20 per cent higher in flocks placed 50 in a pen, giving 4·8 square feet of floor space per bird, than in those of 150 in a pen, with only 3·2 square feet for each. Pens with 100 birds, at 4·8 square feet per bird, were intermediate, indicating that the number of members of the confined group has a marked effect, quite apart from their density on the ground.

21.6. *Similar experiments with rodents*

One other group of vertebrates that have been the subject of many population experiments are the rodents, and it is now well known that they show a series of homeostatic adaptations which are the counterpart of those already described in *Drosophila* and *Tribolium*. Crew and Mirskaia (1931) showed, with a population of laboratory mice (*Mus musculus*), that as crowding increased the litter-size dropped, the number of pregnancies dropped and the death-rate mounted, following a pattern with which we are now familiar. They were the first to notice that in rodents these effects are not uniformly suffered by the whole population: even under extremely crowded conditions there were still some individuals that continued to enjoy a full normal life-span and well-sustained fertility. Evidently this is characteristic of the social organisation of rodents: Clarke (1955, p. 81) found in the vole *Microtus agrestis*, for instance, three classes among his males; there were (i) large, glossy-furred, relatively unscarred individuals that could roam with impunity all over the enclosure, (ii) similarly large, free-roaming individuals, but with tattered fur and scars on their hind quarters, and (iii) light-weight, unaggressive, tattered, lack-lustre individuals with restricted home ranges. Such differences as these depend on social dominance, and once again it will be readily apparent, both in the remainder of this chapter and in the next, how deeply the hierarchy is involved in the homeostatic process.

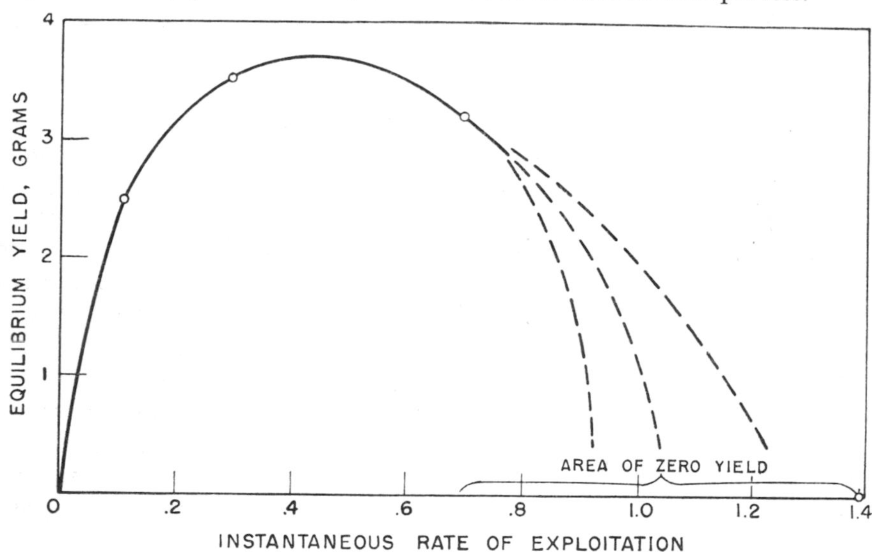

Fig. 44. Relation between exploitation rate and sustainable yield in the guppy fishing experiments. The optimum point comes about the 35-40 per cent exploitation rate and any rate exceeding about 50 per cent leads to extinction and zero yield. (From Silliman and Gutsell, 1958.)

Confined populations of voles, mice and rats build up in typical fashion to a ceiling, even in the presence of superabundant food—a ceiling which is, however, not quite so predictable as those we have previously studied; duplicate stocks in identical enclosures, or the same stock in successive years, do not necessarily attain exactly the same numbers (*cf.* Clarke, loc. cit.). ' Temperamental ' differences are largely responsible (*cf.* Crew and Mirskaia, loc. cit.; Brown, 1953; Southwick, 1955a). The homeostatic mechanism is complex, but many if not most of the component factors can at any rate be identified. In house-mice, for instance, longevity, the duration of breeding condition, the pregnancy-rate, intra-uterine resorption, litter-size (or birth-rate), and neonatal mortality are all capable of density-dependent variation.

In experiments where the food-supply was limited, Strecker and Emlen (1953, p. 380) found that the effective factor in curbing reproductive output initially, when the growing population first began to respond to the food-stress, was a drastic mortality of young mice: before the ' food crisis ' 89 per cent of the young born had survived; after it, in three litters totalling 13 individuals, all but one died before reaching the age of five weeks. By this timely response a food emergency was no doubt averted. It was soon followed by a deeper-seated adjustment—the cessation of breeding altogether; for the next six months, from September to January, the biomass of mice stayed more or less constant, while the actual number of mice slowly declined.

Southwick (1955a, p. 221) had in part a similar experience: ' the major mechanism of population control in each pen ', he wrote, ' was a high litter mortality, apparently related to population pressure. None of the pens was actually limited by failure of litters to appear '. The mortality sprang from cannibalism, desertion, encroachment by adults, or actual destruction of the nest. Though this kind of regulation is properly the subject of the next chapter, we may specially note the importance of aggressive fighting amongst the adults in causing the decline of parental care. By the time a population had built up until fighting involved each mouse on the average in one aggressive encounter per hour, litter-survival had fallen so low that population growth was slowed or ceased altogether (Southwick, 1955b, p. 630).

Southwick's experiments differed from those of Strecker and Emlen in that the mice had unlimited food, and were restricted only in space and cover. It is probably significant that they did actually continue to mate and breed under these highly unnatural conditions, and thereby perpetuated a situation calling for persistent mortality of the new-born. Nutritional stress may be necessary to bring about a downward adjustment of the pregnancy rate: thus, in quite different circumstances, McClure (1958) reported that, by starving and feeding mice in a series of alternating 48-hour periods, the appearance of litters could be completely prevented.

The third combination of environmental conditions, where food was limited but not space, was investigated by Strecker (1954) in a laboratory basement. Food in measured quantity and water were supplied. The mice were caught and individually marked, and, since they could come and go quite freely, traps were set in other parts of the building. A ceiling of numbers was reached in the basement and a steady population maintained; but no food shortage ever developed, and there was no indication of any change in the rate of reproduction (loc. cit., p. 251). Emigration, which had been almost negligible during the months of population growth, suddenly shot up as soon as a ceiling had been established. Crowcroft and Rowe (1958) recently contributed a demonstration that females whose fertility had been inhibited by confined crowding could become fertile immediately they were allowed to disperse into a large pen.

These various experiments with house-mice show that homeostatic restraints can be invoked either by overcrowding *per se*, in the presence of sufficient food, or by a limitation in the food-supply alone. When there was no such stress, on the contrary, a high reproductive output was the rule and recruitment to the population continued at a rapid pace. Experimental conditions are obviously artificial, and either purposely or by force of circumstances unnatural; under wild conditions (if this is the proper expression to use for house-mice) the density-tolerance or space mechanism is perhaps not often likely to be set in action in the absence of any accompanying nutritional stress. As we shall see, however, individual encounters and fights provide the proximate factor that sets the homeostatic process in motion; and crowding beyond a certain density, even in the presence of superabundant food, evidently exerts a disquieting stress and interferes, in a way that we can appreciate from human experience, with the privacy essential for the proper fulfilment of domestic life.

There is a very interesting and important point to be made in the fact that no adult mortality was ever caused in these experiments by actual starvation. In Strecker and Emlen's group on limited rations, after population growth and reproduction had ceased, the authors particularly noticed that the adults differed from those in the other group serving as controls ' in possessing more fat and being reproductively quiescent. All mice in pen *B* had heavy deposits of subcutaneous and abdominal fat while those in pen *C* [population still growing] had no fat or only slight amounts present' (Strecker and Emlen, p. 381). Regulation of the population, it seems safe to infer, had taken place in pen *B* at a ceiling level which assured a sufficient diet to all its occupants.

The physiological changes that bring about the regulation of reproductive output are mediated by the mammal's endocrine system. Selye himself (1939), who first recognised and named the ' general adaptation syndrome ', early showed that various damaging forms of stress could affect the female rat's sex organs. His later work, in revealing the vast significance of human adaptation to stress—a fuller discussion of which is deferred to the next chapter—has demonstrated the intimate relation between stress and the adrenal glands (*cf.* Selye, 1950). Christian (1950, 1959), Richter (1952), and more recently Barnett (1958), have followed up his lead, and shown conclusively that tough sewer-rats (*Rattus norvegicus*) have on the average much larger adrenals than their placid laboratory cousins, and that socially dominant rats generally have smaller adrenals than subordinate ones; Clarke (1953) found the same in the short-tailed vole (*Microtus agrestis*). An ' interloper ' male rat, introduced into a cage occupied by an established colony, and subject as a result to persecution, quickly develops an enlargement or hypertrophy of the adrenal cortex, accompanied by a great increase in its secretory activity. Males, as we might expect, are more susceptible than females, for it is the emotional stress of threat and fighting that induces the physiological response. Christian (1959, p. 10) made this quite clear by the administration of reserpine, a tranquilising drug, which effectively prevented both adrenal enlargements and associated changes in other organs.

The significance of the stress syndrome will appear even more forcibly in the next chapter, in connection with induced mortality. Cortical hormones are specially concerned with the body's ability to adapt itself to new situations, with its reaction and defence against injury, with general reproductive activity—and, most significantly, with preserving a steady state *within the body* (physiological homeostasis). It may be strongly suspected that the adrenals, coupled with the closely-linked pituitary, play a key part in guiding the individual animal's responses, in accordance with the homeostatic needs of the social unit to which it belongs. Our knowledge of the endocrine side of the response to stress still extends hardly at all beyond man and the laboratory mammals; but at present there seems no reason to doubt that the adreno-pituitary system functions similarly in all mammals, as well as in birds and other vertebrates generally. It may indeed have evolved in this group primarily to serve as a physiological mediator, sensitive to the varying stimulus of social pressure and epideictic excitation, and leading to a concordant homeostatic response by the individual concerned. Some such system must necessarily exist, not only in vertebrates but in all other groups of animals where the same kinds of social conventions are found.

21.7. *Field evidence*

It is desirable to make sure that the correlation so clearly established in the laboratory, between population-density and recruitment rate, obtains equally generally in the field; but we shall continue as far as possible to set aside evidence concerned with mortality (including neonatal mortality) for consideration in Chapter 22, and confine ourselves primarily to matters relating to fertility. As might be expected, much of the information available is circumstantial or incomplete.

An example in insects of an apparent correlation between recruitment rate and population-density has been reported by Wallace (1957), in the collembolan *Sminthurus viridis*, known as the lucerne flea. In parts of Western Australia and elsewhere it has become a pest of clover crops. In a 10-acre square marked out in a grid, the population-density was found to vary widely from point to point at the beginning of the winter reproductive season (May). Seven censuses were made in all over a four-month period, and it gradually appeared that a strong compensation for the initial variation was taking place: so far did this go that by September the areas originally carrying the sparsest population had tended to become the densest, and *vice versa* (*see* figs. 45 and 46). There was an inverse correlation between initial and final densities, with a negative correlation coefficient between them of —0·426 (significant at $P < 1/1000$). Similar statistically significant correlations were obtained during the next three years also. It can be assumed that the changes in the density pattern must have been caused by local differences in the birth and survival rates, since the distances to be covered are too great for the effects to have been produced by movement, in such minute flightless insects. It was possible to rule out the most obvious extrinsic factors that might have influenced the changes, including soil, micro-climate and predation-pressure.

It is possible to infer that density-dependent governing factors were responsible for the correlation observed, and there is a strong presumption that these factors were intrinsic or autobiotic—the more so since MacLagan

(1932*b*, p. 216) had shown in laboratory experiments with confined populations of the same species that, above a certain density, egg-production varied inversely as population-density. MacLagan also noticed, as a point of interest, an 'individual distance' phenomenon in *S. viridis* when the insects were crowded in a glass jar.

The best-documented case in birds is Kluyver's study (1951) of the populations of titmice (*Parus* spp.) in the Netherlands, to which we referred previously in discussing the correlation between habitat-productivity and population-density. Here the data are exceptionally full, and cover a very long series of years. They show that there is a highly significant negative correlation (−0·73) between density and the number of second broods laid (loc. cit., p. 71). In four out of five woodland areas analysed there was an additional negative correlation between density and clutch-size (the fifth, a positive one, being near zero). However, although the density itself sometimes varied as two to one, a difference of only about half an egg separated the higher and lower density years; and none of the correlations is statistically significant (Lack, 1952, p. 169). Kluyver's graphs of these two variables in relation to density, and the summation of both of them combined (called 'fecundity' by him), are reproduced in fig. 47.

There was still another woodland—Aardenburg—that showed considerably stronger density-effects, both in the percentage of pairs having second broods and in average clutch-size. Between 1926 and 1929 there were only a few nest-boxes in this wood, but after additional ones had been erected the breeding population at once increased threefold. Over the next five years the average percentage of second broods fell from 63 to 16, and the average

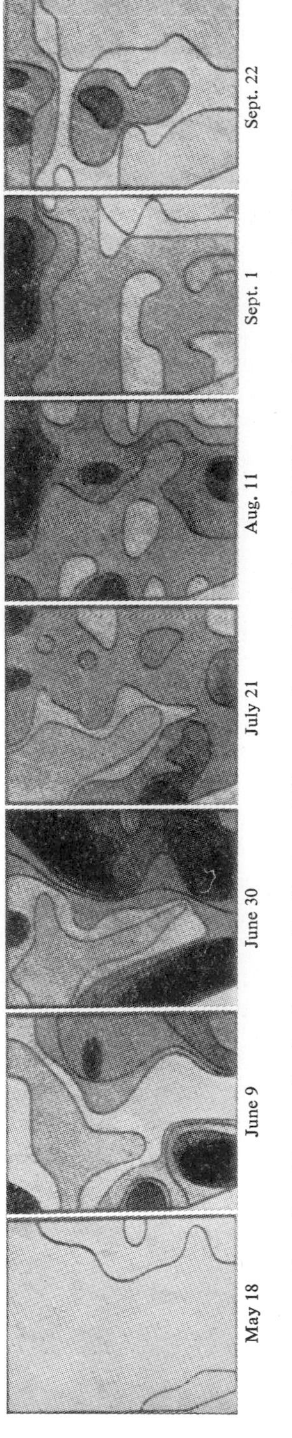

FIG. 45. Changes in the density-distribution of 'lucerne fleas' (*Sminthurus viridis*) on a clover crop in Western Australia during a single season. Note that areas of high and low density in early June had been approximately reversed by September. (From Wallace, 1957.)

clutch-size by nearly two eggs (*cf.* Lack, loc. cit.). Combining these together, the average annual fertility per breeding female decreased from 14·2 at low density to 8·9 at high density (ratio 1:0·63). Kluyver makes it clear that factors other than absolute population-density also enter into the determination of clutch-size—especially the earliness or lateness of the spring; productivity of the habitat, that is to say, being variable from year to year, exerts the important influence that might be anticipated.

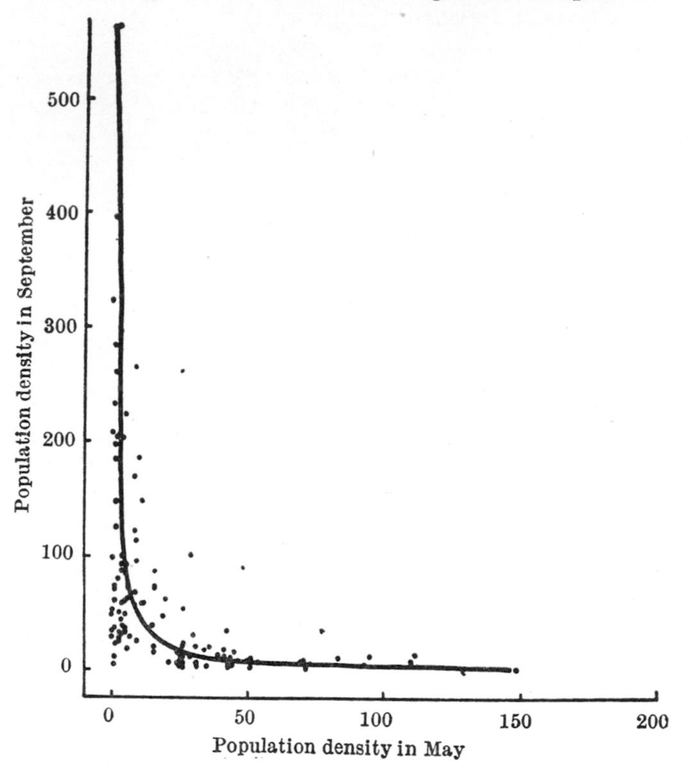

FIG. 46. Graph showing an inverse correlation between density in May and density in September on the study area shown in fig. 45. (From Wallace, 1957.)

Figures for the same three species of titmice obtained in England have been analysed for comparison by Lack (1955); the occurrence of second broods is generally much less common there than in the Netherlands, and on this count the figures available are as yet statistically inadequate; but a similar inverted correlation was shown to occur between density and clutch-size (loc. cit., p. 61).

A third illustration may be chosen from the mammals: it is Kalela's work in Finnish Lapland on the natural reproductive rate of the vole *Clethrionomys rufocanus*, in three successive years (Kalela, 1949). Here the principal inverse density effect was found in the proportion of young animals attaining sexual maturity and breeding in the summer in which they were born. In 1954, a year in which the season started with a low population, all or almost all of the early-born juveniles of both sexes became mature and bred before the season was out. In 1955, the year of high population, the great majority of young males remained immature, though a considerable proportion of the females became fecund. In one special area, however, where the density was

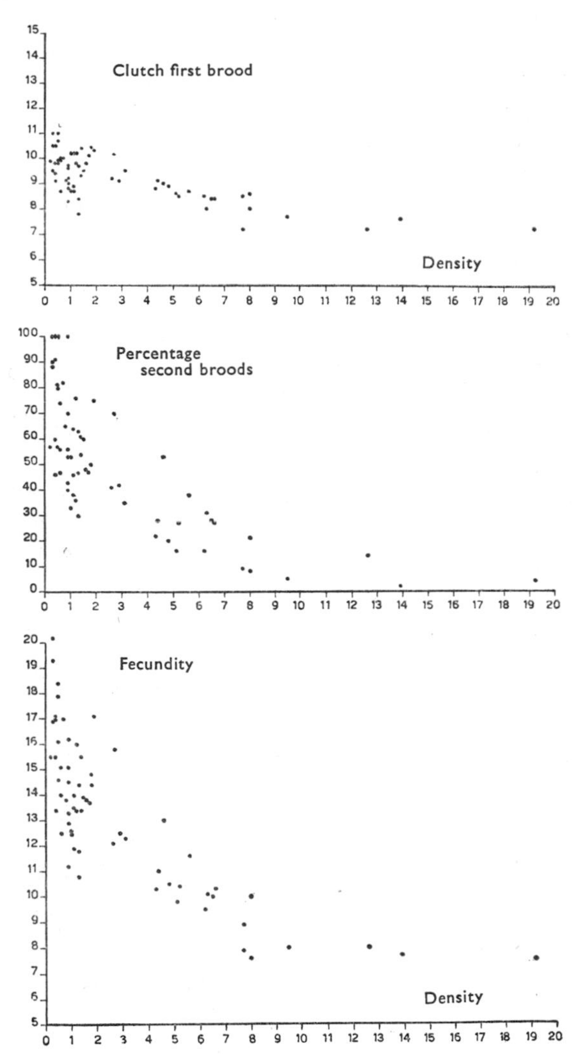

FIG. 47. Correlation between population-density in the great tit (*Parus major*), in pairs per 10 ha, and (i) clutch-size, (ii) percentage of second broods, and (iii) the summation of i and ii, or ' fecundity '. (From Kluyver, 1951.)

about twice as high as anywhere else in 1955, practically all the young females also failed to breed. In 1956, after a severe mortality during the preceding winter, with a population even lower than in 1954, the majority of early-born young of both sexes once again became mature and bred in the later part of the summer. It is significant that Clarke (1955, p. 71), in his experiments with *Microtus agrestis*, found exactly the same difference between a denser and a less dense population: in the less dense each young produced an average total of 10·2 young in the year of her birth, compared with only 1·2 in the denser. Recruitment regulation in both these voles seemed thus to be predominantly a matter of the age at which breeding commenced.

Illustrations as well worked out as these are of course very uncommon; but we have plenty of more or less fragmentary circumstantial evidence pointing to the same conclusion, namely, that the rate of recruitment through reproduction is one of the commonest density-dependent variables to be exploited in securing an optimum population level.

One of our general postulates is that, under favourable conditions, there is no need for the population to exercise its full reproductive power in order to make good current losses and provide the small surplus normally desirable for export; all that is required is some fraction of the possible output. This at once explains the otherwise incredible inefficiency of the ordinary reproductive process as we see it in so many animals—an inefficiency often so gross as to be inexplicable on the hypothesis that the animals are always reproducing at the highest rate allowed by their physiological resources, or by their powers of caring and providing for the young. We have already noted in a different connection that as a general rule in birds there is a proportion of adults excluded from breeding at all, by the stringent conventions of property-tenure—even when, as in so many sea-birds, the crowded colony is limited by nothing but a conventional boundary. Without the status conferred by the possession of a satisfactory site a bird is inhibited from breeding, whatever the state of its physiological resources.

In a scattered colony of American brant geese (*Branta bernicla hrota*) on Southampton Island, a graded scale of reproductive rates was found among the site-holders, apparently depending on the degree of crowding in different parts of the colony. Barry (1956, p. 198) sampled three sections of the 4½-mile breeding strip and found, where the nesting was densest at the east end, an average clutch-size of 3·77; towards the west the colony spread out, and the average rose to 4·21 and finally to 4·41 in the most westerly sample area.

A notorious instance of what is often wrongly assumed to be reproductive inefficiency is the destruction of eggs, embryos or newly-produced young by the parents or other members of the same species. Where we saw this developed as a conventional practice in primitive human races, we followed Carr-Saunders in accepting its function as being homeostatic (p. 492). In all the main types of social insects, including honey-bees, bumble-bees, wasps, ants and termites, egg-eating is known to take place, and is likewise admissible there as an adaptation: ‘ it is . . . certain ’, says Richards (1953, p. 55), ‘ that the regulation of the number of eggs is essential to the well-being of the colony, and egg-eating might be regarded as a form of birth-control ’.

The habit has scarcely ever been regarded in this light, however, when observed in other animals; neither have the equally common phenomena of fratricide among the new-born, or parental neglect leading to their death, which are detailed in the following chapter. Sinclair (1895, p. 39) describes how he watched a male of the common centipede (*Lithobius forficatus*) rush at a female, that had just laid an egg and still held it in her clasping organs, seize the egg and devour it. Brown (1936, p. 136) describes seven cases of European kingfishers (*Alcedo atthis*) destroying the eggs of their own species in Renfrewshire, Scotland: whether the parents, neighbours or strangers were responsible could not be ascertained; the eggs were always rolled out of the nest-burrow and no tracks were revealed in the dry sandy nest-entrance other than those of kingfishers. Nice (1937, p. 145) noted, under the heading of ' parental failures ', eight cases of disappearance of single eggs from the nests of the song-sparrow (*Melospiza melodia*). Ratcliffe (1958) reported a still larger number of instances of nests of the peregrine (*Falco peregrinus*) where one or more eggs were broken or disappeared; and after considering the evidence and alternative explanations, he found it ' difficult to avoid the conclusion that the majority of these broken eggs were eaten by one or other of the owners ' (loc. cit., p. 26). Disappearances were specially common in 1951, occurring in four out of nine nests, compared with a total of one out of thirty-five nests over the preceding five years. The habit of puncturing eggs was one of the principal causes of nesting loss in the arctic tern colony (*Sterna macrura*) studied by Pettingill (1939, p. 426) on Machias Seal Island in Maine; and in the end only 16 per cent of the original eggs produced fledglings.

It has been customary to regard this kind of behaviour as a negative and unnatural perversion, injurious to the species concerned; but of course it is inconceivable that it could survive as a common practice were the results not beneficial to the welfare of the population as a whole. It seems much more reasonable to accept it instead as a positive act of intervention, carried out at a particular point in the reproductive cycle, cutting down the effective output in accordance with the latest revised homeostatic information.

The writer has been struck for many years by the low output of young often achieved in thriving colonies of sea-birds. Keighley and Lockley (1948) recorded that of 50 eggs of the razorbill (*Alca torda*) they had under observation at Skokholm (representing 50 or possibly more pairs of adults), only 18 survived to hatch and grow into sea-going chicks. Assuming the latter were evenly divided between the two sexes, this implies a life-expectancy of $50 \div 9$ or $5\frac{1}{2}$ years for each chick at the time of leaving the colony. (It must be realised, however, that statistically this is not a large sample.) At the great colony of Fowlsheugh in north-east Scotland, where there are several thousand razorbills and perhaps twenty or thirty thousand guillemots (*Uria aalge*), the figures appear just as low or lower. For instance, on 7 July 1957—early enough to be sure that few if any young had gone down to the water—two ledges with a total of ninety-nine adult guillemots held only two visible chicks. This must be below average, but it supports the impression of the low reproductive coefficient required to maintain a flourishing, stable, and nowadays virtually unmolested colony like this of long-lived birds.

We must expect the reproductive coefficient within each species to vary from season to season and place to place, in accordance with what is economically optimal in the prevailing circumstances. A good illustration of geographical variation is provided by the studies of H. Marshall (1947) and Paludan (1951), and the analysis made by the latter (loc. cit., pp. 113-123), which showed that the herring-gull (*Larus argentatus*) has only about half the life-expectancy in the United States that it has in Denmark: it thus requires twice the recruitment rate to maintain its numbers. The balance of recruitment and loss is achieved at these two levels, as Paludan demonstrated, not by any difference in the number of eggs in the clutch, but by differences in the rate of destruction of chicks (mostly done by the adults, *see* p. 532); in his colony in Denmark this allowed only about fifty fledglings to leave the colony for every hundred breeding pairs (loc. cit., p. 99).

A striking example of reproductive output varying from season to season, in response to changing economic conditions, is seen in the effect that myxomatosis, in virtually destroying the abundant rabbit population of the British Isles, had secondarily on the rabbit's specialist bird-predator, the common buzzard (*Buteo buteo*). '1955 showed a great decrease in breeding-activity of buzzards in all regions where rabbits had become rare or extinct. Many, perhaps most, pairs did not breed at all. It was normal, *i.e.* comparable with 1954, in local areas where the rabbit population was not affected, and where rabbits had never been abundant' (Moore, 1957, p. 190). Earlier Wendland (1952) had shown that, in the same species in Germany, above-average fertility occurred in years of vole outbreaks, notably 1933, when the output of fledged young reached 2·6 per nesting pair, compared with an average of 1·41 over the period 1940-51. A somewhat similar relationship between the pomarine jaeger (*Stercorarius pomarinus*) and the lemming (*Lemmus trimucronatus*) in Alaska was mentioned in another connection on p. 150.

In wild mammals indications can also be found of reproductive rates being adjusted or altered in relation to current survival rates, and to varying economic conditions. In North Wales, Brambell and his associates made the unexpected discovery in the rabbit (*Oryctolagus cuniculus*) that an average of 64 per cent of the embryos conceived perish before birth, very largely by the total loss of litters about the twelfth day of pregnancy. The arrested embryos are not aborted, but instead their tissues are broken down and resorbed by the uterus, finally leaving nothing to be seen but an impermanent scar. Losses on this scale are not thought to be abnormal for wild European rabbits in general (this information is cited from Thompson and Worden, 1956, pp. 112-13). The same process of inter-uterine resorption is now being widely detected in rodents, carnivores (*e.g. Alopex*), artiodactyls (*e.g. Cervus elaphus*) and probably other groups of mammals. Regarded as a form of birth-control it becomes intelligible: once again the less satisfactory alternative is to regard it, along with cannibalism and egg-destruction, as a kind of adaptational malfunction.

There are indications in the rabbit that litter-size is variable, and that it may be correlated with current population-density and economic conditions. Litter-size in Caernarvonshire was found to average 4.89 in 1941, and 5·64 in 1942. On Skokholm, where the rabbits at the time were small and, in winter, undernourished, Lockley in 1939 found an average of about four all through the year. In another part of South Wales near Tenby, where rabbits had been allowed to multiply excessively on a fertile area of some 1,200 acres while it was being used for military purposes, and had remained unmolested for nine years, litter-sizes were reported to be remarkably small, usually only 2-3 young to a doe (Thompson and Worden, loc. cit., pp. 56 and 110). The presumption here is that the rabbits had almost or quite reached a stable ceiling of population, and it could be predicted, in the light of the experiments described earlier on exploited populations, that as soon as someone disturbed the colony and started to trap and shoot them, their average litter-size would have mounted again to the sort of figures found in Caernarvonshire.

It is relevant here to recall the traditional practice by British shepherds of flushing their ewes for about three weeks before putting them to the rams, in order to increase the conception of twins. Flushing consists of bringing the ewes from a lower to a much higher plane of nutrition, so that they are in good and still-improving condition at the time of tupping. The practice has been fully investigated (originally by F. H. A. Marshall, 1905), and has been shown to lead to a highly significant rise in the incidence of multiple ovulations: many more of the ewes consequently conceive twins, some (differing in proportion with the breed, but often around 10 per cent) triplets, and occasional ones quadruplets. In comparison with an average production of possibly 100-120 lambs by 100 unconditioned ewes, flocks of flushed ewes may bear as many as 150-195 lambs per cent (*cf.* Fraser and Stamp, 1957, p. 176). This striking link between a steeply-rising standard of living and an enhanced rate of twinning is not shared, so far as we know, by the cow, nor for that matter by man; but it clearly resembles the correlation between available food-supply and litter- or clutch-size indicated in the rabbit and various birds. Many analogous examples could be found in mammals: the lion (*Felis leo*), for instance, is a species in which litter-size is greatly affected by the abundance of game (Stevenson-Hamilton, 1947, p. 130).

In the mouse experiments, it will be remembered, recruitment was curbed chiefly by two factors—an immediate effect on infant mortality, and a more delayed reduction or cessation of pregnancies. The musk-rat (*Ondatra zibethica*) provides an example of the latter effect in a mammal living under wild conditions: the females, so Errington (1956, p. 307) showed in Iowa, when ' subject to the damping effects of crowding past their toleration limits, just quit breeding early in the season after giving birth to a litter or two '. In wild Norway rats in the city of Baltimore, Davis (1951) found it was the general pregnancy-rate that was depressed by crowding; during the spring breeding season in a stable (static) population it was only 14·4 per cent, compared with 41·6 per cent in a population that had previously been ' exploited ' by heavy trapping and was then increasing again.

In two further cases, exploitation by man has already been advanced by the original investigators as a possible explanation for an observed increase in the rate of reproductive output. The first concerns the notorious African quelea bird (*Quelea quelea*), which exists in such incalculable numbers (*see* p. 360) that it has become a serious agricultural scourge. In Senegal, Morel and Bourlière (1956, p. 108) found the average clutch-size rose significantly from 2·72 in 1953, and 2·74 and 2·78 in two colonies in 1954, to 3·09 in 1955. Though it can be no more than a faint suspicion, ' one cannot completely exclude the hypothesis of an increase in fecundity due to a decrease in the effective total of the species (as a result of the repeated hecatombs of the bird-control department), which assures the survivors a more favourable food-situation '.

The second case, considerably less tenuous, originated with a report by N. A. Mackintosh (1942, pp. 223-4) on the *Balaenoptera* whales of the southern ocean. ' A preliminary examination of the data indicates that the percentage of adult females which are pregnant has been increasing in a remarkable degree year by year, as if the actual rate of breeding were becoming faster. . . . No doubt the interval [between pregnancies] varies, and if the rate of breeding has increased in recent years, conceivably as a reaction to whaling, it would probably mean that fewer whales fail to become pregnant after two years from the last pregnancy '. Further research has confirmed this deduction: the pregnancy rate in *B. physalus*, for instance, has risen by an eighth, and it is probable that the females now tend to start breeding at an earlier age than formerly (R. M. Laws, verbally). It appears, therefore, that recruitment is regulated here to an important extent by varying the pregnancy rate, very much as in the Baltimore rats.

It begins to seem fairly certain that the intrinsic control of fertility demonstrated by population experiments operates just as widely under natural conditions—a conclusion that has already been rather widely accepted by population-ecologists. It should be remembered that we have been concerned so far (with rare exceptions) only with evidence for the control of recruitment through reproduction, and not through subsequent mortality. The distinction is purely one of convenience in dealing with the very extensive material. We shall see in Chapter 22 that in many animals the principal modulation of the annual recruitment rate takes the form of juvenile mortality: in terms of anthropology, it is the counterpart of infanticide instead of abstention or abortion.

It seems likely enough that the annual reproductive enterprise is entered upon in many species without accurate fore-knowledge of what the season will bring forth in food-supplies at the time the new generation reaches the stage of peak demand, and without any fore-knowledge either of the extrinsic casualties that are going to befall the population. The accumulated providence that is distilled into adaptive behaviour therefore tends to call at the start for a fecundity distinctly on the generous side, to take care of emergencies: it can always be trimmed down later on to meet contingencies as they develop. In fact the evidence suggests that in all the higher animals such measured homeostatic adjustments can and do take place, at one, or more than one, stage of the reproductive sequence. The farther that ontogenesis is allowed to proceed, however, the more costly is likely to be the wastage of flesh and blood if it is terminated: and this helps us to understand why salvage operations, either by the resorption of foetal embryos or the consumption as food of surplus eggs and young, have commonly and repeatedly evolved.

21.8. *Polygamy and related phenomena*

This is the most appropriate place to return to the subject of polygamy, to which passing references have been made in some of the earlier chapters (pp. 216 and 250). It is well known that polygyny, and much less commonly polyandry, have arisen repeatedly in the higher animals; yet the subject seems never to have been reviewed, as it affects the animal kingdom as a whole, nor is there any general and acceptable explanation of its usefulness or survival value. Some zoologists who have given it passing thought have probably concluded that, because individuals mated to several consorts are likely to leave more progeny than those that are monogamous, selection must tend to favour polygamy. While this may contain a small grain of truth, it is evidently far from being the whole, or in any way the prime, evolutionary cause of the phenomenon. The alternative possibility, that polygamy is a secondary consequence arising from an unbalanced sex-ratio, is equally unsatisfying: for often it is visibly at variance with the facts, where we find a non-breeding surplus of the polygamous sex; or, if not, it poses us with the ulterior problem of explaining why the sex-ratio should have become unequal.

The subject is too large to explore here as thoroughly as it deserves, and a full review must await another occasion. It concerns us, however, because of the possibility, touched upon earlier, that polygamy could have evolved as a refinement of the homeostatic apparatus, effectively placing the control of fertilisation—the first step in actual progeniture—within the power of relatively few individuals. In the commoner of the two alternative forms of polygamy, where it is the dominant males that provide repeated matings (polygyny), it may be postulated that the pregnancy-rate could in this way more easily be held down to the optimal ceiling: the handful of participating males, being fully informed either of their own personal total or, where there is a communal lek, of the group-total of matings performed to date, could be conditioned to respond when the appropriate complement had been reached by becoming sexually inert. Where polygamy does not occur, and the population is dispersed for instance in monogamous pairs, this hypothetical kind of quota control could not be applied; but, as we have seen, the same end could be attained in a dozen other ways—one of them being through the control of breeding density by a territorial system. What is suggested here is that polygamy may be still another of the numerous adaptations contributing to the efficient regulation of reproductive output.

It will be recalled that dispersionary activities of many kinds are commonly controlled by the male sex, and only exceptionally by the female. In the breeding season in the majority of sexually-dimorphic animals it is the males that compete for property-rights (when they exist), and determine the pattern of spacial dispersion of the breeding population; it is they who develop the secondary sexual weapons and adornments, and take part in epideictic displays; they are the ones predominantly involved in the hierarchy (where this develops), their mates assuming social status only as their dependents; theirs also is the active part in courtship, which endows them with the initiative in actual fecundation (*see* p. 250). If polygamy is still another variation on the dispersionary theme, it is clear that at any rate it follows the same pattern as these others, in conferring on the male (with rather rare exceptions) whatever measure of control it affords.

It is useful to examine briefly the possible types of marital relationship between males and females that can be adopted for the purpose of sexual reproduction. There is first the agamic condition known as obligate parthenogenesis, in which the succeeding generations consist solely of females: here males play no part and can be done without. Like many of the asexual forms of budding and reproduction, this is especially effective in species that have to exploit brief periods of success and superabundance, when it is desirable to expand the population at top speed before inimical conditions return; it is found, for instance, among freshwater Cladocera, aphids (Hemiptera), many gall-wasps (Cynipidae) and some other Hymenoptera. In both the insect groups named the life-histories are inclined to be complex; though there is not infrequently an assumption of sedentary habits during rapid build-up of numbers through parthenogenesis, alternating with sexual reproduction and the appearance of winged adults (in the aphids at this stage generally only the males are winged). In the Cladocera sexual reproduction is similarly associated with resistant over-wintering eggs. It is a fair generalisation to say that this type of parthenogenesis is typically associated with populations growing rapidly if not exponentially—with little or no restraint. It is so prolific a method of multiplication chiefly because all adults produce progeny, and not merely half the adults as in dioecious populations; but it suffers of course from the grave disadvantage of genetic invariability, all the progeny of each (haploid) ancestral individual being genotypically identical. Usually sexual reproduction intervenes at least once in every annual cycle, allowing genetical recombinations to take place. Parthenogenesis is evidently another of these sporadically-occurring phenomena whose comparative rarity in the animal kingdom can be taken to imply that they tend to be evolutionary blind alleys.

The second condition, hermaphroditism, resembles the last in that here again all adults are ovigerous; but cross-fertilisation between individuals can of course proceed in a normal manner. Some kinds of hermaphrodites undergo progressive or alternating changes from one sex to the other; others produce ova and sperms simultaneously, as in the earthworms (Lumbricidae) and most of the common slugs and snails. In the absence of cross-fertilisation some of the latter, moreover, can be self-impregnating. Thus hermaphroditism appears to offer an exceptionally efficient kind of sexual regime, capable of combining the genetical advantages of cross-fertilisation, the ability to resort to agamic reproduction when required, and the enhanced fecundity that goes with progeny-production by all adults. That this is a reasonable inference is confirmed by the fact that a closely analogous form of hermaphroditism is by far the commonest condition to be found through the whole range of higher plants.

Where it occurs in animals it tends likewise to be associated with vegetative, sedentary or slow-moving habits and is largely confined to the lower phyla. It is common, for instance, in flat-worms, annelids, cirripedes and molluscs. It occurs most generally, perhaps, in groups that appear not yet to have evolved the common secondary attributes of sex—the differentiation of sexual castes, conventional weapons and nuptial adornments, and the epideictic and courtship displays for which these are used: where, in fact, the more elaborate conventional developments of social organisation seem not to exist. This is well illustrated by the molluscs, which, apart from some of the cephalopods, show practically no secondary sexual developments, whether they are hermaphrodite or dioecious; to the great majority of them the cephalopods stand in rather sharp contrast, with their highly-developed locomotion and vision, separate sexes, and propensity for conventional forms of display and behaviour.

Only rarely and sporadically has hermaphroditism intruded among the higher animals—as, for instance, in the fish *Serranus cabrilla*. We are led to conclude, as usual, that it lacks some valuable characteristics which in the long run gives the separation of the sexes a marked superiority, at least among highly-organised, mobile animals.

The commonest marital condition generally, and more especially so in the higher types, is a one-to-one mating between males and females. Males are usually full-sized individuals, capable of producing enormously greater quantities of sperm than are required to fertilise the eggs of a single female; they make only the minutest material contribution to the production of progeny, however, and may be considered as biologically extravagant from the standpoint of reproduction. We are at least familiar with a variety of economies that it has been found possible to make in the production of males—witness the minute ' complemental ' or parasitic male types that have evolved sporadically in different groups, such as the echiuroid worm *Bonellia*, various parasitic copepods, the barnacle *Scapellum* and certain other cirripedes, and the deep-sea angler-fishes (Ceratioidea), all of which are primarily benefitting from the same physical advantages that would be inherent in self-fertilisation; similar size-reduction in males without loss of potency is found in the male argonaut (*Argonauta*) and the male salmon parr. Another equally interesting economy of the same kind, we may suspect, is the eating of males by the females as soon as their sexual function has been discharged; this occurs in some scorpions, praying mantes (*e.g. Mantis religiosa*), and the ground-beetle *Carabus auratus* (Fabre, V, p. 330 and X, p. 248). On the other hand we do not regard it as in any way abnormal to find that males are even bigger than females: certainly this is more common in the higher animals than the reverse condition.

I think we are justified in concluding from such premises as these that there is some very important secondary division of labour arising from the differentiation of the sexes, quite distinct from the primary one of producing sperms or eggs: that the male, giving so little to posterity in terms of his own substance, discharges some alternative function that has come through natural selection to be scarcely less essential to the welfare and survival of the vast majority of species.

In the highest classes of animals, such as the birds for instance, it is easy to see how great an amount of energy the male expends in dispersionary activities—in territorial forms of self-assertion and defence, and in courtship. At the height of the sexual season, social duties frequently appear to occupy most of the time that he can spare from providing the bare necessities of food and rest. At the birds' advanced level of social evolution it seems clear that this is where the main secondary-sexual division of labour lies. To the male has accrued the primary burden of population-control in its widest sense, whilst the female is left to supply the material resources of reproduction. There is, however, a strong tendency for the mated pair to support one another, or substitute for one another in emergency, in their respective secondary tasks; and not infrequently the sharing of these is developed to a high degree. In this way something sufficiently approaching identity of secondary sexual roles can be achieved for selection to swing the balance occasionally the other way, imposing the dispersionary function on the female and the domestic one on the male; but this, as we saw in Chapter 12, is another of these sporadic anomalies, to be regarded as an evolutionary sidetrack (*see* p. 240).

In a great many species of animals, an individual breeds only once in its life: this applies particularly to those with adult life-spans of less than a year, though there are many longer-living exceptions such as the Pacific salmon (*Oncorhynchus* spp.) and the freshwater eels (*Anguilla* spp.). In this situation a female may require only a single service from one male, in order to achieve her life's reproductive potential. It is not uncommon, nevertheless, for progeny-production by the female to be extended over a period of time, still within a single breeding season; the eggs or young are then usually produced in batches, clutches, or litters, or sometimes in a more continuous succession. In these circumstances it is usual for the services of a male to be required afresh for each batch, or repeated at intervals. This we can see in many small mammals, notably rodents, or in birds with successive broods: earlier in the chapter we noted similarly that female flour-beetles (*Tribolium*) had a higher fertility rate when they received frequent recopulations than when they did not (p. 496).

This does not result from an insuperable lack of vitality in spermatozoa generally. The queens of bees and wasps for instance can be fertilised one summer and retain the living sperms for use in the next; queen ants and honey-bees are fertilised once for life. Donisthorpe (1936) recorded a queen ant of *Stenamma westwoodi* that lived certainly sixteen and probably eighteen years, and Lubbock (1888) one of *Formica fusca* that lived almost fifteen; both were fertile till the end. Cases arise occasionally in zoos of long-isolated females unexpectedly producing viable offspring, for instance in the tropical American colubrid snake *Leptodeira* which laid four fertile clutches of eggs during five years of isolation (Haines, 1940). There are similar instances of fertility retained by captive females up to four years after copulation in the American box-turtle (*Terrapene carolina*) (*cf*. Carr, 1952, p. 146). Cases like these, however, are very exceptional, the normal rule being that, once the male has shed them, sperms are rather short-lived. In animals surviving to breed in successive seasons, it is almost invariably true that mating must be consummated afresh in each year.

This is, of course, consistent with the hypothesis that the male exercises the principal controls in the homeostatic machine. Where the control is most efficient, in the higher groups, the female is virtually awarded a license to breed by the male that woos and mates with her. Even then her commission is brief and limited: it has to be renewed in the same manner as before for each subsequent episode. Indeed there are cases where a separate copulation is required for every single ovum, not only in the many uniparous vertebrates such as man, but even in certain insects, for instance the poplar weevil (*Byctiscus populi*) and the capricorn beetle (*Cerambyx heros*), in which the male unremittingly serves and supervises the female as long as egg-laying proceeds (Fabre, VII, pp. 155-6). It would be difficult to explain these situations on the hypothesis that selection always promotes the highest practicable fecundity, considering the apparently feasible alternative of fertilisation for life.

The outstanding exception to the rule of repeated fertilisation is found in the social Hymenoptera, where, in a series of parallel instances, there has evolved an identical culmination of using a single female to produce offspring for the entire social unit.

It is recognised, as we have noted several times before, that the value of the queen system, and the primary cause for evolving this specialised type of society, lies in the precise control of reproduction it affords. The community of foraging workers provides a very literal kind of feed-back to their oviferous mother, and her fertility-rate in return responds to their integrated needs and efforts. Throughout the Hymenoptera also there runs a propensity towards what is called facultative parthenogenesis. It has been exploited by all the social bees and wasps, if not generally also by the ants (*cf.* Imms, 1957, p. 676), and allows the queen the additional power of determining the primary sex-ratio of her progeny, by either fertilising from her sperm sac those eggs that are to produce females, or laying unfertilised male-producing eggs, to develop parthenogenetically. As a result, workers (infertile females) can be made to predominate in the community.

It may be suspected that even in some of these species the males retain the dispersionary control of community-establishment over a wider area. Congregations of drone honey-bees at selected sites have been recorded (*cf.* Ribbands, 1953, p. 143); and recently I noticed a congregation of hundreds of drone wasps (*Vespula norvegica*), attended by a few workers, on the rocky summit of a hill on the Aberdeenshire-Banffshire border, at a height of 2,360 feet—an incident that has all the appearance of an epideictic gathering. As in so many insects, the males have significantly larger sensory organs (eyes and/or antennae) and brains than the queens and workers.

We discussed in Chapter 11 the common habit in various animals (especially in insects) of males collecting together to participate in communal displays or leks, often at a set place and hour each day; to one of these the female repairs in due course, and finds a mate. The assemblies, as we know, are very varied in character; and in a dancing swarm of male gnats, for instance, there is no obvious competition leading to a male hierarchy, and matings with females appear to involve any male more or less at random in the group. In birds such as Gould's manakin (*Manacus vitellinus*) or the black grouse (*Lyrurus tetrix*), some males have higher status than others and their advances are more often successful. In the sage grouse (*Centrocercus urophasianus*), one male usually dominates all the others and performs all or nearly all the matings. In effect these seem to be scarcely more than procedural differences, as far as mating-control is concerned. In every case we can regard the lekking males as a corporate unit, collectively polygamous; all present are interested witnesses to the matings that occur and can presumably be conditioned by them in a cumulative way: so that whether the actual matings are all performed by one male or variously shared among the group, the same collective impression and result can probably be produced.

Seeing that males are potentially so enormously fecund, and that this kind of polygamy is evidently successful in procuring matings, it would seem to be an obvious economy to reduce the number of males carried in the population, limiting them to only what are actually required to fertilise the females' eggs, with a sufficient reserve allowance to meet an emergency. It is certainly not easy to answer the question of why the sex-ratio should so generally remain near equality when one male could in many cases fertilise a hundred females, given suitable conditions. The immediate answer, that primary sex-determination depends on a simple one-to-one genetical mechanism, is no doubt too facile, since the ratio is regularly modified in many animals, chiefly through differential mortality during development or immaturity (*cf.* Lack, 1954, p. 107). It seems probable that selection would have led, in most cases at least, to whatever proportion was optimal for the species concerned, even if this had been very far from equality. Thus it is reasonable to assume that the surplus males in polygamous species are not just ' dead wood ', otherwise selection would have eliminated them.

The question may be asked whether there is some genetical disadvantage in an unequal sex-ratio, or in assigning the matings of numerous females to one male. Without discussing this at length, it can be stated with little fear of contradiction that any differences in long-term genetical effects arising from polygamous matings, on the scale that occurs in nature, compared with those from monogamous matings, would almost certainly be negligible.

Before drawing any conclusions, it is desirable to indicate how widespread polygamy is known to be in the animal kingdom. In insects it is perhaps uncommon: an example is found in the ambrosia-beetle *Xylephorus*, which lives under tree-bark (Imms, 1957, p. 816); *a priori* it might be expected to occur in the extravagantly-horned beetles (*see* p. 265), though it has not in fact been detected. Among fishes it may be relatively frequent: there are examples in the sponge-blenny (*Paraclinus marmoratus*), where the male guards and tends a mass of eggs laid by a number of females (Breder, 1941); and in the 3-spined stickleback (*Gasterosteus aculeatus*), where again the male is in complete control, claiming and guarding the territory, building the nest, and soliciting one female after another to deposit her eggs in it (*cf.* Tinbergen, 1953, p. 12). Polygamy almost certainly occurs also in the Atlantic salmon (*S. salar*). In birds it is especially prevalent in the Galliformes, although by no means universal: many phasianids are somewhat polygamous, as no doubt are all the species of grouse that hold leks; but other grouse, including the *Lagopus* species and *Bonasa*, are essentially monogamous, and this applies equally to many types of partridge and quail. Another group in which polygamy is common is the Ploceidae (weavers and bishop-birds)—for instance in *Ploceus philippinus* (*cf.* Whistler, 1941, p. 206), *Euplectes hordeacea* (Lack, 1935), *Coliuspasser albonotatus* (Vincent, 1949, p. 503) and *Bubalornis albirostris* (Crook, 1958, p. 176). Many other avian examples can be found. In the rheas and ostriches (*Struthio camelus*) it has long been known: here, as in the stickleback, the male attends and incubates the eggs laid in a single nest by a number of females. It occurs (generally as the exception rather than the rule) in various kinds of wrens, including *Troglodytes aedon*, *T. troglodytes*, in the European corn-bunting (*Emberiza calandra*), the domestic canary (*Serinus canarius*) and many more.

In the Mammalia, polygyny is best authenticated in the highly gregarious types, particularly among the seals, ungulates and primates. In the first of these groups, the Pinnipedia, there are three families; in one of them, the Otariidae, comprising the six genera of sea-lions and fur-seals, polygamy is without exception highly developed; the breeding males secure harems averaging between fifteen and forty females, depending on the species. The second family (Odobenidae) contains only one species, the walrus, in which, if it occurs at all, polygamy is probably on quite a small scale. In the third and largest family, the Phocidae, comprising the thirteen genera of hair-seals, there are only two instances—in the sea-elephants (*Mirounga*) and the grey seal (*Halichoerus*), where the average harem-sizes are about twenty and ten respectively (facts and figures taken from Scheffer, 1951, p. 24). In the southern elephant-seal (*M. leonina*), Laws (1956, p. 46) found the sex-ratio at birth to be actually 1:0·822 in favour of the males, and there is no evidence in any of the other species that, either at birth or maturity, the ratio departs much farther than this from equality, in either direction, or that polygyny results from an unbalanced sex-ratio. Indeed, it is very characteristic of the breeding colonies of polygamous pinnipedes to contain conspicuous numbers of bachelor males. The polygamous habit has probably arisen independently in each of three pinnipede groups in which it now occurs.

In the Perissodactyla, polygamy is probably a common feature of all the Equidae, including horses, asses and zebras; but it appears not to occur in the rhinoceros family. Many of the larger game animals are extremely secretive in their nuptial habits, and few comprehensive life-history studies have ever been made, especially in tropical countries. It is often difficult to ascertain from published accounts whether polygamy occurs or not: but in the absence of actual observations of rutting ceremonies, in which males secure harems of females and repel other males, tentative conclusions can be drawn from the social organisation of the species concerned—whether it is gregarious or not—and from the existence of solitary or bachelor males that are excluded from the main herds.

The great order of Artiodactyla contains three main groups, the Suiformes (pigs and hippos), Tylopoda (llamas and camels) and the Ruminantia (deer, giraffes and hollow-horned ruminants), and in all of these polygyny almost or quite certainly occurs. It is not universal, however, but rather, as in the previous groups, sporadic in its development. Among the deer, for instance, the red deer (*Cervus elaphus*) is renowned for polygamous habits, which are typical of the whole genus *Cervus*, and also of *Rangifer*; but the roe (*Capriolus*) and moose (*Alces*) are thought to be generally, if not quite always, monogamous. In the hollow-horned ruminants polygyny is especially prevalent, alike in buffaloes and other cattle, sheep, goats or antelopes; but still there are apparently monogamous exceptions, such as the African klipspringer (*Oreotragus*) and its near relative the bushbuck (*Tragelaphus*), and the Asiatic four-horned antelope (*Tetraceros*).

Turning to another gregarious group, even more difficult to study in life, the Cetacea, we find that the whalebone whales (Mystacoceti) seem all to be monagamous; but the sperm-whale (*Physeter*) is strongly polygynous. Thomas Beale (1839, p. 51), a trained surgeon who had longer opportunities of studying the behaviour of this species than any more recent author, says that schools of sperm-whales are of two kinds, the one consisting of females, and the other of males not fully grown. Occasionally the schools contain (or did in his time) numbers up to as many as five or six hundred. ‘ With each herd or school of females are always from one to three large " bulls "—the lords of the herd, or, as they are called, the " schoolmasters ". The males are said to be extremely jealous of intrusions by strangers, and to fight fiercely to maintain their rights. The [remaining] full-grown males or " large whales ", almost always go alone in search of food. . . .' Clarke (1956), whose excellent modern account in the main confirms Beale's, adds, however, that sperm-whales seem to form very large herds only on migration. Bachelor schools do not necessarily contain growing animals, but sometimes comprise large, mature whales; while the lone bulls, though they may be sexually inert, are by no means necessarily aged nor sexually effete, as has often been assumed by whalers and others (loc. cit., p. 278). This is of course the species unique among whales for the great size-difference between the sexes, the male growing ultimately to four or five times the bulk of the adult female—a situation closely parallel to what is found in some polygamous seals, such as the sea-elephants and fur-seals (*see* p. 269). In none of the other toothed whales, porpoises or dolphins is sexual dimorphism developed to anything like the same extent; but the great majority of these species are more or less gregarious —far more so than the whalebone whales—and polygyny is very likely just as common among them as it is among the gregarious ungulates. Bachelor schools are known to occur in the common porpoise (*Phocaena*) (Møhl-Hansen, 1954).

According to Zuckerman (1932, p. 208) ' most if not all wild primates are polygynous or tend to polygyny '. The sex-ratios probably do not depart greatly from parity, and breeding males have to maintain their dominant status in competition with bachelor males. There are no doubt many primate species in addition to man, however, in which polygamy is far from universal, and practised, if at all, on a numerically small scale. Precise knowledge of marital relationships is extremely scanty in these as in most other mammals, including the lagomorphs and rodents, which likewise tend commonly towards sociability. In the European rabbit (*Oryctolagus*) Southern (1940) found that bucks belonging to a particular warren maintain a strong hierarchy, and only a few of the dominant ones take an active part in social affairs. These are generally big and old, and they have been seen to chase other males away from the does they are attending: but polygyny is nevertheless only presumptive (*cf.* Thompson and Worden, 1956, p. 82). This seems equally true of even the most familiar rodents, such as house-mice and brown rats; in the domestic laboratory stocks it is, however, usual to mate up one male with several emales in tfhe same cage.

Though the examples so far given may conceal some cases of promiscuous relations between the sexes, certainly none of them offers any hint of polyandry. Polygyny is obviously by far the commoner form of polygamy. The few examples of polyandry that can be found all seem to occur where the normal status of the sexes has been reversed and the female has become the dominant sex. This is true of the cases already detailed in Chapter 12, relating to the bustard-quail (*Turnix suscitator*) and water-pheasant (*Hydrophasianus chirurgus*) (*see* pp. 237-8).

Admittedly this sketch of the incidence of polygamy has left many wide gaps, but it is probably sufficient to allow us to draw certain broad conclusions. Evidently the habit is particularly well developed in the mammals, and characteristic above all of those that are gregarious in their social organisation and dispersion. Where, in contrast, the males are individually and permanently territorial in their behaviour, monogamy is the usual marital condition, and the simultaneous attachment of more than one adult female to a male only occasional at most. The evidence suggests that polygyny requires the existence of a male hierarchy in which some males dominate others on a graded scale; and it is illuminating that in a bird-colony, where the same condition of gregariousness exists, but social competition appears to distinguish only between tenants and non-tenants of nest-sites and not to lead to continuous gradations of dominance and submission, monogamy is practically universal. Outside the mammals, polygyny is especially associated with the lek phenomenon, in which a corporation of males, whether they develop a graded hierarchy or not, act as a single copulatory unit.

The survey confirms, moreover, that polygyny is quite irregular in its incidence. As one of the small number of possible alternative types of marital relationship, it belongs to the great complex of conventional behaviour. It has arisen many times and at different levels in the course of animal evolution, and as an adaptation appears most commonly valuable in gregarious species, though it is not by any means universal among them. We have repeatedly seen the same sort of sporadic incidence in other conventional adaptations, for instance in the various types of dispersionary patterns, or social structures, or sexual adornments and weapons; and the conclusion appears fairly plain, and not unexpected, that conventional characteristics are much more fickle and liable to evolutionary change—probably because of their arbitrary, substitutive quality—than are the more clean-cut, unequivocal adaptations of behaviour, physiology and anatomy that come under the direct influence of extrinsic selection.

It cannot be disputed that polygyny generally places the control of fertilisation in the hands of a fraction of the mature male population. If it is assumed that the sex-ratio is evenly balanced in the adult populations concerned, the figures for harem-size, where these are available, allow us to estimate that the fraction of males that breed is frequently as low as a tenth, and occasionally a fortieth or less. In the lek type of collective polygyny, figures of the same order are found in the different species of grouse (Tetraonidae), where from ten to fifty or even more males usually act in concert in fertilising the local females, effectively forming a single co-ordinated unit. In the latter case, as we saw earlier, the lek continues in daily session throughout the breeding season. In the Atlantic salmon (*S. salar*), while they do not appear to have a well-marked lek, the males in numbers also tend to stay in position on the spawning beds right to the end of the season, by which time they greatly outnumber the females. In spite of this it is well known that some females, called baggots, never mate and shed their eggs, but hang on occasionally for several months, in a functionally ripe condition (Menzies, 1931, pp. 39 and 44): possibly they constitute a genuine surplus of fecundity, over and above what is required to maintain the local stock, and consequently fail to interest the males; if so, were there not such a great toll taken of the salmon by netsmen and anglers before and after they enter our rivers, many more baggots would presumably be found.

The close association between polygyny and the graded type of male hierarchy, frequently supported by some sort of lek-display, is in itself a confirmation of our earlier suspicion that polygyny is a dispersionary or homeostatic adaptation. What we surmised at the beginning of this section appears to have been supported by the evidence presented, namely that polygyny belongs to the series of closely inter-related conventions that commonly operate through the male sex, in which the females passively comply—conventions that are all, so far as we have been able to discover, adapted to achieve intrinsic population control and optimum dispersion. The conclusion is indicated that, where selection has promoted polygyny, it has secured the same kind of advantage that lies in the male's assumption of the initiative in courtship, or in his being obliged to compete with other males for the conventional goals of property and status that qualify him to breed: they all contribute to the machinery by which the output per unit area can be regulated.

Competition among males for success in reproduction is no doubt density-dependent like other forms of social competition; and, other things being equal, its general intensity will be stepped up in polygynous species, because fewer can attain the goal. If we seek to understand why a balanced sex-ratio, resulting in the presence of a non-breeding surplus of bachelor males, is usually retained in these species, it appears possible that part of its functional value lies in maintaining the high level of this competition, and providing a sensitive index of population-density to feed back into the machinery controlling the output. Quite a different value may arise, however, from the fact that conditions are not always stable and optimal, and sometimes, as a result of some heavy disaster or where new habitats are being colonised, survival of the stock may depend on a handful of survivors or prospectors; the odds that these might all be females would increase, the farther the sex-ratio departed from equality, and the preservation of an even balance can therefore be of great importance wherever populations become sparse.

Most populations, whatever their density happens to be, have no difficulty at all in achieving an adequate reproductive rate, and they can therefore easily afford to carry a number of bachelor males for the sake of providing social feed-back and giving an insurance against the non-availability of males should population-densities become greatly reduced. We may judge that polygyny is not a convention that carries universal or overwhelming advantages, and, while it is very likely that it has additional values not detected here, these rather modest features may be sufficient to account in the main for its appearance, and for the retention of the surplus males. In the next chapter we shall be taking note of the existence in various species of a large proportion of adolescents in the population, similarly carried as ' passengers ' that consume but do not produce, since they are inhibited from breeding until they reach a relatively advanced age. In both cases the condition is one which could not arise under the all-against-all regime of individual selection, and can only have been introduced by selection at the group level.

21.9. *Summary of Chapter* 21

1. The chapter starts with an analysis of the equation of recruitment and loss, which is the condition of stability in a population. When recruitment and loss vary together and remain equal, the question arises, which is the dependent and which the independent variable? A break-down shows that the equation can be more easily understood when expanded into the following form:

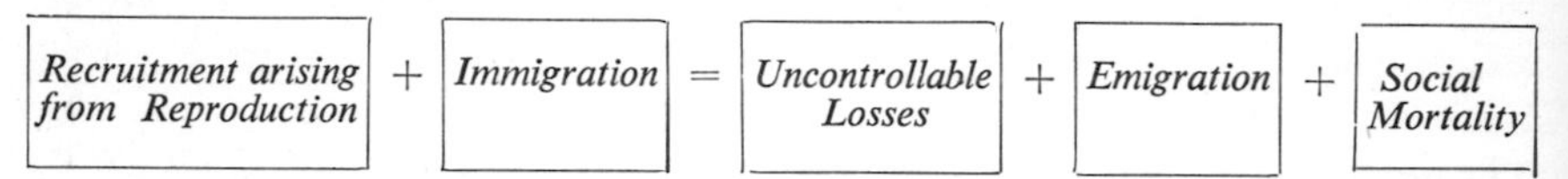

Of these five components, only one, namely ' Uncontrollable losses ', is incapable of being influenced from within through the action of the social machine, so that it is the true independent variable. Of the remaining four, Emigration and Immigration formed the subject of the last chapter, Recruitment arising from Reproduction is the subject of this, and Social Mortality that of the next chapter.

2. The next question is whether the potential fecundity level of each species is adjusted to the sum of the dangers to which the species is exposed, or whether it represents the maximum rate that inflationary selection has been able to induce. The evidence supports the former alternative. For instance, in birds and mammals there is often a broad correlation between body-size and longevity, the bigger forms tending on the whole to live longer and have lower fecundities (the last two variables necessarily go together); but the larger birds and mammals produce egg-clutches and litters that are relatively small in proportion to the mother's size, when compared with those of the smaller species. Since it is the smallest species that produce relatively the heaviest clutches and litters, we may infer that the fecundity of larger, longer-lived, more slowly-reproducing forms does not impose anywhere near a maximal physiological burden on them. These large species not only have eggs or young that are relatively small and few, but they also tend to breed less often; and there are indeed some birds and mammals with restrictive adaptations so stringent that even their maximum breeding efforts produce considerably less than one offspring per adult female per year.

3. ' Fertility ' is taken to mean the actual or realised production of

living young, in contrast to ' fecundity ', which is used here for the potential maximum fertility of which the species is capable. Assuming that recruitment through reproduction can be regulated density-dependently, fertility would be expected to vary, being high when the population is low, and *vice versa*: within the ordinary range of economic conditions something less than the full fecundity is all that is required each year to replenish the population.

Thus in order to control the rate of recruitment a density-dependent brake generally needs to be applied, and this can be done not only, as just indicated, (i) by holding down the number of ova fertilised per female to some point below the possible maximum, but also by restricting (ii) the number of females breeding and (iii) the frequency of successive broods, and (iv) by increasing the losses due to parental neglect, cannibalism, social stress and similar intrinsic causes. A population may launch on a seemingly full-scale reproduction, but, as a result of subsequent socially-regulated elimination, finish up by recruiting to their number as adolescents only a fractional residue of the eggs originally available for fertilisation. As an illustration of this, mention is made of the conventional limitations universally imposed by primitive human tribes, chiefly through abstention, abortion and infanticide, in order to prevent their populations exceeding the optimum.

4. In 1922 Carr-Saunders formulated ' the principle of the optimum number ', as it affects primitive human races. It very closely resembles the present theory, in postulating that primitive tribes controlled their own population-density, by means of a social organisation that imposed a system of land-tenure and conventional restraints on reproduction. He recognised (i) the need for group-selection in evolving social adaptations, (ii) the effects of social pressure in securing compliance with conventional rules, and (iii) the unacceptability of starvation as a means of limiting numbers. He thought the principle applied only to man and dated from his supposed innovation of a social system, possibly in the Lower Palaeolithic era. Apart from this circumscribed application, and from his non-recognition of the overfishing danger, Carr-Saunders' theory almost perfectly anticipates our own.

5. Many experiments have been made in the laboratory on population growth and regulation: all have yielded fundamentally similar results. In *Drosophila* or the flour- and grain-beetles, for instance, experimental populations reach definite and predictable ceilings, imposed by a variety of density-dependent factors that react on fertility and mortality. Crowding *per se* can have a depressing effect even in the presence of abundant food.

When experimental populations such as these are exploited or cropped, by regularly destroying a proportion of their members in a manner that parallels predation or fishing, the recruitment rate is found to respond (as predicted by our theory) to the increased demand by rising. In the sheep blow-fly *Lucilia cuprina*, even when 99 per cent of the flies emerging from pupation each day were being destroyed, the vastly accelerated rate of recruitment, although it could not maintain the population at its pre-exploitation density, was still capable of indefinitely staving off its complete extinction.

Guppies (*Lebistes*), living in aquaria on a superabundant diet, likewise showed an increased rate of recruitment in response to cropping: it was found that removing a 30-40 per cent ' catch ' out of each successive generation produced the heaviest sustainable crop that the population could be induced to provide; but that if the experimental catch were to exceed 60 per cent by weight of each generation, the recruitment rate could not be stepped up far enough to balance the losses, and the stock would consequently decline to extinction. This gives an experimental demonstration both of density-dependent rates of recruitment and of overfishing, and shows the relation between them.

6. In laboratory mice, rats and voles, experimental populations are also invariably self-limiting, just as are those mentioned above; the space factor (*i.e.* population-density) alone, or nutritional stress alone, are again each separately effective in curbing population growth. Social-dominance phenomena and individual temperament have come to influence rather strongly the ceiling densities actually attained by rodents. Their recruitment rate can be regulated at various stages, for instance by altering the rates of pregnancy, inter-uterine resorption, or neonatal mortality; and the adult mortality rate can also be influenced (*see* Chapter 22).

In none of the experiments reported did mortality ever result from starvation: regulation of numbers always took place in advance of such an emergency.

The control of reproductive output is internally mediated in rodents by the endocrine system; and the latter in turn is profoundly influenced by the degree of external stress presently being experienced by the individual: gross changes in the adrenal cortex can easily be induced in rats and short-tailed voles by crowding, associated with social persecution or nutritional stress. It may be suspected, in fact, that Selye's ' general adaptation syndrome ' plays a great part in population homeostasis; and it is significant that the adreno-pituitary system, so closely associated with the physiological homeostasis of the internal environment within the body, should appear to be no less involved in maintaining the external social and ecological equilibrium.

7. It is not difficult to find field evidence indicating a similar density-dependent regulation of recruitment rates in wild populations. Three illustrative examples given in detail refer to the collembolan insect *Sminthurus viridis*, to Kluyver's tits (*Parus* spp.), and to the vole *Clethrionomys rufocanus*. In further support of the hypothesis outlined in §3 above, that the reproductive process often starts by making provision for progeny far in excess of what events subsequently prove to be desirable, evidence is given of cannibalism and destruction by the parents of eggs after laying, and of the termination of development of mammalian foetuses within the uterus followed by their resorption. On any other hypothesis such very common events could only be attributed to wasteful perversion or malfunction. (Similar but post-natal homeostatic controls, such as infanticide, are reserved for Chapter 22.) A number of additional instances are given also of

reproductive rates varying from time to time and place to place, apparently in accordance with changing resource-levels or in response to exploitation.

8. The chapter concludes with a discussion of polygamy and marital relationships. The evidence is considered in relation to the hypothesis that polygamy is yet another homeostatic adaptation, still further concentrating the control of fertilisation within the power of a few individuals. Almost always we have to do with polygyny rather than polyandry, so that polygamy is at least in this respect consistent with the long series of recognised homeostatic adaptations that confer ' dispersionary ' functions on the male sex.

From a discussion of parthenogenesis and hermaphroditism, it would appear that the latter is reproductively very efficient, since it allows all adults to bear progeny and still enjoy the advantages of genetic recombinations. It is the method adopted by most plants. By contrast, one-to-one matings between male and female animals appear very extravagant, especially where the males (which contribute so little materially to posterity) are as big or bigger than the females, equally numerous, and excessively fecund. The conclusion is reached that there must be some major secondary division of labour between the sexes to justify such an apparent functional imbalance, prevailing as it does almost throughout the middle and higher animals. The justification is at once seen to reside in making the male bear the immense burden of population-control, while the female is left to provide the material resources of reproduction.

Reproductive control is in fact so tightly vested in the male that normally the female is commissioned to breed only on his initiative; nor is this commission an extended one, since courtship and impregnation are commonly required anew for each batch of ova, or even for every one singly.

Where the males gather into a lek to which females must come for fertilisation, all copulations may be performed by one or a few dominant males; but whether this is so or not a lek is effectively a collective polygynous unit.

A survey of the occurrence of polygamy reveals that it is irregular in its incidence, but apparently confined to the highest animals (insects and vertebrates); it is frequently associated with the lek phenomenon, and occurs chiefly in gregarious species, including a good many mammals. It is not found in the absence of a male hierarchy, and is typically associated with highly-developed secondary-sexual adornments, or enhanced male size. The evidence in fact strongly supports the initial hypothesis that it belongs to the great complex of conventional homeostatic adaptations.

Possible reasons are discussed for the maintenance of a 1:1 sex-ratio in polygamous species, since it results in a ' surplus ' of bachelors: there seem to be several, the most important being that polygamy is normally density-dependent in its intensity; and at the tenuous densities likely to be found at critical moments of extreme adversity, or of new colonisation, the presence of representatives of both sexes is vital to survival and the achievement of posterity.

Chapter 22

Socially-induced mortality

22.1. *Juvenile mortality—birds*

The remaining term in the equation of recruitment and loss (*see* p. 486) is the mortality that some animal populations contrive to inflict upon themselves. Theory leads us to expect that the exercise of the faculty (where it is developed at all) will be density-dependent, and that the mortality itself will be conditioned by some form of social or conventional stress.

Many of the examples given in the last chapter suggested that the actual fertility realised by a population at any given period, in terms of ontogenesis, is frequently less than the full fecundity of which the population is capable. We saw the brakes being applied at practically all stages along the production line, which resulted in the potential capacity being only partially or incompletely utilised. What is produced in an average breeding season is sufficient to make good the losses in the stock, or to take advantage of prevailing economic conditions, but the population ordinarily withholds a reserve of reproductive power, to be exercised fully only under exceptional conditions of understocking and regeneration.

The process of curbing or tapering off the production of recruits, until what remains just matches prospective requirements, may continue after hatching or birth, as the case may be, just as it did during embryonic development. The young of many of the higher animals remain for some time dependent on their parents; and over the population as a whole the total amount of care devoted to the new-born may vary with circumstances, thus raising or lowering the proportion that die of neglect. Not very rarely, as mentioned earlier, parents intervene directly by killing some of the offspring; in certain cases also the offspring kill each other.

One of the species in which this intrinsic type of infantile mortality has been closely studied is the white stork (*Ciconia ciconia*). This is a bird that does not wait for the completion of its clutch before starting to brood, but ncubates the first egg from the time it appears. Four or five eggs are commonly laid, at intervals of one or, more usually, several days: hatching, in consequence, is spaced out at corresponding intervals, so that the successively-appearing young are graded in size—stepped like organ-pipes, as the German authors describe it. Especially in the first few weeks of the nestling period the strength of the smallest is much less than that of the largest, and their respective chances of surviving to reach the flying stage are influenced by this factor. The smallest nestling not infrequently falls a defenceless victim to fighting among others going on in the nest; or it may be quite purposely thrown overboard. One of the leading authorities, Dr. F. Hornberger, informed Schüz (1957, p. 7) that ejections of nestlings were so commonplace that his observers had not thought it worth recording the individual instances. More dramatically still, individual nestlings are sometimes struck down and thereafter devoured by one of their parents—a phenomenon for which Schüz (loc. cit.) proposes the term 'kronism', in allusion to Kronos, the mythical Titan who killed and ate all his children but one as soon as they were born. Schüz (1943, 1957), on whose valuable recent review and discussion I have relied for much of the material presented here, on different occasions has photographed adult storks in all stages of the act.

There is enough information to suggest that deliberate killings are generally the work of the male stork, and that they are most likely to occur when the perpetrator is a very young parent, for instance a male only three years old, although it has been recorded also of one five-year-old and one six-year-old male. By the same token, casualties are commonly associated with newly-established nests. There are always liable to be numbers of non-breeding adults of both sexes wandering round in the nesting season, and these are given to pestering established nest-holders and provoking bitter defensive battles: they pay scarcely any attention to deserted nests (Haverschmidt, 1949, p. 35). 'Of course these fights are much more numerous in regions with a dense population than . . . where nests are few and far between' (the same, p. 37). All are agreed that the battles result in substantial casualties to eggs and young. Brinckmann (1954) found a record number of young thrown out (actually eleven) in a year when very few nest-sites were occupied (only twenty-two out of forty-six in the area of his survey): at this time there was presumably an unusually high proportion of non-breeders present. Hornberger (1954), moreover, showed that the average number of young ones reared increases progressively, as the age of the parents advances from three to four and four to five years—an increase not correlated with any rise in clutch size, but only with increased 'efficiency' and fewer casualties (*cf.* Schüz, 1957, pp. 7-8).

Schüz (loc. cit., p. 12) concludes that nestlings thus eliminated by their parents are surplus to requirements, and that their loss must enter into the population-balance of the species; but beyond this there remains for him an outstanding unresolved problem, alike in ethology and population-dynamics.

In the light of the dispersion theory, however, kronism presents no paradox; on the contrary, it conforms with the general pattern of adaptations evolved to curb recruitment; even the eating of the young could be regarded as a point of efficiency. There are clear indications in the evidence that social status enters into the situation; and whereas the most junior and subordinate members of the adult population may be completely inhibited by social pressure from breeding at all, those a little higher in the social scale may manage to achieve a nest, eggs perhaps, and even young; but, because of their humbler status, when the communal economy requires a curtailment of the number of young being reared, the stress presumably falls more tellingly on them than it does on better-established, more senior members of the community. This would be the naturally expected consequence of the hierarchic system.

Remembering how widely we have found dispersionary responsibilities to be vested in the male sex, we may read some significance into the fact that the evidence gathered so far has in each case identified the father as the executioner.

Examples were given in the last chapter of birds of other kinds destroying and eating their eggs, which amounts to almost the same thing as kronism; and some further and even closer parallels to what has been described in the stork can now be appended. The most notorious case concerns the herring-gull (*Larus argentatus*), in which prolicide and cannibalism have been recorded over and over again. The earliest record appears to have been made by H. L. Ward (1906) who saw young birds in a breeding colony being struck by adults with the bill, mainly on the back of the head, and afterwards grasped and worried till they were dead. Other observers have seen the victims being eaten, and according to Goethe (1956, p. 80) this may be done not only by strangers into whose neighbourhood the youngsters have wandered, but by the actual parents. Paludan (1951, p. 97) is another observer to have recorded substantial losses of chicks from infanticide. When the youngsters have grown a bit larger, fratricide may augment the mortality (*cf.* Strong, 1914, p. 19).

The same practice is shared to some extent by the terns; Pettingill (1939, p. 426) mentions the kidnapping of young arctic terns (*Sterna macrura*) as one of the causes of post-natal loss; and Sprunt (1948, p. 14) has recorded young sooty terns (*S. fuscata*) falling victims to adults, though on the whole this mortality appeared to be rather light.

Beyond these there are only scattered instances. A brown pelican (*Pelecanus occidentalis*) that killed and ate a chick is mentioned by Murphy (1936, p. 816); and man-o'-war birds (*Fregata magnificens*) have also been seen to snatch and eat unprotected nestlings of their own species (*cf.* Bent, 1922, p. 307). Schüz (1957) collected a number of similar cases from the literature in his review including a heron (*Ardea cinerea*) striking its four young; a female kestrel (*Falco tinnunculus*) that tore up and fed to the remaining nestlings one of their number that had succumbed (cause of death not determined); a white wagtail (*Motacilla alba*) swallowing one of its brood of six; a pair of magpies (*Pica pica*) that hacked a backward youngster, whose remains were afterwards found in the stomach-contents of its brothers; and finally a male red-backed shrike (*Lanius collurio*) that removed and impaled the bodies of its two smaller chicks, afterwards feeding them to the four remaining larger ones. Later in this chapter analogous examples of cannibalism in other groups of animals are given (*see* p. 541).

There are several other forms of intrinsic suppression of nestlings that to some extent overlap what has already been described, and together make up a complex of inter-related sources of mortality. For descriptive purposes it is desirable to try to separate them, although it is clear that this cannot be completely achieved. Broadly, one can attempt to separate the cases where young nest-mates kill one another, from those in which parental care is in some degree deficient, or is prematurely terminated, with the result that some or all the brood fail to survive. There is one basic feature common to many species in which this kind of regulatory mortality occurs; this is the successive hatching of the eggs, already described in the stork, that leads to the nestlings being graduated in size. It can reasonably be regarded as an adaptation to facilitate this particular form of homeostatic control that has arisen independently in a number of separate groups of birds.

Fratricide among nestlings is known to be rather common in the diurnal birds of prey (Raptores) and owls (Striges). Juvenile cannibalism in these two groups has recently been reviewed and discussed by Ingram (1959), who concludes in the first place that the graduations in nestling size so common in these two groups have two advantages: first, that of spreading the parents' task of feeding the young over a more prolonged period; 'and secondly, as it is eugenically preferable to rear, let us say, one or two healthy well-nourished progeny rather than six or seven weaklings, this "staggering" is of the highest importance since it offers perhaps the only satisfactory way—namely by controlled cannibalism—of numerically reducing the family to a feedable size when this has become vitally necessary to a food shortage. . . . Without a marked disparity in the age and size of the fledglings, fratricide would be virtually impossible. . . .' A list is provided of sixteen species of raptors, including several buzzards (*Buteo*), various eagles and falcons, the goshawk (*Astur gentilis*), and the harriers (*Circus*), in all of which fratricide (and almost invariably cannibalism) has been reported. A seventeenth, the honey-buzzard (*Pernis*), is excluded on what seems to us an unwarranted assumption, namely that, because the species is largely insectivorous, cannibalism could at most only be an exceptional phenomenon, and consequently even the record itself cannot be regarded as convincing. Like most other raptors, however, the honey-buzzard possesses the 'enabling adaptation', laying its eggs at minimum intervals of three days and beginning to sit with the first egg (Witherby, 1939, III, 97); and it would not be surprising if natural selection had demanded here as elsewhere that eliminated nestlings be consumed for the benefit of the survivors; among the mammals this happens even in certain herbivores. The list also names five owls, namely two species of barn-owls (*Tyto*), the long- and short-eared owls (*Asio*) and the great horned owl (*Bubo virginianus*).

On the question of how nest cannibalism could be controlled, Ingram suggests (loc. cit., p. 222) that in the short-eared owl (*A. flammeus*), which he has observed closely, it might depend on whether or not the parents leave extra 'control food' lying beside the nest. In this species surplus food is frequently placed a short distance from the rim of the nest, at such a distance that very possibly only the strongest young would be bold enough to retrieve it, and then only when sufficiently famished; but its presence might be made to serve as a controllable deterrent against cannibalism.

Our explanation of the phenomenon would differ from Ingram's only in one important respect, namely that there is certain to be a conventional mechanism that governs the intervention of the parents, and this would as usual be substituted for the ultimate factor—a sufficiency of food. There are a good many instances, as we shall see, of birds that regularly produce two eggs but hardly ever rear more than one young. This is very characteristic of various eagles. In the lesser spotted eagle (*Aquila pomarina*), for example, none of the fifty nests studied by Wendland (1951) produced more than a single eaglet, though twenty-seven out of the twenty-eight nests in which the original clutch could be checked started off with two eggs. In the golden eagle (*A. chrysaetos*), at least in Scotland, two young are not very infrequently raised, but from the first week after hatching the older one persecutes the younger without remorse; and if the parents do not feed the small one generously it is certain to die. If it is fed, it faces the ill-usage stoically, and puts up an increasing defence (*cf.* Gordon, 1955, p. 142). There seems little room for doubt that the aggressor's motive is not primarily to secure its nest-mate as food, but rather to bully and dominate it. The attacks are suspended during engorgement (Gilbert and Brook, 1925, p. 22), and gradually diminish as the victim grows in strength and vigour. Thus it appears that feeding, and especially the feeding of the smaller chick, is the critical factor, and consequently the control depends directly on the parents' actions.

Something of the same kind occurs in the rockhopper and macaroni penguins (*Eudyptes chrysocome* and *E. chrysolophus*). According to Gwynne (1953), the most remarkable fact emerging from his study of the former on Macquarie Island was that the first egg (much smaller than the second, or than the infrequent third egg) is normally wasted. 'Its fate is extremely varied, but only when the second egg is lost does it produce a viable chick, and as the larger egg always receives preferential treatment, this rarely happens under natural conditions. Usually the second egg is the earliest to hatch as a result of the preference in the care devoted to it; thereafter the first egg is generally discarded; but should both hatch, the puny chick that emerges from it usually survives only briefly. Only twice did Gwynne see a rockhopper with two recently hatched chicks of equal size and vigour.

It may be significant that at Herd Island and in some of the rockhopper nesting areas at Macquarie Island, a much higher proportion of first eggs were prematurely ejected during incubation. This seems to have been the general rule everywhere in the macaroni penguins, all the first small eggs being kicked out of their nests quite early, to be harvested by the sheathbills and skuas (loc. cit., p. 8).

It may be suspected that this adaptation of producing eggs of two sizes—apparently unique to the genus *Eudyptes*—has some deeper significance than that merely of enabling those of one size to be raised as alternatives or successors to those of the other. In that case, why the difference in size? Why the ejection of the small egg during incubation, before the large egg has even hatched? More likely it is simply a variant of the commoner adaptation of hatching the eggs in protracted succession in order to ensure an initial size-difference in the young; and it seems likely that in a greatly reduced population, requiring to restore its numbers as rapidly as possible to full strength, these penguins would succeed in hatching both their eggs and rearing a high proportion of twins.

The early decease of the second of two nestlings is characteristic of a number of other birds, including the Sarus crane (*Antigone*) (Baker, 1929, VI, 56), the Sandwich tern (*Sterna sandvicensis*) (Goethe, 1956) the white-tailed black cockatoo (*Calyptorhynchus baudinii*) (Serventy and Whittell, 1951), and some of the boobies, especially *Sula dactylatra*. The boobies and gannets are particularly interesting on account of the series of gradations to be found in the different species, relating to clutch-size and survival of young. In the blue-footed and Peruvian boobies (*S. nebouxii* and *S. variegata*), while the typical clutch is two eggs, one and three are not uncommon, and all the eggs are quite likely to produce viable chicks. Murphy (1936, p. 830) quotes figures for Lobos de Tierra Island, Peru, showing an average clutch of 2·1 eggs in thirty-nine nests of *S. nebouxii*; he himself noticed there that in families of well-grown chicks, two was a commoner number than three, implying that three are sometimes reared (loc. cit., p. 836). As usual in these cases, the young were notably different in size. At the Chincha Islands he found that in *S. variegata* likewise three fledgling chicks were not uncommon, and two were exceedingly numerous.

In the middle of the scale come the brown and masked boobies (*S. leucogaster* and *S. dactylatra*). In the former, three eggs are in fact laid by a small proportion of the birds and two by most of the others—at intervals of about a week—but it is very uncommon for them to rear more than one chick: at the Cay Verde colony in the Bahamas, Chapman (1908, p. 146) records that only two out of 640 nests contained twin youngsters. In the latter species, the clutch almost invariably contains two eggs, although more than one chick is rarely or never reared (Murphy, loc. cit., p. 847). Finally, the red-footed booby (*S. sula*), and the gannets (placed in the separate sub-genus *Morus*), normally lay only one egg.

The evolutionary and adaptive significance of these differences between the species of *Sula* have been briefly discussed by Murphy (loc. cit., p. 844) and by Stresemann (1953). Murphy suggests that the higher reproductive rate in the species of the first group may be related to the ready availability of plenteous food in the Humboldt Current waters where they feed off the coast of Peru. The more strictly tropical, ' blue water ' species in the middle group, he thinks, are ' living beyond their means ' in producing two or three eggs, because of the comparatively poor resources of their habitat; and this leads in practice to the suppression of all but one chick. Stresemann suggests that the second egg laid by members of the middle group has become redundant and functionless, and that a thousand years hence it may have been atrophied altogether, so that the clutch will have shrunk to one, as in the third group.

It is, however, quite possible to reconcile all these stages with a perfectly functional state of adaptation. The potential fecundity of the various species provides them with a certain reserve of reproductive power beyond their normal requirements: in some species the reserve is higher, in some lower. In some species it may take the form of extra eggs, and in other species (like *Sula bassana*) of a surplus of non-breeding adults. The average net productive output is apparently higher in *S. variegata* and *S. nebouxii* than in the others, and if so this must of course imply a shorter expectation of life. Certainly it has nothing to do with the richness or poverty of their respective specific habitats: this factor will govern the population-density of the adults, but it can play little or no direct part in the evolution of basic fecundity in the various species concerned. This appears to be valid as a generalisation, applicable to the animal kingdom as a whole.

Another group of birds given to protracted laying, and brooding from the laying of the first egg, is the heron and bittern family (Ardeidae). The intervals between ovulations may not always be longer than the minimal period of twenty-four hours, though at times three or four days can elapse between successive eggs; moreover, incubation does not invariably start with the first egg in any given species. But the adaptation is there, clearly enough, and completely functional. Of the most exhaustively studied species, the European grey heron (*Ardea cinerea*), Lowe (1954, p. 96) says: ' when incubation starts after the first egg is laid the last chick to hatch is very much smaller than its eldest nest-mate; Holstein and other observers contend that these weaker nestlings frequently fall victims to bullying '. Lowe himself never saw signs of death of nestlings by violence, however, but small young were found dead, and ' cases of cannibalism undoubtedly sometimes occur ': he found a heron pellet containing the remains of a young one, and quotes two other cases, where parts of one nestling were found in another and where one fledgling suffocated in trying to swallow one of its brothers (loc. cit., p. 97).

Our list of birds in this category is still far from complete, but about some of them very little detailed information is available. (For a more complete list, *see* Stresemann, 1927-34, p. 381.) Among the familiar Corvidae, most, perhaps all, the species of *Corvus* sometimes or regularly begin to incubate with the laying of the first egg; and their nesting is not unexpectedly characterised by substantial losses during the fledgling period. In the rook (*C. frugilegus*) the average clutch-size in southern England is between four and five (Lack, 1948, p. 38; Owen, 1959, p. 237). Yeates (1934) and Burkitt (1935) estimated the average number of young raised per nest as between 2·2 and 2·5 and Wontner-Smith (1935) as 1·94. The last author's combined figures for three years show 31·6 per cent that produced only one fledgling, 44 per cent with two, 23·3 per cent with three and 1 per cent with four (in all, 193 broods): it also emerges quite clearly that brood-size tends to be somewhat larger, in any given year, in big colonies than in small ones. Van Koersveld (1958, p. 61) obtained similar figures for young birds per nest in Holland (32 per cent with 1, 50·5 per cent with 2, 17 per cent with 3, and 0·5 per cent with 4): the average production rate for 931 nests in three years was 1·88 nestlings. For our purpose here it is significant to notice that partial losses, amounting to some 60 per cent of the eggs originally laid, are spread over the whole rookery, and there is no suggestion of an all-or-nothing distribution of success in nesting, such as results in many passerines for instance, through predation or desertion. Incidentally I have noticed in arctic Canada that even small birds such as redpolls (*Carduelis flammea* and *hornemanni*) incubate from the first egg, because freezing temperatures are still likely to prevail in the early days of nesting; the young consequently tend to vary in size, and casualties of the smaller ones may occur (Wynne-Edwards 1952*b*, p. 382).

From all the evidence presented it seems difficult to resist the conclusion that there are species of birds in which the parents can exercise an intrinsic control over the quota of young that is to survive until fledging. Frequently it has been made clear how closely nestling survival has been linked with the availability of food; but in some cases at least we can be certain that chick mortality is not due to the parents' simple inability to supply food: witness the rejection of unhatched eggs, or of young killed but not eaten, and the ancillary development of fraternal hostility among the new-born. It is highly probable instead that the parents are protected in the normal way by social conventions from over-exploiting the food-supply and thus jeopardising future resources, merely in order to meet the short-term demands of their nestlings: the question is not one of compulsory or inevitable starvation leading to mortality, but rather of the exercise of prudent restraint in terms of present and future economic conditions. In corroboration of this it may be recalled (*see* for example p. 150) that birds may be completely inhibited from even attempting to breed when food supplies are low. There is little doubt also that devices facilitating the immediate establishment of a hierarchy among the new-born, such as protracted hatching, and different-sized eggs, are adaptations primarily evolved to aid in the exercise of this control. It is interesting to notice that they must have arisen independently many times in the course of avian evolution.

The ' fledging success ' of a bird population can be intrinsically influenced in another way, through the complete desertion of the nest and its contents by the parents. In general it is known that the majority of birds are especially liable to desert during the early stages of nesting, as a result of almost any kind of disturbance; later on they are less easily discouraged, and as their nestlings approach the final stages of fledging, they become extremely tenacious and bold. Sometimes, though perhaps not very commonly, this normal climax of parental devotion appears to be arrested, or over-ridden, and the nest is spontaneously forsaken, in circumstances that demand some different explanation. The causes of desertion are often impossible to ascertain, and in any case the subject is one that has been neglected by most ornithologists, so that only a fragmentary picture can be given here. Occasionally a nest is discovered long afterwards containing dead or mummified young, and it is impossible to say whether perhaps they were a very late brood, abandoned by parents overcome by the urge to be off on their southward migration, or one whose parents had met with an accident. The classical instances are those of the swallow (*Hirundo rustica*) and house-martin (*Delichon urbica*), which sometimes desert quite a high proportion of their broods. Blackwall (1873, pp. 86-98) examined thirty-six disused house-martins' nests in the month of November, in 1825 and 1826, and found that eighteen contained dead nestlings or eggs. The fact that a few even had un-incubated eggs makes one question whether the universal presumption that all were late broods is justified: there is nothing whatever to show that some them of were not forsaken much earlier, long before the final migration date.

Some light is thrown on spontaneous desertion by the discovery of Coulson and White (1958) in the kittiwake (*Rissa tridactyla*) that the parental drive tends to be weaker and more often abortive in younger than in older birds. ' In several instances birds breeding for the first time failed to incubate their eggs satisfactorily, and a number of others were deserted before the incubation period was completed. There were also three records of newly hatched chicks dying because they were never fed, and in each of these cases the parent birds were breeding for the first time ' (loc. cit., p. 47). In general, the oldest birds laid the largest clutches and had the highest breeding success. There is no doubt that the situation bears a close resemblance to what we found earlier in the case of the white stork, culminating there in kronism by junior breeding males (*see* p. 531). Where, as here, the incidence of desertion is correlated with seniority in the breeding community, there is a very strong probability that it is capable of being instigated by social pressure. Notice that this offers a positive or adaptive explanation of the phenomenon ; the only alternative being to ascribe it to some nebulous negative cause, or maladaptation such as inexperience or imperfect attainment of the appropriate instincts.

Another striking instance of nest-desertion in colonial birds has been described by the van Somerens (1945), and relates to two African species of weavers, *Ploceus spekei* and *P. intermedius*. The authors were initially interested in two other ploceids, the silver-bill (*Euodice malabarica*) and the cut-throat (*Amadina fasciata*), which had taken over a disused colony of weavers' nests; and this led them to the discovery of hundreds of deserted eggs, and some dead young, belonging to the original occupants. The weavers had built a mixed colony in an acacia tree, which the authors estimated to contain 1,388 nests; about five-sixths (84 per cent) could be attributed to *P. spekei* and the remainder to *P. intermedius*. Several hundreds were examined and their contents recorded, and it was found that 36 per cent belonging to the former, and 32 per cent belonging to the latter, contained clutches of deserted eggs or (in a few cases) young. The mean sizes of the deserted clutches work out at 2·4 and 2·3 respectively, but as these are probably about normal for the species concerned they afford no clue to the situation. The authors consider what causes could have led to such a tremendous wastage, and suggest a number of negative and somewhat remote possibilities, such as wholesale defects in the embryos, or in the females' maternal instincts; or the death of the females concerned, and so on. But they also refer to a possible social explanation, arising from the intense gregariousness of the weavers, which they think might render a small minority of breeding birds disinclined to stay on and finish their job, after the main bulk of the inhabitants had reared their broods and left.

That this was not a unique or isolated event may perhaps be inferred from Ritter's account (1938, p. 289) of a colony of one of the masked weavers, closely related if not identical to *P. intermedius*, seen by him in an aviary in the zoological park at Sydney. He was told by the director of the park that the birds thrived there excellently, although they never succeeded in bringing any young to maturity. They built nests, mated, laid eggs, incubated and hatched them quite normally, but then abandoned them and let them die. If, as seems possible, the nestling mortality both in the wild and in captive colonies is induced by social stress, then it could be presumed that the aviary birds were overcrowded, and that had they been under less constraint, either from having smaller numbers, or a larger aviary or more satisfying food, some at least of them would have gone right through with the whole process of reproducing and perpetuating their group. Male weavers are known to bird-fanciers to be intensely keen on nest-building, and one can reasonably infer from this that the initial steps in reproduction have come to play some essential, but purely conventional, part in establishing and maintaining their societies.

It seems probable, in the light of other forms of intrinsic neo-natal mortality already discussed in this section, that all three of these cases of nest-desertion—in the house-martin, kittiwake and weavers—could have arisen from similar or identical causes; all relate to colonial species, in which the reproductive instincts of individual members could easily become inhibited as soon as an adequate reproductive target for the colony as a whole had been assured. Over-recruitment would thus be prevented. The inhibition could be expected to fall earliest on the most junior members of the colonial hierarchy; or, alternatively, breeding might stop more or less generally as soon as the quota was assured. Desertion in these circumstances would be a social adaptation, benefitting the group at the expense of the individual deserters, and as such it would require, of course, to be evolved by group-selection.

It may be that this particular regulatory adaptation—if that is what it is —is especially typical of colonial breeders. Something of the same sort happens from time to time in various species of terns; and while the juvenile mortality may in most of these cases be attributed to a seasonal failure of the food-supply, sometimes ' conventional ' features reveal themselves in the situation, for instance when the juvenile mortality is much greater in one part of the colony than another (*cf.* Galloway and Thomson, 1914, p. 272).

In rather different circumstances from any hitherto mentioned, a huge post-natal mortality has been described in the velvet scoter (*Melanitta fusca*), a species of duck breeding abundantly in the Baltic archipelagos (Koskimies, 1955). In this habitat, as a rule, ' at least 90-95 per cent of the young have died during the first five to ten days of life '. Koskimies attributes this to the maladaptation to maritime conditions of a species that normally nests inland on lakes. The ducklings tend to concentrate, and to overcrowd the most sheltered and richest nursery waters; the family bond is ' peculiarly loose '; and whereas he believes this may not matter on a relatively small freshwater lake, on the sea where there is no limit to how far the young can roam it can be fatal; and finally, surface water temperatures in July are much lower in the Baltic, and much more changeable, than in inland lakes, and the heaviest mortality has always been associated with bad weather. Although the velvet scoter populations in the whole outer archipelago of Finland have steadily increased during the last fifteen years, in Koskimies' opinion the rate of native recruitment is so low that there is probably a regular population influx taking place from some distant, presumably inland, source. The breeding females, however, are known to be very long-lived and extremely *ortstreu*, returning usually to exactly the same islet in successive years.

The figures are available (Koskimies and Routamo, 1953, pp. 67, 94) to show that, if the average female breeds for at least five years (and it was found that over 80 per cent survived and returned from year to year), and lays 8·4 eggs annually, of which 73 per cent (*i.e.* 6·1 eggs) hatch, she will produce a total of 30·5 ducklings. To maintain the population, only two of these need survive to adult life, so that 28·5 ducklings are expendable during the period between hatching and maturity: this is 93 per cent of the total. It will be seen that this figure is almost the same as the estimate given by Koskimies for the losses in the first ten days of life; and, provided there was virtually no further mortality until after maturity, even if 90 per cent of the ducklings were to perish the adult population could still increase at some 3 per cent per annum. I have made desultory observations of the eider-duck (*Somateria mollissima*), a related and ecologically similar species, for over twenty-five years, and suspect that it has a not dissimilar mortality regime, but to what extent the mortality in either case is intrinsic and controllable is another question, since most of the ducklings seem either to be eaten by gulls or seals, or to die of disease. To some extent, of course, the extrinsic agents may be density-dependent in their action, and thus auxiliary to the homeostatic machine.

We are accustomed to accept a high mortality during the initial period of independent life as being a normal occurrence in almost any part of the animal kingdom. Confining the discussion in this section to birds, the rook (*Corvus frugilegus*) may be cited as a final illustration. As we have seen, a pair of rooks starts off on the average with a clutch of five eggs, and raises about two young. The death rate in July and August is so heavy, however, that instead of the young birds still equalling their parents in number, or comprising at least a third of the total population when allowance is made for the presence of non-breeding adults, the proportion has dropped by early autumn to one in ten or less (Burkitt, 1935, p. 324). This was the position in Ireland in an area where young rooks were not shot. Rooks frequent their breeding places for much of the year, and Burkitt found that by the time the trees in his rookery at Enniskillen were bare of leaves, there were only six young birds still present in the local group, although the rookery had contained fifty-four nests in the spring. Some juveniles may have left to join other colonies, but the opposite is equally possible, namely that immigrants had been received from outside; and thus the figures, small though the sample is, point to a very high rate of mortality, of the order of 95 per cent. The immediate cause of death of so high a proportion is not known, but we can be sure that predation, anywhere in this country, where there are so few predatory birds large enough to take rooks, can account at most for a very small fraction of it. Corvids are notably aggressive towards other members of their own societies (*cf.* Lockie, 1956), so that it seems not at all unlikely that, in rooks, social stress may take a considerable share in the process.

22.2. *Juvenile mortality in other groups*

The experiments with laboratory rodents discussed in Chapter 21 showed, amongst other things, that litter mortality is generally density-dependent, playing an important part in curbing the rate of recruitment and in bringing the increase of population to a halt as the ceiling density is neared. Litter mortality is wholly intrinsic, depending on such factors as the failure of milk-secretion in the mother, desertion, destruction of the nest during aggressive fighting among the adults, and cannibalism (for references *see* p. 504): it sometimes, and perhaps generally, involves the social status of the mother, those lower in the scale incurring much heavier losses than the dominant females (*cf.* Calhoun, 1950, p. 1115); in the vole *Microtus agrestis* at least, males in captivity are so prone to eating new-born young that females with litters must be isolated from them (Ranson, 1934, p. 74). Lastly, in appropriate circumstances the mortality can be seen to make a highly effective contribution towards the stabilisation of numbers. Indeed, laboratory experience appears to conform so closely with what we are led to expect on theoretical grounds, that we can feel considerable assurance in accepting the mortality as adaptive.

In the domestic rabbit (*Oryctolagus cuniculus*), litter-mortality is likewise common. Sawin and Crary (1953) found that, in four different domestic races, between 8 and 15 per cent of the litters suffered from cannibalism, and between 11 and 23 per cent more from ' scattering ', meaning that the young were left strewn about the floor of the cage outside the nest. The race with the fewest postpartum failures lost one-fifth of the litters, and the one with the most, almost two-fifths. Cannibalism usually occurred shortly after birth, at the time the mother ate the placentas, but in rarer instances it followed after a day or two of normal care.

Precise details about this kind of behaviour are difficult enough to obtain in captive animals; in the wild they are not obtainable at all, on account of the extreme secretiveness of nesting mammals while they are nursing their young. It seems reasonable, *prima facie*, to expect that the close confinement of caged animals would tend to aggravate post-natal losses, and that in nature these would not normally be as heavy as they are in captivity, but it must be remembered that laboratory mothers are quite often isolated in order to maximise their breeding success, in contrast to the wild European rabbit, for instance, which is an inmate of a strongly hierarchical and sometimes congested colonial society. We saw earlier that almost two-thirds of the litters conceived in a British population of wild rabbits were resorbed before birth, so that a fairly heavy rate of control over the survival of the new-born cannot be regarded as at all improbable.

In laboratory rats, Wiesner and Sheard (1935) found that mortality at the nursling stage completely eliminated about two litters out of five before the young were weaned, but, very interestingly, immediately after weaning expectation of further survival reached the highest level attained at any stage of the life-cycle, during the period of youth and early maturity. Sixty per cent of those weaned were still alive two years later, but by that time the death-rate had begun to mount, and rather less than 2 per cent surpassed an age of three years. It would appear in this case that control of recruitment had been completed by the weaning stage, after which all members of the group continued to survive until old age began to take its increasing toll.

About mammals under wild conditions we have only fragmentary information. We saw that a good many primitive races of men, which can properly be included in this category, were adapted by tradition to practice infanticide as a means of family limitation. Primitive man, in fact, applied the brakes to his fecundity at every stage, though not all methods were customary in every tribe; the most important of these were the deferment of marriage, abstention from intercourse, abortion, and infanticide. The notable feature in his case, however, is the comparatively small part played by physiological, and especially endocrine, adaptations in controlling fertility, and the correspondingly enormous importance of traditional tribal customs. These are presumably more flexible in relation to varying local conditions than are hereditary adaptations: they are also open to insight modification by the tribal authorities. In our world, almost all traces of limiting traditions have been lost, and, more seriously, the vital necessity of reimposing intrinsic controls as rapidly as possible in order to avert imminent and overwhelming overpopulation is not widely or properly understood.

In the lion (*F. leo*) almost every litter contains one or more cubs weaker than the rest; 'and, when times are hard and the mother has not a great deal of milk, infant mortality is undoubtedly high' (Stevenson-Hamilton, 1949, p. 133). Losses continue after weaning during the long period that the cubs accompany their mother; the same authority goes on to say that 'a great number of young lions between nine months and two years fall victims to their own kind'.

In the first year of life of the northern fur-seals (*Callorhinus ursinus*) that breed on the Pribilof Islands, the death-rate is about 60 per cent. The part of this mortality that falls on the unweaned pups before they take to the sea is known to have increased with the build-up of the herds to their present ceiling. When, under international treaty, the United States finally gained control of the Pribilof herds in 1911, only a remnant consisting of 132,000 seals still resorted to the islands, after half a century of ruthless over-exploitation. As is well known, subsequent good management allowed the herds to recover steadily, until by 1940 they had reached a total exceeding one and a half million. Since then, however, there has been no further increase. In the early days, before 1917, pup-mortality was extremely small, varying from one rookery to another, but of the order of only 1, 2 or 3 per cent. At the present time it averages 12-14 per cent, but attains 20 per cent in some individual rookeries. The losses are not due to predation, and the immediate cause of death has not been ascertained. (This information all comes from Dr. V. B. Scheffer, orally, 1957.) The *per capita* rate of recruitment of the adult population has, of course, declined to a constant average figure since the population settled down to the present ceiling, and it seems quite probable that the observed rise in pup-mortality is adaptive, contributing to the natural limitation now imposed on reproductive output.

Turning next to the fishes, we have already encountered (p. 499) another excellent laboratory demonstration of intrinsic mortality adapted to population-control in the guppy (*Lebistes reticulatus*). Breder and Coates (1932), in their classical experiments, founded two colonies of guppies in identical aquaria, the first with a single gravid female, and the second with a mixed group of fifty fish. Food was supplied in excess. As related earlier, in the course of about six months both populations reached the same asymptote, namely nine fish. In the first tank, survival of the initial brood of young after birth was 100 per cent: in succeeding broods it was progressively less, until, when the fourth brood appeared, all were eaten by the mother. By this time the biomass of her earlier broods had grown considerably, and she began to thin them also. In the second tank, no young were ever allowed to survive, however briefly, after birth: all were immediately eaten by the group of fish already present in the tank. Moreover, the latter consumed one another by degrees, until here again a final equilibrium population of nine was reached.

The authors were able to conclude unequivocally that ' infanticide in *Lebistes* is directly proportional to population concentration ' (loc. cit., p. 153); and this, taken in conjunction with their other findings, enables us to state that recruitment through reproduction is limited here, at once intrinsically and density-dependently, by the intervention of cannibalism.

Another case resembling this in some respects has been reported in a crustacean, the common spider-crab (*Hyas araneus*), by Schafer (1952). Here, however, we are not dealing strictly with juvenile mortality or cannibalism of the new-born, but rather the destruction of grown or growing individuals by one another. Spider-crabs introduced in numbers from outside into an aquarium live in a state of mutual hostility, but they are proof against each other's weapons and can come to no serious harm, except during the four days immediately succeeding a moult, while the new shell is still soft and they are physically feeble and defenceless. Unless they are isolated at this time they are invariably torn to pieces and devoured by their fellows. At the year's end, consequently, there is only one left—the one that by chance is the last to moult. This is what occurs in a tank of two cubic metres capacity: in one of six cubic metres, two or three crabs will survive. From these and similar observations Schafer concludes that there is a ' critical volume ' for species such as this, providing the mechanism for the natural regulation of its population-density—a conclusion in exact concordance with our own. One of the curious features of the artificial conditions created by overcrowding spider-crabs in an aquarium, where they cannot spread out to an acceptable individual distance, and where cannibalism is consequently unnaturally aggravated, is that the eater almost inevitably ingests a dose of moulting hormone, still present in the blood or tissues of its victim, sufficient to precipitate its own moult and immediately provide another victim for the survivors: indeed, once this cannibalism starts in an aquarium it induces a moulting epidemic among the remaining inmates.

In nature, Shafer concludes, the moult in *Hyas* and other cannibalistic Crustacea is a crucial period, and they must then be assured of adequate space in which to conceal themselves from the sensory perceptions of other crabs, otherwise they are unlikely to survive. Lobsters, he observes, become particularly aggressive for some days before they cast their shells, threatening their neighbours with their claws for hours on end, and whenever possible driving them away; in this way they clear a sufficient territory to give themselves a fair chance of completing the moult safely. In decapods like these the female can only be fertilised when in the fresh-moulted state, and Schafer states that the males of *Hyas* will defend females at that time, though other females (in an aquarium at any rate) will devour them if they can.

Some of Crombie's (1944) experiments on insects infesting grain are highly relevant to this subject. The larvae of the beetle *Rhizopertha dominica* and of the moth *Sitotroga cerealella* both habitually bore their way into cereal grains, feeding and completing their individual larval and pupal lives within a single seed. In crowded conditions, females will oviposit without discrimination on grains already infested, and ' superinfestations ', with several larvae to one grain, occur. The same situation can be induced experimentally in various ways that need not concern us here. While it is not impossible for two adults to emerge from a single grain in either species, the mean of all the superinfestation experiments gave figures of only 1·24 emerging adults per grain in *Rhizopertha* and 0·96 in *Sitotroga*: in other words, considerable elimination of supernumerary larvae goes on during development. Crombie found that, in both insects, two larvae of the same species meeting together

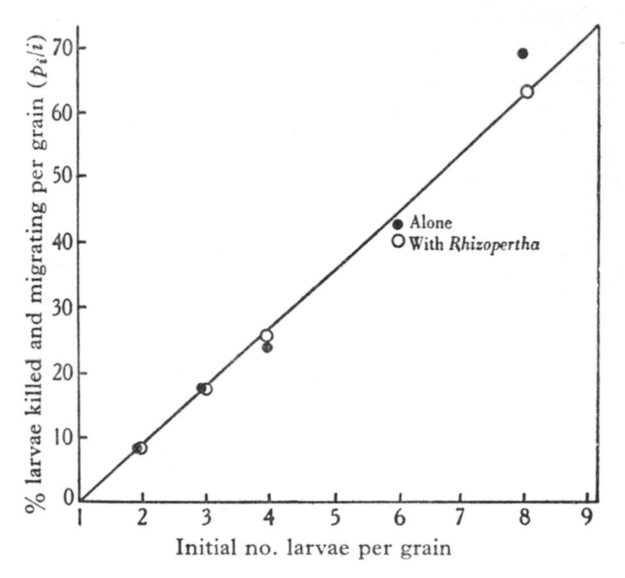

FIG. 48. Relationship between initial density of *Sitotroga* larvae and the proportion subsequently lost through death or migration. The line indicates the theoretical density-dependent relationship, and the points show the experimental results. (From Crombie, 1944.)

always threaten or attack one another with their mandibles. Sometimes their contests result in mortal wounds, and at other times in the emigration of one individual: in either case the victor is left in possession and the loser eliminated. The proportion lost in this way through emigration or death was found to vary directly with the degree of superinfestation, that is to say the elimination of surplus larvae is density-dependent. This is illustrated in figs. 48 and 49, taken from Crombie's 1944 paper, presenting the same experimental results in two different ways: fig. 48 shows that the percentage of larvae eliminated bears a linear proportion to the initial population-density; and in fig. 49 it can be seen that the probability of survival of any individual varies inversely with the density. In the latter the solid line is drawn through the theoretical values, supposing perfect density-dependence, and it can be seen how closely the experimental results accorded with it.

It is worth recording that cannibalism is well known among spiders, especially perhaps in the period succeeding their emergence from the nursery cocoon. In the purse-web spider *Atypus affinis*, Enock (1885) long ago described how, apparently under the stress imposed by inclement weather, the mother spider would devour some or all of her newly emerged young (*cf.* Warburton, 1912, p. 93).

It is interesting, and perhaps unexpected, to find as a result of reviewing the known incidence of prolicide, fratricide and cannibalism in the animal kingdom, as we have done in sections 1 and 2 of this chapter, that these have to be regarded as very advanced adaptations, seemingly almost confined to the highest animals, found in the arthropod and vertebrate classes. That they are adaptations of homeostatic function cannot any longer be seriously doubted.

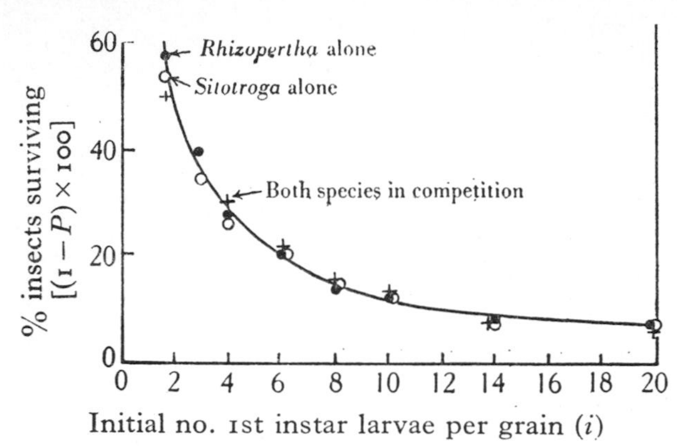

FIG. 49. Relationship between initial density and survival for *Sitotroga* and *Rhizopertha* larvae, alone and mixed together. The line shows the theoretical density-dependent relationship when the average survival-rate is 1·2 insects per grain. The experimental results agree closely with this, showing that the probability of survival varies inversely with total density. (From Crombie, 1944.)

22.3 *Mortality from predation and disease*

In paying so much attention to intrinsic sources of mortality we have run some danger of forgetting or discounting the heavy casualties that can fall on the larval, juvenile or adolescent age-groups as a result of external agencies. In long-lived higher animals, such as man himself, special dangers to the young normally arise from their want of physical strength or stamina, and from inexperience. Deaths attributable to these dangers must certainly be regarded as partly intrinsic, to the extent that adequate parental care, had it been forthcoming, could have prevented them; but in part they are truly extrinsic. The relative proportions of the intrinsic and extrinsic elements in the sum total of juvenile mortality must depend to a considerable extent on the level to which parental care and responsibility has evolved in the species concerned. The free-floating eggs and planktonic larvae of many marine fish, for example, are completely abandoned by the parents and practically defenceless from predators appreciably larger than themselves; and it is to be expected that extrinsic losses will enormously predominate. No doubt these are very variable from place to place and year to year, and tend to make the recruitment rate irregular. The great advantage achieved through the evolutionary development of parental care seems to lie in minimising this unpredictable element in juvenile mortality, with a corresponding gain in control of the recruitment rate. Young individuals lost through uncontrollable causes can be regarded as waste from the reproductive point of view: it would be more economical if they did not have to be produced. But there is of course a compensating factor on the other side of the ledger, in aquatic and parasitic organisms especially, where swarming larval stages, analogous to the seeds of plants, perform a major service in disseminating the species.

Predation, whether on adults or young, is frequently assumed to be density-dependent, at any rate to a significant extent, and at first sight this may appear a very reasonable assumption. It is, unfortunately, almost impossible to test by experiment or to confirm by direct observation in nature. The evidence of earlier chapters has clearly indicated that populations tend to expel surplus individuals; and a sizeable proportion of such outcasts, at least in the smaller species, are likely to fall victims to predators. In so far as this occurs, predation can certainly be expected to be density-dependent; but the mortality being incurred in such circumstances as these is not fundamentally extrinsic at all, but intrinsic, because it is the initial expulsion that governs the mortality, regardless of the form the actual execution may happen to take.

Ecologists have, perhaps, tended to be somewhat uncritical in rather generally accepting the assumption that predation in its own right, as an extrinsic, independent, self-seeking activity, is likely to exert a density-dependent pressure on the prey. Most predators are found on investigation to take a rather small crop off their food species, measured as a percentage of the whole population. When the prey is specially abundant the population-density of predators certainly tends to rise, and *vice versa*; but it is not clear that the percentage the predators take for food is normally higher when the prey is abundant than when it is scarce. Indeed, unless the predators' population-density responds very rapidly to sudden changes in abundance of the prey, the exact opposite is likely to happen, so that they come to bear more heavily on the prey under stress of hunger at times when the latter suddenly becomes scarcer, and are powerless to make any appreciable inroads into the great surplus available to them at a time of sudden plenty. (This is essentially the 'Volterra effect', earlier discussed on p. 389.)

On quite different theoretical grounds we should expect the social adaptations of the predator population to protect it from any tendency to over-exploit a plentiful prey, and thus force prey numbers down below the existing carrying-capacity of the habitat. Theory, in other words, appears to support the probability that predation is not in its own right a density-dependent process, independently capable of controlling a prey population from outside: the 'co-operation' of the prey population, in insuring that its surplus members are specially vulnerable to predators through the operation of the social machine, is almost sure to be the indispensable condition underlying whatever density-dependent, homeostatic influence predation may be found to have. The density-dependence of predation losses, that is to say, may well prove to be a completely secondary effect, regulated by the prey themselves and not by the predators at all.

Predators obviously perform a very desirable function in disposing of surplus members of the prey population, but selection will tend to make sure that the predators confine their catch, as far as this can be contrived, to the expendable surplus, so that they do not seriously interfere with the optimum density of breeding stock, nor with the need for producing a certain small output of pioneers (*see* p. 463).

It is generally recognised that the total death-rate from all causes is relatively high in the very young age-groups in most kinds of animals (*cf.* Lack, 1954, chap. 8); in middle life there is often a more or less sustained period of low mortality, to be succeeded finally by a sharply rising rate in old age. In some types of potentially long-lived species, including many of the larger birds and fishes, a steady drain of mortality year after year throughout the long ' low ' period of middle life may leave only a minute remnant to last out their full potential span, and die of ' old age '; but in man, on the other hand, or in many shorter-lived species, such as the small mammals that have annual or biennial generations, the senile death-rate can be conspicuously high. In the most perfectly adapted species it is to be expected that as much as possible of the mortality, including practically all of it in youth and old age, will be intrinsically controlled through the social organisation, and conform as nearly as possible to the homeostatic ideal; and this is in keeping with the observation that many species in the highest groups are almost wholly immune from predation once their earliest youth is past.

Death from old age is of course another primarily intrinsic phenomenon, arising from an innate adaptive, senile debility, or lowering of the body's defences against the whole range of potentially lethal agents. The relatively constant feature in senile mortality is not the cause of death, which may be any one of a variety of possible alternatives, but the mean age at which it takes place. The age itself is a hereditary characteristic, evidently adapted to the needs of the species concerned, and adjusted by group-selection to fit in with the other parameters of its population-turnover.

It seems likely that infective disease is equally incapable of exerting an independent controlling effect on an animal population. This was the conclusion reached by Lack (1954, chap. 15) after he had undertaken an extensive review of mortality attributable to disease in birds. When they are in good condition, such birds as the red grouse (*Lagopus lagopus scoticus*) can carry a considerable burden of internal parasites without injury, but an adverse change in the quality of the available food, which can result from a partial winter kill of their staple food-plant, *Calluna vulgaris*, or from extensive damage by the heather beetle (*Lochmaea suturalis*), apparently lowers the birds' threshold of resistance, so that they appear to die of parasitic disease. The threshold of resistance is an intrinsic phenomenon, depending no doubt indirectly on the social status as well as directly on the economic well-being of the subject. A surplus individual that has been squeezed out of a saturated habitat may be eliminated by parasitic disease as effectively as by predation or by exhaustion of its food-reserves.

Mortality from communicable disease is by no means a conspicuous everyday phenomenon, taking the animal kingdom as a whole. In man, who may not be a typical species in this respect, there are certainly many communicable diseases that can result in an appreciable percentage of fatal cases, and in the recent past the death-rates from most of them have been higher than they are now. Infants and children often run the greatest danger, because they are meeting the infections for the first time and lack the protection or immunity that may be acquired and in later life retained by having overcome an earlier attack of the same organism. A similar special liability in the young to die of infections is presumably shared by other animals having like powers of acquiring immunity to dangerous parasites.

In summing up this brief discussion of the parts played by predation and infectious disease in the general equation of recruitment and loss, our tentative conclusion should probably be that, in most species of animals, both of these are liable to inflict a certain amount of mortality that must be assigned to the category of ' uncontrollable losses ' (*see* p. 486). Both are also, however, frequently accepted as agents and convenient executioners for mortality that is socially initiated in the ordinary run of homeostatic adjustment.

In the course of attaining the optimum number of recruits at the stage when they pass into the adult class, we normally expect the number originally born to have been whittled down considerably. At birth a sufficient margin of numbers must be allowed to meet all uncontrollable losses during adolescence, no matter whether these happen to be large or small on any given occasion. Intrinsic mortality needs to be very flexible, therefore, so that it can as far as possible make up whatever deficiency is necessary to complement the extrinsic losses, and bring the final quota of recruits down to the figure that will keep the population-density in balance with economic conditions. The greater the potential death-rate from extrinsic causes, the greater the need for the resourcefulness of the population in disposing of an unwanted surplus of adolescents, if sufficient extrinsic mortality fails to occur. The fact that uncontrollable losses tend to be highest in the youngest stages, therefore, is no doubt fundamental to the very conspicuous development of homeostatic adaptations in this juvenile section of the population. Not only do its members supply most of the victims in cases of direct cannibalism and intraspecific slaughter generally, but theirs is usually the principal time of life for emigration and colonising new ground: this they are driven to do largely in response to overproduction and consequent social pressure at the point of origin. Further, in many long-lived animals the duration of immaturity itself is variable, so that recruits can either be allowed to go forward as early as possible to join the breeding stock, or some or all of them can be held back for another year by reason of their supernumerary status in the social hierarchy, if economic conditions so dictate. This last interesting type of adaptation, hitherto unconsidered, is discussed in the final chapter.

22.4. *Adult mortality promoted by stress*

Reference was made in Chapter 21 (p. 502) to experiments showing that social stress can have depressing and injurious effects on the animal body just as severe as those produced by the stresses of disease, hunger, fatigue or exposure to physical hardship. Research on this subject has up to the present time been concentrated almost entirely on mammals, although effects of the same kind might reasonably be looked for in the other highly-adapted groups. More or less severe social stress is likely to be induced in individuals that are bullied or under constant threat from which there is no escape—a condition that can very easily arise in captivity. The attacks of the victim's companions may be physically hurtful: even the hierarchical pecking of domestic hens can be vicious enough to draw blood. In captive voles, as we noted earlier, the free-ranging glossy-coated dominant males are easily distinguished from the hard-bitten, tattered subordinates at the other end of the social scale (Clarke, 1955). Unbearable antagonisms among mice may end in the stronger party killing the weaker. This happened after many weeks of mutual friendship between two adult field-mice (*Apodemus sylvaticus*) that I kept in a cage some years ago. They used habitually to sleep curled up together, until one morning I came down to find signs of a tremendous running fight all round the cage during the night: the loser had not only been killed but completely eaten, skull, teeth and all, except for about a square inch of skin and fur. It was noted on an earlier page that guppies and spider-crabs also freely attack and eat one another, even when adult, if confined at densities greater than they are prepared to tolerate.

This kind of fratricide is probably a fairly common last resort in reducing oppressive crowding in a confined space. Even among large mammals such as the red deer (*Cervus elaphus*), if they are kept artificially at too high a density, fratricidal ' accidents ' are common. ' Stags fight in season and out, and stags will kill calves and strange hinds under such conditions '. Overcrowding thus results in ' behaviour which in itself is one type of check to the further increase of a cramped population ' (Darling, 1937, p. 35), but in fact it seems to go further than this, by actually decreasing the population of the surviving herd.

Social stress can of course have more or less profound inward consequences for the victim, quite distinct from the external injuries inflicted by his companions. In rats, mice and voles, physiological derangements can affect organs in many parts of the body, including the adrenal cortex (the references were given on p. 505), the thymus (Selye, 1950; Clarke, 1953; Christian,

1959), spleen (Clarke, loc. cit.; Dawson, 1956) and other lymphatic tissues, the red-cell content of the blood (Dawson, loc. cit.), the mammary glands, the reproductive system in both males and females (numerous authors), and and the fat-reserves (Strecker and Emlen, 1953).

In man it appears that most or all of these organs and many others besides can be affected by prolonged exposure to severe stress arising from manifold external causes. According to Selye's (1950) general theory, the organism reacts to such conditions not only by specific responses in the parts of the body that form the direct ' targets ' of the stress, but also in a variety of general, non-specific ways. Thus a person in the course of becoming inured to marching long distances develops and strengthens his muscles and skeletal frame, and finds the horny layer progressively thickening on the soles of his feet: these are specific responses. The non-specific responses on the other hand concern targets not at all obviously connected with the particular stress, such as the organs named in the preceding paragraph; and no matter which one of all the many possible forms the specific stress may take, it is characteristic that these non-specific responses tend to follow the same range of common patterns. Constantly recurring responses in the same series of non-specific targets are especially typical of Selye's ' general adaptation syndrome '.

Under prolonged severe systemic stress the general adaptation syndrome ' evolves in three successive stages, namely, first the " alarm reaction ", then the " stage of resistance " and finally the " stage of exhaustion " ' (Selye, loc. cit., p. 13). The alarm reaction may elicit more or less serious shock and injury; but in the typical case this phase is overcome; the body enlists its defences and succeeds in adapting itself to the stress, and resisting its effects. If the stress continues long enough, however, and is sufficiently severe, adaptation in the end wears out, and the lesions produced in the initial alarm stage reappear.

It is known that the endocrine glands, and particularly the close-knit system of pituitary corticotrophins and the adrenal cortex, are especially sensitive to stress, and play a central part in the general adaptation syndrome. Secretion of the cortical hormones is intimately connected with the body's resistance to infection, and to injury generally. The enlargement of the rat's adrenal cortex, for example, in the experiments mentioned in the last chapter, is related to the second or resistance stage of the syndrome. The succeeding, exhaustion stage, correspondingly, is brought on by the breakdown of the defensive endocrine mechanism and the consequent submission of the body to systemic injury. An individual that has thus succumbed to stress must, directly or indirectly, have impaired its chances of survival.

That the stress need not have a physical origin, but can arise in the mind from causes real or imagined, is well known as far as man is concerned. Primitive and superstitious people are particularly susceptible to mental stress, and it is apparently possible for individuals to become ill and die simply from a conviction that they have been afflicted by the evil eye, or some other unbreakable spell (*cf.* Carr-Saunders, 1922, p. 182). In Borneo, for example, according to Hose (1926, p. 251), there can be no doubt that, if a man learns that he has been made the object of a magical attempt against him, he does in many cases suffer in health, ' and it is probable that in some cases such knowledge has proved fatal '. In the Kikuyu race of East Africa ' a person who is under *thahu* [an imaginary curse] becomes emaciated and ill or breaks out into eruptions or boils, and if the *thahu* is not removed, will probably die ' (Hobley, 1922, p. 103). Primitive people often have little if any conception of the natural causation of disease, and are strongly inclined to ascribe illnesses to magic and witchcraft. By intuition they often attribute the cause to the malevolence of some other person; and, convincing themselves and their companions that they must be protected from further afflictions from the same source, they may proceed to put the accused person to death. A good deal of socially induced mortality in primitive and medieval human communities can be laid to the score of witchcraft, especially in troubled times; it is obviously a double-edged weapon that can not only kill people who believe themselves bewitched, but can also lead to the fanatical destruction of unpopular persons or outcasts on whom blame comes to be fixed.

Our knowledge of any analogous situation in animals, leading to the elimination of adults as a result of mental stress, is virtually confined to captive populations. These usually offer the only adequate chance of observing what is going on; but captivity of course frequently tends to exaggerate the effects of dense crowding, in an environment from which escape is being effectively prevented.

In captive ruffed grouse (*Bonasa umbellus*), A. A. Allen (1934, p. 182) observed a definite ' intimidation display ', confined in its complete form to the males although females had something very similar. It seemed to have little or nothing to do with mating, but was intended rather to establish the dominance of the performer over some other, usually weaker, bird. ' A bird that has been completely subjugated . . . is subject to attack from every other bird in the enclosure. He has developed an *inferiorism* and usually, unless removed, he remains in a corner until he dies. His resistance completely shattered, he dies, not from mechanical injury nor from starvation, but from some sort of nervous shock, and death is likely to occur within twenty-four hours ' (Allen, loc. cit., p. 183). This sounds very like the final stage of Selye's syndrome, arising from mental rather than physical stress. Allen states that if such a broken-down individual is isolated he usually recovers, but it may take weeks before he can offer resistance again to other grouse.

The best-known case of a large-scale mortality in mammals, from what appears to be a non-specific reaction to stress, is the one described by Green and Larson (1938) in the snowshoe hare (*Lepus americanus*): it resulted from what the authors called ' shock disease '. During a period when natural hare populations were progressively declining in the state of Minnesota, they found that hares trapped and held in captivity tended to die almost at once, although placed in conditions known from previous experience to be favourable for extended survival. 'The hares appeared at ease, hopped around interestedly, and ate a variety of foods. However, a hare appearing normal would suddenly spring into the air in convulsions or sink to the floor in coma. In either case death usually followed from a few minutes to an hour after the onset of symptoms'. In March 1936 when this investigation was begun, the average survival in captivity was 4·2 days, and all the 204 hares taken died within 20 days; in April survival was somewhat better, averaging 12·3 days (67 hares) (details from Green, Larson and Bell, 1939). The seizures were found to be hyperglycaemic, the most consistent pathological symptom observed being the shrunken and histologically atrophic state of the liver. This could be correlated with a failure to store glycogen and provide the body's normal carbohydrate reserve; and in fact investigation showed that hares in shock had liver glycogen values of 0·02-0·18 per cent, compared with an average of 5·5 per cent in normal hares (Green and Larson, 1938, p. 201). Most of the diseased hares had the spleen contracted, and about a third of those whose adrenals and thryoids were sectioned showed lesions in these glands.

We need not be concerned with the authors' conclusion that shock disease is the principle agent reducing the numbers of snowshoe hares from maximum to minimum over a period of years: in their view it plays a leading part in bringing about the well-known periodic fluctuations in numbers in this species, to which we have earlier referred in another context (*see* p. 469). The deaths recorded by them appear to have arisen from stress brought to a head by the unnatural confinement of either box-traps or a receiving room 20+28 feet in area, but this might of course have happened to any hares brought in from the wild, whether they had been previously under stress or not. The point of importance here is that, under certain conditions at anyrate, hares can suffer a tremendous mortality from a 'non-specific' disease, in circumstances resembling those of the general adaptation syndrome.

A case apparently presenting certain parallels is that of the common toad (*Bufo bufo*), when brought into the laboratory. Toads will survive for years in captivity when kept singly or under such conditions that individuals do not see much of each other, but if they are crowded there is invariably a high mortality, which consequently renders the toad unsuitable as a laboratory animal. In fact they actually starve to death in the presence of an abundance of food (Elkan, 1960). All that succumb are alike in the total regression or absence of the fat-body, showing that they have been unable to assimilate food for some time before death. The study of 100 males that had died in such circumstances failed to show any similar correlation with the incidence of possible pathogenic organisms, or with gross lesions elsewhere in the body; and in Elkan's opinion, 'of all the possible explanations for this phenomenon the effect of crowding seems to be the most likely' (loc. cit., p. 181).

It might be objected that such death-rates as these could never arise except under the highly unnatural conditions of captivity. There is not very much evidence of similar fatality resulting from stress in nature, although Green and Larson (1938, p. 195) describe in detail two instances of apparently the same disease occurring in wild hares. But at least we are justified in concluding that some mammals, and birds and amphibia possess adaptations similar to those recognised in man, causing, under sufficient stress, internal disorders of a non-specific nature that are potentially fatal. Social stress in a natural environment may lead as a rule to an attempt to escape, most commonly by emigrating in search of a situation less competitive; but if this is impracticable, or fails to ease the strain because overcrowding is everywhere the same and equally inescapable, then the general adaptation syndrome presumably runs its course, and effectively removes the surplus.

It seems reasonable on these grounds, and consistent with our previous findings, to suggest that the general adaptation syndrome, as yet known only in vertebrates, is another of the numerous group-adaptations evolved in the higher animals to aid in population homeostasis. Like the others it is intrinsic and density-dependent in its action. This particular one provides a device for suppressing surplus adults, if necessary on the spot, when it benefits the population to do so.

22.5. *Summary of Chapter* 22

1. The first section, comprising roughly half the chapter, is concerned with intrinsic sources of juvenile mortality in birds. It seems clear that the adaptive process of trimming down the output of recruits continues after hatching, just as it did during the embryonic period. In the intensively-studied white stork, nestling mortality is often heavy, and individual chicks may be deliberately killed and sometimes eaten by one of their parents (perhaps always the father). This is most likely to happen where the parents are beginners or young adults, and presumably of low social status. Dr. Schüz has called this habit ' kronism ', and he believes it to be adaptive, serving to eliminate surplus offspring. Kronism and violent prolicide have been reported from a number of other birds, including certain gulls and terns, frigate-birds and pelicans, herons, and a few passerines.

In some of these, and most of the other species considered in this section, the parents start to incubate as soon as the first egg is laid: egg-laying is frequently protracted, and hatching the same, so that the nestlings at any moment tend to be graduated in size. This establishes a nestling hierarchy at the outset, and seems to be in the nature of an enabling adaptation for controlling nestling survival. Penguins of the genus *Eudyptes* lay two eggs of different sizes, an interesting variant of the same device. The hierarchy may lead to fratricide, as for example in many raptors and owls, but the parents can prevent this by supplying adequate amounts of food.

The decease of one of the embryos or nestlings where two eggs are normally laid is a regular occurrence in various other birds, notably some of the boobies (*Sula* spp.). This again is interpreted as representing a reserve of breeding potential that could be called upon whenever the population needed to be built up, but not ordinarily required in stable near-optimal conditions.

Some instances are given of nest-desertion, especially in kittiwakes, weaver-birds and swallows—all colonial breeders. In the first of these the social status of the parents appears again to be involved, and all may represent the operation of another device to check recruitment. Examples are also given to show the huge scale of post-fledging mortality in various species of birds: some at least of this may turn out to be intrinsic, promoted by social stress.

2. Prolicide and cannibalism are known to occur quite widely in mammals, for instance in rodents, lions and primitive man. Cases of cannibalism are also found in fish (*Lebistes*), spider-crabs (*Hyas*), and spiders (*Atypus*); and of fratricide, in insects (larvae of *Sitotroga* moth and *Rhizopertha* beetle). In all cases experimentally investigated, the mortality has been found to be density-dependent, and to cease below a certain critical population-density.

3. Death from *extrinsic* causes may also be high in the young stages, the level depending largely on the degree of parental care developed in the species concerned. Extrinsic mortality tends to be irregular and unpredictable, and one of the main advantages of evolving parental care appears to be the control it affords at this stage over death and recruitment rates. Mortality from predation is examined: it appears likely to be density-dependent, not in its own right, but only to the extent that the prey co-operates by making its surplus members especially vulnerable to predators. The density-dependent element in predation, that is to say, seems to arise on the side of the prey and not on that of the predators.

Natural selection tends to bring all mortality under intrinsic homeostatic control, as far as this can be achieved. This applies of course to senile mortality, which occurs at the age that suits the species concerned, more or less independently of the particular agency that happens to discharge the death sentence.

Infective diseases seem equally incapable of exerting an independent limiting effect, although, because of the varying and intrinsically-controlled resistance of the host, disease like predation can sometimes be used effectively in reducing a surplus—in this case a surplus of individuals predisposed to injury by their depressed physiological state. Both disease and predation do, naturally, inflict a variable amount of ' uncontrollable mortality ', and this tends to fall most heavily on the young, as yet unprotected by an acquired immunity. Uncontrollable mortality is chancy and fluctuating, and, since it is liable to strike most heavily in the early age-groups, especially versatile and flexible adaptations are required at this stage to complement it, so as still to come out at the end with the optimal quota of recruits. In addition to the direct and indirect forms of infanticide, other checks on recruitment employed are expulsion, and the suspension of adult status for a longer or shorter period.

4. Social stress can lead to casualties at all ages, both through direct and mortal combat, and through stress-induced disease. The victim of severe stress is likely to develop physiological disorders, affecting many organs, especially the lymphatic apparatus (including the spleen and thymus), the nervous system, circulatory, digestive and generative organs. The endocrine glands, and above all the adrenal cortex, appear to serve an intermediary role between the ' stresser ' and the ' target ', the latter being the organ affected.

The whole complex is known as the ' general adaptation syndrome ' of Selye. Under sustained severe stress, the individual typically suffers first an alarm reaction and shock damage, next a stage of resistance and adaptation to the stress, and finally a collapse and submission to injury.

Social stress is sometimes partly physical, as when the exercise of peck-order rights leads to the infliction of wounds or withholding food and shelter; but it may be largely mental, as we know in man, who, in his simpler-minded states, may die from the conviction that he has been bewitched. Single cases are given in birds, mammals and amphibia, namely in the ruffed grouse, snowshoe hare, and common toad, of animals similarly dying from non-specific injuries apparently induced by social stress.

It is concluded that Selye's syndrome is another of the many group-adaptations evolved in the highest forms of animals, serving to control population-density, in this case especially by eliminating surplus adults.

Chapter 23

Deferment of growth and maturity

23.1. *Density-dependent rates of individual growth in aquatic animals*

Three chapters have been devoted to surveying the many methods devised for controlling population-density and balancing the equation of recruitment and loss, and still we have not covered all the types of adaptation that have been evolved for the purpose. Of the two that remain to be discussed, the first is the limitation of populations not in terms of numbers but of biomass, whereby a high density has the effect of stunting individual growth; the second centres round the control of recruitment, by holding back a reserve of full-grown adolescents or virgin adults and admitting to breeding status only the number actually justified at the time by circumstances. Of these the second is much the commoner, having developed above all in those species among the higher animals in which there is a long span of individual life.

Many people who have kept goldfish in a small bowl know from experience that, in spite of regular feeding and unremitting care, years can pass without the fish showing any perceptible growth; their size limit appears to be set by the size of the bowl. Over the last century and more, numerous experiments bearing on this phenomenon have been carried out, and these have already been reviewed by Allee (1931, chap. 6) and Richards (1958, p. 147). Only the main facts need be outlined here.

Taking first the effects observed in amphibia, where our knowledge is farthest advanced, itwas demonstrated by Bilski (1921) that when toad tadpoles (*Bufo*) are reared in the laboratory on superabundant food, their growth-rates show a strong positive correlation with the size of the vessel they are living in; and in aquaria of uniform size, the larger the number of tadpoles the slower the growth. Goetsch (1924) extended this work, using among other things axolotl larvae (*Ambystoma tigrinum*) and frog tadpoles (*Rana esculenta*), and found that the primary limiting factor seemed to be the amount of space per individual (total volume divided by number of larvae), whether the individual was in a mixed group or confined alone in a restricted space; the accumulation of soluble metabolites appeared to have only a secondary effect, small in comparison, especially in the case of the axolotl. Later workers confirmed these results.

Understanding of the subject has recently been greatly extended by Richards (1958), exploiting her initial discovery that large tadpoles (*R. pipiens*), when sufficiently crowded, produce something that powerfully inhibits or arrests the growth of small tadpoles. She found this inhibitory substance could be transferred in ' crowded water ', and quickly assayed by measuring its effects on the growth in weight of a standard group of small tadpoles. The centrifuging of ' crowded water ' showed that the substance was present in the residue and absent from the supernatant: it was therefore presumed not to be soluble, but to be either a fairly large particle or attached to such a particle. Passing the water through filters of known pore-size showed the particle to be larger than 10-15 μ and smaller than 40-60 μ—larger, that is to say, than most bacteria: moreover it was not destroyed by penicillin or streptomycin. Next, direct experiment showed that the inhibitor was present in the tadpoles' faeces; that it could be ' killed ' by heat (10 min. at 60° C.), freezing, sonification, ultraviolet light, etc., as if it were something living; and that it was resistant to a number of enzymes, as if it were an intact cell. Microscopic examination revealed the actual presence of peculiar rounded vacuolated cells of the right size in enormous quantities in the guts and faeces of crowded tadpoles; and also their complete absence in non-crowded tadpoles. When fifteen non-crowded tadpoles were assembled together, in order to create crowding, a sample killed and dissected showed that within one and a half hours the guts were full of these cells; and within five hours the faeces of the live tadpoles contained them. Tadpoles released from crowded conditions lost them, and concomitantly started to grow at a normal rate.

The origin of the cells was not traced: they could conceivably have been organisms such as yeasts, fungi, algae or protozoa, but it seemed rather more likely, in view of their rapid spontaneous appearance and the fact that no dividing or double cells were ever found, that they were produced from the gut wall of the tadpole itself. The more they accumulated in the water and bottom deposit (from which the tadpoles re-ingested them along with food), the greater the growth-inhibitory activity of the water became. But the mode of their action—whether they merely occupied bulk space in the gut to the exclusion of digestible food, whether they gave off a short-lived toxic material, or whether they deprived the tadpoles of an essential growth factor, could not be decided.

Analogous inhibitory effects are known to arise in similar circumstances in other kinds of aquatic animals, though so far this type of cellular inhibitory substance has been found only in *Rana pipiens*. Church (1927, unpublished, quoted by Allee, p. 113) and Shaw (1932) demonstrated the stunting effect in confined populations of the small tropical fish *Platypoecilus maculatus*, and showed that it resulted from crowding, but was independent of food-supply. In *Lebistes*, also, when either the size of the containers in which isolated individuals are kept, or the plane of nutrition, is varied, correlated effects are observed on the maximum size the fish attain (Comfort, 1956, p. 76).

Rose (1959) found in two other fish, *Tanichthys albonubes* and *Barbus tetrazona*, that some metabolite is produced which acts ' in feed back fashion ' on the growth of smaller individuals. In a 15-litre aquarium a female of *T. albonubes* could be expected to produce many more eggs than could develop in such a confined space, but no matter how many hatched—even as many as 200—never more than twenty reached a 1-cm size. ' Shortly after feeding begins, differences in size appear. The larger fish continue to grow: the smaller ones stop eating and die in spite of an abundance of food '. There is nothing inherently wrong with the small fish when feeding ceases, because they proceed to grow quite normally if they are either isolated in another aquarium or have their larger brothers removed.

Barbus tetrazona gave even more striking results. (At all times a slight excess of food was being provided.) From a spawning of over 200 in a 15-litre aquarium, no more than fifteen ever survived to the 1-cm size. ' The survivors were always the most rapid growers '. However, by replacing half the water in the aquarium twice, three times, and latterly four times a day, the survival rate was stepped up from fifteen or less to 174: from which it may be concluded that the growth-inhibiting activity of the metabolite depends on its concentration.

The cessation of growth leads eventually, in all these cases, to the death and elimination of surplus individuals: so that, as a mechanism protecting these particular amphibia and teleosts from overpopulation, the device can be made effectively equivalent to many of the other induced-mortality adaptations considered in the preceding chapter.

Corresponding experiments have been done many times with freshwater snails (*Lymnaea*), for instance by Semper (1881) and De Varigny (1894). De Varigny thought he had demonstrated that stunting could be produced purely by a space-effect—a ' psychological ' factor—independent of the accumulation of metabolites, but other workers showed that the accumulation of faeces and excretory products, and, of course, deficiencies in food, oxygen and minerals, could all certainly reduce the rate of growth. Richards (loc. cit., p. 148) therefore attaches some doubt to the validity of De Varigny's conclusions.

The other classical instance of growth stunted by crowding is in the freshwater cladoceran *Daphnia*. Warren (1900) established that, in *D. magna*, crowding results in the progressive atrophy of the spine at the back of the carapace, a dwarfing of size, and a slowing down of reproduction in the rapidly succeeding generations. Pratt (1943) found that, at 18° C., cultures of *Daphnia* built up to a ceiling and later stabilised themselves somewhat below this maximum density level: the sequence of events is essentially similar to what we saw in most of the experimental populations of insects and vertebrates discussed in Chapter 21: the final equilibrium density was very similar in each of a series of replicate cultures. Regulation was achieved almost entirely through adjusting the birth-rate (loc. cit., p. 127). (At 25° C., it is interesting to note, the cultures underwent a series of oscillations, apparently due to the birth-rate reacting too slowly to stabilise numbers—under high-temperature conditions that induced rapid metabolism, short life and a high steady death-rate. Some of the cultures became extinct after one cycle.)

Frank, Boll and Kelly (1957), using *Daphnia pulex*, have confirmed and extended these earlier results. They found that ' numerical density has significant effects on deaths, births and growth '. Experimental populations, seeded with new-born parthenogenetic females, were set up at fixed densities of 1, 2, 4, 8, 16, 24 and 32 *Daphnia* per millilitre; and these showed that, once the females had started to breed (which they did a little more than a week after birth), the births per female per day consistently declined with increasing crowding. To keep the numbers constant, all neonates were systematically removed, and all casualties replaced by individuals from similarly conditioned experimental stocks.

It may be noted that the age at which breeding began increased significantly with density (loc. cit., p. 293); this provides a link with the subject to be discussed in the third and later sections of this chapter. The size attained by *Daphnia* is extremely variable (*see* fig. 50): those growing up in the least crowded cultures, at 1, 2 or 4 per ml, all attained about the same maximum size, but at higher densities the maximum attained became less and less, falling to about a quarter of the normal uncrowded size in the densest cultures (at 32 per ml). Above the critical density of 4 per ml, in fact, the growth-rate of the individual is closely proportional to the population density, and the total biomass of all the cultures increases at roughly the same rate, and attains roughly the same maximum volume, regardless of whether there are 8, 16, 24 or 32 individuals per ml present.

These experiments eliminate the factor of population growth through reproduction: the increment in biomass that takes place is regulated solely through the growth in size of a constant number of individuals. The experimenters have succeeded, therefore, in isolating and emphasising in a very striking way the absolute density-dependence of growth-rate in *Daphnia*, and its efficacy in biomass control.

There is another homeostatic mechanism which also shows up in the *Daphnia* experiments, namely that crowding is associated with a general reduction in the ' rate of living ' of the crowded individuals, and a corresponding increase in the life-span of survivors. Though we have seen that crowding can frequently induce mortality of surplus members of the population, there is nevertheless a complementary tendency for those that survive under conditions that delay normal growth and development to compensate by living longer. This is sometimes known as Lee's phenomenon, having first been postulated by Rosa Lee (1912) as a possible explanation for her discovery that the older fish in any given sample—of herrings, haddock or trout, for instance—always appear from the rings on their scales to have had a slower history of growth than the younger members of the sample. This suggests that the fastest growing individuals in any age-group are the first to die, so that the older age-groups come progressively to be left with more and more retarded veterans. As a general adaptation, Lee's phenomenon appears to give the stock an increased chance of surviving whenever

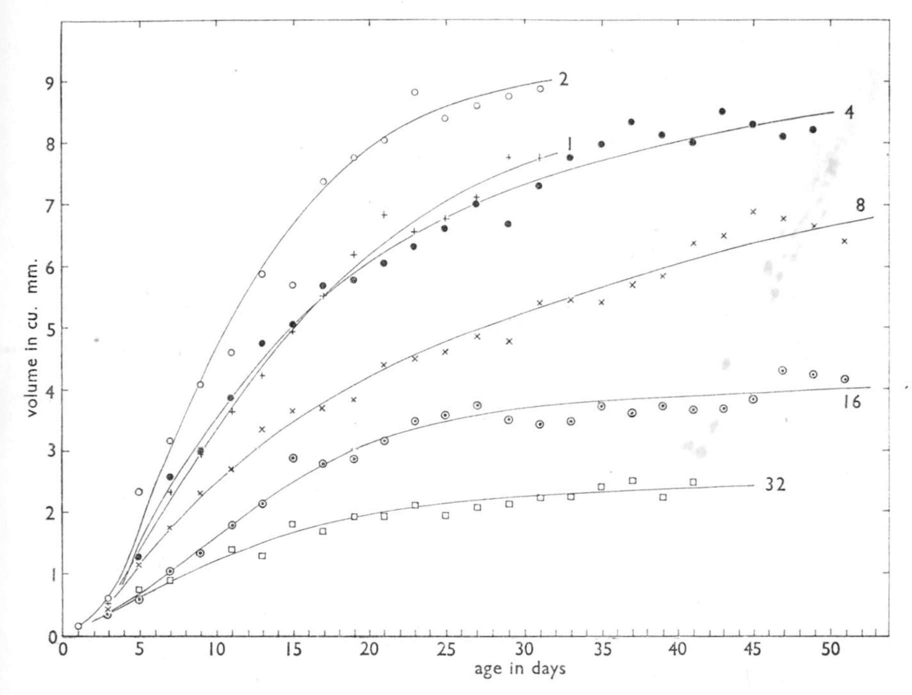

FIG. 50. Growth-curves for *Daphnia* developing at different culture-densities, showing the effect of crowding. Mean individual volume (in cu mm) is plotted against age. At densities of 1, 2 and 4 *Daphnia* per ml, individuals grow to nearly the same mean maximum size, but at higher densities their growth is progressively stunted. In the most crowded culture, at 32 per ml, they attain barely three-tenths of the size normally reached at the optimum density of 2 per ml. (Redrawn from Frank, Boll and Kelly, 1957.)

there is a period of conditions inimical to growth or reproduction; and in the fish it also spreads the age of first maturity, conferring an advantage to be discussed in the next section.

The general concept of homeostasis in aquatic populations through growth-inhibitory substances, such as we have outlined here, is complicated by the fact that, in some of the very same species, growth-*promoting* metabolites have also been discovered. In the *Daphnia* experiments last described it was actually found that significantly the fastest growth-rate occurred, not in the culture with 1 per ml, but in the one with 2 per ml (Frank *et al.*, loc. cit., p. 298). This particular case was not further investigated, but the facts closely parallel what had been found by Allee and his collaborators in certain fishes and amphibian larvae. Reverting to the tropical aquarium fish *Platypoecilus*, Shaw (loc. cit.) discovered that water in which the freshwater mussels *Anodonta* and *Lampsilis* had been living favoured the growth of fish more than plain water. Mussels were left for twenty-four hours in artificially standardised water, which was then said to be ' heterotypically conditioned ': the control medium was water of the same composition, standing alongside in an identical aquarium with no mussels in it for the same period. Each day half the water in the fish tanks was changed, the experimentals receiving conditioned water and the controls plain water. Growth differences between the two were highly significant. Exactly the same results were obtained with salamander larvae (*Ambystoma tigrinum*). Further experiments in the same laboratory by Allee, Bowen, Welty and Oesting (1934) made similar use of homotypically conditioned water—that is, water in which fishes of the same species had been living for twenty-two hours; and here again a marked growth-promoting effect was demonstrated in the conditioned water. This was found alike in *Platypoecilus*, black catfish (*Ameiurus melas*), and goldfish (*Carassius*).

This growth-promoting effect is realised, of course, only in conditions of low density and non-crowding, and is not therefore incompatible with the inhibitory effect discussed earlier. Goldfish, *Platypoecilus* and *Daphnia*, for example, all show both effects in the respectively appropriate circumstances. The nature of the growth-promoting substances is obscure; Livengood (1937) found he could simulate the effect by using, as the conditioned medium for goldfish, water which had been filtered after food juices had merely been squeezed into it; yet Allee and his associates succeeded in extracting a growth-promoting concentrate from the skin of the goldfish, effective in dilutions of as little as 1 part in 800,000: this appeared to be a protein, and was traced to the slime secreted from the dermal glands (Allee, 1938, p. 78). It appears premature as yet to try to draw any conclusions as to the adaptive significance of the phenomenon.

23.2. *Density-dependent rates of growth in terrestrial animals*

Growth-inhibiting (and no doubt growth-promoting) effects are evidently rather widespread in aquatic organisms: the latter live in a medium that lends itself particularly easily to being conditioned by metabolites. It may be suspected also that this kind of regulatory brake is specially effectual and easy to impose in species that maintain the power of growth throughout life. Something of the same kind, however, certainly occurs here and there among terrestrial animals; and it is quite familiar in the insects. As a homeostatic adaptation, indeed, we first came across the ' conditioning of the medium ' in the flour-beetle *Tribolium*, which has gas glands adapted for the purpose (*see* p. 497); and these of course exert their effect on the survival of larval stages, as well as on the reproductive rate of adults.

In many species of insects a considerable variabiltiy in size is found among the adults. We saw an illustration of this earlier (p. 266), in the striking case of the males of some of the horned Coleoptera such as the stag-beetles (*Lucanus*); there it is associated with a higher or lower develop ment of secondary sexual weapons and adornments. Size is notoriously variable also among the workers of some of the social Hymenoptera, for instance the bumble-bees (*Bombus* spp.); in *B. lucorum* workers in a single nest have been found to range from less than 60 to more than 300 mg in weight (Cumber, 1949, p. 13). In some of the common syrphids such as the drone flies (*Eristalis* spp.), and in many butterflies, individual size variation is also marked: similar examples could be found in Hemiptera, Isoptera, and probably every order. In *Bombus* all the workers in one nest are sisters, and it has been established that individual size and growth-rate depend essentially on the conditions of development—especially nutrition—and are not primarily genetical; the same is likely to be true of most if not all the other cases.

In the termites it is known that the rate of development, and the number of moults required before full size is reached, may be correlated with the size of the colony. In *Zootermopsis*, for instance, an adult soldier will be produced after 5, 6, 7 or 8 moults, according to whether the colony consists of about 25, 50, 100 or 500 individuals, corresponding to ages of 1, 2, 2-3 and 4-5 years, respectively (Richards, 1953, p. 182, citing Grassé's experiments). Presumably in younger colonies it is desirable to increase numbers rapidly, at the expense of individual size. The same is true also in the ant *Myrmica rubra*, in which individual worker-size is very small in new colonies, but increases rapidly with colony-growth in colonies up to 300 individuals, above which it remains constant however large the colony may ultimately become (Brian, 1957, p. 178).

A possibly analogous adaptation occurs in mammals, especially in the ungulate orders, leading to a reduction in adult size in, for instance, insular or otherwise confined habitats that preclude the possibility of emigration. The red deer (*Cervus elaphus*) is particularly well known for the manner in which adult size is influenced both by economic conditions and by insulation. The largest red deer stags are raised in parks, where the plane of nutrition is artificially high, and in rich natural forests abroad: Harris in the Outer Hebrides, on the other hand, produces the smallest stags in the British Isles, if not in Europe, weighing little more than half what their cousins do in the better deer-forests on the mainland of Scotland (*cf.* Millais, 1906:III, p. 105). A more striking instance is the miniature white-tailed deer (*Dama virginiana clavia*) native to some of the Florida Keys; a buck Key deer weighs only fifty pounds (Peterson and Fisher, 1956, p. 133), compared with the normal average on the North American mainland of some 200 lb.

The horse (*Equus caballus*) has much the same propensity. Diminutive island breeds from Shetland, Corsica and Sardinia, and from barren hilly regions like Dartmoor, are very well known. Darwin (1868) mentions a stock of small Shetland-like ponies, believed to have originated through exposure to unfavourable conditions, that formerly existed on an island off the coast of Virginia. Other small breeds are or were established in some of the East Indian islands, in the South American cordillera, and elsewhere.

In some breeds of horses small size is of course congenital, and not due to any actual stunting of the growth-rate in response to contemporary stress. The Shetland pony, averaging only ten hands (1 m) in height at the shoulder (compared with 14-16 hands in the larger breeds), comes of an ancient stock that has been domesticated probably from neolithic times; and the same is true of the other small ponies of the British Isles, Scandinavia, Iceland, or Corsica. Their small size is essentially hereditary, fixed by selection over a long period. On the other hand it is well known to horse-breeders that neither the rigorous climates of high latitudes and mountains, nor the humid heat of the tropics, are conducive to raising large horses; new blood introduced from temperate regions quickly 'deteriorates' under these conditions, whether in the Falkland Islands (Darwin, 1845) or in India (Hayes, 1922), for example.

In the same way, feral horses that have gone wild in recent times and are descended from normal domestic stock are liable to become small under adverse conditions. Thus in the Grand Canyon country of Arizona there are feral horses only 10½-13 hands high when full grown (confirmed by the Chief Park Naturalist of Grand Canyon National Park in 1960); and on Sable Island, 100 miles off the Atlantic coast of Nova Scotia, the feral horses, comprising a herd of some 250, are said to attain at a maximum only about fourteen hands (*The Times*, London, 27 May 1960, p. 13).

There is some suggestion, though unfortunately not well documented, that when red deer were introduced into New Zealand from Scotland and elsewhere in the 19th Century, they increased in average stature during the colonisation period, but subsequently declined again in the more congested areas. In an attempt to confirm this, I corresponded with, among others, Mr. R. I. Kean, of the New Zealand Forest Service; and he gave me information about another more recently introduced mammal, the Australian opossum *Trichosurus vulpecula*, which appears to show a similar effect. Thus, in a reserve of indigenous forest with good food-resources, the opossums average at least 7 lb in weight, and 9 lb animals are not uncommon; whereas in another forest study-area where the best foods have been permanently depleted, the average weight is 5-7 lb. 'Decrease in size is general', he writes, 'after quality of food supplies falls and population numbers rise'; and this is amply confirmed by the experience of commercial trappers.

There seems no reason to doubt that this phenomenon can be regarded as a constructive, functional adaptation. The fact is that many mammals, including man himself, have to a greater or less extent developed a capacity for producing perfectly normal healthy adults, varying in size according to the degree of nutritional (and possibly other) stress under which they grew up. Severer stress does not necessarily produce starved or ailing adults, but merely smaller ones, still perfectly capable of reproduction. It seems likely enough to be a group-adaptation, capable of increasing the chances of survival of the stock in any small isolated habitat like an island, by permitting a larger population (of smaller individuals) to be carried in each generation; and perhaps capable also of usefully reducing the biomass, and economising on food-resources, in species like deer that tend to multiply beyond carrying capacity of their habitat whenever they lose their attendant predators.

It is perhaps significant that birds, with their greater mobility, and consequently much smaller risk of being isolated and barred from emigrating for the purpose of relieving population pressure, either do not show this faculty at all, or have it relatively little developed; but there is apparently an exception—where indeed we might have anticipated it—in some of the large flightless birds. Both the ostrich (*Struthio*) and emu (*Dromaius*) are exceptionally variable in size: in the former ' some birds are very much larger than others ' (Cronwright-Schreiner, 1925, p. 97), and in the latter the weight of full-grown adults in good condition ranges from 70-120 lb (Serventy and Whittell, 1951). However, I have been unable to find out whether this is due to a capacity for size-accommodation in response to stress during development, or to an unusual amount of genetic variability.

23.3. *Deferment of maturity in birds*

A parameter that enters strongly into the vital statistics of every species is the age at which breeding commences. Rapid breeders among the Crustacea, insects and mammals, for instance, can run through two or more generations in a year; yet there are other species, even in the same classes, that require five, ten or even twenty years to emerge from adolescence. On the untenable assumption that animals must always reproduce at the fastest rate that natural selection can evoke, it is just as difficult to explain this tremendous spread in the turnover-rate of succeeding generations as it is to explain the analogous differences that we find in fecundity.

We should probably do well again, therefore, to seek an explanation by considering primarily the interests and survival of the stock as a whole, rather than the effect on particular individuals of mutual differences in prolificity. Not always, though indeed very often, an extended adolescence is associated with very low reproductive rates after adult life is reached (Wynne-Edwards, 1955, p. 543); and the two conditions are in these circumstances almost certainly interrelated. Both in fact appear to have been evolved in unison, through the process of group-selection, in order to enhance the homeostatic control of recruitment and loss in populations of long-living species: or so it may at any rate be suspected before the intricacies of the situation have been fully examined.

What we should first ask, therefore, is whether anything can be gained, towards achieving still greater control over population-density, by abandoning early maturation and delaying it in favour of a longer term of sterile adolescence. We shall find this leads to consideration of the ulterior question also—what are the merits of long life and perennial breeding as compared with a swifter succession of, say, annual 'monotelic' (only-once maturing) generations?

As we saw in the last chapter, breeding success in various birds, including white storks (p. 531) and kittiwakes (p. 538), is distinctly lower in the younger adults, particularly those breeding for the first time, compared with older-established individuals. It is quite likely that this is a general phenomenon in birds; in the blackbird (*Turdus merula*), for instance, Snow (1958) showed that yearling females not only lay on the average slightly smaller clutches of eggs than older females, but their hatching success is significantly lower also: in pairs in which the male is a yearling too, the success is lowest of all, and even the survival of the fledglings continues to be below average for two or three weeks after they leave the nest. In the yellow-eyed penguin (*Megadyptes antipodes*), the same effect is very pronounced: Richdale (1957, chap. 7) showed that two-year-old females starting to breed laid smaller eggs, and had a mean clutch-size of 1·62 compared with 1·99 in the old-established groups; they hatched only 32 per cent of the eggs they laid compared with about 90 per cent in maturer birds, and reared a lower proportion of the chicks they did hatch. Performance progressively improved among the third, fourth and fifth year-classes, after which they could be considered fully efficient.

It has been usual to attribute this failure of birds to achieve in their first breeding season the success they acquire later in life to a mixture of physiological impotence, imperfectly developed instincts, and want of experience. It can scarcely be doubted that at any rate the last of these really exerts some influence, but there are indications that here, as elsewhere, fertility is to an important extent governed by the interplay of social rank and economic conditions, with the result that newcomers to the breeding caste are liable to be handicapped by their inferior social position. Pitelka, Tomich and Treichel (1955, p. 93) found at Barrow, Alaska, that in the summer of 1953, when the lemmings were at peak abundance and the pomarine jaegers (*Stercorarius pomarinus*) nested at a very high density in consequence, a small percentage of birds actually bred before they had acquired the characteristic twisted tail-feathers that denote adult status: they were probably only a year old. In 1952, or indeed at any other stage of the lemming cycle, such a thing was never observed. The same economic factor probably explains why certain species of birds tend to start breeding younger in some districts than they do in others. The majority of white storks (*Ciconia ciconia*), for example, first breed successfully when they are four years old or older, but in Jutland they have occasionally been proved to start at the age of two (Lange, 1940, cited from Haverschmidt, 1949). In East Prussia no case has been discovered of successful nesting before the age of three, and even this is uncommon (7 instances in 201 cases of ringed birds of known age starting to breed, given by Hornberger, 1943, p. 351).

It is apparently characteristic of birds having prolonged periods of adolescence that not all the members of any given population start to breed at the same age. Lange's figures for the white stork, relating to fifty-one birds whose age at first breeding was known, were actually distributed as follows:

2 years	2
3 years	12
4 years	22
5 years	15
	51

Hornberger (loc. cit.) found a generally similar, though longer, spread: he continued to come across ringed breeding birds, not previously detected nesting, up to even eleven and fourteen years old (one each).

Another example is provided by the common tern (*Sterna hirundo*). Austin (1945, p. 24) found in Massachusetts that 1·6 per cent bred at one year old, 15·7 per cent at two, and the remaining 82·7 per cent not before they reached three years.

The satin bower-bird (*Ptilonorhynchus violaceus*) is a very slowly maturing species. A. J. Marshall (1954, p. 28) records that, in captivity, a young male acquired a few of the dark adult feathers in its fourth year and completed the transition into glossy blue-black adult plumage by the age of four years and two months: another bird had been actually in captivity for six years and three months before becoming completely dark, and must have been at least five months old when first caught. This is not very much to go on, but it is specially interesting in the present context because, under certain conditions, males in the greenish mottled ' immature ' plumage are known to breed successfully. There is evidence that males in this plumage are attacked and dominated by blue males (loc. cit., p. 68). In a case where a mated blue male was removed for experimental purposes, his bower and female were annexed by, presumably, ' the most aggressive green male in the area '. This bird was then collected and found to have completely developed and fertile testes. His place at the bower was assumed by a second green male; and when this was in turn removed, by a third, ' it was almost certainly with him that the female eventually bred '. A repetition of the experiment, in which a mated blue male was removed from a different bower, similarly resulted in a green male taking possession within a week, and eventually breeding (Marshall, loc. cit., pp. 43-4). The impression conveyed is certainly one of a population dominated by full-plumaged males but carrying a considerable reserve of potential, but usually inhibited, breeders belonging to a sub-adult caste.

In the yellow-eyed penguin (*Megadyptes antipodes*), Richdale (1957) showed a considerable difference between the sexes in the commencement of breeding. They acquired adult plumage at the age of 14-18 months, and about half the females (51 per cent) first bred, on the Otago Peninsula, when two years old: another 45 per cent first bred in their third season, and the remainder (3·5 per cent) not till their fourth. In the males the onset of breeding was spread all the way from two to ten years: 11 per cent first bred at two, 36 per cent at three, 33 per cent at four, 13 per cent at five, and the other 8 per cent at various ages up to ten (loc. cit., p. 134). In explanation of this, Richdale points out that young males have to compete with older males, which always outnumber breeding females. 'In assessing the onset of the breeding urge in males I would claim, therefore, that the delay in breeding . . . is not entirely due to the lack of a breeding urge, but is partially an effect of competition with older males' (loc. cit., p. 108).

There is no need to emphasise that with slight modifications the same general explanation could be applied to the females, or to the white storks, just as we have already applied it in the case of the satin bower-bird. An alternative explanation has been suggested by Lack (1954, p. 63), namely that breeding imposes a strain on the parents, which may prove too heavy for the younger individuals in these slowly maturing species, so that they tend to succumb if they attempt it too soon; natural selection has consequently retarded the onset of maturity, giving them time to develop the stamina required to cope with it. This theory does not enable us to account for the very wide spread in the age of first breeding that we have just seen to occur in different individuals of the same species; presumably the same kind of selection would work about as strongly against individuals that were handicapped by coming to maturity a year or two behind the average, and would result in fixing the age of puberty within rather narrow limits. Nor does it account in any way for the fact that in so many cases there is a reserve of sexually potent, but actually non-breeding, individuals, ready to step in and take the place of any breeding bird that is killed. As Lack was aware, it fails again to suggest why the males should often be retarded longer than the females, when, if anything, the former have the lighter reproductive task. Finally, there is no evidence to support the basic proposition that mortality is higher among breeders than among non-breeders in the younger year-classes.

Our own interpretation is based on the premise that the habitat has a certain carrying capacity, varying from year to year; and the homeostatic machinery allows only the appropriate number of individuals to breed in any given area in any one year. Usually this number is somewhat less than the total population of potential breeders, so that there is a surplus of non-breeders that cannot for the time being win the property and status that entitle them to nest. This surplus we know exists in many—perhaps most—breeding populations of birds. It consists of those that have been unsuccessful in social competition—those, that is to say, standing low in the hierarchy. The majority of them are probably generally junior in age: those that have bred before, we know, often possess a virtually inalienable right to return to and occupy their former sites or territories. It is the new recruits that have never yet won themselves a stake that are, naturally, most likely to be shut out and inhibited from breeding. In long-lived species their chance may sometimes be very slow in coming—especially in males if they tend to survive longer than females or to be polygamous. When they do start to breed they are still likely to be more or less grudgingly tolerated by their established senior neighbours, and to be the ones to give way if the breeding output of the population later turns out to be somewhat too high and requires curtailing: it is they who in the stress of these circumstances are most likely to become the ' bad ' parents, as the evidence has shown us.

It is likely at times, however, that not all members of the excluded non-breeding surplus are maiden individuals that have never yet bred. Circumstances might conceivably arise in which the surplus of potential breeders had reached such proportions that not only could no new recruits be admitted to the breeding ranks, but even some of those that had previously bred would be compelled to stand down. Long ago I found that, among the great host of non-breeding fulmars (*Fulmaris glacialis*) that spend the summer months along the coast of Labrador several hundred miles from the nearest breeding colony, some individuals gave overwhelming evidence of being aged birds and of having bred on previous occasions (Wynne-Edwards, 1939). The evidence was of two kinds. Firstly, in a sample of sixteen birds, some had long, needle-sharp claws, and others blunted stumps: in length, their middle claws varied from 10·6-16 mm. Fulmars cannot wear their claws down except by going to a breeding colony, since at other times they never make contact with anything harder than ice. The presumption was that those with their original sharp points still intact had not been back to land since their nestling days, whereas those with heavily worn claws had, first and last, spent much time there. If the sample is subdivided according to claw-length into 1-mm groups, almost the same number of birds fall into every group, indicating a continuous gradation among members of the sample in the amount of wear and tear sustained.

The second kind of evidence came from a histological examination of the ovaries of four females taken from the same sample; this exactly corroborated the evidence of the claws. In a bird's ovary there are large numbers of oocytes, or potential ova, only some of which are in due time required to develop into actual eggs. Their development necessitates a great enlargement of the original ovum, to contain the enormous supply of yolk—an enlargement that takes place in stages. Many ova go through a preliminary ripening process that results in a yolky cell about 2 mm in diameter (in the fulmar), surrounded by a follicle of nurse cells, and they may remain in that condition for an indefinite period. There is a tendency for these to be further enlarged or sustained at the expense of one another, so that, as the years go on, the ovary comes to contain a progressively increasing number of scars of old follicles, whose ova have been resorbed. This phenomenon is called atresia. Eggs that have actually been completed and laid sometimes leave ovarian scars that still show in after years—the remains of the follicle opening outwards, where it freed its ovum; atretic follicles, on the other hand, tend to be closed and covered over by the tissues of the ovary.

As may be plainly seen in plate X, the ovary of one of these females (*No*. 1360) contained very few atretic scars; the tissues were in an extremely healthy state, with numerous 2-mm ova. In great contrast to this, the ovary of bird *No*. 1373 was full of old scars, leaving comparatively little room for yolky oocytes. The number of resorbed follicles was counted in each of the four females, and compared with the measurements of the claws, with the following result:

Specimen number	*Middle claws (in mm)*		*Resorbed ovarian follicles*
	Left	*Right*	
1364	15·8	15·8	29
1360	15·6	14·8	28
1355	13·0	13·0	50
1373	10·6	10·6	101

The number of follicles given in the last column is believed to be correct within 5 per cent. It will be seen that the first two birds, the youngest on the evidence of claw-length, had the fewest follicular scars: the next bird is intermediate in both respects, and the last is on both counts the oldest. Moreover, the last bird showed clear evidence of having laid an egg in a previous season (plate XC).

The only weak link in this chain of evidence is the last statement. Can it be certainly shown that this was a genuine non-breeder, and not just a bird that had already laid its egg, lost it at once and left the colony, flying south to join the Labrador non-breeders as early as July 26? This would be very difficult to prove. The evidence shows that fulmars lay their eggs at the Cape Searle colony in Baffin Island about the second week of June, so that a period of roughly seven weeks could have elapsed between egg-laying and the time the bird was shot. The massive follicle, large enough to have enveloped an egg-yolk over 30 mm in diameter (roughly the size of the yolk in the largest of hen's eggs), would have had to contract and heal in this time to a minute quiescent scar smaller than a pin-head. Such a thing can only be regarded as very improbable.

This has been related at some length, because the facts presented have in the past always appeared anomalous, or difficult to explain. In Britain where the fulmar is rapidly multiplying and extending its range, we know that most adults breed annually, but in the light of the present context it is easy to see that in arctic Canada, where the fulmar population is static or even slowly declining, the demand for breeding stock may be much less, and the surplus of non-breeders on a far larger scale. It is perfectly possible, when the circumstances are so completely different in these two distant parts of the fulmar's range, that whereas in Canada there are considerably more adults than are required to make up the breeding quota, this is not so in western Europe, where the quota is annually expanding before our eyes.

A.

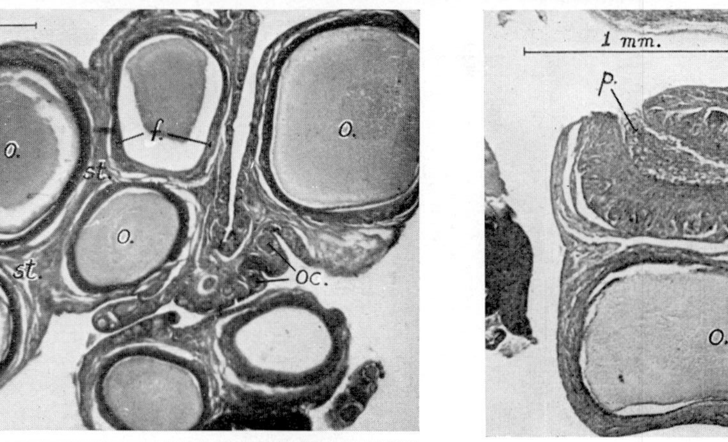

B.

C.

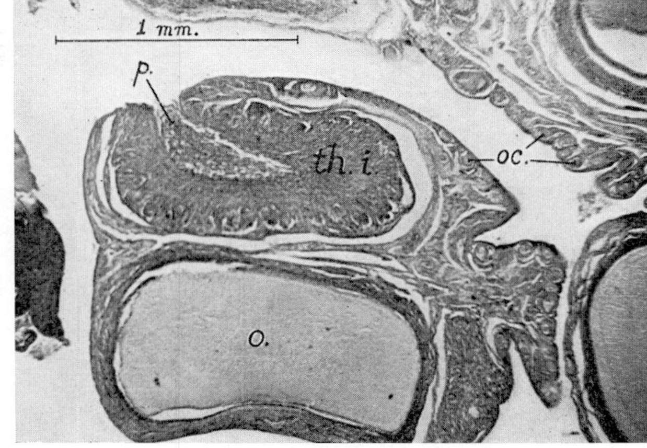

In *A* and *B* the contrast is vividly shown between a young adult (*No*. 1360), with numerous follicles containing small ova 1-2 mm in diameter and few or no scars of old follicles; and an aged bird (*No*. 1373), with relatively fewer, smaller ova and abundant scars of resorbed follicles. *C* shows the contracted remains of an ovulated follicle in *No*. 1360, from which it appears that an egg erupted into the body-cavity (and passed to the oviduct). This indicates that *No*. 1360 has at some time bred, perhaps in the previous year. *f*. follicle; *o*, ovum; *oc*, small oocytes; *p*, plug of cells filling lumen of ovulated follicle; *r.f.*, scars of resorbed follicles; *st*., stroma; *th.i.*, theca interna. (From Wynne-Edwards, 1939.)

PLATE X. Sections of ovaries from two non-breeding fulmars obtained near Cape Chidley Labrador, on 26 July 1937.

The gannetry at Cape St. Mary, Newfoundland, seen from the lighthouse. The gannets' rock, in the middle distance, is a completely detached stack. Note the large congregation of 'unemployed' birds on the nearer slope, above the adjacent mainland cliff. (From Wynne-Edwards, 1935; photographed 23 June 1934.)

Once this possibility has been put on a rational basis, that adults may sometimes be temporarily suspended from breeding by the action of the social guillotine, evidence to support it may come in from many sources. Richdale (1957, p. 133), for example, found a small number of males in the yellow-eyed penguin that passed a season without being mated, although they had previously bred. Oordt and Bruyns (1938) found in the oystercatcher (*Haematopus ostralegus*) that non-breeders may be either yearlings or adults. The existence of large numbers of presumptive non-breeders in adult plumage has been especially noted and commented upon in different species of penguins, gannets, and petrels, but the present-day tendency had generally been to account for them as failed-breeders rather than to admit the possibility of some kind of exclusion or inhibition for which there has seemed to be no rational explanation. In a recent study of the Australian gannet (*Sula serrator*), however, Warham (1958, p. 349) showed that the ' unemployed ' birds at his colony were, in fact, neither failed-breeders nor the off-duty mates of nesting birds; although in adult plumage, he concluded they were ' non-breeders through immaturity '—approximating to what we have described earlier as maiden birds.

At many of the great colonies of the Atlantic gannet (*S. bassana*) hundreds of unemployed birds are to be seen, usually occupying a traditional standing-ground (*see* pl. XI); and it is interesting that Gurney (1913, p. 342) suspected that some of those at the Bass Rock ' may have been merely practising a forced abstention, owing to the particular nest-sites on which they had set their wishes being already occupied by others '. By the 1930s, however, the likelihood of finding barren birds at a breeding colony had come to look rather improbable and insufficiently supported by fact, and was consequently rejected (Wynne-Edwards, Lockley and Salmon, 1936, p. 265). Seeing the matter now in a new light, our conclusion would be that not only gannets, but perhaps even birds generally, do not normally need to employ their full ' manpower ' in order to produce the year's recruitment quota; they can afford to carry a non-breeding adult reserve, absorbing it completely only on occasions when economic circumstances, or mere survival, demand an all-out effort. This is a simple extension of what we saw happening in the case of the individual female or family: these also commonly possessed a considerable reserve of potential fecundity, drawing upon it only to the partial extent required in the average breeding season or generation in order to sustain the population. A non-breeding reserve of adults, exerting a steady pressure to gain admittance to the breeding class, is likely to reflect each minor change in survival rates and economic conditions, and thus to be a sensitive, efficient buffer and stabiliser of breeding numbers.

Kendeigh (1941, p. 45) observed of the house wren (*Troglodytes aedon*) that ' the presence of the non-breeding, but potentially breeding, population seems to have an effect on territorial behavior of nesting birds and to cause them to be continually alert in the defense of their possessions '. Such external pressure, feeding back into the breeding population, may be

Table of durations of immaturity in birds.
(An asterisk * denotes an observation on a captive bird.)

Species	*Age at first breeding, etc.*	*Authority*
Struthio camelus (Ostrich)	Ad. plumage not later than about 4 years.*	Cronwright-Schreiner, 1925, p. 92
	3½ years (♀)	Mayaud, 1950, p. 539
Apteryx (Kiwi)	4¾ years to 1st. (infertile) egg* 5-6 years to maturity*	Robson, 1947, p. 1
	3-5 years (♀)	Mayaud, 1950, p. 539
Aptenodytes patagonica (King Penguin)	5½ years*	Gillespie, 1932, p. 128
Diomedea epomophora (Royal Albatross)	♂, 9 and 11 years, ♀, 8 or 9 years	Richdale, 1957, p. 108
Puffinus tenuirostris (Short-tailed Shearwater)	6 years	Serventy, 1956
Sula bassana (Gannet)	5 years to laying 1st egg*	Saunders, 1884-5, p. 159 (quoting E. T. Booth)
	Estimated 3 years to adult plumage	Gurney, 1913, p. 484
	Estimated 6 years to adult plumage	Saunders, loc. cit., p. 160
Phalacrocorax carbo (Cormorant)	Adult plumage in 5th year*	Bonhote, 1909.
Ciconia ciconia (White Stork)	(2) 4-5 (and more)	(*See* p. 566 for ref.)
Gymnogyps californianus (California Condor)	Assumed to be 5 years at least	Kofoid, 1953, p. 5
	12 years*	*cf.* Bent, 1937, p. 3
Vultur gryphus (Andean Condor)	9 years	Richdale, 1952, p. 134 from K. C. Lint, 1950 *Zoonooz*, 23:1-8
Aquila chrysaetos (Golden Eagle)	3½-4½ years to adult plumage	Jollie, 1947
Otis tarda (Great Bustard)	3 years to reach full size and plumage	La Touche, 1921, p. 16
Stercorarius parasiticus (Arctic Skua)	3-6 years	Davis, 1961, p. 164
Larus argentatus (Herring-Gull)	3 years (refers to one bird only, still with traces of dark band on tail)	Paludan, 1951, p. 119
	3½ years to adult plumage	Witherby, 1941, V, p. 93
Sterna hirundo (Common Tern)	1-4 years (mostly 3 or 4)	(*See* p. 567 for ref.)
Sterna macrura (Arctic Tern)	Minimum 3 years	Grosskopf, 1957 p. 66
Calyptorhynchus banksii (Red-tailed Black Cockatoo)	Mature in 4th year	Serventy and Whittell, 1951, p. 222
Ptilonorhynchus violaceus (Satin Bower-bird)	4 and 7 years to reach full plumage in ♂ *	(*See* p. 567 for ref.)

presumed to find its reflection in the quota of recruits they produce, so that the size of the non-breeding reserve will automatically be regulated.

The reserve is likely to form a valuable insurance against an emergency, enabling colonists to be mustered rapidly after a catastrophe, or advantage to be taken of new opportunities as they open up. The long-lived species will be the ones best able to afford it, because in them the average recruitment rate demanded is low, and correspondingly easy to meet.

Though the possibility of intermittent breeding by mature adults has been especially stressed, it seems altogether probable, as stated earlier, that the great majority of individuals inhibited from breeding by conventional social pressure are maiden birds. Much evidence has already been given, all pointing to the same conclusion that, in general, junior members of the adult hierarchy occupy a subordinate rank, and are therefore unlikely to be successful in winning nest-sites or territories in competition with previously established birds. In human populations, deferment of marriage, especially in women, is well recognised as having a restrictive influence on reproductive output; and the additional faculty this contributes to the marvellously versatile machinery of population-control is certainly one of the adaptive advantages of prolonging adolescence.

In order to give an indication of the extent to which adolescence can be prolonged in birds, a number of cases in which the period of prematurity is on the average three years or longer, as shown either by plumage changes or breeding records of birds of known age, have been assembled in the accompanying table. Though the list is not exhaustive, it is surprising to find how fragmentary our knowledge still is regarding the age of first breeding in long-lived birds. Half a century of bird-ringing has made it clear that this important vital statistic is very difficult to obtain. One of the complicating factors, of course, is this same fact that individual birds do not necessarily start to breed as soon as their reproductive organs are ready to function, and consequently the length of adolescence is individually variable, sometimes over a considerable range.

The final point to be made in this section is that, in so far as the length of adolescence is variable and governed by social competition, it must automatically also be density-dependent.

23.4. *Deferment of maturity in fishes*

Even small birds generally have a potential life-span of several years, though few individuals actually survive all accidents long enough to see it through to the end. Senile death therefore tends to be uncommon, and the expectation of life often rather short (*cf.* Lack, 1954, chap. 9): nevertheless, birds clearly belong to the category of what we have called perennial breeders; that is to say, once they become adult they are physiologically and bionomically adapted to go on reproducing season after season as long as they live. This is a common situation in the animal kingdom, just as it is among the plants, but there is of course a widespread alternative, in which the individual undergoes a period of growth and development that culminates in a single reproductive event, after which it dies. Though the life-span of animals in this second category is quite frequently longer than a year, the analogy with annual (and other monocarpous) species of plants is very close, and some at least of the advantages gained—such as overwintering by means of resistant zygotes—are in certain cases the same.

It is convenient to have a technical term to designate the second situation, and for this 'monotelic'—literally meaning once-reaching-maturity or perfection—seems appropriate. Monotely can be contrasted where necessary with the more perennial condition of 'polytely'.

None of the warm-blooded animals are strictly monotelic. It is true that some species of small rodents and insectivores come very near it, having life-spans only slightly exceeding twelve months, and including only a single breeding season, but the adults usually undergo a succession of oestrous periods, and their adaptations are immediately derived by contraction from the ordinary polytelic type of cycle characteristic of the majority of mammals. Primary monotelic adaptations are, of course, well represented amongst insects.

On the old hypothesis that natural selection must always favour the genes that enable the individual to leave the largest legacy of offspring to posterity, it would be expected that perennation would have a clear advantage over monotely, wherever conditions permitted it. The longer an individual lived and remained fertile, the greater could be its total fecundity. The fact that monotely is so common, and that evolution can obviously take place freely in either direction between it and the perennial state, is additional illustration of the inadequacy of the hypothesis.

For example, most fishes, including many Salmonidae, are perennial breeders; but the Pacific salmon, comprising six species of the genus *Oncorhynchus*, are all monotelic. Like the Atlantic salmon (*Salmo salar*), they ascend rivers to spawn; but having done so once they die. The Atlantic salmon is in an intermediate or borderline condition: most individuals spawn but once, though some survive as many as four or even five excursions into fresh water for the purpose of spawning.

It will repay us to study the case of *Salmo salar* in some detail. The information quoted here has been taken either from Menzies' two books (1931, 1949) or from the numerous papers by Menzies and MacFarlane on Scottish salmon rivers, published between 1915 and 1938, a bibliography of which may be found in Pyefinch's (1955) review of the literature of the biology of the Atlantic salmon.

The proportion of repeating spawners in the adult population differs from one salmon river to another, although, broadly speaking, in any given river it remains about the same level from year to year. Information on this point can be obtained from the scales, which reveal not only the age of the fish, and the number of years it spent in fresh water as a parr before going down to the sea as a smolt, but also whether it has ever previously returned to fresh water as a mature adult: in that case the 'spawning mark' on the scale is unmistakable. Of the Scottish rivers, those on the east coast contain comparatively few fish spawning for a second time: in the Aberdeenshire Dee, in the four years 1922-25, the figure ranged between 1·0 and 2·8 per cent, among samples of many thousands: 0·1 per cent had two previous spawning marks. In the Spey, not far away, the proportion was distinctly higher, averaging 4·9 per cent; and 0·2 per cent were spawning for a third time. But on the west coast, for example in the River Add in Argyll, 34 per cent of the breeding salmon were found to have spawned once already, 6·5 per cent twice, 2·9 per cent three times; one fish was found with four spawning marks, indicating that it was making its fifth visit to fresh water. In the Kirkcudbright Dee, the corresponding figures were 12·3, 1·3 and 0·15 per cent. The vast majority of previously spawned fish are everywhere females.

The diminishing figures for the successive returns on the whole suggest a geometric progression, and this implies a roughly constant survival rate. In the last case given, the sample breaks down as follows.

	Maiden fish	*Previously spawned*		
		once	*twice*	*3 times*
Percentage	86·25	12·3	1·3	0·15
Survival rate	—	1:7·0	1:9·5	1:8·7

In the stocks belonging to different rivers the survival rates of adults undoubtedly differ, due presumably to differences in fishing pressures (especially by commercial netsmen), and in natural hazards (arising for instance from the length and difficulty of the upstream migration entailed). Growth-rates also differ between different stocks, some rivers like the Tay being noted for the greater size of their fish, compared with salmon of the same age from other waters.

The eggs deposited in the gravel-beds of the spawning streams hatch in early spring, and in due course the larvae become parr. Parr ultimately metamorphose into silvery smolts, 10 cm or rather more in length, and descend to the sea about May, but this can occur at the age of $1\frac{1}{4}$, $2\frac{1}{4}$, $3\frac{1}{4}$ years or even later (especially in far northern latitudes), depending on growth-conditions in the river concerned. Not all the parr of one spawning or year-class are as a rule ready to metamorphose at the same time. The newly laid eggs vary in size, and even from the time of hatching some fish are larger and probably grow faster than their companions: consequently the most precocious may be ready to leave the river one, two, or more years ahead of the slowest developers.

The smolts grow rapidly in the sea until they are approaching maturity, when they return to the river for the first time, entering fresh water at any season between early spring and late autumn. Virtually none come back until they have spent a year in the sea, but their first return can take place after any interval, from one up to four years (and occasionally longer), after their original departure as smolts. In other words, the age of first breeding is very variable indeed. Hutton (1924, p. 52) worked out, in a representative case, that the progeny of the 1913 spawning in the River Wye, hatched in 1914, returned to breed over a span of no less than six different years, from 1916 to 1921. The yearly showings were as follows: 1916, 0·03 per cent; 1917, 7·3 per cent; 1918, 54·2 per cent; 1919, 35·7 per cent; 1920, 2·7 per cent; 1921, 0·07 per cent. Moreover, when in turn he analysed the age-composition of their parents—the 1913 spawning stock—he found they were similarly drawn from eight different year-classes, hatched between 1904 and 1911.

Fish like the salmon that individually lay thousands of eggs and leave them to their fate generally have less control over the numbers of recruits produced than have the more slowly reproducing species. Salmon parr are known to exhibit an individual-distance reaction to one another, and certainly have considerable control over their own population-densities (*cf.* Kalleberg, 1958), but in practice the recruitment of smolts is not very constant from year to year, and there are consequently good smolt years and poor ones.

It is easy to see that the variable age of maturity in the salmon is an excellent buffer or stabiliser against such annual irregularity: instead of a bumper smolt year producing, a fixed number of years later, an exceptional spawning stock, or a disastrous smolt year resulting in almost no spawners at all, the divided return of each year-class has the same kind of effect as we employ, when making a smoothed curve in a graph, by taking a series of running averages over, say, three years at a time. By the device of recruiting the breeding stock from five to six year-classes at once, extreme fluctuations must be considerably damped. Over the period 1914-1946, for example, the largest annual catch of salmon in Scotland was barely three times the smallest (Menzies, 1949, p. 79). The year's catch, it shoud be added, is not necessarily a reliable index of the ultimate spawning stock, although it probably serves well enough to show the general magnitude of annual fluctuations in the numbers of incoming migrants.

The smoothing effect on recruitment resulting from variability in the age of individuals attaining maturity would appear to be a general advantage, of adaptive significance to many kinds of animals, but especially valuable in monotelic species, such as the Pacific salmon.

The remaining homeostatic processes regulating the populations of salmon are still very imperfectly known, and apparently complex. There are indications that late-developing smolts tend on the average to return sooner to the rivers, after a shorter sojourn in the sea; also, on the average, small female salmon are less fecund than large ones, though exceptions occur. It has long been known that some male parr become fertile before they ever go to sea at all, and, weighing as little as $\frac{1}{2}$ oz (15 g), may shed their ripe sperm over the eggs of spawning females. The proportions of grilse to larger salmon vary much from river to river, and records show that in some rivers they have undergone long-term changes. Some of the fish spawning for a second time come back in the season following their first spawning, some stay away longer and miss a year. Among all this confusion, however, adaptations are discernible (*e.g.* the gradations in egg-size, in rate of development, and in body-size) resembling those that elsewhere have been shown to contribute to the regulation of numbers, and to be controlled through the social organisation. Certainly there is no indication at all that selection in the salmon has pursued the narrow path of inducing the maximum possible fecundity in the individual.

There are of course species of fish that mature more slowly than the salmon. The ammocoete larvae of the lampreys live and grow in fresh water for five or six years before metamorphosis into adult lampreys, both in *Lampetra* and *Petromyzon* (MacDonald, 1959). European eels (*Anguilla anguilla*), developing from eggs spawned in the Sargasso Sea area, spend about three years as larvae on their journey to the coasts of Europe: having metamorphosed and entered fresh water, they take up their abode there, and feed and grow for several years. All freshwater eels are monotelic, and by the time they ultimately develop the silver livery and migrate back to the sea as maturing adults the males (in *A. anguilla*) are in the 8th to 14th years of life, and the females (much larger) in the 10th to 18th (Bertin, 1956, p. 94). Individual differences in the length of adolescence are no doubt mainly due to differences in environment, but the smoothing effect on the annual production of spawning eels must be the same as it is in the salmon.

A still more variable period of adolescence occurs in the halibut (*Hippoglossus hippoglossus*). Various workers have shown that males may first reach maturity at any age between five and seventeen years, and females between seven or eight and eighteen (*cf.* Rae, 1959, p. 7). This variability is probably a general phenomenon in long-lived fishes; even the herring (*Clupea harengus*), which matures more quickly than any of the species so far mentioned, may take from two to eight years to attain its first spawning; though in any one spawning-stock the vast majority of new recruits usually come from only one or at most two or three year-classes. The herring is a typical polytelic breeder, and this in itself tends to smooth out the annual variations in recruitment, by bringing together members of a number of year-classes in every spawning group.

We should expect from our theory that promotion from adolescence to the status of breeder would be governed by the social organisation. Where the stock of senior fish in the population is adequate to provide the reproductive output required to meet the needs of the population in any particular year, junior fish will presumably be inhibited from maturing, just as we found in the birds. One of the expected consequences of exploiting a fishery, which lowers the proportion of big, old fish in the stock, will consequently be to bring down the age or size at which maiden fish are allowed to mature. Intensive fishing will, in other words, tend to mobilise the reserve of young potential breeders, in an attempt to keep up the recruitment rate in the face of increased mortality.

23.5. *Deferment of maturity in mammals and reptiles*

Prolonged adolescence, usually coupled with indeterminateness in the age of first breeding, is also known in mammals, where, as in birds and fishes, it is to a greater or less extent characteristic of the longer-lived species. Captive female Indian elephants, for example, generally bear their first calf at the age of fifteen or sixteen years, but there are cases in the literature of calves being born to mothers eight and nine years old: males similarly are usually ready to reproduce at fourteen to fifteen, but in exceptional cases as early as nine (for references, *see* Frade, 1955, p. 766). There is a single old report of an Indian rhinoceros (*Rh. unicornis*)—a male—that exhibited the first symptoms of puberty in its tenth year (*cf.* Gee, 1953a, p. 344). Humpback whales (*Megaptera nodosa*) in Australian waters reach puberty between three and seven years of age, the majority being four or five: five is the commonest age in the North Pacific (*cf.* Chittleborough, 1960). In the female harp seal (*Pagophilus groenlandicus*) maturity, as marked by the first ovulation, is reached between the ages of five and eight, with a peak at six (Fisher, 1956, p. 514). Walrus (*Odobenus rosmarus*) must also be five or six years old before they can mate (Freuchen, 1935, p. 219; Scheffer, 1958, p. 26); but in some of the other pinnipedes, including even the southern elephant-seal (*Mirounga leonina*), the females are mature at two. In the polygamous seals the males do not of course take much part in reproduction for some years after they become mature; in *Mirounga*, for instance, the bulls do not normally begin to intercept females on the beaches in South Georgia until their sixth year (Laws, 1956, p. 46). As a result of sealing operations at South Georgia, younger bulls have now come to play a more important part than they would do if older bulls were allowed to survive; which illustrates the point just made regarding the effect of intensive fishing and shows clearly the value of having a non-breeding reserve to be called into service when mortality from uncontrollable sources becomes severe. Probably in most of these mammals, as in the last case, once puberty is reached the actual age of first breeding is decided for each individual by the social pressures existing at the time.

Among the reptiles, turtles lend themselves especially to close study and experimental marking, and many of them are of course notoriously long-lived; in fact it can reasonably be assumed that the longest life-spans to be found anywhere among the present-day higher animals occur in this group. Risley (1933, p. 706) found that, in the American stinkpot turtle *Sternotherus odoratus*, whereas the males were adult by their third or fourth year, the females did not mature until their ninth to eleventh. This occurred in a population with a decidedly unbalanced sex-ratio—possibly in no way abnormal for the species—of one male to 2·3 females; and it might be suspected that the presence of such a large surplus of females in a monogamous species could have something to do with the sexual difference in the age of entering the breeding caste.

The diamondback terrapin (*Malaclemys terrapin*), a brackish-water species much in demand as a table-delicacy, has been artificially bred and raised in pens; and under these conditions, at least, the sex-ratio has also proved very uneven, namely one male to 5·9 females, among 1,442 adults (Hildebrand, 1932, p. 562). The males, much smaller than the females, scarcely ever attain marketable size, but both sexes nevertheless appear to reach maturity at about the same age in this case : in the females usually at seven years, although individual first egg-layings were recorded from all age-groups between four and eight (the same, p. 561).

In the box-turtle (*Terrapene carolina*) Nichols (1939, p. 20) obtained evidence that the lower limit of sexual maturity in males (on Long Island) was about twelve to fourteen years. Long ago Louis Agassiz (1857, p. 491) reported that females of the painted turtle (*Chrysemys picta*) did not contain mature ova until they were eleven years old, although they had been seen to copulate at seven; and that the eleventh to the fourteenth year is about the normal age at which most species of freshwater turtles in North America lay their eggs for the first time (loc. cit., p. 496). His age-determinations were based on the annually-ringed scales of the carapace, which in the strongly seasonal climate of the north-eastern United States are fairly reliable in early life at least, for a considerable variety of species in this, the world's richest fauna of freshwater Chelonia.

It will occasion little surprise to find that some of these slowly maturing species lay very small clutches of eggs—five to seven in the case of *C. picta* at Cambridge, Massachusetts, according to Agassiz; usually four to five in the box-turtle; two to four in the Old World Grecian tortoise (*Testudo graeca*); and only one in the curious soft-shelled North African species *Testudo tornieri* (*cf.* Carr, 1952, p. 13). Like the long-lived birds and mammals mentioned in Chapter 21 (p. 488), such well-protected and perennially-breeding chelonians require only a relatively low fertility to sustain their numbers, even though in this case they do not stand guard over their eggs.

It is safe to say that in the majority of vertebrates maturity is reached within one or two years of birth, but it is clear from the examples given in this and the two preceding sections that there are many exceptions in which adolescence has been secondarily prolonged. In each class that we have considered there have indeed been one or more extreme instances in which normal immaturity lasts ten years or over.

We may with advantage go back to our question at this point and see what further light can be shed on the function of this retardment. Thinking especially of the mammals, which cover such a wide range in this respect, it might appear superficially that it is little more than a consequence of increase in size, and that prolongation of development becomes a physical necessity in large animals because they have got so much farther to grow. We could hardly expect that the body of an elephant, for example, ' could be built up like that of a mouse in three weeks ' (Weissmann, 1899, p. 7), and it is in general true that there are more slow-maturing species among the large mammals than among the small; but any primary connexion breaks down on close examination (*cf.* Weissmann, loc. cit.; Mitchell, 1912, p. 47). Among the larger mammals the ruminants, hippopotamus, large cats and many of the seals and Cetacea, are full grown in a fifth to a half the time it takes an elephant or a man to reach adult stature. Just the same sort of loose correlation exists between body-size and total life-span, but no one would want to argue from this that large physical size *per se* necessitates a long life.

Youth is rightly regarded as being a period of education and preparation for adult life; and it might seem an advantage therefore to prolong it, especially in the very long-lived vertebrates that are generally governed predominantly by tradition, and have consequently so much to learn on that score alone before they are qualified to tread without deviating the paths of convention, and conditioned to respond correctly to every change in the state of their economy. But that this can be the sole, or even the main reason for prolonging youth, is nevertheless difficult to accept. Among the large but quick-growing mammals just mentioned there are several noted for their sagacity and complex traditional conservatism, that nevertheless achieve adulthood in two or three years. It seems impossible to resist the conclusion that, by comparison with a lion, an antelope, or a whale, man is exceptionally retarded, not only in physical growth but also in mental development. It is not difficult to recognise, on detached introspection, how surprisingly his helplessness in infancy persists, and later his childish irresponsibility and slowness to learn, compared with the rapid, efficient progress of discipline in these more precocious mammals. True, he has more to learn; but this argument applies with less force to the elephant, which shares with man the trait of taking four or five times as long over its apprenticeship as a horse or a buffalo.

It may be noticed next that growth generally continues in mammals for some time after puberty; and because of their inferior size, the youngest adults automatically merit a lower social status than those in the prime of life. We have repeatedly seen in the foregoing pages, either that the fecundity of these young adults is not exploited at all, or that reproductive success is lower with them than with adults already advanced to senior status. In the same way, in man, although pubescent members of both sexes are capable of reproduction, to consummate marriage at the first opportunity has in many societies and ages besides our own been rigidly prevented by custom, and often expressly forbidden by law. Nowadays we may seem to perpetuate this restriction mainly, perhaps, to prevent rash marriages that riper wisdom would later repent, but it really derives from earlier times when the chief value to the community of postponement of marriage lay in the limitation it imposed on the number of householders, and on the birth-rate. This is something especially to be desired in agricultural races whose smallholdings cannot be any further subdivided or multiplied without over-exploiting the land and leading to degradation and want: races, that is to say, already living at or above the optimum density (*cf.* Carr-Saunders, 1922, p. 264). The exact converse was, until very recent times, to be seen in the great American West (*e.g.* Texas), where the rapidly expanding pioneer communities, far from imposing any restrictions, tended rather to extol the manly precocity of those who entered into teen-age marriages.

Our conclusion from what we saw in birds, that there was frequently a non-breeding ' floating reserve ' of sub-adults and young adults ready to be called into action on demand, appears to apply more or less widely to the remaining vertebrate groups. We have noted also, wherever the facts have been available, the correlated tendency for variation in the age of first breeding, in species with long adolescent periods; and there seems no reason to doubt the additional value of this in smoothing out irregularities between good breeding years and poor ones, when it comes to the stage of adult recruitment.

We have still not managed to explain, however, what there is to be gained by the great prolongation of the earlier, ' pulline ' period of life, before the age of puberty is reached at all, and when reproduction is not in any circumstances possible. There can be no doubt that this stage is secondarily prolonged in many slowly-maturing species—especially in the mammals—and that there must be some positive adaptive advantages in it. The possibility that it simply allows more time for education has to be set aside on the grounds that mental development is often as much retarded as physical, at least in the mammals, where education is of most importance. It may be significant, however, that the pulline mammal is first and foremost a dependent, and that considerable responsibility for its welfare falls on the parental generation. Even though weaned, the burden of one or more followers, depending on their mother as leader and protector, is probably one of the factors impairing her ability to achieve another pregnancy. The alternative would be for her to default in her duties as a guardian, and, by abandoning her dependents, prejudice their survival.

This is a stage when life is still cheap and expendable in the other vertebrates also. It seems not illogical, therefore, as a tentative possibility, to guess that the longer the period of dependence is extended, the more will be the scope allowed for regulating ultimate recruitment to the breeding caste, either by conditioning the adults against too high a fertility rate when there are already plenty of dependents in the population, or by allowing a still longer period in which to thin out the recruits by means of controlled, socially-induced mortality.

23.6. *Deferred maturity, or prolonged larval life, in arthropods*

We have often found in earlier chapters that no sooner had we identified some dispersionary adaptation, developed and highly perfected in the vertebrates, than we turned to find the same device quite independently evolved in the arthropods. The prolongation of pre-adult life provides another such parallel; and in a phylum where body-size tends on the whole to be rather small, and the turnover of generations annual if not even faster, examples can be found in most if not all the major classes, including the Crustacea, arachnids and insects, of adolescent or larval life extended up to seven or more years. Factual data are scantier here than in the vertebrates, and most of them are derived from animals in captivity or under other experimental conditions, which may not always faithfully reflect those of a wholly natural environment.

Among the larger Crustacea, even in commercially valuable species like the European lobster (*Homarus vulgaris*) and edible crab (*Cancer pagurus*), our knowledge of the age of first maturity rests only on circumstantial evidence, being in the first case probably seven or eight years, and in the second, three (Orton, 1936, p. 686), four or five (*cf.* Graham, 1956, pp. 183, 190; Pearson, 1908, p. 187). Both these species are perennial breeders, and may live several decades. The initial planktonic larval stage is relatively short, so that by far the greater part of their early lives are spent as adolescents, resembling adults in all except size and the relative development of the great claws, which very likely determine their social status. Growth, fastest in the earlier years, probably continues with each moult throughout life, though latterly at a decreasing rate.

Not all decapods are as long-lived as these, or as slow to develop. In the common prawn (*Palaemon serratus*), for example, ' the bulk of the female population [at Plymouth] spawn before their first birthday ', and few live into a third year (Forster, 1959, p. 626; in North Wales the tempo is somewhat slower, *cf.* Cole, 1958). In another, smaller prawn, *Spirontocaris lilljeborgii*, the usual life-span for males appears to be eighteen months, with a possible maximum of thirty months; and for females, twenty-four months, although a few appear to live thirty-six months (Pike, 1954, p. 743). The shore crab (*Carcinus maenas*) and squat-lobster (*Galathea squamifera*) both likewise mature at about twelve months old; the latter appears to live two years, or rarely three (Orton, 1936, p. 680 and Pike, 1947, p. 120).

Turning to the arachnids, the spiders and probably the scorpions also again include species with a secondarily prolonged pre-adult stage. This lasts about four years in the purse-web spider *Atypus affnis* ; and the fact that in this species ' eighteen months normally elapse between mating and the emergence of the young from their mother's nest ' (Bristowe, 1958, p. 78) rather suggests that here too, as in many of the mammals, what we there called the pulline stage, when the young are dependent on their parents, is correspondingly lengthened. *Atypus* is not strictly monotelic: Bristowe states that it may survive to breed a second time, in which case the whole life span can extend to seven or eight years.

The adolescent period is longer still in the tarantulas (*Aviculariidae*). In a species common in Arkansas (by inference *Dugesiella hentzii*), the male takes eight to ten years and sometimes more to reach maturity. It appears normally to enjoy one season of fertility before dying, though in the laboratory males have been kept alive as long as twenty months after they became adult. Females have matured in ten years in the laboratory, but in the field they probably take eleven or twelve according to Baerg (1945, p. 106), by whom these vital data were worked out. Unlike the males, adult females continued to moult at intervals and thrive more or less indefinitely: when he wrote, Baerg had one aged twenty, still in excellent health and vigour, but in the laboratory his females produced eggs only for one or two years, and afterwards remained barren.

These developmental time-scales are again evidently much above average for the class. A good many spiders appear to be typical annuals, and others, after a rapid development, survive their maturity by only one year or two (Bristowe, loc. cit., p. 68). Bristowe observes that, while there is a tendency for larger forms to grow up more slowly and live longer, this is not an invariable rule, any more than it was in the vertebrates we considered earlier.

In their social and dispersionary adaptations the arachnids and Crustacea have evolved to a stage roughly parallel with that of many vertebrates; and it seems therefore not improbable that the same kinds of selective advantage as were suggested earlier on p. 576 may have contributed to the evolution of their adaptations.

The insects at first sight present a rather different picture. Here we have what is in some evolutionary respects the most advanced and successful class of animals, almost entirely commited to the monotelic habit. Obviously one of the keys to their success, except in the lowest groups, is the power of flight, to which their morphology and life-history tend very largely to be subordinated. As we have had reason to notice earlier (p. 457), insect wings are generally delicate unprotectable structures, easily damaged and impossible to repair; nor have any insects acquired the power of regenerating wings at a subsequent moult. Only the Ephemeroptera have two winged stages in the life-cycle, the subimago and imago, and in their case the transformation is achieved by merely sloughing off a thin outer wrapping, which at first invests the wings, at the time of the final moult. Elsewhere the ability to fly invariably pertains to the last instar alone, and is no doubt generally of special value in facilitating pair-formation or mating, and in allowing the female to seek the appropriate nidus for her eggs. More than anything else, it seems to be the wings that deny the majority of the Pterygota the chance of prolonging adult life beyond a single season.

This conclusion is at once confirmed by considering the few exceptions to the rule of having only one season of adult life. The most notable of all occur in the most advanced social insects, where polytely in the royal caste has evolved three times independently, in the ants, honey-bees and termites.

The queen honey-bee retains her wings intact to the end of her life, which may extend to five years or longer (Butler, 1954, p. 74); and she is consequently always capable of flying off with a swarm of workers to establish headquarters elsewhere. But once her original nuptial flight is over she will not take wing again unless the colony swarms.

The queen ant and the royal termites each go a stage farther (presenting in this one of their many remarkable instances of parallel evolution) as in both cases they get rid of their wings immediately the wedding flight is over. Thereafter they may live to a very advanced age, known to be not less than fifteen years in the ant *Formica fusca*, and from fifteen possibly to fifty or more in some of the higher termites (*cf.* Imms, 1957, p. 384).

Neither termites nor ants possess and make use of wings except for mating and colony-formation. Highly successful as they are, they have become secondarily flightless in their everyday life, and no limitation is therefore imposed on individual survival by dependence on frail wings. Even the sterile castes may live two to four years, in some of the higher termites at least (Imms, loc. cit.).

In all three groups the queens, once they are established in their colonies, live in the equable environment of the nest, nourished, tended and guarded by the workers in order that their fecundity may be sustained. Conditions are thus exceptionally favourable for survival and in fact the oldest royal individuals in the ants and termites are thought to outlive all other adult insects.

Apart from these highly specialised social insects, perennation of the imago stage appears to be confined, so far as is known, to some of the Coleoptera and Thysanura. The latter, commonly known in this country as silverfish, are primitively wingless, and retain the capacity for moulting every few weeks or months throughout life, long after maturity is attained. *Ctenolepisma longicaudata* reaches maturity at 2½-3 years, and in Melbourne, Australia, normally breeds for at least four more years, making a life-span of seven years (Lindsay, 1940, p. 75). The beetles, comprising the largest and at the same time one of the most advanced orders of insects, are also in general the most heavily armoured and indestructible. Many of those that have wings use them very seldom; others have lost the hind wings and power of flight completely, their elytra in that case often being fused and immovable. It is one of these last, the cellar beetle *Blaps gigas*, that at present holds the longevity record for adult Coleoptera, namely nine years five months (Labitte, 1916, p. 110). This was achieved in a terrarium, into which eight individuals had been introduced in the summer of 1905; the seven others also all survived until they were unfortunately devoured by rats after 5½ years. Labitte obtained longevity records of three other species of *Blaps*, in which the oldest survivors lived 4½, 3¼, and 3 years respectively. 'Planned survival' in this genus, therefore, evidently has a potential maximum of several years. Whether the adults remain potentially fertile as long as they live is not known, but at least it may be expected that this would be so, otherwise their survival would be a useless burden to the population. Labitte, who kept hundreds of beetles under observation over a period of twenty years, lists eleven other species with adult life-span maxima extending into a third summer, including *Carabus auratus*, *Dytiscus marginalis*, *Copris hispanus*, *Geotrupes stercorarius* and two species of *Akis*. (Additional records, some of them pertaining to the same or closely related species, have been assembled by Comfort, 1956.)

Very much commoner in insects than the perennation of adult life is the lengthening of the larval stages. It must be made clear that in some cases this is probably principally due to the harshness or poverty of the environment, which limits the growth-rate so drastically that it is impossible to attain adult stature within a single season. For instance, in central Baffin Island in the summer of 1950 we found the woolly-bear caterpillars of the lymantriid moth *Gynaephora rossi* very common and active on warm days: they were present in all stages of growth right from the time the snow melted, and we concluded that it takes them at least three of the brief summers in that high latitude to get through all their larval instars and pupate. They are hardy enough to be frozen in their tracks, as it were, and quite capable of resuming the digestion of the same meal when they thaw out again, presumably nine months later when necessary. In a very different environment, some of the woodworm larvae of timber beetles continue an even slower development in dry wood, on a very meagre plane of nutrition, sometimes for immense periods; the emergence of adults of *Eburia quadrigeminata* from American oak timber, for example, up to twenty and even forty years after it has been sawn, seasoned and made into furniture, is well authenticated (*cf.* Hunter, 1959). Though not perhaps to such an extreme extent as this, an element of the same kind of nutritional stringency may enter into the deferment of maturity in some of the other insects to be mentioned below.

A good many of the Coleoptera have their larval stages prolonged for three or more years, particularly perhaps in species spending the growing period underground. Familiar examples are the larger cockchafers and June-bugs (Melolonthidae), most of which take three years from egg to pupa, the adults emerging in their fourth summer of life; *Melolontha hippocastani* sometimes requires five years (*cf.* Richter, 1958, p. 321). Species of *Pleocoma*, another scarabaeid genus, require still longer, perhaps even eight or more years (*ibid.*). Not unexpectedly, most of the longer life-cycles that have been closely studied are those of destructive pests, among them the wireworms—the larvae of click-beetles (Elateridae). These also live underground, and this makes their habits and behaviour difficult to observe. In the common genera *Agriotes* and *Athous* the eggs are laid in the spring, and the larvae feed and grow several years, finally pupating and transforming into beetles, which overwinter before they reproduce. Authorities differ on the exact length of larval life; according to Stapley (1949) it is four years, the adults emerging as a rule the following (fifth) spring. Roberts (1919, 1921) concluded that the length of the larval stage is, or may be, six years (in

Agriotes); and Thomas (1940, p. 10), referring to the species in the eastern United States, states that most require from three to six years to complete their whole cycle. Though in part these differences relate to different species, there is some evidence at least that the duration of larval life within a single species may differ from one individual to another even in the same habitat (*cf.* Roberts, loc. cit., p. 197): this would be in accordance with the parallel situation we have already found in the vertebrates.

Similar indefinite periods of development probably occur in several of the other insect orders. Among the dragonflies (Odonata), Corbet (1959, p. 243) found that the large *Aeshna cyanea* usually takes three years from egg to maturity. The eggs overwinter before they hatch, and the larvae then have two summers in which to grow (and two winters to survive), before they emerge as adults in July or August. Various related species, including *Aeshna grandis* and *Brachytron pratense*, take as long or longer. ' Probably no species in Britain regularly requires more than five years to complete development, although there are indications that sometimes *Cordulegaster* and *A. grandis* may be exceptions ' (Corbet, Longfield and Moore, 1960, p. 82). As these authors point out, environmental temperatures and food-supply can be expected to influence the rate of development, but it should be noted that other equally large dragonflies reach adult status in similar (sometimes the same) environments in two years (*e.g. Anax imperator*) or only one (*e.g. Aeshna mixta*, and a minority of *Anax imperator—see* Corbet *et al.*, p. 140); so that the speed of development would appear to be adapted in part at any rate in response to selective pressures quite independent of food-supply, water-temperature or body-size.

Another group with long-lasting freshwater juvenile stages is the Ephemeroptera (mayflies): in some species the nymphs are believed to take three years to grow up to the stage of metamorphosis, in sharp contrast to their spans of adult life which may be measured in hours rather than days (Imms, 1957, p. 284) and are thus among the shortest known in the higher animals.

Next we may cite two more wood-boring larvae of different orders namely the great wood-wasp (*Sirex gigas*) among the Hymenoptera, and the goat moth (*Cossus cossus*) in the Lepidoptera. Both have larval periods varying between one and three years (even four in *Cossus*), and it is suspected that, as with the timber beetles, the duration is conditioned in part by the water-content of the wood on which the larva feeds: the drier it is the slower the development (*cf.* Chrystal, 1937, p. 105). A smaller relative of *Cossus*, the leopard moth (*Zeuzera pyrina*), which also has a wood-eating larva, completes its development in either two or three years.

It is clearly impossible to decide on present information how much of the slowing down of larval growth is due to deficiencies of food and water, and how much of it has been ' cultivated ' by natural selection, on account of the contribution it may make, as earlier suggested, to stabilising numbers. Undoubtedly one of the great disadvantages inherent in the monotelic, annual sequence of generations is the erratic fluctuation in the annual crops of adults it is certain to produce, in all but the most equable environments. The phenomenon we have customarily referred to as recruitment can hardly be said to occur in an annual species: there is no standing population, but instead it is reborn each season, its size depending partly on the fecundity of the previous generation and partly on current environmental conditions. Insects certainly appear liable to fluctuate in numbers more sharply from year to year than any other higher animals familiar to us; and it seems not improbable that for this their short life-cycles and rapid turnover are in no small measure responsible. As a class they lack the population stability that characterises long-lived perennial animals to be found in the higher vertebrate classes, or even in the Crustacea and Mollusca.

If this is true, it may be expected that larval periods—extended in the first place, perhaps, because the habitat offered no choice—may become purposely variable in extent within a single brood of progeny, so that members of one age-group are prevented from all emerging as adults the same year: in this way the annual crop of parents can be drawn from the progeny of several antecedent generations. It may be anticipated that larval growth-rates will prove in some cases to be density-dependent, so that life-cycles are completed more quickly when numbers are low, but are more spun out and more variable in duration when they are high. The latter would produce a smoothing effect on adult emergences from year to year, analogous to what was recognised earlier in this chapter in the case of the salmon (p. 576).

There is another advantage in having the duration of pre-adult life variable if it has in any case been forced to extend over two or more years, on account of the coldness or barrenness of the habitat in which the species is adapted to live. This is to avoid the danger of allowing the species to become broken up into different breeding stocks completely isolated from each other by a time factor: odd-year stocks and even-year stocks, for instance, in a biennial species. It could lead to a species being particularly abundant every second or every third or fourth year, depending on the length of adolescence and the size of the parental population from which it came. This kind of reproductive isolation is on general grounds undesirable. Where it occurs, the habitat will be required to support two (or more) parallel populations, depending for their continued survival on one common set of resources, yet regulating their population-densities in more or less complete isolation, and unable to reinforce one another in conditions of adversity. They might easily become, indirectly, mutual competitors. One year-stock may die out locally, or over a wider area, leaving the resources of the habitat unclaimed in the leap years that resulted, and allowing other competing species to encroach uncontested on the use of them. Lastly, gene-flow is even more to be desired between populations partially isolated in time than it is between those isolated in space, since they all require to be adapted to identically the same environment.

The need for some degree of overspill from one year-group to another in populations that tend to be temporarily isolated has of course been recognised by previous authors, notably by Dunbar (*cf.* 1957) in his studies of such arctic plankton organisms as *Themisto libellula* and *Sagitta elegans*, which require two years for their life-cycles in cold northern waters. In temperate latitudes they generally have annual or even more frequent generations. There are certainly some cases where temporal isolation really is almost complete. One that has come to my knowledge concerns the noctuid moth known as the northern dart (*Agrotis hyperborea*), which also has a biennial life-cycle: it is a Scottish rarity, much sought by collectors, but ' the pupal stage is reached and the imagines found only in the even-numbered years ' (Tod, 1953, p. 12). Only two specimens collected in odd-numbered years were known to Tod, both taken long ago (1839 and 1873), and all recent attempts to discover even a limited odd-year emergence have been unsuccessful. A better-known case is that of the sockeye salmon (*Oncorhynchus nerka*) in the Fraser River in British Columbia, where, until the disastrous blockages of the Canyon section in 1912-13 largely cut off the upper spawning grounds and put an end to the cycle, there was a remarkable four-yearly recurrence of a dominant generation (the class of 1901, 1905, etc.), estimated to be over a hundred times as numerous on the average as the off-cycle runs in the same upper tributaries: here in the big years over 99 per cent of the spawners were four years old (Ricker, 1950). The age-spread generally typical of spawning salmon had locally become so reduced as to be incapable of rapidly restoring a balance between the year-classes.

There is of course one famous insect, or genus of insects, which appears to have specialised and exaggerated this trait of intermittent appearance, namely the periodical cicada (*Magicicada septendecim* and its two relatives) of eastern North America. The four presently recognised forms all spend their nymphal life underground, mostly in hardwood forest habitats, where they sometimes occur in prodigious numbers. In typical *M. septendecim* and *M. cassini* development takes seventeen years, but each species has also a 13-year race with a generally more southern distribution, although there is a considerable overlap in latitude: one of these shorter-period forms is called *M. septendecim tredecim*, and the other, recently discovered, is as yet apparently unnamed. Their life-cycles are maintained with an extraordinary degree of precision. Although under certain conditions odd individuals may be accelerated or retarded, and appear in the season just before or just after the main swarm, the integrity of the particular ' brood ' in question seems not to be thereby impaired. In Marlatt's classical paper (1907) thirty broods were recognised, each identified by what one might describe as their 'golden number ' (giving the sequence of emergence-years) and by the geographical area occupied. Brood XIV, for example, appears to have maintained its integrity since the Pilgrims first observed it in 1634, and its next appearance in 1974 can be confidently predicted. It is the same with the other broods except that history does not trace them back so far. There must clearly be very effective adaptations inhibiting successful reproduction by individuals that emerge out of step.

The next point to note is that, in spite of the vast swarms characteristic of emergence years—at least until modern civilisation began to whittle down the size of many of the broods—the winged adult cicadas are more or less sedentary, and show no tendency to travel far from their natal place or colonise new regions. Each brood has indeed a more or less sharply circumscribed territory; and it is rather typical that different broods may share a common boundary without overlapping, so far at least as the typical *septendecim* is concerned. This may be true also as between *septendecim* and *tredecim* in regions where both occur: Dr. Monte Lloyd, who has given me valued instruction on the subject of periodical cicadas, tells me that the Middle West and Mississippi valley will see in 1963 the simultaneous emergence of 17-year brood III and 13-year brood XXIII, an event that recurs only at intervals of 221 years, but, as far as is known before the event, the territories of these two broods will be mutually exclusive although exactly contiguous.

This strange set-up contains yet another remarkable feature. Over parts of the range *M. septendecim* a smaller, closely related species lives with it, *M. cassini.* There has been much controversy in the past regarding its taxonomic status, but its specific distinctness is now fully established (*cf.* Moore and Alexander, 1958): it is not only smaller, but has differences in the genitalia and a completely distinctive song of its own; it also associates with others of its own kind, and, in Indiana, showed different preferences for the kind of trees it collected in (Jacobs, 1953). The earlier reluctance on the part of some entomologists to accord it specific rank arose from the fact that it can also have either a 17-year or a 13-year cycle, which is always exactly in step with that of the *septendecim* brood whose territory it shares. The two species are rigidly interlocked in time.

In trying to understand the survival value of these extraordinary adaptations we must not neglect first to observe that, even among the examples that have filled the preceding pages of this section, *Magicicada* finds no equal for the length of its life-cycle. It has the slowest turnover of generations we have yet found among monotelic animals. Secondly, there must be a special advantage in high synchronisation, since we find two species being kept in perfect step by natural selection. Third, no intermediate appearances are tolerated—no intrusions to break the long succession of blank years by emergents coming from other broods, whether sympatric or immigrant: for natural selection has also provided the machinery—whatever it is—for keeping the broods geographically apart. It is still the current view, Dr. Lloyd informs me, that the primary goal achieved is emancipation from all specialist parasites and predators: no exploiting species is ever likely to succeed in locking its time-scale to theirs. Their emergence is invariably therefore a surprise to potential ' enemies '; and the vast horde that suddenly appears on the scene completely floods the capacity of all predators and parasites to take serious advantage of them, or endanger their reproductive success.

This explanation, first put forward by Marlatt (loc. cit., p. 13; *see* also Beamer, 1931) is not necessarily the final one; nor is it in some ways wholly plausible, but it is still the best that offers.

Magicicada presents us with another example of natural selection having made capital, as it were, out of inescapable difficulties and defects. We thought we detected a somewhat analogous case—and perhaps a more successful venture—in the migratory locusts, where, instead of struggling against a fickle environment to maintain a balance between density and resources, huge surplus swarms are 'deliberately' built up, and sent out as expeditionary forces. In *Magicicada* the basic evils of temporal isolation have just as evidently been turned to account. But here it seems a particularly improvident path to have followed. There are in fact apparently-suitable areas of North America, which have never to our knowledge had periodical cicadas at all, standing out as empty gaps in the distribution. On these it seems rather probable that some former brood dwindled to extinction before the era of recorded history began (*cf.* Young, 1958*b*, p. 165). If our deductions have been made correctly, the ordinary machinery for pioneering and the recolonisation of lost ground has been largely suppressed in the interests of brood isolation; and in that case the long-term prospects for survival in *Magicicada* must be poor indeed.

23.7. *The length of the life-span*

Our attention has often been drawn, especially in the present chapter, to adaptations in some way connected with the duration of individual life. Though some species have a still more rapid turnover, the majority of monotelic animals have generations that succeed one another annually. A very great number of other animals have life-spans longer than this, potentially attaining, in the longest-lived species of tortoise, a maximum of over a hundred years.

We are familiar also with the general characteristics of survivorship, and recognise that a tremendous early mortality in the immature sections of the population is a normal, acceptable outlay, to an important extent intrinsic in origin in the higher animals, automatically regulated, and advantageous to the population concerned. After the adult stage is reached, two divergent (though intergrading) patterns of survivorship can be recognised in polytelic species. Either (as in man) mortality tends to drop to a rather low level during the long prime of life, and a substantial proportion of individuals live to attain senility, at which time mortality mounts very steeply once more, so that quite soon afterwards they are all dead from one cause or another. Or, often in species potentially just as long-lived, mortality continues at a steady high level even in adult life, and few if any individuals survive to become senile. The latter applies to many large birds, as we saw earlier (p. 548); for instance, although under the protection of captivity a herring gull (*Larus argentatus*) may live for more than forty years, in the wild its expectation of further life—which remains practically the same at any age after it is a year old and does not appreciably diminish with the passing years—is only about three years in the north-eastern United States, or six years in Denmark (*cf.* Paludan, 1951, p. 125 for data: life-expectancies worked out here by Lack's formula—*see* Lack, 1954, in which the whole of Chapter 9 is devoted to a discussion of survivorship.)

In these latter circumstances each year-class undergoes a roughly exponential decline in numbers as it grows older, and the remnant—if any—surviving to ultimate senility is so small that we often have only the vaguest notion of how long the potential life-span may be in the species concerned. It is not even safe to assume that degenerative senility would be reached at any even roughly constant age in animals like these. In the longest-lived perennial plants, for instance, the best known of which are mostly trees, it is scarcely possible to say more than that some species have a longer average survival than others. Oaks in one habitat may pass their prime and start to decline at a hundred years, and in another keep going for a thousand. For these species there is no universal life-span corresponding to our ' three score years and ten ', that can be accepted as characteristic of the species as a whole: indeed, in some cases the durability and resistance of the dead heart-wood seem to be the critical limiting factors, rather than the true viability of living cells.

It has been widely assumed that natural selection will normally promote the longevity of adult organisms, since those that survive longest are likely to make the largest total contribution to posterity, and their genes must consequently increase in frequency with succeeding generations. We ought, however, to regard this with the same suspicion as the various other similar assumptions that take no account of the superior force of group-selection. It seems very much more probable that, in so far as the average potential life-span in any species is physiologically fixed, its duration has been arrived at by selection, in the process of permitting the stock as a whole to survive; and that it is more or less completely integrated with the other basic parameters of population balance, such as fecundity. Important group-characters like these could theoretically be protected from disturbance, due to selection at the individual level, by coming to involve not one or two but a great many genes: this would make them highly conservative and resistant to change as a result of single mutations or selection affecting particular individuals. Some at least of the other fundamental characters of species, for instance body-size and intelligence in man, are known to depend in this way on polygenic systems.

From the mass of information available it is easy to see that there are plenty of thriving species, families and even orders of animals that have evolved relatively short individual life-spans, although others make out just as successfully with long ones. Within the world of living organisms there appear to be some whose generations, given the opportunity, proliferate and die as rapidly as possible, and others that retain the spark of individual life as long as it can be cherished. From this it may be concluded that there are advantages and disadvantages on both sides of the scales, and that the balance arrived at is a compromise between conflicting benefits and evils, that vary in weight with the particular circumstances and way of life of the species concerned.

Most habitats in which animals (and plants) live are affected by the annual climatic cycle, and there is usually, even in the Tropics, an alternation of seasons more and less favourable for activity and growth. That so many organisms should have evolved life-spans geared to this prime period, therefore, with annually-succeeding generations, seems readily understandable. The calendar period suitable for reproduction may be quite short, perhaps just at the beginning of the growing season; and in that case, once the adult has completed its reproductive duties, further survival would be pointless. Thus in several orders of insects, including the mayflies (Ephemeroptera), the big silkworm moths (Saturniidae), and the warble-flies (Oestridae), the adults have even dispensed with the means of ingesting food, having sufficient reserves laid up in the body to carry them through the very brief period required for successful reproduction. Adult life is equally truncated in the Pacific salmon (*Oncorhynchus* spp.), which do not feed again once they have entered fresh water from the sea, and die almost immediately they have spawned.

Typical annuals among animals tend on the whole to breed either at the beginning or the end of the growing season, but many pass through a succession of generations (sometimes parthenogenetic) within a single growing season, the last of which overwinters or produces resistant eggs; and these, like the seeds of plants, inertly carry the stock through the succeeding unfavourable months. Those that breed only at the beginning of the growing season generally complete their growth from the egg not long before the inclement period arrives: this they survive in a more or less inactive state, ready to emerge and breed as soon as the next growing season commences. Neither deposited eggs nor torpid adults can defend themselves or make any escape if danger threatens, but on general grounds it is probably more efficient for sub-adults or adults to survive the winter (or dry season, etc., as the case may be), and be present in spring to select sites for laying the eggs where they can start to develop right away, rather than consigning them the previous autumn to what must in the first instance be a suitable hibernaculum: some degree of parental care may even develop in the former circumstances, as in many spiders. Epideictic ceremonies among adults are likely to give a more direct reflection of existing economic conditions in the spring than they could ordinarily do in the preceding autumn. Certainly it is true of the majority of the higher insects that they overwinter (or aestivate) as pupae or adults, rather than as eggs.

Advantages from a rapid turnover in generations presumably arise through the resulting speed and intensity of the genetic and selective machinery. In a given term of years the number of individuals it takes to sustain a given biomass depends not only on their size but on their life-span; and the larger the number of generations, the faster the progress of selection. Gene-recombinations will be more frequent and numerous, and favourable genes will be more rapidly spread. Short life and small size are partially correlated; but though there is no doubt often a strict limit to the growth in size that could be achieved in a brief span, this is seldom the primary determining factor, as we noted earlier (p. 580). In actual fact, large animals and plants tend to grow more rapidly than small ones, regardless of the size of the egg or seed from which they come, which suggests that growth-rate is in general adapted to the ultimate size of the organism, rather than that size is limited by the time available for all-out growth.

Small size can have tremendous advantages in its own right—for example in exploiting special habitats of diminutive extent. Without any need for impossibly great individual powers of locomotion, breeding populations of small organisms of sufficient numerical strength can be successfully carried by isolated colonies of particular food-plants, or by small bodies of water with particular characteristics for instance, as long as these persist. Many of these innumerable micro-habitats would be quite untenable, and frequently inaccessible, to larger species.

Another advantage of rapid turnover is the facility with which population-density can be matched to the variable or ephemeral resources available at any given time. A short-lived species can manage to exist much more from hand to mouth than a long-lived perennial, and be far less committed to the future. A good year or a good site allows a dense population to spring up, but it will inevitably have to start up again the next season, perhaps in a new place, when conditions may be quite different, and a new optimum desirable. This provides a great element of opportunism, typically exploited among plants by the annual weeds of cultivation, and among animals by some of the species commonly designated as pests—many of which have annual or biennial cycles—and by the short-term occupiers of transitory sites generally.

Whether or not there are other undetected benefits in the balance, that ought to be added to these, there can be no doubt that rapid turnover—often correlated with small size—confers immense evolutionary vitality on many of its adherent species. What then are its disadvantages; or, looking at the question the other way round, what is there to be gained by increasing longevity?

In practice it often appears to us that annual organisms are not after all particularly efficient in exploiting their resources: we may be justified in suspecting that the conspicuous fluctuations in numbers that they sometimes undergo from year to year do not solely reflect changes in their available food-supply, but are due in part to an incompetence or failure of their homeostatic and dispersive machinery. A high proportion of each individual life tends to be spent in growth, before the strength, defence and locomotor powers of the adult have been fully attained, and while the individual is still in consequence relatively vulnerable. Compare the average bird, which attains almost its full stature and powers of flight within a few weeks of hatching (even before it leaves the protection of its parents) and then lives on for perhaps three to five years, with the typical twelve-month insect, able to use its wings for only perhaps the last twentieth or fiftieth of its total existence as an individual. The ' uncontrollable losses ' term in our equation of recruitment and loss tends consequently to be relatively much larger, and the maintenance of anything like an optimum density that much more difficult to achieve.

Large size *per se* offers a number of independent advantages, including the strength, and the speed and range of locomotion, that generally go with it. Long life, on the other hand, seems to confer all its primary benefits in the sphere of social homeostasis, contributing especially towards the better use and conservation of the resources of permanent types of habitat. Most of these benefits have already been separately discussed in the preceding sections, but they may with advantage be recapitulated.

Prolonging adult life to include more than one annual breeding cycle implies the existence at all times of a standing adult population—in itself a powerful factor in maintaining stability of numbers. The resulting breeding caste will be drawn not from one but from several or many year-classes, and chance irregularities in juvenile survival from year to year tend consequently to be smoothed out. Even the total failure of a breeding season is not so insuperable a disaster as it might in some cases be for an annual species. The way is clear for the development of parental care, and the control of recruitment through juvenile survival; and indeed these developments are much commoner in polytelic than in monotelic animals.

One of the most important results of the perennial way of life is that offspring grow to maturity in the presence of their parents, and, among other advantages, opportunity is created for the transfer of traditional information from generation to generation. So valuable is this faculty, as we saw in Chapter 19, that even some apparently annual, monotelic insects have contrived to develop it: though it will be recalled that we found it difficult in some of the instances cited there to avoid the suspicion that at least a few parental survivors mingled with each daughter generation. Individual learning and adoption of precedent is the only possible means of adapting a social convention to a particular topographical setting; and it can be safely assumed that the benefit of access to tradition, above all as an aid to conventional behaviour, exercises a tremendous influence in the evolution of polytely and the multi-annual span of life.

Advantages different again spring from the extension of juvenile or larval life. We have found that in species where the developmental and pre-mature stages cover several years, there is generally a considerable individual variation in the age at which breeding actually begins. The number of recruits admitted each year to the breeding caste can in many animals be varied according to current requirements, and recruiting can draw if necessary on several year-classes at the same time, so that great flexibility becomes possible in varying the intake. Many monotelic species, living only for one season as adults, nevertheless have prolonged juvenile lives and consequently reap these advantages; and it ought to be noted that even the annuals tend to be similarly protected against the failure of one year's crop of progeny, by evolving eggs or seeds that vary individually in their length of dormancy. Thus the germination of seeds from a crop of annual plants—and this includes some of the commonest garden weeds—is generally spread over several years after the initial seed-fall, even when conditions are continuously favourable, so that by the strictest definition they are not pure annuals. Similarly Moscona (1950, p. 201) records that in the stick-insect *Bacillus libanicus*, a phasmid with an annual cycle, a proportion of the eggs do not start to develop until the second year after laying.

Monotelic species, even when their full development extends over two or more years, run the risk of splitting into separate broods, successively producing generations of adults each time their turn comes round—at intervals of two, three, or any other constant number of years—while between one year's brood and the next complete reproductive isolation results. This is a danger that can equally well be avoided if the population possesses (as a group-character) variability in developmental rate sufficient to ensure that not all members of the same year-class are ready to breed in the same year.

Very long life of course implies a correspondingly great measure of independence from year-to-year fluctuations in the environment, and no doubt it is achieved most easily in habitats that are by nature stable, extensive and permanent. It implies a general supremacy over the outer world and a perfection of inward control over dispersion, recruitment and loss. Average rates of recruitment and loss are always necessarily correlated inversely with the average life-span; and to ensure long average life, the sources of 'uncontrollable mortality' must all have been minimised by successful adaptations acquired against them. It is a sign of the highest accomplishment in this direction when, as in some contemporary races of man, as many as half the individuals that reach maturity are still living at seventy (*cf.* Comfort, 1956, p. 16), having enjoyed their life's full span.

This prompts one to ask what decides the optimum duration of life? If the species is to be polytelic at all, what are the merits of living ten years rather than twenty?

Judging by the rarity of maximum life-spans exceeding fifty years, very special circumstances are necessary to justify them, in terms of evolutionary survival. In plants, with their great powers of vegetative regeneration and complete lack of social organisation as understood here, it is quite different: fifty years cannot be at all unusual for the life of a vascular perennial. It is usually assumed that the life-spans at least of some animal species are limited by the durability of certain permanent structures or organ-systems indispensable to life; and we have already tentatively accepted the conclusion that the possession of non-repairable wings imposes a binding limitation on the extension of adult life in insects. It is sometimes suggested similarly that mammal can live only as long as their teeth last, but the elephants have evolved a system of molar replacement that allows them one of the longest potential life-spans known among animals (*i.e.* some seventy years in *E. indicus*); and the rodents have dispensed with tooth-replacement altogether, having only a single set that grow ' by the yard ' as required, as long as the animal lives. Where they are advantageous and have therefore been promoted by selection, life-spans approaching a century appear to have evolved in a number of very diverse animal groups, and thus to lie within the general range of attainability for animal tissues.

Naturally we have a fuller knowledge and experience of the general biology of senescence in man than in any other species; and, whether or not his is a widely typical case, there can be very little doubt that man is adapted to die from intrinsic causes when he reaches the age of seventy or eighty years. It is not one outworn organ or system that is responsible for his failure, but a general physiological degeneration; with the result that his actual death may be produced by a wide variety of causes. The time of life at which people die from old age is inherent, determined and relatively predictable, in complete contrast to the chance agency that happens to deliver the mortal stroke.

Since he acquired the power of speech and the ability it confers to make the fruits of experience available to the rising generation, man's potential life-span has probably increased, through the agency of group-selection, by some twenty-five or thirty years. None of our dumb relatives among the primates, and especially the great apes, appear to live to be more than thirty or forty (*cf.* Comfort, 1956, p. 46). But man, in contrast, as pointed out already in Chapter 13 (p. 248), undergoes at a corresponding age a special transformation in outward appearance, no less striking than the one at puberty: the hair of his head whitens, and in some races he may go bald on the crown, signifying that he is now an elder member of society, to whom respect is due for his ripe wisdom and valued counsel; indeed it is on him that the chief burden of government has come to devolve.

This increase in human life-span has involved no overall rise in fecundity, since the woman completely abandons her fertility on entering the elder caste; and although it has presumably added almost 50 per cent to the length of service demanded of the other bodily organs, there is no reason to think that evoking this extra endurance has entailed any fundamental revolution in primate physiology. Man has probably taken the increase in his stride.

It seems reasonable to conclude, with Weissman (1899), and consistent with what we found earlier in relation to maximum fecundity, that the potential span of life in the higher animals has been fixed by natural selection at a point that is optimal for the species, being one of the parameters that has contributed in all presently existing animal societies to their hitherto unbroken record of survival.

Long life in our case makes it possible for us to interest ourselves in slowly maturing plans and projects, and to devote, for example, twenty or thirty years to the upbringing and education of children. It enables us to acquire an unparalleled amount of traditional learning, and, looking far ahead, still to have time to employ it in long-term constructive and creative work. Our mastery over the forces of nature, and ability to preserve the complex social organisation on which our lives depend, rely overwhelmingly on accumulated experience, and this makes it easy to appreciate the special advantages of long living; but when we turn to the Indian elephant, the land-tortoises or the lobster we are admittedly at a loss to explain the slow, protracted tempo of their lives.

The commonest range of life-spans in polytelic animals lies, of course, very much below these; and it may reasonably be questioned whether the retardment of successive generations down to three or four a century does not slow the genetic and selective machines to a dangerous extent, and render the species refractory and inadaptable to permanent secular changes arising in the world in which it lives. Is not the chance of survival enhanced, perhaps, by keeping the genetic wheels turning faster? It is not easy to reach a conclusion on this point: certainly the rate of evolutionary advance has been influenced by many powerful factors besides the turnover rate of successive generations. Highly conservative groups, such as the ' primitive ' apterygote insects or the archaic bony fishes, for instance, that have changed comparatively little since palaeozoic times, have turnover rates not very obviously different from those of the most recent and progressive members of their respective classes.

Two generalisations can be made, however, that seem to throw some light on this point. First, when we consider the animal kingdom at large, long life—over thirty years, say—is seen on the whole to be a sporadic phenomenon; and this implies that in the long run of evolution its adherents have been more likely to become extinct than to found the orders and classes of the future. Secondly, we may single out the insects for their unrivalled success in peopling the world of land and water with numberless species surpassing those of all other animals put together: in a lesser degree the Passeres—the principal order of small birds—have come easily to outnumber in species the sum total of all the twenty-odd other orders of living birds (Mayr and Amadon, 1951); and in the mammals it is the rodents that account for more than a third of the living genera, and are by far the largest single order (Simpson, 1945). It can scarcely be questioned that these are three of the most successful groups of animals living in the contemporary world, and all are characterised by average life-spans inclining well towards the lower end of the scale existing in the groups with which they can most appropriately be compared.

Longevity, therefore, is probably seldom determined by the inability of selection to evoke from the body any greater endurance, although this may apply in the case of winged insects. In the great majority of animals it should probably be regarded instead as one of the cardinal factors in population balance, fixed by group-selection and co-ordinated with fecundity, dispersal-rate and other parameters of the same kind, in the interests of keeping every population as near as possible to its optimum dispersion. On general grounds it seems probable that in those animal species where a high proportion of individuals survive to be overtaken by senility (as in many insects and mammals), the potential life-span is most likely to be precisely determinate and invariable.

23.8. *Summary of Chapter 23*

1. Two further types of homeostatic response form the main subject of this chapter. The first is the density-dependent accommodation of the rate of growth and maximum body-size of individual animals. In many aquatic species, including various amphibia, fish, Crustacea and gastropods, crowding has an inhibitory effect on growth; and when, in certain of the experiments, the number of individuals present is large their body-size is correspondingly reduced, so that the total biomass per unit volume reaches a constant ceiling regardless of the number of individuals present. The nature of some of the growth-inhibiting substances has been partially investigated. Growth-promoting metabolites may also occur, and are typically produced at the other extreme, when densities are very low.

2. Analogous effects can be seen in certain terrestrial animals: *Tribolium*, as noted earlier, conditions its medium (flour) with secretions from special gas-glands, and this affects, among other things, larval growth and survival. Individual size variation is widespread in adult insects, and often—perhaps usually—not determined genetically, but dependant instead on economic conditions and especially on larval nutrition. In certain social Hymenoptera (namely bumble-bees and ants) large or small workers are produced according to the economic needs of the colony; and exactly the same is true of some of the termites.

In the mammals, body-size is often much influenced by nutrition during the growing period; and various ungulates, for instance, tend to perpetuate small body-size in stocks living continuously in restricted or adverse habitats. Various deer and horses provide good illustrations of this. On the contrary in birds, which can easily relieve congestion by flying away, this kind of facultative size-accommodation is not seen: only in the flightless ostrich and emu do we find any indication of it.

3. The second theme to be developed concerns the deferment of the age at which breeding commences. Some species have a rapid turnover of generations, whereas in others, because of the time required for development and adolescence, it may be slowed down to once in many years. Does the delayed development achieve any greater measure of population control? Indeed, what are the merits of long life and perennial breeding generally, as compared with a swifter succession of monotelic (only-once-breeding) generations? The task of answering these questions is spread over the remainder of the chapter.

In birds, breeding success is sometimes—and perhaps generally—lower in young adults than in older ones, and there is evidence that the degree of success, and even the admission of recruits to the breeding caste, is dependent on social standing and economic conditions. The age of first breeding in species with long adolescent periods is individually variable (*see* table on p. 572): under very good conditions, or when there are few adults present, even birds in sub-adult plumage sometimes breed.

There is considerable evidence for the existence of a non-breeding reserve, perhaps as a normal phenomenon, consisting of potentially mature individuals inhibited for want of the necessary qualification of property or social rank. At times even some of the birds that have bred successfully in earlier seasons may find themselves temporarily excluded by the social guillotine from nesting—resulting in the phenomenon of ' intermittent breeding ', which has been recorded in Canadian fulmars, and elsewhere. But most of the non-breeders are probably junior birds, as a rule; and one of the advantages of deferment of the age of first breeding, at any rate beyond the minimum age of puberty, is no doubt that it provides such a reserve, which can be mobilised and brought into action whenever a greater reproductive output is called for.

4. The differences between monotely and polytely are well illustrated in fishes. Pacific salmon are all monotelic, whereas the Atlantic salmon is polytelic. All salmon have variable rates of development in fresh water, even in the same river; and after they leave it they spend a variable number of years in the sea. Consequently each year's incoming breeding stock is drawn from many year-classes; and, although breeding success and parr production vary very much from year to year, this diversity of age among first breeders produces a ' running-average effect ', and keeps the breeding population comparatively steady. The effect is probably one of the main advantages resulting from variability in the age of first breeding, which seems so universal among slowly maturing animals: it is especially valuable to those that are monotelic. In polytelic species, where younger and older breeders mingle inevitably, the same smoothing effect is produced in an alternative way.

There are extreme cases in fishes where first breeding is delayed for between five and seventeen or eighteen years (*e.g.* European eel, halibut).

Where, as already seen in birds, the age of admission to breeding status in fish depends on the social hierarchy, it may be expected that an active fishery, by removing a disproportionate number of large adults, will tend to lower the average age of maiden breeders.

5. In mammals deferred maturity and variability in first-breeding age are no less common and closely correlated. Female Indian elephants will bear first calves at any age between eight and sixteen. In the polygamous seals it is obvious how much the onset of breeding is governed in males by social status; and recent commercial exploitation of the elephant-seal, by removing the largest bulls, has in fact resulted in lowering their average first-breeding age.

Among reptiles, age determinations are possible in various freshwater turtles, and they all show similar phenomena. Postponement of maturation extends in a number of species to ages exceeding ten years.

In considering further the general functions of retardment it should be pointed out that, while larger animals tend in fact to take longer to mature, the exigencies of growth and large size *per se* are not the primary cause of delay. Man himself, though not among the largest mammals, is one of the slowest growers; and it seems clear that by comparison with the ungulates, say, not only his physical but also his mental development is slowed down. This and other evidence suggests that slow development cannot be serving primarily as an extended apprecticeship either, merely allowing more time for education.

In addition to confirming the conclusions already drawn, it is tentatively suggested that the extension of the so-called pulline period, when the offspring is still a dependent on its parents, may significantly condition the latters' fertility. It may also allow a still more extended opportunity for thinning out recruits through socially-induced mortality.

6. When we turn to the arthropods we find, not for the first time, almost exactly the same adaptations. Some Crustacea and arachnids take seven or more years to mature, apparently partaking, as would be expected, in the usual individual variability. The insects present at first sight a rather different picture, because they are almost wholly committed to monotely. This may to some extent be forced on them by their dependence on frail and non-renewable wings. Certainly where polytely has evolved among them it has usually been in species that do not fly continually—among the social Hymenoptera (ants, honey-bees), termites, beetles and Apterygota.

Lengthening of insect larval stages to cover several years is fairly common. Primitively this may sometimes be imposed by the harshness or poverty of the habitat, but it is often secondarily developed and adapted in exactly the same ways as we found in vertebrates. Individual variability in age of emergence is authenticated in a number of orders, as usual enabling adults to be recruited from two or more year-classes at once. Another important advantage of variability is that it prevents temporal isolation in, for instance, biennial species. This isolation does occur sometimes: the northern dart-moth in Scotland, for example, has in recent decades had no known emergences in odd-numbered years.

A most remarkable and exceptional case is that of the periodical cicadas (*Magicicada* spp.) of North America. Here the nymphal life occupies seventeen or thirteen years, depending on the stock. In any one locality almost no emergences occur in intervening years. The thirty separate ' broods ' that are known are all allopatric, each appearing in their exclusive years and areas, with the apparent result that reproductive isolation between them is virtually complete. There are two distinct seventeen-year species (*M. septendecim* and *M. cassini*) which occur together in some regions, always exactly in step.

It is generally supposed that these remotely spaced swarms always come as a 'surprise' to potential predators and parasites, whose powers of inflicting damage are completely swamped by the vast glut of emergents. From an evolutionary standpoint, however, *Magicicada* appears to have embarked on a dangerous mode of existence.

7. As to longevity, it would be fallacious to assume that selection necessarily favours longer life, any more than it does higher fecundity. These are group characteristics, closely integrated together, giving the group as a whole the chance of survival: this was pointed out by Weissmann in 1882.

Some very successful animals have a rapid turnover of generations, others a slow one; and it is clear that selection must strike an appropriate compromise between the various benefits and evils inherent in either extreme. A short life has the advantage of giving more frequent gene-recombinations, a faster flow of genes, and an ability to compass swift on-the-spot changes in numbers in transient or fluctuating habitats. It tends to be correlated with small size, which allows prolific populations to flourish in all sorts of specialised micro-habitats. Short-lived species, however, tend broadly to be far less stable or well-regulated in numbers than long-lived ones.

Long life makes for stability in numbers: for instance, (i) it mixes breeders from different year-classes, (ii) parents are associated with offspring and can supervise recruitment, and (iii) tradition can be used with tremendous advantage, especially in conventions involving local topography. Other benefits have already been pointed out. Long life tends to be associated with permanent, extensive, stable environments.

Maximum life-spans approaching 100 years are apparently within the range of general attainability by animals, but anything exceeding even a third of this is rare and sporadic in practice. In evolutionary terms, therefore, specialising in very long life is apt to be a dead end. Longevity is probably not generally limited by the inability of selection to elicit further endurance from tissues and organs. For instance, man, without any accompanying increase in fecundity, has probably added the twenty-five years or so that comprise his post-reproductive 'elder' stage of life, since he acquired the power of speech, and without entailing any particular physiological revolution in tissue and organ endurance.

Most animal life-spans are much shorter than ours; and for what it is worth it is pointed out that (i) the largest class in the animal kingdom—the insects, (ii) the largest order of birds—the Passeres, and (iii) the largest order of mammals—the rodents, all tend to have life-spans towards the lower end of the range existing in the groups from which we have singled them out.

References

AGASSIZ, L. 1857. Contributions to the natural history of the United States of America First Monograph. 2 vols. Boston, Mass.

AIYAR, R. G., and N. K. PANIKKAR. 1937. Observations on the swarming habits and lunar periodicity of *Platynereis* sp. from Madras harbour. *Proc. Indian Acad. Sci.*, **5** (B): 245-60.

ALCOCK, A. W. 1902. *A naturalist in Indian seas.* London.

ALEXANDER, H. G. 1923. The migrations of ducks. *Brit. Birds*, **16**: 330-1.

ALEXANDER, P., and D. H. R. BARTON. 1943. The excretion of ethylquinone by the flour beetle. *Biochem. J.*, **37**: 463-5.

ALLARD, H. A. 1930. The first morning song of some birds of Washington, D.C.; its relation to light. *Amer. Nat.*, **64**: 436-69.

ALLEE, W. C. 1931. *Animal aggregations.* Chicago.

ALLEE, W. C. 1934. Recent studies in mass physiology. *Biol. Rev.*, **9**: 1-48.

ALLEE, W. C. 1938. *The social life of animals.* New York, London and Toronto.

ALLEE, W. C. 1940. Concerning the origin of sociality in animals. *Scientia*, 1940: 154-60.

ALLEE, W. C., E. S. BOWEN, J. C. WELTY and R. OESTING. 1934. The effect of homotypic conditioning of water on the growth of fishes, and chemical studies of the factors involved. *J. exp. Zool.*, **68**: 183-213.

ALLEE, W. C., A. E. EMERSON, O. PARK, T. PARK and K. P. SCHMIDT. 1949. *Principles of animal ecology.* Philadelphia and London.

ALLEN, A. A. 1934. Sex rhythm in the ruffed grouse (*Bonasa umbellus* L.) and other birds. *Auk*, **51**: 180-99.

ALLEN, G. M. 1939. *Bats.* Cambridge, Mass.

ALLEN, M. DELIA. 1956. The behaviour of honeybees preparing to swarm. *Brit. J. Anim. Behav.*, **4**: 14-22.

ALTUM, B. 1868. *Der Vogel und sein Leben. See* E. Mayr (1935).

AMADON, D. 1943. Birds weights and egg weights. *Auk*, **60**: 221-34.

ANDERSON, A. R. 1894. Note on the sound produced by the Ocypode crab. *J. Asiatic Soc. Bengal.*, **63**: 138-9.

ANDREWARTHA, H. G. 1959. Self-regulatory mechanisms in animal populations. *Aust. J. Sci.*, **22**: 200-5.

ANTHONY, A. 1955. Behavior patterns in a laboratory colony of prairie dogs, *Cynomys ludovicianus*. *J. Mammal.*, **36**: 69-78.

ANTHONY, H. E. 1928. *Field Book of North American mammals.* New York and London.

ARROW, G. J. 1951. *Horned beetles, a study of the fantastic in nature.* The Hague.

AUDUBON, J. J. 1831-9. *Ornithological biography*, etc. 5 vols. Edinburgh.

AUSTIN, O. L. 1945. The role of longevity in successful breeding of the common tern (*Sterna hirundo*). *Bird-banding*, **16**: 21-8.

AUSTIN, O. L. 1949. Site tenacity, a behavior trait of the common tern (*Sterna hirundo* L.). *Bird-banding*, **20**: 1-39.

AZARA, F. DE. 1838. *The natural history of the quadrupeds of Paraguay.* Vol. 1. Edinburgh.

BAERG, W. J. 1945. The black widow and the tarantula. *Trans. Conn. Acad. Arts Sci.*, **36**: 99-113.

BAILEY, V. 1905. Biological survey of Texas. *North Amer. Fauna*, No. 25. U.S. Dept. Agric. (Biol. Surv.), Washington: 222 pp.

BAINBRIDGE, R. 1952. Underwater observations on the swimming of marine zoo plankton. *J. mar. biol. Ass. U.K.*, **31**: 107-12.

BAKER, E. C. STUART. 1922-9. *The fauna of British India: Birds.* 2nd Ed., 6 vols. London.

BAKER, J. R., and Z. BAKER. 1936. The seasons in a tropical rain-forest (New Hebrides).—Pt. 3. Fruit-bats (Pteropidae). *J. Linn. Soc. (Zool.)*, **40**: 123-41.

BAKER, J. R., and T. F. BIRD. 1936. The seasons in a tropical rain-forest (New Hebrides).—Pt. 4. Insectivorous bats (Vespertilionidae and Rhinolophidae). *J. Linn. Soc. (Zool.)*, **40**: 143-61.

BAKUS, G. J. 1959. Observations on the life history of the dipper in Montana. *Auk*, **76**: 190-207.

BANG, B. G. 1960. Anatomical evidence for olfactory function in some species of birds. *Nature, Lond.*, **188**: 547-8.

BARKOW, H. C. L. 1846. *Der Winterschlaf nach seinen Erscheinungen im Thierreich.* Berlin.

BARNETT, S. A. 1955. Competition among wild rats. *Nature, Lond.*, **175**: 126.

BARNETT, S. A. 1958. Physiological effects of " social stress " in wild rats.—I. The adrenal cortex. *J. psychosomat. Res.*, **3**: 1-11.

BARRETT, C. G. 1895. *The Lepidoptera of the British Islands.* Vol. 2. *Heterocera, Sphinges, Bombyces.* London.

BARRETT-HAMILTON, G. E. H. 1910-21. *A history of British mammals.* Vols. 1 and 2, pt. 1 and 2. London.

BARRINGTON, R. M., and C. B. MOFFAT. 1901. Wasps in County Wicklow. *Irish Nat.*, **10**: 197-301.

BARRY, T. W. 1956. Observations of a nesting colony of American brant. *Auk*, **73**: 193-202.

BATES, G. L. 1908. Observations regarding the breeding-seasons of the birds in Southern Kamerun. *Ibis*, 1908: 558-70.

BATES, H. W. 1863. *The naturalist on the river Amazons.* 2 Vols. London.

BATESON, W., and H. H. BRINDLEY. 1892. On some cases of variation in secondary sexual characters, statistically examined. *Proc. zool. Soc. Lond.*, 1892: 585-94.

BAUMGARTNER, F. M. 1939. Territory and population in the great horned owl. *Auk*, **56**: 274-82.

BAXTER, E. V., and L. J. RINTOUL. 1953. *The birds of Scotland.* 2 vols. Edinburgh.

BEALE, T. 1839. *The natural history of the sperm whale. . . . To which is added a sketch of a South-Sea whaling voyage.* 2nd Ed. London.

BEAMER, R. H. 1931. Notes on the 17-year cicada in Kansas. *J. Kansas ent. Soc.*, **4**: 53-8.

BEEBE, W. 1917. *Tropical wild life in British Guiana.* New York.

BEEBE, W. 1925. Studies of a tropical jungle. *Zoologica*, N.Y., **6**: 5-193.

BEEBE, W. 1935. *Half mile down.* London.

BEEBE, W. 1944. The function of secondary sexual characters in two species of Dynastidae (Coleoptera). *Zoologica*, N.Y.: 29: 53-8.

BEIRNE, B. P. 1944. The causes of occasional abundance or scarcity of wasps (*Vespula* spp.). *Ent. mon. Mag.*, **80**: 121-4.

BELT, T. 1874. *The naturalist in Nicaragua.* London.

BENHAM, W. B. 1896. Polychaet worms. In *Cambridge Natural History*, **2**: 241-344.

BENT, A. C. 1922. Life histories of North American petrels and pelicans and their allies. *Bull. U.S. nat. Mus.*, **121**: 343 pp.

BENT, A. C. 1937. Life histories of North American birds of prey, Pt. 1. *Bull. U.S. nat. Mus.*, **167**: 409 pp.

BERRY, J. 1939. *The status and distribution of wild geese and wild duck in Scotland.* Cambridge.

BERTIN, L. 1956. *Eels, a biological study.* London.

BETTS, MONICA W. 1955. The food of titmice in oak woodland. *J. Anim. Ecol.*, **24**: 282-323.

BEVERTON, R. J. H., and S. J. HOLT. 1957. On the dynamics of exploited fish populations. *Fishery Invest., Lond.*, ser. 2, **19**: 533 pp.

BIGELOW, H. B., and W. W. WELSH. 1925. Fishes of the Gulf of Maine. *Bull. U.S. Bur. Fish.*, **40**, pt. 1: 567 pp.

BILSKI, F. 1921. Über den Einfluss des Lebensraumes auf das Wachstum der Kaulquappen. *Pflüg. Arch. ges. Physiol.*, **187**: 254-72.

BIRGE, E. A. 1897. The vertical distribution of the limnetic Crustacea of Lake Mendota. *Biol. Zbl.*, **17**: 371-4.

BLACKWALL, J. 1873. *Researches in zoology.* 2nd Ed. London.

BLAIR, K. G. 1921. Insects in winter. *Proc. S. Lond. ent. and N.H. Soc.*, 1920-1: 26-42.

BLANFORD, W. T. 1888-91. *The fauna of British India: Mammalia.* London.

BONHOTE, J. L. 1909. Changes of plumage in the cormorant. *Field*, **114**: 974.

BOYD, H. 1955. The role of tradition in determining the winter distribution of pinkfeet in Britain. *7th ann. Rep. Wildfowl Trust, Lond.*: 107-22.

BRADLEY, J. C. 1908. Gregarious sleeping habits among aculeate Hymenoptera. *Ann. ent. Soc. Amer.*, **1**: 127-30.

BRAUER, A. 1908. Die Tiefsee-Fische. II. Anatomische Teil. *Wiss. Ergeb. Deutsch Tiefsee Exp.* (' *Valdivia* '), **15**: 266 pp.

BRECKENRIDGE, W. J. 1953. Night rafting of American golden-eyes on the Mississippi river. *Auk*, **70**: 201-4.

BREDER, C. M., Jr. 1941. On the reproductive behavior of the sponge blenny, *Paraclinus marmoratus* (Steindacher). *Zoologica*, N.Y., **26**: 233-5.

BREDER, C. M., Jr., and C. W. COATES. 1932. A preliminary study of population stability and sex ratio of *Lebistes*. *Copeia*, 1932: 147-55.

BREHM, A. 1922-6. *Brehms Tierleben.* 4th Ed. 13 vols. Leipzig.

BREWSTER, W. 1890. Summer robin roosts. *Auk*, **7**: 360-73.

BREWSTER, W. 1906. The birds of the Cambridge region of Massachusetts. *Mem. Nuttall orn. Club*, no. 4: 426 pp.

BREWSTER, W., and F. M. CHAPMAN. 1895. Trinidad birds. *Auk*, **12**: 207-8.

BRIAN, ANNE D. 1949. Dominance in the great tit *Parus major*. *Scot. Nat.*, **61**: 144-55.

BRIAN, M. V. 1956. Segregation of species of the ant genus *Myrmica*. *J. Anim. Ecol.*, **25**: 319-37.

BRIAN, M. V. 1957. The growth and development of colonies of the ant *Myrmica*. *Insectes sociaux*, **4**: 177-90.

BRIDGE, T. W. 1904. Fishes (exclusive of the systematic account of the Teleostei). In *Cambridge Natural History*. London.

BRINCKMANN, M. 1954. Vom Weissen Storch in Bezirk Osnabrück. *Vogelwelt*, **75**: 194-200.

BRINDLEY, H. H. 1910. Further notes on the procession of *Cnethocampa pinivora*. *Proc. Camb. phil. Soc.*, **15**: 576-87.

BRISTOWE, W. S. 1958. *The world of spiders.* London.

BROOK, G. 1886. Report on the herring fishery of Loch Fyne and the adjacent districts during 1885. *Fish. Bd. Scotl., 4th ann. Rep. for* 1885: 47-61.

BROWN, H. 1954. *The challenge of man's future.* New York.

BROWN, L. H. 1952-3. On the biology of the large birds of prey of the Embu district Kenya Colony. *Ibis*, **94**: 577-620, and **95**: 74-114.

BROWN, L. H. 1955. *Eagles.* London.

BROWN, R. L. 1936. The egg-breaking habit of the kingfisher. *Scot. Nat.*, 1936: 135-8.

BROWN, R. Z. 1953. Social behavior, reproduction and population changes in the house mouse. *Ecol. Monogr.*, **23**: 217-40.

BRUES, C. T. 1903. On the sleeping habits of some aculeate Hymenoptera. *J. N.Y. ent. Soc.*, **11**: 228-30.

BUDGETT, J. S. 1907. *The work of John Samuel Budgett.* Edited by J. Graham Kerr. Cambridge.

BÜNNING, E. 1958. *Die physiologische Uhr.* Berlin.

BURKITT, J. P. 1935. Notes on the rook: with special reference to the proportion of young in flocks, etc. *Brit. Birds,* **28**: 322-6.

BURT, W. H. 1937. Territorial behavior and population densities in mammals. *Bull. ecol. Soc. Amer.,* **18**: 53.

BURT, W. H. 1943. Territoriality and home range concepts as applied to mammals. *J. Mammal.,* **24**: 346-52.

BURT, W. H. 1949. Territoriality. *J. Mammal.,* **30**: 25-7.

BUTLER, C. G. 1954. *The world of the honeybee.* London.

BUXTON, P. A. 1923. *Animal life in deserts.* London.

CALDERWOOD, W. L. 1907. *The life of the salmon.* London.

CALHOUN, J. B. 1950. The study of wild animals under controlled conditions. *Ann. N.Y. Acad. Sci.,* **51**: 1113-22.

CARPENTER, C. R. 1934. A field study of the behavior and social relations of howling monkeys (*Aloatta palliata*). *Comp. Psych. Monogr.,* **10**, no. 48: 168 pp.

CARPENTER, G. D. H. and E. B. FORD. 1933. *Mimicry.* London.

CARR, A. 1952. *Handbook of turtles.* Ithaca, N.Y.

CARR-SAUNDERS, A. M. 1922. *The population problem: a study in human evolution.* Oxford.

CARTER, G. S. 1951. *Animal evolution.* London.

CAULLERY, M., and F. MESNIL. 1898. Les formes épitoques et l'évolution des Cirratuliens. *Ann. Univ. Lyon,* **39** (quoted from Hempelmann, 1911).

CHAPIN, J. P. 1954. The calendar of Wideawake Fair. *Auk,* **71**: 1-15.

CHAPIN, J. P., and L. W. WING. 1959. The Wideawake calendar, 1953 to 1958. *Auk,* **76**: 153-8.

CHAPMAN, F. M. 1908. A contribution to the life histories of the booby (*Sula leucogastra*) and man-o-war bird (*Fregata aquila*). *Carneg. Instn., Papers from Tortugas Lab.,* **2**: 139-151.

CHAPMAN, F. M. 1928. *Handbook of birds of eastern North America.* Revised Ed. New York.

CHAPMAN, F. M. 1935. The courtship of Gould's manakin (*Manacus vitellinus vitellinus*) on Barro Colorado Island, Canal Zone. *Bull. Amer. Mus. nat. Hist.,* **68**: 471-525.

CHAPMAN, R. N. 1928. The quantitative analysis of environmental factors. *Ecology,* **9**: 111-22.

CHITTLEBOROUGH, R. G. 1960. Marked humpback whale of known age. *Nature, Lond.,* **187**: 164.

CHITTY, D. H. 1952. Mortality among voles (*Microtus agrestis*) at Lake Vyrnwy, Montgomeryshire in 1936-9. *Philos. Trans.,* B, **236**: 505-52.

CHRISTIAN, J. J. 1950. The adreno-pituitary system and population cycles in mammals. *J. Mammal.,* **31**: 247-59.

CHRISTIAN, J. J. 1959. Control of population growth in rodents by interplay between population density and endocrine physiology. *Wildlife Disease,* no. 2: 38 pp. (microcard).

CHRYSTAL, R. N. 1937. *Insects of the British woodlands.* London and New York.

CLARKE, C. H. D. 1944. Biological reconnaissance of the Alaska military highway with particular reference to the Yukon Territory. Ottawa: duplicated typescript privately circulated, 44 pp.

CLARKE, G. L. 1933. Diurnal migration of plankton in the Gulf of Maine and its correlation with changes in submarine illumination. *Biol. Bull. Wood's Hole,* **65**: 402-36.

CLARKE, G. L. 1934. Further observations on the diurnal migration of copepods in the Gulf of Maine. *Biol. Bull. Wood's Hole,* **67**: 432-55.

CLARKE, J. R. 1953. The effect of fighting on the adrenals, thymus and spleen of the vole (*Microtus agrestis*). *J. Endocrin.,* **9**: 114-26.

Clarke, J. R. 1955. Influence of numbers on reproduction and survival in two experimental vole populations. *Proc. roy. Soc.*, B, **144**: 68-85.

Clarke, R. 1956. Sperm whales of the Azores. '*Discovery*' *Rep.*, **28**: 237-98.

Clarke, S. F. 1891. The habits and embryology of the American alligator. *J. Morph.*, **5**: 181-214.

Cloudsley-Thompson, J. L. 1952. The relationship between reproductive rate and environmental resistance in British birds. *Ann. Mag. nat. Hist.*, ser. 12, **5**: 989-1019.

Coates, C. W., M. Altamirano and H. Grundfest. 1954. Activity in electrogenic organs of knifefishes. *Science*, **120**: 845-6.

Cole, H. A. 1958. Notes on the biology of the common prawn *Palaemon serratus* (Pennant). *Fishery Invest., Lond.*, ser. 2, **22**, no. 5; 22 pp.

Cole, H. A., and E. W. Knight-Jones. 1949. The setting behaviour of larvae of the European flat oyster, *Ostrea edulis* L., and its influence on methods of cultivation and spat collection. *Fishery Invest., Lond.*, ser. 2, **17**, no. 3 : 59 pp.

Cole, L. C. 1954. Some features of random population cycles. *J. Wildlife Mgt.*, **18**: 2-24.

Collett, R. 1895. *Myodes lemmus*, its habits and migrations in Norway. *Christiania Vid.-Selsk. Forh.*, no. 3: 1-63.

Colyer, C. N., and C. O. Hammond. 1951. *Flies of the British Isles.* London.

Comfort, A. 1956. *The biology of senescence.* London.

Common, I. F. B. 1952. Migration and gregarious aestivation in the bogong moth, *Agrotis infusa*. *Nature, Lond.*, **170**: 981-2.

Common, I. F. B. 1954. A study of the ecology of the adult bogong moth, *Agrotis infusa* (Boisd.), with special reference to its behaviour during migration and aestivation. *Aust. J. Zool.*, **2**: 223-63.

Conder, P. J. 1949. Individual distance. *Ibis*, **91**: 649-55.

Coombes, R. A. H. 1950. The moult migration of the sheld-duck. *Ibis*, **92**: 405-18.

Corbet, P. S. 1959. The larval development and emergence of *Aeshna cyanea* (Müll.) (Odon. Aeshnidae). *Ent. mon. Mag.*, **95**: 241-5.

Corbet, P. S., C. Longfield and N. W. Moore. 1960. *Dragonflies.* London.

Corbet, P. S., and A. T. Tjønneland. 1955. Rhythmic flight activity of certain East African Trichoptera. *Nature, Lond.*, **175**: 1122-3.

Corney, B. G. 1922. Abstract of a paper on the periodicity of swarming of pálolo (*Eunice viridis* Gr.). *J. Torquay nat. Hist. Soc.*, 1922: 5 pp.

Cott, H. B. 1929. Observations on the natural history of the racing-crab *Ocypoda ceratophthalma*, from Beira. *Proc. zool. Soc. Lond.*, 1929: 755-65.

Cott, H. B. 1940. *Adaptive coloration in animals.* London.

Couch, J. 1863. *A history of the fishes of the British Islands.* Vol. 2. London.

Coues, E. 1876. Habits of the prairie hare. *Bull. Essex Inst.*, **7** (1875): 73-85 (quoted from T. S. Palmer, 1896, *U.S. Dept. Agric., Div. Orn. and Mamm., Bull.* no. 8).

Coulson, J. C., and E. White. 1958. The effect of age on the breeding biology of the kittiwake *Rissa tridactyla*. *Ibis*, **100**: 40-51.

Coutière, H. 1899. Les ' Alpheidae ', morphologie externe and interne, formes larvaires, bionomie. *Ann. Sci. Nat. Zool.*, (8) **9**: 1-559.

Crane, J. 1941. Crabs of the genus *Uca* from the west coast of Central America. *Zoologica*, N.Y., **26**: 145-207.

Crawshay, L. R. 1935. Possible bearing of a luminous syllid on the landfall of Columbus. *Nature, Lond.*, **136**: 559-60.

Crew, F. A. E., and L. Mirskaia. 1931. The effects of density on an adult mouse population. *Biol. gen.*, **7**: 239-250.

Crombie, A. C. 1942. The effect of crowding upon the oviposition of grain-infesting insects. *J. exp. Biol.*, **19**: 311-40.

Crombie, A. C. 1944. On intraspecific and interspecific competition in larvae of graminivorous insects. *J. exp. Biol.*, **20**: 135-66.

CRONWRIGHT-SCHREINER, S. C. 1925. *The migratory springbucks of South Africa (the Trekbokke). Also an essay on the ostrich*, etc. London.

CROOK, J. H. 1958. Études sur le comportement social de *Bubalornis a. albirostris* (Vieillot). *Alauda*, 1958: 161-95.

CROWCROFT, P. 1954. The daily cycle of activity in British shrews. *Proc. zool. Soc. Lond.*, **123**: 715-29.

CROWCROFT, P. 1955. Notes on the behaviour of shrews. *Behaviour*, **8**: 63-80.

CROWCROFT, P., and F. P. ROWE. 1958. The growth of confined colonies of the wild house-mouse (*Mus musculus*): the effect of dispersal on female fecundity. *Proc. zool. Soc. Lond.*, **131**: 357-65.

CULLEN, J. M. 1957. Plumage, age and mortality in the arctic tern. *Bird Study*, **4**: 197-207.

CUMBER, R. A. 1949. The biology of humble-bees, with special reference to the production of the worker caste. *Trans. R. ent. Soc., Lond.*, **100**: 1-45.

CUMMINS, H. 1920. The role of voice and coloration in the spring migration and sex recognition of frogs. *J. exp. Zool.*, **30**: 325-43.

CUSHING, D. H. 1951. The vertical migration of planktonic Crustacea. *Biol. Rev.*, **26**: 158-92.

DAHL, E. 1959. The amphipod, *Hyperia galba*, an ectoparasite of the jellyfish, *Cyanea capillata*. *Nature, Lond.*, **183**: 1749.

DAHLGREN, U., and C. F. SILVESTER. 1906. The electric organ of the stargazer, *Astroscopus* Brevoort. *Anat. Anz.*, **29**: 387-403.

DALQUEST, W. W. 1947. Notes on the natural history of the bat *Corynorhinus rafinesquii* in California. *J. Mammal.*, **28**: 17-30.

DARLING, F. F. 1937. *A herd of red deer: a study in animal behaviour*. London.

DARLING, F. F. 1938. *Bird flocks and the breeding cycle*. Cambridge.

DARLINGTON, P. J. 1938. Experiments on mimicry in Cuba. *Trans. R. ent. Soc., Lond.*, **87**: 681-95.

DARNTON, I. 1958. The display of the manakin, *M. manacus*. *Ibis*, **100**: 52-8.

DARWIN, C. 1845. *Journal of researches. . . .* London. (Page references are given to 10th Ed., 1891.)

DARWIN, C. 1868. *The variation of animals and plants under domestication*. 2 vols., London. (Page references are given to the popular Ed., 1905.)

DARWIN, C. 1874. *The descent of man*. 2nd Ed. 2 vols., London. (Page references are given to John Murray's new Ed., 1901.)

DASMANN, R. F., and R. D. TABER. 1950. Columbian black-tailed deer with reference to population ecology. *J. Mammal.*, **37**: 143-64.

DAVIS, D. E. 1940*a*. Social nesting habits of the smooth-billed ani. *Auk*, **57**: 179-218.

DAVIS, D. E. 1940*b*. Social nesting habits of *Guira guira*. *Auk*, **57**: 472-84.

DAVIS, D. E. 1946. A seasonal analysis of mixed flocks of birds in Brazil. *Ecology*, **27**: 168-81.

DAVIS, D. E. 1951. The relation between level of population and pregnancy of normal rats. *Ecology*, **32**: 459-61.

DAVIS, P. 1957. The breeding of the storm petrel. *Brit. Birds*, **50**: 85-101, 371-84.

DAVIS, P. 1961. The arctic skua colony, 1960. *Fair Isle Bird Obs. Bull.*, **4**: 162-5.

DAVIS, T. A. W. 1934. Notes on display in the humming-birds. *Ibis*, 1934: 732-8.

DAVIS, T. A. W. 1958. The displays and nests of three forest humming-birds of British Guiana. *Ibis*, **100**: 31-9.

DAWSON, JANET. 1956. Splenic hypertrophy in voles. *Nature, Lond.*, **178**: 1183-4.

DEANESLY, RUTH. 1944. The reproductive cycle of the female weasel (*Mustela nivalis*). *Proc. zool. Soc. Lond.*, **114**: 339-49.

DEEGENER, P. 1918. *Die Formen der Vergesellschaftung im Tierreiche*. Leipzig.

DEHORNE, A. 1924. Multiplication asexuée chez Dodecaceria du Portel par émiettement métamérique, ou processus de cténodrilisation. *C.R. Acad. Sci., Paris*, **178**: 143-5.

DEHORNE, A. 1927. Le cycle reproducteur annuel de *Dodecaceria concharum* au Portel. La schizométamétrie. *C.R. Acad. Sci., Paris,* **184**: 547-9.

DELMÉE, E. 1954. Douze années d'observations sur le comportement du pigeon columbin (*Columba oenas* L.). *Gerfaut,* **44**: 193-259.

DE VARIGNY, H. 1894. Recherches sur le nanisme expérimentale. Contribution à l'étude de l'influence du milieu sur les organismes. *J. Anat., Paris,* **30**: 147-88.

DEWAR, J. M. 1933. Northumberland rook roosts. *Brit. Birds,* **27**: 103-4.

DIAKONOV, D. M. 1925. Experimental and biometrical investigations on dimorphic variability of *Forficula. J. Genet.,* **15**: 201-32.

DIJKGRAAF, S. 1955. Lauterzeugung und Schallwahrnehmung bei der Languste (*Palinurus vulgaris*). *Experientia,* **11**: 330-1.

DITMARS, R. L. 1907. *The reptile book.* New York. (Page references are from the 1931 impression.)

DIVER, C. 1940. The problem of closely related species living in the same area. In *The new systematics,* edited by J. Huxley, London : 303-28.

DOBRIN, M. B. 1947. Measurements of underwater noise produced by marine life. *Science,* **105**: 19-23.

DOBZHANSKY, T. 1941. *Genetics and the origin of species.* New York.

DONISTHORPE, H. ST. J. K. 1927. *British ants, their life-history and classification.* 2nd Ed., London.

DONISTHORPE, H. 1936. The oldest insect on record. *Ent. Rec.,* **48**: 1-2.

DONKER, J. K. 1959. Migration and distribution of the wigeon, *Anas penelope* L., in Europe, based on ringing results. *Ardea,* **47**: 1-27.

DOWNES, J. A. 1955. Observations on the swarming flight and mating of *Culicoides. Trans. R. ent. Soc., Lond.,* **106**: 213-36.

DRESSER, H. E. 1871-90. *A history of the birds of Europe.* 9 vols. London.

DUGMORE, A. R. 1913. *The romance of the Newfoundland caribou.* London.

DUNBAR, M. J. 1956. The *Calanus* expeditions in the Canadian Arctic, 1947 to 1955. *Arctic,* **9**: 178-90.

DUNBAR, M. J. 1957. The determinants of production in northern seas: a study of the biology of *Themisto libellula* Mandt. *Canad. J. Zool.,* **35**: 797-819.

DUNNET, G. M. 1955. The breeding of the starling *Sturnus vulgaris* in relation to its food supply. *Ibis,* **97**: 619-62.

EDWARDS, T. G. 1910. On the procession and pupation of the larva of *Cnethocampa pinivora. Proc. Camb. phil. Soc.,* **15**: 431-6.

EGGELING, W. J. 1955. The breeding birds of the Isle of May. *Scot. Nat.,* **67**: 72-89.

EGGERS, F. 1927. Nähere Mitteilungen über das Johnstonsche Sinnesorgan und über das Ausweichvermögen der Taumelkäfer. *Zool. Anz.,* **71**: 136-56.

EHLERS, E. 1898. Ueber Palolo (*Eunice viridis* Gr.). *Nachr. K. Ges. Wiss., Göttingen, Math.-phys. Kl.,* 1898: 400-15.

ELKAN, E. 1960. The common toad (*Bufo bufo L.*) in the laboratory. *Brit. J. Herpet.,* **2**: 177-82.

ELMHIRST, R. 1912. Some observations on the glow-worm (*Lampyris noctiluca*). *Zoologist,* ser. 4, **16**: 190-2.

ELTON, C. S. 1924. Periodic fluctuations in the numbers of animals: their causes and effects. *Brit. J. exp. Biol.,* **2**: 119-63.

ELTON, C. S. 1942. *Voles, mice and lemmings.* Oxford.

ELTON, C., E. B. FORD, J. R. BAKER and A. D. GARDNER. 1931. The health and parasites of a wild mouse population. *Proc. zool. Soc., Lond.,* 1931: 657-721.

ELTON, C. S., and M. NICHOLSON. 1942. The ten-year cycle in numbers of the lynx in Canada. *J. Anim. Ecol.,* **11**: 215-44.

ELTRINGHAM, H. 1933. *The senses of insects.* London.

ELTRINGHAM, S. K., and A. R. HOCKLEY. 1958. Coexistence of three species of the wood-boring isopod *Limnoria* in Southampton Water. *Nature, Lond.,* **181**: 1659-60.

ENOCK, F. 1885. The life-history of *Atypus piceus*. *Trans. ent. Soc., Lond.*, 1885: 394.

ERRINGTON, P. L. 1954. On the hazards of overemphasizing numerical fluctuations in studies of 'cyclic' phenomena in muskrat populations. *J. Wildlife Mgt.*, **18**: 66-90.

ERRINGTON, P. L. 1956. Factors limiting higher vertebrate populations. *Science*, **124**: 304-7.

ESPINAS, A. 1878. *Des sociétés animales*. Paris.

ESTERLY, C. O. 1912. The occurrence and vertical distribution of the Copepoda of the San Diego region. *Univ. Calif. Publ. Zool.*, **9**: 253-340.

ESTERLY, C. O. 1914. A study of the occurrence and manner of distribution of the Ctenophora of the San Diego region. *Univ. Calif. Publ. Zool.*, **13**: 21-38.

ESTERLY, C. O. 1919. Reactions of various plankton animals with reference to their diurnal migrations. *Univ. Calif. Publ. Zool.*, **19**, no. 1: 1-83.

EVANS, H. MUIR. 1940. *Brain and body of fish: A study of brain pattern in relation to hunting and feeding in fish*. Philadelphia.

EVANS, L. T. 1938. Cuban field studies on territoriality of the lizard, *Anolis sagrei*. *J. comp. Psych.*, **25**: 97-125.

EVANS, W. 1922. Edinburgh rookeries in 1921. *Scot. Nat.*, 1921: 9-12.

EWART, J. C. 1888. The electric organs of the skate. *Philos. Trans.*, **179** B: 399-416, 539-56.

FABRE, J. H. 1879-1909. *Souvenirs entomologiques*. Series I-X. (The series are quoted by roman numerals, and the pages from the standard Delagrave edition.)

FAGE, L., and R. LEGENDRE. 1923. Les danses nuptiales des quelques Néréidiens. *C.R. Acad. Sci. Paris*, **177**: 1150-2.

FARRAN, G. P. 1947. Vertical distribution of plankton (*Sagitta*, *Calanus* and *Metridia*) off the south coast of Ireland. *Proc. R. Irish Acad.*, **51**: 121.

FERRAR, M. L. 1934. Daily flighting of flying foxes (*Pteropus giganteus* Brünn.). *J. Bombay nat. Hist. Soc.*, **37**: 214-5. [Not seen: quoted by Allen, 1939, p. 212.]

FISHER, H. D. 1956. Utilization of Atlantic harp seal populations. *Trans. 20th N. Amer. Wildl. Conf.*, 507-18. [Quoted from Scheffer, 1958.]

FISHER, J. 1952. *The fulmar*. London.

FISHER, J., and R. A. HINDE. 1949, 1951. The opening of milk bottles by birds. *Brit. Birds*, **42**: 347-57; **44**: 393-6.

FISHER, R. A., and E. B. FORD. 1928. The variability of species in the Lepidoptera, with reference to abundance and sex. *Trans. ent. Soc., Lond.*, **79**: 367-79.

FITZSIMONS, F. W. 1919-20. *The natural history of South Africa: Mammals*. 4 vols. London.

FLOWER, W. H., and R. LYDEKKER. 1891. *An introduction to the study of mammals, living and extinct*. London.

FOLK, G. E., M. R. MELTZER and R. E. GRINDELAND. 1958. A mammalian activity rhythm independent of temperature. *Nature, Lond.*, **181**: 1598.

FORBES, H. O. 1896. *A hand-book to the Primates*. 2 vols. London.

FORD, E. B. 1945. *Butterflies*. London.

FORSTER, G. R. 1959. The biology of the prawn, *Palaemon* (= *Leander*) *serratus* (Pennant). *J. mar. biol. Ass. U.K.*, **38**: 621-7.

FOX-WILSON, G. 1946. Factors affecting populations of social wasps, *Vespula* species, in England (Hymenoptera). *Proc. R. ent. Soc. Lond.* (A) **21**: 17-27.

FRADE, F. 1955. Proboscidiens. In Grassé's *Traité de Zoologie*, **17** (1): 715-83.

FRANK, A. 1941. Eigenartige Flugbahnen bei Hummelmännchen. *Z. vergl. Physiol.*, **28**: 467-84.

FRANK, P. W., C. D. BOLL and R. W. KELLY. 1957. Vital statistics of laboratory cultures of *Daphnia pulex* De Geer as related to density. *Physiol. Zool.*, **30**: 287-305.

FRASER, A., and J. T. STAMP. 1957. *Sheep husbandry and diseases*. London.

FRASER, F. C. 1937. On the development and distribution of the young stages of krill (*Euphausia superba*). '*Discovery*' *Rep.*, **14**: 1-192.

FRASER, F. C., and P. E. PURVES. 1953. Fractured earbones of blue whales. *Scot. Nat.*, **65**: 154-6.

FRAZER, J. F. D., and A. ST. G. HUGGETT. 1959. The breeding seasons and length of pregnancy in four species of large whales. *Proc. XV Int. Congr. Zool.* (*London*): 311-12.

FREE, J. B., and C. G. BUTLER. 1959. *Bumblebees.* London.

FREUCHEN, P. 1935. Field notes and biological observations. *Rep. 5th Thule Exped.* 1921-24, 2, Zoology, pt. 1, Mammals: 68-278.

FRIEDLAENDER, B. 1898. Ueber den sogennanten Palolowurm. *Biol. Clbt.*, **18**: 337-57.

FRISCH, K. VON 1936. Über den Gehörsinn der Fische. *Biol. Rev.*, **11**: 210-46.

FRYER, J. C. F. 1913. An investigation by pedigree breeding into the polymorphism of *Papilio polytes* Linn. *Philos. Trans.*, **204** B: 227-54.

FUGGLES-COUCHMAN, N. R. 1943. A contribution to the breeding ecology of two species of *Euplectes* (bishop-birds) in Tanganyika Territory. *Ibis*, **85**: 311-26.

GADOW, H. 1901. Amphibia and reptiles. In *Cambridge Natural History*, vol. 8. London.

GALLOWAY, A. R., and A. L. THOMSON. 1914. Notes on high mortality among young common terns in certain seasons. *Scot. Nat.*, 1914: 271-8.

GALLOWAY, T. W., and P. S. WELCH. 1911. Studies on a phosphorescent Bermudan annelid, *Odontosyllis enopla* Verrill. *Trans. Amer. micr. Soc.*, **30**: 13-39.

GAMULIN, T., and J. HURE. 1956. Spawning of the sardine at a definite time of day. *Nature, Lond.*, **177**: 193-4.

GARDINER, A. C. 1933. Vertical distribution in *Calanus finmarchicus*. *J. mar. biol. Assoc. U.K.*, **18**: 575-610.

GARTEN, S. 1914. Die elektrischen Erscheinungen an elektrischen Organen der Zitterfische. *Wintersteins Handbuch der vergleichenden Physiologie*, 3, pt. 2: 170-212.

GAULD, D. T. 1953. Diurnal variations in the grazing of planktonic copepods. *J. mar. biol. Assoc. U.K.*, **31**: 461-74.

GAUSE, G. F. 1934. *The struggle for existence.* Baltimore, Md.

GEE, E. P. 1953*a*. The life-history of the great Indian one-horned rhinoceros (*R. unicornis* Linn.). *J. Bombay nat. Hist. Soc.*, **51**: 341-8.

GEE, E. P. 1953*b*. Further observations on the great Indian one-horned rhinoceros. *J. Bombay nat. Hist. Soc.*, **51**: 765-72.

GIBB, J. 1950. The breeding biology of the great and blue titmice. *Ibis*, **92**: 507-39.

GIESBRECHT, W. 1892. Pelagische Copepoden. *Fauna und Flora des Golfes von Neapel*, **19**: 831 pp. and 54 pl. Berlin.

GILBERT, H. A., and A. BROOK. 1925. *The secrets of the eagle and of other rare birds.* London.

GILLESPIE, T. H. 1932. *A book of king penguins.* London.

GILLIARD, E. T. 1956. Bower ornamentation versus plumage characters in bower-birds. *Auk*, **73**: 450-1.

GOETHE, F. 1937. Beobachtungen und Untersuchungen zur Biologie der Silbermöwe (*Larus a. argentatus* Pontopp.) auf der Vogelinsel Memmertsand. *J. Orn., Lpz.*, **85**: 1-119.

GOETHE, F. 1956. *Die Silbermöwe.* Wittenberg Lutherstadt.

GOETHE, F. 1961. The moult gatherings and moult migrations of shelduck in north-west Germany. *Brit. Birds*, **54**: 145-61.

GOETSCH, W. 1924. Lebensraum und Korpergrösse. *Biol. Zbl.*, **44**: 529-60.

GOODE, G. W., and T. H. BEAN. 1896. Oceanic ichthyology. *Mem. Mus. comp. Zool. Harvard*, **22**: 1-533. Reprinted in *Smithsonian misc. Coll.*, **30** (1895), xxxv+529 pp., 123 pl.

GOODSIR, H. D. S. 1843. On the maidre of the fishermen. *Edin. new Philos. J.*, ser. 1, **35**: 102-4.

GOODWIN, D. 1952. Notes and display of the magpie. *Brit. Birds*, **45**: 113-22.

GORDON, SETON. 1955. *The golden eagle, king of birds.* London.

GOSSE, P. H. 1840. *The Canadian naturalist.* London.

GRAF, W. 1956. Territorialism in deer. *J. Mammal.*, **37**: 165-70.

GRAHAM, M. 1956. *Sea fisheries, their investigation in the United Kingdom.* London.

GRANT, E. C., and M. R. A. CHANCE. 1958. Rank order in caged rats. *Brit. J. Anim. Behav.*, **6**: 183-94.

GRAY, J. E. 1847. An account of Palolo, a sea worm eaten in the Navigator Islands, by the Rev. J. B. Stair, with a description by J. E. Gray. *Proc. zool. Soc. Lond.*, **15**: 17-18.

GREEN, R. G., and C. L. LARSON. 1938. A description of shock disease in the snowshoe hare. *Amer. J. Hyg.*, **28**: 190-212.

GREEN, R. G., C. L. LARSON and J. F. BELL. 1939. Shock disease as the cause of the periodic decimation of the snowshoe hare. *Amer. J. Hyg.*, **30**, sec. B: 83-102.

GREENE, C. W. 1924. Physiological reactions and structure of the vocal apparatus of the California singing fish. *Amer. J. Physiol.*, **70**: 496-9.

GREENE, C. W. 1930. The smelts of Lake Champlain. In 'A biological survey of the Champlain watershed'. *19th ann. Rep. N.Y. State Cons. Dept.*, suppl.: 105-29.

GROSKIN, H. 1945. Chimney swifts roosting at Ardmore, Pennsylvania. *Auk*, **62**: 361-70.

GROSS, A. O. 1951. The herring-gull—cormorant control project. *Proc. Xth Int. orn. Congr. Uppsala:* 532-6.

GROSSKOPF, G. 1957. Das Durchschnittsalter der auf Wangerooge nistenden Küsten-seeschwalben (*Sterna macrura*). *J. Orn., Lpz.*, **98**: 65-70.

GUDMUNDSSON, F. 1951. The effects of the recent climatic changes on the bird life of Iceland. *Proc. Xth Int. orn. Congr. Uppsala:* 502-14.

GUNN, D. L. 1940. The daily rhythm of activity of the cockroach, *Blatta orientalis* L. *J. exp. Biol.*, **17**: 267-77.

GURNEY, J. H. 1913. *The gannet, a bird with a history.* London.

GUTHRIE, MARY J. 1933*a*. Notes on the seasonal movements and habits of some cave bats. *J. Mammal.*, **14**: 1-19.

GUTHRIE, MARY J. 1933*b*. The reproductive cycles of some cave bats. *J. Mammal.*, **14**: 199-216.

GWYNN, A. M. 1953. The egg-laying and incubation periods of rockhopper, macaroni and gentoo penguins. *ANARE Reps.*, B, **1**: 1-29.

HAARTMAN, L. VON 1955. Clutch-size in polygamous species. *Acta XI Congr. Int. Orn., Basel*, 450-3.

HAARTMAN, L. VON 1956. Territory in the pied flycatcher *Muscicapa hypoleuca. Ibis*, **98**: 460-75.

HAAS, A. 1946. Neue Beobachtungen zum Problem der Flugbahner bei Hummelmänn-chen. *Z. Naturf.*, **1**: 596-600.

HAAS, A. 1952. Die Mandibeldrüse als Duftorgan bei eigenen Hymenopteren. *Naturw.*, **39**: 484.

HADDOW, A. J. 1956. Rhythmic biting activity of certain East African mosquitoes. *Nature, Lond.*, **177**: 531-2.

HAINES, T. P. 1940. Delayed fertilisation in *Leptodeira annulata polysticta. Copeia*, **2**: 116-18.

HALE, W. G. 1956. The lack of territory in the redshank *Tringa totanus. Ibis*, **98**: 398-400.

HALME, E. 1937. Eine neue Methode zur Bestimmung der relativen Wanderungsintensität des Zooplanktons. *Acta Soc. Fauna Flora fenn.*, **60**: 347-73.

HAMERSTROM, F. and F. 1955. Population-density and behavior in Wisconsin prairie chickens. *Acta XI Congr. Int. Orn., Basel*, 459-66.

HAMILTON, J. STEVENSON-. 1947. *Wild life in South Africa.* London.

HAMM, A. H. 1908. Observations on *Empis livida. Ent. mon. Mag.*, **44**: 181-4.

HANEDA, Y. 1955. Luminous organisms of Japan and the Far East. In *The luminescence of biological systems*, ed. by F. H. Johnson: 335-85. Washington.

HARDY, A. C. 1936. Observations on the uneven distribution of oceanic plankton. '*Discovery*' *Rep.*, **11**: 511-38.

HARDY, A. C. 1938. Change and choice: a study in pelagic ecology. In *Evolution, essays . . . presented to Professor E. S. Goodrich . . .*: 139-59. Oxford.

HARDY, A. C. 1956. *The open sea.* London.

HARDY, A. C., and R. BAINBRIDGE. 1951. Effect of pressure on the behaviour of decapod larvae (Crustacea). *Nature, Lond.*, **167**: 354-5.

HARDY, A. C., and E. R. GUNTHER. 1935. The plankton of the South Georgia whaling ground and adjacent waters 1926-7. '*Discovery*' *Rep.*, **11**: 1-456.

HARKER, JANET E. 1954. Diurnal rhythms in *Periplaneta americana. Nature, Lond.*, **173**: 689-90.

HARKER, JANET E. 1955. Control of diurnal rhythms of activity in *Periplaneta americana* L. *Nature, Lond.*, **175**: 733.

HARPER, F. 1931. Amphibians and reptiles of the Athabaska and Great Slave Lakes region. *Canad. Field-Nat.*, **45**: 68-70.

HARRIS, J. E., and U. K. WOLFE. 1955. A laboratory study of vertical migration. *Proc. roy. Soc.*, B, **144**: 329-54.

HARTLEY, P. H. T. 1947. The natural history of some British freshwater fishes. *Proc. zool. Soc. Lond.*, **117**: 129-206.

HARTLEY, P. H. T. 1949. The biology of the mourning chat in winter quarters. *Ibis*, **91**: 393-413.

HARVEY, E. N. 1940. *Living light.* Princeton, N.J.

HARVEY, E. N. 1952. *Bioluminescence.* New York.

HASKELL, P. T. 1953. The stridulation behaviour of the domestic cricket. *Brit. J. Anim. Behav.*, **1**: 120-1.

HASLE, GRETE R. 1950. Phototactic vertical migration in marine dinoflagellates. *Oikos*, **2**: 162-74.

HASLER, A. D., and W. J. WISBY. 1951. Discrimination of stream odors by fishes and its relation to parent stream behavior. *Amer. Nat.*, **85**: 223-38.

HAUSMAN, L. A. 1925. On the utterances of the kingbird, *Tyrannus tyrannus* Linn., with reference to a recently recorded song. *Auk*, **42**: 320-6.

HAVERSCHMIDT, F. 1949. *The life of the white stork.* Leiden.

HAYES, M. H. 1922. *Points of the horse.* 4th Ed. New imp. London.

HEAPE, W. E. 1931. *Emigration, migration and nomadism.* London.

HEDIGER, H. 1941. Biologische Gesetzmässigkeiten in Verhalten von Wirbeltieren. *Mitt. Naturf. Ges. Bern* (1940). [Not seen.]

HEDIGER, H. 1955. *Studies of the psychology and behaviour of captive animals in zoos and circuses.* London.

HEMPELMANN, F. 1911. Zur Naturgeschichte von *Nereis dumerilii* Aud. and Edw. *Zoologica, Stuttgart*, **25**: 135 pp.

HEMPELMANN, F. 1931. Archiannelida und Polychaeta. In *Kükenthals Handbuch der Zoologie*, 2 Bd, 2 Hälfte, (7): 212 pp.

HENRY, G. M. 1955. *A guide to the birds of Ceylon.* London.

HENSHAW, H. W. 1876. Report on the ornithology . . . of California during the field-season of 1875. *Ann. Rep. geog. Surv. W. of 100th Merid.*, Appendix JJ: 224-78.

HENSHAW, H. W. 1921. Storage of acorns by the California woodpecker. *Condor*, **23**: 109-18.

HENSLEY, M. M., and J. B. COPE. 1951. Further data on removal and reproduction of the breeding birds in a spruce-fir forest community. *Auk*, **68**: 483-93.

HERRICK, C. J. 1924. *Neurological foundations of animal behavior.* New York.

HILDEBRAND, S. F. 1932. Growth of diamond-back terrapins—size attained, sex ratio and longevity. *Zoologica, N.Y.*, **9**: 551-63.

HILDEBRAND, S. F., and W. C. SCHROEDER. 1927. Fishes of Chesapeake Bay. *Bull. U.S. Bur. Fish.*, **43**, pt. 1: 366 pp.

Hindwood, K. A. 1956. Clustering of wood-swallows. *Emu*, **56**: 165-6.

Hingston, R. W. G. 1933. *The meaning of animal colour and adornment.* London.

Hobley, C. W. 1922. *Bantu beliefs and magic.* London.

Hoffmann, A. 1949. Über die Brutpflege des polyandrischen Wasserfasans, *Hydrophasianus chirurgus* (Scop.). *Zool. Jb.* (*Syst.*), **78**: 367-403.

Holme, N. A. 1950. Population-dispersion in *Tellina tenuis* da Costa. *J. mar. biol. Ass. U.K.*, **29**: 267-80.

Holmquist, A. M. 1926. Studies in arthropod hibernation. *Ann. ent. Soc. Amer.*, **19**: 395-428.

Holstein, V. 1942. Duehøgen—*Astur gentilis dubius* (Sparrmann). *Biol. Stud. over Danske Rovfugle*, **1**: 155 pp. Copenhagen.

Hornberger, F. 1943. Einige Ergebnisse zehnjähriger Planarbeit im ' Storchforschungskreis Insterburg ' der Vogelwarte Rossitten. *J. Orn., Lpz.*, **91**: 341-55.

Hornberger, F. 1944. Reifealter und Ansiedlung beim Weissen Storch. *Vogelwarte*, **17**: 114-49. [Not seen.]

Horst, R. 1893. On the habits of *Thalassina anomala* Herbst. *Notes from the Leyden Museum*, **15**: 314-15.

Hose, C. 1926. *Natural man: a record from Borneo.* London.

Howard, H. E. 1920. *Territory in bird life.* London.

Howlett, M. 1907. Note on the coupling of *Empis borealis*. *Ent. mon. Mag.*, **43**: 229-32.

Hudson, W. H. 1892. *The naturalist in La Plata.* London. (Pages quoted from New (4th) Ed., 1922.)

Hudson, W. H. 1915. *Birds and man.* London. (Pages quoted from New Readers Library Ed., 1927.)

Hunter, F. A. 1959. *Eburia quadrigeminata* Say (Col., Cerambycidae) a further British record of delayed emergence from furniture. *Ent. mon. Mag.*, **95**: 261.

Huntsman, A. G. 1948. *Odontosyllis* at Bermuda and lunar periodicity. *J. Fish. Res. Bd. Can.*, **7**: 363-9.

Husson, R. 1952. Attaque des pinèdes de la Sarre par l'insecte *Neodiprion sertifer* Geoffroy (= *Lophyrus rufus*, Hymenoptera). *Ann. Univ. Saraviensis, Sciences*, **1**: 71-9.

Hutton, J. A. 1924. *The life-history of the salmon.* Aberdeen.

Huxley, J. S. 1932. *Problems of relative growth.* London.

Huxley, J. S. 1934. A natural experiment on the territorial instinct. *Brit. Birds*, **27**: 270-7.

Hyman, L. H. 1955. *The Invertebrates: Echinodermata.* New York.

Ihering, R. von. 1930. Sur la voix des poissons d'eau douce. *C.R. Soc. Biol., Paris*, **103**: 1327-8.

Ilse, D. R. 1955. Olfactory marking of territory in two young male loris, *Loris tardigradus lydekkerianus*, kept in captivity in Poona. *Brit. J. Anim. Behav.*, **3**: 118-20.

Imms, A. D. 1931. *Recent advances in entomology.* London.

Imms, A. D. 1947. *Insect natural history.* London.

Imms, A. D. 1957. *A general textbook of entomology.* 9th Ed., revised by O. W. Richards and R. G. Davies. London.

Ingram, C. 1959. The importance of juvenile cannibalism in the breeding biology of certain birds of prey. *Auk*, **76**: 218-26.

Jacobs, Merle. 1953. Observations on the two forms of the periodical cicada. *Proc. Indiana Acad. Sci.*, **63**: 177-9.

Jacobs, M. E. 1955. Studies on territorialism and sexual selection in dragonflies. *Ecology*, **36**: 566-86.

Jenkins, D. W. 1944. Territory as a result of despotism and social organisation in geese. *Auk*, **61**: 30-47.

Jerdon, T. C. 1862-4. *The birds of India.* 3 vols. Calcutta.

JERDON, T. C. 1874. *The mammals of India*. London. (Reprint of 1867 Ed.)

JESPERSEN, P. 1924. The frequency of birds over the high Atlantic Ocean. *Nature, Lond.*, **114**: 281-3.

JESPERSEN, P. 1930. Ornithological observations in the North Atlantic Ocean. *Danish 'Dana'-Exped.* 1920-22, *Oceanogr. Rep.* no. 7: 1-36.

JESPERSEN, P. 1946. The breeding birds of Denmark with special reference to changes during the last century. Danish Section, Int. Ctee. Bird Preserv., Copenhagen.

JOHNSON, N. W., F. A. EVEREST and R. W. YOUNG. 1947. The role of snapping shrimp (*Crangon* and *Synalpheus*) in the production of underwater noise in the sea. *Biol. Bull. Wood's Hole*, **93**: 122-38.

JOLLIE, M. 1947. Plumage changes in the golden eagle. *Auk*, **64**: 549-76.

JONES, F. M. 1931. The gregarious sleeping habit of *Heliconius charithonea* L. *Proc. ent. Soc., Lond.*, **6**: 4-10.

JONES, J. W. 1959. *The salmon*. London.

JOURDAIN, F. C. R. 1927. Review of 'How Birds Live'. *Brit. Birds*, **21**: 71-2.

JUDD, W. W. 1956. Red-headed woodpeckers (*Melanerpes erythrocephalus*) feeding on Carolina locusts. *Auk*, **73**: 285-6.

JUMBER, J. F. 1956. Roosting behavior of the starling in central Pennsylvania. *Auk*, **73**: 411-26.

JUST, E. E. 1914. Breeding habits of the heteronereis form of *Platynereis megalops* at Wood's Hole. *Biol. Bull. Wood's Hole*, **27**: 201-12.

KALABUKHOV, N. I. 1939. [Some ecological peculiarities of closely-allied species of rodents.] *Probl. Ecol. Biocen. Leningr.*, **7**: 92-112. (In Russian. Not seen; ref. copied from Southern, 1954*a*, p. 98.)

KALELA, O. 1949. Über Fjeldlemming-Invasionen und andere irreguläre Tierwanderungen. *Ann. zool. Soc. 'Vanamo'*, **13**, no. 5: 1-90.

KALELA, O. 1954. Über den Revierbesitz bei Vögeln und Säugetieren als populationsökologischer Faktor. *Ann. zool. Soc. 'Vanamo'*, **16**, no. 2: 48 pp.

KALLEBERG, H. 1958. Observations in a stream tank of territoriality and competition in juvenile salmon and trout (*Salmo salar* L. and S. *trutta* L.). *Inst. freshw. Res. Drottningholm*, **39**: 55-98.

KALMBACH, E. R. 1932. Winters tarling roosts of Washington. *Wilson Bull.*, **44**: 65-74.

KEENLEYSIDE, M. H. A. 1955. Some aspects of the schooling behaviour of fish. *Behaviour*, **8**: 183-247.

KEIGHLEY, J., and R. M. LOCKLEY. 1948. The incubation and fledging-periods of the razorbill. *Brit. Birds*, **41**: 113-14.

KELLOGG, W. N. 1958. Echo ranging in the porpoise. *Science*, **128**: 982-8.

KELLOGG, W. N., R. KOHLER and H. N. MORRIS. 1953. Porpoise sounds as sonar signals. *Science*, **117**: 239-43.

KENDEIGH, S. C. 1941. Territorial and mating behavior of the house wren. *Ill. biol. Monogr.*, **18**: 1-120.

KENDEIGH, S. C. 1952. Parental care and its evolution in birds. *Ill. biol. Monogr.*, 22, nos. 1-3: 356 pp.

KENDEIGH, S. C., and S. P. BALDWIN. 1937. Factors affecting yearly abundance of passerine birds. *Ecol. Monogr.*, **7**: 91-124.

KENNEDY, P. G., R. F. RUTTLEDGE and C. F. SCROOPE. 1954. *The birds of Ireland*. Edinburgh and London.

KETTLEWELL, B. D. H. 1946. Female assembling scents with reference to an important paper on the subject. *Entomologist*, **79**: 8-14.

KEY, K. H. L. 1950. A critique on the phase theory of locusts. *Quart. Rev. Biol.*, **25**: 363-407.

KIRBY, W., and W. SPENCE. 1858. *An introduction to entomology*. 7th Ed. London. (1st Ed., 1817, 2 vols.)

KLEERECOPER, H., and K. SIBAKIN. 1956*a*. An investigation of the electric 'spike' potentials produced by the sea lamprey (*Petromyzon marinus*) in the water surrounding the head region. *J. Fish. Res. Bd. Canad.*, **13**: 375-83.

KLEEREKOPER, H., and K. SIBAKIN. 1956*b*. Spike potentials produced by the sea lamprey (*Petromyzon marinus*) in the water surrounding the head region. *Nature, Lond.*, **178**: 490-1.

KLOTS, A. B. 1951. *A field guide to the butterflies of North America, east of the Great Plains*. Boston, Mass.

KLUYVER, H. N. 1933. Bijdrage tot de biologie en de ecologie van den Spreeuw (*Sturnus vulgaris vulgaris* L.) gedurende zijn voortplantingstijd. *Versl. Plziekt. Dienst Wageningen*, 69.

KLUYVER, H. N. 1951. The population ecology of the great tit, *Parus m. major* L. *Ardea*, **39**: 1-135.

KLUYVER, H. N. 1955. Das Verhalten des Drosselrohrsängers, *Acrocephalus arundinaceus* (L.), am Brutplatz, mit besonderer Berücksichtung der Nestbautechnik und der Revierbehauptung. *Ardea*, **43**: 1-50.

KLUYVER, H. N., and L. TINBERGEN. 1953. Territory and the regulation of density in titmice. *Arch. néerl. Zool.*, **10**: 265-89.

KNIGHT-JONES, E. W., and S. Z. QASIM. 1955. Responses of some marine plankton animals to changes in hydrostatic pressure. *Nature, Lond.*, **175**: 941-2.

KNIGHT-JONES, E. W., and J. P. STEVENSON. 1951. Gregariousness during settlement in the barnacle *Elminius modestus* Darwin. *J. mar. biol. Ass. U.K.*, **29**: 281-97.

KNUDSEN, V. O., R. S. ALFORD and J. W. EMLING. 1948. Underwater ambient noise. *J. mar. Res.*, **7**: 410-29.

KOERSVELD, E. VAN. 1958. A few data on the reproduction of the rook *Corvus f. frugilegus* L. *Ardea*, **46**: 58-62.

KOFOID, C. B. 1953. The California condor. *National Audubon Society, Research Rep.* no. 4: xiii+154 pp. New York.

KORRINGA, P. 1957. Lunar periodicity. *In* 'Treatise on marine ecology and palaeontology', Chap. 27. *Geol. Soc. Amer., Mem.* 67, **1**: 917-34.

KOSKIMIES, J. 1955. Juvenile mortality and population balance in the velvet scoter (*Melanitta fusca*) in maritime conditions. *Acta XI Congr. Int. Orn., Basel* (1954): 476-9.

KOSKIMIES, J. 1949. Some methodological notes concerning the waterfowl census in the archipelago. *Papers on Game Research* (*Helsinki*), no. 3: 18 pp.

KOSKIMIES, J., and E. ROUTAMO. 1953. Zur Fortpflantzungsbiologie der Samtente *Melanitta f. fusca* (L.). I. Allegemeine Nistökologie. *Papers on Game Research* (*Helsinki*), no. 10: 105 pp.

KROPOTKIN, P. 1902. *Mutual aid, a factor in evolution.* London.

KUHL, W. 1928. Die Variabilität der abdominalen Körperanhänge bei *Forficula auricularia* L. unter Berücksichtigung ihrer normalen und abnormen Entwicklung, nebst einem Anhang über die Geschlechtsbiologie. *Z. Morph. Oek. Tiere*, **12**: 299-532.

KULLENBERG, B. 1947. Sound emitted by dolphins. *Nature, Lond.*, **160**: 648.

LABITTE, A. 1916. Longévité de quelques insectes en captivité. *Bull. Mus. Hist. nat. Paris*, **22**: 105-13.

LACK, D. 1935. Territory and polygamy in a bishop-bird, *Euplectes hordeacea hordeacea* (Linn.). *Ibis*, 13th Ser., **5**: 817-36.

LACK, D. 1939. The display of the blackcock. *Brit. Birds*, **32**: 290-303.

LACK, D. 1943. *The life of the robin.* London.

LACK, D. 1943-44. The problem of partial migration. *Brit. Birds*, **37**: 122-30, 143-50.

LACK, D. 1948. The significance of clutch-size. Part III.—Some interspecific comparisons. *Ibis*, **90**: 25-45.

LACK, D. 1950. Breeding seasons in the Galapagos. *Ibis*, **92**: 268-78.

LACK, D. 1952. Reproductive rate and population density in the great tit: Kluijver's study. *Ibis*, **94**: 167-73.

LACK, D. 1954*a*. *The natural regulation of animal numbers.* Oxford.

LACK, D. 1954*b*. The stability of the heron population. *Brit. Birds*, **47**: 111-19.

LACK, D. 1955. British tits (*Parus* spp.) in nesting boxes. *Ardea*, **43**: 50-84.

LACK, D. 1956. *Swifts in a tower.* London.

LACK, D., and H. N. SOUTHERN. 1949. Birds on Tenerife. *Ibis*, **91**: 607-26.

LANG, H., and J. P. CHAPIN. 1917. Notes on the distribution and ecology of Centra African Chiroptera. *Bull. Amer. Mus. nat. Hist.*, **37**: 479-96.

LANYON, W. E. 1956*a*. Ecological aspects of the sympatric distribution of meadowlarks in the north-central states. *Ecology*, **37**: 98-108.

LANYON, W. E. 1956*b*. Territory in the meadowlarks, genus *Sturnella*. *Ibis*, **98**: 485-9.

LA TOUCHE, J. D. D. 1921. Notes on the birds of north-east Chihli, in north China. *Ibis*, 11th ser., **3**: 3-48.

LA TOUCHE, J. D. D. 1934. *A handbook of the birds of eastern China.* Vol. 2. London.

LAWS, R. M. 1953, 1956. The elephant seal (*Mirounga leonina* Linn.). I. Growth and Age. *Falkland Is. Dep. Surv., Sci. Rep.*, no. 8: 62 pp. II. General, special and reproductive behaviour. Ibid., no. 13: 88 pp.

LEBOUR, M. V. 1928. The larval stages of the Plymouth Brachyura. *Proc. zool. Soc. Lond.*, 1928: 473-560.

LEE, ROSA M. 1912. An investigation into the methods of growth determination in fishes. *Cons. perm. Int. Expl. Mer., Publ. Circonst.*, no. 63: 35 pp.

LEGG, K., and F. A. PITELKA. 1956. Ecologic overlap of Allen and Anna hummingbirds nesting at Santa Cruz, California. *Condor*, **58**: 393-405.

LEHTONEN, L. 1951. Zur herbstlichen Ethologie des Ziegenmelkers, *Caprimulgus e. europaeus*. *Ornis Fennica*, **28**: 89-109.

LEITCH, I., F. E. HYTTEN and W. Z. BILLEWICZ. 1959. The maternal and neonatal weights of some Mammalia. *Proc. zool. Soc. Lond.*, **133**: 11-28.

LEVICK, G. M. 1914. *Antarctic penguins, a study of their social habits.* London.

LHOSTE, J. 1942. Les cerques des Dermaptères. *Bull. biol.*, **76**: 192-201.

LILLIE, F., and E. E. JUST. 1913. Breeding habits of the heteronereis form of *Nereis limbata* at Wood's Hole, Mass. *Biol. Bull. Wood's Hole*, **24**: 147-60.

LINDROTH, A. 1955. Distribution, territorial behaviour and movements of sea trout fry in the River Indälsalven. *Rep. Inst. freshw. Res., Drottningholm*, **36**: 104-19.

LINDSAY, E. 1940. The biology of the silverfish, *Ctenolepisma longicaudata* Esch., with particular reference to its feeding habits. *Proc. roy. Soc. Victoria*, **52**: 35-83.

LISSMANN, H. W. 1951. Continuous electrical signals from the tail of a fish *Gymnarchus niloticus* Cuv. *Nature, Lond.*, **167**: 201-2.

LISSMANN, H. W. 1958. On the function and evolution of electric organs in fish. *J. exp. Biol.*, **35**: 156-91.

LIVENGOOD, W. F. 1937. An experimental analysis of certain factors affecting growth of goldfishes in homotypically conditioned water. *Copeia*, 1937: 81-8.

LLOYD, L. 1867. *The game birds and wild fowl of Sweden and Norway*, etc. London.

LOCKIE, J. D. 1955. The breeding habits and food of short-eared owls after a vole plague. *Bird Study*, **2**: 53-69.

LOCKIE, J. D. 1956. Winter fighting in feeding flocks of rooks, jackdaws and carrion crows. *Bird Study*, **3**: 180-90.

LOCKLEY, R. M. 1942. *Shearwaters.* London.

LONG, D. B. 1953. Effects of population density on larvae of Lepidoptera. *Trans. R. ent. Soc., Lond.*, **104**: 543-84.

LONG, D. B. 1955. Observations on sub-social behaviour in two species of lepidopterous larvae, *Pieris brassicae* L., and *Plusia gamma* L. *Trans. R. ent. Soc., Lond.*, **106**: 421-37.

LONGFIELD, C. 1949. *The dragonflies of the British Isles.* London.

LORENZ, K. 1931. Beiträge zur Ethologie sozialer Corviden. *J. Orn., Lpz.*, **79**: 67-127.

LOWE, F. A. 1954. *The Heron.* London.

LOWE, P. R. 1931. On the relations of the Gruimorphae to the Charadriimorphae and Rallimorphae, etc. *Ibis*, 1931: 491-534.

LOWERY, G. H., Jr. 1939. Vaux swift in Louisiana. *Wilson Bull.*, **51**: 200.

LUBBOCK, J. 1888. Observations on ants, bees, and wasps. Part XI. *J. Linn. Soc.* (*Zool.*), **20**: 118-36.

LUCAS, C. E. 1938. Some aspects of integration in plankton communities. *J. Cons. int. Explor. Mer*, **13**: 309-22.

LUCAS, C. E. 1947. The ecological effects of external metabolites. *Biol. Rev.*, **22**: 270-95.

LUCAS, C. E. 1956. External metabolites in the sea. In *Papers in Marine Biology and Oceanography*, London, pp. 139-48.

LYDDEKER, R. 1893-4. *The royal natural history.* 6 vols. London.

MCBRIDE, A. F., and D. O. HEBB. 1948. Behavior of the captive bottle-nose dolphin, *Tursiops truncatus. J. comp. physiol. Psych.*, **41**: 111-23.

MCCLURE, T. J. 1958. Temporary nutritional stress and infertility in mice. *Nature, Lond.*, **181**: 1132.

MACDONALD, T. H. 1959. Estimates of the length of larval life in three species of lamprey found in Britain. *J. Anim. Ecol.*, **28**: 293-7.

MACGILLIVRAY, J. 1860. Zoological notes from Aneiteum, New Hebrides. *Zoologist*, **18**: 7133.

MACGILLIVRAY, W. 1837-52. *A history of British birds.* 5 vols. London.

MACGINITIE, G. E. and N. 1949. *Natural history of marine animals.* New York.

MCINTOSH, W. C. 1885. Report on the Annelida Polychaeta collected by H.M.S. Challenger during the years 1873-76. '*Challenger*' *Rep.*, vol. 12.

MCINTYRE, A. D. 1956. The use of trawl, grab and camera in estimating marine benthos. *J. mar. Ass. U.K.*, **35**: 419-29.

MACKINTOSH, N. A. 1942. The southern stocks of whalebone whales. '*Discovery*' *Rep.*, **22**: 197-300.

MACKINTOSH, N. A., and J. G. F. WHEELER. 1929. Southern blue and fin whales. '*Discovery*' *Rep.*, **1**: 257-540.

MACLAGAN, D. S. 1932*a*. The effect of population density upon rate of reproduction with special reference to insects, *Proc. roy. Soc. B.*, **111**: 437-54.

MACLAGAN, D. S. 1932*b*. An ecological study of the 'lucerne flea' (*Smynthurus viridis*, Linn.). *Bull. ent. Res.*, **23**: 101-90.

MACLAGAN, D. S., and E. DUNN. 1936. The experimental analysis of the growth of an insect-population. *Proc. roy. Soc. Edin.*, **55**: 126-39.

MARCHANT, S. 1942. Some birds of the Owerri Province, S. Nigeria. *Ibis*, 14th ser., **6**: 137-96.

MAREY, E. J. 1879. Nouvelles récherches sur les poissons électriques: charactères de la décharge du Gymnotus, effets d'une décharge de Torpille lanceé dans une telephone. *C.R. Acad. Sci., Paris*, **88**: 318-21.

MARINE BIOLOGICAL ASSOCIATION. 1957. *Plymouth marine fauna.* 3rd Ed. Plymouth.

MARLATT, C. L. 1907. The periodical cicada. *U.S. Dept. Agri., Div. Ent., Bull.* 71: 181 pp.

MARPLES, B. J. 1934. The winter starling roosts of Great Britain. *J. Anim. Ecol.*, **3**: 187-203.

MARSHALL, A. J. 1954. *Bower-birds.* Oxford.

MARSHALL, A. J. 1957. On the function of 'clustering' in wood-swallows. *Emu*, **57**: 53-4.

MARSHALL, F. H. A. 1905. Fertility in Scottish sheep. *Proc. roy. Soc.*, B, **77**: 58-62.

MARSHALL, H. 1947. Longevity of the American herring gull. *Auk*, **64**: 188-98.

MARSHALL, N. B. 1954. *Aspects of deep sea biology.* London.

MARSHALL, S. M., and A. P. ORR. 1955. The biology of a marine copepod *Calanus finmarchicus* (Gunnerus). Edinburgh.

MARSHALL, S. M., A. P. ORR and C. B. REES. 1953. *Calanus finmarchicus* and related forms. *Nature, Lond.*, **171**: 1163-4.

MARTIN, M. 1698. *A late voyage to St. Kilda*, etc. London.

MASURE, R. H., and W. C. ALLEE. 1934*a*. The social order in flocks of the common chicken and the pigeon. *Auk*, **51**: 306-27.

MASURE, R. H., and W. C. ALLEE. 1934*b*. Flock organisation of the shell-parakeet, *Melopsittacus undulatus* Shaw. *Ecology*, **15**: 388-98.

MATTHEWS, L. H. 1929. The birds of South Georgia. '*Discovery*' *Rep.*, **1**: 561-92.

MATTHEWS, L. H. 1935. The oestrous cycle and intersexuality in the female mole (*Talpa europaea* Linn.) *Proc. zool. Soc. Lond.*, 1935, pt. 2: 347-83.

MATTHEWS, L. H. 1939. The bionomics of the spotted hyaena, *Crocuta crocuta*. *Proc. zool. Soc., Lond.*, Ser. A, **109**: 43-56.

MAYAUD, N. 1950. Biologie de la Reproduction. In *Traité de Zoologie*, tome 15, *Oiseaux*: 539-653. Paris.

MAYNARD, C. J. 1888. Notes on the breeding habits of the American flamingo. *Oologist*, **5**: 108-10. (Quoted from A. C. Bent (1926) Life histories of North American marsh birds. *U.S. nat. Mus. Bull.*, **135**: 4.)

MAYR, E. 1935. Bernard Altum and the territory theory. *Proc. Linn. Soc. N.Y.*, nos. 45, 46: 24-38.

MAYR, E. 1942. *Systematics and the origin of species*. New York.

MAYR, E., and D. AMADON. 1951. A classification of recent birds. *Amer. Mus. Novit.*, no. 1496: 42 pp.

MEEK, A. 1916. *The migrations of fish*. London.

MEIKLEJOHN, M. F. M. 1937. Adult pied wagtails using a roost in June. *Brit. Birds*, **31**: 85 and 124.

MEINERTZHAGEN, R. 1954. *Birds of Arabia*. Edinburgh.

MEINERTZHAGEN, R. 1956. Roost of wintering harriers. *Ibis*, **98**: 535.

MEIXNER, J. 1933-36. Coleopteroidea. In *Kükenthals Handbuch der Zoologie*, 4 Bd., 2 Hälfte, 1 Pt.: 1037-1348. Berlin.

MENZIES, R. J., and T. M. WIDRIG. 1955. Aggregation by the marine wood-boring isopod, *Limnoria*. *Oikos*, **6**: 149-52.

MENZIES, W. J. M. 1931. *The salmon: its life story*. New Ed. Edinburgh and London.

MENZIES, W. J. M. 1949. *The stock of salmon* . . . being the Buckland Lectures for 1947. London.

MENZIES, W. J. M., and W. M. SHEARER. 1957. Long-distance migration of salmon. *Nature, Lond.*, **179**: 790.

MICHAEL, E. L. 1911. Classification and vertical distribution of the Chaetognatha of the San Diego region. *Univ. Calif. Publ. Zool.*, **8**: 21-186.

MICHENER, C. D. and M. H. 1951. *American social insects*. New York.

MIERS, E. J. 1883. On the species of *Ocypoda* in the collection of the British Museum. *Ann. Mag. nat. Hist.*, 5th ser., **10**: 376-88.

MILLAIS, J. G. 1892. *Game birds and shooting sketches*. London.

MILLAIS, J. G. 1895. *A breath from the veldt*. London.

MILLAIS, J. G. 1904-6. *The mammals of Great Britain and Ireland*. 3 vols. London.

MILLER, A. H. 1955. Breeding cycles in a constant equatorial environment in Colombia, South America. *Acta XI Congr. Int. Orn.*, 495-503.

MILLER, R. S. 1955. Activity rhythms in the wood mouse, *Apodemus sylvaticus*, and the bank vole, *Clethrionomys glareolus*. *Proc. zool. Soc. Lond.*, **125**: 505-19.

MITCHELL, P. CHALMERS. 1912. *The childhood of animals*. London.

MOFFAT, C. B. 1903. The spring rivalry of birds: some views on the limit to multiplication. *Irish Naturalist*, **12**: 152-66.

MOFFAT, C. B. 1931, 1932. A pied wagtail roost in Dublin. *Brit. Birds*, **24**: 364-66; **26**: 93-4.
MØHL-HANSEN, U. 1954. Investigation on reproduction and growth of the porpoise (*Phocaena phocaena* (L.)) from the Baltic. *Vidensk. Medd. Dansk naturh. Forening*, **106**: 369-96.
MOHR, C. E. 1933. Observations on the young of cave-dwelling bats. *J. Mammal.*, **14**: 49-53.
MÖHRES, F. P. 1957. Elektrische Entadungen in Dienste der Revierabgrenzung. *Naturwissenschaften*, **44**: 431-2.
MOORE, N. W. 1952. On the so-called ' territories ' of dragonflies (Odonata-Anisoptera). *Behaviour*, **4**: 85-100.
MOORE, N. W. 1953. Population density in adult dragonflies (Odonata-Anisoptera). *J. Anim. Ecol.*, **22**: 344-59.
MOORE, N. W. 1957. The past and present status of the buzzard in the British Isles. *Brit. Birds*, **50**: 173-97.
MOORE, T. E., and R. D. ALEXANDER. 1958. The periodical cicada complex (*Homoptera: Cicadidae*). *Proc. X. int. Congr. Ent.*, **1**: 349-55.
MOORE, W. G. 1948. Bat caves and bat bombs. *Turtox News* (Chicago), **26**: 262-65.
MOREAU, R. E. 1936. Breeding seasons of birds in East African evergreen forest. *Proc. zool. Soc. Lond.*, 1936: 631-53.
MOREAU, R. E. 1950. The breeding seasons of African birds. *Ibis*, **92**: 223-67, 419-33.
MOREAU, R. E., and W. M. MOREAU. 1938. The comparative breeding ecology of two species of *Euplectes* (bishop birds) in Usambara. *J. Anim. Ecol.*, **7**: 314-27.
MOREL, G., and F. BOURLIÈRE. 1956. Recherches écologiques sur les *Quelea quelea quelea* (L.) de la basse vallée du Sénégal. II. La reproduction. *Alauda*, **24**: 97-122.
MORISON, G. D. 1941. Notes on earwigs in northern Scotland. *Ent. mon. Mag.*, **77**: 128-30.
MORRISON, T. F. 1929. Observations on the synchronous flashing of fireflies in Siam. *Science*, **69**: 400-1.
MOSCONA, A. 1950. Studies of the egg of *Bacillus libanicus* (Orthoptera, Phasmidae). *Quart. J. micr. Sci.*, **91**: 183-203.
MOSELEY, H. N. 1879. Notes by a naturalist on the ' Challenger '. London.
MOULTON, J. M. 1956. Influencing the calling of sea robins (*Prionotus* spp.). *Biol. Bull. Wood's Hole*, **111**: 393-8.
MOULTON, J. M. 1957. Sound production in the spiny lobster *Panulirus argus* (Latreille). *Biol. Bull. Wood's Hole*, **113**: 286-95.
MOUNTFORT, G. 1956. The territorial behaviour of the hawfinch *Coccothraustes coccothraustes*. *Ibis*, **98**: 490-5.
MUIRHEAD-THOMSON, R. C. 1956. Communal oviposition in *Simulium damnosum* Theobald (Diptera, Simuliidae). *Nature, Lond.*, **178**: 1297-8.
MÜLLER, F. 1869. *Facts and arguments for Darwin*. London.
MÜLLER, JOH. 1857. Ueber die Fische welche Töne von sich geben. *Arch. Anat. Physiol. Lpz.*, 1857: 249-79. [Not seen.]
MUNRO, J. H. B. 1948. Rook roosts in the Lothians, winter 1946-47. *Scot. Nat.*, **60**: 20-9.
MURPHY, R. C. 1936. *Oceanic birds of South America*. 2 vols. New York.
MURRAY, J., and J. HJORT. 1912. *The depths of the ocean*. London.
MYERS, J. G. 1929. *Insect singers: a natural history of the cicadas*. London.
MYERS, K., and R. MYKYTOWYCZ. 1958. Social behaviour in the wild rabbit. *Nature, Lond.*, **181**: 1515-16.
NAUMANN, J. A. 1822. *Naturgeschichte der Vögel Deutschlands*. Edited by J. F. Naumann. Pt. 1. Leipzig.
NEAL, E. 1948. *The Badger*. London.

NELSON, E. W. 1909. The rabbits of North America. *N. Amer. Fauna*, no. 29, *U.S· biol. Surv., Wash.*: 314 pp.

NEWMAN, E. 1869. *Illustrated natural history of British moths*. London.

NEWMAN, H. H. 1917. A case of synchronic behavior in Phalangidae. *Science*, **45**: 44.

NEWTON, A. 1864. *Ootheca Wolleyana*. Vol. 1, pt. 1. London.

NEWTON, A. 1893-6. *A dictionary of birds*. London.

NICE, M. M. 1935. Some observations on the behavior of starlings and grackles in relation to light. *Auk*, **52**: 91-2.

NICE, M. M. 1937. Studies in the life history of the song sparrow. I. A population study of the song sparrow. *Trans. Linn. Soc. N.Y.*, **4**: 1-247.

NICE, M. M. 1939. 'Territorial song' and non-territorial behavior of goldfinches in Ohio. *Wilson Bull.*, **51**: 123.

NICE, M. N. 1941. The role of territory in bird life. *Amer. Midl. Nat.*, **26**: 441-87.

NICHOLLS, A. G. 1933. On the biology of *Calanus finmarchicus*. III. Distribution and diurnal migration in the Clyde sea-area. *J. mar. biol. Ass. U.K.*, **19**: 139-64.

NICHOLS, J. T. 1939. Data on size, growth and age in the box turtle, *Terrapene carolina*. *Copeia*, 1939: 14-20.

NICHOLSON, A. J. 1927. A new theory of mimicry in insects. *Australian Zoologist*, **5**: 10-104.

NICHOLSON, A. J. 1947. Fluctuation of animal populations. *Aust. N.Z. Ass. Adv. Sci,.* Presidential Address, sec. D: 1-14.

NICHOLSON, A. J. 1955*a*. Compensatory reactions of populations to stresses, and their evolutionary significance. *Aust. J. Zool.*, **2**: 1-8.

NICHOLSON, A. J. 1955*b*. An outline of the dynamics of animal populations. *Aust. J. Zool.*, **2**: 9-65.

NICHOLSON, E. M. 1931*a*. Communal display in humming-birds. *Ibis*, 1931: 74-83.

NICHOLSON, E. M. 1931*b*. *The art of bird-watching*. London.

NICHOLSON, E. M., and B. D. NICHOLSON. 1930. The rookeries of the Oxford district. *J. Ecol.*, **18**: 51-66.

NICOL, J. A. C. 1953. Luminescence in polynoid worms. *J. mar. biol. Ass. U.K.*, **32**: 65-84.

NIELSEN, J. 1961. Contributions to the biology of the Salmonidae in Greenland. *Medd. Grønl.*, **159** (8): 76 pp.

NOBLE, G. K. 1931. *The biology of the Amphibia*. New York.

NOBLE, G. K. 1938. Sexual selection among fishes. *Biol. Rev.*, **13**: 133-58.

NOBLE, G. K., M. WURM and A. SCHMIDT. 1938. Social behavior of the black-crowned night heron. *Auk*, **55**: 7-40.

NORMAN, J. R. 1936. *A history of fishes*. 2nd Ed. London.

ODUM, E. P. 1942. Annual cycle of the black-capped chickadee—3. *Auk*, **59**: 499-531.

OLIVER, W. R. B. 1955. *New Zealand birds*. Wellington, N.Z.

OORDT, G. J. VAN, and M. F. M. BRUYNS. 1938. Die Gonaden übersommernder Austernfischer (*Haematopus ostralegus* L.). *Z. Morph. Ökol. Tiere*, **34**: 161-72.

ORTON, J. H. 1936. Experiments in the sea on the rate of growth of some Crustacea Decapoda. *J. mar. biol. Ass. U.K.*, **20**: 673-89.

OSSIANNILSSON, F. 1949. Insect drummers. *Opuscula entomologica*, Suppl. 10 (Lund Entom. Sällsk.): 146 pp.

OWEN, D. F. 1959. The breeding season and clutch-size of the rook *Corvus frugilegus*. *Ibis*, **101**: 235-39.

OWEN, R. 1868. *On the anatomy of vertebrates*. Vol. 3. Mammals. London.

PALMEN, E. 1955. Diel periodicity of pupal emergence in natural populations of some chironomids (Diptera). *Ann. zool. Soc. Vanamo*, **17**: 30 pp.

PALMGREN, P. 1935. Über Tagesrhytmus der Vögel im arktischen Sommer. *Ornis Fennica*, **12**: 107-21.

PALMGREN, P. 1949. On the diurnal rhythm of activity and rest in birds. *Ibis*, **91**: 561-76.

PALUDAN, K. 1951. Contributions to the breeding biology of *Larus argentatus* and *Larus fuscus*. *Vidensk. Medd. Dansk naturh. Foren.*, **114**: 1-128.

PARK, T. 1933. Studies in population physiology. II. Factors regulating initial growth of *Tribolium confusum* populations. *J. exp. Zool.*, **65**: 17-42.

PARK, T. 1934. Studies in population physiology. III. The effect of conditioned flour upon the productivity and population decline of *Tribolium confusum*. *J. exp. Zool.*, **68**: 167-82.

PARK, T. 1949. See Allee, Emerson, Park, Park and Schmidt (1949), Chapters 18-22.

PEARL, R. 1932. The influence of density of population upon egg production in *Drosophila melanogaster*. *J. exp. Zool.*, **63**: 57-84.

PEARL, R., and F. M. SURFACE. 1909. A biometrical study of egg production in the domestic fowl. I. Variation in annual egg-production. *U.S. Dept. Agric., Bur. Anim. Indust. Bull.*, **110**, pt. 1: 1-80.

PEARSALL, W. H. 1957. Report on an ecological survey of the Serengeti National Park, Tanganyika. *Oryx*, **4**: 71-136.

PEARSE, A. S. 1914. Habits of fiddler-crabs. *Ann. Rep. Smithson. Inst., Wash.* for 1913: 415-28.

PEARSON, J. 1908. Cancer. *L.M.B.C. Memoirs*, no. 16. London.

PEARSON, O. P. 1946. Scent glands of the short-tailed shrew. *Anat. Rec.*, **94**: 615-29.

PEARSON, O. P., MARY R. KOFOID and ANITA K. PEARSON. 1952. Reproduction of the lump-nosed bat (*Corynorhinus rafinesquei*) in California. *J. Mammal.*, **33**: 273-320.

PETERLE, T. 1955. Notes on the display of the red grouse. *Scot. Nat.*, **67**: 61-4.

PETERSON, R. T., and J. FISHER. 1956. *Wild America*. London.

PETTINGILL, O. S. 1939. History of one hundred nests of arctic tern. *Auk*, **56**: 420-8.

PHILIPSON, W. R. 1933. The rook roosts of south Northumberland and the boundaries between their feeding territories. *Brit. Birds*, **27**: 66-71.

PHILIPSON, W. R. 1937. Two contrasting seasons at a redwing roost. *Brit. Birds*, **30**: 343-5.

PICKENS, A. L. 1935. Evening drill of chimney swifts during the late summer. *Auk*, **52**: 149-53.

PIERCE, G. W. 1948. *The songs of insects*. Cambridge, Mass.

PIKE, R. B. 1947. Galathea. *L.M.B.C. Memoirs*, no. 34: 179 pp. Liverpool.

PIKE, R. B. 1954. Notes on the growth and biology of the prawn *Spirontocaris lilljeborgii* (Danielssen). *J. mar. biol. Ass. U.K.*, **33**: 739-47.

PINCHIN, R. D., and J. ANDERSON. 1936. The nocturnal activity of Tipulinae (Diptera) as measured by a light trap. *Proc. R. ent. Soc., Lond.*, (A) **11**: 69-78.

PITELKA, F. A. 1942. Territoriality and related problems in North American hummingbirds. *Condor*, **44**: 189-204.

PITELKA, F. A. 1951. Ecologic overlap and interspecific strife in breeding populations of Anna and Allen hummingbirds. *Ecology*, **32**: 641-66.

PITELKA, F. A. 1957. Some characteristics of microtine cycles in the Arctic. *18th Biology Colloqium, Oregon State College, Corvallis, Oregon:* 73-88.

PITELKA, F. A., P. Q. TOMICH and G. W. TREICHEL. 1955. Ecological relations of jaegers and owls as lemming predators near Barrow, Alaska. *Ecol. Monogr.*, **25**: 85-117.

POCOCK, R. I. 1939-41. Mammalia. 2 vols. *The Fauna of British India*. London.

POTTS, F. A. 1913. Stolon formation in certain species of *Trypanosyllis*. *Quart. J. micr. Sci.*, **58**: 411-46.

POULTON, E. B. 1890. *The colours of animals*. London.

POWELL, T. 1882. Remarks on the structure and habits of the coral-reef annelid, *Palolo viridis*. *J. Linn.Soc. Lond., Zool.*, **16**: 393-6.

PRAED, C. W. MACKWORTH, and C. H. B. GRANT. 1952. *Birds of eastern and north eastern Africa*. Vol. 1. London.

PRATT, D. M. 1943. Analysis of population development in *Daphnia* at different temperatures. *Biol. Bull. Wood's Hole*, **85**: 116-40.

PRÉVOST, J. 1953. Formation des couples, ponte et incubation chez le manchot empereur. *Alauda*, **21**: 141-56.

PRÉVOST, J., and J. BOURLIÈRE. 1957. Vie sociale et thermorégulation chez le manchot empereur *Aptenodytes forsteri*. *Alauda*, **25**: 167-73.

PRYER, H. 1884. An account of a visit to the birds'-nest caves of British North Borneo. *Proc. zool. Soc., Lond.*, 1884: 532-8.

PUKOWSKI, E. 1933. Ökologische Untersuchungen an *Necrophorus* F. *Z. Morph. Ökol. Tiere*, **27**: 518-86.

PUMPHREY, R. F. 1950. Hearing. *Symposia Soc. exp. Biol.*, no. 4: 3-18.

PUNNETT, R. C. 1915. *Mimicry in butterflies*. Cambridge.

PYEFINCH, K. A. 1955. A review of the literature on the biology of the Atlantic salmon. *Sci. Invest. Freshw. Fish, Scotland*, no. 9: 44 pp.

QUAINTANCE, C. W. 1938. Content, meaning and possible origin of male song in the brown towhee. *Condor*, **40**: 97-101.

RAE, B. B. 1959. Halibut—observations on its size at first maturity, etc. *Mar. Res., Scotland*, 1959 (4): 19 pp.

RAGGE, D. R. 1955. A note on female stridulation in the British Acrididae. *Bri . J. Anim. Behav.*, **3**: 70.

RAND, A. L. 1945. Mammals of Yukon. *Nat. Mus. Canada, Bull.* no. 100: 93 pp.

RANSON, R. M. 1934. The field vole (*Microtus*) as a laboratory animal. *J. Anim. Ecol.*, **3**: 70-6.

RATCLIFFE, D. A. 1958. Broken eggs in peregrine eyries. *Brit. Birds*, **51**: 23-6.

RATCLIFFE, F. 1932. Notes on the fruit-bats (*Pteropus* spp.) of Australia. *J. Anim. Ecol.*, **1**: 32-57.

RAU, P. 1932. Rhythmic periodicity and synchronous flashing in the firefly, *Photinus pyralis*, with notes on *Photurus pennsylvanicus*. *Ecology*, **13**: 7-11.

RAU, P., and N. RAU. 1916. The sleep of insects; an ecological study. *Ann. ent. Soc. Amer.*, **9**: 227-74.

REGNART, H. C. 1931. The lower limits of perception of electrical currents by fish. *J. mar. biol. Ass. U.K.*, **17**: 415-20.

REIGHARD, J. 1920. The breeding behaviour of the suckers and minnows. *Biol. Bull. Wood's Hole*, **38**: 1-32.

RIBBANDS, C. R. 1953. *The behaviour and social life of honeybees*. London.

RICHARDS, CHRISTINA M. 1958. The inhibition of growth in crowded *Rana pipiens* tadpoles. *Physiol. Zool.*, **31**: 138-51.

RICHARDS, O. W. 1927. Sexual selection and allied problems in insects. *Biol. Rev.*, **2**: 298-360.

RICHARDS, O. W. 1953. *The social insects*. London.

RICHDALE, L. E. 1942. Supplementary notes on the royal albatross. *Emu*, **41**: 169-84, 253-64.

RICHDALE, L. E. 1952. *Post-egg period in albatrosses*. Dunedin.

RICHDALE, L. E. 1957. *A population study of penguins*. Oxford.

RICHTER, C. P. 1952. The effect of domestication on the steroids of animals and man. *Ciba Foundation Colloquia on Endocrinology*, 3: 89-107.

RICHTER, H. 1953. Zur Lebensweise der Wasseramsel. *J. Orn., Lpz.*, **94**: 68-82.

RICHTER, P. O. 1958. Biology of the Scarabaeidae. *Ann. Rev. Entom.*, **3**: 311-34.

RICKER, W. E. 1950. Cycle dominance among the Fraser sockeye. *Ecology*, **31**: 6-26.

RISLEY, P. L. 1933. Observations on the natural history of the common musk turtle, *Sternotherus odoratus* (Latreille). *Papers Mich. Acad. Sci. Arts Lett.*, **17**: 685-711.

RITTER, W. E. 1938. *The California woodpecker and I*. Berkeley.

ROBERTS, A. W. R. 1919, 1921. On the life history of ' wireworms ' of the genus *Agriotes*, Esch., etc. *Ann. appl. Biol.*, **6**: 116-35; **8**: 193-215.

ROBERTSON, F. W., and J. H. SANG. 1944. The ecological determinants of population growth in a *Drosophila* culture. I. Fecundity of adult flies. *Proc. roy. Soc.*, B, **132**: 258-77.

ROBSON, F. D. 1947. Kiwis in captivity, as told to Robert Gibbings. *Hawkes Bay Art Gallery and Museum, Napier, N.Z.*, 7 pp.

ROEBUCK, A. 1933. A survey of the rooks in the Midlands. *Brit. Birds*, **27**: 4-23.

ROLLINAT, R., and E. TROUESSART. 1896. Sur la reproduction des chauve-souris. *Mém. Soc. Zool. Fr.*, **9**: 214-40.

ROSE, M. 1933. Copépodes pélagiques. *Faune de France*, **26**: 374 pp.

ROSE, S. M. 1959. Failure of survival of slowly growing members of a population. *Science*, **129**: 1026.

ROTH, L. M. 1943. Studies on the gaseous secretion of *Tribolium confusum* Duval. II. The odiferous glands of *Tribolium confusum*. *Ann. ent. Soc. Amer.*, **36**: 397-424.

ROTH, L. M., and RUTH B. HOWLAND. 1941. Studies on the gaseous secretion of *Tribolium confusum* Duval. I. Abnormalities produced . . . by exposure to a secretion given off by the adults. *Ann. ent. Soc. Amer.*, **34**: 151-72.

ROULE, L. 1935. *Fishes: their ways of life*. London.

ROWAN, M. K. 1955. The breeding biology and behaviour of the redwinged starling *Onychognathus morio*. *Ibis*, **97**: 663-705.

RUDEBECK, G. 1955. Some observations at a roost of European swallows and other birds in the south-eastern Transvaal. *Ibis*, **97**: 572-80.

RUSSELL, F. S. 1925. The vertical distribution of marine macroplankton. An observation of diurnal changes. *J. mar. biol. Ass. U.K.*, **13**: 769-809.

RUSSELL, F. S. 1927*a*. The vertical distribution of plankton in the sea. *Biol. Rev.*, **2**: 213-62.

RUSSELL, F. S. 1927*b*. The distribution of animals caught in the ring-trawl in the daytime in the Plymouth area. *J. mar. biol. Ass. U.K.*, **14**: 557-608.

RUSSELL, F. S. 1928*a*. The vertical distribution of marine macroplankton. Further observations on diurnal changes. *J. mar. biol. Ass. U.K.*, **15**: 81-103.

RUSSELL, F. S. 1928*b*. Observations on the behaviour of *Calanus finmarchicus*. *J. mar. biol. Ass. U.K.*, **15**: 429-54.

RUSSELL, F. S., and J. S. COLMAN. 1934. The zooplankton. *Great Barrier Reef Exped. 1928-9, Sci. Rep.*, **2**: 159-276.

RUTTNER, F. 1905. Über das Verhalten des Oberflächenplanktons zu verschiedenen Tageszeiten im Grossen Plöner See und in zwei nordböhmischen Teichen. *Forschungsber. Biol. Sta. Plön*, **12**: 35-62.

SÁLIM ALI. 1953. The Keoladeo Ghana of Bharatpur (Rajasthan). *J. Bombay nat. Hist. Soc.*, **51**: 531-36.

SALOMONSEN, F. 1950-51. *The birds of Greenland*. Copenhagen.

SALOMONSEN, F. 1955. The evolutionary significance of bird migration. *Dansk. biol. Medd.*, **22**, no. 6: 62 pp.

SAND, A. 1937. The mechanism of the lateral sense organs of fishes. *Proc. roy. Soc. B.*, **123**: 472-95.

SAND, A. 1938. The function of the ampullae of Lorenzini with some observations on the effect of temperature on sensory rhythms. *Proc. roy. Soc. B.*, **125**: 524-553.

SAUER. F., and E. SAUER. 1960. Orientation of nocturnal bird migrants by the stars. *Proc. XII Congr. Int. Orn. Helsinki*: 645-8.

SAUNDERS, H. 1884-5. *A history of British birds*, by W. Yarrell. 4th Ed., Vol. 4, rewritten by H. Saunders. London.

SAWIN, P. B., and D. D. CRARY. 1953. Genetic and physiological background of reproduction in the rabbit. II. Some racial differences in the pattern of maternal behaviour. *Behaviour*, **6**: 128-46.

SCHÄFER, W. 1952. Der 'Kritische Raum', Masseinheit und Mass für die mögliche Bevölkerungsdichte innerhalb einer Art. *Zool. Anz.*, 16 Supplementband: 391-95.

SCHEFFER, V. B. 1958. *Seals, sea lions and walruses.* Stanford and London.

SCHEVILL, W. E., and BARBARA LAWRENCE. 1949. Underwater listening to the white porpoise. *Science,* **109**: 143-4.

SCHJELDERUP-EBBE, T. 1922. Beiträge zur Sozialpsychologie des Haushuhns. *Z. Psychol.*, **88**: 226-52.

SCHMIDT, J. 1922. The breeding places of the eel. *Philos. Trans.*, **211**: 179-208.

SCHORGER, A. W. 1955. *The passenger pigeon, its natural history and extinction.* Madison, Wis.

SCHREIBER, B. 1935. Experimentie considerazioni sul significato funzionale dell'apparato di Weber degli Ostariofisi. *Boll. Zool.*, **6**: 67-70.

SCHROTTKY, C. 1922. Soziale Gewohnheiten bei solitären Insekten. *Z. wiss. Insekt Biol.*, **17**: 49.

SCHÜZ, E. 1943. Über die Jungenaufzucht des Weissen Storches. *Z. Morph. Ökol, Tiere*, **40**: 181-237.

SCHÜZ, E. 1957. Das Verschlingen eigener Junger ('Kronismus') bei Vögeln und seine Bedentung. *Vogelwarte*, **19**: 1-15.

SCHWAN, A. 1920. Vogelsang und Wetter, physikalischbiologisch untersucht. *Pflug. Arch. ges. Physiol.*, **180**: 341-7.

SCOTT, H. 1916. Note on the swarming of Chloropid flies, Psocidae, etc., in houses. *Ent. mon. Mag.*, **52**: 18-21.

SCOTT, J. P., and E. FREDERICSON. 1951. The causes of fighting in mice and rats. *Physiol. Zoöl.*, **24**: 273-309.

SCOTT, J. W. 1942. Mating behavior of the sage grouse. *Auk*, **59**: 477-98.

SCOTT, J. W. 1950. A study of the phylogenetic or comparative behavior of three species of grouse. *Ann. N.Y. Acad. Sci.*, **51**: 1062-73.

SEELEY, H. G. 1886. *The fresh-water fishes of Europe.* London.

SELOUS, E. 1906-7. Observations tending to throw light on the question of sexual selection in birds, including a day-to-day diary on the breeding habits of the ruff (*Machetes pugnax*). *Zoologist*, 4th ser, **10**: 201, etc.; **11**: 65, etc.

SELOUS, E. 1909. An observational diary on the nuptial habits of the blackcock (*Tetrao tetrix*) in Scandinavia and England. Part 1, Scandinavia. *Zoologist*, 4th ser., **13**: 400, etc.

SELYE, H. 1939. The effect of adaptation to various damaging agents on the female sex organs in the rat. *Endocrinology*, **25**: 615-24.

SELYE, H. 1950. *The physiology and pathology of exposure to stress: a treatise . . .* Montreal.

SEMON, R. 1899. *In the Australian bush.* London.

SEMPER, K. 1881. *Animal life as affected by the natural conditions of existence.* New York. [Not seen: quoted from Allee, 1931.]

SERGEANT, D. E. 1951. Ecological relationships of the guillemots *Uria aalge* and *Uria lomvia. Proc. Xth Int. Orn. Congr., Uppsala*: 578-87.

SERVENTY, D. L. 1956. Age at first breeding of the short-tailed shearwater. *Ibis*, **98**: 532-3.

SERVENTY, D. L., and H. M. WHITTELL. 1951. *A handbook of the birds of Western Australia.* 2nd Ed. Perth, W.A.

SERVICE, R. 1902. The adder in Solway. *Ann. Scot. nat. Hist.*, 1902: 153-62.

SETON, E. T. 1909. *Life-histories of northern animals, an account of the mammals of Manitoba.* 2 vols. New York.

SHARP, D. 1895. Insects. Part. I. In *Cambridge Natural History.* London.

SHARP, D. 1899. Insects. Part II. In *Cambridge Natural History.* London.

SHAW, G. 1932. The effect of biologically conditioned water upon the rate of growth in fishes and amphibia. *Ecology*, **13**: 263-78.

SHOEMAKER, H. H. 1939. Social hierarchy in flocks of the canary. *Auk*, **56**: 381-406.

SHORTRIDGE, G. C. 1934. *The mammals of south west Africa.* 2 vols. London.

SICK, H. 1959. Die Balz der Schmuckvögel. *J. Orn., Lpz.*, **100**: 269-302.

SIIVONEN, L. 1957. The problem of short-term fluctuations in numbers of tetraonids in Europe. *Papers on Game Research, Helsinki*, no. 19: 1-44.

SILLIMAN, R. P., and J. S. GUTSELL. 1958. Experimental exploitation of fish populations. *U.S. Fish and Wildlife Service Fishery Bull.*, **58** (no. 133): 214-52.

SIMMONS, K. E. L. 1951. Interspecific territorialism. *Ibis*, **93**: 407-13.

SIMPSON, G. G. 1945. The principles of classification and a classification of mammals. *Bull. Amer. Mus. nat. Hist.*, **85**: 1-350.

SINCLAIR, F. G. 1895. Myriapods. In *The Cambridge Natural History*. Vol. 5. London.

SKAIFE, S. H. 1954. *African insect life*. London, Cape Town and New York.

SKUES, G. E. M. 1921. *The way of a trout with a fly*. London.

SLADEN, F. W. L. 1912. *The humble-bee*. London.

SLUD, P. 1957. The song and dance of the long-tailed manakin, *Chiroxiphia linearis*. *Auk*, **74**: 333-9.

SMITH, G. 1905. High and low dimorphism. With an account of certain Tanaidae of the Bay of Naples. *Mitt. zool. Sta. Neapel*, **17**: 312-40.

SMITH, G. 1909. Crustacea. In *The Cambridge Natural History*. Vol. 4. London.

SMITH, H. N. 1935. Synchronous flashing of fireflies. *Science*, **82**: 151-2.

SMITH, M. 1951. *The British amphibians and reptiles*. London.

SNODGRASS, R. E. 1925. Insect musicians, their music, and their instruments. *Smiths' Inst. Ann. Rep.* for 1923: 405-52.

SNOW, D. W. 1958. The breeding of the blackbird *Turdus merula* at Oxford. *Ibis*, **100**: 1-30.

SNOW, D. W. 1959. *A study of blackbirds*. London.

SOLLAS, W. J. 1924. *Ancient hunters and their modern representatives*. London.

SORBY, H. C. 1906. Notes on some species of *Nereis* in the district of the Thames estuary. *J. Linn. Soc. Lond.*, **29**: 434-9.

SØRENSEN, W. E. 1884. *Om Lydorganer hos Fiske; en physiologisk og comparativ anatomisk Undersøgelse*. Copenhagen.

SOUTHERN, H. N. 1940. The ecology and population dynamics of the wild rabbit, *Oryctolagus cuniculus*. *Ann. appl. Biol.*, **27**: 509-26.

SOUTHERN, H. N. 1954*a*. *Control of rats and mice*. Vol. 3, *House mice*. Oxford.

SOUTHERN, H. N. 1954*b*. Tawny owls and their prey. *Ibis*, **96**: 384-410.

SOUTHERN, R., and A. C. GARDINER. 1932. The diurnal migrations of the Crustacea of the plankton in Lough Derg. *Proc. R. Irish Acad.*, **40**: 121-59.

SOUTHWICK, C. H. 1955*a*. The population dynamics of confined house mice supplied with unlimited food. *Ecology*, **36**: 212-25.

SOUTHWICK, C. H. 1955*b*. Regulatory mechanisms of house mouse populations: social behaviour affecting litter survival. *Ecology*, **36**: 627-34.

SPENCER, K. G. 1953. *The lapwing in Britain*. London and Hull.

SPRUNT, A., Jr. 1948. The tern colonies of the Dry Tortugas Keys. *Auk*, **65**: 1-19.

STANFORD, J. K. 1947. Birds parties in forest in Burma. *Ibis*, **89**: 507-9.

STAPLEY, J. H. 1949. *Pests of farm crops*. London.

STEGMANN, B. 1956. Über die Herkunft des flüchtigen rosroten Federpigments. *J. Orn., Lpz.*, **97**: 204-5.

STEWART, R. E., and J. W. ALDRICH. 1951. Removal and repopulation of breeding birds in a spruce-fir forest community. *Auk*, **68**: 471-82.

STEWART, W. 1928. Studies of some Lanarkshire birds. The carrion crow. *Scot. Nat.*, 1928: 19-23.

STONEHOUSE, B. 1953. The emperor penguin. I. Breeding behaviour and development. *Falkland Is. Depend. Surv., Sci. Rep.* no. 6: 33 pp.

STONEHOUSE, B. 1956. The kind penguin of South Georgia. *Nature, Lond.*, **178**: 1424-6.

STONER, D. 1941. Homing instinct in the bank swallow. *Bird-banding*, **12**: 104-8.

STONOR, C. R. 1940. *Courtship and display among birds*. London.

STORER, T. I. 1925. A synopsis of the Amphibia of California. *Univ. Calif. Publ. Zool.*, **27**: 1-342.

STRECKER, R. L. 1954. Regulating mechanisms in house mouse populations: the effect of limited food-supply on an unconfined population. *Ecology*, **35**: 249-53.

STRECKER, R. L., and J. T. EMLEN. 1953. Regulating mechanisms in house mouse populations. *Ecology*, **34**: 375-85.

STRESEMANN, E. 1927-34. Sauropsida: Aves. In *Kükenthals Handbuch der Zoologie*, 7 Bd., 2 Hälfte. Berlin and Leipzig.

STRESEMANN, E. 1950. Interspecific competition in chats. *Ibis*, **92**: 148.

STRESEMANN, E. 1953. So sind Töpfel. *Orion*, **8**: 772-80. [Not seen, quoted from Schüz, 1957.]

STRONG, R. M. 1914. On the habits and behavior of the herring gull. *Auk*, **31**: 22-49, 178-99.

STUART, T. A. 1953. Spawning migration, reproduction and young stages of loch trout (*Salmo trutta* L.). *Scottish Home Dept., Freshw. & Salmon Fish. Res.* no. 5: 39 pp.

STUART, T. A. 1957. The migrations and homing behaviour of brown trout (*Salmo trutta* L.). *Scottish Home Dept., Freshw. & Salmon Fish. Res.*, no. 18: 27 pp.

STUBBS, F. J. 1910. Ceremonial gatherings of the magpie. *Brit. Birds*, **3**: 334-6.

STURTEVANT, A. H. 1938. Essays on evolution. II. On the effects of selection on social insects. *Quart. Rev. Biol.*, **13**: 74-6.

SUND, O. 1920. Peneides and Stenopides. *Rep. Sci. Res. ' Michael Sars '*, **3** (ii): 32 pp.

SUOMALAINEN, H. 1939. Regelbundene Tagesrhythmik beim Gryllteist, *Uria g. grylle* (L.). *Ornis Fennica*, 16. (Quoted from Koskimies, 1949.)

SUTTER, E. 1946. Das Abwehrverhalten nestjunger Wiedehopfe. *Orn. Beob.*, **43**: 72-81.

SUTTER, E. 1958. Frequenz und Ablauf des Nachtzuges nach Radar-Beobachtungen. (Unpublished paper given at XII Int. orn. Congr., Helsinki.)

SVÄRDSON, G. 1957. The ' invasion ' type of bird migration. *Brit. Birds*, **50**: 314-43.

SVIHLA, A. 1931. Habits of the Louisiana mink. *J. Mammal.*, **13**: 366-8.

SWANBERG, O. 1951. Food storage, territory and song in the thick-billed nutcracker. *Proc. X Int. Orn. Congr.*: 545-54.

TEMPLEMAN, W. 1948. The life history of the caplin (*Mallotus villosus* O. F. Müller) in Newfoundland waters. *Bull. Nfld. Govt. Lab.*, no. 17: 151 pp.

THOMAS, C. A. 1940. *The biology and control of wireworms.* State College, Penna.

THOMPSON, E. NETHERSOLE. 1951. *The Greenshank.* (The New Naturalist Monographs.) London.

THOMPSON, H. V., and A. N. WORDEN. 1956. *The Rabbit.* (The New Naturalist Monographs.) London.

THOMSON, A. L. 1923. The migrations of some British ducks: results of the marking method. *Brit. Birds*, **16**: 262-76.

TICEHURST, C. B. 1938. On the birds of northern Burma. Part 1. By J. K. Stanford assisted by C. B. Ticehurst. *Ibis*, 14th ser., **2**: 65-102.

TINBERGEN, N. 1953. *Social behaviour in animals.* London and New York.

TINBERGEN, N. 1957. The functions of territory. *Bird Study*, **4**: 14-27.

TOD, K. 1953. The distribution of the northern dart, *Agrotis hyperborea* (Zetterstedt), on the mainland of Scotland. *Scot. Nat.*, **65**: 11-18.

TOMPKINS, GRACE. 1933. Individuality and territoriality as displayed by three passerine species. *Condor*, **35**: 98-106.

TOWNSEND, C. H. T. 1891. Notes on Acrididae of Michigan. *Proc. ent. Soc. Wash.*, **2**: 43-4.

TRAUTMAN, M. B. 1940. The birds of Buckeye Lake, Ohio. *Misc. Publ. Mus. Zool. Univ. Mich.*, no. 44: 466 pp.

TRISTRAM, H. B. 1884. *The fauna and flora of Palestine.* London.

TUCKER, B. W. 1949. Remarks on a seasonal colour change in the bill and legs of herons. *Brit. Birds*, **42**: 46-50.

TUCKER, D. W. 1959. A new solution to the Atlantic eel problem. *Nature, Lond.*, **183**: 495-501.

TURNER, E. L. 1920. Some notes on the ruff. *Brit. Birds*, **14**: 146-53.

TURNER, L. M. 1886. Contributions to the natural history of Alaska. *U.S. Army Signal Service, Arctic Series*, no. 2: 226 pp. Wash., D.C.

UHRICH, J. 1938. The social hierarchy in albino mice. *J. comp. Psychol.*, **25**: 373-413. [Not seen].

URQUHART, F. A. 1960. *The monarch butterfly*. Toronto.

UVAROV, B. P. 1928. *Locusts and grasshoppers*. London.

VALVERDE, J. A. 1955-6. Essai sur l'aigrette garzette (*Egretta g. garzetta*) en France. *Alauda*, **23**: 145-71, 254-79; **24**: 1-36.

VAN SOMEREN, V. D. 1940. The factors conditioning the rising of trout (*Salmo trutta*) in a small freshwater lake. *J. Anim. Ecol.*, **9**: 89-107.

VAN SOMEREN, V. G. L. and G. R. C. 1945. Evacuated weaver colonies. . . . *Ibis*, **87**: 33-44.

VARLEY, G. C. 1947. The natural control of populations balance in the knapweed gall-fly (*Urophora jaceana*). *J. Anim. Ecol.*, **16**: 139-87.

VENABLES, L. S. V. 1943. Observations at a pipistrelle bat roost. *J. anim. Ecol.*, **12**: 19-26.

VERWEY, J. 1930. Einiges über die Biologie ost-indischer Mangrovekrabben. *Treubia*, **12**: 167-261.

VESEY-FITZGERALD, D. F. 1958. Notes on breeding colonies of the red-billed quelea in S.W. Tanganyika. *Ibis*, **100**: 167-74.

VEVERS, H. G. 1953. Photographic survey of certain areas of the sea floor near Plymouth. *J. mar. biol. Ass. U.K.*, **31**: 215-21.

VINCENT, A. W. 1949. On the breeding habits of some African birds. *Ibis*, **91**: 483-507.

VLEUGEL, D. A. 1941. Sociale roestgewoonten bij vogels, inzonderheit bij Vink (*Fringilla coelebs*) en Keep (*Fringilla montifringilla*). *Ardea*, **30**: 89-106.

VOLTERRA, V. 1928. Variations and fluctuations of the number of individuals in animal species living together. (Translated by M. E. Wells.) *J. Cons. int. Explor. Mer*, **3**: 3-51.

VOOUS, K. H. 1950. The breeding seasons of birds in Indonesia. *Ibis*, **92**: 279-87.

WAGNER, G. 1958. Die Brutvögel von Röst (Lofoten). *Sterna*, **3**: 59-72.

WALLACE, A. R. 1869. *The Malay Archipelago*. 2 vols. London.

WALLACE, M. M. H. 1957. Field evidence of density-governing reaction in *Sminthurus viridis* (L.). *Nature, Lond.*, **180**: 388-90.

WALLS, G. L. 1942. *The vertebrate eye and its adaptive radiation*. Bloomfield Hills, Mich.

WALTON, I. 1653. *The compleat angler*. London.

WARBURTON, C. 1912. *Spiders*. Cambridge.

WARD, F. 1919. *Animal life under water*. London.

WARD, H. L. 1906. Why do herring gulls kill their young? *Science*, **24**: 593-4.

WARHAM, J. 1958. The nesting of the Australian gannet. *Emu*, **58**: 339-69.

WARREN, E. 1900. On the reaction of *Daphnia magna* to certain changes in its environment. *Quart. J. micr. Sci.*, **43**: 199-224.

WATERMAN, T. H., R. F. NUNNEMACHER, F. A., CHACE and G. L. CLARKE. 1939. Diurnal vertical migrations of deep-water plankton. *Biol. Bull. Wood's Hole*, **76**: 256-79.

WEISSMANN, A. 1889. The duration of life. In *Essays upon heredity and kindred biological problems*. Oxford.

WELCH, P. S. 1938. Diurnal rhythms. *Quart. Rev. Biol.*, **13**: 123-39.

WELSH, J. H., and F. A. CHACE, Jr. 1937. Eyes of deep sea crustaceans. II. Sergestidae. *Biol. Bull. Wood's Hole*, **74**: 364-75.

WELSH, J. H., F. A. CHACE, Jr., and R. F. NUNNEMACHER. 1937. The diurnal migration of deep-water animals. *Biol. Bull. Wood's Hole*, **73**: 185-96.

WELSH, W. W., and C. M. BREDER. 1924. Contributions to life histories of Sciaenidae of the eastern United States coast. *Bull. U.S. Bur. Fish.*, **39**: 141-201.

WENDLAND, V. 1951. Zwanzigjährige Beobachtungen über den Schreiadler *Aquila pomarina*. *Vogelwelt*, **72**: 6-11. (Quoted from Schüz, 1957.)

WENDLAND, V. 1952. Populationsstudien an Raubvögeln. I. Zur Vermehrung des Maüsebussards (*Buteo b. buteo* (L.)). *J. Orn., Lpz.*, **93**: 144-53.

WERTH, IRENE. 1947. The tendency of blackbird and song-thrush to breed in their birth-places. *Brit. Birds*, **40**: 328-30.

WESSELL, J. P., and H. LEIGH. 1941. Studies of the flock organisation of the white-throated sparrow. *Wilson Bull.*, **53**: 222-30. [Not seen.]

WHEELER, W. M. 1910. *Ants, their structure, development and behavior*. New York.

WHISTLER, H. 1941. *Popular handbook of Indian Birds*. 3rd Ed. London.

WHITE, C. M. N. 1951. Weaver birds at Lake Mweru. *Ibis*, **93**: 626-7.

WHITMEE, S. J. 1875. On the habits of *Palolo viridis*. *Proc. zool. Soc. Lond.*, 1875: 496-502.

WHITTLE, C. L., and L. B. FLETCHER. 1924. Further observations on the group habit among birds. *Auk*, **41**: 327-33.

WIESNER, B. P., and N. M. SHEARD. 1935. The duration of life in an albino rat population. *Proc. roy. Soc. Edin.*, **55**: 1-22.

WIGGLESWORTH, V. B. 1953. *The principles of insect physiology*. 5th Ed. London.

WILDE, J. DE. 1941. Contribution to the physiology of the Johnston organ and its part in the behaviour of *Gyrinus*. *Arch. néerl. Physiol.*, **25**: 381-400.

WILLEY, A. 1904. Crows and flying foxes at Barberyn. *Spolia Zeylanica*, **2**: 50-1.

WILLIAMS, C. B. 1935. The times of activity of certain nocturnal insects, chiefly Lepidoptera, as indicated by a light-trap. *Trans. R. ent. Soc. Lond.*, **83**: 523-56.

WILLIAMS, C. B. 1958. *Insect migration*. London.

WILLIAMSON, F. S. L. 1957. Hybrids of the Anna and Allen hummingbirds. *Condor*, **59**: 118-23.

WILLIAMSON, K. 1948. *The Atlantic islands*. London.

WILLIAMSON, K. 1959. Changes of mating within a colony of arctic skuas. *Bird Study*, **6**: 51-60.

WILSON, D. P. 1952. The influence of the nature of the substratum on the metamorphosis of the larvae of marine animals, especially the larvae of *Ophelia bicornis* Savigny. *Ann. Inst. océan.*, **27**: 49-156.

WILSON, D. P. 1953. Notes from the Plymouth aquarium, II. *J. mar. biol. Ass. U.K.*, **32**: 199-208.

WINTERBOTTOM, J. M. 1943. On woodland bird parties in Northern Rhodesia. *Ibis*, **85**: 437-42.

WINTERBOTTOM, J. M. 1949. Mixed bird parties in the Tropics, with special reference to Northern Rhodesia. *Auk*, **66**: 258-63.

WITHERBY, H. F. 1927. Wigeon ringed in Cumberland reported in north-east Russia. *Brit. Birds*, **21**: 97-8.

WITHERBY, H. F. (editor). 1938-40. *A handbook of British Birds*. 5 vols. London.

WONTNER-SMITH, C. 1935. Number of young reared by the rook. *Brit. Birds*, **29**: 26-7.

WOODBURY, A. M., and R. HARDY. 1940. The dens and behaviour of the desert tortoise. *Science*, **92**: 529.

WORTHINGTON, E. B. 1931. Vertical movements of freshwater macroplankton. *Int. Rev. Hydrobiol.*, **25**: 394-436.

WRIGHT, A. H. 1914. North American Anura. Life-histories of the Anura of Ithaca, New York. *Publ. Carneg. Instn.*, no. 197.

WRIGHT, S. 1938. Size of population and breeding structure in relation to evolution. *Science*, **87**: 430-1.

WYNNE-EDWARDS, V. C. 1929*a*. The behaviour of starlings in winter. *Brit. Birds*, **23**: 138-53, 170-80.

WYNNE-EDWARDS, V. C. 1929*b*. On the waking-time of the nightjar (*Caprimulgus e. europaeus*). *J. exp. Biol.*, **7**: 241-7.

WYNNE-EDWARDS, V. C. 1931. The behaviour of starlings in winter. II. Observations in Somerset, 1929-30. *Brit. Birds*, **24**: 346-53.

WYNNE-EDWARDS, V. C. 1932. The breeding habits of the black-headed minnow (*Pimephales promelas* Raf.). *Trans. Amer. Fish. Soc.*, **62**: 382-3.

WYNNE-EDWARDS, V. C. 1939. Intermittent breeding of the fulmar (*Fulmarus glacialis* (L.)), with some general observations on non-breeding in sea-birds. *Proc. zool. Soc. Lond.*, A., **109**: 127-32.

WYNNE-EDWARDS, V. C. 1952*a*. Geographical variation in the bill of the fulmar (*Fulmarus glacialis*). *Scot. Nat.*, **64**: 84-101.

WYNNE-EDWARDS, V. C. 1952*b*. Zoology of the Baird expedition. The birds observed in central and south-east Baffin Island. *Auk*, **69**: 353-91.

WYNNE-EDWARDS, V. C. 1952*c*. Freshwater vertebrates of the Arctic and Subarctic. *Fish Res. Bd. Canada, Bull.* **94**: 28 pp.

WYNNE-EDWARDS, V. C. 1955. Low reproductive rates in birds, especially sea-birds. *Acta XI Congr. Int. Orn., Basel:* 540-7.

WYNNE-EDWARDS, V. C., R. M. LOCKLEY and H. M. SALMON. 1936. The distribution and numbers of breeding gannets (*Sula bassana* L.). *Brit. Birds*, **29**: 262-76.

YARRELL, W. 1859. *A history of British fishes*. 3rd Ed. 2 vols. London.

YARRELL, W. 1871-85. *A history of British birds*. 4th Ed., revised by A. Newton and H. Saunders. 4 vols. London.

YEATES, G. K. 1934. *The life of the rook*. London.

YONGE, C. M. 1930. *A year on the Great Barrier Reef*. London and New York.

YONGE, C. M. 1949. *The sea shore*. London.

YOUNG, F. N. 1958*a*. A large aggregation of larval millipeds, *Zinaria butleri* (McNeill), in Brown County, Indiana. *Proc. Indiana Acad. Sci.*, **67**: 171-2.

YOUNG, F. N. 1958*b*. Some facts and theories about the broods and periodicity of the periodical cicada. *Proc. Indiana Acad. Sci.*, **68**: 164-70.

YOUNG, J. Z. 1950. *The life of vertebrates*. London.

ZUCKERMAN, S. 1932. *The social life of monkeys and apes*. London.

Index